Y0-AAY-461

MOLECULAR BIOLOGY AND PHYSIOLOGY OF INSULIN AND INSULIN-LIKE GROWTH FACTORS

ADVANCES IN EXPERIMENTAL MEDICINE AND BIOLOGY

Recent Volumes in this Series

MOLECULAR BIOLOGY AND PHYSIOLOGY OF INSULIN AND INSULIN-LIKE GROWTH FACTORS

Edited by

Mohan K. Raizada
University of Florida
Gainesville, Florida

and

Derek LeRoith
National Institutes of Health
Bethesda, Maryland

PLENUM PRESS • NEW YORK AND LONDON

Library of Congress Cataloging-in-Publication Data

International Symposium on Molecular and Cellular Biology of Insulin and IGFs (3rd : 1990 : Gainesville, Fla.)
Molecular biology and physiology of insulin and insulin-like growth factors / edited by Mohan K. Raizada and Derek LeRoith.
p. cm. -- (Advances in experimental medicine and biology ; v. 293)
"Proceedings of the Third International Symposium on Molecular and Cellular Biology of Insulin and IGFs, held October 12-14, 1990, in Gainesville, Florida"--T.p. verso.
Includes bibliographical references and index.
ISBN 0-306-43928-X
1. Somatomedin--Mechanism of action--Congresses. 2. Insulin-like growth factor-binding proteins--Congresses. 3. Insulin--Mechanism of action--Congresses. I. Raizada, Mohan K. II. LeRoith, Derek, 1945- . III. Title. IV. Series.
[DNLM: 1. Central Nervous System--physiology--congresses. 2. Insulin--physiology--congresses. 3. Insulin-Like Growth Factor I--physiology--congresses. 4. Insulin-Like Growth Factor II--physiology--congresses. 5. Receptors, Insulin--congresses. W1 AD559 v. 293 / WK 820 I611m 1990]
QP552.S65I58 1991
599'.031--dc20
DNLM/DLC
for Library of Congress 91-24015
CIP

Proceedings of the Third International Symposium on
Molecular and Cellular Biology of Insulin and IGFs,
held October 12-14, 1990, in Gainesville, Florida

ISBN 0-306-43928-X

A Division of Plenum Publishing Corporation
233 Spring Street, New York, N.Y. 10013

Printed in the United States of America

PREFACE

This volume addresses a fundamental puzzle in biology and medicine, namely, how does tissue develop, repair and replace itself. The answer appears to lie in growth factors and their regulation. To thrive and survive we need growth factors and this book concentrates on two factors that are related to growth hormone. Growth hormone does not act directly on all tissues, but mediates many of its actions through the release of insulin-like growth factors from the liver. The growth factors were originally called somatomedins by McConaghey and Sledge [1], who discovered that they mediated growth-like effects of growth hormone. However, the factors were purified on the basis of their insulinomimetic actions on fat and muscle and it is their relationship to the insulin family of peptides that now gives them their name [2,3] of insulin-like growth factors (IGFs). They mediate the actions of growth hormone on the proteoglycan synthesis of cartilage and produce mitogenic effects in fibroblast cultures. It is now recognized that IGF-I and IGF-II have local cellular actions and that they are the key to many essential functions in the daily routine lives of cells of every tissue. They are not only factors for linear growth in skeletal tissue, as was first thought, but they also stimulate DNA synthesis and cause cell division in diverse cell populations, including human cell lines. One cannot yet make a hard and fast rule about the different actions of IGF-I compared to IGF-II but we do know they have separate receptors. There is consensus that both factors are fetal growth promoters and have both short-term metabolic effects and long-term growth effects. In the adult IGF-I is mostly synthesized in the liver and secreted into the circulation, whereas as IGF-II is synthesized in nonhepatic tissues. It appears that sustained and balanced growth requires the endocrine and paracrine functions of both factors.

In humans, the importance of IGFs is well illustrated by the finding that pygmies have a genetic inability to produce IGF-I, particularly at puberty, which results in their shorter stature [4]. The lack of IGF-I secretion appears to be due to a GH receptor deficiency. The human IGF-I and IGF-II gene have been mapped [5,6] but the expression is complex due to the different stages of development in different tissues. There are promoters for early development and other promoters for later development. It is this interplay of genetics, hormone secretion, regulation, and the molecular receptors for the growth factors and insulin that is the focus of this book. The volume collects together the most recent findings and insights into the molecular physiology of IGFs. It is a timely harvesting of a rapidly growing field of research.

The first section deals with the structural and regulatory aspects of IGFs. This includes the human genes which encode them, the expression in somatic cell hybrids, and the application of recombinant EGF-I. The EGFs are characterized by their biological activity and the complexity of biosynthesis and points at which EGFs can be regulated are discussed. The molecular biology of the insulin-like growth factors are covered in these chapters, bringing together the world's most preeminent researchers in growth hormone and growth factor research.

In the second section, new areas which have blossomed in this field including studies on insulin-like growth factor binding proteins are reviewed. The IGFs are present in the serum in other biological fluids and although they are water soluble, they bind to large proteins. These insulin growth factor binding proteins (IGFBP) are unrelated to be the IGF receptor proteins. Four IGFBPs have been cloned from human and rat tissues. It appears likely that more may be discovered in the future. It is still a mystery what functional role these binding proteins play. They may protect tissue from the insulin activity of the insulin-like peptides or they may enhance the actions of IGFs and their binding ability to receptors. Whatever role they play, binding proteins are upregulated in the many conditions which cause growth retardation. The latest data is reviewed in these chapters and the molecular analysis of the binding proteins are published here for a variety of different tissues, including those found in the brain ventricles.

Insulin resistance as a clinical entity has stimulated intense interest in insulin and insulin-like receptors. The importance of receptors for binding to insulin and IGFs is covered in the third section of this book. Molecular biology has improved the picture that we now have of membrane receptors and the structural maps of these receptors have been revealed. Several differences appear between the maps of the insulin, IGF-I and the IGF-II/cation independent mannose-6-phosphate receptors. Also, the highly relevant glucose transporter gene family is included in this book. The importance of these receptors to muscle development, vascular tissue in normal and diabetic states, and during embryogenesis is analyzed by the chapters in this section.

The final section is devoted to insulin and IGF in the central nervous system. Like nerve growth factor, IGFs appear to have neurotrophic actions, particularly on cholinergic cells. It is suggested that since these cells are impaired in Alzheimer's disease, IGF-I may have a previously unsuspected role in the disease. IGF-II mRNA is reported to be expressed on choroid plexus and the leptomeninges of the CNS. IGF-I is expressed in glial and neuronal cultures. The two components of brain tissue, glia and neurons have proved to be accessible in cell culture for the study of insulin and IGF receptors and the early response genes, also known as oncogenes. The early response genes appear to be the key to understanding how the peptide hormones act, after stimulating the membrane surface receptors and activating second messengers. Through these genes they influence the promoter sequence of different genes which lead to the suppression or enhancement of transcription, and thus regulate gene expression. Both transient and prolonged effects of expression in neurons in glia can be controlled by the super family of oncogenes. The unfolding picture of IGF action in the central nervous system is covered in this section and represents the most complete source of work in this area that is currently available.

For an up-to-date volume covering all the latest developments in insulin-like growth factors, Mohan K. Raizada and Derek LeRoith have put together this book from the third of three conferences they have organized in the mild winter climate of Gainesville, Florida. The first conference in 1987, was on insulin and insulin-like peptides, emphasizing the role of insulin receptors in the brain [7]. The second conference in 1989, was on the molecular and cellular biology of insulin-like growth factors and their receptors [8]. The present book is based on the 1990 conference. Between each conference more advances have been made and

"*Molecular Biology and Physiology of Insulin and Insulin-like Growth Factors*", brings insulin and insulin-like growth factors back together again in a most comprehensive review of their molecular biology and physiology. As an observer of the field, I am full of admiration for the achievements of all the authors in this volume and the incredible progress that has been made in the last few years. This book will be an essential source for physicians, nutritionists, and scientists, dealing with the molecular and cellular physiology of growth and growth retardation. This includes a widening circle of clinical states in which EGFs are implicated Among these are diabetes, renal failure, steroid excess, lung and breast cancer, disorders of nutrition and metabolism and disturbances of the brain which may be involved in Alzheimer's disease, Down's syndrome and megalencaphaly.

M. Ian Phillips, D.Sc.
Professor and Chairman
Department of Physiology
University of Florida

REFERENCES

1. McConaghey, P. and Sledge, C.G. Production of sulphation factor by the perfused liver. *Nature* 225:1249, 1970.

2. Rinderknecht, R. and Humbel, R.E. The amino acid sequence of human insulin-like growth factor I and its structural homology to proinsulin. *J. Biol. Chem.* 253:2769-2776, 1978.

3. Rinderknecht, E. and Humbel, R.E. Primary structure of human insulin growth factor II. *FEBS Letter* 89:283-286, 1978.

4. Merimee, T.J., Zapf, J. and Froesch, E.R. Insulin-like growth factors in pygmies and subjects with pygmy tract: Characterization of the metabolic actions of IGF-I and IGF-II in man. *J. Clin. Endocrinol. and Metab.* SS:1081-1087, 1982.

5. Jansen, M., Van Schaik, F.M.A., Ricker, A.T., Bullock, B, Woods, K.H., Gabbay, D.E., Nussbaum, A.L. Sussenbach, J.S. and Van den Brande, J.L. Sequence of cDNA^R encoding human insulin-like growth factor I precursor. *Nature* 306:609-611, 1983.

6. Jansen, M., Van Schaik, F.M.A., Van Tol, H. Van den Brande, J.L. and Sussenbach, J.S. Nucleotide sequences of cDNAs encoding precursors of human insulin-like growth factor II (IGF-II) and a variant of IGF-II. *FEBS Letter* 179:243-246, 1988.

7. Raizada, M., Phillips, M.I. and LeRoith, D. Insulin, insulin-like growth factors and their receptors in the central nervous system. Plenum Press, New York, NY, 1987.

8. LeRoith, D. and Raizada, M. Molecular and cellular biology of insulin-like growth factors and their receptors. Plenum Press, New York, 1989.

CONTENTS

SECTION I: INSULIN-LIKE GROWTH FACTORS I AND II

SECTION II: INSULIN-LIKE GROWTH FACTOR BINDING PROTEINS

SECTION III: RECEPTORS FOR INSULIN, IGF-I AND II

SECTION IV: INSULIN, IGFs AND THE CENTRAL NERVOUS SYSTEM

STRUCTURAL AND REGULATORY ASPECTS OF THE HUMAN GENES ENCODING IGF-I AND -II

J.S. Sussenbach, P.H. Steenbergh, E. Jansen, P. Holthuizen, D. Meinsma+, M.A. van Dijk, and T. Gloudemans

Laboratory for Physiological Chemistry and +Dept. of Pediatrics, State University of Utrecht, Utrecht, The Netherlands

INTRODUCTION

Insulin-like growth factors (IGFs) play an important role in regulation of growth. Of the two major forms of IGFs, IGF-I is important for postnatal growth, whereas IGF-II is involved in fetal growth and development. These factors are synthesized in many different tissues and exert their physiological functions probably through an endocrine as well as an autocrine/paracrine mechanism of action. During the last years it has become clear that the autocrine/paracrine mechanism of action is probably more important than originally foreseen. Elegant studies of Han and others have revealed that connective tissues play an important role in production of IGF-I, but that the peptide subsequently accumulates in the adjacent tissues[1,2]. The hypothesis has been formulated that IGF levels in plasma regulate the overall growth status, while locally produced IGFs are important for cell proliferation in specific tissues (for reviews see 3,4). If this assumption is correct, balanced growth requires that the local production of IGFs is delicately controlled in a tissue-specific manner. We have established the gene structure for human IGF-I and -II and investigated the mechanisms involved in the expression of these genes.

STRUCTURE OF THE IGF-I GENE

The first step in the characterization of the human IGF-I gene was the isolation of an IGF-I cDNA from an human adult liver cDNA library by Jansen et al.[5]. Nucleotide sequence analysis revealed that this cDNA (pIGF-Ia) coded for an IGF-I precursor consisting of a signal peptide (25 or 48 amino acids long), mature IGF-I (70 amino acids) and a trailer peptide (35 amino acids). The structure of the precursor indicates that release of IGF-I requires extensive post-transcriptional processing. Rotwein isolated from a human adult liver cDNA library an IGF-I cDNA (pIGF-Ib) which was for the main part identical to the Jansen cDNA, but differed at the 3'-end[6]. This cDNA encoded an IGF-I precursor with a different trailer peptide (77 amino acids). Analysis of the IGF-I gene revealed that pIGF-Ia and pIGF-Ib correspond to mRNAs, which arise by alternative splicing. The gene consists of 5 exons numbered 1 through 5 of which exons 4 and 5 are alternatively expressed. Employing pIGF-Ia as a probe the expression of the IGF-I gene at the RNA level was investigated in detail. Northern blotting revealed the existence of mRNAs of 7.6, 1.4 and 1.1. kilobases (kb), respectively[7,8]. Also minor species of various sizes could be detected.

Molecular Biology and Physiology of Insulin and Insulin-Like Growth Factors
Edited by M.K. Raizada and D. LeRoith, Plenum Press, New York, 1991

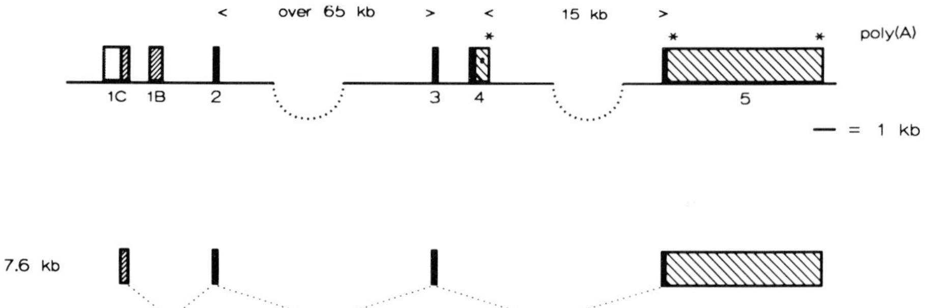

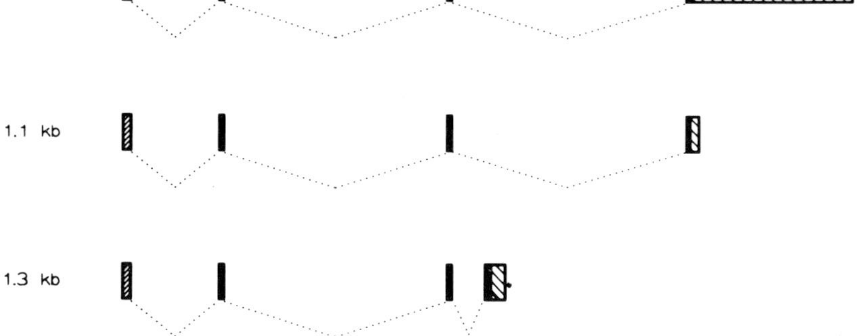

Fig.1. Structure of the human IGF-I gene and corresponding mRNAs. The black boxes indicate the coding regions. The asterisks represent polyadenylation sites. The exons are numbered.

Since the length of the cDNA was around 1 kb, it was obvious, that an important part of the IGF-I gene was still not identified. Lund and coworkers have shown, that the major missing part of the gene must be located at the 3′-end of the gene[9]. Therefore, we isolated fragments from an exon 5 containing cosmid and hybridized these to polyA$^+$ enriched RNA derived from human liver. For the isolation of polyA$^+$ enriched RNA, total liver RNA was first incubated with biotinylated oligodT. Subsequently the mRNA-biotinylated oligodT was isolated with the help of avidin-coupled paramagnetic particles. The mRNA obtained is free of rRNA and hybridizes very well in Northern blot analyses. Hybridization of the Northern blots with ^{32}P-labeled IGF-I cDNA showed only mRNAs of 7.6, 1.4 and 1.1 kb, respectively, while hardly any minor species could be detected. These latter minor RNAs probably represent degraded IGF-I mRNAs, which are lost during the purification procedure. Analysis of the IGF-I mRNA revealed that the missing 3′-part of the gene is contained within exon 5. Nucleotide sequence analysis of this part of the gene revealed that this exon is about 6000 bp long. In the nucleotide sequence several putative polyadenylation signals are found. Probably only two signals are actually used leading to the production of 7.6 and 1.1 kb mRNA, respectively.

As mentioned above exons 4 and 5 can be employed alternatively. A genomic probe encompassing exon 4 was hybridized to liver mRNA. This probe only hybridized to 1.4 kb mRNA indicating that this exon is only present in that particular mRNA and not in the 7.6 and/or 1.1 mRNA species. This is in agreement with the data of Höppener et al.[10] and Gloudemans et al. (personal communication), but differs from the results of Rotwein[6].

As for the 5′-end of the gene, analysis of the rat IGF-I gene has indicated that the gene contains multiple leader exons designated 1A, 1B and 1C, yielding mRNAs with different 5′-ends[11,12]. In rat liver, the major leader 1C is found in about 80% of the IGF-I messengers. The remaining 20% contains leaders 1A and 1B. Nucleotide sequence analysis of rat IGF-I cDNAs has revealed that exon 1A is linked to the 3′-terminal part of leader 1B. We have investigated the structure of the

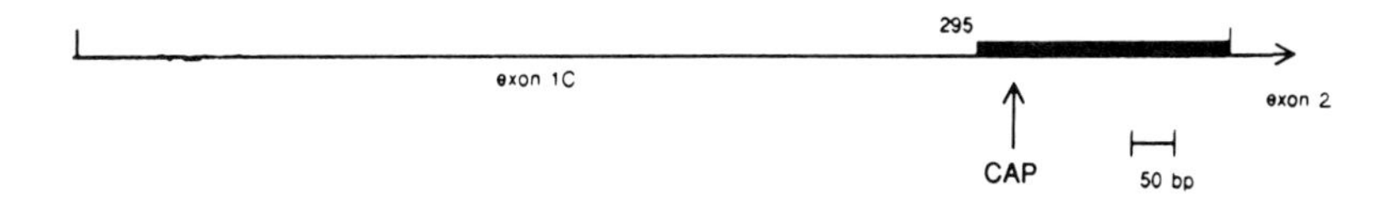

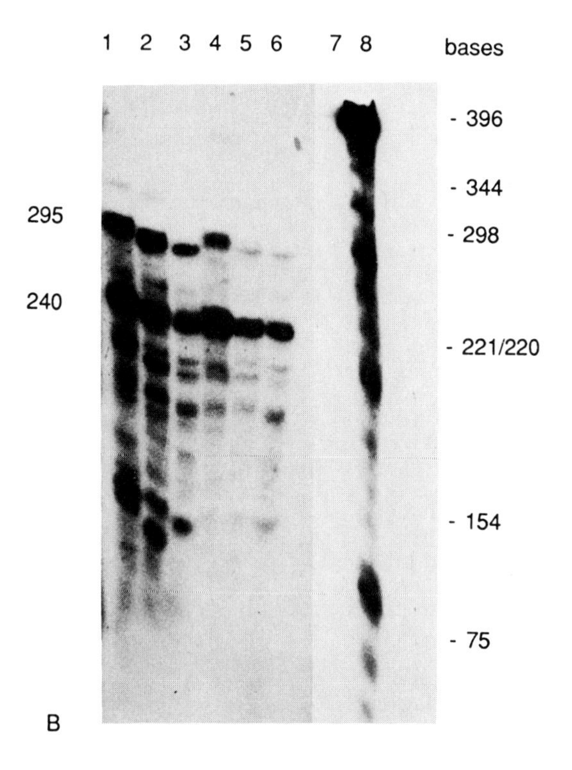

Fig.2. RNase protection assay of the 1C leader exon. A. Schematic presentation of the 1C leader exon. The probe is indicated by the thick line. RNase protection of 50 μg total RNA from human adult liver (lanes 1-3) and 5 μ total RNA from uterus (lanes 4-6). Lane 7, probe alone; lane 8, probe without RNase. The blot was hybridized with 200,000 cpm ^{32}P-labeled Riboprobe. RNase digestion was performed with 10 U/ml RNase T1 and 1, 10 or 40 μg/ml RNase A (lanes 1-3 and 4-6, respectively).

human IGF-I leader exons employing the polymerase chain reaction (PCR) and RNase protection assays.

Analysis of different cDNAs isolated from a human adult liver cDNA library has revealed that exon 1C is at least 227 nucleotides long. To determine the precise position of the cap site, a ^{32}P-labeled RNA fragment was hybridized to total RNA isolated from human liver and uterus and subsequently subjected to RNase A and T1 digestion. The products were characterized by gel electrophoresis (Fig.2). Interestingly two protected fragments can be detected, one is 240 nucleotides long, while a fainter band in the autoradiograph represents the completely protected probe of 295 nucleotides. From the smaller protected fragment it can be derived, that exon 1C is 240 nucleotides long. The presence of the other protected fragment of 295 nucleotides suggests that also a longer version of exon 1C exists. The identification of the upstream 5′-end is hampered by the low abundance of the corresponding RNA. However, employing PCR we were able to map the upstream start site between 1161 and 1144 nucleotides from the 3′-end of exon 1C. The results described indicate that exon 1C might have two different cap sites located about 900 nucleotides apart. Starting from the downstream cap site (cap 2), calculation of the theoretical length of IGF-I mRNA from the total length of the different exons plus a polyA tail of 200 residues yields values of 7.5, 1.3 and 1.1 nucleotides long, depending on the polyadenylation signals used. This fits very well with the observed IGF-I mRNA species of 7.6, 1.4 and 1.1 kb long. Employment of the upstream cap site (cap 1) would yield mRNAs of 8.5, 2.3 and 2.0 kb long, respectively. However, mRNAs of these lengths have not been observed yet. This raises the question whether cap 1 is actually used for the production of IGF-I mRNA. For the rat, in addition to the 1C exon, two other leaders have been detected: 1A and 1B. To establish whether similar exons are also present in the human gene Southern blot analysis was performed with a cosmid containing a 40 kb human genomic DNA insert encompassing exons 1C and 2. As a probe a cDNA fragment encoding the rat 1B exon was used. This experiment revealed that indeed homologous sequences were present in the cosmid. Using digestions with different restriction endonucleases the homologous region was mapped more precisely and subsequently subjected to nucleotide sequence analysis. This analysis revealed that the rat and human sequences from this region show a high degree of homology. The next question was whether this region is indeed expressed at the transcriptional level. One of the major obstacles to answer this question is the low abundance of IGF-I mRNA in human liver. In comparison, in rat liver the IGF-I mRNA levels are at least ten times higher than in human liver. Since the analysis of rat IGF-I mRNA had already shown that the exon 1B containing species represent only a few percent of total IGF-I mRNA, it was expected that the detection of the human counterpart by Northern blot analysis would almost be impossible. This indeed appeared to be the case. Therefore, again we employed PCR to show the presence of exon 1B in human mRNA. Total human liver RNA was incubated with reverse transcriptase employing an exon 2 primer to make cDNA copies of the 5′-ends of IGF-I mRNAs. These single-stranded cDNA copies were used in a PCR reaction using a second more upstream exon 2 primer and different oligonucleotides derived from upstream parts of the putative 1B exon. Indeed, products of the expected length can be detected indicating the presence of exon 1B in human IGF-I RNA. With the help of different primers from the putative 1B exon region we have established that human exon 1B is between 750 and 900 nucleotides long. To obtain an impression of the relative amount of exon 1B RNAs we have performed a semi-quantitative experiment in which the amounts of PCR products with exon 1B primers were compared with the amount of products obtained with exon 1C primers employing different amounts of polyA$^+$ RNA. This analysis indicates that probably in only 3% of total human IGF-I mRNA exon 1B is present.

In rat liver, exon 1A represents about 20% of total IGF-I RNA. Unfortunately, the sequence established for rat exon 1A is too short to investigate whether a similar exon is present in the human gene. To investigate the possibility of the existence of a human analog of exon 1A, we have established the relative amounts of exon 1C and non-exon 1C containing IGF-I mRNAs in human liver employing RNase protection assays. The result of that analysis indicates that the non-exon 1C fraction amounts to about 20% of total IGF-I mRNA. Since the exon 1B containing fraction is only about 3% of total IGF-I RNA, it is likely that indeed a human exon 1A-like exon exists.

STRUCTURE OF THE HUMAN IGF-II GENE

From the analysis of different IGF-II cDNAs we originally proposed a structure for the human IGF-II gene consisting of 8 different exons (in Fig.3 exons 1-3 and 5-9) and three different promoters (assigned as P1, P3 and P4)[13,14,15]. We have shown that these promoters are expressed in a tissue-specific and development-dependent way. Promoter P1 is only expressed in adult human liver and has no counterpart in the

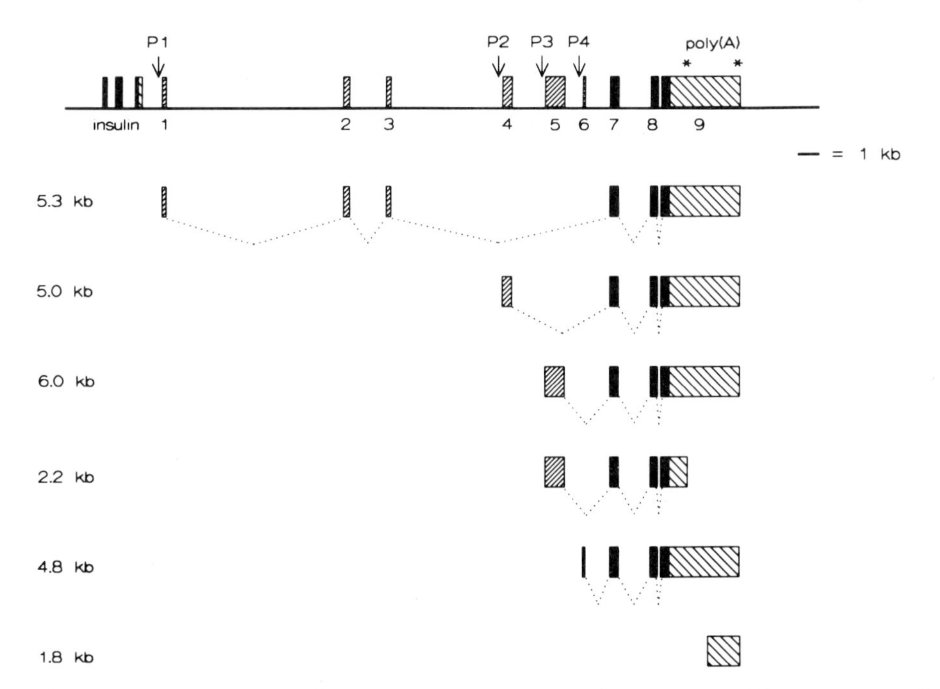

Fig.3. Structure of the human IGF-II gene and corresponding mRNAs. The black boxes indicate the coding regions. The asterisks represent polyadenylation sites. The exons are numbered.

rat IGF-II gene. Promoters P3 and P4 are expressed in all fetal tissues tested and in adult non-hepatic tissues. Yamamoto and coworkers showed that in rat analogs of P3 and P4 exist and they detected an additional promoter designated rat P1[16]. Employing Southern blotting and nucleotide sequence analysis we have established that an analog of rat P1 (P2) and of the corresponding exon (exon 4) exists in the human gene located between exons 3 and 5 (Holthuizen et al., in press) (Fig.3). Northern blotting of total RNA from fetal liver and leiomyosarcoma RNA has shown that exon 4 is indeed expressed in IGF-II mRNA. Probes derived from exon 4 hybridize specifically with an mRNA species of 5.0 kb. These results bring the total number of human IGF-II exons to nine and the number of promoters to four. Attempts to detect promoter activity in transient experiments employing hepatoma cell lines of DNA fragments upstream of exon 4 were not successful until now. Yamamoto and coworkers also were not able to show promoter activity in their cell lines[17]. Since the corresponding mRNA can be detected in both species, the failure to show promoter activity is probably due to the cell lines employed for transient expression and the extremely low promoter activity.

During our analysis of the human IGF-II gene we have detected a polyadenylated 1.8 kb RNA which consists only of the 3'-terminal region of exon 9[14,18] (Fig.3). The presence of one of the leader exons 1, 4, 5 or 6 could not be detected in this RNA. One of the obvious questions concerning the 1.8 kb RNA is its origin. In this respect, it is noteworthy that evidence was obtained for co-expression of this RNA species with the IGF-II encoding mRNAs. Two possible routes of production of the 1.8 kb RNA were considered: (i) a distinct promoter is present in front of the region encoding the 1.8 kb RNA which is coregulated with the IGF-II promoters, and (ii) the 1.8 kb RNA arises as a cleavage product of IGF-II mRNA. To discriminate between the two possibilities, constructs were made in which truncated IGF-II genes were placed under control of cytomegalovirus (CMV) immediate early enhancer-promoter elements. These constructs were used for transfection

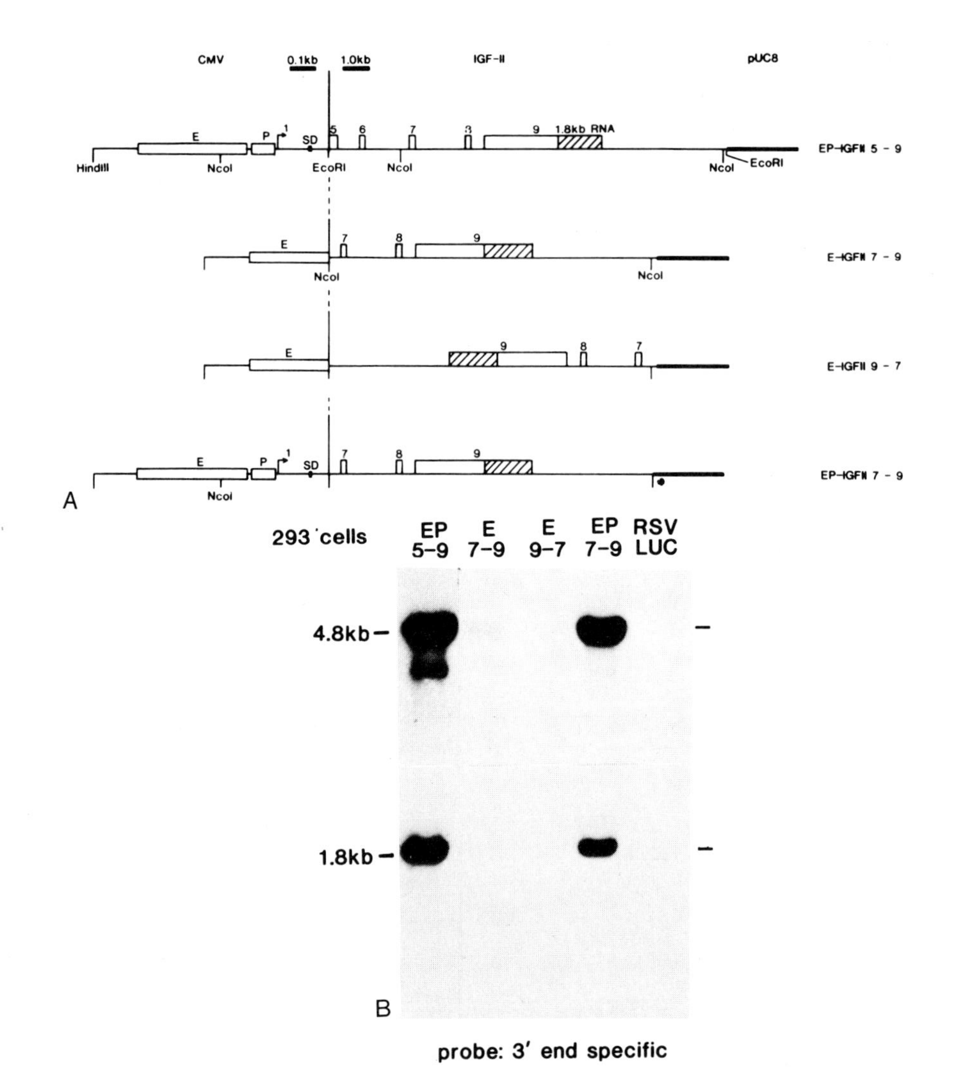

Fig.4. A. Expression of 1.8 kb RNA. A. Constructs used for the expression of 1.8 kb RNA. The IGF-II exons present in each construct are numbered. The region representing the 1.8 kb RNA is shaded. E and P indicate the CMV enhancer and promoter regions, respectively. SD, splice donor site in the CMV DNA. B. Northern blot analysis of total RNA isolated from 293 transfected cells. The cells were transfected with the indicated plasmids. RSV LUC, an expression plasmid containing the luciferase gene under control of the RSV promoter. The blot was hybridized with a ^{32}P-labeled probe derived from the 3′ end of exon 9.

of human 293 cells, and the expression of the 1.8 kb RNA and IGF-II mRNAs was analyzed by Northern blotting. A CMV enhancer linked to an IGF-II gene fragment containing exons 7,8 and 9 could not stimulate expression of the 1.8 kb RNA (Fig.4, E-IGF-II 7-9, E-IGF-II 9-7). However, when the complete CMV enhancer-promoter combination was present, strong expression of the 1.8 kb RNA and of a 4.8 kb mRNA containing CMV exon 1 spliced to the IGF-II exons 7, 8 and 9 (EP-IGF-II

7-9) was obtained. Similar observations were made when a larger IGF-II fragment containing exons 5-9 was coupled to the CMV enhancer-promoter in the two possible orientations. The sense construct (Fig.4, EP-IGF-II 5-9) yielded the expected 4.8 kb mRNA and the 1.8 kb RNA, while the antisense construct EP-IGF-II 9-5 gave no detectable RNA expression (result not shown). These experiments suggest that the 1.8 kb RNA does not arise from transcriptional activation by a distinct promoter. Instead, expression of the 1.8 kb RNA seems to depend on the presence of a larger IGF-II transcript from which it is most likely cleaved off in a specific way. Analysis of actinomycin-treated Hep3B cells showed that the typical IGF-II mRNAs (6.0 kb originating from P3 and 4.8 kb derived from P4) as well as the 1.8 kb RNAs have half-lives longer than 10 hours. An important question is whether the 1.8 kb RNA also is translated into protein. In an *in vitro* translation system it was found that 1.8 kb RNA may yield a 8.3 kDa protein. However, the nucleotide sequence of the 1.8 kb RNA does not contain an open reading frame encoding a protein of this size. It is known, that in rat cells a similar 1.8 kb RNA associated with the IGF-II gene is present[18]. Comparison of the human and the rat 1.8 kb RNAs does not show homologous open reading frames[19]. These data make a protein encoding function of the 1.8 kb RNA rather doubtful. Currently, we favor the hypothesis, that specific cleavage of IGF-II mRNAs plays a role in regulating the stability of these mRNAs. As shown above mRNAs containing the 3′-part of exon 9 are very stable. It is feasible, that removal of the 3′-end leads to destabilization of IGF-II 5′- and coding regions. In this way, by regulating the specific cleavage of IGF-II mRNAs yielding the accumulation of the 1.8 kb RNA the cell is able to regulate the levels of IGF-II mRNAs. Cleavage might be controlled by the secondary and tertiary structure of the 3′-end. The rat and human IGF-II RNAs may have similar tertiary structures of the 3′-ends.

ANALYSIS OF THE IGF-II PROMOTERS

To gain more insight in IGF-II gene expression, an analysis of promoter elements and nuclear factors involved in the transcriptional regulation of the gene was initiated, employing transient expression of IGF-II promoter constructs and DNase I footprinting. To establish which elements in the promoter regions are involved in transcriptional regulation we made a series of constructs carrying different IGF-II promoter fragments fused to the bacterial chloramphenicol acetyltransferase (CAT) gene or the luciferase (Luc) gene and tested these in transient expression studies. Hep3B cells or HeLa cells were cotransfected with pSV2Apap encoding the placental alkaline phosphatase (PAP) gene or a ß-galacosidase (ßgal) expression plasmid to measure the transfection efficiency. After 48 h cell extracts were prepared and tested for CAT, Luc, PAP or ßgal activity depending on the plasmids employed in the transfection experiments (TABLE 1). The results indicate that in P1 negatively cis-acting sequences are present upstream of -184. The region between -159 and -147 appears to be very important for promoter activity.

The proximal region of promoter P3 (-184/+134) contains a TATA-box, a CCAAT-box and several Sp1 sites. Our analysis indicates that this proximal region does not contain sufficient information for maximal transcription and that regions located further upstream are required. The areas responsible for this increase are located downstream of -1300. The additional 1000 bp in P3 has little effect on the activity of P3 in HeLa cells, but in Hep3B cells the addition of this region lowers promoter activity approximately 4-fold. The stimulatory elements of the upstream region between -1300 and -184 on P3 are distributed over the entire region. The observed differences in promoter activity in Hep3B and HeLa cells indicates that P3 contains ubiquitous as well as tissue-specific components. When in construct -565/+134 a 55 bp SacII fragment (-98/-44) containing the CCAAT-box is deleted, the activity of the promoter is severely reduced in all cell lines, indicating the importance of this element for the activity of promoter P3.

TABLE 1. Relative promoter activity

Promoter fragment	Relative activity	
	Hep3B	HeLa
P1		
-889/+52	100	100
-338/+52	98	79
-204/+52	128	121
-184/+52	154	101
-159/+52	108	93
-147/+52	60	17
-54/+52	32	5
P3		
-2300/+134	100	100
-1300/+134	397	76
-565/+134	45	28
-290/+134	18	28
-184/+134	1	3
-565/+134 without -98/-44	6	17
P4		
-1075/+103	100	100
-677/+103	51	48
-450/+103	39	52
-52/+103	28	2

Analysis of promoter P4 revealed a straight decrease of activity upon truncation. Interestingly construct -52/+103, which contains only the first 52 nucleotides of promoter P4 including the TATA-box, shows hardly any activity in HeLa cells, but is active in Hep3B. Maximal promoter activity, however, requires the presence of upstream elements. Evans et al. (13) have shown that rat promoter P3 (homologous to human P4) only requires the first 128 bp 5′ of the cap site containing the TATA-box and several GC-motifs for maximal promoter activity. The human promoter P4 displays slightly different properties. Starting from the EcoRI-site at -1075 deletion to -450 results in a 2-fold decrease in CAT activity, indicating that this region contributes to the activity of P4. In promoter P3 of the rat this region is not required for maximal promoter activity. In the human promoter P4 the region between -450 and -52 contributes considerably to the activity of this promoter; it cannot be excluded that, as in the rat, the essential elements are located between -128 and -52.

To extend the identification of the nucleotide sequences involved in transcriptional regulation of the IGF-II promoters we investigated the binding of nuclear factors to promoter sequences employing DNase I footprint analysis. For this analysis, fractionated nuclear extracts obtained from HeLa cells were incubated with DNA fragments spanning the proximal regions for each of the three promoters as probes. Using a probe covering the first 188 basepairs of promoter P1 we detected two distinct footprints with HeLa nuclear extracts. These were named Proximal Element 1-1 and 1-2 (PE 1-1 and PE 1-2) (result not shown). PE 1-1 covers approximately 26 basepairs close to the cap-site of exon 1 and contains a potential Sp1 binding site. PE 1-2 covers a 38 bp region located upstream of PE 1-1. Its sequence does not show strong homology to any known transcription factor binding site.

Footprint analysis of P3 employing HeLa cell nuclear extracts showed footprints over the CCAAT-box (PE 3-1: -89/-64) and the Sp1-sites (PE 3-2:-126/-102) as well as two additional footprints located further upstream (PE 3-3:-143/-126, and PE 3-4: -190/-169) (Fig.5).

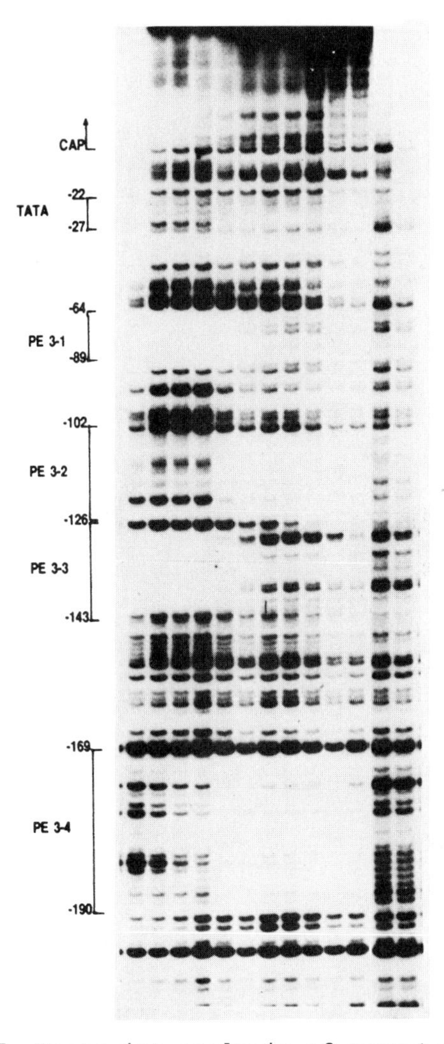

Fig.5. Footprint analysis of promoter P3.

In promoter P4 two footprints were detected (PE 4-1: -28/15, and PE 4-2: -351/-336). In the region between PE 4-1 and PE 4-2 we could not detect additional footprints.

In conclusion, the structure of the gene and the heterogeneity of its mRNAs suggests a complex mechanism of regulating IGF-II gene expression. Further characterization of the components involved in the transcriptional regulation will provide more insight in this mechanism and its role in the development of an organism. Currently we are investigating in an *in vitro* transcription system the binding proteins which are required for transcription.

ROLE OF IGFs IN TUMORIGENESIS

Polypeptide growth factors not only play an important role in

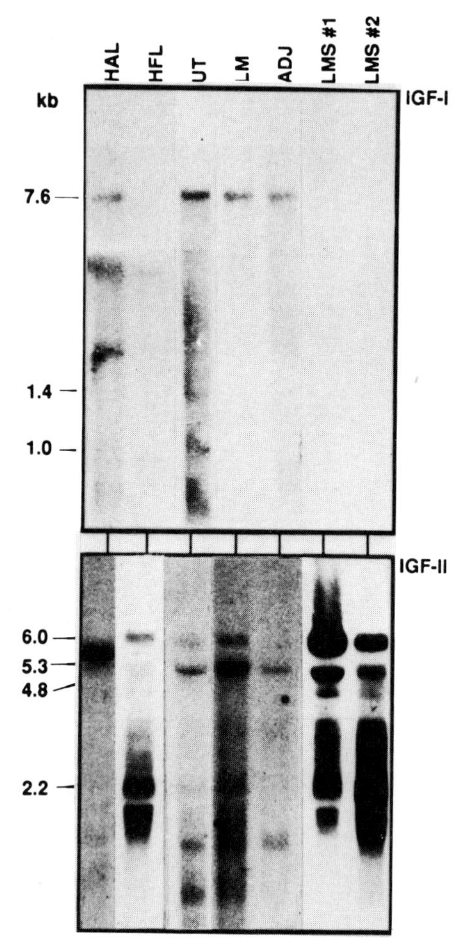

Fig.6. Northern blot analysis of total RNA from human adult liver (HAL), human fetal liver (HFL), normal uterus myometrium (UT), leiomyoma (LM), non-tumorous myometrium directly adjacent to the leiomyoma (ADJ), two leiomyosarcomas from small intestine (LSM#1) and the abdominal wall (LSM#2), respectively. Top panel, hybridization with ^{32}P-labeled IGF-I cDNA. Exposure time 5 days. Bottom panel, hybridization with ^{32}P-labeled IGF-II cDNA. Exposure time 9 days, except for the lanes HFL, LSM#1 and LSM#2, which were exposed for 16 h. The sizes of the major hybridizing RNAs are indicated in kb.

growth and development but probably also in tumor formation. The findings that the oncogene v-*sis* codes for a growth factor (PDGF) or that v-*erb*-B codes for a truncated form of the EGF receptor support this. Several observations indicate that IGFs also play an important role in tumorigenesis. Autocrine and paracrine growth stimulation of cell growth in several tumors has been reported for IGF-I and IGF-II.

An interesting model system to study the role of IGFs in tumorigenesis is constituted by smooth muscle tumors since (i) benign (leiomyoma) and malignant (leiomyosarcoma) smooth muscle tumors can be distinguished by the pathologist and (ii) in an initial study higher levels of IGF-I and IGF-II mRNAs were detected in this type of tumors than in normal smooth muscle tissue[10].

The levels of IGF-I or IGF-II RNA in smooth muscle tumors were investigated by RNA blot analysis. Total RNA was isolated from a panel of smooth muscle tissues: normal uterus myometrium (16 samples),

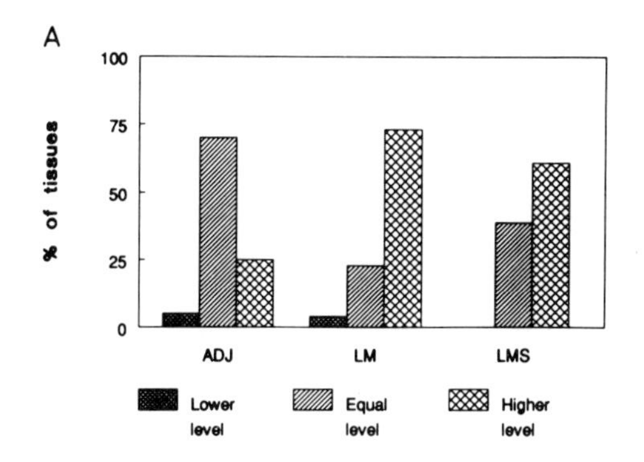

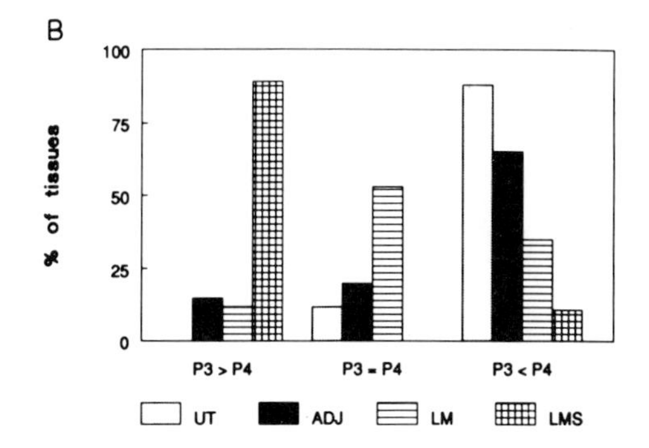

Fig.7. Expression of IGF-II RNA. A. Abundance of IGF-II 6.0 kb mRNA in non-tumorous myometria adjacent to leiomyoma (ADJ)(n=20), leiomyomas (LM)(n=26), and leiomyosarcomas (LMS)(n=18) compared to normal uterus myometrium. The percentage of samples in which the level of IGF-II 6.0 kb mRNA is higher, equal, or lower, respectively than in normal myometrium is indicated. B, Abundance of IGF-II 6.0 kb mRNA (P3) compared to IGF-II 4.8 kb mRNA (P4) in the same tissues. The percentage of samples in which the level of 6.0 kb mRNA (P3) is higher, equal, or lower, respectively, than 4.8 kb mRNA (P4) is shown.

leiomyomas from the uterus (26 samples) and in 20 cases myometrium directly adjacent to the leiomyoma, and from leiomyosarcomas (20 samples) from other smooth muscle tissues. These RNAs were subjected to RNA blot analysis and probed with IGF-I and IGF-II specific cDNAs. Fig.6 shows an example of an RNA transfer blot containing some of these RNAs.

IGF-I. Fig.6 shows that in normal myometrium and in leiomyoma the levels of the 7.6 kb mRNA are higher than in adult liver. In myometrium adjacent to a leiomyoma the level is lower, whereas in leiomyosarcoma and in fetal liver the 7.6 kb mRNA is not detectable. Comparing the levels of IGF-I mRNA in all normal smooth muscle tissues, leiomyomas, and leiomyosarcomas studied it was found that in almost every leiomyosarcoma the IGF-I gene was not expressed, whereas in leiomyomas and normal myometrium the IGF-I gene was expressed at approximately the same level. This indicates that the IGF-I gene is repressed in malignant smooth muscle tumors.

IGF-II. The IGF-II mRNA species in smooth muscle tissues are the same as in fetal liver. In normal myometrium and leiomyomas the IGF-II gene is expressed at a low level and the 4.8 kb mRNA is the dominant messenger, while in leiomyosarcomas the IGF-II gene is expressed at very high levels and the 6.0 kb mRNA is the most abundant species as is the case for fetal liver. Comparison of the IGF-II mRNA levels in all smooth muscle tissues and tumors studied to those in normal uterus shows that the IGF-II gene becomes activated in malignant smooth muscle tumors (Fig.7). When in malignant smooth muscle tumors the levels of the IGF-II mRNAs transcribed from P3 and P4 were compared, this revealed that P3 is activated to a higher extent than P4 (Fig.7).

So, the IGF-I gene is expressed in normal smooth muscle tissue, while the IGF-II gene promoters P3 and P4 are only active at a very low level. The same expression pattern has been reported for other adult tissues[20,21]. In malignant tumors of smooth muscle a decrease of IGF-I gene expression is found. IGF-II gene expression increases in malignant smooth muscle tumors by activation of P3 and P4. In normal adult smooth muscle tissue P4 is more active than P3, while in fetal tissues P3 is by far the most active promoter. From our experiments it is clear that in tumors of smooth muscle P3 is activated predominantly with increasing malignancy. The cause for the shift from P4 to P3 usage is not clear yet, but at least two explanations are feasible. First, the activity of the promoters during tumorigenesis is altered due to changes in the levels of specific transcription factors or second, the IGF promoters might be mutated in the tumors. The second alternative seems unlikely since in many cases where IGF-II gene expression is enhanced the IGF-II gene itself appears to be normal and never aberrant IGF-II gene transcripts have been detected[22,23,24]. The methods used, however, could only detect gross gene rearrangements. The first possibility seems more likely, which is stressed by our demonstration that the human IGF-II promoters are able to bind different sets of nuclear factors. Probably the shift in IGF-II gene expression is mediated by differential expression of genes encoding regulatory proteins binding to the promoters of the IGF-II gene. How the expression of the genes encoding these factors is regulated is still obscure.

In conclusion, a role for IGF-I in smooth muscle tumor formation is not likely, since in malignant tumors IGF-I gene expression is repressed. On the other hand, the strong activation of the IGF-II gene in malignant smooth muscle tumors suggests that IGF-II plays a role in tumorigenesis. It is tempting to speculate that the shift in promoter usage in malignant tumors reflects a fetal state of differentiation of these cells.

REFERENCES

1. V. K. M. Han, A. J. D'Ercole and P. K. Lund, Cellular localization of somatomedin (insulin-like growth factor) messenger RNA in the human fetus, Science 236:193 (1987).

2. V. K. M. Han, P. K. Lund, D. C. Lee and A. J. D'Ercole, Expression of somatomedin/insulin-like growth factor messenger ribonucleic acids in the human fetus: Identification, characterization and tissue distribution, J. Clin. Endocrinol. Metab. 66:422 (1988).

3. W. H. Daughaday and P. S. Rotwein, Insulin-like growth factors I and II. Peptide, messenger ribonucleic acid and gene structures, serum, and tissue concentrations, Endocrine Reviews 10:68 (1989).

4. J. L. Van den Brande, Somatomedins on the move, Horm. Res. 32:58 (1990).

5. M. Jansen, F. M. A. Van Schaik, A. T. Ricker, B. Bullock, K. H. Woods, D. E. Gabbay, A. L. Nussbaum, J. S. Sussenbach and J. L. Van den Brande, Sequence of cDNA encoding human insulin-like growth factor I precursor, Nature 306:609 (1983).

6. P. S. Rotwein, Two insulin-like growth factor I messenger RNAs are expressed in human liver, Proc. Natl. Acad. Sci. USA 83:77 (1986).

7. M. Jansen, P. De Pagter-Holthuizen, F. M. A. Van Schaik, J. S.

Sussenbach and J.L.Van den Brande, Characterization and organization of the somatomedin/insulin-like growth factor genes, in: "Somatomedins and other peptide growth factors: Relevance to Pediatrics," R. L. Hintz and L. E. Underwood, eds. Report Eighty-Ninth Ross Conference on Pediatric Res. Ross Laboratories, Columbus,Ohio, pp 12 (1985).

8. M. Jansen, P. De Pagter-Holthuizen, J. W. M. Höppener, J. S. Sussenbach and J. L. Van den Brande, Analysis of cDNA and mRNA encoding the somatomedin precursors, in: "Diabetes 1985'" M. Serrano-Rios and P. J. Lefebvre,eds., Elsevier Science Publishers B.V., Amsterdam, pp 289 (1986).

9. P. K. Lund, E. C. Hoyt and J. J. Van Wyk, The size heterogeneity of rat insulin-like growth factor-I mRNAs is due primarily to differences in the length of 3'-untranslated sequence, Mol.Endocrinol.3:2054 (1989).

10. J. W. M. Höppener, S. Mosselman, P. J. M. Roholl, C. Lambrechts, R. J. C. Slebos, P. De Pagter-Holthuizen, C. J. M. Lips, H. S. Jansz and J. S. Sussenbach, Expression of insulin-like growth factor-I and -II genes in human smooth muscle tumours, EMBO. J. 7:1379 (1988).

11. C. T. Roberts Jr, S. R. Lasky, W. L. Lowe and D. LeRoith, Rat IGF-I cDNAs contain multiple 5'-untranslated regions, Biochem. Biophys. Res. Commun. 146:1154 (1987).

12. W. L. Lowe, C. T. Roberts Jr, S. R. Lasky and D. LeRoith, Differential expression of alternative 5' untranslated regions in mRNAs encoding rat insulin-like growth factor I, Proc. Natl. Acad. Sci. USA 84:8946 (1987).

13. P. De Pagter-Holthuizen, F. M. A. Van Schaik, G. M. Verduijn, G. J. B. Van Ommen, B. N. Bouma, M. Jansen and J. S. Sussenbach, Organization of the human genes for insulin-like growth factors I and II, FEBS Lett. 195:179 (1986).

14. P. De Pagter-Holthuizen, M. Jansen, R. A. van der Kammen, F. M. A. Van Schaik and J. S. Sussenbach, Differential expression of the human insulin-like growth factor II gene. Characterization of the IGF-II mRNAs and an mRNA encoding a putative IGF-II-associated protein, Biochim. Biophys. Acta 950:282 (1988).

15. P. Holthuizen, The structure and expression of the human insulin-like growth factor genes, in: "Molecular and cellular biology of insulin-like growth factors and their receptors," D. LeRoith and M. K. Raizada, eds., Plenum Press, New York, pp 97 (1989).

16. T. Ueno, K. Takahashi, T. Matsuguchi, H. Endo and M. Yamamoto, A new leader exon identified in the rat insulin-like growth factor II gene, Biochem. Biophys. Res. Commun. 148:344 (1987).

17. T. Matsuguchi, K. Takahashi, K. Ikejiri, T. Ueno, H. Endo and M. Yamamoto, Functional analysis of multiple promoters of the rat insulin-like growth factor II gene, Biochim. Biophys. Acta 1048:165 (1990).

18. L. Chiariotti, A. L. Brown, R. Frunzio, D. R. Clemmons, M. M. Rechler and C.B.Bruni, Structure of the rat insulin-like growth factor II transcriptional unit: Heterogeneous transcripts are generated from two promoters by use of multiple polyadenylation sites and differential ribonucleic acid splicing, Mol. Endocrinol. 2:1115 (1988).

19. J. S. Sussenbach, The gene structure of the insulin-like growth

factor family, Progr. Growth Factor Res. 1:33 (1989).

20. J. Scott, J. Cowell, M. E. Robertson, L. M. Priestley, R. Wadey, B. Hopkins, J. Pritchard, G. I. Bell, L. B. Rall, C. F. Graham and T. J. Knott, Insulin-like growth factor II gene expression in Wilms' tumour and embryonic tissue, Nature 317:260 (1985).

21. A. E. Reeve, M. R. Eccles, R. J. Wilkins, G. I. Bell and L. J. Millow, Expression of insulin-like growth factor II transcripts in Wilms' tumour, Nature 317:258 (1985).

22. D. Yee, K. J. Cullen, S. Paik, J. F. Perdue, B. Hampton, A. Schwartz, M. E. Lippman and N. Rosen, Insulin-like growth factor II mRNA expression in human breast cancer, Cancer Res. 48:6691 (1988).

23. T. S. Su, W. Y. Liu, S. H. Han, M. Jansen, T. L. Yang-Fen, F. K. P'eng and C. K. Chou, Transcripts of the insulin-like growth factors I and II in human hepatoma, Cancer Res. 49:1773 (1989).

24. W. L. Lowe, C. T. Roberts Jr, D. LeRoith, M. T. Rojeski, T. J. Merimee, S. T. Fui, H. Keen, D. Arnold, J. Mersey, S. Gluzman, D. Spratt, R. C. Eastman and J. Roth, Insulin-like growth factor-II in nonislet cell tumors associated with hypoglycemia: Increased levels of messenger ribonucleic acid, J. Clin. Endocrinol. Metab. 69:1153 (1989).

STRUCTURAL AND FUNCTIONAL CHARACTERIZATION OF IGF-I RNA 3' VARIANTS

P. Kay Lund, Eileen C. Hoyt[+], Judson J. Van Wyk[+]
Department of Physiology and Department of Pediatrics[+] University of North Carolina at Chapel Hill, North Carolina 27599

INTRODUCTION

In rat and other mammalian species a single insulin-like growth factor I (IGF-I) gene gives rise to a heterogeneous and complex family of IGF-I mRNAs (1,2). There are two levels to the rat IGF-I mRNA heterogeneity :

1. Size heterogeneity - There are multiple IGF-I mRNAs with a given coding sequence that differ in size.

2 Coding sequence heterogeneity - All rat IGF-I mRNAs encode the same mature IGF-I but differ in coding sequences for amino-terminal or carboxyl-terminal precursor peptides that flank the mature IGF-I sequence (1,2).

This chapter will focus on studies in the rat to (i) determine the structural basis of IGF-I mRNA size heterogeneity and (ii) analyze the functional relevance of rat IGF-I mRNAs that differ in size.

SIZE HETEROGENEITY OF IGF-I mRNAs

In adult rat liver, the major site of IGF-I mRNA expression (3), IGF-I mRNAs exist as two predominant forms with estimated sizes of 7.5 -7.0 kilobases (kb) and 1.2 - 0.9 kb (3-5). There are also minor rat liver IGF-I mRNAs that are intermediate in size. Similar size heterogeneity is observed in other mammalian species (1,2). Characterized rat IGF-I cDNA and genomic sequences account for only 2 kb of the large molecular weight 7.5 -7.0 kb IGF-I mRNAs and therefore more than 5 kb of the sequence of the large IGF-I mRNAs is unknown. Elucidation of the unknown sequence in the large

Molecular Biology and Physiology of Insulin and Insulin-Like Growth Factors
Edited by M.K. Raizada and D. LeRoith, Plenum Press, New York, 1991

molecular weight mRNAs is essential for complete characterization of the transcribed portion of the rat IGF-I gene. Since the regulatory elements that control gene expression are located 5' or 3' to the transcribed portion of a gene, characterization of the unknown sequence in the 7.5 - 7.0 kb rat IGF-I mRNAs is also important for identification and characterization of the promoters and regulatory elements that control expression of the IGF-I gene.

RNase H Mapping

It was not clear if the unknown sequence in the large 7.5 - 7.0 kb rat IGF-I mRNAs lay 5' or 3' to the known sequence. A useful strategy to map the location of unknown sequences in large mRNAs is oligomer directed RNase H mapping as shown in Figure 1.

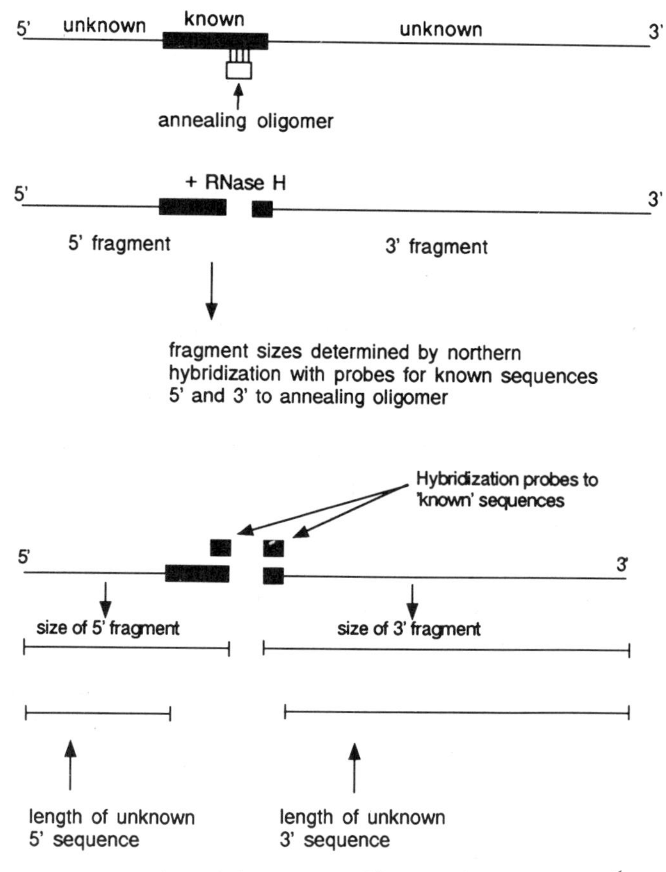

Figure 1. A schematic of the RNase H mapping strategy (see text for explanation).

A synthetic oligodeoxyribonucleotide probe (oligomer) complementary to "known" mRNA sequence is annealed with the RNA. This oligomer is termed the annealing oligomer (Figure 1). The reaction is then treated with RNase H which hydrolyzes the RNA strand of an RNA: DNA hybrid. This results in two RNA fragments, one corresponding to the entire RNA sequence located 5' to the annealing oligomer, and one corresponding to the entire RNA sequence located 3' to the annealing oligomer. The RNA fragments can then be analyzed on northern blots by hybridization with probes complementary to *known* sequences that lie 5' or 3' to the annealing oligomer. By subtracting the length of "known" 5' sequence from the length of the RNase H generated 5' fragment, the length of unknown 5' sequence is determined. By subtracting the length of "known" sequence from the size of the RNase generated 3' fragment, the length of the unknown 3' sequence is determined.

RNase H mapping of 7.5 - 7.0 kb rat IGF-I mRNAs

Using RNase H mapping we established that the large rat IGF-I mRNAs have more than 6 kb of sequence 3' to the IGF-I precursor coding sequence

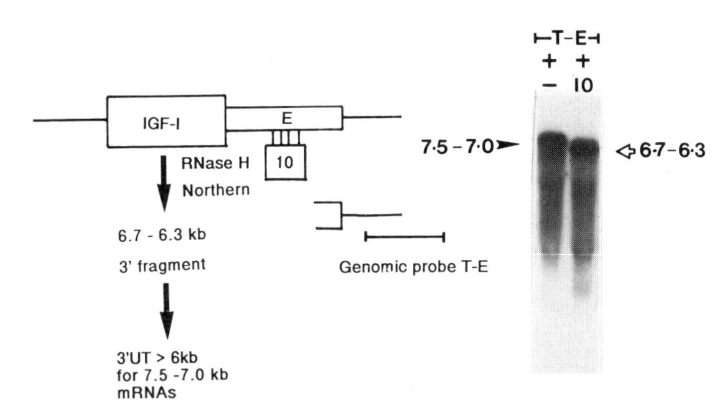

Figure 2. RNase H mapping of the 7.5 - 7.0 kb rat IGF-I mRNAs. In the schematic (left) the numbered box shows that annealing oligomer #10 is complementary to E domain coding sequences close to the translation stop codon in characterized rat IGF-I cDNA sequences. This oligomer was annealed with rat liver poly A^+ RNAs. RNase H was then used to hydrolyze the complementary RNA sequence in the 7.5 - 7.0 kb rat IGF-I mRNAs. A genomic probe (T-E) that lies 3' to the annealing probe was then used to analyze the resultant 3' fragment on a northern blot. As shown in the autoradiogram (right) RNase H cleavage of the 7.5 - 7.0 kb rat IGF-I mRNA at annealing oligomer #10 sequences generated a 6.3 - 6.7 kb 3' fragment.

(Figure 2, 8) and thus, that most of the unknown sequence in the 7.5 - 7.0 kb IGF-I mRNAs is 3' untranslated sequence (3' UT). These RNase H mapping studies also demonstrated that less than 600 bases of sequence lies 5' to IGF-I precursor coding sequence in the predominant rat IGF-I type 1/ class C mRNAs (8).

Structure of the long 3' UT in the 7.5 - 7.0 k rat IGF-I mRNAs

We isolated 3' rat IGF-I genomic and cDNA probes. By a combination of northern blot hybridization and RNase protection, we mapped the polyadenylation site(s) that results in the long 7.5 - 7.0 kb rat IGF-I mRNAs. By nucleotide sequencing, we determined the sequence of the 6354 base long 3' exon of the rat IGF-I gene. This sequence, composed of primarily 3' UT, exhibits some features of interest: The sequence is (i) AU rich, composed of 61% AU residues (ii) contains multiple $(U)_nA$ motifs and (iii) contains multiple inverted repeats.

PHYSIOLOGICAL SIGNIFICANCE OF THE SIZE HETEROGENEITY OF IGF-I mRNAs

All mammalian species studied to date exhibit size heterogeneity of IGF-I mRNAs. Our findings above demonstrate that in the rat, this size heterogeneity is due to different lengths of 3' UT. Similar studies suggest that human IGF-I mRNAs also exhibit size heterogeneity and that large human IGF-I mRNAs have long 3' UT's as in the rat (Dr. John Sussenbach, University of Utrecht, personal communication). Evolutionary conservation of size heterogeneity indicates functional significance.

Rat IGF-I mRNAs that differ in length of 3' UT exhibit different half-lives in vitro and in vivo

Accumulating evidence suggests that 3' UTs influence mRNA stability and that mRNA stability is a key determinant of the rate of synthesis of growth factors and oncogenes (9-12). Sequences implicated in mRNA destabilization or in regulating mRNA stability include AUUUA, $(U)_nA$ sequences and inverted repeats (9-12). The long 3' UT of the 7.5 - 7.0 kb rat IGF-I mRNAs contains such sequence motifs raising the possibility that the large rat IGF-I mRNAs differ in stability from the smaller mRNAs. We tested this possibility in vitro and in vivo (13). We used the reticulocyte lysate cell-free translation system to analyze the decay of rat liver IGF-I mRNAs in vitro in the absence of ongoing transcription. We used a hypophysectomized rat model to examine the in vivo half life of rat liver IGF-I mRNAs. In this model, liver IGF-I mRNAs are very low in abundance but are rapidly induced in response to growth hormone (GH) (13). We therefore gave hypophysectomized rats a single injection of GH and assayed liver IGF-I mRNAs at 1-32 h after the injection. The half-life of IGF-I mRNAs was calculated as the time taken to decay from maximum abundance induced by the GH injection. In both the in vitro and in vivo systems we found that the half life of the 7.5 - 7.0 kb IGF-I mRNAs was much shorter than that of the smaller IGF-I mRNAs (13). These findings suggest that the 7.5 - 7.0 kb IGF-I mRNAs are less stable than the smaller forms and that the size heterogeneity of IGF-I mRNAs provide a mechanism for post-transcriptional down regulation of IGF-I synthesis at the level of mRNA stability.

It is noteworthy that using a different approach Lowe et al (14) found no evidence for differences in half life of 7.5 - 7.0 and 1.2 - 0.9 kb rat IGF-I mRNAs. Lowe et al (14) treated rat GH_3 cells with the transcription inhibitor actinomycin D and observed parallel decay of large and small IGF-I mRNAs. Recent studies indicate however, that transcription inhibitors may actually stabilize short lived mRNAs such as c fos and may therefore not be optimal for studies of mRNA half life (12). The mechanism by which transcription inhibitors stabilize mRNAs is not known but may involve inhibition of the synthesis of proteins involved in mRNA destabilization or non-specific effects such as intercalation into RNA and disruption of sequences involved in destabilization. In preliminary studies we have obtained evidence for the latter mechanism since addition of actinomycin D to the cell-free reticulocyte lysate translation system resulted in prolongation of the half-life of the 7.5 - 7.0 kb rat IGF-I mRNAs (unpublished observations).

Physiological relevance of the different half-lives of 7.5 - 7.0 and1.2 - 1.0 kb rat IGF-I mRNAs

Recent studies reveal that protein deprivation induces preferential down regulation of the 7.5 - 7.0 kb rat liver IGF-I mRNAs compared with smaller IGF-I mRNAs probably by a post-transcriptional mechanism (15-17). Since protein deprivation results in decreased synthesis of IGF-I, these studies suggest that preferential destabilization of the 7.5 - 7.0 kb IGF-I mRNAs may represent a mechanism for post-transcriptional down regulation of IGF-I synthesis. Some non-hepatic tissues express primarily the large molecular weight IGF-I mRNAs (3,5). Post-transcriptional regulation of mRNA stability may therefore represent a particularly important mechanism for down regulation of IGF-I synthesis in these tissues.

Acknowledgements

This work was supported by NIH grant AM1022. We gratefully acknowledge the work and input of Jessie Hepler.

References

1. Hepler JE, Lund PK (1990) Molecular biology of the insulin-like growth factors: Revelance to nervous system function. Molecular Neurobiology: In Press.

2. Daughaday WH, Rotwein P (1989) Insulin-like growth factors I and II. Peptide, messenger ribonucleic acid and gene structures, serum, and tissue concentrations. Endocrine Reviews 10: 68-91.

3. Lund PK, Moats-Staats BM, Hynes MA, Simmons JG, Jansen M, D'Ercole AJ, Van Wyk JJ (1986) Somatomedin-C/insulin-like growth factor-I and insulin-like growth factor-II mRNAs in rat fetal and adult tissues. Journal of Biological Chemistry 261: 14539-14544.

4. Casella S, Smith E, Van Wyk JJ, Joseph D, Hynes M, Hoyt E, Lund P (1987) Isolation of rat testis cDNA encoding an insulin-like growth factor I precursor. DNA 6: 325-330.

5. Hoyt E, Van Wyk JJ, Lund P (1988) Tissue and development specific regulation of a complex family of insulin-like growth factor I messenger ribonucleic acids. Molecular Endocrinology 2: 1077-1086.

6. Berger SL (1987) Direct mapping of rare mRNAs by means of oligomer-directed ribonuclease H cleavage. Anal Biochem 161: 272-279.

7. Irminger J, Rosen KM, Humbel RE, Villa-Komaroff L (1987) Tissue specific expression of insulin-like growth factor II mRNAs with distinct 5' untranslated regions. Proc Natl Acad Sci USA 84: 6330-6334.

8. Lund PK, Hoyt E, Van Wyk JJ (1989) The size heterogeneity of rat insulin-like growth factor-I mRNAs is due primarily to differences in the length of 3'-untranslated sequence. Molecular Endocrinology 3: 2054-2061.

9. Wreschner D, Rechavi G (1988) Differential mRNA stability to reticulocyte ribonucleases correlates with 3' non-coding $(U)_n$A sequences. Eur J Biochem 172: 333-340.

10. Brawerman G (1987) Determinants of messenger RNA stability. Cell 48: 5-6.

11. Shaw G, Kamen R (1986) A conserved AU sequence from the untranslated region of GM-CSF mRNA mediates selective mRNA degradation. Cell 46: 659-667.

12. Shyu A, Greenberg M, Belasco J (1989) The c-fos transcript is targeted for rapid decay by two distinct mRNA degradation pathways. Genes & Development 3: 60-72.

13. Hepler JE, Van Wyk JJ, Lund PK (1990) Different half-lives of insulin-like growth factor mRNAs that differ in length of 3' untranslated sequence. Endocrinology 127: 1550-1552.

14. Lowe W, Adamo M, LeRoith D, Roberts C (1989) Expression and stability of insulin-like grwoth factor-I (IGF-I) mRNA splicing variants in the GH3 rat pituitary cell line. Biochem Biophys Res Comm 162: 1174-1179.

15. Straus DS, Takemoto CD (1990) Effect of dietary protein deprivation on insulin-like growth factor (IGF)-I and II, IGF binding protein-2 and serum albumin gene expression in rat. Endocrinology 127: 1849-1859.

16. Thiessen JP, Underwood LE (1991) Translational status of the insulin like growth factor I mRNAs in liver of protein restricted rats. Mol. Endocrinol. submitted.

17. Kato H, Miura Y, Okoshi A, Umezawa T, Takahashi S, Noguchi T (1989) Dietary and hormonal factors affecting the mRNA level of IGF-I in rat liver in vivo and in primary cultures of rat hepatocytes. In: LeRoith D, Raizada MK (eds) Molecular and Cellular Biology of Insulin-Like Growth Factors and Their Receptors. Plenum Press New York 125-128.

CHARACTERIZATION OF THE BIOLOGICAL ACTIVITY OF IGF I ANALOGS WITH REDUCED AFFINITY FOR IGF RECEPTORS AND BINDING PROTEINS

Margaret A. Cascieri, Gary G. Chicchi, and Marvin L. Bayne

Departments of Biochemical Endocrinology and Growth Factor Research
Merck Sharp & Dohme Research Laboratories
Rahway, N.J.

We have prepared a series of IGF I analogs in order to determine the structural requirements for binding to the types 1 and 2 IGF receptors and to IGF binding proteins (1-5). These analogs were designed by substituting regions of IGF I with analogous regions of insulin, by deleting or replacing regions of IGF I that are absent in insulin, or by replacing specific residues of IGF I. The binding data for these analogs is summarized in Table 1.

RECEPTOR BINDING DOMAIN

Replacement of residues 28 to 37 of hIGF I with four glycines results in a 30-fold loss in affinity for the type 1 IGF receptor (1). This analog, [1-27,gly^4,38-70] IGF I, has normal affinity for the type 2 IGF receptor, the insulin receptor and human serum binding proteins. These data suggest that specific binding determinants reside within the C-region of IGF I.

Replacement of the tyrosine residues at positions 24, 31 and 60 results in 18-, 6-, and 20-fold loss in type 1 receptor affinity, respectively (2,3). Tyrosine 24 is homologous to the phenylalanine at B25 in the B-chain of insulin, which was previously shown to be important in maintaining high affinity binding of insulin to its receptor (6,7). Tyrosine 31, which is absent in hIGF II and insulin, is within the C-region deletion and its loss may account for some of the reduced binding affinity of [1-27, gly^4, 38-70] IGF I. [Leu 24] IGF I and [Ala 31] IGF I have normal affinity for the type 2 receptor and human serum binding proteins (2,3). In contrast, [Leu 60] IGF I has reduced affinity for the rat liver type 2 IGF receptor (3). The structural model of IGF I proposed by Blundell et al. (8) predicts that tyrosine 60 and isoleucine 43 maintain a Van der Waals interaction that is analogous to the interaction between isoleucine A2 and tyrosine A19 of insulin. These data suggest that the replacement of tyrosine 60 results in a conformational change or destabilization leading to loss in affinity for both types 1 and 2 IGF receptors. Replacement of two or more tyrosines results in a dramatic loss in type 1 receptor affinity. Thus, [Leu 24, Ala 31] IGF I, [Leu 24, Leu 60] IGF I, [Ala 31, Leu 60] IGF I and [Leu 24, Ala 31, Leu 60] IGF I have 240-,>1200-, 520-, and 1200-fold lower affinity than IGF I, respectively (3).

Molecular Biology and Physiology of Insulin and Insulin-Like Growth Factors
Edited by M.K. Raizada and D. LeRoith, Plenum Press, New York, 1991

Table 1. Summary of the Binding Activity of IGF I and Analogs. Binding affinity to IGF-BP1, 2, and 3 was determined in collaboration with Dr. David Clemmons. R1=type 1 receptor, R2=type 2 receptor, hs= acid-stable human serum binding protein, BP1=human, BP2=bovine.

Peptide	Receptors		Binding Proteins				Antibodies	
	R_1	R_2	hs	BP 1	BP 2	BP 3	Polyclonal	Sm 1.2
B-chain mutant	↔	↓	↓↓	↓↓	↓↓	↓↓	↓↓	↓↓
[$Q^3A^4Y^{15}L^{16}$] IGF I	↔	↔	↓	↓	↓	↓	↓	↓
[Q^3A^4] IGF I	↔	↔	↓	↓	↓	↔	↓	↔
[$Y^{15}L^{16}$] IGF I	↔	↔	↓	↔	↓	↔	↓	↓
A-chain mutant	↔	↓	↔	↓	↓	↔	n.d.	↑
[$T^{49}S^{50}I^{51}$] IGF I	↔	↓	↔	↓	↓	↔	↓	↑
[$Y^{55}Q^{56}$] IGF I	↔	↑	↔	↔	↔	↔	↓	↑
[F^{23}, F^{24}, Y^{25}] IGF I	↔	↔	↔	n.d.	n.d.	n.d.	n.d.	n.d.
[S^{24}] IGF I	↓	↔	↔	n.d.	n.d.	n.d.	↔	↔
[L^{24}] IGF I	↓	↔	↔	n.d.	n.d.	n.d.	n.d.	n.d.
[L^{60}] IGF I	↓	↓	↔	n.d.	n.d.	n.d.	n.d.	n.d.
[A^{31}] IGF I	↓	↔	↔	↔	↔	↔	↔	↔
[I-62] IGF I	↔	↔	↔	↔	↔	↔	↔	↔
[Leu^{24}] [I-62] IGF I	↓	↔	↔	↔	↔	↔	↔	n.d.
[1-27, G^4, 38-70] IGF I	↓	↔	↔	n.d.	n.d.	n.d.	↑	↔
[1-27, G^4, 38-62] IGF I	↓	↔	↔	↔	↔	↔	n.d.	n.d.

↔ = no change in affinity
↓ = decreased affinity
↑ = increased affinity
n.d. = not determined

In contrast to these data, large changes in the A-, B- and D-regions of IGF I have minimal effects on binding to the type 1 IGF receptor. Replacement of residues 1-16 of IGF I with residues 1-17 of the B-chain of insulin (B-chain mutant) or of residues 42-56 with residues 1-15 of the A-chain of insulin (A-chain mutant) has a minimal effect on type 1 IGF receptor affinity but results in a loss in type 2 IGF receptor affinity (4,5). In addition, deletion of residues 63-70 (D-region) of IGF I has no effect on type 1 receptor binding but results in slightly increased affinity for insulin receptors (1). These data strongly suggest that the important type 1 receptor binding domain is within residues 24-37 of IGF I.

The data described above were generated using human placental membranes as a source of type 1 IGF receptors. In order to determine if there are species differences in the binding specificity of the analogs, placental membranes were prepared from sheep, rat and dog. Type 1 receptor binding was determined by incubating labelled IGF I in the presence or absence of IGF I, analog or insulin. Insulin (10 uM) inhibits >90% of the specific binding of labelled IGF I to placental membranes from human, sheep and rat, but only 15% of the specific binding to dog membranes. Addition of disuccinimidyl suberate to the reaction mixture results in covalent modification of the alpha-subunit of the dog type 1 receptor (Mr=135 kD) and of a 46 kD IGF binding protein (Figure 1). Addition of up to 10 nM [1-62] [Leu 24] IGF I (2) results in decreased covalent modification of the 46kD IGF-BP and increased covalent modification of the receptor alpha-subunit (Figure 1). Modification of the alpha-subunit is decreased to control levels by addition of 1 uM [1-62] [Leu 24] IGF I. Therefore, this analog can be utilized to differentially block binding of labelled IGF I to IGF-BP to determine the affinity of analogs for the dog placental type 1 IGF receptor.

The binding specificity of the type 1 IGF receptor in human, sheep, rat and dog placental membranes is nearly identical (Table 2). Thus,

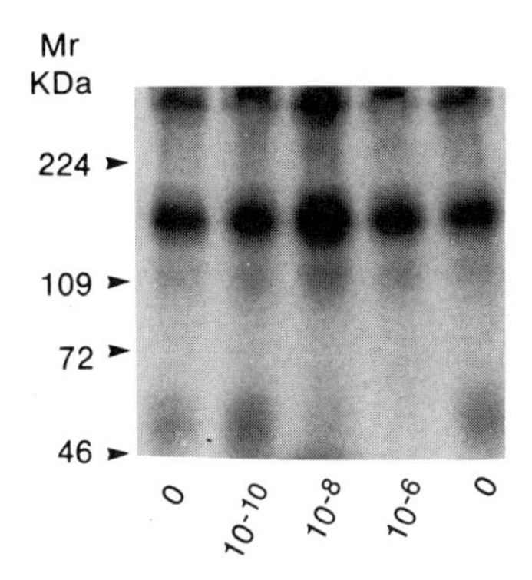

Figure 1. Disuccinimidyl suberate crosslinking of labelled IGF I to dog placental membranes in the presence of absence of [I-62] [Leu 24] IGF I.

analogs with insulin-like replacements in the B-region and A-region or with the D-region deleted have nearly normal affinity. However, replacement of tyrosine 24, 31 or 60 or replacement of residues 28-37 with four glycines results in dramatic loss in receptor affinity.

IGF-BP BINDING DOMAINS

The B-chain mutant, [Gln 3, Ala 4, Tyr 15, Leu 16] IGF I, [Gln 3, Ala 4] IGF I and [Tyr 15, Leu 16] IGF I have 1000-, 600-, 4- and 4-fold reduced affinity for the acid-stable IGF binding proteins in human serum, respectively (4). In contrast, analogs with insulin-like replacements in the A-region (5), with deletions or replacements in the C- and D-regions (1) or with replacements of tyrosines 24, 31 or 60 (2,3) have nearly normal affinity for binding proteins in this preparation. In collaboration with Dr. David Clemmons, we have measured the affinity of these analogs for purified human IGF-BP1 (9), bovine IGF-BP2 and bovine IGF-BP3 (Table 1). The B-chain mutant and [Gln 3, Ala 4, Tyr15, Leu 16] IGF I have > 30-fold and 15-30-fold, respectively, reduced affinity for all three proteins. However, [Gln 3, Ala 4] IGF I has > 20-fold reduced affinity for IGF-BP1 and 2 but normal affinity for IGF-BP3.

Table 2. Relative potency of selected IGF analogs for the type 1 receptor from human, sheep, rat and dog. Relative potency is the ratio of the IC50 for IGF I/IC50 for the analog. The IC50 for IGF I was 4.9nM, 4.2nM, 6.9nM, and 8.1nM for human, sheep, rat, and dog membranes, respectively.

Peptide	Relative Potency IGF-R1 Placental Membranes			
	Human	Sheep	Rat	Dog
IGF-I	1	1	1	1
B-chain mutant	0.5	1.1	0.9	0.9
A-chain mutant	1.5	2.0	1.4	1.2
[Y^{55}, Q^{56}] IGF I	1.0	1.0	0.4	0.8
[T^{49}, S^{50}, I^{51}] IGF I	0.7	0.8	0.3	0.5
[L^{24}] IGF I	0.03	0.03	0.03	0.04
[A^{31}] IGF I	0.1	0.06	0.06	0.09
[L^{60}] IGF I	0.05	0.03	0.03	0.03
[l-62] IGF I	1.2	1.0	0.9	1.0
[l-27, G^4, 38-70] IGF I	0.03	0.03	0.03	0.03
[l-27, G^4, 38-62] IGF I	0.02	0.01	0.02	0.02
Insulin	0.01	0.02	0.01	0.008

In addition, analogs with insulin-like replacements at residues 49-51 (A-chain mutant and [Thr 49, Ser 50, Ile 51] IGF I) have >40-fold reduced affinity for IGF-BP1 and 2 and normal affinity for IGF-BP3 (9, Table 1). These data suggest that the B-region helix (residues 1-16) of IGF I contains the IGF-BP3 binding domain and that residues 3 and 4 and residues 49-51 of the first A-region helix contain the IGF-BP1 and IGF-BP2 binding domains. Thus distinct regions of IGF I are responsible for binding to these proteins.

ANTIBODY BINDING DOMAINS

The domains of IGF I that are required to maintain high affinity binding to polyclonal antibodies are similar to those required for IGF-BP binding (10). The B-chain mutant, [Tyr 15, Leu 16] IGF I and [Gln 3, Ala 4, Tyr 15, Leu 16] IGF I have >100-, 100- and 25-fold reduced affinity, respectively, for the polyclonal antibody, UB286 (gift of Drs. J.J. Van Wyk and L.E. Underwood, obtained from the National Hormone and Pituitary Program) (10). In addition, [Thr 49, Ser 50, Ile 51] IGF I and [Tyr 55, Gln 56] IGF I have 30- and 50-fold reduced affinity (10). Thus, the B-region helix and the surface residues of the two helices in the A-region are involved in antibody binding.

The human IGF I-specific, monoclonal antibody SM1.2 (gift of Drs. J.J. Van Wyk and L.E. Underwood, obtained from the Developmental Studies

Hybridoma Bank) has reduced affinity for the B-chain mutant and the two analogs in which glutamine 15 and phenylalanine 16 are substituted with Tyr-Leu (10). Both UB286 and SM1.2 are neutralizing antibodies in that they inhibit the binding of IGF I to the type 1 receptor. The biological activity of analogs with high affinity for SM1.2 is attenuated by SM1.2 while the antibody does not affect the activity of [Tyr 15, Leu 16] IGF I or the B-chain mutant.

BIOLOGICAL ACTIVITY OF ANALOGS

The B-chain mutant and [Gln 3, Ala 4, Tyr 15, Leu 16] IGF I have a 4-fold reduced half-life compared to IGF I in rats after intravenous administration (11). However, despite their reduced half-life, these analogs are 4-fold and 2-fold, respectively, more potent than IGF I in stimulating glucose incorporation into glycogen after intraperitoneal administration (11). Thus, binding to serum binding proteins appears to attenuate the interaction of IGF I with the type 1 receptor on muscle cells.

We have measured the potency of various IGF analogs in three cell types, rat A10 cells, mouse Balb C/3T3 cells and mouse L7 cells. In general, the biological potency is well correlated to the potency at the type 1 IGF receptor (Figure 2). However, analogs with reduced affinity for IGF-BP 1 and 2 have increased relative potency as compared to IGF I in Balb C/3T3 cells (12). Thus, the B-chain mutant, [Gln 3, Ala 4, Tyr 15, Leu 16] IGF I, [Thr 49, Ser 50, Ile 51] IGF I and the A-chain mutant have 5-fold, 5-fold, 7.5-fold and 3.5-fold increased relative potency, respectively.

Conditioned medium from Balb C/3T3 cells inhibits the activity of IGF I, but not [Gln 3, Ala 4, Tyr 15, Leu 16] IGF I in A10 cells (12). A ligand blot of this media shows that it contains the IGF-BP3 doublet (Mr=46 and 43 kD), 35.5, 30 and 27 kD binding proteins (Figure 3). The 30 kD band co-migrates with IGF-BP1 from HepG2 cells. The conditioned medium of L7 cells contains similar amounts of the IGF-BP3 doublet but reduced levels of the other three proteins (Figure 3). Since the biological potencies of the A-chain mutant and [Thr 49, Ser 50, Ile 51] IGF I, two analogs with normal affinity for IGF-BP3, are not inhibited by conditioned medium from Balb C/3T3 cells, it is unlikely that IGF-BP3 is responsible for the inhibition.

In collaboration with Drs. Sherida Tollefsen, Ellen Heath-Monnig and William Daughaday, we have studied the activity of [Gln 3, Ala 4, Tyr 15, Leu 16] IGF I ([QAYL] IGF I) in fibroblasts isolated from a child of short stature. These fibroblasts have previously been shown to be significantly less responsive to IGF I (13). Stimulation of AIB uptake and thymidine incorporation in these cells required 3-fold higher levels of IGF I than in normal human fibroblasts. In contrast, the patient's cells were normally responsive to [QAYL] IGF I (14). Binding of labelled IGF I to normal fibroblasts and the patient's fibroblasts is inhibited 50% and $< 20\%$ by α IR3, respectively. However, binding of labelled [QAYL] IGF I to normal and to the patient's fibroblasts is 60% and 95% lower, respectively, than binding of labelled [QAYL] IGF I, and is $> 90\%$ inhibited by αIR3 (14). These data suggest that a cell surface binding protein may be responsible for the decreased sensitivity to IGF I in these abnormal cells. In fact, they have detected a 10-fold increase in the cell surface expression of a 32 kD IGF-BP in the patient's fibroblasts (14).

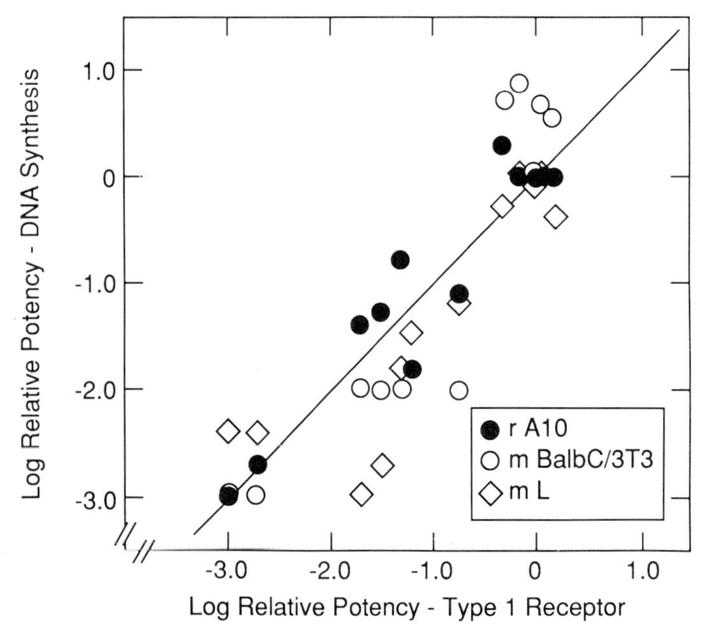

Figure 2. Correlation between type receptor potency and biological potency in three cell lines.

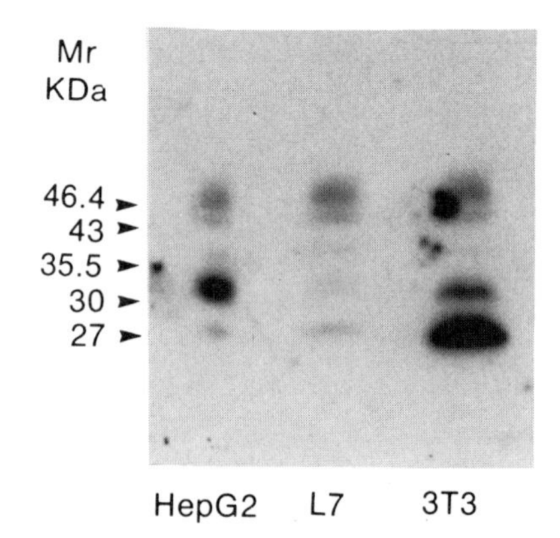

Figure 3. Ligand blotting of conditioned medium from three cell lines.

The above data suggests that the activity of IGF I in many in vitro and in vivo systems is modulated by inhibitory IGF-BP. David Clemmons has previously demonstrated that IGF-BP1 potentiates the biological activity of IGF I in many systems (15, 16). He has recently shown that IGF-BP1 potentiates the ability of IGF I and [1-62] [Leu 24] IGF I, but not the B-chain mutant, to stimulate DNA synthesis in porcine smooth muscle cells (9). In addition, IGF I and the type 1 receptor-inactive analog, [1-62] [Leu 24] IGF I stimulate the secretion of a 31 kD IGF-BP from humal fetal fibroblasts whereas analogs with poor IGF-BP affinity, such as the B-chain mutant, [QAYL] IGF I and [Thr 49, Ser 50, Ile 51] IGF I, do not (9). These data suggest that some biological responses are potentiated or mediated by IGF-BP.

Dr. Robert Bar has shown that partially purified endothelial cell IGF-BP stimulate 2-deoxglucose uptake and AIB uptake in cultured microvessel endothelial cells (17). He now has shown that this IGF-BP preparation is active when covalently crosslinked (ie., utilizing disuccinimidyl suberate) with either IGF I or the receptor-inactive analog, [1-27, gly^{4}, 38-70] IGF I (18). These data suggest that the IGF-BP preparation is active irrespective of activation of the type 1 IGF receptor.

CONCLUSIONS

We have demonstrated that distinct domains of IGF I mediate binding to the types 1 and 2 IGF receptor, insulin receptor, IGF-BP1, 2, and 3 and antibodies. In the absence of IGF-BP, the biological activity of these analogs correlates with their type 1 receptor binding affinity. However, the role of IGF-BP in regulating IGF I activity is complex, and may be binding protein-and/or cell type-specific. The IGF analogs that we have prepared and characterized that selectively bind to receptors or IGF-BP have proven to be useful reagents in understanding the complex regulation of IGF I biological activity.

ACKNOWLEDGEMENTS

We gratefully acknowledge the assistance of our other associates in these studies, Ms. Joy Applebaum, Ms. Nancy S. Hayes, and Ms. Barbara G. Green.

REFERENCES

1. Bayne, M.L., Applebaum, J., Underwood, D., Chicchi, G.G., Green, B.G., Hayes, N.S. and Cascieri, M.A. (1989) J. Biol. Chem., 264, 11004-11008.
2. Cascieri, M.A., Chicchi, G.G., Applebaum, J., Hayes, N.S., Green, B.G. and Bayne, M.L. (1988) Biochemistry, 27, 3229-3233.
3. Bayne, M.L., Applebaum, J., Chicchi, G.G., Miller, R.E. and Cascieri, M.A. (1990) J. Biol. Chem., 265, 15648-15652.
4. Bayne, M.L., Applebaum, J., Chicchi, G.G., Hayes, N.S., Green, B.G. and Cascieri, M.A. (1988) J. Biol. Chem., 263, 6233-6239.
5. Cascieri, M.A., Chicchi, G.G., Applebaum, J., Green, B.G., Hayes, N.S. and Bayne, M.L. (1989) J. Biol. Chem., 264, 2199-2202.
6. Tager, H., Thomas, N., Assoian, R., Rubenstein, A., Saekow, M., Olefsky, J. and Kaiser, E.T. (1980) Proc. Natl. Acad. Sci. USA, 77, 3181-3185.

7. Kobayashi, M., Ohgaku, S., Iwasaki, M., Maegawa, H., Shigeta U. and Inouye, K. (1982) Biochem. J., 206, 597-603.
8. Blundell, T.L., Bedarker, S. and Humbel, R.E. (1983) Fed. Proc., 42, 2592-2597.
9. Clemmons, D.R., Cascieri, M.A., Comacho-Hubner, C., McCusker, R.H. and Bayne, M.L. (1990) J. Biol. Chem., 265, 12210-12216.
10. Cascieri, M.A., Bayne, M.L., Ber, E., Green, B.G., Men, G.W. and Chicchi, G.G. (1990) Endocrinology, 126, 2773-2777.
11. Cascieri, M.A., Saperstein, R., Hayes, N.S., Green, B.G., Chicchi, G.G., Applebaum, J. and Bayne, M.L. (1988) Endocrinology, 123, 373-381.
12. Cascieri, M.A., Hayes, N.S. and Bayne, M.L. (1989) J. Cell. Physiol., 139, 181-188.
13. Heath-Monnig, E., Wohlmann, H.J., Mills-Dunlop, B. and Daughaday, W.H. (1987) J. Clin. Endocrinol. Metab., 64, 501-507.
14. Tollefsen, S.E., Heath-Monnig, E., Cascieri, M.A., Bayne, M.L. and Daughaday, W.H. (1990) J. Clin. Invest., in press.
15. Elgin, R.G., Busby, W.H. and Clemmons, D.R. (1987) Proc. Natl. Acad. Sci. USA, 84, 3254-3258.
16. Busby, W.H., Klapper, D.G., and Clemmons, D.R. (1988) J. Biol. Chem., 263, 14203-14210.
17. Bar, R.S., Booth, B.A., Boes, M., and Dake, B.L. (1989) Endocrinology, 125, 1910.
18. Booth, B.A., Bar, R.S., Boes, M., Dake, B.L., Bayne, M. and Cascieri, M.A. (1990) Endocrinology, in press.

INSULIN-LIKE GROWTH FACTOR II: COMPLEXITY OF BIOSYNTHESIS AND RECEPTOR BINDING

Steen Gammeltoft, Jan Christiansen, Finn C. Nielsen, and Sten Verland

Department of Clinical Chemistry, Bispebjerg Hospital
DK 2400 Copenhagen NV, Denmark.

SUMMARY

Insulin-like growth factor II (IGF-II) belongs to the insulin family of peptides and acts as a growth factor in many fetal tissues and tumors. The gene expression of IGF-II is initiated at three different promoters which gives rise to multiple transcripts. In a human rhabdomyosarcoma cell line IN 157 IGF-II mRNAs of 6.0-kb, 4.8-kb, and 4.2-kb are present. Fractionation of cellular extracts on sucrose gradients and Northern blot analysis showed that only the 4.8-kb mRNA was associated with polysomes, whereas the other transcripts co-sedimented with monosomal particles. This suggests that only the 4.8-kb mRNA is translated to IGF-II. The cell line secretes two forms of immunoreactive and bioactive IGF-II to the medium of molecular size 10 kd and 7.5 kd which may be involved in autocrine control of cell growth. IGF-II binds to two receptors on the surface of many cell types: the IGF-I receptor and the mannose-6-phosphate (Man-6-P)/IGF-II receptor. There is consensus that the cellular effects of IGF-II are mediated by the IGF-I receptor via activation of its intrinsic tyrosine kinase. The Man-6-P/IGF-II receptor is involved in endocytosis of lysosomal enzymes and IGF-II. In selected cell types, however, Man-6-P induces cellular responses. We have studied rat brain neuronal precursor cells where Man-6-P acted as a mitogen suggesting that phosphomannosylated proteins may act as growth factors via the Man-6-P/IGF-II receptor. In conclusion, the gene expression and mechanism of action of IGF-II is very complex suggesting that its biological actions can be regulated at different levels including the transcription, translation, posttranslational processing, receptor binding and intracellular signalling.

INTRODUCTION

Insulin-like growth factor II (IGF-II) is a member of the insulin family of peptides and shows 50-70% amino acid sequence homology with insulin and insulin-like growth factor I (IGF-I) (Humbel, 1984). Computer modeling has shown that their three-dimensional structure is also similar (Blundell et al., 1978). IGF-I is identical with somatomedin C which stimulates skeletal growth during adolescence under growth hormone control whereas IGF-II stimulates growth of many tissues in the fetus (Froesch et al., 1985). Following the cloning of IGF-I and IGF-II cDNAs and genes (Sussenbach, 1989), studies of the distribution of IGF-I and IGF-II mRNAs in fetal and adult mammals revealed that both transcripts are widely occurring during fetal and postnatal life (Daughaday and Rotwein, 1989). In the fetus IGF-I and IGF-II genes are expressed in many tissues indicating that both peptides may act as fetal growth promoters. In the adult mammal IGF-I is predominantly synthesized in the liver and secreted to the circulation, but synthesis of IGF-I has also been shown in connective tissue, muscle and cartilage. The main site of IGF-II synthesis in adults is the choroid plexus epithelium of the brain ventricles, but IGF-II is also synthesized in the pituitary gland, adrenal medulla and skin. Thus, IGF-I and IGF-II may act as endocrine and paracrine growth regulators (Gammeltoft, 1989).

IGF-I and IGF-II induce their actions by interaction with two types of receptors, the IGF-I receptor and the IGF-II receptor (Gammeltoft, 1989). The IGF-I receptor is a disulphide-linked heterotetrameric protein composed of two extracellular α-subunits which bind IGF-I and two β-subunits with transmembrane domain and cytoplasmic domain which has tyrosine kinase activity. The IGF-I receptor is similar to the insulin receptor, but insulin and IGF-I show only 1% cross-reactivity with each others receptors. The IGF-I receptor mediates the intracellular signals of both IGF-I and IGF-II (Czech, 1989). The IGF-II receptor is a monomeric protein and is identical with the mannose-6-phosphate (Man-6-P) receptor. The Man-6-P/IGF-II receptor is a multifunctional protein involved in endocytosis of IGF-II and intracellular transport of lysosomal hydrolases from the Golgi or plasma membrane to the lysosomes (Dahms et al., 1990). At present a role of the Man-6-P/IGF-II receptor in intracellular signalling of IGF-II has only been demonstrated in specific cell lines (Roth, 1988).

The cloning of IGF-I, IGF-II and their receptors and studies of their cellular biology has raised many questions concerning the gene regulation, biosynthesis and mechanism of action of IGF-I and IGF-II. These include the presence of three IGF-II gene promoters, the alternative splicing giving two IGF-I precursors, the posttranslational processing of IGF-I and IGF-II, the release of E-peptide fragments, the signalling mechanism of IGF-I receptor tyrosine kinase, the signal function of the Man-6-P/IGF-II receptor and the multiple actions of IGF-I and IGF-II. In the following we will discuss the translational regulation of IGF-II biosynthesis, the processing of prepro-IGF-II and the signalling function of IGF-I and Man-6-P/IGF-II receptors.

TRANSLATIONAL DISCRIMINATION OF IGF-II mRNAs

The characterization of the human and rat chromosomal IGF-II genes and studies of their transcription revealed complex patterns of the IGF-II gene expression in different tissues and developmental stages (Sussenbach, 1989). The IGF-II gene in man spans about 30 kilobase pairs (kbp) of chromosomal

DNA and comprises eight exons (Fig. 1). Exons 5 and 6, and 234 nucleotides of exon 7 provide the coding region for the prepropeptide consisting of 180 amino acids, while exons 1-4 and 4B are not coding for preproIGF-II. The IGF-II gene has been mapped on the short arm of chromosome 11 where it is located very close to the insulin gene at a distance of 1.4-kbp.

Analysis of mRNAs from fetal and adult man has shown that in fetal tissues and non-liver adult tissues two major mRNAs of 6.0- and 4.8-kb are present (Gray et al., 1987; de Pagter-Holthuizen et al., 1988). The 6.0-kb species is derived from promoter P2, consists of exons 4, 5, 6 and 7. The 4.8-kb mRNA is derived from promoter P3. This mRNA consists of exons 4B, 5, 6 and 7. In this case, the translational apparatus is presented with two mRNAs with different 5′ noncoding regions. In adult liver, P2 and P3 are inactive and transcription is controlled by promoter P1. The corresponding mRNA has a length of 5.3-kb and consists of exons 1, 2, 3, 5, 6 and 7. Thus, the expression of the IGF-II gene is regulated in a tissue- and development-specific way. P1 is only active in the adult liver while promoters P2 and P3 are active in the fetal stage and in adult non-liver tissues. The expression of the IGF-II gene in other mammals like rat and pig is also developmentally regulated (Frunzio et al., 1986; Hedley et al., 1989).

The simultaneous use of two promoters in one cell leading to multiple transcripts has been a puzzle with respect to translation and biosynthesis of IGF-II. In order to adress the question we have examined the translational status of IGF-II mRNAs using sucrose gradient and Northern blot analysis. Our studies suggest that only the minor 4.8-kb mRNA is actively engaged in protein synthesis while the predominant 6.0-kb mRNA is present in a cytoplasmic

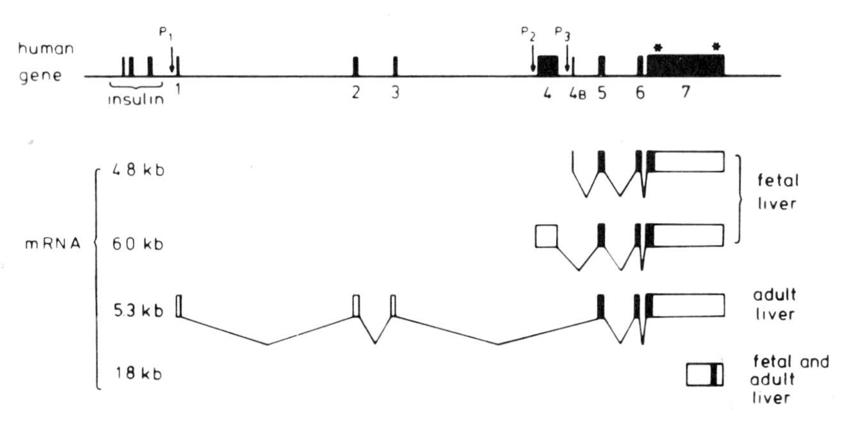

Fig. 1 Map of the human IGF-II gene (top) and mRNAs (bottom). P_1, P_2 and P_3 indicate promoters, and polyadenylation sites are indicated by asterisks. In the mRNAs the coding regions are indicated as black boxes. (Reproduced from Sussenbach, 1989)

ribonucleoprotein particle that is stable in EDTA (Nielsen et al., 1990). The study was performed on a human rhabdomyosarcoma cell-line IN 157 which produces IGF-II. Comparison with normal human placenta showed similar transcript patterns on Northern analysis with a major transcript of 6.0-kb and two less abundant transcripts of 4.8-kb and 4.2-kb. The 6.0-kb and 4.8-kb transcripts were poly(A)$^{+}$ whereas the 4.2-kb transcript lacked the poly(A)tract. Hybridization with exon-specific probes indicated that the 4.8-kb transcripts originated from exon 4B, whereas the 6.0- and 4.2-kb transcripts were detected with the exon 4 probe.

The translational status of the different IGF-II mRNAs was determined by isolation of polysomes from detergent-lysed IN 157 cells. Only the 4.8-kb transcript sedimented with polysomes whereas the predominant 6.0-kb transcript and the 4.2-kb transcript co-migrated with the monosome fraction. Furthermore, the 4.8-kb transcript was associated with membrane-bound and not "free" polysomes, whereas the other transcripts were free in the cytoplasm. Finally, the sedimentation of the monosomal 6.0- and 4.2-kb transcripts was not sensitive to EDTA indicating that they are not associated with ribosomes or their subunits, but are present in stable cytoplasmic particles not directly engaged in protein synthesis. The functional significance of the ribonucleoprotein particle with the 6.0-kb transcript is unknown. It may be speculated that the mRNA could be activated leading to translation of IGF-II or another peptide since exon 7 has an open reading frame encoding a 113 amino acid protein (de Pagter-Holthuizen et al., 1988).

POSTTRANSLATIONAL PROCESSING OF IGF-II

The cloning and sequencing of cDNAs encoding IGF-II revealed that prepro-IGF-II is synthesized as a 20-kd precursor composed of signal peptide, mature IGF-II and a carboxyl-terminal extension (E-domain). Thus, formation of IGF-II from its precursor requires proteolytic processing at both ends whereby the signal peptide and E-domain are removed. Inspection of the amino acid sequence shows that the signal peptidase acts on alanine in the IGF-II precursor like in IGF I and insulin precursors (Steiner et al., 1980). The mature peptide is produced from pro-IGF-I by proteolytic cleavage at an arginine residue whereby the E-domain is removed. Cleavage at single basic residues during posttranslational processing of hormones has been described for a number of peptides (Schwartz, 1986).

The presence of several pairs of basic residues and single basic residues within the E-domain suggests that proteolytic cleavages may occur at multiple sites (Fig. 2). Various classes of IGF-II-like molecules (Mr ~ 9.000-19.000) are secreted by cultured BRL-3A rat liver cells (Marquardt et al., 1981; Yang et al., 1985) and two forms of pro-IGF-II (Mr ~ 10.000-15.000) are found in serum (Zumstein et al., 1985) and cerebrospinal fluid (Haselbacher and Humbel, 1982).

Recent studies in our laboratory have focused on the nature of IGF-II-like peptides secreted from the rhabdomyosarcoma cell-line IN 157 (Nielsen et al., 1991). Using a specific radioimmunoassay we have identified two molecular forms of IGF-II with Mr 7.500 and 10.000 in the conditioned medium under serum-free conditions. These forms were also bioactive in cellular assays of receptor binding and DNA synthesis. Based on their molecular size determined on gel filtration it is suggested that they correspond to mature IGF-II and partially processed pro-IGF-II.

It is not known whether the various IGF-II precursors which are present in biological fluids have specific functions which are distinct from IGF-II. Furthermore, it is possible that the carboxylterminal E-fragments of pro-IGF-II may exert unique effects either alone or in synergy with IGF-II. Preliminary data suggest that E-peptides may potentiate the mitogenic activity of IGF-II (Hylka et al., 1985).

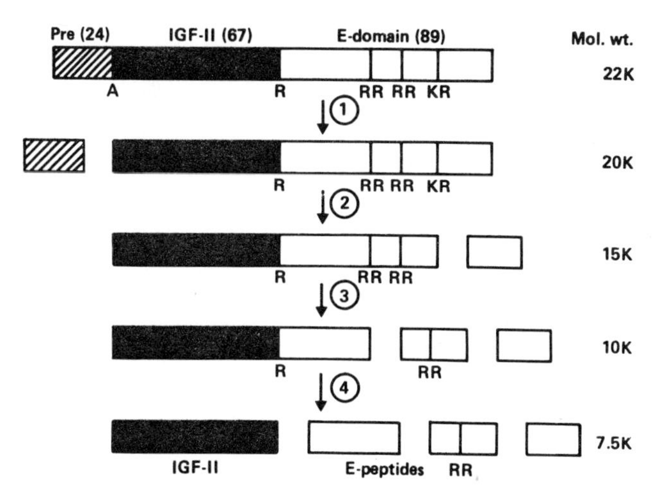

Fig. 2 Model of the posttranslational processing of prepro-IGF-II. The IGF-II precursor (22 kd) is initially cleaved by the signal peptidase (1) at an alanine residue whereby the signal peptide is removed. Pro-IGF-II (20 kd) is sequentially cleaved by processing enzymes (2 and 3) at dibasic residues to yield two intermediate forms of pro-IGF-II (15 and 10 kd). Finally, clevage at a single basic residue (4) leads to mature IGF-II (7.5 kd). E-peptides of variable length are produced. (Reproduced from Gammeltoft, 1989)

IGF-I RECEPTOR TYROSINE KINASE IN GROWTH STIMULATION OF IGF-II

The IGF-I receptor belongs to the family of receptors with protein tyrosine kinase activity (Ullrich and Schlessinger, 1990). The amino acid sequence of the IGF-I receptor is similar to that of the insulin receptor (Fig. 3) (Ullrich et al., 1985, 1986). This structural conservation suggests that the mechanism of action and responses of the two receptors are also similar. Both insulin and IGF-I receptors mediate acute responses like glucose and amino acid transport as well as long-term effects like protein synthesis and cell division in vitro (Czech, 1989). The binding sites of insulin and IGF-I receptors are probably different as reflected by the low level of cross-reactivity (1%) between the two receptors and

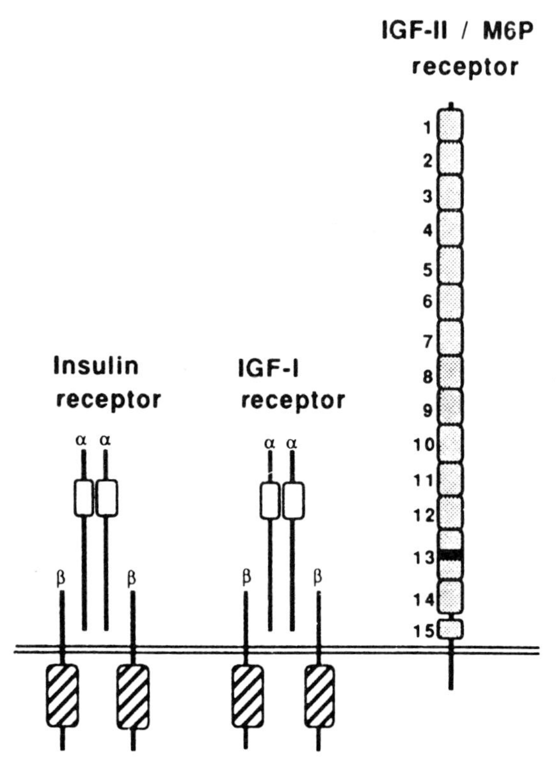

Fig. 3 *Schematic comparison of insulin, IGF-I and Man-6-P/IGF-II receptors. The structures are based on published primary sequences and drawn to the same scale with carboxyl terminals inside. Extracellular cysteine-rich regions and protein tyrosine kinase domains in insulin and IGF-I receptors are indicated by open boxes and striped boxes, respectively. In the Man-6-P/IGF-II receptor the stippled boxes represent repeat sequences numbered 1-15. The band in repeat 13 sequences the type II fibronectin homology region. (Reproduced from Morgan et al., 1987)*

each others ligands (Gammeltoft et al., 1985). Furthermore, the sequence similarity between the extracellular domains varies between 41 and 67% (Ullrich et al., 1986). Finally, a chimeric receptor consisting of the extracellular insulin receptor and the N-terminal 313 amino acids of the IGF-I receptor binds IGF-I with much higher affinity than insulin (Andersen et al., 1990).

The role of the insulin receptor tyrosine kinase in insulin action has been studied extensively regarding its autophosphorylation, enzyme kinetics, structure-function relationship, cellular substrates and dephosphorylation (Gammeltoft and Van Obberghen, 1986; Rosen, 1990). In contrast, only few studies have adressed these questions for the IGF-I receptor kinase. It has been shown that the insulin and IGF-I receptors are regulated by similar mechanisms (Yu et al., 1986). On the other hand, expression of cDNA constructs expressing receptor chimeres with the cytoplasmic domain of insulin and IGF-I receptors suggests that the IGF-I receptor confers a higher mitogenic potential than the insulin receptor whereas both are potent stimulators of glucose transport (Lammers et al., 1989).

The presence of two types of IGF receptors, the IGF-I receptor and the Man-6-P/IGF-II receptor in many cell types has raised the question about their role in signal transduction of IGF-I and IGF-II (Czech, 1989). The IGF-I receptor binds IGF-I and IGF-II with almost equal affinity whereas the Man-6-P/IGF-II receptor binds IGF-II with 10-50 times higher affinity than IGF-I. Insulin interacts with the IGF-I receptor, but not the Man-6-P/IGF-II receptor. Two lines of evidence have suggested that the IGF-I receptor mediates the actions of both IGF-I and IGF-II whereas the Man-6-P/IGF-II receptor is not involved. First, the dose-response relationship of ligand binding to the IGF-I receptor correlates with that of DNA synthesis and amino acid transport induced by IGF-I, IGF-II, and insulin in several cell types (Ewton et al., 1987; Ballotti et al., 1987; Nielsen and Gammeltoft, 1988; Verland and Gammeltoft 1989; Nielsen et al., 1990). Secondly, antibodies to the IGF-I receptor inhibit binding as well as actions of IGF-I and IGF-II like glucose transport, amino acid uptake and DNA synthesis whereas antibodies to the Man-6-P/IGF-II receptor had no such inhibitory effects (Flier et al., 1986; Kiess et al., 1987; Nielsen et al., 1990). Thus, in many cells the IGF-I receptor mediates the growth-promoting actions of IGF-I and IGF-II (Fig. 4).

FUNCTION OF MAN-6-P/IGF-II RECEPTOR

The cloning of the mammalian IGF-II receptor revealed its identity with the Man-6-P receptor (Morgan et al., 1987). The main function of the Man-6-P/IGF-II receptor is to transport newly synthesized lysosomal enzymes from the Golgi apparatus to the lysosomes. A small fraction of the receptors are expressed on the cell surface where they mediate endocytosis of lysosomal enzymes and IGF-II (Von Figura and Hasilik, 1986; Gammeltoft, 1989). A long standing question has been the role of Man-6-P/IGF-II receptors in intracellular signalling of IGF-II (Fig. 4). Some exceptions to the model that only the IGF-I receptor

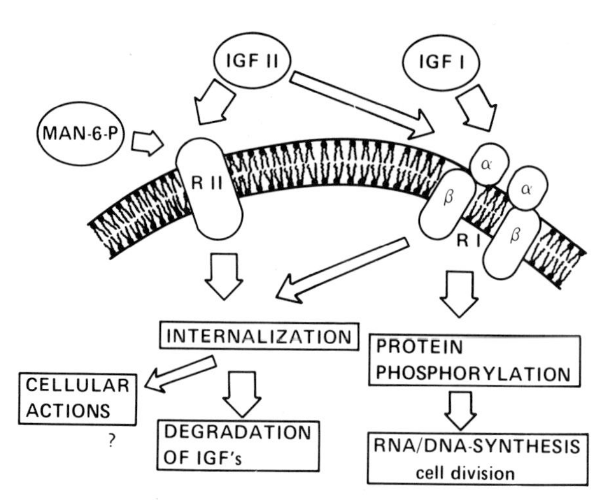

Fig. 4 *Model of the cellular mechanism of IGF-I and IGF-II action in mammalian tissues. Two types of receptors are present in the cell. IGF-I and IGF-II interact with the α-subunit of the IGF-I receptor (R I) and activate the β-subunit tyrosine kinase. Protein phosphorylations mediate effects on nucleotide synthesis and cell division. IGF-II binds to the Man-6-P/IGF-II receptor (R II). Internalization of IGFs via both receptors leads to their degradation. The role of the Man-6-P/IGF-II receptor in cellular actions of its ligands is not yet clear. (Reproduced from Gammeltoft, 1989)*

tyrosine kinase mediates actions of IGF-II have been reported. In selected cell lines IGF-II elicits responses which seem to be mediated by the Man-6-P/IGF-II receptor. These include stimulation of calcium influx in Balb/c 3T3 fibroblasts (Nishimoto et al., 1987), glucose and amino acid uptake in L6 myoblasts (Beguinot et al., 1985), and glycogen synthesis in HepG2 hepatoma cells (Hari et al., 1987). These results suggest that Man-6-P/IGF-II receptors may be linked to cellular proteins involved in signal transduction and it has been reported that the Man-6-P/IGF-II receptor functionally interacts with a G_i-like protein. Analysis of the small cytoplasmic domain of the receptor revealed that it contains a segment which activates the GTP-binding rate of G_{i-2} protein (Okamoto et al., 1990).

The notion that the Man-6-P/IGF-II receptor may be involved in intracellular signalling was supported by observations that Man-6-P induces cellular effects. In fibroblasts Man-6-P stimulated expression of its receptor on the cell surface by a pertussis toxin-sensitive mechanism involving G-proteins (Braulke et al., 1989). Studies in our laboratory of neuronal precursor cells showed that Man-6-P stimulated cell proliferation by interaction with the Man-6-P/IGF-II receptor (Nielsen and Gammeltoft, 1990). The response showed a sugar specificity which correlated with the potency of various phosphorylated carbohydrates in inhibiting binding and endocytosis of Man-6-P-containing proteins and oligosaccharides. Furthermore, the response was blocked by an antibody to the Man-6-P/IGF-II receptor. These observations suggest that secreted Man-6-P-containing glycoproteins may be involved in regulation of cell division. Possible candidates are proliferin, a prolactin-related glycoprotein secreted by placenta (Lee and Nathans, 1988), uteroferrin, a progesterone-induced secreted acid phosphatase (Baumbach et al., 1984), renin (Faust et al., 1987), thyroglobulin (Herzog et al., 1987) and TGF-β precursor (Purchio et al., 1988).

In addition to endocytosis of lysosomal hydrolases, the Man-6-P/IGF-II receptor internalizes IGF-II (Gammeltoft, 1989). Internalization of IGF-II is followed by its rapid degradation (Oka et al., 1985; Nielsen et al., 1990). The Man-6-P/IGF-II is recycled back to the plasma membrane via the Golgi network and the movement of receptors occurs constitutively in the absence and presence of its ligands with a half-time at 3-5 min (Von Figura and Hasilik, 1986). The amount of Man-6-P/IGF-II receptors on the cell surface is regulated by various factors and conditions. In cultured fibroblasts the receptor number is increased 1.5- to 2-fold within 10-15 min by insulin, IGF-I, IGF-II, Man-6-P, and EGF (Braulke et al., 1989). Insulin increases the number of Man-6-P/IGF-II receptors in the adipocyte membrane and decreases their phosphorylation (Corvera and Czech, 1985). As a consequence, the internalization and degradation of IGF-II is increased. Man-6-P/IGF-II receptors are also increased during regeneration or hyperplasia of various tissues such as liver following partial hepatectomy (Scott and Baxter, 1990), the small intestine after 50% resection (Grey et al., 1991), the thyroid gland during propylthiouracil treatment (Polychronakos et al., 1986), thymocytes during activation with concanavalin A and interferon Γ (Verland et al., 1991), and astrocytes during proliferation (Auletta et al., 1991). These studies suggest that Man-6-P/IGF-II receptors may be active in the endocytosis of Man-6-P-containing proteins and IGF-II as well as generation of intracellular signals during cell proliferation (Fig. 4).

CONCLUSION

The biosynthesis and actions of IGF-II are characterized by a high degree of complexity and diversity. Biosynthesis of IGF-II can be regulated at several levels including the transcriptional level by initiation at three different promoters, the translational level by discrimination of multiple mRNAs, and the posttranslational level by differential processing of the IGF-II precursor. Regulation at these different steps may account for the development- and tissue-specific biosynthesis of IGF-II which occurs in different cell types under various physiological and pathological conditions.

The actions of IGF-II can be regulated at several levels including the receptor level by interaction with three different receptors, the level of intracellular signal transduction by activating different pathways, and the level of biochemical effects by stimulation of different cellular responses. The diversity in IGF-II actions in different cell types during development and postnatal life may be related to the existence of multiple signalling pathways like stimulation of protein tyrosine kinase activity and activation of G-protein-coupled enzymes.

In conclusion, the spectrum of regulatory levels and the complex relationship between biosynthesis, receptor binding and signalling of IGF-II agrees well with the concept that IGF-II acts in an autocrine or paracrine manner and that IGF-II exerts diverse cellular actions under a variety of conditions. At present our understanding is limited and further studies are needed to elucidate the role of these regulatory events in IGF-II functions.

REFERENCES

Andersen, A. S., Kjeldsen, T., Wiberg, F. C., Christensen, P.M., Rasmussen J. S., Norris, K., Møller, K. B., and Møller N. P. H., 1990, Changing of the insulin receptor to possess insulin-like growth factor I ligand-specificity. Biochemistry, in print.

Auletta, M., Nielsen, F. C., and Gammeltoft, S., 1990, Receptor-mediated endocytosis and degradation of insulin-like growth factors in neonatal rat astrocytes. J. Neuroscience Research, submitted.

Ballotti, R., Nielsen, F. C., Pringle, N., Kowalski, A., Richardson, W. D., Van Obberghen, E., and Gammeltoft, S., 1987, Insulin-like growth factor I in cultured rat astrocytes: expression of the gene, and receptor tyrosine kinase. EMBO J., 6:3633.

Baumbach, G. A., Saunders, P. T. K., Bazer, F. W., and Roberts, R. M., 1984, Uteroferrin has N-asparagine-linked high-mannose-type oligosaccharides that contain mannose-6-phosphate. Proc. Natl. Acad. Sci. USA, 81: 2985.

Beguinot, F., Kahn, C. R., Moses, A. C., and Smith, R. J., 1985, Distinct biologically active receptors for insulin, insulin-like growth factor I, and insulin-like growth factor II in cultured skeletal muscle cells. J. Biol. Chem., 260:15892.

Blundell, T. L., Bedarkar, S., Rinderknecht, E., and Humbel,R. E., 1978, Insulin-like growth factor: a model for tertiary structure accounting for immunoreactivity and receptor binding. Proc. Natl. Acad. Sci. USA, 75:180.

Braulke, T., Tippmer, S., Neher, E., and Von Figura, K., 1989, Regulation of the mannose 6-phosphate/IGF II receptor expression at the cell surface by mannose 6-phosphate, insulin like growth factors and epidermal growth factor. EMBO J., 8:686.

Corvera, S., and Czech, M. P., 1985, Mechanism of insulin action on membrane protein recycling: a selective decrease in the phosphorylation state of insulin-like growth factor II receptors in the cell surface membrane. Proc. Natl. Acad. Sci. USA, 82:7314.

Czech, M. P., 1989, Signal transmission by the insulin-like growth factors. Cell, 59:235.

Dahms, N. M., Lobel, P., and Kornfeld, S., 1989, Mannose 6-phosphate receptors and lysosomal enzyme targeting. J. Biol. Chem., 264:12115.

Daughaday, W. H., Rotwein, P., 1989, Insulin-like growth factors I and II. Peptide, messenger ribonucleic acid and gene structures, serum and tissue concentrations. Endocrine Rev., 10:68.

de Pagter-Holthuizen, P., Jansen, M., van der Kammen, R. A., van Schaik, F. M. A., and Sussenbach, J. S., 1988, Differential expression of the human insulin-like growth factor II gene. Characterization of the IGF-II mRNAs and mRNA encoding a putative IGF-II-associated protein. Biochim. Biophys. Acta, 950:282.

Ewton, D. Z., Falen, S. L., and Florini, J. R., 1987, The type II insulin-like growth factor (IGF) receptor has low affinity for IGF-I analogs: Pleiotypic actions of IGFs on myoblasts are apparently mediated by the type I receptor. Endocrinology, 120:115.

Faust, P. L., Chirgwin, J. M., and Kornfeld, S., 1987,Renin, a secretory glycoprotein, acquires phosphomannosyl residues. J. Cell Biol., 105:1947.

Flier, J. S., Usher, P. A., Moses, A. C., 1986, Monoclonal antibody to the type I insulin-like growth factor (IGF-I) receptor blocks IGF-I receptor-mediated DNA synthesis: clarification of the mitogenic mechanisms of IGF-I and insulin in human skin fibroblasts. Proc. Natl. Acad. Sci. USA, 83:664.

Froesch, E. R., Schmid, C., Schwander, J., and Zapf, J., 1986, Actions of insulin-like growth factors. Ann. Rev. Physiol., 47:443.

Frunzio, R., Chiariotti, L., Brown, A. L., Graham, D. E., Rechler, M. M., and Bruni, C. B., 1986, Structure and expression of the rat insulin-like growth factor II (rIGF-II) gene. J. Biol. Chem., 261:138.

Gammeltoft, S., 1989, Insulin-like growth factors and insulin: gene expression, receptors and biological actions,in: "Peptide hormones as prohormones", Martinez, ed., Ellis Horwood Limited, pp. 176-210.

Gammeltoft, S., and van Obberghen, E., 1986, Protein kinase activity of the insulin receptor. Biochem. J., 235:1.

Gammeltoft, S., Haselbacher, G. K., Humbel, R. E., Fehlmann, M., and Van Obberghen, E., 1985, Two types of receptor for insulin-like growth factors in mammalian brain. EMBO J., 4:3407.

Gray, A., Tam, A. W., Dull, T. J., Hayflick, J., Pintar, J., Cavanee, W. K., Koufos, A., and Ullrich, A., 1987, Tissue-specific and developmentally regulated transcription of the insulin-like growth factor 2 gene. DNA, 6:283.

Grey, V., Rouyer-Fessard, C., Gammeltoft, S., Bourque, M., Morin, C., and Laburthe, M., 1991, IGF-II/Man-6-P receptors are transiently increased in the rat distal intestinal epithelium after resection. Mol. Cell. Endocrinol., in print.

Hari, J., Pierce, S. B., Morgan, D. O., Sara, V., Smith, M. C., and Roth, R. A., 1987, The receptor for insulin-like growth factor II mediates an insulin-like response. EMBO J., 6:3367.

Haselbacher, G. K., and Humbel, R., 1982, Evidence for two species of insulin-like growth factor II (IGF-II and "big" IGF-II) in human spinal fluid. Endocrinology, 110:1822.

Hedley, P. E., Dalin, A. M., Engströmm, W., 1989, Developmental regulation of insulin like growth factor II gene expression in the pig. Cell Biol. Int. Reports, 13:857.

Herzog, V., Neumüller, W., Holzmann, B., 1987, Thyroglobulin, the major and obligatory exportable protein of thyroid follicle cells, carries the lysosomal recognition marker mannose-6-phosphate. EMBO J., 6:555.

Humbel, R. E., 1984, Insulin-like growth factors, somatomedins, and multiplication stimulating activity: chemistry, in: "Hormonal proteins and peptides", C.H. Li, ed., Academic Press, New York, pp. 57-59.

Hylka, V. W., Teplow, D. B., Kent, S. B. H, and Straus, D. S.,1985, Identification of a peptide fragment from the carboxyl-terminal extension region (E-domain) of rat proinsulin-like growth factor-II. J. Biol. Chem., 260:417.

Kiess, W., Haskell, J. F., Lee, L., Greenstein, L. A., Miller, B. E., Aarons, A. L., Rechler, M. M., and Nissley, S. P., 1987, An antibody that blocks insulin-like growth factor (IGF) binding to the type II IGF receptor is neither an antagonist nor an inhibitor of IGF-stimulated biologic responses in L6 myoblasts. J. Biol. Chem., 262:12745.

Lammers, R., Gray, Alane, Schlessinger, J., and Ullrich A.,1989, Differential signalling potential of insulin- and IGF-1-receptor cytoplasmic domains. EMBO J., 8:1369.

Lee, S.-J., and Nathans, D., 1988, Proliferin secreted by cultured cells binds to mannose 6-phosphate receptors. J. Biol. Chem., 263:3521.

Marquardt, H., Todaro, G. J., Hendersen, L. E., and Oroszlan, S., 1981, Purification and primary structure of a polypeptide with multiplication-stimulating activity from rat liver cell cultures. J. Biol. Chem., 256:6859.

Morgan, D. O., Edman, J. C., Standring, D. N., Fried, V. A., Smith, M.C., Roth, R. A., and Rutter, W. J., 1987, Insulin-like growth factor II receptor as a multifunctional binding protein. Nature, 329:301.

Nielsen, F. C., and Gammeltoft, S., 1988, Insulin-like growth factors are mitogens for rat pheochromocytoma PC 12 cells. Biochem. and Biophys. Res. Com., 154:1018.

Nielsen, F. C., and Gammeltoft, S., 1990, Mannose-6-phosphate stimulates proliferation of neuronal precursor cells. FEBS Letters, 262:142.

Nielsen, F. C., Gammeltoft, S., and Christiansen, J., 1990,Translational discrimination of mRNAs coding for human insulin-like growth factor II. J. Biol. Chem., 265: 13431.

Nielsen, F. C., Wang, E., and Gammeltoft, S., 1990, Receptor binding, endocytosis, and mitogenesis of insulin-like growth factors I and II infetal rat brain neurons. J. Neurochem., in press.

Nielsen, F. C., Haselbacher, G. K., and Gammeltoft, S., 1991, Biosynthesis of insulin-like growth factor II in a human rhabdomyosarcoma cell line. Mol. Cell. Endocrinol., submitted.

Nishimoto, I., Hata, Y., Ogata, E., and Kojuma, I., 1987, Insulin-like growth factor II stimulates calcium influx in compentent BALB/c 3T3 cells primed with epidermal growth factor. J. Biol. Chem., 262:12120.

Oka, Y., Rozek, L. M., and Czech, M. P., 1985, Direct demonstration of rapid insulin-like growth factor II receptor internalizatioand recycling in rat adipocytes. J. Biol. Chem., 260:9435.

Okamoto, T., Katada, T., Murayama, Y., Ui, M., Ogata, E., and Nishimoto, I, 1990, A simple structure encodes G protein-activating function of the IGF-II/Mannose 6-phosphate receptor. Cell, 62:709.

Polychronakos, C., Guyda, H. J., Patel, B., and Posner B. I., 1986, Increase in the number of type II insulin-like growth factor receptors during propylthiouracil-induced hyperplasia in the rat thyroid. Endocrinology, 119:1204.

Purchio, A. F., Cooper, J. A., Brunner, A. M., Lioubin, M. N., Gentry, L. E., Kovacina, K. S., Roth, R. A., and Marquardt, H., 1988, Identification of Mannose 6-phosphate in two asparagine-linked sugar chains of recombinant transforming growth factor-β1 precursor. J. Biol. Chem., 263:14211.

Rosen, O. M., 1990, Insulin-receptor approaches to studying protein kinase domain. Diabetes Care, 13:1990.

Roth, R. A., 1988, Structure of the receptor for insulin-like growth factor II: the puzzle amplified. Science, 239:1269.

Schwartz, T. W., 1986, The processing of peptide precursors. Proline-directed arginyl cleavage and other monobasic processing mechanisms. FEBS Lett., 200:1.

Scott, C. D., and Baxter, R. C., 1990, Insulin-like growth factor-II/Mannose-6-phosphate receptors are increased in hepatocytes from regenerating rat liver. Endocrinology, 126:2543.

Steiner, D. F., Quinn, P. S., Chan, S. J., Marsh, J., and Tager, H. S., 1980, Processing mechanisms in the biosynthesis of proteins. Ann. NY Acad. Sci., 343:1.

Sussenbach, J. S., 1989, The gene structure of the insulin-like growth factor family. Progress in Growth Factor Res., 1:33.

Ullrich, A., Bell, J. R., Chen, E. Y., Herrera, R., Petruzelli, L. M., Dull, T. J., Gray, A., Coussens, L., Liao, Y.-C., Tsubokawa, M., Mason, A., Seeburg, P. H., Grunfeld, C., Rosen, O. M., and Ramachandran, J., 1985, Human insulin receptor and its relationship to the tyrosine kianse family of oncogenes. Nature, 313:756.

Ullrich, A., Gray, A., Tam, A. W., Yang-Feng, T., Tsubokawa, M., Collins, C., Henzel, W., Le Bon, T., Kathuria, S., Chen, E., Jacobs, S., Francke, U., Ramachandran, J., and Fujita-Yamaguchi, Y., 1986, Insulin-like growth factor I receptor primary structure: comparison with insulin receptor suggests structural determinants that define functional specificity. EMBO J., 5:2503.

Ullrich, A., and Schlessinger, J., 1990, Signal transduction by receptors with tyrosine kinase activity. Cell, 61:203.

Verland, S., and Gammeltoft, S., 1989, Functional receptors for insulin-like growth factors I and II in rat thymocytes and mouse thymoma cells. Mol. Cel. Endocrinol., 67:207.

Verland, S., Nielsen, F. C., and Gammeltoft, S., 1990, Gamma-interferon induces expression of mannose-6-phosphate/insulin-like growth factor II receptor gene in rat thymocytes. Endocrinology, submitted.

Von Figura, K., and Hasilik, A., 1986, Lysosomal enzymes and their receptors. Ann. Rev. Biochem., 55:167.

Yang, Y. W. H., Romanus, J. A., Liu, T. Y., Nissley, S. P., and Rechler, M. M., 1985a, Biosynthesis of rat insulin-like growth factor II. I. Immunochemical demonstration of A ~ 20-kilodalton biosynthesis precursor of rat insulin-like growth factor II in metabolically labelled BRL-3A rat liver cells. J. Biol. Chem., 260:2570.

Yu K.-T., Peters, M. A., and Czech, M. P., 1986, Similar control mechanisms regulate the insulin and type I insulin-like growth factor receptor kinases. J. Biol. Chem., 261:11349.

Zumstein, P. P., Lüthi, C., and Humbel, R. E., 1985, Amino acid sequence of a variant pro-form of insulin-like growth factor II. Proc. Natl. Acad. Sci. USA, 82:3169.

DEVELOPMENT OF A SPECIFIC RADIOIMMUNO ASSAY FOR E DOMAIN CONTAINING FORMS OF INSULIN-LIKE GROWTH FACTOR II

James F. Perdue, Linda K. Gowan*, W. Robert Hudgins, Joan Scheuermann, Beverly Foster, and Sharron Northcutt Brown

Laboratory of Molecular Biology
Jerome H. Holland Laboratory for the Biomedical Sciences
American Red Cross
15601 Crabbs Branch Way
Rockville, Maryland 20855

*Electro-Nucleonics Inc.
7101 Riverwood Drive
Columbia, Maryland 21046

INTRODUCTION

The analysis of cDNA clones for human (1,2) and rat (3) insulin-like growth factor-II (IGF-II)[1] has led to the prediction that the processed forms of the growth factors, i.e. M_r = 7422 (67 amino acids) for human IGF-II are synthesized as precursors with an extension of 89 amino acids at the carboxyl terminus. This extension is termed the E domain. Moses et al (4) identified two precursor forms of rat IGF-II (originally designated multiplication-stimulating activity) in the conditioned medium of Buffalo rat liver, i.e. BRL-3A, cells with appM_rs = 16,270 (MSA-1) and 8,700 (MSA-II). Human serum, spinal fluid and tissue extracts also contain high M_r forms of IGF-II (5-8). Zumstein et al., (5) have purified a M_r = 10,000 variant form of IGF-II from serum that contained Cys-Gly-Asp for Ser^{33} in the C domain and an E domain extension of 21-amino acids. This 10 kDa IGF-II may be similar or identical to the "big IGF-II" that was reported to be present in human serum and in spinal fluid (6). We have isolated a still larger form of IGF-II from normal human serum (7). N-terminal amino acid sequence analysis through the first 28 residues and RRAs using rat placental membranes established it as a form of IGF-II. As evidenced from its mobility during SDS-PAGE, it has an appM_r = 15,000. Very recently, Hudgins et al., (8) established that normal human serum contains several forms of precursor IGF-II with acidic isoelectric points, i.e. pI's. The mass and acidic nature of one of these molecules with an apparent M_r=15,000 was contributed, in part, by polysaccharides and sialic acids, respectively (8). These results may explain the observations of several investigators that extracts from the tissues and serum of patients with malignant tumors contain a broad size range of IGF-II (2,9,10).

RIAs that use polyclonal or monoclonal antisera prepared against M_r = 7,500 IGF-II have been used to quantitate the amounts of M_r = 10,000 IGF-II (5), big IGF-II (6) and the large M_r forms of IGF-II in extracts of the plasma from patients with human tumors (9,10). Values of 50-

120, 20-30, and 1000- 2000 ng/ml have been calculated for serum (6) and cerebrospinal fluids (6), and plasma from patients with tumors (10), respectively. Quantitation of the large M_r forms of IGF-II in the different tissue and serum samples by RIA using polyvalent serum (5,6 and 9) are at best estimates since they gave non-parallel displacement curves compared to serum-derived, completely processed M_r = 7,500 IGF-II. In the studies with the Amano mAb (10), the contribution of the E-domain to its specificity (11) has not been evaluated. Moreover, since the epitopes recognized by both Abs are present on M_r = 7,500 IGF-II, the different forms of the growth factor must be separated by chromatography prior to the determination of IGF-II concentrations by RIA. We have, therefore, undertaken to develop a RIA for the higher M_r forms of IGF-II that does not require chromatography. This has been achieved by preparing antiserum to a chemically synthesized peptide E(1-21) corresponding to the twenty-one amino acids of the E domain beginning at Arg^{68} and employing IgG from this antiserum to develop a specific and sensitive RIA. In this report, we describe the application of this antisera and RIA for the isolation and characterization of higher M_r forms of IGF-II from human plasma.

MATERIALS AND METHODS

Cohn fraction IV_1 paste was obtained from Hyland Laboratories, and single donor samples of serum were obtained from staff members of the Holland Laboratory of the American Red Cross (Rockville, MD).

Purification of IGFs

Human IGF-I, M_r = 7,500 IGF-II, and the higher M_r forms of IGF-II or MSA-III, MSA-II and MSA-III, i.e. rat IGF-II with M_rs = 15,000, 9842 and 7468 were purified from Cohn fraction IV_1 paste or the conditioned media of cultured BRL cells, respectively, by procedures described in references 7,12, and in the text of this manuscript. Purity of the preparations was assessed by SDS-PAGE in the Laemmli buffer system (13) and from their N-terminal amino acid sequence and composition; protein concentration was determined by amino acid analysis (12).

Synthesis of Peptide and Preparation of Polyclonal Antiserum

The peptide Arg-Asp-Val-Ser-Thr-Pro-Pro-Thr-Val-Leu-Pro-Asp-Asn-Phe-Pro-Arg-Tyr-Pro-Val-Gly-Lys, i.e. the E(1-21) peptide, that corresponded to the amino acid sequence in the E-domain from Arg^{68} to Lys^{88} was chemically synthesized using an Applied Biosystems Model 430 A automated peptide synthesizer and purified by reverse-phase HPLC on a 10 x 250 mm Synchrom RP-P C18 column. Purity was assessed by amino acid composition. The peptide was chemically coupled to bovine thyroglobulin (TRG; Sigma) or keyhole limpet hemocyanin (KLH; Sigma) at a molar ratio of 85:1 with the coupling reagent 1-ethyl-3-(3'-dimethylaminopropyl) carbodiimide hydrochloride (Pierce) in 10 mM sodium phosphate, pH 6.5, 0.15 M NaCl at room temperature for 72 h. The efficiency of coupling to TRG and KLH were 19% and 17%, respectively. Rabbits (New Zealand White female) were immunized intradermally with 500 μg of conjugate in Freund's complete adjuvant and subsequently boosted with 250 μg of conjugate in Freund's incomplete adjuvant at intervals of 2, 4, and 7 weeks. The results reported in this paper were obtained using antiserum drawn 17 days after the final boost. Immune IgG was isolated on a column of Protein A-Sepharose CL4B (Sigma) and stored in 2 ml aliquots at -20°C. A molar absorption coefficient of 1.34 ml/mg at 280 nm was used to calculate the concentration of the IgG.

Iodination

One μg of the synthetic peptide E(1-21) was iodinated by the chloramine T method (14), diluted with 2 ml of 0.1% (v/v) trifluoracetic acid (TFA), and absorbed to a C_{18} Sep-Pak

(Waters). Following a wash with 20 ml of 0.1% TFA, the radiolabeled peptide was eluted in 2 ml of 50% (v/v) acetonitrile, 0.1% TFA. Bovine serum albumin (Sigma, RIA grade) was added to a concentration of 1 mg/ml and the radiolabeled peptide was stored at 4°C. IGFs were iodinated by a lactoperoxidase technique as previously described (15).

Solution Phase RIA

Polyclonal antisera raised against E(1-21) peptide or IgG isolated from this antisera was employed in a solution phase RIA using PEG to precipitate the immune complex (16). Briefly, 100 μl of antiserum or IgG that had been diluted from 1:400 to 1:3000 fold in 100 mM sodium phosphate, pH 7.4, 0.1% w/v ovalbumin (RIA buffer), was incubated overnight at 24°C with buffer, standards or unknown samples and 20,000 cpms of ^{125}I-E-peptide or appM_r = 15,000 IGF-II in a final volume of 0.4 ml (equilibrium assay). Immune complexes were precipitated by the addition of 100 μl of 3% (w/v) bovine gamma globulin (Sigma) and 500 μl of 25% (w/v) PEG-8000 (Baker), chilled to 4°C and centrifuged at 2500 rpm for 30 min at 4°C to form pellets that were counted in a LKB 1272 Clini Gamma Counter. In a nonequilibrium assay, the samples were incubated overnight in the absence of ^{125}I-labeled ligand followed by a 4 h incubation with 50,000 cpms of the ^{125}I-labeled ligand prior to the addition of PEG. The data is expressed as B/B_o, i.e. ligand bound in the presence of competing unlabeled ligand divided by the quantity of ligand bound in the absence of competing ligand after correction for non-specific binding (NSB). NSB was calculated from the cpm obtained in the presence of 1 μg of E(1-21) peptide or in the absence of antibody.

Sephadex G-50 Gel Filtration

Serum samples (typically 100 μl) were acidified with one volume of 2 M HAc, incubated for 30 min at 24°C, and the IGFs separated from binding proteins by chromatography on a 0.7 x 27 cm column of Sephadex G-50 superfine (Sigma) in 1 M HAc at a flow rate of 0.1 ml/min. The elution positions of M_r = 7,500 IGF-II and appM_r = 15,000 IGF-II were determined by chromatographing radioactivity-labelled growth factors individually in the presence of 100 μl of acidified serum.

Fractions to be assayed for high M_r IGF-II were dried (Savant Speed Vac), dissolved in RIA buffer, and assayed in triplicate. If dried fractions were to be tested for IGF-binding protein (17), they were dissolved in 50 mM sodium phosphate, pH 6.5, 0.25% fatty-acid free BSA, and assayed in duplicate.

TSK-2000 Gel Filtration

Serum samples (typically 100 μl) were acidified by the addition of a 0.25 volume of 1 M HAc, 0.5 M trimethylamine (TMA), pH 2.8, and incubated for 30 min at 24°C. Samples were microfuged to remove precipitated proteins, injected onto a 0.75 x 60 cm TSK-2000 column (LKB), and eluted in 0.2 M HAc-0.1 M TMA at a flow rate of 0.5 ml/min to separate the IGFs from the binding proteins (17). The column was calibrated for the elution of M_r = 7,500 IGF-II, appM_r = 15,000 and IGF-II binding protein as described above for the Sephadex G-50.

C-18 Cartridge Reversed Phase Chromatography

Serum samples (typically 150 μl) were acidified by the addition of an equal volume of 0.5 M HCl and incubated for 2 h at 24°C. Acid-dissociated IGFs were separated from binding proteins by (i) incubating the acidified serum for 5 min with an activated C-18 resin (Waters Sep-Pak cartridge, Millipore Corporation, Milford, MA); (ii) removing loosely bound proteins from the resin by washing with 10 ml of 4% HAc; and (iii) eluting bound IGFs with 4 mls of abs MeOH (18). The percentage recovery of M_r = 7,500 IGF-II and appM_r = 15,000 IGF-II was determined by measuring the levels of these growth factors in serum samples that had been spiked with ^{125}I-labeled or 37.5 ng of unlabeled IGFs prior to the acidification step.

Assays for IGF-II and Binding Protein

A rat placental membrane RRA (15), an RIA using mAb to IGF-II (Amano International, Troy, VA) and procedures similar to those described for the solution phase RIA were used to quantitate IGF-II. The quantity of IGF binding protein in the serum of fractions separated by chromatography and in the eluate from the Sep-Pak was determined by a charcoal absorption procedure (19).

Isoelectric Focusing of High M_r Forms of IGF-II

High M_r forms of IGF-II that had been dissolved in 25 μl of 9.5 M Urea, 20% (v/v) NP-40, 2% (v/v) pH 4 to 6.5 ampholine (Pharmacia, Piscataway, NJ) were separated based on their isoelectric points (pIs) on 190 x 3 mm in diameter tube gels of 4% (w/v) polyacrylamide that had been prerun for 3 h at 800 volts. Anodic and cathodic buffers were 100 mM H_3PO_4 and 20 mM NaOH, respectively. After washing of the gels to remove ampholines, the proteins were transferred to nitrocellulose and identified by immunostaining using the anti E(1-21) peptide Ab, goat biotinylated antirabbit IgG and alkaline phosphatase coupled to Avidin D (Vector Laboratories, Inc., Burlingame, CA). The approximate pI of each protein was determined by slicing a control gel into 1 cm in length segments and measuring their pH following 3 h of incubation in distilled water.

RESULTS

Characterization of the E (1-21) antisera and the development of a RIA

Six rabbits were immunized with E(1-21) peptide coupled to KLH or TRG. Serum from two of the rabbits that had been immunized with the former precipitated ^{125}I-E(1-21) peptide at titres of sera of between 1:200 to 1:3200 in a solution phase equilibrium assay. IgG isolated from one of the serums precipitated 87 to 20% of the ^{125}I-labeled peptide at concentrations of 11 to 0.01 μg of protein (data not presented). Using concentrations as low as 20 ng of this IgG and 20,000 to 50,000 cpm's of radioactivity, a sensitive RIA was developed. The majority of the data have been obtained using the RIA under equilibrium conditions. As is illustrated in Figure 1A, unlabeled E(1-21) peptide competed with ^{125}I-E(1-21) peptide for binding to the Ab at concentrations of between 0.05 to 6 ng. Unlabeled peptides displaced 50% of the bound ^{125}I-E(1-21) peptide at a concentration of 0.38 $\pm$ 0.06 ng (mean $\pm$ Sd of four experiments). The sensitivity of the assay could be increased approximately 3-fold by using non-equilibrium conditions of incubation, e.g. B/B_o = 0.5 occurs at a concentration of 0.14 $\pm$ 0.02 ng of unlabeled ligand. AppM_r = 15,000 IGF-II competed with the ^{125}I-E(1-21) peptide for binding to the Ab. Under equilibrium incubation conditions, 50% of the labeled peptide was displaced by about 3 ng of unlabeled appM_r = 15,000 IGF-II. As illustrated in Figure 1B, assays can also be carried out using ^{125}I-appM_r = 15,000 IGF-II as the labeled antigen and unlabeled E(1-21) peptide as the competing antigen.

Specificity of the E (1-21) peptide antibody

The Ab made against the E(1-21) peptide is specific for forms of IGF-II that have E domain sequences but it does not recognize hIGF-II of M_r = 7,500, rIGF-I or insulin (Figure 1A). Although there are four amino acid differences predicted between the amino acid sequence of the E-domains of human IGF-II and MSA-II, i.e. rat IGF-II with M_r = 9842 (3,4), the latter competed with ^{125}I-appM_r = 15,000 IGF-II for binding to the antibody at a half-maximal concentration of 2.5 ng (Figure 1B). In this same experiment, MSA-III (i.e., rat IGF-II with M_r = 7468 which lacks an E domain) did not compete with ^{125}I-appM_r = 15,000 IGF-II for binding to the antiserum at concentrations as high as 50 ng. The specificity of the antisera for the different M_r forms of IGF-II has been confirmed in dot blot experiments (unpublished results).

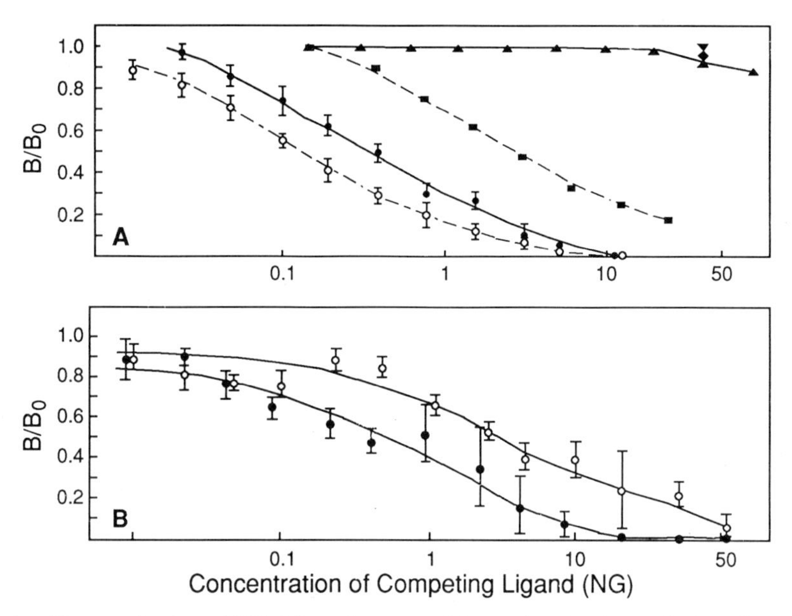

Figure 1. Characterization of E(1-21) antiserum and its application in a solution phase RIA.

Panel A: Competition for the formation of ^{125}I-E(1-21)-antibody complexes by unlabeled ligands. ^{125}I-E(1-21) peptide was incubated in RIA buffer under equilibrium conditions containing anti E(1-21) peptide IgG and the unlabeled ligands, E(1-21) peptide (●), appM_r = 15,000 IGF-II (■), M_r = 7,500 IGF-II (▲), insulin (▼), or IGF-I (♦) by procedures described in Materials and Methods. The Ab was also incubated with radiolabeled and unlabeled E(1-21) peptide (○) under non-equilibrium conditions. The results of these experiments demonstrated that anti E(1-21) peptide antibody interacts specifically and with moderate affinity to ligands containing the E domain sequence of IGF-II. The concentration of E(1-21) peptide that displaced 50% of the specificality bound ^{125}I-E(1-21) peptide from the Ab under equilibrium and non-equilibrium conditions were 0.35 and 0.14 ng, respectively.

Panel B: Competition for the formation of ^{125}I-appM_r = 15,000 IGF-II-antibody complexes by unlabeled E(1-21) peptide (●) and MSA-II (rIGF-II M_r = 9842) (○). Half-maximal inhibition of the formation of the antigen-antibody complex occurred with 1 ng of E(1-21) peptide or 2.5 ng of MSA-II.

Gel Filtration Chromatography of the IGFs

The quantitative determination of IGF's in serum requires additional manipulations because of their association with carrier proteins (10,16,17) and the frequent interferences that these binding proteins cause in RIAs and RRAs. Gel filtration has proven to be a generally applicable method for the separation of serum binding protein from IGF's with M_r = 7500. It had not been established, however, if these procedures separate binding protein from the high M_r forms of IGF-II. Sephadex G-50 gel chromatography and TSK-2000 HPLC (17), therefore, were evaluated for their ability to separate high M_r IGF-II from adult serum binding protein. For this evaluation, each chromatography procedure was calibrated for the elution positions of (a) ^{125}I-labeled appM_r = 15,000 IGF-II and M_r = 7,500 IGF-II; (b) binding proteins; and (c) high M_r IGF-II forms as evidenced by their ability to compete with ^{125}I-appM_r = 15,000 IGF-II for binding to the E(1-21) antibody.

In the Sephadex G-50 gel system, the majority of the applied serum proteins eluted in the void volume of the column (Figure 2C). Binding proteins (Figure 2B) were separated from M_r = 7,500 IGF-II (Figure 2A), however, they overlapped slightly with the eluting appM_r = 15,000 IGF-II (Figure 2A) and other high M_r forms of IGF-II (Figure 3C). During chromatography on TSK-2000, the proteins present in serum can be separated into two peaks near the void volume (Figure 3C). Binding protein was present in the second of these peaks (Figure 3B). There were smaller differences between the elution positions of appM_r = 15,000 and M_r = 7,500 IGF-II on the TSK-2000 (Figure 3A) than on the Sephadex G-50 (Figure 2A). This is not a liability since the latter does not cross react with the E(1-21) antibody. However, this gel filtration system did not completely separate the high M_r forms of IGF-II from the binding protein (Figure 4C).

C-18 Cartridge Reverse Phase Chromatography of the IGFs

Due to (i) the incomplete separation of large M_r forms of IGF-II from binding proteins by Sephadex G-50 or TSK-2000 exclusion gel chromatography procedures, (ii) the use of the noxious reagent TMA in TSK-2000 HPLC, and (iii) the inordinate amount of time to chromatographic individual serum samples on Sephadex G-50, a Sep-Pac C-18 reverse phase separation procedure (18) was evaluated. The recoveries of the different IGFs from sera were estimated from the % recovery of ^{125}I-labeled growth factors or of unlabeled growth factors that had been added to a sample of serum prior to acidification and C-18 absorption-desorption chromatography.

The recoveries of ^{125}I-labeled IGF-II of M_r = 7,500, M_r = 7,800 variant (12) or appM_r = 15,000 in the absolute MeOH eluted fractions was about 60%. This is about 10% less than that of recovered ^{125}I-IGF-I (Table 1). These results are consistent with the observed greater hydrophobicity of all three forms of IGF-II compared to IGF-I and of the iodinated compared to the uniodinated forms of the growth factors during rp HPLC on C-4 and C-18 columns (7,8, and 12). In agreement with the observations of Davenport, et al. (18), acidified IGF binding proteins are not absorbed to the C-18 resin under acidic conditions but are found almost exclusively in the flow through and wash fractions. The small quantity of absorbed binding protein could be eluted, in part, from the Sep-Pac along with the IGFs by abs MeOH. Based on an analysis of four samples, it was equal to 0.7% of the quantity of binding protein in the initial 150 μl of serum. Decreasing the MeOH concentration to 70% or using acetonitrile or acidified MeOH or acetonitrile increased the quantity of eluted ^{125}I-IGFs but it also increased the quantity of eluted binding protein in these fractions by 4- to 7-fold (data not presented).

An estimate of the recovery of appM_r = 15,000 IGF-II that had been in sera isolated by absorption to and desorption from the C-18 resins was obtained by measuring the concentration of the growth factor in control sera and in sera that had been "spiked" with 250 ng/ml of the purified polypeptide. In these assays, dried MeOH eluted fractions were suspended in buffer and

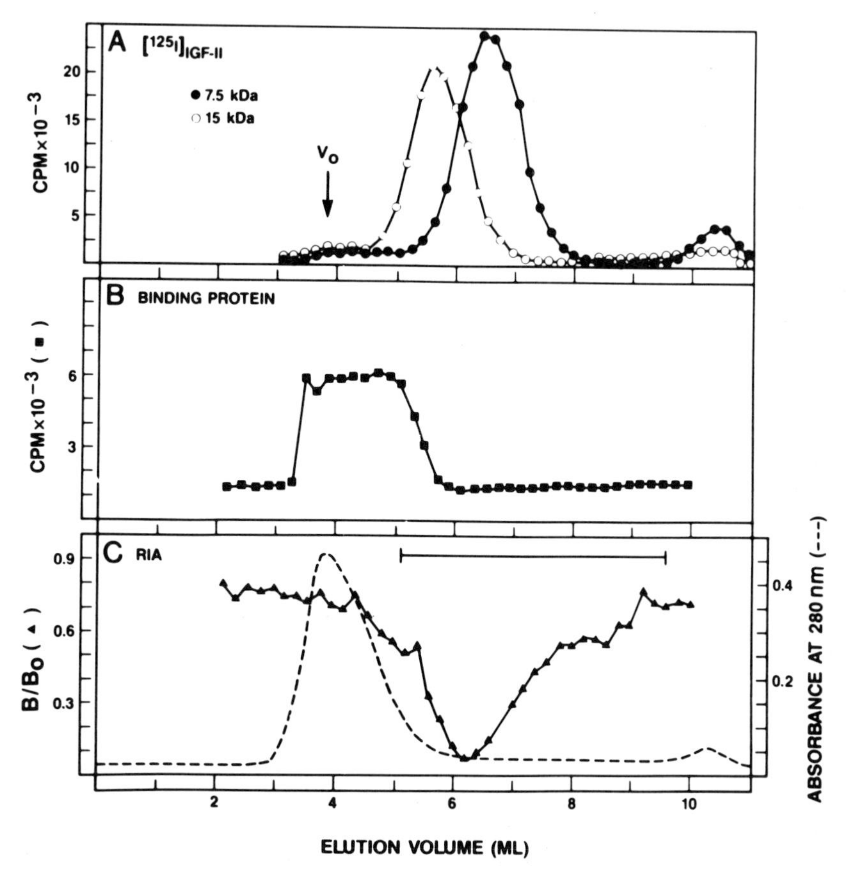

Figure 2. Evaluation of the separation of IGF's from binding protein by Sephadex G-50 gel filtration chromatography. ^{125}I-labeled IGF-IIs and 0.1 ml of acidified human serum were chromatographed on a column of Sephadex G-50 as described in Materials and Methods.

Panel A: The elution of [^{125}I]-labeled 7.5 kDa (●) and 15 kDa IGF-II (○).

Panel B: The distribution of IGF-binding protein.

Panel C: The distribution of protein (---) and high M_r forms of IGF-II (▲-▲). The solid line indicates the fractions which were combined for the quantification of total IGFs and residual binding protein.

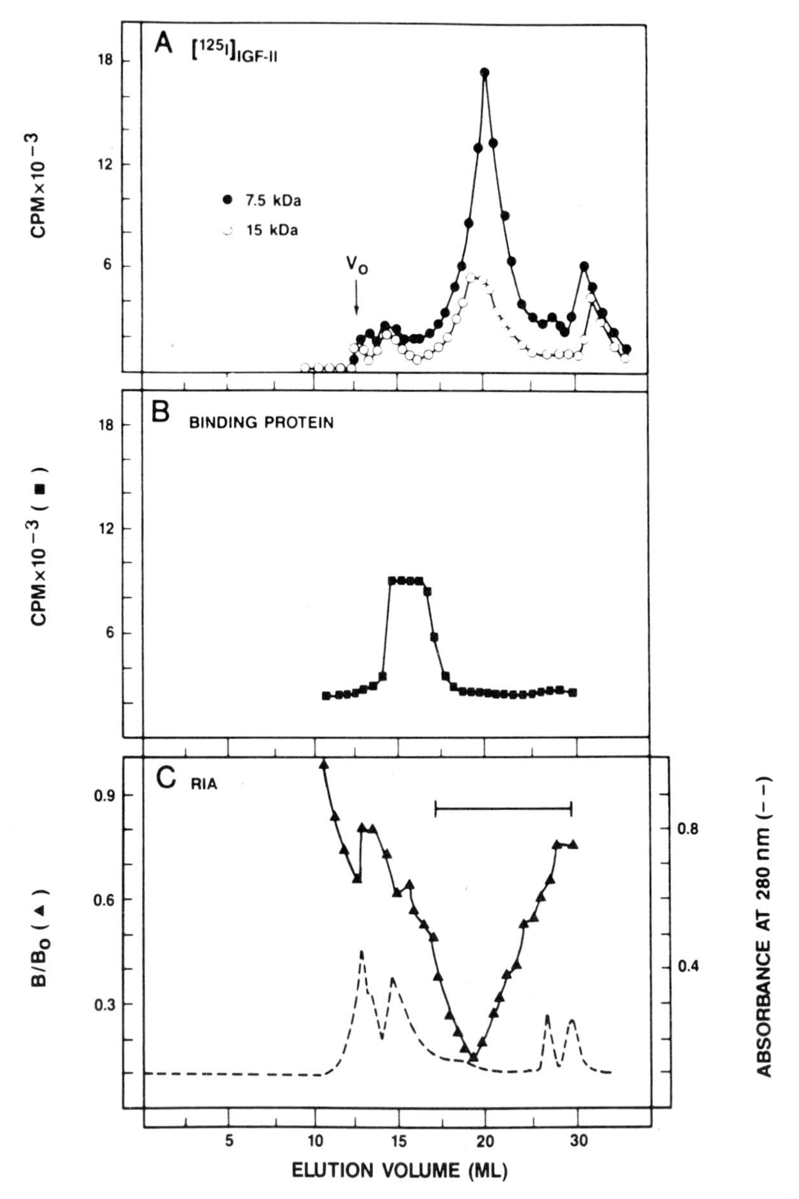

Figure 3. Evaluation of the separation of IGF's from binding protein by TSK-2000 gel filtration chromatography. ^{125}I-labeled IGF-IIs and 0.1 ml of acidified human serum were chromatographed on a TSK-2000 SW column as described in Material and Methods.

Panel A: The elution of [^{125}I]-labelled 7.5 kDa, (●) and 15 kDa IGF-II (○).

Panel B: The distribution of IGF-binding protein.

Panel C: The distribution of protein (----) and high M_r forms of IGF-II (▲-▲).

aliquots corresponding to 15 μl of untreated serum or 2 μl of "spiked" serum were serially diluted to provide a range of unknown IGF-II concentrations. The E-domain competing activity at each dilution of the reconstituted preparation was determined by the non-equilibrium RIA described in Figure 1A. These values were used to calculate the concentration of higher M_r forms of IGF-II in the serum and the percent recovery of appM_r = 15,000 IGF-II that had been added to the serum prior to its isolation from the Sep-Pac. The E(1-21) peptide equivalent competing activity in the serum of two normal adults was found to be equal and was 28 $\pm$ 3 (S.d.) ng/ml. The mean percent recovery of added appM_r = 15,000 IGF-II in four experiments was 90 $\pm$ 8 (S.d.).

Applications Using the Anti E(1-21) Antibody

We have identified anti E(1-21) Ab-reactive forms of IGF-II with appM_rs of 10,000, 13,000, 15,000 and 16,000 and IGF-II variant (12) with appM_rs of 10,000 and 13,000 in extracts of Cohn fraction IV_1 that had been purified to near homogeneity (20). These molecules have a pI that is more acidic than M_r = 7,500 IGF-II. This is illustrated in Figure 4 which shows an anti E(1-21) peptide Ab immunoblot of a mixture of precursor IGF-II forms that had been electrofocused between pH 6.5 and 4.0. Two forms with pIs of 5.1 and 5.4 and a minor component with a pI of 4.9 were identified. Subsequent analysis by SDS-PAGE and additional studies established that the pI 5.1 and 5.4 forms have appM_r of 15,000 and 11,500, respectively, and have polysaccharide additions through an O-glycosidic linkage at Thr^{75} contributes four to five kDa of mass to the appM_r = 15,000 form of IGF-II (8).

DISCUSSION

The presence of large M_r forms of IGF-II in human serum, CSF and extracts of human brain and pituitary (and presumably other tissues), requires the development of more specific procedures for their quantification than the currently available RIAs that recognize all M_r forms of IGF-II. We have developed such a method using antisera that is directed against the E(1-21) peptide portion of the E domain of IGF-II beginning at Arg^{68} and ending at Lys^{88}. This antisera is specific since it interacts only with the E(1-21) peptide, the appM_r = 15,000 IGF-II that contains an identical E domain sequence and with a rat IGF-II of M_r = 10,000 (Figures 1A and B). It did not interact with human or rat IGF-IIs that lack an E domain or insulin or IGF-I. This antibody has been employed in a sensitive solution-phase assay under equilibrium and non-equilibrium conditions. The latter modification increased the sensitivity of the assay about 3-fold from a B/B_o = 0.5 of 0.38 to 0.14 ng.

This assay has been used to follow the distribution and the recoveries of the higher M_r forms of IGF-II during the fractionation of HAc extracts of Cohn fraction IV_1 (20). The Ab has been employed also in the detection by Western blotting of the higher M_r forms of IGF-II that had been separated by isoelectric focusing. The finding that the precursor forms of IGF-II have acidic pIs led to experiments that established that they contianed sialic acid and other sugars that were chemically linked to the polypeptide at Thr^{76} (8).

Three methods were evaluated for separating acid-dissociated IGFs from their binding proteins. Of the approaches that had been evaluated, i.e. Sephadex G-50 or TSK-2000 gel filtration chromatography (19) and C-18 cartridge reverse phase chromatography (18), only the latter separated the binding proteins from the growth factors cleanly and quickly. Finally, using this method we have determined that the quantity of E peptide equivalents in the serum of two normal adults was 31 ng/ml. A recovery of 90% was assumed for this calculation. Since the mass of the E(1-21) peptide is about one-fourth that of appM_r = 15,000 IGF-II, a value of 120 ng/ml is a more realistic estimate of the quantities of precursor forms of IGF-II in their blood.

TABLE 1

Distribution of ^{125}I-Labeled rIGFs During C-18 Absorption Chromatography

Fraction	rIGF-I	hIGF-II		
		M_r=7,500	appM_r=15,000	M_r=7,800 variant
	%	%	%	%
Number of experiments	4	5	3	5
Flow through and wash	13.2 ± 1.7*	19.6 ± 6	20.4 ± 2.8	18.2 ± 2
MeOH Elution	72.5 ± 2.8	62.1 ± 13	56.0 ± 9	61.2 ± 8
C-18 Cartridge	15.5 ± 1.7	18.4 ± 8	18.6 ± 10.1	19.6 ± 10
Total recovery	101.2	100.1	95.3	99.6

*Mean ± S.d.

Acidified sera from normal individuals was "spiked" with 100,000 cpms of ^{125}I-labeled IGF-I or the different M_r forms of IGF-II and the unlabeled and labeled growth factors separated from IGF-binding proteins by absorption to and elution from C-18 Sep-Pac cartridges by procedures (18) described in Materials and Methods. Approximately 60% of the IGF-II and 70% of IGF-I can be eluted from the cartridges with Abs MeOH. The contamination of these preparations with IGF-binding protein was less than 1%.

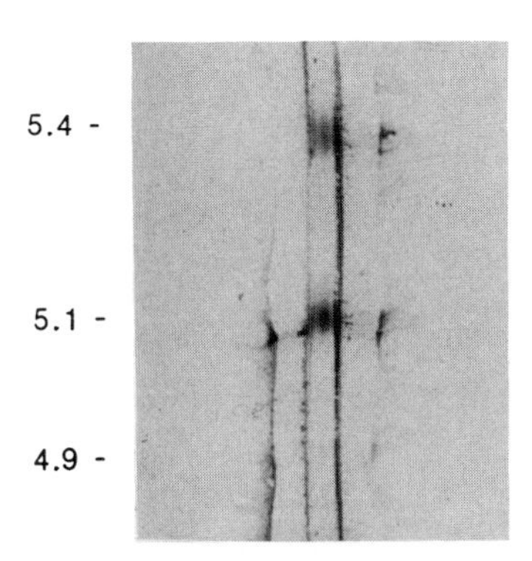

Figure 4. Separation of precursor forms of IGF-II by isoelectric focusing - Enriched preparations of precursor forms of IGF-II were separated according to their pIs on 4% (w/v) acrylamide gels containing ampholines in the pH 4.0 to 6.5 range and electrotransfered to nitrocellulose. Higher M_r forms of IGF-II were identified by immunoblotting with anti E(1-21) peptide Ab and their pIs by procedures described in Materials and Methods. Precursor forms of IGF-II contain two major species with pIs of 5.4 and 5.1 and a minor species with a pI of 4.9. SDS-PAGE of the pI 5.1 form established that it had an appM_r = 15,000.

This value agrees with the quantities of M_r = >7,500 IGF-II observed by Daughaday et al. (21), for normal serum, i.e. 142 ± 11 ng/ml, following its separation from M_r = 7,500 IGF-II by chromatography on Biogel P-60 and quantification by the Amino RIA.

ACKNOWLEDGEMENTS

The authors gratefully acknowledge Drs. Akira Komoriya and Yvonne Yang for the synthesis of the E(1-21) peptide, Dr. Wilson Burgess, Tevie Mehlman and Brian Hampton for the isolation and analysis of the peptide, Kitty Wawzinski for typing assistance and Etana Finkler, Kristine Thompson and Joe Watson for graphic and photographic assistance. This work was supported, in part, by grants to J.F.P. from the National Institutes of Health (CA 47150) and the Donaldson Charitable Trust.

REFERENCES

1. G.I. Bell, J.P. Merryweather, R. Sanchez-Pescador, M.M. Stempien, L.P. Priestley, J. Scott, L.B. Rall, Sequence of a cDNA clone encoding human preproinsulin-like growth factor II, Nature (London) 310:775 (1984).
2. W.H. Daughaday, and P. Rotwein, Insulin-like Growth Factors I and II Peptide, Messenger Ribonucleic Acid and Gene Structures, Serum, and Tissue Concentrations Endocrine Rev. 10:68 (1989).
3. H.J. Whitfield, C.B. Bruni, R. Frunzio, J.E. Terrell, S.P. Nissley, M.M. Rechler, Isolation of a cDNA clone encoding the rat insulin-like growth factor-II precursor, Nature (London) 312:277 (1984).
4. A.C. Moses, S.P. Nissley, P.A. Short, M.M. Rechler, J.M. Podskalny, Purification and characterization of multiplication-stimulating activity, Eur. J. Biochem. 103:387 (1980).
5. P.P. Zumstein, C. Luthi, R. Humbel, Amino acid sequence of a variant pro-form of insulin-like growth factor II, Proc. Natl. Acad. Sci. (USA) 82:169 (1985).
6. G. Haselbacher, R. Humbel, Evidence for two species of insulin-like growth factor II (IGF-II and "big" IGF-II) in human spinal fluid, Endocrinology 110:1822 (1982).
7. L.K. Gowan, B. Hampton, D.J. Hill, R.J. Schlueter, and J.F. Perdue, Purification and characterization of a unique high M_r form of insulin-like growth factor II, Endocrinology 121:449 (1986).
8. W.R. Hudgins, B. Hampton, W.H. Burgess, and J.F. Perdue, 2nd International IGF Symposium, San Francisco, CA, January 1991. Abstract, in press
9. G.K. Haselbacher, J.C. Irmilnger, J. Zapf, W.H. Ziegler, and R.E. Humbel, Insulin-like growth factor II in human adrenal pheochromocytomas and Wilms' tumors: Expression at the mRNA and protein level, Proc. Natl. Acad. Sci., (USA) 84:1104 (1987).
10. W.H. Daughaday, M. Kapadia, Significance of abnormal serum binding of insulin-like growth factor II in the development of hypoglycemia in patients with non-islet-cell tumors, Proc. Nat. Acad. Sci. 86:6778 (1989).
11. H. Tanaka, O. Asami, T. Hayano, I. Sasaki, Y. Yoshitake, and K. Nishikawa, Identification of a family of insulin-like growth factor II secreted by cultured rat epithelial-like cell line 18, 54-SF: Application of a monoclonal antibody, Endocrinology 124:870 (1989).
12. B. Hampton, W.H. Burgess, D.R. Marshak, K.J. Cullen, and J.F. Perdue, Purification and characterization of an insulin-like growth factor II variant from human plasma, J. Biol. Chem. 264:19155 (1989).
13. U.K. Laemmli, Cleavage of structural proteins during the assembly of the head of bacteriophage T4, Nature (London) 227:680 (1970).

14. W.M. Hunter, and F.C. Greenwood, Preparation of iodine-131-labeled human growth hormone of high specific activity, Nature 194:495 (1962).
15. J.F. Perdue, J.K. Chan, C. Thibault, P. Radaj, B. Mills, W.H. Daughaday, The biochemical characterization of detergent-solubilized insulin-like growth factor II receptors from rat placenta, J. Biol. Chem. 258:7800 (1983).
16. R.W. Furlanetto, L.E. Underwood, J.J. Van Wyk, A.J. D'Ercole, Estimation of somatomedin-C levels in normals and patients with pituitary disease by radioimmunoassay, J. Clin. Invest. 60:648 (1977).
17. C.D. Scott, J.L. Martin, R.C. Baxter, Production of insulin-like growth factor I and its binding protein by adult rat hepatocytes in primary culture, Endocrinology 116:1094 (1985).
18. M.L. Davenport, M.E. Svobolda, K.L. Koerber, J.J. VanWyk, D.R. Clemmons, and L.E. Underwood, Serum concentrations of insulin-like growth factor II are not changed by short term fasting and re-feeding, J. Clin. Endo. Metab. 67:1231 (1988).
19. J.L. Martin, and R.C. Baxter, Insulin-like growth factor binding protein from human plasma: Purification and characterization, J. Biol. Chem. 261:8754 (1986).
20. W.R. Hudgins, and J.F. Perdue, High Molecular Weight Forms of Insulin-Like Growth Factor-II, 72nd Annual Meeting of the Endocrine Society, Atlanta, Georgia, June 1990, Abs. P. 306.
21. W.H. Daughaday, J.-G. Wu, S.-D. Lee, and M. Kapadia, Abnormal Processing of pro-IGF-II in Patients with Hepatoma and in Some Hepatitis B Virus Antibody-Positive Asymptomatic Individuals, J. Lab. Clin. Med. 115:555 (1990).

THE ROLE OF THE IGFs IN MYOGENIC DIFFERENTIATION

K. A. Magri, D. Z. Ewton, and J. R. Florini

Biology Department, Syracuse University
Syracuse, NY 13244

INTRODUCTION - Myogenesis is an important aspect of developmental biology that has been actively investigated in cultured cells for nearly three decades. Interest has been heightened and activity more sharply focussed during the past few years by the discovery of a family of myogenesis controlling genes, which appear to play major roles in the aspect of myogenesis to be considered here: the terminal differentiation of proliferating myoblasts to form post-mitotic myotubes in which a number of muscle-specific genes are expressed. Most attention on the control of this process has been concentrated on its negative regulation by medium components initially characterized loosely as "mitogens," but subsequently identified as Fibroblast Growth Factor (FGF) and Transforming Growth Factor-beta (TGF-ß). This laboratory has shown that the Insulin-like Growth Factors (IGFs), in contrast, are stimulators of myogenesis, and recent work suggests that autocrine/paracrine actions of the IGFs may play an important role in differentiation in low serum medium even when the exogenous growth factors are not added. Thus, initiation of myogenic differentiation *in vitro* requires not only the removal of inhibitors present in serum, but also the secretion or addition of *stimulators* of this process.

We emphasize that myogenesis as considered here is only a small portion of the extensive developmental pathway by which functioning skeletal muscle is formed. Furthermore, virtually all of the studies considered here have been done on cultured cells, and there are obvious limitations on interpretation of such results. Nevertheless, the aspects of differentiation that do occur in cultured cells closely parallel that portion of *in vivo* differentiation, so it seems likely that many of the conclusions presented here will be relevant to physiological processes as they occur in the intact organism.

Muscle Cell Lines - Immortalized cell lines offer the substantial advantages of lack of contamination by nonmuscle cells and a relatively high (but by no means absolute) degree of reproducibility. (For instance, we have been surprised by the differences between sublines of C2 and L6 myoblasts from different sources.) The L6 line was isolated from the thigh muscle of neonatal rats (Yaffe, 1968) and has been the most widely used. In our experience, it provides a useful combination of properties, exhibiting a convenient rate of differentiation that is readily amenable to manipulation by changes in medium components. Its major disadvan-

Molecular Biology and Physiology of Insulin and Insulin-Like Growth Factors
Edited by M.K. Raizada and D. LeRoith, Plenum Press, New York, 1991

tage is its lack of responsiveness to FGF (Florini et al., 1986). The C2 line (Yaffe and Saxel, 1977) is from satellite cells of adult mouse muscle, and it has been widely used in recent years. It differentiates more rapidly than L6 cells (which can be an advantage or disadvantage, depending upon the study being done), and it is responsive to FGF. Its relatively high level of expression of the IGF-II gene (Tollefsen et al., 1989a) makes it less sensitive than L6 cells to exogenous IGFs added to the medium. The BC_3H1 line is unusual in several ways. It was initially isolated (Schubert et al., 1974) from a mouse pituitary tumor, and it expresses at least one smooth muscle protein (α-actin) in addition to a number of skeletal muscle proteins. Its most striking property is its failure to fuse to form postmitotic myotubes; it has recently been elegantly demonstrated by Brennan et al. (1990b) that this deficiency is attributable to its failure to express the MyoD1 gene. This separation of biochemical and morphological differentiation allows investigation of the reversibility of processes involved in myogenesis. The MM14 mouse muscle line (Linkhart et al., 1982) has been very useful in studies of FGF action, but it has not yet been very widely studied beyond the laboratory where it was developed. A cell line of slow muscle origin, designated sol 8, has recently been derived from the week-old mouse soleus muscle (Mulle et al., 1989), and we have found some of its properties promising, although it too has not yet been used in many laboratories.

In spite of their convenience, all of these cell lines have decided limitations. For instance, the L6 line has been out of the animal for more than 20 years, and its properties may have diverged from those of skeletal muscle in undetected ways in addition to the loss of responsiveness to FGF (Florini et al., 1986) and failure to express cardiac α-actin early in differentiation (Hickey et al., 1986). Indeed, the fact that none of these cells exhibit senescence demonstrates that they are different from normal diploid cells in at least one major property. Thus we believe that it is essential that all major conclusions be checked in primary cultures of skeletal muscle cells. Two approaches are used for such studies; in most cases, skeletal muscle cells are cultured from late embryos or early postnatal (or post-hatching) animals, while others (in particular, R. Allen and his associates) have done the more difficult preparation of proliferating myogenic cells from the satellite cell population of adult muscles. In spite of these misgivings about cell lines, we know of no major conclusion concerning actions of hormones and growth factors on terminal myogenic differentiation reached with cell lines that could not be confirmed in studies on primary muscle cell cultures.

The Myogenesis Determination Genes - MyoD1 (Davis et al., 1987; Tapscott et al., 1988) and *myd* (Pinney et al., 1988) were the first genes described as controlling determination to the myogenic lineage. Myogenin was initially identified as a controller of terminal differentiation in both rat (Wright et al., 1989) and mouse (Edmondson and Olson, 1989) muscle cell lines. (The distinction about controls of these two aspects of myogenesis has not remained so clear as more has been learned about these and related genes.) Subsequently, Braun et al. (1989a) described human counterparts of murine MyoD1 and myogenin, and they also cloned an additional gene, Myf-5, that had not previously been recognized in animal studies. The essentially identical genes MRF4 (Rhodes and Konieczny, 1990), herculin (Miner and Wold, 1990), and Myf-6 (Braun et al., 1990) have recently been added to this family of putative muscle determination genes, all of which are capable of converting nonmuscle cells to the myogenic lineage. These factors have been shown to be expressed exclusively in skeletal muscle, and they appear to positively autoregulate their own expression as well as to stimulate expression of other members of the family (Braun et al., 1989b; Thayer et al., 1989). A significant role for MyoD1 in myogenesis is indicated by reports that its expression is blocked by inhibitors of myogenic differentiation such as transforming growth factor-beta (TGF-ß) and fibroblast growth factor (FGF) (Vaidya et al., 1989) or expression of the H-*ras* oncogene (Konieczny et al., 1989). MyoD1 is not expressed in L6 or BC3H1 myoblasts (Braun et al., 1989b), and MRF4/herculin/myf-6 is not expressed in any myogenic cell line (Rhodes and Konieczny, 1990; Miner and Wold, 1990; Braun et al., 1990), so it seems that not all of these genes are essential for expression of the myogenic program in all muscle cells. Alternatively, it is possible that some factors may act earlier or later in development and not at those stages of development that are being observed *in vitro*. In contrast, myogenin is expressed in every cell undergoing myogenic differentiation (Wright et al., 1989; Edmondson and Olson, 1989; Braun et al., 1989b; Florini et al., unpublished). The interrelationships among these genes (some of which can induce expression of the others) are not yet clear, and our present level of understanding is insufficient to explain the physiological significance of each of the members of the myogenic gene family.

All of these myogenic factors are members of a class of DNA binding proteins (characterized by Murre et al., 1989a) that possess a distinctive secondary structure in which two amphipathic helices are separated by an intervening loop. This helix-loop-helix (HLH) motif has extensive homology with a region of the c-*myc* proto-oncogene family. The myogenic factor proteins also share a conserved highly basic domain adjacent to the HLH region. This ~70 amino acid conserved basic/HLH region is both necessary and sufficient to convert 10T1/2 cells to the myogenic lineage (Tapscott et al., 1988) and for DNA binding and activation of target genes (Davis et al., 1990). MyoD1 (Murre et al., 1989b) and myogenin (Brennan and Olson, 1990) dimerize through the HLH domain with ubiquitous proteins, designated E12 and E47, which are also members of the HLH family of proteins and were initially identified through their binding to immunoglobulin gene enhancers. Myogenin and MyoD1 have been shown to bind to a specific target sequence, termed the E box (CANNTG), that has been found upstream of several target genes such as the muscle creatine kinase (Brennan and Olson, 1990; Lassar et al., 1989), MLC 1/3 (Rosenthal et al., personal communication) and the acetylcholine receptor (Wang et al., 1990) genes. This rapidly emerging field has recently been reviewed by Olson (1990).

Mitogenesis *vs*. Myogenesis - For many years, most investigators in this field accepted the suggestion by Konigsberg (1971) that myogenic differentiation is negatively controlled by mitogens, which force cells in G1 phase of the cell cycle to enter S phase and to continue proliferation rather than to undergo terminal differentiation. This view was strongly supported by the widespread observation that myogenic differentiation is suppressed under conditions in which rapid growth of the cells occurs (high serum and chick embryo extract concentrations) and is triggered by incubation in low serum. It was further supported by a number of elegant studies from the Hauschka and Glaser laboratories on the inhibitory effects of FGF, which is a potent mitogen in the presence of serum. In general, the possibility that serum might contain non-mitogenic inhibitors of differentiation was not considered, and repeated reports, going back more than two decades, that insulin (mitogenic at the concentrations used) *stimulated* differentiation (de la Haba, 1966) were generally ignored. In spite of the clear understanding that mitogenicity and inhibition of differentiation by FGF are unrelated (Linkhart et al., 1982; Spizz et al., 1986), many investigators still write (and say) that they are stimulating differentiation by removing mitogens or by incubating cells in low mitogen medium. We (Florini et al., 1986) and others (Olson et al., 1986; Massague et al., 1986) have found that TGF-ß, which is not a mitogen for muscle cells, inhibits myogenic differentiation even in confluent cultures at high cell density. TGF-ß and FGF have both been demonstrated to act at specific loci unrelated to cell proliferation. For example, both of these growth factors are known to block MyoD1 (Vaidya et al., 1989) and myogenin (Heino and Massague, 1990) expression. Thus it is clear that TGF-ß and FGF inhibit myogenic differentiation by mechanisms unrelated to cell proliferation.

EFFECTS OF THE IGFs ON MUSCLE CELLS

General Anabolic Actions - The IGFs exert a number of anabolic actions on skeletal muscle cells (reviewed by Florini, 1987), as they do on many other kinds of cells (reviewed by Baxter, 1988). Increased rates of amino acid uptake and cell proliferation in L6 cells - as well as decreased protein degradation - have been shown by a number of groups (Beguinot et al., 1985; Ballard et al., 1986; Roeder et al., 1986; Ewton et al., 1987; Roeder et al., 1988). Florini et al. (1984) also showed stimulation of leucine, thymidine, and uridine incorporation in response to IGF-II. Increased protein synthesis and decreased degradation in rat L8 cells has been reported by Gulve and Dice (1989). DeVroede et al. (1984) showed increased amino acid and glucose uptake by IGFs and insulin in mouse BC_3H1 cells. These observations in cell lines have been extended to cultures prepared directly from animal muscles. In primary chick muscle cell cultures, increased proliferation and differentiation (Schmid et al., 1983) in response to IGFs, and decreased proteolysis in myotubes treated with IGF-II (Janeczko and Etlinger, 1984) has been reported. Allen's group (Dodson et al., 1985; Allen and Boxhorn, 1989) demonstrated that IGFs and insulin stimulate proliferation and differentiation in satellite cells from adult rat muscle. IGF-I and insulin stimulate protein synthesis in fetal sheep muscle myoblasts (Harper et al., 1987), but with a lower potency than in primary rat muscle cells or L6 myoblasts. In primary human fetal myoblasts, a concentration-dependent stimulation of amino acid uptake by IGF-I, IGF-II, and insulin was found (Hill et al., 1986), but higher levels of each growth factor were required to stimulate ^3H-thymidine incorporation. (This is in agreement with the observation (Ewton et al., 1987) that the stimulation of amino acid uptake and inhibition of proteolysis in L6 cells were approximately an order of magnitude more sensitive to IGFs and insulin than were the stimulation of proliferation and of differentiation.) Shimizu et al. (1986) reported that IGF-I and IGF-II were equally active in stimulating amino acid uptake in myoblasts, but IGF-I was more potent in myotubes. Unlike FGF (Linkhart et al., 1982), IGFs do not require the presence of other serum components to exert their mitogenic activity (Florini et al., 1986).

Thus it appears that a pleiotypic set of anabolic actions are exerted by IGFs on all skeletal muscle cells. Indeed, we have never seen a report of any negative results in studies on effects of IGFs on such cells, although we (Florini et al., unpublished) have found relatively low sensitivity of the sol 8 and C2 cell lines to exogenous IGFs. This may be attributable to their production of high levels of these growth factors (see below).

Stimulation of Differentiation by the IGFs - It was reported by de la Haba et al. (1966) that high levels of insulin stimulate differentiation in chick muscle cells, and Mandel and Pearson (1974) and others subsequently found similar stimulatory effects of insulin on L6 cells (Ball and Sanwall, 1980), and chick cells (Kumegawa et al., 1980). The latter two reports also demonstrated synergism with glucocorticoids and thyroxine, respectively, but these relationships have not been explored in subsequent studies. Ewton and Florini (1981) were the first to show that relatively low levels of IGFs stimulated L6 cell differentiation measured either as morphological (fusion) or biochemical (CK activity) changes, and suggested that the previous observations on insulin represented cross-reaction with the IGF-I receptor (Florini and Ewton, 1981). This stimulation is also observed in primary muscle cell cultures; similar stimulation by IGFs was observed in chick embryo muscle cells (Schmid et al., 1983), and in rat satellite cells (Allen and Boxhorn, 1989). Spira et al. (1988) reported a 20-fold enhancement of myogenesis by insulin and the IGFs in a population of skeletal muscle cells derived from adult rat pituitaries. Thus the stimulation of differentiation by the insulin-like hormones appears to be a general phenomenon, even in the case of cell lines such as the C2 and sol 8 lines that secrete substantial amounts of these growth factors themselves (Tollefsen et al., 1989a,1989b; Florini et al., manuscript in preparation).

In L6 cells, the concentration dependency curve for stimulation of differentiation is strikingly biphasic in nature, returning almost to the baseline at high concentrations of IGF-I, IGF-II, or insulin (Florini et al., 1986). For IGF-I, the stimulation exhibits the expected logarithmic character from approximately 0.5 to 40 ng/ml, with a rather sharp peak followed by lower levels of differentiation down to near-control levels at very high concentrations (640 ng/ml). For IGF-II and insulin, the curves are displaced to the right, corresponding to the generally lower potencies of these agents. The biphasic response was observed at both the biochemical (CK) and morphological (fusion) levels, as well as in the elevation of myogenin mRNA (see below). The decrease at higher concentrations was not due to toxicity; parallel incubations *with the same hormone solutions* gave the usual increase to a flat plateau when other responses to IGF-I (such as stimulation of amino acid uptake and inhibition of protein degradation) were measured. Such decreased responses at high levels are not uncommon for steroid hormones, but we have not seen other examples of them for peptide hormones or growth factors. (This may result in part from the fact that they have only recently become available in quantities that permit experiments with higher concentrations.) Recent experiments by a collaborating laboratory (Foster et al., unpublished) have shown that there is a similar biphasic response to IGF-I in a quite different system - the stimulation of expression of the elastin gene in aortic smooth muscle cells. The mechanism underlying this biphasic response is not obvious, but it cannot be explained simply as a result of cross-reaction with the IGF-II receptor by IGF-I, as IGF-II exhibits a similar effect.

Muscle is not the only tissue in which the IGFs stimulate differentiation. For example, porcine granulosa cells (Adashi et al., 1985) and rat epiphysial osteoblasts (Schmid et al., 1984) both differentiate in response to IGFs. Other growth factors, including TGF-ß (Roberts and Sporn, 1990) and FGF (Gospodarowicz et al., 1987) have also been shown to stimulate differentiation in some target tissues. As all of these agents are mitogens under some conditions, we must conclude that the generalization that mitogens block differentiation is simply not true.

Mechanism of IGF Stimulation of Differentiation

Initial Possibilities - Some obvious mechanisms were eliminated soon after the discovery of the stimulation of myogenesis by IGF-I. A likely possibility was that increased differentiation was a result of the greater culture density resulting from the mitogenic effects of IGF-I; dense cultures of myoblasts differentiate more rapidly than sparse cultures. This was eliminated by two kinds of experiments - incubations with an inhibitor of DNA synthesis (cytosine arabinoside) and demonstration of stimulation under conditions in which ^{3}H-thymidine incorporation was abolished (Turo and Florini, 1982), and initial plating of the cells over a wide range of densities and demonstration that the effect of IGF-I was much greater than the effect of cell density (Ewton and Florini, 1981). The possibility that the cells differentiated more rapidly upon treatment with IGF-I because of the anabolic actions of the hormone (the "happy cell" theory) was eliminated by incubating cells with a series of serum concentrations (rather than our usual serum-free conditions) and demonstration that differentiation was stimulated by IGF-I and IGF-II even in serum concentrations that supported maximal rates of proliferation of the cells (Florini et al., 1986).

Another possible mechanism was suggested by our observation that an inhibitor of polyamine formation, difluoromethylornithine (DFMO), inhibits myogenesis in L6 cells (Ewton et al., 1984). Like most growth factors, IGF-I is a potent inducer of ornithine decarboxylase, which is generally described as the rate-limiting enzyme on the pathway to polyamine synthesis. Ewton et al. (1984) showed that addition of exogenous polyamines to cells in which differentiation was inhibited by DFMO reversed that inhibition. Thus one possible mechanism for stimulation of differentiation by IGF-I would be the induction of elevated levels of polyamines in the cells. However, addition of polyamines rather than IGF-I to L6 myoblasts did not induce differentiation, so we concluded that elevated polyamine levels were necessary but not sufficient for induction of differentiation. This conclusion must be tempered by the report (Multhauf and Lough, 1986) that addition of much higher levels (100 µM) of polyamines did stimulate differentiation of mouse myoblasts.

Induction of Myogenin - At present, the most likely mechanism for the stimulation of differentiation by the IGFs is substantially increased expression of myogenin, which appears to be the most widely expressed of the myogenesis determination genes (Olson, 1990). We (Florini et al., manuscript in preparation) have recently demonstrated that treatment of L6A1 cells with IGF-

I, IGF-II, or high levels of insulin gives a very large (up to 60-fold) increase in myogenin mRNA content at 24 to 40 hours - substantially before the elevation of creatine kinase at 48 to 72 hours. Strong evidence for an essential role of myogenin gene expression in the stimulation of differentiation by IGF-I was provided by experiments in which an antisense oligodeoxyribonucleotide complementary to the first 15 nucleotides in the translated portion of the myogenin mRNA was a potent inhibitor of IGF-induced differentiation (Florini and Ewton, 1990). This inhibition was specific for differentiation; other processes stimulated by IGF-I were unaffected. It was also highly sequence specific; changes in as few as two of the nucleotides substantially decreased inhibition by the antisense oligo. In addition, we have recently shown (unpublished data) that IGF-I, IGF-II and insulin stimulate the expression of bacterial chloramphenicol acetyl transferase (CAT) in C2 muscle cells stably transfected with a 1500 bp myogenin-CAT construct, suggesting that the IGFs induce transcription of myogenin mRNA.

As promising as we find these observations, in a sense they merely point to a new question: "What is the mechanism by which IGF-I induces expression of the myogenin gene?" Obviously, experiments are under way to investigate this question. The relatively long time required for the elevation of myogenin mRNA appears to eliminate the possibility that a rapid single step is involved - i.e., that a substrate of the IGF-I receptor tyrosine kinase activity acts directly on an enhancer associated with the myogenin gene. That the receptor tyrosine kinase plays *any* role in stimulation of differentiation has been questioned by Beguinot et al., (1988), who found that the 3- to 5-fold increase in phosphorylation of the beta-subunit of the IGF-I receptor (and associated appearance of a 175 K phosphoprotein) in response to IGF that was observed in myoblasts did not occur in myotubes. They concluded that phosphorylation of this protein was involved in the stimulation of proliferation but not of differentiation. We (Magri and Florini, unpublished observations) found no detectable difference in the population of phosphoproteins under conditions in which differentiation was blocked by TGF-ß, an observation consistent with the suggestion that the IGF-I receptor tyrosine kinase does not play a role in stimulation of differentiation.

Role of IGF Receptors - The identity of the receptor mediating the stimulation of differentiation was a matter of interest to us. Three different receptors might be involved - the IGF-I receptor, the IGF-II receptor, and the insulin receptor (properties of these receptors are summarized by Rechler and Nissley (1985) and by Roth (1988)). The presence of all three kinds of receptors in isolated muscle has been demonstrated (Yu and Czech, 1984), and on BC3H1 cells (De Vroede et al., 1984) as well as L6 myoblasts (Beguinot et al., 1985; Ballard et al., 1986; Ewton et al., 1987) and chick embryonic myoblasts (Bassas et al., 1987). Early experiments with rat IGF-II (Florini et al., 1984) showed that stimulation of amino acid uptake and suppression of protein degradation were much more sensitive to IGF-II than were the stimulation of proliferation and differentiation. At that time, this suggested to us that the first two processes were mediated by the IGF-II receptor and the latter by the IGF-I receptor, which binds IGF-II with lower affinity. However, when increased availability of IGF-I made detailed concentration dependency studies possible, we (Ewton et al., 1987) found that all four actions exhibited the same relative potency (IGF-I > IGF-II > insulin); this hierarchy corresponds to the affinities of the IGF-I receptor. Further evidence for the primary role of the IGF-I receptor came from experiments in which we showed that a recombinant DNA-produced analog of IGF-I that did not exhibit detectable binding to the IGF-II receptor was as active as IGF-I purified from human plasma. Ballard et al. (1986, 1988) similarly concluded that the stimulation of protein and DNA synthesis in L6 cells was mediated by the IGF-I receptor, and also demonstrated lack of binding of two forms of bovine IGF-I to the IGF-II receptor in L6 myoblasts. Conclusive evidence for an active role of the IGF-I receptor was provided by Kiess et al. (1987), who used an antibody that specifically blocked the binding of IGF-II to the IGF-II receptor without affecting the binding of IGF-I to its receptor. This antibody did not alter the effects of IGF-II on amino acid or glucose uptake or leucine incorporation into proteins in L6 cells, and it was concluded that the IGF-II receptor did not mediate these actions of the IGFs.

As mentioned above, the observation that IGF-II was much more potent in stimulating amino acid uptake and inhibiting proteolysis than it was in stimulating cell proliferation and myogenesis initially let Florini et al. (1984) to suggest that the first two were mediated by the IGF-II receptor and the latter two by the IGF-I receptor, but the observations summarized in the preceding paragraph clearly disagreed with that view. How could the same receptor mediate actions that exhibit more than an order of magnitude difference in sensitivity? A theoretical paper by Loeb and Strickland (1987) demonstrates clearly that such observations could result from differences in affinities of second messengers for their intracellular receptors independent of receptor-ligand affinity; among other things, this analysis explains the "spare receptor" phenomenon so widely observed with peptide hormones. According to Loeb and Strickland, direct correlations of activity with receptor occupancy occur only when the intracellular message has the same affinity for its "receptor" that the plasma membrane receptor has for its ligand.

Significance of the IGF-II Receptor - The role of the IGF-II receptor in myogenesis remains unclear; in L6 cells, the relative potencies of the IGFs are consistent with IGF-II acting through the IGF-I receptor. The IGF-II receptor is present in very large quantities in L6 cells (Beguinot et al., 1985), and it is strikingly elevated during differentiation in C2 cells (Tollefsen et al., 1989b), but no function for it has been shown in muscle cells. Its well-established identity to the mannose-6-phosphate receptor (associated with the targeting of lysosomal enzymes) (Morgan et al., 1987) indicates that it might be involved in degradation of IGF-II, a possibility supported by the observation (Kiess et al., 1987) that an antibody to the IGF-II receptor inhibited IGF-II degradation by 90% in L6 myoblasts.

There have been reports of actions mediated by the IGF-II receptor in nonmuscle cells (Mellas et al., 1986; Rosenfeld et al., 1987), but the general role of the IGF-II receptor is unclear. Kovacina et al. (1989) found that antibodies to the IGF-II/mannose-6-phosphate receptor blocked the insulin and IGF-I induced decrease in protein catabolism of CHO cells; they suggested that this might result from a disruption of movement of lysosomal enzymes to the lysosomes. Unfortunately for our purposes, this study does not include a demonstration that IGF-II is more potent than IGF-I in suppressing proteolysis (as would be expected if it is acting through the IGF-II receptor), and we found the converse to be true in L6 cells (Ewton et al., 1987).

Effects of IGF Binding Proteins - Unlike other peptide hormones, the IGFs circulate bound to specific binding proteins (Baxter, 1988; Baxter and Martin, 1989), similarly to the steroids and thyroid hormones. In recent years, there has been a great deal of interest in the properties and actions of these proteins. For some time, the general view was that the binding proteins functioned to decrease metabolism and regulate the concentration of free IGFs, and most early studies showed inhibitory effects of the binding proteins on IGF actions (Zapf et al., 1979). More recently, stimulatory effects of binding proteins have been reported (Elgin et al, 1987), and a number of different binding proteins have been described (Baxter and Martin, 1989). Although this rapidly growing area is beyond the scope of this review, observations that human fetal myoblasts (Hill et al., 1989) as well as L6 and BC_3H1 myoblasts and porcine smooth muscle cells (McCusker et al., 1989; McCusker and Clemmons, 1988) all secrete IGF binding proteins, and this secretion is regulated by insulin and IGF-I. Tollefsen et al. (1989a) found a 30-fold increase in binding protein concentration in the medium during differentiation of C2 cells. The binding proteins may play an important role in regulating the growth and differentiation of skeletal and smooth muscle, but the specifics of this action are not yet clear.

Actions of the IGFs on Proto-oncogene Expression - It is logical to assume that early effects (increased proto-oncogene expression) might be on the pathway to subsequent actions (induction of myogenin gene expression). Increased proto-oncogene expression is a frequently reported early effect of growth factors in nonmuscle cells. In L6 cells, a fourfold increase in c-*fos* mRNA by IGF-I was found by Ong et al. (1987); a number of other proto-oncogenes were not affected. However, the finding that elevated c-*fos* expression in transfected L6 myoblasts inhibited differentiation (Rahm et al., 1989) makes this a rather unlikely candidate for the mediator of the IGF stimulation of myogenesis. Induction of c-*myc* by high levels (1 µg/ml) of IGF-I, and even higher levels of insulin (30 µg/ml) has been observed (Endo and Nadal-Ginard, 1986), but at these levels differentiation of myoblasts is not stimulated (see above discussion of biphasic responses). Rat-1 fibroblasts induced to overexpress H-*ras* showed an enhanced response to IGF-I and insulin (Burgering et al., 1989), but Gossett et al. (1988) have demonstrated that *ras* mimics TGF-ß and FGF in inhibiting myoblast differentiation. (A listing of proto-oncogenes that change during myogenic differentiation is presented in our earlier review (Florini and Magri, 1989), and Schneider and Olson (1988) have considered this subject in detail.) All told, the available evidence does not yet reveal any proto-oncogene that is involved in the stimulation of myogenesis by IGF, although they may well be involved in proliferative and other anabolic responses.

We conclude that we have taken a major step toward understanding of the stimulation of differentiation with the discovery that IGFs stimulate expression of the myogenin gene, but a number of unknowns remain concerning events between occupancy of the IGF receptor and increased myogenin mRNA levels that are associated with terminal myogenic differentiation.

ROLE OF AUTOCRINE/PARACRINE IGFS IN "SPONTANEOUS" DIFFERENTIATION

There is growing evidence for synthesis of the IGFs by skeletal muscle cells, and this process may play a role in control of differentiation as well as in wound healing. Local production of IGF mRNA and immunoreactive IGFs have been reported for muscle at several stages of development - IGF-II in embryonic (Stylianopoulou et al., 1988), and neonatal rats (Beck et al., 1988), as well as in fetal human muscle (Han et al., 1987, 1988). Synthesis of IGF-I is detectable in satellite cells for only three days after birth in the rat (Jennische and Olivecrona, 1987), but it can be stimulated in several ways. Growth hormone administration to hypophysectomized rats led to marked increases in IGF-I mRNA in skeletal muscle (Isgaard et al., 1989; Murphy et al., 1987) and both IGF-I and IGF-II mRNAs increased in skeletal muscle of adult rats implanted with GH-secreting cells (Turner et al., 1988). Work-induced hypertrophy of soleus or plantaris muscle results in a marked increase in IGF-I and -II transcripts in the affected muscle of both intact and hypophysectomized rats (DeVol et al., 1990). Regeneration of injured skeletal muscle is associated with increased IGF-I gene expression in that muscle (Jennische and Hansson, 1987; Jennische et al., 1987) and IGF-I mRNA is elevated in proliferating muscle cells (Edwall et al., 1989). There is a substantial increase (in both mRNA and protein) of both IGF-I (Tollefsen et al., 1989a) and IGF-II (Tollefsen et al., 1989b) in C2 myoblasts as they undergo terminal differentiation. These are accompanied by similar increases in the receptors and binding proteins for the IGFs. It seems likely to us that the relatively low sensitivity of C2 cells to exogenous IGF-I (Florini et al., manuscript in preparation) is attributable to the induction of the IGF-I and -II genes resulting in a relatively high rate of secretion of IGFs by these cells when transferred into low-serum differentiation medium.

In collaboration with P. Rotwein's laboratory , we (Florini et al., manuscript in preparation) have recently undertaken a more extended study of the role of autocrine/paracrine secretion of the IGFs in "spontaneous" (i.e., not stimulated by exogenous IGFs) differentiation. We compared several sublines of L6, C2, and sol 8 cells and found a striking correlation between rate of differentiation and accumulation of IGF-II mRNA. Experiments with antisense oligomers to the IGFs (similar to the anti-myogenin experiments described above) indicate that these oligos are potent inhibitors of differentiation in low serum medium, thus supporting the view that secretion of the IGFs stimulate differentiation even when they are not added to the medium. Our observation that the antisense oligomers do not block the stimulation of differentiation by exogenous IGF-II eliminates a number of possible alternative explanations for the antisense effects, and strongly supports the view that autocrine/paracrine secretion of the IGFs is essential for myogenesis in myoblasts shifted to low serum medium to stimulate differentiation.

Rosen et al. (1991) have recently completed a study of autocrine secretion of the IGFs that confirms and extends the observations reported in the preceding paragraph. They found that C_2C_{12} cells (which differentiate very rapidly in 2% horse serum) showed a very rapid increase (within 1 hour) in IGF-II mRNA. These cells expressed the full array of different IGF-II mRNAs, but individual transcripts appeared and disappeared with different kinetics. In contrast, the much more slowly differentiating L6 cells took a much longer time before they began to express IGF-II mRNAs, and the amounts of those mRNAs never reached the levels they did in C_2 cells. IGF-I appears to play little if any role in this "spontaneous" differentiation. Rosen et al. found low expression of IGF-I mRNAs during pre-differentiation and early post-differentiation time points in C_2 cells, but the IGF-I mRNA disappeared as differentiation continued; it could not be detected at any time in L6 cells. These striking observations are further evidence of the importance of autocrine/paracrine IGF-II in myogenic differentiation.

Another example of a paracrine effect involving IGFs has recently been reported by Ishi (1989), who found that IGF-II gene expression appears to be correlated with the formation of neuromuscular synapses in fetal rat hind limb muscles. IGF-II mRNA levels are very high prior to synapse formation and decrease postnatally at the same time that superfluous synapses are eliminated. IGF-II mRNA was found to increase dramatically in response to transection of the sciatic nerve. The author postulated that innervation generates a signal that down-regulates IGF-II gene expression. This is supported by the finding that the IGF-II gene is expressed at a remarkably high level in myoblast cultures (Tollefsen et al., 1989b) where innervation,of course, does not occur.

CHANGES IN GROWTH FACTOR RECEPTORS WITH DIFFERENTIATION

On the basis of a substantial decrease in the binding of EGF during differentiation of mouse MM14 myoblasts, Lim and Hauschka (1984) suggested that loss of growth factor receptors might be the basis of the postmitotic state of differentiated myotubes. Unfortunately, MM14 cells do not respond to EGF, so any functional consequences of this decline in receptor number could not be evaluated. When purified FGF became available, Olwin and Hauschka (1988) showed that FGF receptors also disappear from the surfaces of differentiated MM14 cells. There was a decrease from 2000 receptors/cell in MM14 myoblasts to an undetectable number in differentiated cells (24 hours after removal of FGF). In differentiation defective MM14 cells, the number of FGF receptors remained unchanged at about 9000-12000 per cell.

Not all of the receptors that decrease with differentiation bind mitogenic growth factors. We (Ewton et al., 1988) have demonstrated that TGF-ß receptors virtually disappear from several muscle cell lines as they differentiate, and TGF-ß is *not* mitogenic in these cells. (The parallel decrease in TGF-ß-stimulated amino acid uptake in L6A1 myoblasts showed that the receptors that disappeared were biologically functional.) In contrast, there was only a 50% decrease in binding of TGF-ß as BC_3H1 cells differentiated - perhaps attributable to the fact that their expression of differentiated functions can be reversed by changing the medium. Uncoupling of biochemical and morphological differentiation by EGTA showed that the downregulation of TGF-ß receptors is associated with the latter (Hu and Olson, 1990), and is not an essential biochemical event in myogenesis. There is no decrease in FGF receptors with differentiation in BC_3H1 cells which do not fuse to form myotubes upon differentiating (Olwin and Hauschka, 1988)).

Clearly, any generalization that mitogen receptors decrease with differentiation does not extend to the IGFs, which are active mitogens for muscle cells (Ewton et al., 1987). Under differentiation-inducing conditions in which the loss of TGF-ß receptors was found, we (Ewton et al., 1988) detected little or no decrease in binding of ^{125}I-IGF-I and little decrease in two responses to IGF-I - stimulation of amino acid uptake and suppression of protein degradation. Others (Beguinot et al., 1985)) have reported a decrease (about 70%) in IGF-I receptors with differentiation in a different subline of L6 cells, but this may reflect their normalizing results to protein (which continues to accumulate after myotubes are formed) rather than DNA content of the cultures, as we did. There was no decrease in binding of IGF-I and IGF-II to human muscle cells during differentiation (Shimizu et al., 1986), and a transient increase in binding of IGF-I and a substantial increase in IGF-II binding as C2 cells differentiate have been reported (Tollefsen et al., 1989a,b). There is a significant increase in both insulin receptors and responsiveness with differentiation (De Vroede et al., 1984; Beguinot et al., 1986). Thus it seems clear that the observed loss of EGF and FGF receptors does not explain the postmitotic state of committed or differentiated muscle cells, because myotubes do not respond to IGF-I by increased DNA synthesis in spite of the continued presence of functional IGF receptors.

SUMMARY AND CONCLUSIONS

Of the three families of growth factors/hormones (the FGFs, TGF-ßs, and IGFs) that have major effects on the differentiation of skeletal muscle cells, only the IGFs stimulate the process; indeed, the IGFs are the *only* well-defined agents thus far shown to stimulate myogenesis. All of these agents affect the expression of myogenin, one of the recently discovered family of myogenesis controlling genes, and TGF-ß and FGF inhibit the expression of MyoD1 as well. (L6 cells do not express MyoD1, so we have not looked for an effect of IGFs on it.) At least partly as a result of this action, these agents inhibit or stimulate all aspects of myogenic differentiation - fusion, expression of a set of muscle-specific proteins, and attainment of a postmitotic state - in all tested cell lines and primary muscle cell cultures. It is becoming clear that the myogenic controlling genes are capable of regulating expression of genes for the entire family of muscle specific proteins, so the principal question remaining about actions of these growth factors is the mechanism by which they inhibit or induce expression of the myogenin or MyoD1 genes. In spite of the uncertainty about their interactions, the discovery of the myogenesis controlling genes now provides a much sharper focus for studies on the processes involved in terminal differentiation of skeletal muscle cells. The demonstration that expression of these genes is controlled, both positively and negatively, by specific growth factors that are now readily available opens exciting new possibilities in endocrinology and developmental biology.

REFERENCES

Adashi, E. Y., Resnick, C. E., D'Ercole, A. J., Svoboda, M. E., and Van Wyk, J. J., 1985, Insulin-like growth factors as intraovarian regulators of granulosa cell growth and function, Endocr. Rev., 6:400.

Allen, R. E., and Boxhorn, L. K., 1989, Regulation of skeletal muscle satellite cell proliferation and differentiation by transforming growth factor-beta, insulin-like growth factor-I, and fibroblast growth factor, J. Cell. Physiol., 138:311.

Ball, E. H., and Sanwall, B. D., 1980, A synergistic effect of glucocorticoids and insulin on the differentiation of myoblasts, J. Cell. Physiol., 102:27.

Ballard, F. J., Read, L. C., Francis, G. L., Bagley, C. J., and Wallace, J. C., 1986, Binding properties and biological potencies of insulin-like growth factors in L6 myoblasts, Biochem. J., 233:223.

Ballard, F. J., Ross, M., Upton, F. M., and Francis, G. L., 1988, Specific binding of insulin-like growth factors 1 and 2 to the type 1 and type 2 receptors respectively, Biochem. J., 249:721.

Bassas, L., de Pablo, F., Lesniak, M., A., and Roth, J., 1987, The insulin receptors of chick embryo show tissue-specific structural differences which parallel those of the insulin-like growth factor I receptors, Endocrinology., 121:1468.

Baxter, R. C., 1988, The insulin-like growth factors and their binding proteins, Comp. Biochem. Physiol., 91B:229.

Baxter, R. C., and Martin, J. L., 1989, Binding proteins for the insulin-like growth factors: structure, regulation and function, Progr. Growth. Factor. Res., 1:49.

Beck, F., Samani, N. J., Byrne, S., Morgan, K., Gebhard, R., Brammar, W. J., 1988, Histochemical localization of IGF-I and IGF-II mRNA in the rat between birth and adulthood, Development., 104:29.

Beguinot, F., Kahn, C. R., Moses, A. C., and Smith, R. J., 1985, Distinct biologically active receptors for insulin, insulin-like growth factor I, and insulin-like growth factor II in cultured skeletal muscle cells, J. Biol. Chem., 260:1589 .

Beguinot, F., Kahn, C. R., Moses, A. C., and Smith, R. J., 1986, The development of insulin receptors and responsiveness is an early marker of differentiation in the muscle cell line L6, Endocrinol., 18:446.

Beguinot, F., Formisano, P., Condorelli, G., Tramontano, D., Villone, G., Liquoro, D., Consiglio, E., and Aloj, S. M., 1988, The endogenous substrate for the IGF-I receptor kinase pp175 in the FRTL-5 cell is a cytoskeleton-associated protein, Program. and. Abstracts,. the. Endocrine. Society., 1988:.

Braun, T., Bober, E., Buschhausen-Denker, G. Kotz, S., Grzeschik, K.-H., and Arnold, H. H., 1989a, Differential expression of myogenic determination genes in muscle cells: possible autoactivation by the myf gene products, EMBO. J., 8:3617.

Braun, T., Buschhausen-Denker, G., Bober, E., Tannich, E., and Arnold, H. H., 1989b, A novel human muscle factor related to but distinct from myoD1 induces myogenic conversion in 10T1/2 fibroblasts, EMBO. J., 8:701.

Braun, T., Bober, E., Winter, B., Rosenthal, N.,and Arnold, H. H., 1990, Myf-6, a new member of the human gene family of

myogenic determination factors: evidence for a gene cluster on chromosome 12, EMBO. J., 9:821.
Brennan, T. J., and Olson, E. N., 1990a, Myogenin resides in the nucleus and acquires high affinity for a conserved enhancer element on heterodimerization, Genes. &. Develop., 4:582.
Brennan, T. J., Edmondton, D. G., and Olson, E. N., 1990, Aberrant regulation of MyoD1 contributes to the potentailly defective myogenic phenotype of BC_3H1 cells, J Cell. Biol. 110:929.
Burgering, B. M. T., Snijders, A. J., Maassen, J. A., van der Eb, A. J., and Bos, J. L., 1989, Possible involvement of normal p21 h-ras in the insulin/Insulinlike growth factor 1 signal transduction pathway, Mol. Cell. Biol., 9:4312.
Davis, R. L., Weintraub, H., and Lassar, A. B., 1987, Expression of a single transfected cDNA converts fibroblasts to myoblasts, Cell., 51:987.
Davis, R. L., Cheng, P.-F., Lassar, A. B., and Weintraub, H., 1990, The myoD DNA binding domain contains a recognition code for muscle-specific gene activation, Cell., 60:733.
de la Haba, G., Cooper, G. W., and Elting, V., 1966, Hormonal requirements for myogenesis in vitro: insulin and somatotropin, Proc. Nat. Acad. Sci. USA., 56:1719.
DeVol, D. L., Rotwein, P., Sadow, J. L., Novakofski, J., and Bechtel, P. J., 1990, Activation of insulin-like growth factor gene expression during work-induced skeletal muscle growth, Am. J. Physiol. (Endocrinol. Metab)., 259:E89.
De Vroede, M. A., Romanus, J. A., Standaert, M. L., Pollett, R. J., Nissley, S. P., Rechler, M. M. 1984. Interaction of insulin-like growth factors with a nonfusing mouse muscle cell line: binding, action, and receptor down-regulation. Endocrinol. 114:1917-1929
Dodson, M. V., Allen, R. E., and Hossner, K. L., 1985, Ovine somatomedin, multiplication-stimulating activity, and insulin promote skeletal muscle satellite cell proliferation in vitro, Endocrinol., 117:2357.
Edmondson, D. G., and Olson, E. N., 1989, A gene with homology to the myc similarity region of myoD1 is expressed during myogenesis and is sufficient to activate the muscle differentiation program, Genes. &. Develop., 3:628.
Edwall, D., Schalling, M., Jennische, E., and Norstedt, G., 1989, Induction of insulin-like growth factor I messenger ribonucleic acid during regeneration of rat skeletal muscle, Endocrinol., 124:820.
Elgin, R. G., Busby, W. H., Jr., and Clemmons, D. R., 1987, An insulin-like growth factor (IGF) binding protein enhances the biological response to IGF-I, Proc. Natl. Acad. Sci., 84:3254.
Endo, T., and Nadal-Ginard, B., 1986, Transcriptional and post-transcriptional control of c-myc during myogenesis: its mRNA remains inducible in differentiated cells and does not suppress the differentiated phenotype, Mol. Cell. Biol., 6:1412.
Ewton, D. Z., and Florini, J. R., 1981, Effect of somatomedins and insulin on myoblast differentiation in vitro, Develop. Biol., 86:31.
Ewton, D. Z., Erwin, B. G., Pegg, A. E., and Florini, J. R., 1984, The role of polyamines in somatomedin-stimulated differentiation of L6 myoblasts, J. Cell. Physiol., 120:263.

Ewton, D. Z., Falen, S. L., and Florini, J. R., 1987, The type II IGF receptor has low affinity for IGF-I analogs: pleiotypic actions of IGFs on myoblasts are apparently mediated by the type I receptor, Endocrinology., 120:115.
Ewton, D. Z., Spizz, G., Olson, E. N., and Florini, J. R., 1988, Decrease in transforming growth factor-ß binding and action during differentiation in muscle cells, J. Biol. Chem., 263:4029.
Florini, J. R., 1987, Hormonal control of muscle growth, Muscle. &. Nerve., 7:577.
Florini, J. R., and Ewton, D. Z., 1981, Insulin acts as a somatomedin analog in stimulating myoblast growth in serum-free medium, In. Vitro., 17:763.
Florini, J. R., and Ewton, D. Z., 1990, Highly specific inhibition of IGF-I-stimulated differentiation by an antisense oligodeoxyribonucleotide to myogenin mRNA; no effects on other actions of IGF-I, J. Biol. Chem., Sub:.
Florini, J. R., Ewton, D. Z., Evinger-Hodges, M. J., Falen, S. L., Lau, R.L., Regan, J. F., and Vertel, B. M., 1984, Stimulation and inhibition of myoblast differentiation by hormones, In. Vitro., 20:942.
Florini, J. R., Ewton, D. Z., Falen, S. L., and Van Wyk, J. J., 1986, Biphasic concentration dependency of the stimulation of myoblast differentiation by somatomedins, Am. J. Physiol. (Cell. Physiol)., 250:771.
Florini, J. R., and Magri, K. A., 1989, Effects of growth factors on myogenic differentiation, Am. J. Physiol. (Cell. Physiol. 25)., 256:C701.
Florini, J. R., Roberts, A. B., Ewton, D. Z., Falen, S. B., Flanders, K. C., and Sporn, M. B., 1986, Transforming growth factor-ß. A very potent inhibitor of myoblast differentiation, identical to the differentiation inhibitor secreted by buffalo rat liver cells, J. Biol. Chem., 261:16509.
Gospodarowicz, D., Ferrara, N., Schweigerer, L., and Neufeld, G., 1987, Structural characterization and biological functions of fibroblast growth factor, Endocrine. Rev., 8:95.
Gossett, L. A., Zhang, W., and Olson, E. N., 1988, Dexamethasone-dependent inhibition of differentiation of C2 myoblasts bearing steroid-inducible N-ras oncogenes, J. Cell. Biol., 106:2127.
Gulve, E. A., and Dice, J. F., 1989, Regulation of protein synthesis and degradation in L8 myotubes. effects of serum, insulin, and insulin-like growth factors, Biochem. J., 260:377.
Han, V. K., Hill, D. J., Strain, A. J., Towle, A. C., Lauder, J. M., Underwood, L. E., and D'Ercole, A. J., 1987, Identification of somatomedin/Insulin-like growth factor immunoreactive cells in the human fetus, Pediatr. Res., 22:245.
Han, V. K., Lund, P. K., Lee, D. C., D'Ercole, A. J., 1988, Expression fo somatomedin/insulin-like growth factor messenger ribonucleic acids in the human fetus: identification characterization, and tissue distribution, J. Clin. Endocrinol. Metab., 66:422.
Harper, J. M., Soar, J. B., Buttery, P. J., 1987, Changes in protein metabolism of ovine primary muscle cultures on treatment with growth hormone, insulin, insulin-like growth factor I or epidermal growth factor, J. Endocrinol., 112:87.
Heino, J., and Massague, J., 1990, Cell adhesion to collagen and decreased myogenic gene expression implicated in the

control of myogenesis by transforming growth factor ß, J. Biol. Chem., 265:10181.

Hickey, R., Skoultchi, A., Gunning, P., and Kedes, L., 1986, Regulation of a human cardiac actin gene introduced into rat L6 myoblasts suggests a defect in their myogenic program, Mol. Cell. Biol., 6:3287.

Hill, D. J., Crace, C. J., Strain, A. J., and Milner, R. D. G., 1986, Regulation of amino acid uptake and deoxyribonucleic acid synthesis in isolated human fetal fibroblasts and myoblasts: effect of human placental lactogen, somatomedin-C, multiplication-J. Clin. End. Metab., 62:753.

Hill, D. J., Clemmons, D. R., Wilson, S., Han, V. K., Strain, A. J., and Milner, R. D., 1989, Immunological distribution of one form of insulin-like growth factor (IGF)-binding protein and IGF peptides in human fetal tissues, J. Mol. Endocrinol., 2:31.

Hu, J. S., and Olson, E. N. during Terminal Differentiation, 1990, Functional receptors for transforming growth factor-ß are retained by biochemically differentiated C2 myocytes in growth factor deficient medium containing EGTA, but down-regulated, J. Biol. Chem., 265:7914.

Isgaard, J., Nilsson, A., Vikman, K., Isaksson, O. G., 1989, Growth hormone regulates the level of insulin-like growth factor-I mRNA in rat skeletal muscle, J. Endocrinol., 120:107.

Ishi, D. N., 1989, Relationship of insulin-like growth factor II gene expression in muscle to synaptogenesis, Proc. Natl. Acad. Sci. USA., 86:2898.

Janeczko, R. A., and Etlinger, J. D., 1984, Inhibition of intracellular proteolysis in muscle cultures by multiplication stimulating activity (MSA): comparison of effects of MSA and insulin on proteolysis, protein synthesis, amino, J. Biol. Chem., 259:6292.

Jennische, E., and Hansson, H.-A., 1987, Regenerating skeletal muscle cells express insulin-like growth factor I, Acta. Physiol. Scand., 130:327.

Jennische, E., and Olivecrona, H., 1987, Transient expression of insulin-like growth factor I immunoreactivity in skeletal muscle cells during postnatal development in the rat, Acta. Physiol. Scand., 131:619.

Jennische, E., Skottner, A., and Hansson, H.-A., 1987, Satellite cells express the trophic factor IGF-I in regenerating skeletal muscle, Acta. Physiol. Scand., 129:9.

Jennische, E., Skottner, A., and Hansson, H.-A., 1987, Dynamic changes in isulin-like growth factor I immunoreactivity correlate to repair events in rat ear after freeze-thaw injury, Exper. Mol. Pathol., 47:193.

Kiess, W., Haskell, J. F., Greenstein, L. A., Miller, B. E., Aarons, A. L., Rechler, M. M., and Nissley, S. P., 1987, An antibody that blocks insulin-like growth factor (IGF) binding to the type II IGF receptor is neither an agonist nor an inhibitor of IGF-stimulated biologic responses in L6 myoblasts, J. Biol. Chem., 262:12745.

Konieczny, S. F., Drobes, B. L.,, Menke, S. L., and Taparowsky, E. J., 1989, Inhibition of myogenic differentiation by the H-ras oncogene is associated with the down regulation of the myoD1 gene, Oncogene., 4:473.

Konigsberg, I. R., 1971, Diffusion-mediated control of myoblast fusion, Develop. Biol., 26:133.

Kovacina, K. S., Steele-Perkins, G., and Roth, R. A., 1989, A role of the Insulin-like Growth Factor II/mannose-6-phosphate receptor in the insulin-induced inhibition of protein catabolism, Molec. Endocrinol., 3:901.
Kumegawa, M., Ikeda, E., Hosoda, S., and Takuma, T., 1980, In vitro effects of thyroxine and insulin on myoblasts from chick embryo skeletal muscle, Develop. Biol., 79:493.
Lassar, A. B., Buskin, J. N., Lockshon, D., Davis, R. L., Apone, S., Hauschka, S. D., and Weintraub, H., 1989a, MyoD is a sequence specific DNA binding protein requiring a region of myc homology to bind to the muscle creatine kinase enhancer, Cell., 58:823.
Lassar, A. B., Thayer, M. J., Overell, R. W., and Weintraub, H., 1989b, Transformation by activated *ras* or *fos* prevents myogenesis by inhibiting expression of MyoD1, Cell., 58:659.
Lim, R. W., and Hauschka, S. D., 1984, A rapid decrease in epidermal growth factor-binding capacity accompanies the terminal differentiation of mouse myoblasts in vitro, J. Cell. Biol., 98:739.
Linkhart, T A., Clegg, C. H., Lim, R. W., Merrill, G. F., Chamberlain, J. S., and Hauschka, S. D., 1982, Control of mouse myoblast commitment to terminal differentiation by mitogens. in Pearson, M. L., and Epstein, H. F. (Eds): Molecular and Cellular Biology of Myogenesis. Cold. Springs. Harbor. Conf., 8:877.
Loeb, J. N., and Strickland, S., 1987, Hormone binding and coupled response relationships in systems dependent on the generation of secondary mediators, Mol. Endocrinol., 1:75.
Mandel, J.-L., and Pearson, M. L., 1974, Insulin stimulates myogenesis in a rat myoblast line, Nature., 251:618.
Massague, J., Cheifetz, S., Endo, T., and Nadal-Ginard, B., 1986, Type beta transforming growth factor is an inhibitor of myogenic differentiation, Proc. Natl. Acad. Sci. USA., 83:8206.
McCusker, R. H. and Clemmons, D. R., 1988, Insulin-like growth factor binding protein secretion by muscle cells: effect of cellular differentiation and proliferation, J. Cell. Physiol., 137:505.
McCusker, R. H., Camacho-Hubner, C., and Clemmons, D. R., 1989, Identification of the types of insulin-like growth factor binding proteins that are secreted by muscle cells in vitro, J. Biol. Chem., 264:7795.
Mellas, J., Gavin, J. R., and Hammesman, M. R., 1986, Multiplication-Stimulating Activity-induced alkalinization of canine renal proximal tubular cells, J. Biol. Chem., 261:14437.
Miner, J. H., and Wold, B., 1990, Herculin, a fourth member of the myoD family of myogenic regulatory genes, Proc. Natl. Acad. Sci. USA., 87:1089.
Morgan, D. O., Edman, J. C., Standring, D. N., Fried, V. A., Smith, M. C., Roth, R. A., and Rutter, W. J., 1987, Insulin-like growth factor-II receptor as a multifunctional binding protein, Nature., 329:301.
Multhauf, C., and Lough, J., 1986, Interferon-mediated inhibition of differentiation in a murine myoblast cell line, J. Cell. Physiol., 126:211.
Mulle, C., Benoit, P., Pinset, C., Roa, M., and Changeux, J.-P., 1988, Calcitonin gene-related peptide enhances the rate of desensitization of the nicotinic acetylcholine receptor

in cultured mouse muscle cells, Proc. Natl. Acad. Sci. USA, 85:5728.
Murphy, L. J., Bell, G. I., Duckworth, M. L., and Friesen, H. G., 1987, Identification, characterization, and regulation of a rat complementary deoxyribonucleic acid which encodes insulin-like growth factor-I., Endocrinol., 121:684.
Murre, C., McCaw, P. S., and Baltimore, D., 1989a, A new DNA binding and dimerization motif in immunoglobulin enhancer binding, daughterless, myoD, and myc proteins, Cell., 56:777.
Murre, C., Mccaw, P.S., Vaessin, H., Caudy, M., Jan, L.Y., Cabrera, C.V., Hauschka, S., Lassar, A.B., and Baltimore, D., 1989b, Interactions between heterologous helix-loop-helix proteins generate complexes that bind specifically to a common DNA sequence, Cell., 58:537.
Olson, E. N., 1990, MyoD family: A paradigm for development?, Genes. &. Develop., 4:1454.
Olson, E. N., Sternberg, E., Hu, J. S., Spizz, G., and Wilcox, C., 1986, Regulation of myogenic differentiation by type beta transforming growth factor, J. Cell. Biol., 103:1799.
Olwin, B. B., and Hauschka, S. D., 1988, Cell surface fibroblast growth factor and epidermal growth factor receptors are permanently lost during skeletal muscle terminal differentiation in culture, J. Cell. Biol., 107:761.
Ong, J., Yamashita, S., and Melmed, S., 1987, Insulin-like growth factor I induces c-fos messenger ribonucleic acid in L6 skeletal muscle cells, Endocrinology., 120:353.
Pinney, D. F., Pearson-White, S. H., Konieczny, S. F., Latham, K. E., and Emerson, C. P., Jr., 1988, Myogenic lineage determination and differentiation: evidence for a regulatory gene pathway, Cell., 53:781.
Rahm, M., Jin., P., Sumegi, J., and Sejersen, T., 1989, Elevated c-fos expression inhibits differentiation of L6 rat myoblasts, J. Cell. Physiol., 139:237.
Rechler, M. M., and Nissley, S. P., 1985, The nature and regulation of the receptors for insulin-like growth factors, Ann. Rev. Physiol., 47:425.
Rhodes, S. J., Konieczny, S. F., 1990, Identification of MRF4: A new member of the muscle regulatory factor gene family, Genes & Develop, in press
Roberts, A. B., and Sporn, M. B. The Transforming Growth Factor-betas., 1990, in sporn, M. B.,and roberts, A. B. Heds) peptide growth factors and their receptors. vol 1. Handbook of Exptl Pharm vol 95. Springer Verlag, Heidelberg,., :419.
Roeder, R.A., Thorpe, S. D., Byers, F. M., Schelling, G. T., and Gunn, J. M., 1986, Influence of anabolic agents on protein synthesis and degradation in muscle cells grown in culture, Growth., 50:485.
Roeder, R. A., Hossner, K. L., Sasser, R. G., and Gunn, J. M., 1988, Regulation of protein turnover by recombinant human insulin-like growth factor-I in L6 myotube cultures, Horm. Metabol. Res., 220:698.
Rosen, K.M., Wentworth, B. M., Rosenthal, N., and Villa-Komaroff, L., manuscript in preparation.
Rosenfeld, R. G., Pham, H., James, P., Shsh, R., Diaz, G., and wyche, J, 1987, Demonstration of a autocrine for insulin-like growth factor-II, medicated through the Type II receptor. (Abstract) Endocrinol. 120 (Suppl. 1): A89.

Roth, R. A., 1988, Structure of the receptor for insulin-like growth factor II: the puzzle amplified, Science., 239:1269.
Schmid, Ch., Steiner, Th., and Froesch, E. R., 1983, Preferentiatial enhancement of myoblast differentiation by insulin-like growth factors (IGF-I and IGF-II) in primary cultures of chicken embryonic cells, FEBS. Letters., 161:117.
Schmid, C., Steiner, T., and Froesch, E. R., 1984, Isulin-like growth factor I supports differentiation of cultured osteoblast-like cells, FEBS. Letters., 173:48.
Schneider, M. D., and Olson, E. N., 1988, Control of myogenic differentiation by cellular oncogenes, Molecular. Neurobiol., 2:1.
Schubert, D., Harris, J., Devine, C. E., and Heinemann, S., 1974, Characterization of a unique muscle cell line, J. Cell. Biol., 61:398.
Shimizu, M., Webster, C., Morgan, D. O., Blau, H. M., and Roth, R. A., 1986, Insulin and insulinlike growth factor receptors and responses in cultured human muscle cells, Am. J. Physiol., 251:E611.
Spira, O., Atzmon, R., Rahamim, E., Bar-Shavit, R., Gross, J., Gordon, A., and Vlodavsky, I., 1988, Striated muscle fibers differentiate in primary cultures of adult anterior pituitary cells, Endocrinology., 122:3002.
Spizz, G., Roman, D., Strauss, A., and Olson, E. N., 1986, Serum and fibroblast growth factor inhibit myogenic differentiation through a mechanism dependent on protein synthesis and independent of cell proliferation, J. Biol. Chem., 261:9483.
Stylianopoulou, F., Efstratiadis, A. Herbert, J., and Pintar, J., 1988, Pattern of the insulin-like growth factor II gene expression during rat embryogenesis, Development., 103:497.
Tapscott, S. J., Davis, R. L., Thayer, M. J., Cheng, P.-F., Weintraub, H., and Lassar, A. B., 1988, MyoD1: A nuclear phosphoprotein requiring a myc homology region to convert fibroblasts to myoblasts, Science., 242:405.
Thayer, M. J., Tapscott, S. J., Davis, R. L., Wright, W. E., Lassar, A. B., and Weintraub, H., 1989, Positive autoregulation of the myogenic determination gene myoD1, Cell., 58:241.
Tollefsen, S. E., Lajara, R., McCusker, R. H., Clemmons, D. R., and Rotwein, P., 1989a, Insulin-like growth factors (IGF) in muscle development. expression of IGF-I, the IGF-I receptor, and an IGF binding protein during myoblast differentiation, J. Biol. Chem., 264:13810.
Tollefsen, S. E., Sadow, J. L., and Rotwein, P., 1989b, Coordinate expression of insulin-like growth factor II and its receptor during muscle differentiation, Proc. Natl. Acad. Sci. USA., 86:1543.
Turner, J. D., Rotwein, P., Novakofski, J., Bechtel, P. J., 1988, Induction of mRNA for IGF-I and -II during growth hormone-stimulated muscle hypertrophy, Am. J. Physiol., 255:E513.
Turo, K. A., and Florini, J. R., 1982, Hormonal stimulation of myoblast differentiation in the absence of DNA synthesis, Am. J. Physiol. (Cell. Physiol. 12)., 243:C278.
Vaidya, T. B., Rhodes, S. J., Taparowsky, E. J., and Konieczny, S. F., 1989, Fibroblast growth factor and transforming growth factor ß repress transcription of the myogenic regulatory gene myoD1, Mol. Cell. Biol., 9:3576.

Wright, W. E., Sassoon, D. A., and Lin, V. K., 1989, Myogenin, a factor regulating myogenesis, has a domain homologous to myoD, Cell., 56:607.

Yaffe, D., 1968, Retention of differentiation potentialities during prolonged cultivation of myogenic cells, Proc. Natl. Acad. Sci. USA., 61:477.

Yaffe, D., and Saxel, O., 1977, Serial passaging and differentiation of myogenic cells isolated from dystrophic mouse muscles, Nature. (London)., 270:725.

Yu, K.-T., and Czech, M. P., 1984, Tyrosine phosphorylation of the insulin receptor B subunit activates the receptor-associated tyrosine kinase activity, J. Biol. Chem., 259:5277.

Zapf, J., Schoenle, E., Jagers, E., Sand, I., and Froesch, E. R., 1979, Inhibition of the actions of non-suppressible insulin-like activity on isolated fat cells by binding to its carrier protein, J. Clin. Invest., 63:1077.

EXTINCTION OF HUMAN INSULIN-LIKE GROWTH FACTOR II EXPRESSION IN SOMATIC CELL HYBRIDS

Raffaele Zarrilli°, Vittorio Colantuoni*, Raffaella Faraonio*, Stefano Casola°, Elena Rossi°, and Carmelo B. Bruni°

°Dipartimento di Biologia e Patologia Cellulare e Molecolare and Centro di Endocrinologia ed Oncologia Sperimentale del C.N.R., *Dipartimento di Biochimica e Biotecnologie Mediche, II Facoltà di Medicina e Chirurgia, Università degli Studi di Napoli, Italy 80122.

INTRODUCTION

Insulin-like growth factor (IGF) II is a mitogenic polypeptide (1) that plays an important role in fetal and post-natal development (2). Both in human and rodents, IGF-II is a single copy gene that gives origin to a family of RNA transcripts (3-7). These heterogeneous mRNAs are originated from at least three promoters, functioning in many tissues of the rat during the embryonic and neonatal period (3,4,8). On the contrary, in adult animals, their expression is confined to the choroid plexus and to the leptomeninges (9). The human gene also utilizes three different promoters, two of them functioning in fetal liver and in most fetal tissues (5). The third one is a tissue specific promoter, active in adult liver only (6).

Very little is known about the molecular mechanisms by which IGF-II gene expression is regulated during human and rat development. Analysis of the cis-regulatory sequences of the three rat promoters did not allow the identification of any element able to confer developmental-specific transcription to a reporter gene (10,11). "In vitro" studies on P_2 and P_3 rat promoters by gel-shift and footprinting analysis did not evidence any difference when nuclear extracts from IGF-II-expressing and -non-expressing cells were compared (10,11).

In the present study, somatic cell hybrids (12) were employed. They have been widely used to study the mechanisms regulating tissue-specific as well as developmental-specific gene expression (13,14). Both heterokaryons and stable cell hybrids between cells derived from different embryonic lineages and developmental stages were produced and analyzed to investigate IGF-II expression during ontogeny.

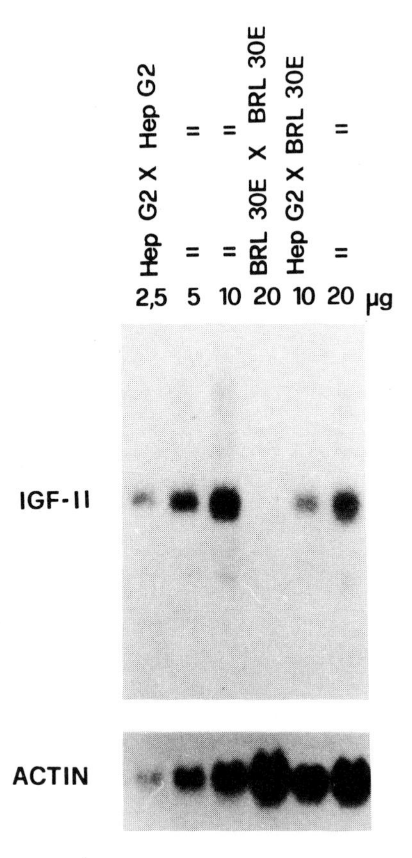

Figure 1. RNA analysis from HepG2 x BRL30E heterokaryons.
A representative Northern blot is shown. RNA isolated from HepG2 x HepG2, HepG2 x BRL30E and BRL30E x BRL30E fusions was size fractionated on a denaturing agarose gel and blotted to a nylon filter. The RNA blot was sequentially hybridized to the different probes shown in the figure. The amounts of total RNA loaded are indicated on the top of each lane.

EXTINCTION OF IGF-II EXPRESSION IN HepG2 x BRL30E HETEROKARYONS

Human hepatoma HepG2 cells (13) that express high levels of fetal 6.0, 4.8 and 2.2 kb IGF-II mRNAs, were fused to IGF-II non-expressing rat hepatocytes BRL30E (Buffalo rat liver), a thymidine kinase-deficient subclone of the BRL3A2 cells (15). Equal amounts of cells were plated together, fused 8-12 hours later using the polyethylene glycol (PEG) method (16) and harvested for RNA extraction after 40 hours. The fusion and RNA extraction were also

TABLE I. Extinction of IGF-II expression in heterokaryons.

Cell fusion products	IGF-II mRNA levels*
HepG2 x HepG2	100
BRL30E x BRL30E	-
HepG2 x BRL30E	30 ± 3.8
HGPRT-HepG2 x HGPRT-HepG2	100
TK-Fibroblasts x TK-Fibroblasts	-
HGPRT-HepG2 x TK-Fibroblasts	33 ± 2.5

* *Values are expressed as percentage of the mRNA levels found in the HepG2 cells, which have been set at 100%. The means ± SD of densitometric scan values of at least three different experiments are reported.*

TABLE II. Chromosome composition of HepG2 x TK-Fibroblasts intertypic intraspecific hybrids.

Cell lines	Chromosome number*
HGPRT-HepG2	49 ± 2.1
TK-Fibroblasts	70 ± 4.3
HGPRT-HepG2 x TK-Fibroblasts cl.1	96 ± 7.3
HGPRT-HepG2 x TK-Fibroblasts cl.2	105 ± 4.8
HGPRT-HepG2 x TK-Fibroblasts cl.3	111 ± 5.3

* *Determined by counting at least 20 metaphases for each cell line. The mean ± SD is shown for each value.*

performed on the same number of HepG2 and BRL30E cells plated alone. These latter samples were duplicated in all the experiments and taken as positive and negative control, respectively.

Figure 1 shows a Northern blot analysis of RNA extracted from HepG2 x HepG2, BRL30E x BRL30E homokaryons and HepG2 x BRL30E heterokaryons. IGF-II mRNA hybridization was detected in the HepG2 x HepG2 homokaryons, the intensity of which was proportional to the amount of the sample loaded on the gel. As expected, no hybridization was detected in BRL30E x BRL30E homokaryons. IGF-II hybridization signal was significantly less intense in the HepG2 x BRL30E heterokaryons. Densitometric analysis of

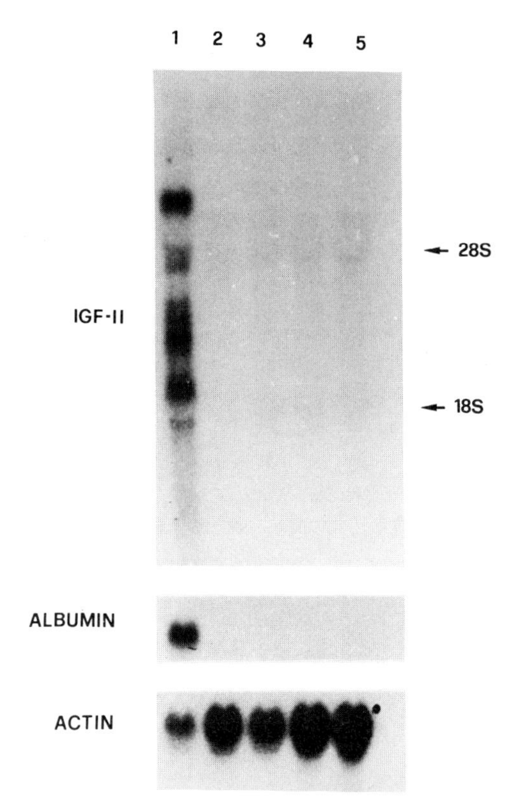

Figure 2. RNA analysis from HGPRT-HepG2 x TK-Fibroblasts intertypic hybrids. *Northern blot analysis of IGF-II, ALBUMIN and β-ACTIN mRNAs of the following parental and hybrids cells: HGPRT-HepG2 (lane 1),TK-Fibroblasts (lane 2), HGPRT-HepG2 x TK-Fibroblasts cl.1 (lane 3), cl.2 (lane 4) and cl.3 (lane 5). Ten micrograms of total RNA was loaded for each lane. The same filter was sequentially hybridized to the different probes shown. Mobility of 28S and 18S ribosomal RNA bands is indicated on the left of the figure.*

the autoradiographs revealed that IGF-II mRNA levels were reduced to a residual activity of 30% in the heterokaryons compared to the positive homokaryons (see also Table I). This reduction was the result of an extinction mechanism. The phenomenon was not complete, presumably because of the heterogeneity of the population. Moreover, cocultivation without fusion of HepG2 and BRL30E cells produced 50% reduction of IGF-II mRNA levels (data not shown). Since all the IGF-II transcripts were comparably reduced in HepG2 x BRL30E heterokaryons, we conclude that they all undergo the same phenomenon.

Down regulation of IGF-II expression was also found in HepG2 x TK-fibroblast heterokaryons (see Table I), supporting the hypothesis that the gene is subject to extinction.

EXTINCTION OF IGF-II EXPRESSION IN HGPRT-HepG2 x TK-FIBROBLASTS HYBRIDS

In order to determine whether extinction of IGF-II expression occurred also in stable cell hybrids, the same cell types used in the heterokaryons experiments were fused to obtain hybrid clones. Hypoxanthine Guanosine Phosphoribosyl transferase deficient HepG2 (HGPRT-HepG2) were fused to thymidine-kinase deficient (TK-) human fibroblasts (17) and selected in HAT medium. After three weeks, single clones were isolated; three of them were further characterized. They showed a chromosome content as expected from intraspecies hybrids (Table II) and complete extinction of some liver-specific genes (Figure 2 and References 19,20).

As shown in Figure 2, IGF-II gene expression was also completely extinguished in the hybrids. None of the mRNA species detected in the positive HGPRT-HepG2 cells (lane 1) was present in the TK-fibroblast cells (lane 2), as well as in any of the hybrid clones isolated (lanes 3-5). On the contrary, β-actin, used as an internal control of the RNA amounts loaded, gave similar hybridization signals in the parental and in the hybrid cells.

CONCLUSIONS

We demonstrate that human IGF-II gene is subject to extinction. The phenomenon is partial or complete, according to the type of fusion analyzed. It appears to be a general event, since it is observed in inter- as well in intratypic heterokaryons. The IGF-II mRNA species extinguished are those transcribed by the two fetal promoters. No mRNA transcribed from the adult liver promoter is detected in the positive HepG2 cells.

Although extinction of IGF-II is comparable to that of differentiated hepatic functions (18-20), the mechanisms underlying the two events may be different. The data presented in this report suggest that negative trans-acting factors may play a role in turning off the gene during normal human development.

REFERENCES

1. E.R. Humbel, Insulin-like growth factors I and II, *Eur. J. Biochem.* 190:445 (1990).
2. T.M DeChiara, A. Efstratiadis, and E.J. Robertson, A growth-deficiency phenotype in heterozygous mice carrying an insulin-like growth factor II gene disrupted by targeting, *Nature* 345:78 (1990).

3. R. Frunzio, L. Chiariotti, A.L. Brown, D.E. Graham, M.M. Rechler, and C.B. Bruni, Structure and expression of the rat insulin-like growth factor II (rIGF-II) gene, *J. Biol. Chem.* 261:17138 (1986).
4. L. Chiariotti, A.L. Brown, R. Frunzio, D.R. Clemmons, M.M. Rechler, and C.B. Bruni, Structure of the rat insulin-like growth factor transcriptional unit: heterogeneous transcripts are generated from two promoters by use of multiple polyadenylation sites and differential ribonucleic acid splicing, *Mol. Endocrinol.* 2:1115 (1988).
5. P. De Pagter-Holthuizen, M. Jansen, F.M.A. van Schaik, R. van der Kammen, C. Oosterwijk, J.L. Van den Brande, and J.S. Sussenbach, The human insulin-like growth factor II gene contains two development-specific promoters, *FEBS Lett.* 214:259 (1987).
6. P. De Pagter-Holthuizen, M. Jansen, R.A. van der Kammen, F.M.A. van Schaik, and J.S. Sussenbach, Differential expression of the human insulin-like growth factor II gene. Characterization of the IGF-II mRNA and an mRNA encoding a putative IGF-II-associated protein, *Biochim. Biophys. Acta* 950:282 (1988).
7. S. Cocozza, S. Garofalo, A. Monticelli, A. Conti, L. Chiariotti, R. Frunzio, C.B. Bruni, and S. Varrone, EcoRI RFLP in the human IGF-II gene, *Nucleic Acid Res.* 16:2737 (1988).
8. A.L. Brown, D.E. Graham, S.P. Nissley, D.J. Hill, A.J. Strain, and M.M. Rechler, Developmental regulation of insulin-like growth factor II mRNA in different rat tissues, *J. Biol. Chem.* 261:13144 (1986).
9. F. Stylianopoulou, J. Herbert, M.B. Soares, and A. Efstratiadis, Expression of the insulin-like growth factor II gene in the choroid plexus and the leptomeninges of the adult rat central nervous system, *Proc. Natl. Acad. Sci. USA* 85:141 (1988).
10. T. Evans, T. DeChiara, and A. Efstratiadis, A promoter of the rat insulin-like growth factor II gene consists of minimal control elements, *J. Mol. Biol.* 199:61 (1988).
11. T. Matsuguchi, K. Takahashi, K. Ikejiri, T. Ueno, H. Endo, and M. Yamamoto, Functional analysis of multiple promoters of the rat insulin-like growth factor II gene, *Biochim. Biophys. Acta* 1048:165 (1990).
12. R.L. Davidson, Gene expression in somatic cell hybrids, *Annu. Rev. Genet.* 8:195 (1974).
13. B.B. Knowles, C.C. Howe, and D.P. Aden, Human hepatocellular carcinoma cell lines secrete the major plasma proteins and hepatitis B surface antigen, *Science* 209:497 (1980).
14. C.A. Peterson, H. Gordon, Z.W. Hall, B.M. Paterson, and H. Blau, Negative control of the helix-loop-helix family of myogenic regulators in the NFB mutant, *Cell* 62:493 (1990).
15. S.P. Nissley, P.A. Short, M.M. Rechler, J. Podskalny, and H.G. Coon, Proliferation of Buffalo rat liver cells in serum-free medium does not depend upon multiplication-stimulating activity (MSA), *Cell* 11:441 (1977).
16. M. Mevel-Ninio, and M.C. Weiss, Immunofluorescence analysis of the time-course of extinction, reexpression, and activaction of albumin production in rat hepatoma-mouse fibroblast heterokaryons and hybrids, *J. Cell Biol.* 90:339 (1981).

17. *A.T.C.C.* CRL 8304.
18. A.M. Killary, and R.E.K. Fournier, A genetic analysis of extinction: trans-dominant loci regulate expression of liver-specific traits in hepatoma hybrids cells, *Cell* 38:523 (1984).
19. V. Colantuoni, A. Pirozzi, C. Blance, and R. Cortese, Negative control of liver-specific gene expression: cloned human retinol-binding protein gene is repressed in HeLa cells, *EMBO J.* 6:631 (1987).
20. R. Faraonio, M. Musy, and V. Colantuoni, Extinction of retinol-binding protein gene expression in somatic cell hybrids: identification of the target sequences, *Nucleic Acid Res.* (In Press).

THE GROWTH HORMONE-INSULIN-LIKE GROWTH FACTOR I AXIS: STUDIES IN MAN DURING GROWTH

Thomas J. Merimee, Suzanne Quinn, Betty Russell, and William Riley

University of Florida Department of Medicine
Division of Endocrinology, Box J-226, JHMHC, Gainesville, FL 32610-0226

INTRODUCTION

An individual's linear growth and development from childhood through adolescence to adulthood is modulated by multiple hormones. In man, linear growth takes place until closure of the epiphyses but distinct periods of growth acceleration or deceleration are clearly discernable within the general pattern. Growth velocity which is greatest at birth, declines steadily until the prepubertal growth spurt, at which time a dramatic increase occurs in height velocity. As a child completes puberty, the growth rate falls once again and eventually growth ceases when the epiphyses close. Growth hormone (GH) and related growth factors, in particular, insulin-like growth factor I (IGF I) are thought to control linear growth in phases of acceleration. Since growth acceleration occurs predominantly during or shortly before puberty, the sex hormones, estrogen and testosterone, may also have an important role. Unfortunately, studies to date have usually involved pooling of subjects of multiple ages to form a broad age range and in no single study has a sufficient number of subjects been studied to yield truly interpretable data. Our current study was conducted in over 3,000 children and adolescents spanning the ages from birth through 16 years. This study was motivated from preliminary data, suggesting there may be two clearly different patterns of interaction between hormones and receptors, one pattern regulating growth in boys, another in girls.

METHODS

Subjects for the current study were healthy children and adolescents attending school in the Tampa-St. Petersburg district. Ambulatory serum samples were collected from each in order to screen for the presence of antibodies to the pancreatic beta cell and through the courtesy of the Department of Pathology, these samples were made available for the present evaluation. Each sample was thawed, an aliquot of serum was obtained and the sample was refrozen. Equal volume aliquots were combined to form serum pools for girls at each year of age between

Molecular Biology and Physiology of Insulin and Insulin-Like Growth Factors
Edited by M.K. Raizada and D. LeRoith, Plenum Press, New York, 1991

3 and 16 years and for boys between 3 and 16 years. Additional sera were also obtained for some individual studies in the ages felt to represent puberty. In all samples we measured GH, IGF I, and testosterone by immunoassay. Growth hormone binding protein (GHBP), a measure of GH-receptors, was quantitated by the Ultrogel system of Daughaday (1) and confirmed by a Sephadex system according to the method of Baumann (2).

SPECIFIC METHODOLOGY

GHBP and GH-Receptors

Direct quantitation of GH-receptors on cellular membranes has not been possible in the past without tissue biopsies. In addition, even when tissues can be obtained, measurement of GH-receptors is difficult and often inaccurate in the best of laboratories. There also appears to be heterogeneity of receptor structure as well as variable biologic response to GH-receptor interaction (3).

The identification of GHBP, a protein found in normal serum, has made it possible to better assess GH receptors. GHBP can bind ^{125}I-GH added to serum samples. Studies to date have shown that its structure represents the extracellular domain of the GH receptor and that it closely reflects the receptor status of multiple tissues (4-6). Two separatory methods are currently available for quantifying GHBP levels. In our current study, all results are given for the Ultrogel separatory system (1), but virtually similar results were obtained by the Sephadex method as well (2). We assayed each serum pool for GHBP 5-7 times with internal controls accompanying each assay. All pooled serum samples were measured for GH in duplicate using a standard immunoassay technique, and IGF I was measured in duplicate at each of two dilutions (1:100 and 1:200). Serum testosterone concentrations were measured using solid phase RIA kits from Diagnostic Products Corporation (Los Angeles, CA).

Assay Methods

To ready serum or plasma for Ultrogel filtration, two assay tubes were prepared. Each tube contained 100 μl of human serum with ^{125}I-human growth hormone (20-25,000 counts/min/tube); one of these tubes also contained 2.5 μg of uniodinated human growth hormone. The samples were brought to a total volume of 250 ml with Tris buffer (Tris 25 Mm, calcium chloride 10 mM, 0.02% NaN_3 and 0.1% bovine serum albumin, PH 7.5), then incubated for two hours either at 20°C or room temperature and transferred to an appropriate column. Two hundred μl of the reaction mixture prepared as above, were added to an Ultrogel AcA 44 column measuring 0.9 x 15-17 cm. The material was then eluted at 0.12 ml/min at 22°C and collected in 10 drop fractions. The same column was used for gel filtering the sample containing only ^{125}I-human growth hormone followed by the duplicate sample containing both ^{125}I-growth hormone and excess unlabeled human growth hormone. Samples for the Sephadex column were prepared in an identical manner: the total volume was increased to 1.1 ml by addition of a phosphate buffer (11.36 g $NaHPO_4$, 2.8 g KH_2PO_4, 5.8 g NaCl, 0.4 g NaN_3 in two liters distilled water, pH 7.4) to insure even distribution across the column surface. Columns measured 1.5 x 75 cm, were eluted at a rate of .08-0.15 ml/min at 4°C, and were collected in 1.2 ml fractions. Because of column length and elution time, samples with and without excess growth hormone were run on separate but identical columns. We did not use a third column to establish an elution profile for ^{125}I-growth hormone alone as described in the initial work (2). Usually 4 to 5 samples were assayed per day, along with the internal standard. This normal reference serum was constructed from 5 pooled adult sera, the growth hormone concentration of which was <1.5 ng/ml.

The mean percentage GH bound in the control sample was 19.8 ± 1.7 ± 0.53% (n=10), (x ± SD ± SEM). Less than 10% of samples required correction for the GH present in serum. This was done as described by Daughaday (1). Variation between Ultrogel assays was 15.2 ± 3.6%, which is similar to that recorded by others with this method. The percent growth hormone bound was calculated by subtracting the free counts in the sample without excess growth hormone from the free counts in the sample containing an excess of unlabeled growth hormone. This figure was similar to that obtained by subtracting the bound fractions from each other. Prior to the above calculations the bound and free fractions were corrected for any variations in total counts between the assay tubes. Results are given as amount bound and as amount bound (%) relative to a standard adult pool of sera. Ultrogel AcA 34 and Ultrogel AcA 44 were obtained from IBF Corporation. Sephadex G100 was obtained from Pharmacia. ^{125}I-GH was obtained from Radioassay Systems Laboratory and had a mean specific activity of 57 mc/mg calculated from multiple iodinations. It was repurified on arrival and weekly on an Ultrogel AcA 34 column as described by Daughaday.

Radioimmunoassay for IGF I

The method for separating free IGF I from its binding proteins prior to immunoassay was based on the method of Zapf (7), but Sep-paks rather than Sephadex columns were used as in earlier studies. Sep-paks were obtained from Morters Association, Millipore. IGF I specific antisera was a gift from E.R. Froesch, M.D. of Zurich. The coefficient of correlation of these two systems, i.e. column vs. Sep-paks when adjusted with a proper standard control is r=0.93.

RESULTS

Figure 1 compares the elution pattern of GHBP from the Ultrogel and Sephadex columns with and without excess unlabeled GH. GHBP is depicted as Peak I on the Ultrogel column and Peak II on the Sephadex column. Peak I on the Sephadex columns represents GH bound to an uncharacterized binding protein which is non-displaceable with GH concentrations up to 200 ng/ml and is unrelated to GHBP (2,8). This binding remains relatively constant from individual to individual and is unrelated to the percent binding by GHBP. As recently measured in our laboratory in 50 individuals and 50 pooled samples, it comprises 5.1% ± 1.43% of total growth hormone and remains constant throughout the age groups studied. Ultrogel peak II and Sephadex peak III represent free growth hormone and the last and smallest peaks are the ^{125}I-salts. Addition of unlabeled GH displaces radioactivity from the GHBP to the free peak.

GHBP Concentrations in Serum

From birth through adolescence and into early adulthood (ages 23-26) the binding of ^{125}I-GH by the specific high affinity GHBP in human serum (as measured by radioactivity in Peak I on the Ultrogel and Peak II on the Sephadex columns) progressively increased. At birth or shortly thereafter, binding of ^{125}I-GH is 2.0 ±

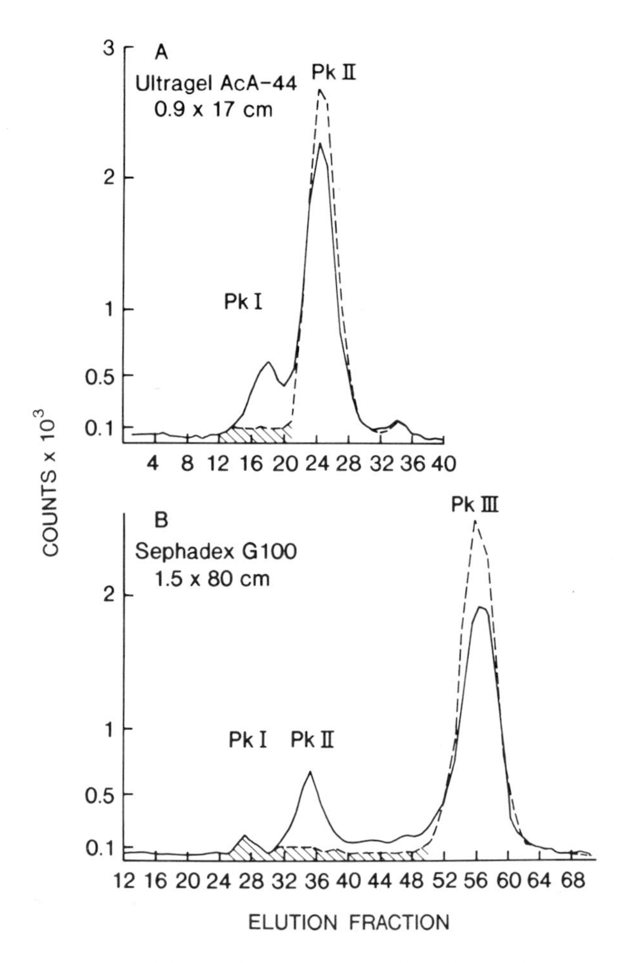

Figure 1. Ultrogel (Top): Peak 1 = GHBP. Peak II = free GH
Sephadex (Bottom): Peak 4 = GHBP. Peak III = free GH
Addition of unlabeled GH displaces radioactivity from the GHBP to the free GH peak (solid to dashed curve).

0.8%; at two years of age, 3.5 ± 0.56% (x ± SEM). These figures are respectively 11 ± 2% and 18 ± 3% of normal adult binding, $p<.001$. Fifty percent of the normal mean binding of young adulthood is achieved by ages 6 to 7 years.

Although the percent of ^{125}I-GH bound by the high affinity binding protein increased with each year of age, the rate at which the increase of binding occurred clearly varied between the years 10 through 16 and the younger age period of 3-10 years (See Fig. 2). For the younger age group, GH-binding increased at 1.2 ± 0.11% yearly until approximately 10 years of age. From ages 10 through 16 years, GH-binding then increased at a rate of 0.33 ± .04% per year. From the mean binding of 19.8% ± 2.2% (x ± SD) of $_{125}$I-GH present at age 23-25, it can be estimated that binding after age 16 increases at approximately 0.38% per year, a figure close to that calculated for the years 10 through 16.

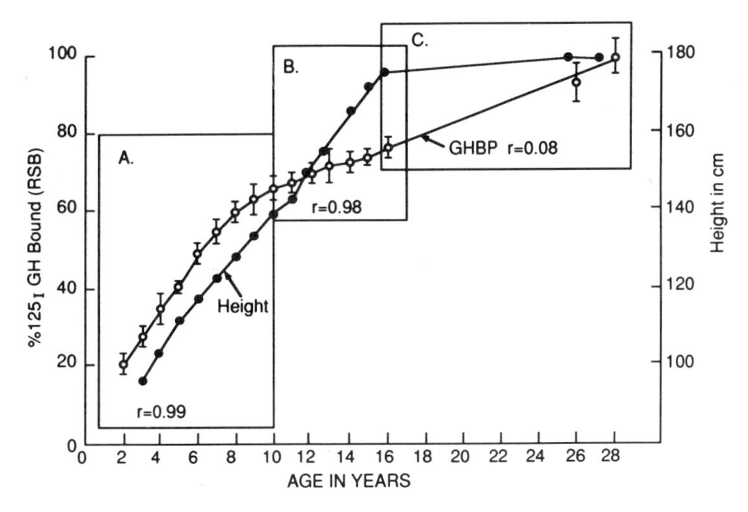

Figure 2. Relative Specific Binding of GHBP with Age: Correlation with Linear Height

The GHBP activity, as estimated in this study, was identical for males and females and this was true at all age periods. There was likewise no difference between males and females or between the age groups in affinity of the GHBP for ^{125}I-GH (data not shown). Despite the limited number of samples at young ages, we were able to obtain 35-40 umbilical cord serum samples. From preliminary data, it appears that GHBP and by inference, GH-receptors are virtually absent at birth, and seldom exceed 10% of the normal adult component. Interestingly, the second protein binder of GH (corresponding to Peak I on the Sephadex column) which is unrelated to GHBP and has remained constant throughout all age groups is also absent to extremely low in concentration at birth. These receptors most likely begin to appear shortly after birth but factors which control their appearance are not known (that GH is not essential for intrauterine growth is obvious from the normal size of anencephalic individuals).

GHBP-Linear Height

Mean GHBP levels correlated strongly with linear heights taken from the 50th centile for U.S. boys and girls (see Figure 2) (9). This was true whether one

TABLE I

GHBP and GH in Relation to Linear Height and Change in Height

	Males						**Females**				
Age Years	GHBP %		Linear Ht Cm	GH μg/l		▴Ht. Cm	GHBP %		Linear Ht Cm	GH μg/l		▴Ht Cm
3	5.5		95.0	3.08		8.0	5.5		94.5	3.99		8.0
4	6.8		103.0	3.40		8.0	6.8		102.0	4.22		7.5
5	8.0		110.0	3.47		7.0	8.0		108.5	4.60		6.5
6	9.6		116.0	2.85		6.0	9.6		115.0	4.36		6.3
7	10.9	r=0.99	122.0	3.62		6.0	10.9	r=0.98	121.0	4.30		6.0
8	11.6		127.0	3.19		5.0	11.6		127.0	4.30		6.0
9	12.5	r=0.96	133.0	4.19		6.0	12.5	r=0.94	132.5	3.42		5.0
10	12.8		137.5	3.80		4.5	12.8		138.0	4.62		6.0
11	13.1		143.0	3.20		5.5	13.1		145.0	5.38		7.0
12	13.5	r=0.98	149.0	3.33		6.0	13.5		151.5	6.10	r=0.86	6.5
13	13.9		156.0	3.19	r=-0.13	7.0	13.9	r=0.92	157.0	4.59		5.5
14	14.0		163.0	3.88		7.0	14.0		160.0	4.19		3.0
15	14.4		169.0	3.98		6.0	14.4		162.0	3.62		2.0
16	14.8		173.0	3.60		4.0	14.8		163.0	4.06		1.0

GH and GHBP are mean values obtained from pooled sera. (See Methods) Linear height and change in height are given for the 50th centile taken from revised tables of the National Center for Health Statistics Percentials. (9)

considered the age span of 2-16 years or two separate age periods, i.e. 2-10 years and 10-16 years. Respective coefficients of correlation for boys were 0.96, 0.99, and 0.98, and for girls very similar: r=0.94, 0.98, and 0.92 (see Table 1). Similarly, GHBP correlated in all groups with IGF I at each of the above age periods (r values between 0.88 and 0.99). In older girls ages 10 through 16, linear height correlated less directly with IGF I levels with an r value of 0.61. The rate at which height increased, i.e. height velocity, showed a negative correlation with GHBP in males and females at all ages as expected. There was also a negative correlation between GH and change in linear height in females ages 10-16. In older males, however, this relationship failed to exist. Linear height in males ages 10-16 was significantly less strongly correlated to GH with an r value of -0.13 compared to +0.86 for the girls (see Table 1).

GH and IGF I

Provocative tests and 24 hour sampling techniques for growth hormone were not possible, but the pooled ambulatory samples from the large number of subjects available for each age group exhibited clear cut variations of GH concentrations based on age and sex. Absolute values for GH levels compared with percent increase over a basal period (ages 3-8) are given in Figure 3 and Table 2.

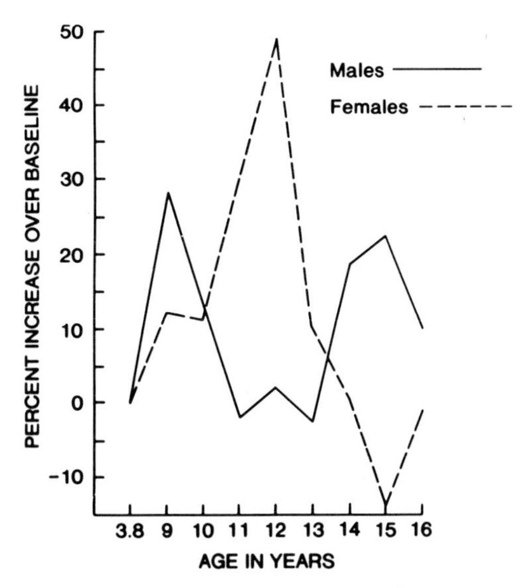

Figure 3. Percent Change in GH with Age in Males and Females

Surprisingly, GH concentrations were greater in females than males at each age period from 3 to 14 years (Chi Square Analysis, $p<.001$); GH levels in girls age 3-10 years were 4.25 ± 0.16 μg/l vs. 3.27 ± 0.12 μg/l for the boys ($p<.05$). At ages 10 through 16 years corresponding values were 4.95 ± 0.28 μg/l in girls and 3.60 ± 0.17 μg/l in boys ($p<.01$). These differences in GH concentrations were quite evident in

TABLE II

Absolute GH Levels and Percent Change with Age

	MALE		FEMALE	
Age Years	Baseline (μg/d)	% Increase	Baseline (μg/l)	% Increase
3-8	3.27		4.16	
9	4.19	28%	4.66	12%
10	3.80	16%	4.62	11%
11	3.20	-2.1%	5.38	29%
12	3.33	18%	6.20	49%
13	3.19	-2.4%	4.59	10%
14	3.88	18.6%	4.19	0.72%
15	4.00	22.3%	3.62	-14%
16	3.60	10%	4.06	-2.4%

ambulatory sampling even without the advantage of provocative stimuli.

Serum IGF I levels were low in children of both sexes before 5 years of age and increased progressively in both males and females (x ± SD in adults = 190 ± 18 ng/ml). From basal values of 88 μg/l in males and 106 μg/l in females (age 3), concentrations rose to peak values in adolescence of 343 ng/ml in boys at age 15, and 328 ng/ml in girls at age 13. Like GH concentrations, IGF I serum concentrations were higher in girls than in boys ($p<.001$ by Chi Square Analysis).

A strong positive correlation existed between IGF I and linear height in males ages 9-14 ($r=0.86$). There was no such correlation in 8-13 year old girls ($r=0.34$) during the period of maximal pubertal growth.

Serum Testosterone

Serum levels of testosterone were undetectable in all pooled male samples before the age of 11. Between 12 and 15 years of age the mean serum concentration of testosterone rose steadily with values of 58, 161, 314, and 403 ng/dl at 12, 13, 14, and 15 years of age, respectively. Serum testosterone correlated with linear height change

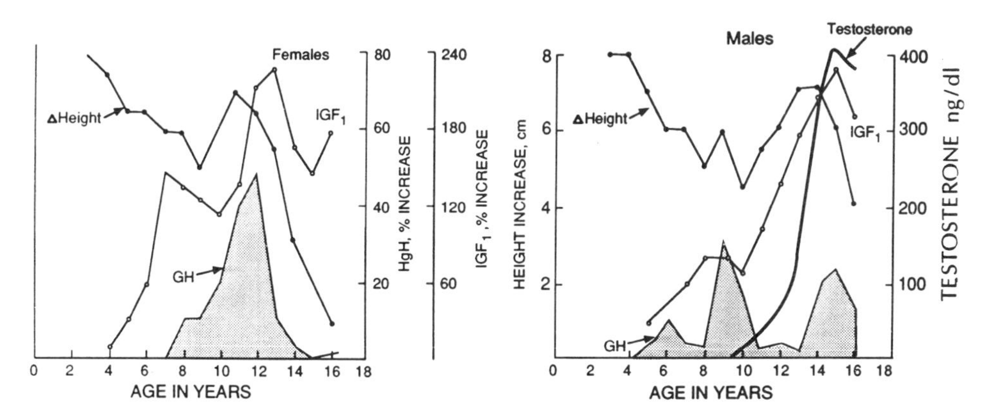

Figure 4. Correlation Between Change in Height and GH, IGF I, and Testosterone in Males and Females

with an r value of 0.88. At the time serum testosterone levels were increasing, IGF I concentrations and linear growth rates were also increasing (see Figure 4). During this same period, GH concentrations did not increase in male subjects.

DISCUSSION

The manner in which linear growth is controlled in man has been studied extensively, but the ability to relate changes of hormones to receptors and to patterns of growth has been compromised. First, only a limited number of subjects have been available for study at any age period. This has resulted in combining subjects from several ages as for example, ages 13 through 15. Secondly, only a small number of subjects have been available over an adequate age span. In most previous studies, our own included, it has been necessary to group subjects into age ranges. Such grouping obscures important differences between the sexes as well as subtle differences related to age. This became obvious in the course of the current study. It can readily be shown by combining results from consecutive age periods near or at puberty that key relationships are obliterated. The close relationship existing between GHBP serum concentrations, GH-receptors, and linear height, for example, is much less evident when age grouping is necessary. A major difference between males and females in the pattern of GH secretion is totally obliterated by studying age-clustered individuals. We were thus fortunate to be able to segregate samples into groups whose ages were within twelve months of each other and thereby identify growth patterns previously not described.

Another problem confounding GHBP studies surrounds the assay itself. As previously mentioned, GHBP is most commonly measured by gel filtration based on the methods of either Daughaday or Baumann; i.e. using Ultrogel or Sephadex columns respectively. Previously, the two methods yielded differing results. After examining four methods of calculating and expressing GHBP values, however, we were able to produce comparable results with both columns. Using the method of calculating binding described under Methods, both techniques yielded similar results, and the interassay variability was maintained at $15.2 \pm 3.6\%$, which as noted earlier is comparable to that recorded by others.

Several investigators are currently developing a radioimmunoassay technique to measure GHBP. While an RIA technique would decrease the time involved in studying multiple samples, early reports appear to offer no improvement in interassay variability (10).

The validity of ambulatory GH measurements for given age periods might conceivably be viewed as a problem. With a small number of subjects, comparison of GH values in ambulatory samples obviously cannot be expected to reflect the overall secretory pattern, whereas, this could be expected with large numbers. The reason for this is fairly obvious, since the larger the sample size, the more it is capable of reflecting a change in either the number of secretory pulses or the amplitude of these pulses. We were able to validate this assumption from a previous study involving normal subjects and diabetics (11).

We believe several important and essentially new observations relating to growth stand out: first, in relation to GHBP. An increase of GHBP and by inference, an increase in cellular receptors for GH correlates strongly with linear height in both sexes, but does not correlate with height velocity. This increase in GHBP as age increases is virtually identical in males and females. Furthermore, the change in the rate in which GHBP concentration increases from infancy to approximately age 10 and then from age 10 to young adulthood can not be related in any way to a change of GH or sex hormone secretion. Secondly, in relation to GH, from age 3 through 14 girls had higher concentrations of GH than males and this was closely reflected in higher serum concentrations of IGF I in the girls. Whether this is related to estrogen concentration was not determined, but the difference existed even in the young where estrogen levels should be low. Finally, and most importantly, the primary period of growth acceleration which occurs at puberty appears to be differently controlled in males and females. The major growth spurt in males is not temporally related to an increase of GH secretion but is temporally related to a rise of serum testosterone. In girls, the major growth spurt is strongly, if not solely linked to GH. The regression coefficients for GH and height velocity make this very evident. The coefficient for correlation of growth hormone vs. height velocity in males between the ages of 10-16 was -0.13 and for girls at the same age +0.86, a highly significant difference ($p<.001$).

There were two other noteworthy variations between boys and girls as related to GH and linear growth velocity. At approximately ages 8-9, boys had a brief period of growth acceleration related to the first of two peaks of GH. No such "early growth spurt" was noted in girls prior to the major peak of pubertal growth. Lastly, the decrease in growth velocity which followed the major period of growth acceleration was more precipitous in girls than in boys.

Examining GH and IGF I

Data from the pubertal years would suggest that the growth period of the male is extended in time by a second peak of GH secretion associated with an additional rise of IGF I once deceleration of the growth phase has begun. Once again, this is totally absent in females.

It is virtually impossible from the current data to describe the precise interaction of testosterone with GH-receptors and IGF I. It is evident, however, that the major growth acceleration phase in males relates to an increase of testosterone, not growth hormone. Furthermore, testosterone's effect must be mediated via the GH-receptor. This is strongly supported by a combination of findings: no correlation between GH and height velocity; yet a strong correlation with testosterone accompanied by a high coefficient of correlation between GHBP and linear height in males.

Studies in African pygmies would strongly reinforce this interpretation. We noted previously that pygmies secrete GH in a normal fashion (12) and that male pygmies secrete testosterone normally (13). However, both male and female pygmies show a blunted to abolished phase of growth acceleration at puberty (14). The reduction of GH-receptors to 25-30% of normal in pygmy subjects would explain this phenomena only if testosterone was acting at the GH, [GH-receptor] IGF I axis.

SUMMARY

Several major differences are noted between males and females in their patterns of growth at puberty. Accelerated pubertal growth in both males and females depends upon the integrity of the GH-receptor system. In males, acceleration of growth results primarily from enhanced sensitivity of the GH-receptor-IGF I system to GH brought about by testosterone. Whether testosterone itself is responsible for this observation is still unclear. Perhaps the initial GH, IGF I peak present in males and absent in females occurs at the time when sleep-related rises of gonadotropins and testosterone begin just prior to puberty.

Though the pygmy data certainly supports a relationship between testosterone and the GH-receptor-IGF I axis, the undisputed tall stature of eunuchs remains a puzzle. It is possible that the maturing male gonad secretes another growth factor and/or growth inhibitor in conjunction with testosterone and that it is this unidentified factor which modulates growth. At any rate, acceleration of growth in males results from sensitization or the GH-receptor-IGF I system while growth acceleration in females results almost solely from increased secretion of GH and not sensitization of the system.

REFERENCES

1. Daughaday, W.H., and Trivedi, B., 1987, Absence of serum growth hormone binding protein in patients with growth hormone receptor deficiency, Proc Soc Natl Acad Sci, USA, 84:4636.
2. Baumann, G., Stolar, M.W., Amburn, K., Barsano, C.P., and DeVries, B.C., 1986, A specific GH-binding protein in human plasma: Initial characterization, J Clin Endocrinol Metab, 62:134.
3. Thomas, H., Green, I.C., Wallis, M., and Aston, R., 1987, Heterogeneity of growth-hormone receptors detected with monoclonal antibodies to human growth hormone, Biochem J, 243:365.
4. Leung, D.W., Spencer, S.A., Cachianes, G., Hammonds, G., Colins, G., Henzel, W.J., Barnard, R., Waters, M.J., and Wood, W.I., 1987, Growth hormone receptor and serum binding protein: Purification, cloning and expression, Nature, 330:537.
5. Spencer, S.A., Hammonds, R.G., Henzel, W.J., Rodriquez, H., Waters, M.J., and Wood, W.I., 1988, The rabbit liver growth hormone receptor and serum binding protein: Purification, characterization and sequence, J Biol Chem, 263:7862.
6. Baumann, G., Shaw, M.A., 1988, Immunochemical similarity of human plasma growth hormone-binding protein and the rabbit liver growth hormone receptor, Biochem Biophys Res Commun, 121:573.
7. Zapf, J., Walter, H., Froesch, E.R., 1981, Radioimmunological determinations of insulin-like growth factors I and II in normal subjects and in patients with growth disorders and extrapancreatic hypoglycemia, J Clin Invest, 68:1321.
8. Baumann, G., Shaw, M.A., and Amburn, K., 1989, Regulation of plasma growth hormone-binding proteins in health and disease, Metabolism, 38:683.
9. Hamill, P.V., Drizd, T.A., Johnson, C.L., Reed, R.B., Rocke, A.F., and Moore, W.M., 1979, Physical growth: National center for health statistics percentiles, Am J Clin Notr, 32:607.
10. Laird, D.M., Melton, M.A., Creely, D.P., Hauser, S.D., Hill, S.R., Glen, K.C., 1990, Expression, purification and characterization of the somatotropin receptor binding domain/binding protein, Endocrine Society, 265A, p. 91.
11. Merimee, T.J., Fitzgerald, C.R., Gold, L.A., McCourt, J.P., 1979, Characteristics of growth hormone secretion in clinically stable diabetes, Diabetes, 28:308.
12. Merimee, T.J., Rimoin, D.L., Penetti, E., and Cavalli-Sforza, L.L., 1972, Metabolic studies in the African pygmy, J Clin Invest, 51:395.
13. Merimee, T.J., Zapf, J., Hewlett, B., and Cavalli-Sforza, L.L., 1987, Insulin-like growth factors in pygmies: The role of puberty in determining final stature, N Engl J Med, 316:906.
14. Meredith, H.V., 1978, Research on the standing height of young children in different parts of the world, Adv Child Dev Behav, 12:1.

RECOMBINANT HUMAN INSULIN-LIKE GROWTH FACTOR I: EFFECTS IN NORMAL SUBJECTS AND IMPLICATIONS FOR USE IN PATIENTS

Hans-Peter Guler*, Katharina Wettstein, Werner Schurr**, Jurgen Zapf, E. Rudolf Froesch

Metabolic Unit, Department of Medicine, University Hospital, 8091 Zurich, Switzerland; *Ciba-Geigy Corp., Summit, NJ 07901; **Ciba-Geigy AG, 4002 Basle, Switzerland

INTRODUCTION

Two major forms of insulin-like growth factors (IGF-I and IGF-II) and their specific binding proteins have been detected in body fluids, including serum, cerebrospinal fluid, breast milk, lymph, and saliva. Extracts of almost any tissue reveal the presence of IGFs. Many cells in culture secret IGFs and various forms of their carrier proteins into the medium (1-3).

IGFs stimulate cell replication and differentiation in vitro (4). The binding proteins are important mediators for these processes (5). Our understanding of the physiological significance of the IGFs in growth and development becomes increasingly clear (6-8). However, it is much less clear how these peptides and their carrier proteins are regulated and what might be their functions (1-3) in adult life. Several factors such as age, nutrition, insulin, growth hormone and IGF-I by itself appear to be crucial (1-3, 9-11).

Abundant quantities of recombinant human IGF-I (rhIGF-I) have made it possible to evaluate pharmacological effects of IGF-I in humans. Here, we present data on the effects of five-day infusions of rhIGF-I on serum levels of two important endocrine actors: insulin and growth hormone.

METHODS

Subjects: Three healthy young men underwent one experiment each. Three additional normal males participated in two experiments each at two different doses of rhIGF-I. The ages ranged from 22 to 26 years.

Protocol: The trial consisted of three phases: A baseline period of three days, a treatment period of five days, and a follow-up period of three days. Each subject served as his own control. The subjects were blinded to the dose of rhIGF-I they received. RhIGF-I was produced by Chiron Corporation, Emeryville, CA, and was prepared for clinical use by CIBA-GEIGY Corporation, Summit, NJ. Three different doses of rhIGF-I were administered by continuous s.c. infusion: 7, 14 or 21 μg/kg/hr. Each dose level was tested in three subjects. The drug was delivered by a portable insulin infusor (MRS3-Infusor, Disetronic AG, Burgdorf, Switzerland). The subjects were on a diet of 2500 kcal/day throughout the study. Blood samples were drawn every morning in the fasting state.

During one night of the baseline period and during the last night of the rhIGF-I infusion, blood samples were drawn every 20 minutes to determine growth hormone and hourly to determine insulin and C-peptide from 10 PM to 6 AM. The overnight measurements are expressed as area under the curve (AUC is the sum of the products of serum levels multiplied by time intervals). Subjects were in the supine position during this time period and were mostly asleep.

RESULTS

Total serum IGF-I levels rose from average baseline values around 250 ng/ml to approximately 600 ng/ml in the low dose group. For the two higher dosing levels, the respective numbers were both in the 900 ng/ml range.

Blood Glucose: Fasting glucose (Table 1) decreased slightly but remained within the normal range at the 14 and 21 μg/kg/hr doses. No subject had clinical symptoms of hypoglycemia, although a few values in the highest dose group were between 3.7 and 4.0 mmol/l.

TABLE 1

Serum-Glucose Levels [mmol/l] in Normal Subjects
Before, During and After Continuous s.c. Infusion of rhIGF-I:
Means ± SD, n = 8 to 15 For Each Value

	Dose of rhIGF-I (μg/kg/hr)		
	7	14	21
Before	4.08 ± 0.48	4.66 ± 0.32	4.30 ± 0.48
During	4.49 ± 0.43	4.12 ± 0.49	4.04 ± 0.53
After	4.41 ± 0.36	4.34 ± 0.29	4.03 ± 0.33

Insulin: Fasting serum insulin levels are presented in Table 2. The serum insulin levels fell considerably at all three doses. At the 14 and 21 μg/kg/hr dose, the decrements were -44.32 ± 10.71 and -31.19 ± 4.08, respectively, and reached statistical significance ($p<0.01$).

TABLE 2

Serum-Insulin Levels [pmol/l] in Normal Subjects
Before, During and After Continuous s.c. Infusion of rhIGF-I:
Samples Drawn in the Morning, After an Overnight Fast
Means ± SD, n = 8 to 15 For Each Value

	Dose of rhIGF-I (μg/kg/hr)		
	7	14	21
Before	73.66 ± 29.41	83.56 ± 22.43	55.92 ± 16.39
During	51.45 ± 32.85	36.70 ± 18.99	26.70 ± 18.37
After	96.91 ± 61.18	71.44 ± 14.60	46.91 ± 15.83

C-Peptide: Fasting C-peptide levels are given in Table 3. A dose dependent suppression of C-Peptide was noted. The decrements from baseline were -323 ± 90 pmol/l (7 μg/kg/hr dose, $p<0.01$), -330 ± 106 pmol/l (14 μg/kg/hr dose, $p<0.05$), and - 363 ± 107 pmol/l (21 μg/kg/hr dose, $p<0.01$), respectively. In the highest dose group the C-peptide levels were roughly 25 % of baseline, whereas insulin levels were only reduced to about 50 %. The ratio of insulin:C-peptide changed from 0.118 (baseline) to 0.219 (21 μg/kg/hr group).

TABLE 3

C-Peptide Levels [pmol/l] in Normal Subjects
Before, During and After Continuous s.c. Infusion of rhIGF-I:
Samples Drawn in the Morning, After an Overnight Fast
Means ± SD, n = 8 to 15 For Each Value

	Dose of rhIGF-I (μg/kg/hr)		
	7	14	21
Before	547.50 ± 154.90	492.50 ± 156.36	472.50 ± 122.21
During	215.33 ± 95.91	175.33 ± 71.60	122.00 ± 80.91
After	445.56 ± 189.02	423.33 ± 160.39	457.78 ± 68.15

Triglycerides: Lower serum triglyceride levels were seen under IGF-I treatment. At the two higher doses, the decrements from baseline reached statistical significance ($p<0.05$).

TABLE 4

Triglyceride Levels [mmol/l] in Normal Subjects
Before, During and After Continuous s.c. Infusion of rhIGF-I:
Samples Drawn in the Morning, After an Overnight Fast
Means ± SD, n = 8 to 15 For Each Value

	Dose of rhIGF-I (μg/kg/hr)		
	7	14	21
Before	1.59 ± 0.88	1.11 ± 0.36	1.12 ± 0.19
During	0.94 ± 0.21	0.70 ± 0.16	0.76 ± 0.14
After	1.32 ± 0.36	0.87 ± 0.31	1.08 ± 0.26

Total Cholesterol: Fasting cholesterol levels (Table 5) decreased under IGF-I treatment. The decrements at the 7 and 14 μg/kg/hr dose were -0.12 ± 0.05 mmol/l ($p<0.05$) and -0.45 ± 0.16 mmol/l ($p<0.05$), respectively. The variability in the highest dose group was too high to reach statistical significance.

TABLE 5

Total Cholesterol Levels [mmol/l] in Normal Subjects
Before, During and After Continuous s.c. Infusion of rhIGF-I:
Samples Drawn in the Morning, After an Overnight Fast
Means ± SD, n = 8 to 15 For Each Value

	Dose of rhIGF-I (μg/kg/hr)		
	7	14	21
Before	4.65 ± 0.52	5.31 ± 0.65	4.85 ± 0.30
During	4.49 ± 0.54	4.77 ± 0.74	4.71 ± 0.56
After	4.62 ± 0.54	4.80 ± 0.69	4.80 ± 0.46

Insulin: Area under the curve (AUC) from 10 PM to 6 AM (Table 6). At the lowest dose of rhIGF-I (7 μg/kg/hr) two subjects did not show any reduction of the AUC. The two higher doses, however, revealed a significant reduction of the AUC. Due to the small sample size, individual data for each subject is presented.

TABLE 6

AUC for Insulin [pmol/1.8 hours]
Before and During Continuous s.c. Infusion of rhIGF-I:
Sampling every hour from 10 PM to 6 AM

	Dose of rhIGF-I (μg/kg/hr)								
	7			14			21		
Subject	2	3	6	1	3	4	1	2	5
Before	231	453	499	706	611	474	365	423	402
During	216	572	253	376	468	145	188	265	227

C-Peptide: Area under the curve (AUC) from 10 PM to 6 AM (Table 7). Two of three subjects showed a large reduction of the AUC at the low dose already. A striking reduction of the AUC in all three subjects at both higher doses was noted.

TABLE 7

AUC for C-Peptide [pmol/1.8 hours]
Before and During Continuous s.c. Infusion of rhIGF-I:
Sampling every hour from 10 PM to 6 AM

	Dose of rhIGF-I (μg/kg/hr)								
	7			14			21		
Subject	2	3	6	1	3	4	1	2	5
Before	3310.0	2480.0	5389.0	4017.5	3125.0	4310.0	4283.0	3690.0	3940.0
During	1495.0	2095.0	3190.0	2660.0	743.0	2530.0	1301.5	597.5	1240.0

Growth Hormone: Area under the curve (AUC) from 10 PM to 6 AM (Table 8). Two of three subjects at the 7 μg/kg/day dose displayed a marked diminuition of the AUC. This effect was dose dependent and became more prominent at the higher doses.

TABLE 8

AUC for Growth Hormone [ng/ml.8 hours]
Before and During Continuous s.c. Infusion of rhIGF-I:
Sampling every 20 minutes from 10 PM to 6 AM

	Dose of rhIGF-I (μg/kg/hr)								
	7			14			21		
Subject	2	3	6	1	3	4	1	2	5
Before	40.13	41.98	17.75	32.22	32.25	58.80	43.68	44.64	24.30
During	18.64	25.05	15.09	18.53	13.65	23.28	9.09	16.88	18.05

DISCUSSION

The data presented in this paper demonstrates that continuous s.c. infusions of rhIGF-I in doses up to 21 μg/kg/hr do not lead to hypoglycemia in normal subjects fed 2500 kcal/day. In contrast, intravenous bolus injections of IGF-I into fasted normal humans (12), or Laron dwarfs (13) do lead to a dose dependent decrease of blood glucose. It was shown that under these circumstances the serum levels of free IGF-I are elevated approximately 30 fold above baseline (12). Furthermore, subcutaneous bolus injections of 120 μg/kg of rhIGF-I lead to depressed blood glucose levels in the range of 2.80 mmol/l ir healthy subjects (14).

Serum levels of insulin and C-peptide were suppressed in a dose dependent fashion in our subjects. This was seen both for levels measured in the fasting state as well as those measured during the night under IGF-I infusion. Our findings are consistent with results from in vitro studies with pancreatic beta cells: the beta cells' response to a glucose challenge was significantly decreased when the culture medium was supplemented with rhIGF-I (15,16). It is possible, therefore, that our results are due to a direct suppressive effect on the beta cell. Our subjects were able to maintain normal to low-normal blood glucose levels with much less than normal insulin. Apparently, IGF-I has partially substituted for the endogenous insulin. The insulin: C-peptide ratio shifted in a way suggesting a relative increase of insulin over C-peptide. This observation was also made in an earlier trial (17). In addition, data from experiments with HepG2 cells showed that the half-life of insulin was prolonged in the presence of high concentrations of IGF-I (18).

The treatment of Type II diabetics with rhIGF-I concomitantly decreases serum insulin levels and blood glucose (personal communication: D. S. Schalch, Madison, WI). There is evidence from the literature that the type I IGF receptor, in contrast to the insulin receptor, may be normal in type II diabetes patients: insulin receptor number and tyrosine kinase activity of muscle was reduced while the type I IGF receptor was normal (19). Furthermore, glucose uptake in insulin resistant BB rats is maintained comparable to non-diabetic BB rats (20). This points to the fact that the type I IGF receptor may in part substitute for the defective insulin receptor in these rats. Contradictory evidence comes from a recent paper by Dohm et al. (21). IGF-I failed to increase in-vitro glucose uptake in muscles prepared from obese subjects with and without type II diabetes. The conclusion was that these patients' skeletal muscles are resistant to IGF-I. However, one has to keep in mind that these patients were very obese (body mass index in the range of 49 to 60 kg/m^2).

Serum levels of triglycerides and total cholesterol were lower during IGF-I treatment. Further investigation will be needed to establish a role of IGF-I as a lipid lowering drug. In particular, patients with hypertriglyceridemia need to be studied.

Finally, we observed a striking effect of IGF-I on the AUC of growth hormone: Pharmacological doses of IGF-I suppressed nocturnal serum growth hormone levels considerably. Earlier work in cell cultures and in-vivo in rats supports our data (22-26). Since GH has been cited as contributing to the dawn phenomenon in diabetic patients (27), suppression of nocturnal GH secretion may be beneficial. Whether IGF-I is able to reduce morning hyperglycemia in diabetic patients remains to be shown.

In conclusion, rhIGF-I leads to reduced serum levels of insulin, C-peptide, triglycerides and cholesterol while blood glucose remains within the normal range. This profile of action makes rhIGF-I a particularly attractive tool for research in patients who are hyperinsulinemic, hyperglycemic, and hyperlipidemic, as frequently observed in type II diabetics.

REFERENCES

1. Daughaday WH, Rotwein P. Insulin-Like Growth Factors I and II Peptide, Messenger Ribonucleic Acid and Gene Structures, Serum, and Tissue Concentrations. Endocrine Rev 10:68-91, 1989.

2. Humbel RE. Insulin-like Growth Factors I and II. Eur J Biochem 190:445-462, 1990.

3. Sara VR, Hall K. Insulin-Like Growth Factors and Their Binding Proteins. Physiol Rev 70:591-614, 1990.

4. Froesch ER, Schmid Chr, Schwander J, Zapf J. Actions of Insulin-Like Growth Factors. Ann Rev Physiol 47:443-467, 1985.

5. Hardouin S, Gourmelen M, Noguiez P, Seurin D, Roghani M, LeBouc Y, Povoa G, Merimee TJ, Hossenlopp P, Binoux M. Molecular Forms of Serum Insulin-Like Growth Factor (IGF)-Binding Proteins in Man: Relationships with Growth Hormone and IGFs and Physiological Significance. J Clin Endocrin Metabol 65:1291-1301, 1989.

6. Schoenle E, Zapf J, Humbel RE, Froesch ER. Insulin-Like Growth factor I Stimulates Growth in Hypophysectomized Rats. Nature 796:252-253, 1982.

7. Van Buul-Offers S, Ueda I, Van den Brande JL. Biosynthetic Somatomedin C (SM-C/IGF-I) Increases the Length and Weight of Snell Dwarf Mice. Pediatric Res 20:825-827, 1986.

8. Guler HP, Zapf J, Scheiwiller E, Froesch ER. Recombinant Human Insulin-Like Growth Factor I Stimulates Growth and Has Distinct Effects on Organ Size in Hypophysectomized Rats. Proc. Natl. Acad. Sci. 85:4889-4893, 1988.

9. Zapf J, Schmid C, Guler HP, Waldvogel M, Hauri C, Futo E, Hossenlopp P, Binoux M, Froesch ER. Regulation of Binding Proteins for Insulin-Like Growth Factors (IGF) in Man: Increased Expression of IGF Binding Protein-2 During IGF-I Treatment of Healthy Adults and in Patients with Extrapancreatic Tumor Hypoglycemia. J Clin Invest; 86:952-962, 1990.

10. Clemmons DR, Thissen JP, Maes M, Ketelslegers JM, Underwood LE. IGF-I infusion into hypophysectomized or protein deprived rats induces specific IGF-binding proteins in serum. Endocrinology 125:2967-2972, 1989.

11. Zapf J, Hauri C, Waldvogel M, Futo E, Hasler H, Binz K, Guler HP, Schmid Ch, Froesch ER. Recombiant human IGF-I its own specific carrier protein in hypophysectmoized and diabetic rats. Proc Natl Acad Sci 86:3813-3817, 1989.

12. Guler HP, Zapf J, Froesch ER. Short-Term Metabolic Effects of Recombinant Human Insulin-Like Growth Factor in Healthy Adults. NEJM 317:137-140, 1987.

13. Laron Z, Klinger B, Erster B, Anin S. Effect of acute administration of insulin-like growth factor I in patients with Laron-type dwarfism. Lancet II:1170-1172, 1988.

14. Takano K, Hizuka N, Asakawa K, Sukegawa I, Shizume K, Demura H. Effects of sc administration of rhIGF-I on normal human subjects. Endocrinol Japon 37:309-317, 1990.

15. Leahy JL, Vandekerkhove KM. Insulin-Like Growth Factor-I at Physiological Concentrations Is a Potent Inhibitor of Insulin Secretion. Endocrinology 126:1593-1598, 1990.

16. Van Schravendijk C, Heylen L, Van Den Brande JL, Pipeleers D. Effects of Insulin and IGF-I on the Secretory Activity of Pancreatic B-Cells. Diabetologia 32:551A, 1989.

17. Guler HP, Schmid Ch, Zapf J, Froesch ER. Effects of recombinant human IGF-I on insulin secretion and renal function in normal human subjects. Proc Natl Acad Sci 86:2868-2872, 1989.

18. Keller S, Schmid Ch, Zapf J, Froesch ER. Inhibition of insulin degradation by insulin-like growth factors I and II in human hepatoma (HepG2) cells. Acta Endocrin 121:279-285, 1989.

19. Livingston N, Pollare T, Lithell H, Arner P. Characterisation of Insulin-Like Growth Factor I Receptor in Skeletal Muscles of Normal and Insulin Resistant Subjects. Diabetologia 31:871-877, 1988.

20. Jacob R, Bowen L, Fryburg D, Fagin K, Tamborlane W, Shulman GI. Increased Metabolic Response to IGF-I in Insulin Resistant Diabetic BB Rats. Ann. Meeting Amer. Diab Assn., Detroit 1989, Abstract #261.

21. Dohm GL, Elton CW, Raju MS, Mooney ND, DiMarchi R, Pories WJ, Flickinger EG, Atkinson M, Caro JF. IGF-I stimulated glucose transport in human skeletal muscle and IGF-I resistance in obesity and NIDDM. Diabetes 39:1028-1032, 1990.

22. Berelowitz M, Szabo M, Frohman LA, Firestone S, Chu L. Somatomedin-C Mediates Growth Hormone Netative Feedback by Effects on Both the Hypothalamus and the Pituitary. Science 212:1279-1281, 1981.

23. Hiromi A, Molitch ME, Van Wyk JJ, Underwood LE. Human Growth Hormone and Somatomedin C Suppress the Spontaneous Release of Growth Hormone in Unanesthetized Rats. Endocrinology 113:1319, 1983.

24. Yamashita S, Melmed S. Insulin-Like Growth Factor I Action on Rat Anterior Pituitary Cells: Suppression of Growth Hormone Secretion and Messenger Ribonucleic Acid Levels. Endocrinology 118:176-181, 1986.

25. Morita S, Yamashita S, Melmed S. Insulin-Like Growth Factor I Action on Rat Anterior Pituitary Cells: Effects of Intracellular Messengers on Growth Hormone Secretion and Messenger Ribonucleic Acid Levels. Endocrinology 121: 2000-2006, 1987.

26. Fagin JA, Brown A, Melmed S. Regulation of Pituitary Insulin-Like Growth Factor-I Messenger Ribonucleic Acid Levels in Rats Harboring Somatomammotropic Tumors: Implications for Growth Hormone Autoregulations. Endocrinology 122:2204-2210, 1988.

27. Perriello G, De Feo P, Torlone E, Fanelli C, Santeusanio F, Brunetti P, Bolli GB. Nocturnal spikes of growth hormone secretion cause the dawn phenomenon in type I (insulin-dependent) diabetes mellitus by decreasing hepatic (and extrahepatic) sensitivity to insulin in the absence of insulin waning. Diabetologia 33:52-59, 1990.

SINGLE sc ADMINISTRATION OF INSULIN-LIKE GROWTH FACTOR I (IGF-I) IN NORMAL MEN

Naomi Hizuka, Kazue Takano, Kumiko Asakawa, Izumi Fukuda, Izumi Sukegawa, Kazuo Shizume*, and Hiroshi Demura

Department of Medicine, Tokyo Women's Medical College, Tokyo, 162, and *Research Laboratory, Foundation for Growth Science, Tokyo, 162, Japan

INTRODUCTION

Insulin-like growth factor I (IGF-I) has been synthesized by recombinant DNA technology (1). With the availability of large quantities of the biosynthetic peptide, the biological effects have been studied in animals in vivo. We have previously reported that IGF-I has growth promoting, anabolic, and insulin-like effects in hypophysectomized and normal rats, and IGF-I inhibits catabolism in fasted rats in vivo (2-5). Recently, the biological effects of IGF-I have been studied in normal men and patients with Laron-type dwarfism (6-9). We have also studied effects of IGF-I single sc administration in normal men (10). In this paper, the results of the single sc administration of IGF-I combined with our previous data (10) are reported.

SUBJECTS AND STUDY PROTOCOL

Subjects

Five healthy young adults (22-24 yrs) participated in this study. None of the subjects had clinical evidence of illness, and all were within 20% of their ideal body weight. Informed concent was obtained from each volunteer and the experimental protocol was approved by the Human Subjects Investigation Committee of Tokyo Women's Medical College.

IGF-I Preparation

Recombinant human IGF-I was kindly provided by Fujisawa Pharmaceutical Co. Ltd., Osaka. The preparation of IGF-I was synthesized by DNA technology as described (1). The amino acid sequence of the IGF-I preparation is the same as that of natural IGF-I. IGF-I was dissolved in physiological saline at a concentration of 0.6% just before use.

Study Protocol

Each subject underwent three experiments (saline, 0.06 mg/kg and 0.12 mg/kg IGF-I administration, respectively). The experimental interval was 7-14 days. Each subject received sc administrations of IGF-I or saline after overnight fasting. Blood samples were taken before and at 1, 2, 3, 4, 5, 6, 8, 10, 12, 24 h after the sc administration. Urine samples were collected for 24 h to measure growth hormone (GH), urea nitrogen, and electrolytes. Subject had lunch, light snack and dinner at 4, 6 and 10 h after IGF-I administration, respectively.

Assays

IGF-I was measured by radioimmunoassay (RIA) as previously reported (11). Plasma total IGF-I levels were measured using acid-ethanol extracted plasma (12). Free form of IGF-I levels were measured using Sep-Pak C_{18} cartridges (13): briefly, Sep-Pak C_{18} cartridges were prepared by washing with 10 ml of methanol followed by 20 ml H_2O, and then 2.5 ml syringe was connected with lower end of the cartridge. The

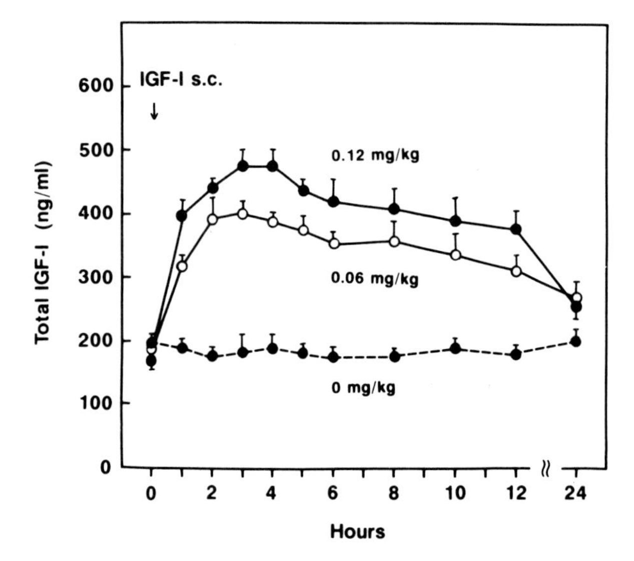

Fig. 1. Changes of total IGF-I levels in plasma after sc administration of IGF-I. Vertical lines indicate mean ± SEM. Redrawn from Takano et al. (10).

plasma sample (0.2 ml) was applied on the upper end of the cartridge, and passed through the cartridge by slowly pulling down of the syringe plunger. The cartridge was washed with 20 ml H_2O and then slowly eluted with 2 ml 75% ethanol-0.01 M HCl. The eluate was dried using a Speed Vac-Concentrator (Savant Inst., Hickville, NY). The sample was reconstituted with assay buffer for measurement of IGF-I RIA.

Plasma IGF-II was measured using acid-ethanol extracted plasma by RIA (14), urinary GH by enzyme immunoassay (15), serum insulin and C-peptide by commercially available RIA kits, and blood glucose, urea nitrogen and electrolytes with "Autoanalyzer".

Statistics

Mann-Whiney test, one-way ANOVA, two-way ANOVA and Kruskal-Wallis test were used for statistical analyses.

TOTAL AND FREE FORM OF IGF-I IN PLASMA

The mean total IGF-I levels in plasma after IGF-I administration are shown in Fig. 1. Plasma total IGF-I levels did not change throughout the study after saline administration. After 0.06 mg/kg IGF-I administration, mean total IGF-I levels in plasma increased from 185±17 ng/ml to a maximal level of 396±21 ng/ml at 3 h and then gradually decreased. However, the level did not return to the basal level even at 24 h after the injection. After 0.12 mg/kg IGF-I administration, the levels rose further; the peak values reached at 4 h had a mean value of 480±27 ng/ml.

The mean values of free form of IGF-I in plasma after IGF-I administration are shown in Fig. 2. The free form of IGF-I values in plasma did not change throughout the study after saline administration. After 0.06 mg/kg or 0.12 mg IGF-I administration, the free form of IGF-I values increased from 2.7±0.3 to 26.1±3.1 ng/ml at 2 h, or from 2.9±0.3 to 60.6±6.3 ng/ml at 2 h, respectively. The values decreased more rapidly thereafter than total IGF-I. The values at 1 - 12 h after 0.12 mg/kg IGF-I administration were greater than those after 0.06 mg/kg ($p<0.05$).

These data indicate that the total and free form of IGF-I in plasma increased after IGF-I administration in a dose dependent manner, and free form of IGF-I levels decreased more rapidly than total IGF-I.

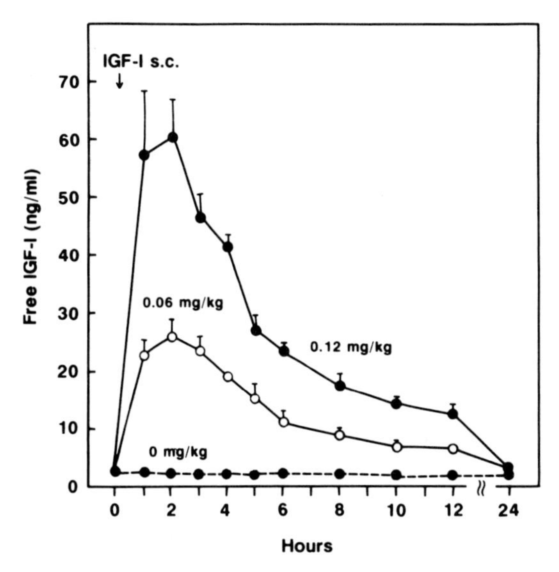

Fig. 2. Changes of free form of IGF-I levels in plasma after sc administration of IGF-I. Vertical lines indicate mean ± SEM.

BLOOD GLUCOSE, SERUM INSULIN AND C-PEPTIDE LEVELS

The blood glucose levels after IGF-I administration are shown in Fig. 3. The blood glucose levels did not change throughout the study after saline administration. The blood glucose levels after 0.06 mg/kg IGF-I administration decreased gradually and reached the minimal value of 73±3 mg/dl (Mean±SEM) at 3 h. The blood glucose levels at 2, 3 and 4 h after the injection were significantly lower than those for saline and the basal values for 0.06 mg/kg IGF-I ($p<0.05$). After 0.12 mg/kg IGF-I administration, the blood glucose levels decreased rapidly; the minimal value was observed at 2 h with a mean of 50±4 mg/dl. Three of the five subjects had hypoglycemia (the blood glucose levels <50 mg/dl). The blood glucose levels at 1, 2, 3, 4 h after the injection were significant lower than those for saline and 0.06 mg/kg IGF-I, and the basal values for 0.12 mg/kg IGF-I ($p<0.05$). These data indicate that the bood glucose levels decrease after IGF-I administration in a dose dependent manner. The decreased blood glucose levels might be due to the increase in free form of IGF-I but not to the increase in total IGF-I according to the results of changes in plasma total and free form IGF-I described above.

The blood glucose levels at 1 and 2 h after lunch (5 and 6 h after injection) for 0.12 mg/kg IGF-I were significantly greater than those for saline and 0.06 mg/kg IGF-I. The reason of the increased values for 0.12 mg/kg IGF-I are not clearly explained, however, the precedence of decreased insulin secretion (Fig. 4) could be related to the phenomenon. After 8 h administration, there were not significant differences in the blood glucose levels among the three groups.

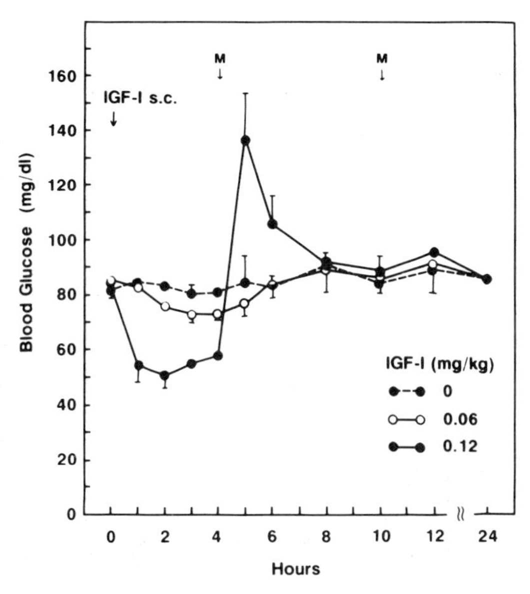

Fig. 3. Changes of blood glucose levels after sc administration of IGF-I. Vertical lines indicate mean ± SEM, and M indicates lunch or dinner. Redrawn from Takano et al. (10).

Serum insulin (IRI) and serum C-peptide levels after administration of IGF-I decreased in a dose dependent manner until 4 h (Fig. 4). The levels increased after food intake. The decreased IRI and C-peptide levels in serum might be due to the decreased blood glucose levels, however, the possibility of direct inhibition of insulin secretion by IGF-I could not be ruled out.

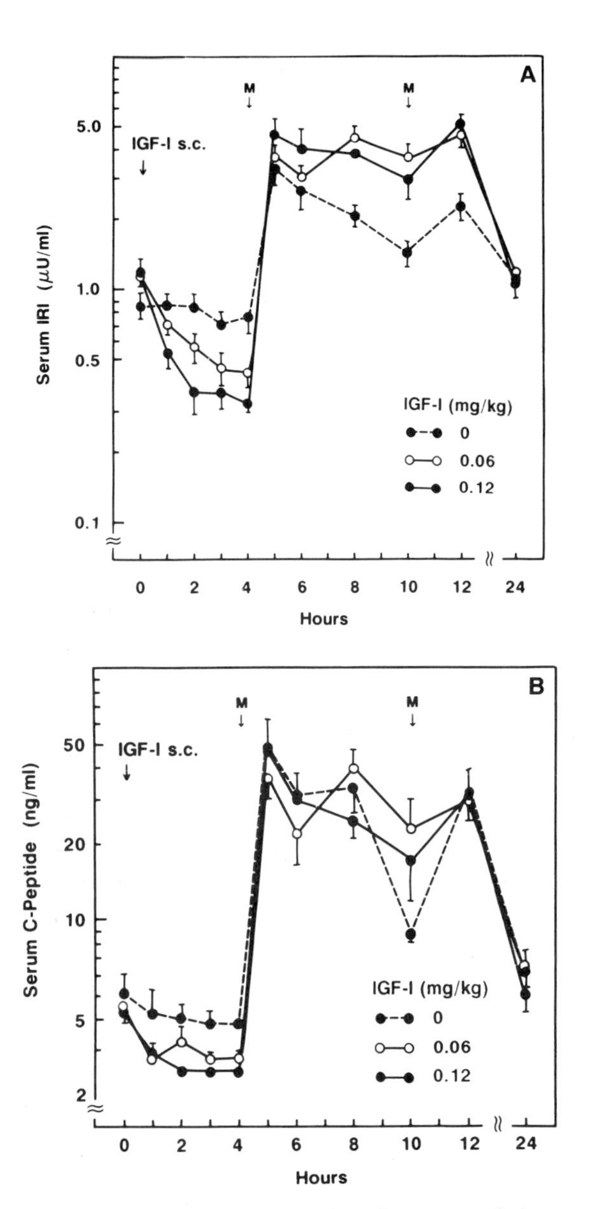

Fig. 4. Changes of serum insulin (IRI) levels (A) and serum C-peptide levels (B) after sc administration of IGF-I. Vertical lines indicate mean ± SEM, and M indicates lunch or dinner. Redrwan from Takano et al. (10).

PLASMA IGF-II

Plasma IGF-II levels after IGF-I administrations are shown in Fig. 5. Plasma IGF-II levels did not change throughout the study after saline administration. Plasma IGF-II values significantly decreased at 4 - 24 h ($p<0.05$) after IGF-I administration; at 0.06 mg/kg of IGF-I, from 660±21 to minimal levels of 513±28 ng/ml at 6 h, and at 0.12 mg/kg, from 661±22 to 495±15 ng/ml at 10 h. The decreased plasma IGF-II after IGF-I administration might be explained by the finding that IGF-I and IGF-II share the IGF binding proteins.

URINARY GH, UREA NITROGEN AND ELECTROLYTES EXCRETION

The urinary growth hormone (GH) values were 13.1±2.4, 4.2±1.2 and 10.6±2.3 ng/day after 0, 0.06 or 0.12 mg/kg IGF-I administration, respectively. The values after 0.06 mg/kg IGF-I administration were significantly lower than those for 0 and 0.12 mg/kg IGF-I administration ($p<0.01$). Guler et al. reported that sc infusion of IGF-I (20 µg/kg/h), where euglycemia was maintained, suppressed spontaneous GH secretion at night (7). Our study comfirms the result that IGF-I suppress the GH secretion in the euglycemic state. Urinary GH values did not decrease after 0.12 mg/kg IGF-I administration, suggesting that hypoglycemia caused by the IGF-I administration induced GH secretion as reported by Guler et al. (6).

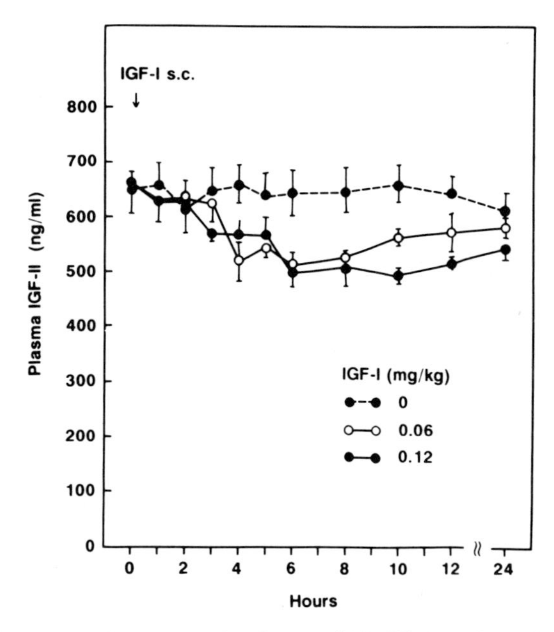

Fig. 5. Changes of plasma IGF-II levels after sc administration of IGF-I. Vertical lines indicate mean ± SEM.

Table 1. Effect of IGF-I administration on urinary excretion of urea nitrogen, Ca, P, Na and K in 5 normal subjects.

	Dose of IGF-I (mg/kg)			
	0	0.06	0.12	
Urea N (g/day)	9.1±0.8	6.7±0.7	7.5±0.6	$p<0.05$
Ca (mg/day)	155±16	149±25	147±15	
P (mg/day)	647±50	576±127	620±91	
Na (mEq/day)	207±14	126±15	142±17	$p<0.05$
K (mEq/day)	32±3	27±4	32±7	

Values given are the Mean±SEM.
Two-way ANOVA was used for statistical analysis.

Urinary urea nitrogen and sodium excretion decreased after IGF-I administration (Table 1). There were no significant changes in other electrolytes after IGF-I administration.

SUMMARY AND CONCLUSION

Total and free form of IGF-I in plasma increased in a dose dependent manner after sc IGF-I administration. Peak values of total IGF-I were obtained at 3 - 4 h after the administration, and then the values decreased gradually. However, peak values of free form of IGF-I were obtained at 2 h, and then rapidly decreased thereafter. The blood glucose, serum insulin and C-peptide levels decreased until 4 h after IGF-I administration in a dose dependent manner. Plasma IGF-II values significantly decreased at 4 - 12 h after IGF-I administration. Urinary urea nitrogen and sodium excretion decreased after IGF-I administration. Urinary GH excretion also decreased after 0.06 mg/kg IGF-I administration.

These data demonstrate that IGF-I may play a role in glucose, protein and electolyte metabolism, and plasma IGF-II levels and GH secretion might be regulated by IGF-I in man.

ACKNOWLEDGEMENTS

The authors are greatly indebted to Fujisawa Pharmaceutical Co. Ltd. (Osaka, Japan), and Sumitomo Pharmaceuticals Co. Ltd. (Osaka, Japan) for supplying biosynthetic IGF-I, and GH enzyme immunoassay kit, respectively. The authors also thank Ms. T. Ohta for her excellent technical assistance. This work was supported in part by a Grant-in-Aid for General Scientific Research (no. 63570548) from the Ministry of Education, Science, and Culture and a research grant from the Intractable Disease Division, Public Health Bureau, Ministry of Health and Welfare.

REFERENCES

1. Niwa M, Sato S, Saito Y, Uchiyama F, Ono H, Yamashita M, Kitaguchi Y, Shiga Y, Notani J, Yamada H, Ishii Y, Ueda I and Takagi Y. Chemical synthesis, cloning, and expression of genes for human somatomedin C (insulin-like growth factor I) and ^{59}Val-somatomedin C. Ann NY Acad Sci 469:31-52 (1986).
2. Hizuka N, Takano K, Shizume K, Asakawa K, Miyakawa M, Tanaka I and Horikawa R. Insulin-like growth factor I stimulates growth in normal growing rats. Eur J Phamacol 125:143-146 (1986).
3. Hizuka N, Takano K, Asakawa K, Miyakawa M, Tanaka I, Horikawa R, Hasegawa S, Mikasa Y, Saito S, Shibasaki T and Shizume K. In vivo effects of insulin-like growth factor I in rats. Endocrinol Japon 34(Suppl):115-121 (1987).
4. Hizuka N, Takano K, Asakawa K, Sukegawa I, Horikawa R, Kikuchi H and Shizume K. Biological action of insulin-like growth factor-I in vivo. In: Bercu BB (ed) Basic and Clinical Aspects of Growth Hormone Plenum Press New York p. 223-231 (1988).
5. Hizuka N, Takano K, Asakawa K, Shizume K and Kikuchi H. Biological effects of insulin-like growth factor I in vivo. Gunma Symposia on Endocrinology 25:137-147 (1988).
6. Guler H-P, Zapf J and Froesch ER. Short-term metabolic effects of recombinant human insulin-like growth factor I in healthy adults. N Engl J Med 317:137-140 (1987).
7. Guler H-P, Schmid C, Zapf J and Froesch ER. Effects of recombinant IGF-I on insulin secretion and renal function in normal human subjects. Proc Natl Acad Sci USA 86:2868-2872 (1989).
8. Guler H-P, Eckardt K-U, Zapf J, Bauer C and Froesch ER. Insulin-like growth factor I increases glomerular filtration rate and renal plasma flow in man. Acta Endocrinol 121:101-106 (1989).
9. Laron Z, Klinger B, Ersner B and Anin S. Effects of acute administration of insulin-like growth factor I in patients with Laron-type dwarfism. Lancet II 1170-1172 (1988).
10. Takano K, Hizuka N, Asakawa K, Sukegawa I, Shizume K and Demura H. Effects of sc administration of recombinant human insulin-like growth factor I (IGF-I) on normal human subjects. Endocrinol Japon 37:309-317 (1990)
11. Miyakawa M, Hizuka N, Takano K, Tanaka I, Horikawa R, Honda N and Shizume K. Radioimmunoassay for insulin-like growth factor I (IGF-I) using biosynthetic IGF-I. Endocrinol Japon 33:795-801 (1986).
12. Daughaday WH, Mariz IK and Blethen SK. Inhibition of access of bound somatomedin to membrane receptor and immunobinding sites: a comparison of radioreceptor and radioimmunoassay of somatomedin in native and acid-ethanol-extracted serum. J Clin Endocrinol Metab 51:781-788 (1980).
13. Hizuka N, Takano K, Asakawa K, Sukegawa I, Fukuda I, Demura H, Iwashita M, Adachi T and Shizume K. Measurement of free form of insulin-like growth factor I in human plasma. Growth Regulation submitted.
14. Asakawa K, Hizuka N, Takano K, Fukuda I, Sukegawa I, Demura H and Shizume K Radioimmunoassay for insulin-like growth factor II (IGF-II). Endocrinol Japon 37:in press (1990).
15. Sukegawa I, Hizuka N, Takano K, Asakawa K, Horikawa R, Hashida S, Ishikawa E, Mohri Z, Murakami Y and Shizume K. Urinary growth hormone (GH) measurements are useful for evaluating endogenous GH secretion. J Clin Endocrinol Metab 66:1119-1123 (1988).

INSULIN-LIKE GROWTH FACTOR BINDING PROTEIN CONTROL SECRETION AND MECHANISMS OF ACTION

David R. Clemmons

Division of Endocrinology
University of North Carolina
Chapel Hill, NC

The insulin-like growth factors are present in extracellular fluids bound to high affinity soluble binding proteins. This localization of IGF-I and II in the pericellular microenvironment in association with IGF binding proteins suggests a role for these proteins in controlling the amount of IGF that is transported to specific cell types and the amount of growth factor that is available to associate with type I and II IGF receptors. This theory has recently been strengthened by the observation that each of the IGF binding proteins that have been purified has an affinity constant that is higher than the type I IGF receptor.[1] Therefore these proteins have the capacity to partition IGF-I and II between receptors and the extracellular fluid compartment. Other proposed functions of the IGF binding proteins include acting as a high affinity carriers in plasma and preventing uncontrolled efflux of the IGFs from the vascular space. In blood IGF-I and II form a 150 kilodalton complex with IGF binding protein-3 (IGFBP-3).[2] The complex is not transported across intact capillaries and forms a reservoir of IGF that is available for utilization by peripheral tissues. Since IGF-I and II are believed to be secreted into blood and utilized by peripheral tissues at a relatively constant rate, the binding protein in the vascular compartment may provide a means for maintaining a sustained amount of growth factor that is available for transport into extracellular fluids.

An additional proposed function for these binding proteins is to mediate the transport of IGF-I and II out of the vascular space. Although it is unknown at this time whether IGF-I and II are transported in a free or bound form, two principals have clearly been established. First, the carrier proteins themselves can cross intact vascular endothelium.[3,4] A second important finding is the observation that when IGF-I is crosslinked to these carrier proteins the complex can also traverse that intact capillaries, suggesting that there is nothing inherent in the IGF-carrier protein complex that results in steric hinderance of its transport.[5] These findings do not exclude the possibility, however that IGF-I and II can leave the vasculature in an unbound state. The evaluation of this hypothesis will require further analysis.

Following their exit from the vascular space the IGF binding proteins have been postulated to direct IGF-I and II to specific tissues, however direct proof that IGF binding protein mediated tissue specific localization occurs has not been provided. Likewise whether each specific binding protein has unique functional properties in addition to their general function as IGF carriers has not been definitively established. The only findings that have been published to date that have any relevance to this point are as follows: 1) the mechanisms that regulate secretion of each form of IGF binding protein are clearly distinct, and 2) each specific cell type appears to secrete a distinct pattern of binding proteins and in some cell types the variables that regulate secretion appear to have distinct effects. However several types of binding proteins are secreted by different cell types and therefore the specificity often resides in the relative ratios of the specific forms that are secreted and only to a lesser extent in the actual species of binding protein that is secreted. The complexity of this system, for example at least four structurally characterized binding proteins, variable patterns of secretion by specific cell types and variable regulation as well as an incomplete understanding of their target cell functions, means that a complete, integrated understanding of how these proteins function cannot be proposed at the present time.

Control of Blood Concentrations

Four specific IGF binding proteins have been isolated and structurally characterized.[6-9] All four have been shown to be present in normal human serum. Several studies have shown that other forms of IGF binding proteins may exist in relatively low concentrations in rat plasma,[10] but it appears that IGFBP-1, 2, 3 and 4 are the major forms in human plasma. Because IGFBP-4 has been isolated and structurally characterized only recently, the control of its plasma concentrations has not been studied. However the variables controlling plasma IGFBP-1, 2 and 3 concentrations have been reported. IGFBP-3 is the major IGF binding moiety in serum.[11] It circulates in concentrations between 2 and 4 ug/ml and has an extremely high affinity for IGF-I and II.[12] It associates with a 88000 dalton non-binding subunit to form a 150,000 dalton complex.[13] Baxter has shown definitively that there is a stoichiometric relationship between the molar concentration of IGF-I and II and the amount of IGFBP-3 in serum. There is an excess of free unbound acid labile subunits, but there is no free IGFBP-3 or IGF-I and II.[14] Any excess IGF-I or II that occurs appears to be bound by IGFBP-1, 2 and 4. Several studies have shown that infusion of IGF-I directly stimulates the secretion of IGFBP-3. Infusion of IGF-I into hypophysectomized rats for six days induces an increase in IGFBP-3 that is equal to that which is induced by growth hormone administration.[15] Likewise if rats are fed a 5% protein diet it results in decreased serum IGF-I and IGFBP-3 (Figure 1). Infusion of IGF-I restores IGFPB-3 to normal levels whereas GH fails to increase IGF-I or IGFBP-3.[15] This suggests that IGF-I directly induces the secretion of IGFBP-3 and that GH mediates its effects through IGF-I. Whether or not growth hormone is working solely through IGF-I cannot be absolutely determined at the present time. However the data suggest that induction of IGF-I by growth hormone is required prior to induction of IGFBP-3. Although the effect of direct infusion of IGF-II on IGFBP-3 has not been reported, IGF-II can induce IGFBP-3 secretion by cultured human fibroblasts.[16] Therefore it appears that the blood IGFBP-3 concentration is dependent upon the sum of the molar concentrations of IGF-I and IGF-II. The importance of these findings is that they would provide a mechanism for maintaining the equimolar ratio between the sum of the IGF-I and II concentration and IGFBP-3 concentration. This could provide a mechanism for tightly regulating the amount of IGF that is available for extravascular transport.

The regulation of plasma IGFBP-1 and 2 concentrations appears to be more complex. Whereas IGFBP-3 concentrations are relatively constant throughout a typical 24 hour day IGFBP-1 concentrations are much more labile.[11] They are markedly increased after an overnight fast and easily suppressible with ingestion of a meal. This meal induced suppression

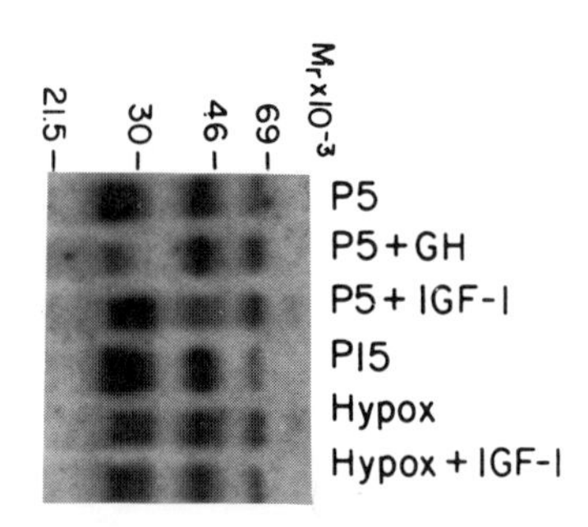

Fig. 1. Induction of IGFBP-3 by IGF-I. Normal rats were fed a 5% protein diet for six days and infused with vehicle (P5) GH, 200 ug/day, or IGF-I, 300 ug/day. Likewise hypophysectomized rats that were fed a 15% protein diet received similar doses. The results show that IGFBP-3 (MW 46-69000 d) was induced with IGF-I but not GH in the P5 animals.

appears to be primarily mediated by changes in insulin secretion. Infusion of insulin into diabetic subjects results in a marked suppression of plasma IGFBP-1 concentrations and the degree of suppression is proportionate to the amount of insulin infused.[17] Likewise withdrawl of insulin and/or glucose results in an acute increase in IGFBP-1 plasma concentrations. Specifically the concentration of IGFBP-1 is increased approximately 3.5 - 12-fold, depending on the age of the subject, after an overnight fast.[18] Continued fasting for 9 days results in an additional 3.5-fold increase.[18] Presumably these changes are due to decreases in insulin secretion, although changes in glucose flux have not been completely excluded. Lewitt and Baxter have shown that addition of glucose to cultured human hepatic explants results in suppression of IGFBP-1 secretion.[19] Recently we infused glucose or fructose into normal volunteers after overnight fast. The glucose infusion resulted in a 2.5-fold increase in plasma c-peptide whereas fructose had no effect.[20] The glucose infusion caused a 61% decrease in IGFBP-1 over 4 hours and this change correlated with the change in c-peptide r=0.74 (Figure 2). However fructose infusion also resulted in a 39% decrease in IGFBP-1. Therefore the glucose associated fall in IGFBP-1 appears to be partially due to glucose stimulated insulin secretion.

The molecular mechanisms accounting for these changes in serum IGFBP-1 are incompletely described, however at least two components appear to be necessary. First in cultured human hepatoma cells it has been shown that exposure of the cells to insulin results in a decrease IGFBP-1 mRNA abundance.[21] Likewise administration of insulin to diabetic rats has been shown to reduce IGFBP-1 mRNA abundance.[22] However simple reduction in biosynthesis of this growth factor is not the sole mechanism accounting for the decrease in its plasma concentrations. Specifically we have shown in collaboration with Dr. Robert Bar that perfusion of the isolated rat heart with radiolabelled IGFBP-1 or IGFBP-2 shows that either protein can cross intact vascular endothelial barrier.[3] Passage appears to be due to traverse across the capillary bed into the heart muscle. When insulin is coperfused with IGFBP-1, there is a 40% increase in the amount of IGFBP-1 that is translocated within the first 5 minutes of the perfusion.[4] In contrast there is no increase in the amount of IGFBP-2 transported. This indicates that insulin selectively stimulates IGFBP-1 translocation out of

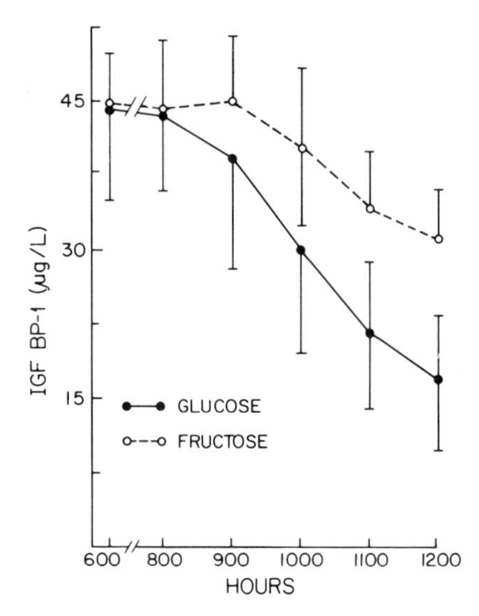

Fig. 2. Glucose Induced Change in IGFBP-1. Six normal volunteers received glucose or fructose 12.5 gm/hr for four hours. Plasma c-peptide rose 2.6-fold in response to the glucose infusion, but did not change during the fructose infusion.

the vascular compartment and this may account of the acute rapid decline in serum IGFBP-1 levels that occurs following an increase in insulin secretion. Undoubtedly suppression of gene transcription also plays a role in this reduction, however, it appears that the rapid effect that occurs in the first hour, e.g. 25% suppression, may be due specifically to a direct effect of insulin upon the efflux of this binding protein from the vascular space. This opens up the possibility that during the normal fasting-refeeding cycle there is a flux of IGF-I and IGF-II out of the vasculature that is meal dependent and this meal dependent flux is dependent upon the ambient rate of IGFBP-1 transport. Conversely, it is possible that this is simply a mechanism for translocating IGFBP-1 itself and that no IGF-I or II is transported bound to IGFBP-1. Resolution of these two alternatives will require direct analysis of simultaneous rates of transport of each growth factor and IGFBP-1 in intact animals.

In contrast to IGFBP-1, IGFBP-2 concentrations fluctuate much less during a 24 hour cycle. Specifically when frequent sampling is undertaken over a 48 hour interval the fluctuations in IGFBP-2 are much less than those in IGFBP-1 and there is no post-prandial decline in IGFBP-2. Likewise when glucose is infused in amounts sufficient to cause a 2.5-fold increase in c-peptide levels, IGFBP-2 concentrations do not change. This suggests that unlike IGFBP-1, physiologic variations in insulin concentrations have no direct effect on plasma IGFBP-2 concentrations in humans. In rats it has been shown that the uncontrolled diabetic state results in an increase in IGFBP-2 mRNA abundance[22] and that serum levels as determined by immunoblotting decrease after insulin administration.[23] Therefore it is possible that the extreme fluctuations in insulin levels that occur in diabetes may result in some change in serum IGFBP-2 levels in humans. These observations are consistent with the observations of Bar that physiologic changes in insulin do not acutely stimulate IGFBP-2 translocation out of the vasculture.

The role of growth hormone in regulating IGFBP-2 is less well-defined. Although IGFBP-2 concentrations are clearly elevated in hypophysectomized animals[24] and some hypopituitary humans,[25] the response to growth hormone by hypopituitary subjects is not uniform. Specifically when growth hormone is administered to normal subjects ingesting a calorically restricted diet there is no change in plasma IGFBP-2 concentrations. When it is administered to GH deficient children who are ingesting a normal diet the response is erratic with some subjects showing no suppression whatsoever. In contrast in well-fed cattle receiving growth hormone there was a 3.5-fold suppression in plasma IGFBP-2 concentrations.[26] This suggests that a normal caloric intake may be required for growth hormone to suppress IGFBP-2 and the response to pharmacologic doses of growth hormone is variable. Interestingly however during phase IV sleep when growth hormone secretion is the highest there is a significant decrease in plasma IGFBP-2, suggesting that physiologic pulsatile secretion of growth hormone may result in some suppression of this binding protein. The most potent inducer of IGFBP-2 secretion into blood appears to be IGF-I. When IGF-I is infused into calorically restricted normal human volunteers there is a 4-fold increase in plasma IGFBP-2 concentrations over 5 days and this response attenuates immediately upon cessation of administration of the growth factor. The role of homologous induction of this high affinity carrier protein by IGF-I is unknown. In neonatal and fetal blood IGFBP-2 is the predominant form of IGF binding protein.[27] This suggests that it may have a major role as a plasma carrier during fetal development although the physiologic significance of this finding is uncertain at the present time. Further studies will be required to delineate the role of IGFBP-2 in mediating IGF transport out of the vascular compartment.

Control of Secretion By Cells

Secretion of IGFBP-1, 2 and 3 by several normal cell types has been extensively delineated. In general each cell type secretes a distinct pattern of binding proteins. For example fibroblasts are a primary source of IGFBP-3 and its secretion is controlled by both cyclic nucleotides[28] and the IGFs themselves.[29] Likewise muscle cells which do not secrete IGFBP-3 produce primarily IGFBP-2 and other forms of IGFBPs and changes in their secretion have been associated with changes in muscle cell differentiation.[30] In addition it appears that at least one muscle cell type, the C2 myoblast, secretes a unique form of IGF binding protein that is not present in smooth muscle cells and is not present in normal serum.[31]

IGFBP-1 secretion by decidual cells has been studied in detail since IGFBP-1 is the most abundant form of IGFBP that is secreted by this cell type. Decidual cells also secrete lesser amounts of IGFBP-2 and 4. IGF-I suppresses IGFBP-1 secretion by these cells by 6.8 fold at concentrations of 50 ng/ml.[32] Insulin also suppresses IGFBP-1 secretion but its maximal effect is only 68% of the maximal IGF-I response.[32] Relaxin and cAMP are potent stimuli of IGFBP-1 secretion resulting in 20 and 6 fold stimulation of secretion over 24 hours (Figure 3). The significance of this major change in IGFBP-1 levels remains to be determined. Similarly cyclic AMP is a potent stimulant whereas agents that regulate protein kinase C activity have no effect.[33] The effects of these agents are relatively specific since they have no effect on IGFBP-2 secretion by this cell type. Insulin induces differential regulation of IGFBP secretion since it stimulates IGFBP-4 while suppressing IGFBP-1. The significance of this observation for decidual cell function is unknown. IGFBP-1 secretion has also been detected by rat hepatocytes during rat liver regeneration. Since IGFBP-1 is rapidly induced during liver regeneration,[34] this suggests that its induction may be linked to cell proliferation.

Recently our laboratory has been interested in delineating the secretion of IGF binding proteins by tumor cells and determining the

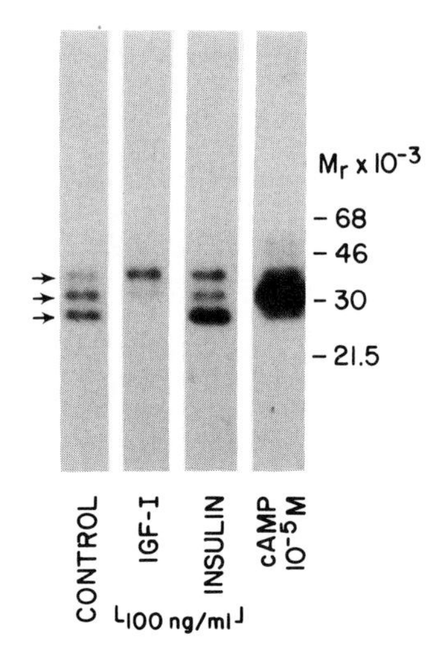

Fig. 3. Differential regulation of IGFBP secretion by Decidual Cells. Decidual cell cultures were exposed to IGF-I, insulin or cAMP and 24 hr conditioned media analyzed by SDS-PAGE with ligand blotting. The results show that insulin and IGF-I suppress IGFBP-1 (30k band) and relaxin is a potent stimulant of its secretion.

response of these target cells to ambient changes in the surrounding IGF binding protein concentrations. Human breast carcinoma cells have been the most extensively studied. We have determined that the types binding proteins that are constitutively released under basal conditions is determined primarily by estrogen receptor status.[35] Cell lines that are estrogen receptor negative secrete IGFBP-3 and IGFBP-1 as well as IGFBP-4. In contrast those cells that are ER-positive secrete no IGFBP-3 or 1 but secrete substantial amounts of IGFBP-2 and IGFBP-4 (Figure 4).[35] IGFBP-2 secretion by ER-positive cells can be stimulated by estrogen. Likewise in the ER-negative cell line MDA-231, cyclic nucleotides appear to increase IGFBP-3 secretion as well as IGFBP-1. IGF-I and insulin suppress IGFBP-1 secretion by this cell type.[36] That IGFBP-1 secretion may be an important modulator of growth is evidenced by two observations. First, tumor cell lines that constitutively secrete high concentrations of IGFBP-1 grow in response to exposure to IGF-I alone. Second, in breast tumor cell lines that secrete lower concentrations of IGFBP-1, the exogenous addition of IGFBP-1 results in potentiation of cell growth rate in response to IGF-I. Therefore variables that enhance IGFBP-1 secretion may potentiate growth by enhancing the secretion rate of this protein. Whether IGFBP-1 alters other functions of these tumor cells has not been determined.

Alterations in IGF-receptor Interaction

Since the IGF binding proteins have affinities for IGF-I and II that are greater than the affinity of the type I IGF receptor they have the potential to significantly alter IGF association with this receptor. At equilibrium the addition of pure IGFBP-1, 2 or 3 results in competition with the receptor for binding radiolabeled IGF-I.[37] If radiolabeled IGF-I is added it binds preferentially to the IGFBPs in solution. However this is not a simple two component equilibrium system since the IGFBPs themselves can bind to cell surfaces. When IGF-I is added at concentrations between 25-100 ug/ml it binds not only to type I receptors but also to cell surface associated IGFBPs. The affinity of the cell surface associated forms of IGFBPs is clearly less than their affinity in solution.[38] Therefore at equilibrium the IGFs will preferentially associate with the solution phase forms of IGFBPs then at higher concentrations they will bind to receptors and cell surface associated IGFBPs. The physiologic consequences of these observations are significant. They indicate that under equilibrium conditions the IGFs are bound to IGFBPs in extracellular fluids and will associate with cell

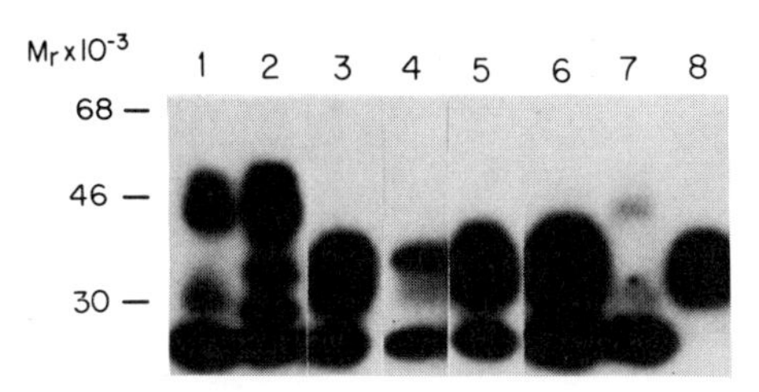

Fig. 4. Ligand Blot Analysis of IGFBPs Secreted by Breast Carcinoma Cell Lines. Conditioned media from estrogen receptor negative cell lines are shown in lanes 1, 2 and 7. They secrete IGFBP-3 (42-48 k band), IGFBP-1 (29-30 k band) and IGFBP-4 (24 k band). Lanes 3, 4, 5, 6 and 8 contain conditioned media from estrogen receptor positive cell lines. They contain IGFBP-2 (34 k band) and IGFBP-4.

surface binding sites only if they are present in higher concentrations than those required to bind to extracellular fluid IGFBPs or if the IGFBP/IGF complex directly associates with the cell surface. Whether direct transfer of IGF off of this cell surface complex onto the receptor occurs has not been determined.

Target Cell Action

Both IGFBP-1 and 2 have been shown to potentiate the cellular growth response to IGF-I in vitro.[39,40] However in other studies IGFBP-1 has been shown to inhibit cell growth[41] and processes linked to growth, such as amino acid transport.[42] Both proteins have interesting structural features which could account for their capacity to alter cellular responsiveness. Both proteins have an Arg-Gly-Asp sequence near their carboxyl terminus which may be responsible for mediating cell adherence. Since cell adherence bears a correlative relationship with the capacity of IGFBP-1 to potentiate the cellular growth response to IGF-I, cellular attachment through this sequence may be an important requirement for biologic activity. An additional property of both proteins is their capacity to form higher order oligomers when stored at high concentrations.[43] Since oligomer formation has been correlated with biologic responsiveness, disulfide exchange with subsequent formation of multimers that are linked by intermolecular disulfide bonds may be a requirement for potentiation of cellular responsiveness to IGF-I. Most recently we have determined that both IGFBP-1 and IGFBP-2 are phosphorylated. Phosphorylation of IGFBP-1 occurs on serine residues.[44] IGFBP-3 has also been reported to be phosphorylated.[45] Phosphorylation appears to result in significant enhancement of the affinity of IGFBP-1 for IGF-I with a 3-7 fold potentiation of affinity being detectable following the phosphorylation reaction. Interestingly the dephosphorylation reaction may also be of biologic significance. Specifically only the dephosphorylated form has been noted to have the property of potentiating the cellular growth response to IGF-I. Therefore the capacity of specific cell types to dephosphorylate this protein may be an important requirement for cells to be able to respond to IGFBP-1 with enhanced biologic responsiveness. Determination of the specific serines that are phosphorylated within IGFBP-1 and whether dephosphorylation occurs constitutively prior to potentiation of DNA synthesis are important questions that need to be addressed. Likewise we have determined that the dephosphorylated form does exist in normal human fetal serum and in amniotic fluid, therefore identifying the cell types that specifically secrete dephosphorylated form of this protein and its secretion rate into physiologic fluids relative to secretion of the phosphorylated form will be important future parameters to determine.

In summary the variables that control the synthesis and secretion of IGF binding proteins as well as their efflux from the vascular compartment in association with the IGFs are important parameters to determine and will continue to be the subject of intensive investigation. Studies to determine how these proteins function at the target cell level to enhance or inhibit cellular responses to the IGFs should lead us to a better understanding of the multiple functional roles of these proteins. Understanding the specific functions of each form of binding protein will be required to form an integrated picture of how they work together to modulate IGF actions.

Acknowledgements

The author gratefully acknowledges Ms. Jennifer O'Lear who prepared the manuscript and Dr. Louis Underwood for his gift of radiolabeled IGF-I. This work was supported by a grant from the National Institutes of Health AG02331.

References

1. Baxter, R.C. and Martin, J.L., Binding proteins for insulin like growth factors: Structure, Regulation and Function, Prog. Growth. Fact. Res. 1:49-68 (1989).
2. Baxter, R.C., Martin, J.L. and Beniac, V.A., High molecular weight insulin-like growth factor binding protein complex: purification and properties of the acid-labile subunit from human serum, J Biol Chem 264:11843-11848 (1989).
3. Bar, R.S., Clemmons, D.R., Busby, W.H., et al, Transcapillary permeability and subendothelial distribution of endothelial and amniotic fluid (IGFBP-1) IGF-binding proteins in the rat heart, Endocrinology 127:1078-1086 (1990).
4. Bar, R.S., Boes, M., Clemmons, D.R., et al, Insulin differentially alters transcapillary movement of intravascular IGFBP-1, IGFBP-2 and endothelial cell IGF binding proteins in rat heart, Endocrinology 127:497-499 (1990).
5. Bar, R.S., Boes, M., Dake, B.L., et al., Tissue localization of perfused endothelial cell IGF bindign protein is markedly altered by association with IGF-I. Endocrinol. 127:3243-3245 (1990).

6. Brewer, M.T., Stetler, G.L., Squires, C.H., et al, Cloning, characterization and expression of a human insulin- like growth factor binding protein, Biochem.Biophys.Res.Comm. 152:1289-1297 (1988).
7. Brown, A.L., Chairiotti, L., Orlowski, C.C., et al, Nucleotide sequence and expression of a cDNA cloning encoding a fetal rat binding protein for insulin-like growth factors, J.Biol.Chem. 264:5148-5154 (1989).
8. Wood, W.I., Cachianes, G., Henzel, W.J., et al, Cloning and expression of the GH dependent insulin like growth factor binding protein, Mol. Endo. 2:1176-1185 (1988).
9. Shimasaki, S., Shimonaka, M., Ui, M., et al, Structural characterization of a follicle-stimulating hormone action inhibitor in porcine ovarian follicular fluid: Its identification as the insulin-like growth factor-binding protein, J. Biol. Chem. 265:2198-2202 (1990).
10. Zapf, J., Bom, W., Chang, J.Y., et al, Isolation and NH2-terminal amino acid sequences of rat serum carrier proteins for insulin-like growth factors, Biochem.Biophys.Res.Comm. 156:1187-1194 (1988).
11. Baxter, R.C. and Martin, J.L., Radioimmunoassay of growth hormone dependent insulin-like growth factor binding protein in human plasma, J.Clin.Invest. 78:1504-1512 (1986).
12. Baxter, R.C., Characterization of the acid labile subunit of the growth hormone dependent insulin like growth factor binding protein complex, J.Clin.Endocrinol.Metab. 67:265-272 (1988).
13. Baxter, R.C. and Martin, J.L., Structure of the Mr 140,000 growth hormone-dependent insulin-like growth factor binding protein complex: determination by reconstitution and affinity labeling, PNAS 86:6898-6902 (1989).
14. Van Wyk, J.J. The somatomedins: Biological actions and physiological control mechanisms. In: *Hormonal proteins and peptides*, edited by Li, C.H. Orlando: Academic Press, 1985, p. 81-125.
15. Clemmons, D.R., Thissen, J.P., Maes, M., et al, Insulin-like growth factor-I (IGF-I) infusion into hypophysectomized or protein-deprived rats induces specific IGF binding proteins in serum, Endocrinology 125:2967-2972 (1989).
16. Thissen, J. P., Underwood, L. E., Maiter, D., et al, Failure of IGF-I infusion to promote growth in protein restricted rats despite normalization of serum IGF-I concnetrations, Endocrinology, in press (1991).
17. Suikkari, A.M., Koivisto, V.A., Koistinen, R., et al, Dose response characteristics for suppression of low molecular weight plasma insulin like growth factor binding protein by insulin, J. Clin. Endocrinol. Metab. 68:135-140 (1989).
18. Busby, W.H., Snyder, D.K. and Clemmons, D.R., Radioimmunoassay of a 26,000 dalton plasma insulin like growth factor binding protein: Control by nutritional variables, J.Clin.Endocrinol.Metab. 67:1225-1230 (1988).
19. Lewitt, M.S. and Baxter, R.C., Inhibitors of glucose uptake stimulate the production of insulin-like growth factor-binding protein (IGFBP-1) by human fetal liver, Endocrinol. 126:1527-1533 (1990).
20. Snyder, D.K. and Clemmons, D.R.,Insulin dependent regulation of IGFBP-1, J. Clin. Endocrinol. Metab. 71:1632-1638 (1990).
21. Conover, C.A.,Regulation of insulin-like growth factor (IGF)-binding protein synthesis by insulin and IGF-I in cultured bovine fibroblasts, Endocrinology 126:3139-3145 (1990).
22. Ooi, G.T., Orlowski, C.C., Brown, A.L., et al,Different tissue distribution and hormonal regulation of messenger RNAs encoding rat insulin-like growth factor binding proteins-1 and 2, Mol. Endo. 4:321-328 (1990).

23. Unterman, T.G., Patel, K., Kumar Mahathre, V., et al, Regulation of low molecular weight insulin-like growth factor binding proteins in experimental diabetes mellitus, Endocrinol. 126:2614-2626 (1990).
24. Orlowski, C.C., Brown, A.L., Ooi, G.T., et al, Tissue, developmental and metabolic regulation of messenger ribonucleic acid encoding a rat insulin-like growth factor-binding protein, Endocrinology 126:644-652 (1990).
25. Hardouin, S., Gourmelen, M., Noguiez, P., et al, Molecular forms of serum insulin-like growth factor (IGF)-binding proteins in man: relationship with growth hormone and IGFs and physiological significance, J. Clin. Endocrinol. Metab. 69:1291-1301 (1989).
26. Cohick, W. S., Busby, W. H., Clemmons, D. R. and Bauman, D. E., Insulin-like growth factor binding proteins (IGFBP) in serum of lactating cows treated with recombinant in-methionlyl bovine somatotropin (BST). Presented to the Annual Meeting of the American Society of Animal Science, Ames Iowa, July, 1990.
27. McCusker, R.H., Campion, D.R., Jones, W.K. and Clemmons, D.R., The insulin-like growth factor-binding proteins of porcine serum: Endocrine and nutritional regulation, Endocrinology 125:501-509 (1989).
28. Martin, J.L. and Baxter, R.C., Insulin like growth factor binding proteins IGF-BP's produced by human skin fibroblasts immunological relationship to other human IGF-BP's, Endocrinology 123:1907-1915 (1988).
29. Hill, D.J., Camacho-Hubner, C., Rashid, P., et al, Insulin like growth factor binding protein secretion by human fibroblasts: Dependence on cell density and IGF peptides, J. Endocrinol. 122:87-98 (1989).
30. McCusker, R.H. and Clemmons, D.R., Insulin-like growth factor-binding proteins (IGF-BPs) secretion by muscle cells: Effect of cellular differentiation and proliferation, J. Cell. Physiol. 137:505-512 (1988).
31. Tollefsen, S.E., Lajara, R., McCusker, R.H., et al, Insulin-like growth factors (IGF) in muscle development, J.Biol.Chem. 264:13810-13817 (1989).
32. Thrailkill, K.M., Clemmons, D.R., Busby, W.H. and Handwerger, S.R., Differential regulation of IGF binding protein (IGFBP) secretion from human decidual cells by IGF-I, insulin and relaxin, J. Clin. Invest. 86:878-883 (1990).
33. Clemmons, D.R., Thrailkill, K.M., Handwerger, S.R. and Busby, W.H., Three distinct forms of insulin-like growth factor binding proteins are released by decidual cells in culture, Endocrinology 127:643-650 (1990).
34. Melby, A. E., Mohn, K. L., Laz, T., et al, The gene encoding the rat low molecular weight IGF binding protein gene is highly expressed immediate early gene in liver regeneration, Mol. Cell. Biol. in press (1991).
35. Clemmons, D.R., Camacho-Hubner, C., Coronado, E. and Osborne, C.K., Insulin-like growth factor binding protein secretion by breast carcinoma cell lines: Correlation with estrogen receptor status, Endocrinology 127:2679-2686 (1990).
36. Camacho-Hubner, C., McCusker, R.H. and Clemmons, D.R., Regulation of insulin-like growth factor binding protein secretion by human tumor cells in vitro, J. Cell. Physiol. submitted.
37. McCusker, R. H., Busby, W. H., Dehoff, M., et al, Insulin-like growth factors (IGF) binding to cell monolayers is modulated by the addition of binding proteins (IGFBPs), Presented to the Endocrine Society, Atlanta, GH, June 22, 1990.
38. McCusker, R.H., Camacho-Hubner, C., Bayne, M.L., et al, Insulin-like growth factor (IGF) binding to human fibroblast and glioblastoma cells: The modulating effect of cell released IGF binding proteins (IGFBPs), J. Cell. Physiol. 144:244-254 (1990).

39. Elgin, R.G., Busby, W.H. and Clemmons, D.R., An insulin-like growth factor binding protein enhances the biogic response to IGF-I, Proc.Natl.Acad.Sci. 84:3313-3318 (1987).
40. Bourner, M.J., Busby, W.H., Siegel, N.R., et al, Purification, structural determination and characterization of biologic and binding activity of bovine insulin-like growth factor binding protein-2, J. Biol. Chem. in press (1991).
41. Busby, W.H., Klapper, D.G. and Clemmons, D.R., Purification of a 31000 dalton insulin like growth factor binding protein from human amniotic fluid, J.Biol.Chem. 263:14203-14210 (1988).
42. Ritvos, O., Ranta, T., Jalkanen, J., et al, Insulin-like growth factor (IGF) binding protein from human decidua inhibits the binding and biological action of IGF-I in cultured choriocarcinoma cells, Endocrinology 122:2150-2157 (1988).
43. Busby, W.H., Hossenlopp, P., Binoux, M. and Clemmons, D.R., Purified preparations of the amniotic fluid IGF binding protein contain multimeric forms that are biologically active, Endocrinol. 125:773-777 (1989).
44. Mukku, V.R., and Chu, H., Phosphorylation of insulin-like growth factor binding protein by mammalian cells, Presented to the Endocrine Society Meeting, Abstract 190, 1990.
45. Jones, J.I., and Clemmons, D.R., The bioactivity of an insulin-like growth factor binding protein is modulated by serine phosphorylation, To Be Presented to the Southern Society for Clinical Investigation Meeting, 1991.

REGULATION AND ACTIONS OF INSULIN-LIKE GROWTH FACTOR BINDING PROTEIN-3*

Robert C. Baxter and Janet L. Martin

Endocrinology Department
Royal Prince Alfred Hospital
Camperdown, NSW 2050, Australia

INTRODUCTION

Binding proteins for the insulin-like growth factors (IGFs), first described 15 years ago, are now recognized as playing important roles as carriers of IGFs in the circulation, and modulators of IGF actions on target cells [1]. Three distinct structural classes of IGF binding protein (IGFBP) have been identified on the basis of cDNA cloning [2-6], and at least two other structurally distinct proteins appear to exist [7-9]. The predominant circulating IGFBP is an acid-stable glycoprotein with the systematic name IGFBP-3. This paper will discuss recent studies, mainly from the authors' laboratory, which relate to the dual roles of IGFBP-3, as part of the growth hormone (GH) dependent, high molecular weight ternary IGF complex in serum, and as local cell regulators of mitogenic actions of the IGFs.

IGFBP-3 was first isolated from Cohn fraction IV of human plasma by IGF-affinity chromatography and reverse-phase HPLC [10]. Analyzed by SDS-PAGE, the purified protein appears as a doublet, with major and minor components of 53 and 47 kDa, respectively, under nonreducing conditions, or 43 and 40 kDa when reduced [10]. Preparations from rat serum [11,12], porcine serum [13], and porcine follicular fluid [14], similarly appear as doublets. Deglycosylation experiments indicate that the two bands are glycosylation variants of a single core protein, with a predicted molecular mass of 28.7 kDa. The primary structure of IGFBP-3 is highly homologous with those of the other well-characterised IGFBPs, IGFBP-1 and IGFBP-2. A total of 18 Cys residues are conserved in the three proteins, clustered into two domains, at the N- and C-termini of the proteins [6]. IGFBP-3 possesses a single high-affinity IGF-binding site, with slightly higher affinity for IGF-II (30 l/nmol) than IGF-I (20 l/nmol)[10].

---

* This work was supported by the National Health & Medical Research Council, Australia

Molecular Biology and Physiology of Insulin and Insulin-Like Growth Factors
Edited by M.K. Raizada and D. LeRoith, Plenum Press, New York, 1991

FORMATION AND REGULATION OF THE TERNARY COMPLEX

IGFBP-3 differs from the other IGFBPs in that it circulates predominantly as part of a ternary complex of ~140 kDa. The other components of this complex are IGF-I or IGF-II, and another, acid-labile protein of ~85 kDa [15]. The acid-labile, binding, and growth factor subunits of the complex have been termed the α-, β-, and γ-subunits, respectively [16]. Like the β-subunit (IGFBP-3), the purified α-subunit also appears on SDS-PAGE as a doublet with bands of 84 and 86 kDa, decreasing to a single 70 kDa band on enzymatic deglycosylation [15]. Since each of the three components of the 140 kDa complex exists in at least two different forms, it is clear that at least eight forms of ternary complex might exist. However, how these different complexes might differ functionally is entirely unknown.

Recently it has become possible to perform binding kinetic experiments on ternary complex formation. In these studies, radioiodinated α-subunit binding to β-γ (IGFBP-3•IGF) is quantitated by precipitating the complex with an IGFBP-3 antiserum [15]. By this means it has been demonstrated that :

(i) α-subunit binding has an absolute requirement for the presence of IGF-I or IGF-II, i.e. binary α-β complexes do not form

(ii) the association constant for α-subunit binding to β-γ, ~0.5 l/nmol, is approximately two orders of magnitude lower than that for IGF binding to IGFBP-3

(iii) the presence of α-subunit has no effect on the kinetics of β-γ complex formation (i.e. IGF binding to IGFBP-3).

The ternary complex appears to play an important role in stabilizing IGFs in the circulation. Free IGFs are estimated to have a circulating half-life of only ~10 min, whereas those in the 140 kDa complex have a half-life of ~10 h [17]. Thus the concentration of unbound IGFs must remain very low. In fact, total serum IGFBP-3, ~110 nmol/l, is very similar to the total serum IGF concentration [18], implying that virtually all circulating IGF-I and IGF-II are associated with IGFBP-3. In contrast, α-subunit circulates at a considerable molar excess, approximately 250 nmol/l, as quantitated by radioimmunoassay [19]. This can also be demonstrated directly, by depleting serum of IGFBP-3 by immunoaffinity chromatography. Although this effectively removes all of the ternary complex, about half of the α-subunit remains, indicating that it was present in the free form, rather than as part of the complex [19]. However, it can be demonstrated simply that this free α-subunit is fully functional, as it is able to convert a covalent IGFBP-3•[^{125}I]IGF-I complex to the 140 kDa form [20]. It may be speculated that the high concentration of α-subunit is important in maintaining most of the IGFs and IGFBP-3 in the high molecular weight form, despite the low association constant for ternary complex formation. This would have the effect of "trapping" the IGFs in the circulation, as the 140 kDa complex is thought to cross the capillary barrier very poorly [21].

The factors known to regulate the circulating concentration of IGFBP-3 and the high molecular weight complex are similar to those that regulate IGF-I production : age, nutritional status, and GH [1,22]. Both IGFBP-3 and α-subunit levels are quite low in neonates, rise steadily through childhood, show highest levels in the pubertal years, and decline steadily throughout adult life [18,19]. The role of nutritional status is indicated by the close association between nitrogen balance and both IGF-I and IGFBP-3 levels in patients undergoing parenteral nutrition [24]. Complex regulation by GH has also been well documented, with IGFBP-3 levels elevated in acromegaly and decreased in GH-deficiency [18,23]. The close association between IGF-I and IGFBP-3 regulation raises the question whether IGF-I is in fact the prime regulator of IGFBP-3. There is some evidence to support this concept : for example, Zapf et al. have demonstrated that IGF-I infusion in hypophysectomized rats strongly induces IGFBP-3 [25]. As discussed below, IGF-I can also be shown to induce IGFBP-3 production in cell culture systems, in some cases acting in synergism with other growth factors.

POTENTIAL ROLE OF THE CIRCULATING COMPLEX

The functional significance of IGFs, and the ternary complex, in the circulation is not well understood. If it is true that IGFs in the high molecular weight form cannot cross the capillary barrier, then a dissociation mechanism must exist to allow circulating IGFs to reach target tissues and exert biological activity. Whether this really occurs *in vivo* is quite controversial, as there is evidence to suggest that IGFs have a significant role as autocrine or paracrine growth factors, i.e. they are generated locally, and act locally [26,27]. If this is the universal mechanism of IGF action, then the circulating complex might simply be a reservoir of peptides on the path to degradation and disposal. However, the very long half-life of IGFs present in the ternary complex makes the complex a very unlikely medium for IGF disposal. Furthermore, experiments in which IGF-I or IGF-II have been infused into rats show clear growth effects over several days [28], despite the fact that exogenous IGF-I is found within hours of administration predominantly in the high molecular weight form [17]. This implies that complexed IGFs are able to perform an endocrine role.

There have been several suggestions as to how IGFs in the circulation might dissociate from the ternary complex to exert actions on target cells. Dilution in the interstitial fluid [29], or the actions of a specific protease or heparin [30], have been proposed to release IGFs from the binding protein. However, an equally significant problem appears to be how the binding protein is released from the α-subunit, i.e. how the ternary complex dissociates. This is of particular importance for two reasons. First, as mentioned above, IGFs, with or without associated IGFBPs, may never gain access to the extracellular fluid surrounding their target cells unless the ternary complex dissociates, since in that form capillary endothelial cells may present an impenetrable barrier [21]. Second, it is not at all certain that IGFs must dissociate from IGFBP-3 to exert their growth promoting activities; indeed, recent studies provide strong support for the concept that the presence of IGFBP-3 might enhance mitogenic activity in some cell systems [31,32].

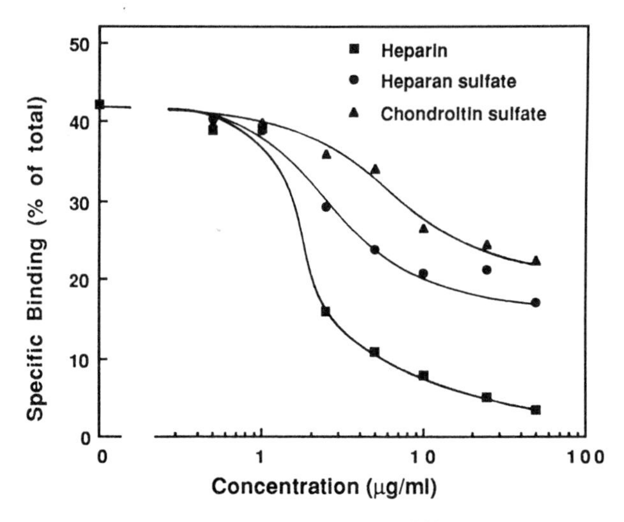

Fig. 1 Inhibition of the binding of ^{125}I-labeled α-subunit to mixtures of IGFBP-3 and IGF-I by increasing concentrations of glycosaminoglycans. Ternary complexes were precipitated using a specific IGFBP-3 antiserum. Experimental conditions were as previously described [33]. Reproduced from Baxter [33] with permission.

The relatively weak binding affinity of α-subunit for IGFBP-3·IGF complexes may be an important factor in ternary complex dissociation. This binding affinity is in fact greatly affected by ionic conditions, implying an important charge-interaction component in the association of α-subunit with the binding protein. In the presence of increasing NaCl concentrations, a progressive reduction in the binding affinity of α-subunit for IGFBP-3·IGF-I is seen [33]. Similarly, polyionic substances appear to be powerful inhibitors of ternary complex formation. Notably, glycosaminoglycans have recently been shown to inhibt this process. Fig. 1 illustrates the inhibitory effect of the glycosaminoglycans heparin, heparan sulfate, and chondroitin sulfate, on the binding of α-subunit to IGFBP-3·IGF-I. It should be noted that in these studies no effect of these substances on binary complex formation, i.e. the binding of IGF-I to IGFBP-3, was observed [33].

The inhibition of α-subunit binding by glycosaminoglycans suggestss how ternary complexes might dissociate, to release IGFBP-3 bearing IGFs which might then cross the capillary barrier. This could occur if the complexes interact with proteoglycans on the surface of capillary endothelial cells; such an interaction would greatly lower the affinity of α-subunit for IGFBP-3. Cell-surface proteoglycans have been shown in other studies to act as weak "receptors" for certain proteins [34]. It may further be speculated that the potential diversity of carbohydrate chains on the ternary complex, resulting from the known glycosylation variants of both IGFBP-3 and α-subunit, may play an important role in regulating this process. However, it must be emphasized that this scheme remains unconfirmed at present.

REGULATION OF CELLULAR IGFBP-3 PRODUCTION

In recent years considerable attention in the authors' laboratory has been directed to the factors that control IGFBP-3 synthesis. These studies have relied on a radioimmunoassay for IGFBP-3 which is not affected by the presence of non-primate proteins, so that the effect of growth factors in fetal calf serum (FCS) could be tested. The main cell system used in these studies has been cultures of neonatal human fibroblasts derived from foreskin explants [35]. These cells produce an IGFBP-3 glycoprotein doublet of ~50 kDa that is indistinguishable by size or binding characteristics from the protein isolated from plasma. When analyzed by immunoprecipitation with an IGFBP-3 antiserum following affinity-labeling, a smaller IGFBP-3-related protein is also detectable in fibroblast culture medium [35]; however, recent experiments with a transformed fibroblast line that does not express the IGFBP-3 gene suggest that this is a distinct protein which shares an epitope with IGFBP-3, but cross-reacts too weakly under the limited-antibody conditions of radioimmunoassay to be detectable by this method [9].

FCS is a strong stimulus to IGFBP-3 production. The stimulatory effect is dose-dependent up to a concentration of at least 10%, and is maintained for three days in culture [35]. Although fibroblast IGFBP-3 normally appears to be ~50 kDa, in the presence of FCS a new peak of ~150 kDa appears. This suggests that synthesis of α-subunit by fibroblasts might be stimulated by the serum [35]. However, this seems unlikely, since the serum itself can be shown by a functional assay to contain α-subunit, whereas fibroblast-conditioned medium, analyzed using a primate-specific radioimmunoassay for α-subunit, contains no activity. Therefore fibroblasts appear not to make α-subunit, and in fact no human cell type has yet been shown to secrete this protein.

We have recently demonstrated that acidification of FCS greatly increases its ability to stimulate IGFBP-3 production by neonatal fibroblasts (Martin JL, Baxter RC, submitted). When the serum was fractionated by gel permeation chromatography, and each fraction tested before and after acification for its ability to increase IGFBP-3 production, the fractions that were potentiated by acidification appeared at ~250 kDa. Since transforming growth factor-β (TGF-β) is known to exist in a latent high molecular weight form, from which the active ~25 kDa form is released by acidification [36], we tested the effect of a neutralizing antiserum to TGF-β on the ability of the active fractions to stimulate IGFBP-3 production. A shown in Table 1, this antiserum entirely abolished the increase in activity resulting from acidification, indicating that active TGF-β was indeed being released by this process.

To characterize the stimulatory effect of TGF-β on fibroblast IGFBP-3 production, the effects of exposure time, cell density, and TGF-β dose, were examined in detail. Exposure of cells to TGF-β for up to three days increasingly stimulated IGFBP-3 production, so that 1 ng/ml TGF-β caused only 2-fold stimulation after 24 h, but 6-fold stimulation after 72 h. The stimulatory effect of TGF-β was very dependent on cell density, with a relatively small effect seen in sparse cells (1×10^4 cell/well), and a greatly increased effect in dense cultures.

Table 1. The effect of TGF-β antiserum on stimulation of fibroblast IGFBP-3 production by fetal calf serum

	IGFBP-3 (ng/well)	
	No addition	TGF-β antiserum
No addition	33.3 ± 2.4	37.2 ± 1.1
+Untreated fraction	72.8 ± 4.9	77.1 ± 4.0
+Acidified fraction	123.2 ± 3.5	69.9 ± 1.4*

Results are means ± SEM for triplicate wells. The "untreated and acidified fractions" are pooled active fractions of fetal calf serum, with or without transient acidification.
* P<0.01 compared to "no antibody".

Stimulation of IGFBP-3 production by TGF-β was half-maximal at a concentration of 0.40 ± 0.05 ng/ml, and maximal at 1-2 ng/ml. IGF-I was found to act synergistically with TGF-β; when added alone, it stimulated IGFBP-3 production no more than 1.5- to 2-fold, but when added in the presence of TGF-β, it was able to double the maximum rate achievable by TGF-β alone.

To test whether fibroblast cultures secreted latent TGF-β similar to that found in FCS, cells were exposed to conditioned medium, and the effect of the neutralizing TGF-β antibody was tested. Conditioned medium stimulated IGFBP-3 production by about 2-fold, this stimulation being unaffected by TGF-β antibody. However, when the conditioned medium was acidified before testing, a further 2-fold stimulation was seen, and this additional activity was reversed in the presence of TGF-β antibody. This indicates that fibroblasts themselves secrete latent TGF-β which is capable of being activated to stimulate IGFBP-3 production.Although the mechanism of activation is unknown, it appears that proteolysis, like acidification, is able to release active TGF-β from the latent form [37]. Thus the possibility exists of a regulatory "cascade" in which latent TGF-β secreted by fibroblasts is activated by an endogenous protease, resulting in stimulation of IGFBP-3 and subsequent modulation of IGF actions.

EFFECTS OF IGFBP-3 AT THE CELLULAR LEVEL

Relatively few studies have addressed the question of the ways in which IGFBP-3 modulates IGF action at the cellular level, and the available studies do not provide an unambiguous answer. Binding proteins have generally been considered to block the activity of their bound ligand, and this has certainly been demonstrated for several impure preparations of IGFBPs [1]. However, Blum et al.[31] reported that in baby hamster kidney fibroblasts, co-incubation of IGFBP-3 with IGF-I greatly enhanced the mitogenic activity of the IGF-I. Similarly, in rat osteoblastic cells, IGF-I-stimulated DNA synthesis and collagen gene expression were shown to increase under conditions where

IGFBP-3 production by the cells was increased, but not if an IGF-I analogue incapable of binding to IGFBP-3 was used [32]. These studies imply a direct enhancing effect of IGFBP-3, either endogenous or exogenous, on IGF action, possibly by protecting the growth factor from degradation, or more specifically, by somehow increasing its accessibility to cell receptors.

Different results have been observed in human and bovine fibroblasts. We have reported that, in neonatal skin fibroblasts, co-incubation of IGF-I with IGFBP-3 resulted in an inhibition of the ability of IGF-I to stimulate DNA synthesis, the effect being maximal when IGFBP-3 was added at a concentration equimolar with that of the IGF-I [38]. In contrast, if the cells were preincubated with IGFBP-3, the ability of subsequently added IGF-I to stimulate DNA synthesis was potentiated. A very similar phenomenon of inhibition by co-incubation, and potentiation by preincubation, has recently been reported in bovine fibroblast cultures, in which aminoisobutyrate uptake rather than DNA synthesis was studied [39]. The mechanism by which the potentiation occurs is unknown, but conceivably preincubation with IGFBP-3, by sequestering endogenous IGFs, might allow up-regulation of IGF receptors, thus making the cells more responsive to subsequently added IGF-I.

Finally, two recent reports indicate that endogenous IGFBP-3 may act as an inhibitor of various cell functions. Blat et al. [40] purified an endogenous 45 kDa inhibitor of DNA synthesis from mouse 3T3 fibroblasts, and found that it had the same N-terminal sequence as rat IGFBP-3. Similarly, an inhibitor of FSH-stimulated steroidogenesis in ovarian granulosa cells was purified from follicular fluid, and shown to be identical to IGFBP-3 [14]. Thus, in different cell types, and depending upon the way in which it is presented to the cells, IGFBP-3 appears capable of acting both as an inhibitor and a potentiator of IGF action.

Which of these roles does IGFBP-3 play in neonatal skin fibroblasts stimulated by TGF-β? Fig. 2 shows TGF-β dose-response curves for IGFBP-3 production and [^{3}H]thymidine incorporation, after 24 or 72 h exposure. As mentioned above, the stimulatory effect of TGF-β on IGFBP-3 production is greatly enhanced over 72 h compared to 24 h. However, the reverse is true for DNA synthesis; the highest rates of thymidine incorporation are seen after 24 h, when IGFBP-3 production is low, and greatly suppressed incorporation is seen after 72 h, when IGFBP-3 production is increased. Therefore, if IGFBP-3 has a regulatory effect on DNA synthesis in these cells, it appears to be an inhibitory one. This implies a self-limiting role for TGF-β in fibroblast growth, whereby initially it stimulates DNA synthesis, and subsequently, by increasing IGFBP-3 production, it suppresses DNA synthesis.

CONCLUDING COMMENT

In this paper we have discussed the two well-recognized roles of IGFBP-3, as the binding component of the major IGF-carrying complex in serum, and as a local regulator of IGF actions. The functional relationship between these roles, which

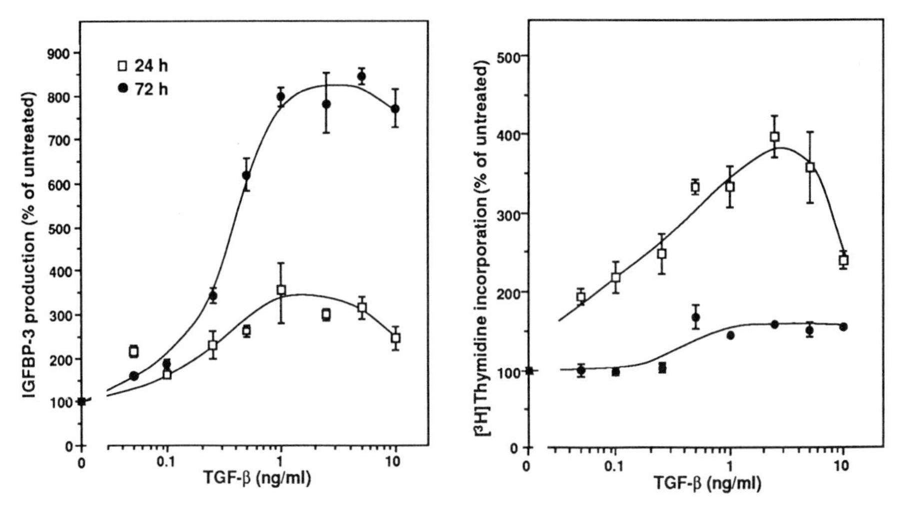

Fig. 2 Left: Stimulation of IGFBP-3 production in neonatal human skin fibroblast cultures by exposure to increasing concentrations of TGF-β for 24 or 72 h, as indicated. Right : Thymidine incorporation in the same cells, measured over the 4 h period following the indicated incubation times with TGF-β.

appear to be quite distinct, is not fully understood. For example, there is little information on the major tissues of origin of IGFBP-3 and α-subunit in the circulation, or the precise way in which they associate with IGFs *in vivo* to form the ternary complex. The biological importance of the multiple potential glycosylation variants of the complex is also unclear. Even more fundamental are the questions whether complexed IGFs are indeed available to target tissues to promote growth-related processes, and whether IGFBP-3 from the circulation, like that produced locally, is able to modulate IGF activity.

At the cellular level, we have demonstrated that fibroblasts produce latent TGF-β which can be activated to stimulate IGFBP-3 production. If this mechanism operates widely, it might provide an explanation for the very widespread actions of TGF-β on growth related processes in a variety of cell types. Finally, when the significance of the paradoxical roles of IGFBP-3 as both an inhibitor and a stimulator of cellular functions is understood, some of the diverse actions of TGF-β might also be explained.

REFERENCES

1. Baxter RC, Martin JL. Binding proteins for the insulin-like growth factors : Structure, regulation and function. Prog Growth Factor Res 1989; 1 : 49-68.
2. Ballard J, Baxter R, Binoux M, Clemmons D, Drop S, Hall K, Hintz R, Rechler M, Rutanen E, Schwander J. On the nomenclature of the IGF binding proteins. Acta Endocrinol (Copenh) 1989; 121 : 751-752

3. Brinkman A, Groggen C, Geurts van Kessel A, Drop SLS. Isolation and characterization of a cDNA encoding the low molecular weight insulin-like growth factor binding protein (IBP-1). EMBO J 1988; 7 : 2417-2423
4. Lee YL, Hintz RL, James PM, Lee PDK, Shively, JE, Powell, DR. Insulin-like growth factor (IGF) binding protein complementary deoxyribonucleic acid from human HEP G2 hepatoma cells : Predicted protein sequence suggests an IGF binding domain different from those of the IGF-I and IGF-II receptors. Molec Endocrinol 1988; 2 : 404-411
5. Binkert C, Landwehr J, Mary J-L, Schwander J, Heinrich G. Cloning, sequence analysis and expression of a cDNA encoding a novel insulin-like growth factor binding protein (IGFBP-2). EMBO J 1989; 8 : 2497-2502
6. Wood WI, Cachianes G, Henzel WJ, Winslow GA, Spencer SA, Hellmiss R, Martin JL, Baxter RC. Cloning and expression of the growth hormone-dependent insulin-like growth factor-binding protein. Molec Endocrinol 1988; 2 : 1176-1185
7. Mohan S, Bautista CM, Wergedal J, Baylink DJ. Isolation of an inhibitory insulin-like growth factor (IGF) binding protein from bone cell-conditioned medium : A potential local regulator of IGF action. Proc Natl Acad Sci USA 1989; 86: 8338-8342
8. Roghani M, Hossenlopp P, Lepage P, Balland A, Binoux M. Isolation from human cerebrospinal fluid of a new insulin-like growth factor-binding protein with a selective affinity for IGF-II. FEBS Lett 1989; 255 : 253-258
9. Martin JL, Willetts KE, Baxter RC. Purification and properties of a novel insulin-like growth factor-II binding protein from transformed human fibroblasts. J Biol Chem 1990; 265 : 4124-4130
10. Martin JL, Baxter RC. Insulin-like growth factor binding protein from human plasma. Purification and characterization. J Biol Chem 1986; 261 : 8754-8760
11. Baxter RC, Martin JL. Binding proteins for insulin-like growth factors in adult rat serum. Comparison with other human and rat binding proteins. Biochem Biophys Res Commun 1987; 147 : 408-415
12. Zapf J, Born W, Chang J-Y, James P, Froesch ER, Fischer JA. Isolation and NH_2-terminal amino acid sequences of rat serum carrier proteins for insulin-like growth factors. Biochem Biophys Res Commun 1988; 156 : 1187-1194
13. Walton PE, Baxter RC, Burleigh BD, Etherton TD. Purification of the serum acid-stable insulin-like growth factor binding protein frim the pig (*Sus scrofa*). Comp Biochem Physiol 1989; 92B : 561-567
14. Ui M, Shimonaka M, Shimasaki S, Ling N. An insulin-like growth factor-binding protein in ovarian follicular fluid blocks follicle-stimulating hormone-stimulated steroid production by ovarian granulosa cells. Endocrinology 1989; 125 : 912-916
15. Baxter RC, Martin JL, Beniac VA. High molecular weight insulin-like growth factor binding protein complex. Purification and properties of the acid-labile subunit from human serum. J Biol Chem 1989; 264 : 11843-11848
16. Baxter RC, Martin JL. Structure of the M_r 140,000 growth hormone-dependent insulin-like growth factor binding protein complex: Determination by reconstitution and affinity-labeling. Proc Natl Acad Sci USA 1989; 86 : 6898-6902

17. Guler H-P, Zapf J, Schmid C, Froesch ER. Insulin-like growth factors I and II in healthy man. Estimations of half-lives and production rates. Acta Endocrinol (Copenh) 1989; 121 : 753-758
18. Baxter RC, Martin JL. Radioimmunoassay of growth hormone-dependent insulinlike growth factor binding protein in human plasma. J Clin Invest 1986; 78: 1504-1512
19. Baxter RC. Circulating levels and molecular distribution of the acid-labile (α) subunit of the high molecular weight insulin-like growth factor-binding protein complex. J Clin Endocrinol Metab 1990; 70 : 1347-1353
20. Baxter RC. Characterization of the acid-labile subunit of the growth hormone-dependent insulin-like growth factor binding protein complex. J Clin Endocrinol Metab 1988; 67 : 265-272
21. Binoux M, Hossenlopp P. Insulin-like growth factor (IGF) and IGF-binding proteins : Comparison of human serum and lymph. J Clin Endocrinol Metab 1988; 67 : 509-514
22. Baxter RC. The somatomedins : Insulin-like growth factors. Adv Clin Chem 1986; 25 : 49-115
23. Blum WF, Ranke MB, Kietzmann K, Gauggel E, Zeisel HJ, Bierich JR. A specific radioimmunoassay for the growth hormone (GH)-dependent somatomedin-binding protein : Its use for diagnosis of GH deficiency. J Clin Endocrinol Metab 1990; 70 : 1292-1298
24. Tan K, Baxter RC. Nutritional regulation of somatomedin-C/insulin-like growth factor-I (IGF-I) and its binding proteins during recovery from surgery. Abstract, 8th Int Congr Endocrinol, Kyoto, Japan, 1988, p. 616
25. Zapf J, Hauri C, Waldvogel M et al. Recombinant human insulin-like growth factor I induces its own specific carrier protein in hypophysectomized and diabetic rats. Proc Natl Acad Sci USA 1989; 86 : 3813-3817
26. D'Ercole AJ, Stiles AD, Underwood LE. Tissue concentrations of somatomedin C : Further evidence for multiple sites of synthesis and paracrine or autocrine mechanisms of action. Proc Natl Acad Sci USA 1984; 81 : 935-939
27. Fant M, Munro H, Moses AC. An autocrine/paracrine role for insulin-like growth factors in the regulation of human placental growth. J Clin Endocrinol Metab 1986; 63 : 499-505
28. Schoenle E, Zapf J, Hauri C, Steiner T, Froesch ER. Comparison of in vivo effects of insulin-like growth factors I and II and of growth hormone in hypophysectomized rats. Acta Endocrinol 1985; 108 : 167-174
29. Daughaday WH, Ward AP, Goldberg AC, Trivedi B, Kapadia M. Characterization of somatomedin binding in human serum by ultracentrifugation and gel filtration. J Clin Endocrinol Metab 1982; 55 : 916-921
30. Clemmons DR, Underwood LE, Chatelain PG, Van Wyk JJ. Liberation of immunoreactive somatomedin-C from its binding proteins by proteolytic enzymes and heparin. J Clin Endocrinol Metab 1983; 56 : 384-389
31. Blum WF, Jenne EW, Reppin F, Kietzmann K, Ranke MB, Bierich JR. Insulin-like growth factor I (IGF-I)-binding protein complex is a better mitogen than free IGF-I. Endocrinology 1989; 125 : 766-77
32. Ernst M, Rodan G. Increased activity of insulin-like growth factor (IGF) in osteoblastic cells in the presence of growth hormone (GH) : Positive correlation with the

presence of the GH-induced IGF-binding protein BP-3. Endocrinology 1990; 127 : 807-814
33. Baxter RC. Glycosaminoglycans inhibit formation of the 140 kilodalton insulin-like growth factor-binding protein complex. Biochem J 1990; 272 (in press)
34. Höök M. Cell-surface glycosaminoglycans. Ann Rev Biochem 1984; 53 : 847-869
35. Martin JL, Baxter RC. Insulin-like growth factor-binding proteins (IGF-BPs) produced by human skin fibroblasts : Immunological relationship to other human IGF-BPs. Endocrinology 1988; 123 : 1907-1915
36. Lawrence DA, Pircher R, Julien P. Conversion of a high molecular weight latent β-TGF from chicken embryo fibroblasts into a low molecular weight active β-TGF under acidic conditions. Biochem Biophys Res Commun 1985; 133 : 1026-1034
37. Lyons RM, Keski-Oja J, Moses HL. Proteolytic activation of latent transforming growth factor-β from fibroblast conditioned medium. J Cell Biol 1988; 106 : 1659-1665
38. De Mellow JSM, Baxter RC.Growth hormone-dependent insulin-like growth factor (IGF) binding protein both inhibits and potentiates IGF-I-stimulated DNA synthesis in human skin fibroblasts.Biochem Biophys Res Commun 1988; 156 : 199-204
39. Conover CA. Biological actions of insulin-like growth factor binding protein-3 in cultured bovine fibroblasts. Abstract, 72nd Annual Meeting, The Endocrine Society, Atlanta, GA, 1990, p. 71
40. Blat C, Bohlen P, Villaudy J, Chatelain G, Golde A, Harel L. Isolation and amino-terminal sequence of a novel cellular growth inhibitor (inhibitory diffusible factor 45) secreted by 3T3 fibroblasts. J Biol Chem 1989; 264 : 6021-6024

REGULATION OF GENE EXPRESSION OF RAT INSULIN-LIKE GROWTH FACTOR BINDING PROTEINS 1 AND 2

Matthew M. Rechler, Alexandra L. Brown, Guck T. Ooi,
Craig C. Orlowski, Lucy Y.-H. Tseng, and Yvonne W.-H. Yang

Growth and Development Section, Molecular, Cellular and Nutritional Endocrinology Branch, National Institute of Diabetes and Digestive and Kidney Disease, National Institutes of Health, Bethesda, Maryland 20892

INTRODUCTION

Virtually all of the insulin-like growth factors (IGFs) in extracellular fluids and cell culture medium occur complexed to specific IGF-binding proteins (IGFBPs).[1,2] The IGFBPs are a family of proteins that bind IGF-I and IGF-II but are unrelated to IGF receptors. Four IGFBPs have been cloned from human and rat sources,[1-8] and partial protein sequence information is available for a fifth IGFBP.[9-11] Other members of the IGFBP family undoubtedly exist,[12] but specific assignment must await amino acid or nucleotide sequencing.

Increasing evidence suggests that the IGFBPs play an important role in regulating the biological actions of IGF-I and IGF-II in different tissues. They have been proposed to determine the bioavailability of IGFs (i.e., half-life in the circulation, transfer from the vascular compartment to tissue space, and tissue distribution)[13-15] and modulate their biological actions on target tissues.[16,17] The major 150 kDa IGF:binding protein complex in adult plasma is a ternary complex consisting of an IGF-binding subunit (IGFBP-3), an acid-labile subunit, and IGF-I or IGF-II.[18] The 150 kDa complex has a long half-life in plasma and may represent an inactive storage form of the IGFs.[13] IGFBP-3 is under-represented in lymph, suggesting that the 150 kDa complex does not cross the capillary barrier.[19] By contrast, IGFBP-1, a low molecular weight IGFBP present in plasma in small amounts and not complexed to the acid-labile subunit, shows dynamic metabolic regulation. Plasma IGFBP-1 decreases rapidly after refeeding fasted patients,[20] or after insulin-treatment of insulin-dependent diabetics,[21] suggesting that it may actively transport IGFs from the vascular compartment to specific tissues. Insulin promotes the exit of IGFBP-1 from the plasma in perfused rat heart,[14] whereupon IGFBP-1 localizes selectively to cardiac muscle rather than to connective tissue.[15]

The approach of our laboratory to understanding the biological role of the IGFBPs has been to study the molecular regulation of IGFBP genes in rat tissues under different physiological conditions and in selected cell model systems. Our working hypothesis is that regulation of

IGFBP mRNA abundance determines the level of IGFBPs in tissues and contributes to the long-term regulation of IGFBP levels in plasma. Differential regulation of the expression of different IGFBPs also should provide clues to the physiological function of these molecules. This paper summarizes studies performed in our laboratory addressing the expression of the rat genes for IGFBP-1 and a second low molecular weight IGFBP, IGFBP-2, the major IGFBP in fetal rat serum.[22]

TISSUE AND DEVELOPMENTAL EXPRESSION OF IGFBP-1 and IGFBP-2 mRNAs.

The abundance of IGFBP-1 and IGFBP-2 mRNAs in liver at different ages was determined by Northern blotting of total liver RNA and hybridization with a cDNA probe for rat IGFBP-2 obtained from BRL-3A cells,[4] and a rat IGFBP-1 cDNA probe obtained from dexamethasone-treated H4-II-E rat hepatoma cells.[23] Both IGFBP-1 and IGFBP-2 mRNAs are expressed at high levels in livers from term-gestation and 1-day old rats and at greatly reduced levels at 21 and 65 days of age[24] (Fig. 1). In term fetal tissues, both IGFBP-1 and IGFBP-2 mRNAs are most abundant in liver.[24] Expression of IGFBP-2 mRNA is moderate in fetal brain, stomach, kidney,

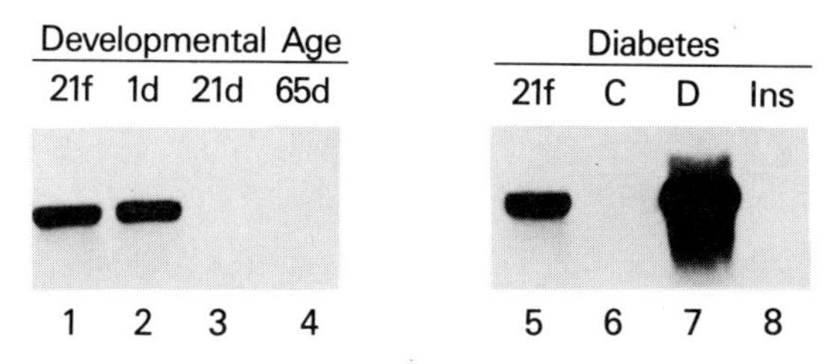

Fig. 1. Expression of IGFBP-1 mRNA in rat liver. Autoradiographs of Northern blots of total rat liver RNA (12 μg/lane) hybridized with a 1450 bp rat IGFBP-1 cDNA probe[23] that had been labeled by nick translation. Lanes 1-4 represent RNA from fetal liver at 21-days gestation (lane 1), or from postnatal liver at 1, 21, and 65 days of age (lanes 2-4, respectively). Lanes 6-8 represent RNA from control (C, lane 6), streptozotocin-induced diabetic (D, lane 7), and insulin-treated diabetic (Ins, lane 8) adult rat liver. Liver from 21-day gestation fetuses is shown in lane 5 for comparison. Similar results were obtained when a human IGFBP-1 probe was hybridized to the same samples.[24]

and lung, and lower in intestine, muscle, heart, and skin.[24] The relative expression of IGFBP-1 and IGFBP-2 mRNAs is similar in most tissues. Two exceptions are kidney and brain which have 8 and 25-fold less IGFBP-1 mRNA, respectively, than IGFBP-2 mRNA. The predominance of IGFBP-2 mRNA in brain suggested that IGFBP-2 may have a special role in the central nervous system.

Unlike liver in which IGFBP-2 mRNA is developmentally regulated, IGFBP-2 mRNA continues to be expressed at high levels in adult rat brain. *In situ* hybridization to coronal sections of the anterior hypothalamus of adult rat brain localized IGFBP-2 mRNA to the choroid plexus.[25] In brains from 13 day-gestation rat embryos, Wood et al.[26] more precisely localized IGFBP-2 mRNA to the epithelial layer of the choroid plexus, whereas IGF-II mRNA was expressed in the underlying mesenchymal layer. This suggested that IGFBP-2 might function to transport IGF-II through the tight-junction epithelial layer into the cerebrospinal fluid. IGFBP-2 was identified immunologically in the cerebrospinal fluid of adult rats[24] and in human cerebrospinal fluid.[27,28] Transport proteins for other hormones and nutrients such as transthyretin (thyroxine and retinol binding protein), transferrin (iron), and ceruloplasmin (copper) also are synthesized in the choroid plexus and transported to the cerebrospinal fluid.

METABOLIC AND HORMONAL REGULATION OF IGFBP-1 AND IGFBP-2 mRNAs IN LIVER

IGFBP-1 and IGFBP-2 mRNAs are increased in adult rat liver in several conditions associated with relative insulin deficiency: hypophysectomy,[24,29,30] fasting,[3,24,29] and diabetes.[24,31] IGFBP-2 mRNA was increased 10-fold after hypophysectomy, but was not decreased after growth hormone treatment.[5,24,29] This is consistent with experiments in cultured rat hepatocytes,[32] which showed that insulin but not growth hormone decreased IGFBP-2 mRNA. Although we saw only a small increase in IGFBP-1 mRNA in the liver of hypophysectomized rats,[24] Seneviratne et al.[30] reported a larger increase in IGFBP-1 mRNA in livers from rats hypophysectomized at different ages and a corresponding increase in IGFBP-1 gene transcription. Fasting for 48 h resulted in 10-fold increases in hepatic IGFBP-2 mRNA[24,29] and IGFBP-1 mRNA[3,23] that were reversed by refeeding.

IGFBP-1 mRNA was increased 100-fold in rats made diabetic and severely ketotic by injection with 140 mg/kg of streptozotocin[24] (Fig. 1); IGFBP-2 mRNA was increased 10-fold.[24,31] In this model, the increase in IGFBP-1 mRNA was readily reversed by insulin treatment, whereas the increase in IGFBP-2 mRNA was not. However, in non-ketotic diabetes induced with 70 mg/kg of streptozotocin, the increase in IGFBP-2 mRNA also was reversible.[23,31] Preliminary results of nuclear runoff assays suggest that increased IGFBP-1 mRNA synthesis contributes to the increase in IGFBP-1 mRNA in diabetic rat liver.[23]

Corresponding increases occur in plasma IGFBP-1 in diabetes[33] and plasma IGFBP-2 in fasting and hypophysectomy,[29] suggesting that the regulation of IGFBP mRNA synthesis and abundance may contribute to (or be responsible for) the alterations in plasma levels of IGFBP-1 and -2.

EXPRESSION AND REGULATION OF IGFBP-1 IN CULTURED CELLS

We next sought to identify cell culture systems in which rat IGFBP genes were expressed and regulated similarly to intact animals. A panel of rat cell lines was screened by ligand blotting, immunoblotting, immunoprecipitation, and Northern blotting to identify cells that preferentially expressed IGFBP-1, -2, or -3.[34] IGFBP-2 was expressed in BRL-3A cells and three other cell lines (Clone 9 and TRL 12-15 cells derived from rat liver, and NRK-52E rat kidney epithelial cells). The H4-II-E cell line, derived from the well-differentiated Reuber H35 rat hepatoma, expresses IGFBP-1, and C6 glial and B104 neuroblastoma cells synthesize IGFBP-3.

Following preliminary screening for hormonal regulation of IGFBP-2 in BRL-3A cells (not shown) and IGFBP-1 in H4-II-E cells, the synthetic glucocorticoid, dexamethasone, was identified as a potent inducer of IGFBP-1 in H4-II-E cells.[35] IGFBP-1 was increased 10-fold in the medium of confluent H4-II-E cells that had been incubated for 48 h with dexamethasone[35] (Fig. 2). IGF-I, growth hormone, progesterone, and testosterone were without effect. Half-maximal stimulation was seen with 6×10^{-9} M dexamethasone, a concentration equivalent to physiological levels of corticosterone. A parallel increase occurred in IGFBP-1 mRNA. The

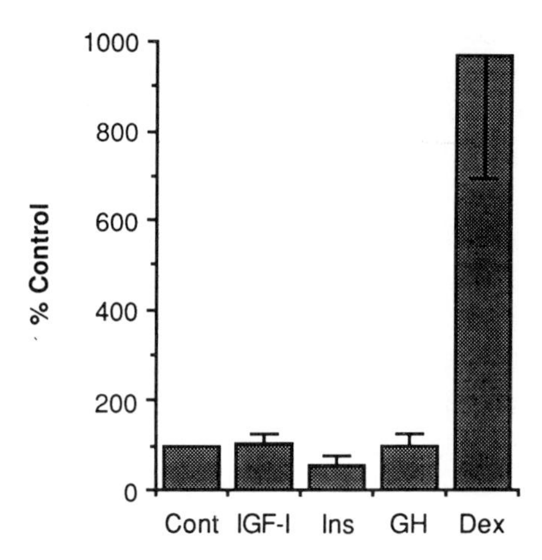

Fig. 2. Hormonal regulation of IGFBP-1 in H4-II-E conditioned medium. Cells were incubated for 48 h without hormone (Cont), or with IGF-I (100 ng/ml; Amgen), insulin (Ins; porcine insulin, Eli Lilly; 1 μg/ml), rat growth hormone (GH, 100 ng/ml; rGH B-11, National Hormone and Pituitary Program and the NIDDK), or dexamethasone (Dex; 10^{-6} M; Sigma). Medium was electrophoresed on sodium dodecyl sulfate-10% polyacrylamide gels under nonreducing conditions, proteins transferred onto nitrocellulose membranes by electroblotting, and IGFBPs identified by incubation with ^{125}I-IGF-I and autoradiography. Signal in the 30 kDa region corresponding to IGFBP-1 was quantitated by Beta scanning. (Modified from reference 35.)

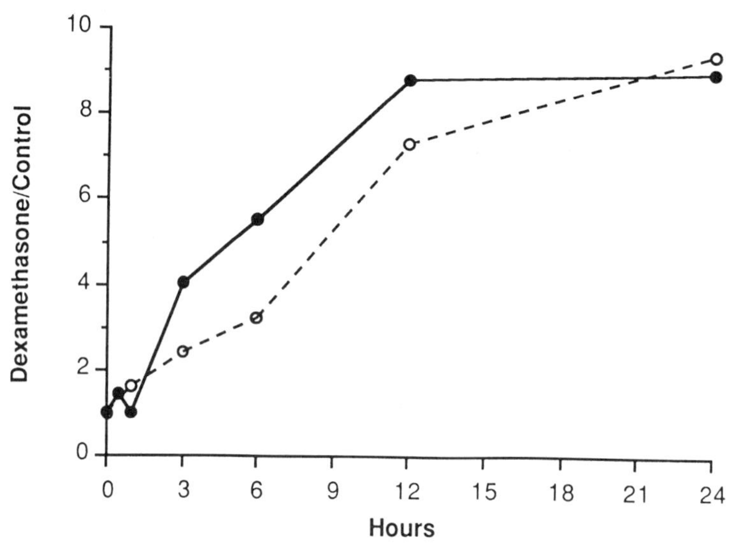

Fig. 3. IGFBP-1 mRNA (*solid circles*) and media IGFBP-1 (*open circles*) at different times after addition of dexamethasone to H4-II-E cells. Dexamethasone (10^{-6} M) was added to confluent H4-II-E cells in serum-free medium at time zero. Cells were harvested for RNA and analyzed by hybridization with a 1450 bp rat IGFBP-1 cDNA probe. IGFBP-1 in the medium was analyzed by ligand blotting as described in Fig. 2. Relative abundance (dexamethasone-treated/control) determined by Beta scanning is plotted at different times. (Reprinted from reference 35 with permission.)

increase in IGFBP-1 mRNA was detected after 3 h, and was maximal after 12 h (Fig. 3). The increase in media IGFBP-1 occurred after a slight delay. Both IGFBP-1 mRNA and, more slowly, IGFBP-1 in the media, decrease following dexamethasone withdrawal.[35] To determine whether dexamethasone increased IGFBP-1 mRNA synthesis, stability, or both, the half-life of IGFBP-1 mRNA was determined in dexamethasone-treated and control cells following addition of actinomycin to block new transcription initiation. The half-life of IGFBP-1 mRNA was similar in the presence or absence of dexamethasone ($t_{1/2}$~2h), suggesting that the principal effect of dexamethasone was on IGFBP-1 mRNA synthesis.[35] Preliminary results of nuclear runoff experiments support this conclusion. Thus, dexamethasone, at physiologically appropriate concentrations, causes a rapid and reversible increase in IGFBP-1 mRNA and, secondarily, in IGFBP-1 protein.

Although glucocorticoids provide the most dramatic hormonal regulation of IGFBP-1 or IGFBP-2 that we have observed to date in cultured cells, it is unlikely that they are important physiological regulators of IGFBP-1. The increase in IGFBP-1 after fasting is blunted in patients with Cushing's disease,[36] but this most likely reflects changes in insulin sensitivity. IGFBP-1 mRNA is increased in liver following administration of glucocorticoids to rats,[37] but this may be a pharmacological effect. By contrast, insulin, acting through the insulin receptor, rapidly decreases IGFBP-1 and IGFBP-1 mRNA produced by H4-II-E cells (C. C. Orlowski and G. T. Ooi, unpublished results). This effect is more obviously related to the effects of insulin administration *in vivo*, suggesting that transcriptional regulation may contribute significantly to the observed regulation of plasma levels of IGFBP-1.

CHARACTERIZATION OF THE RAT IGFBP-2 GENE

The rat IGFBP-2 gene consists of 4 exons, each of which contains protein-coding regions[38] (Fig. 4). It is organized similarly to the human IGFBP-1 gene.[39,40] The human IGFBP-3 gene differs in having a fifth exon consisting entirely of 3' untranslated sequences.[41] Exon-intron boundaries are conserved for the 3 genes. Exons 1, 3, and 4 show 30-40% homology, and the position of the 18 cysteine residues (12 in exon 1, 1 in exon 3, and 5 in exon 4) are conserved. Exon 2 shows no significant homology and does not contain cysteine.

The transcription initiation site of the IGFBP-2 gene was identified by reverse transcription-primer extension and ribonuclease protection experiments[38] (Fig. 5). Results from both approaches suggest a major mRNA start site at nucleotide -151 (with respect to the ATG), and probable utilization of multiple transcription start sites from nucleotide -151 to -116. Reverse transcription terminated prematurely at nucleotide -88 using more 3' primers, presumably because of secondary structure of the mRNA in this region.

The presence of a functional promoter was demonstrated by transient transfection assays in BRL-3A cells using a construct in which the 5' flanking region of the IGFBP-2 gene (1144 bp) was coupled to a luciferase reporter gene[38] (Fig. 6). The construct in which the 5' flanking region was in the same orientation as the sense strand of the luciferase gene was expressed with approximately 10% the efficiency of constructs containing the RSV promoter. Control vectors without the promoter region or with the IGFBP-2 5' flanking region in the opposite orientation were without activity. A construct containing 430 bp of 5' flanking region also was active, whereas a construct consisting of the region between nucleotides -189 and -35 was inactive (A. L. Brown, unpublished results). These results further support the assignment of nucleotide -151 as the transcription initiation site. Utilization of the rat IGFBP-2 promoter is cell-type specific

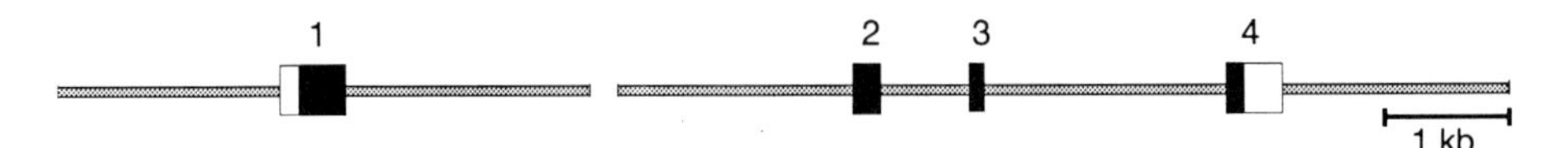

Fig. 4. Structure of the rat IGFBP-2 gene. Schematic diagram of the rat IGFBP-2 gene as determined by restriction mapping and nucleotide sequencing.[38] Boxes indicate the 4 exons. Solid areas are protein coding regions; open areas in exons 1 and 4 are the 5' and 3'-untranslated regions, respectively. The map is drawn to scale. Intron 1 is ~35 kb (J. B. Margot, personal communication).

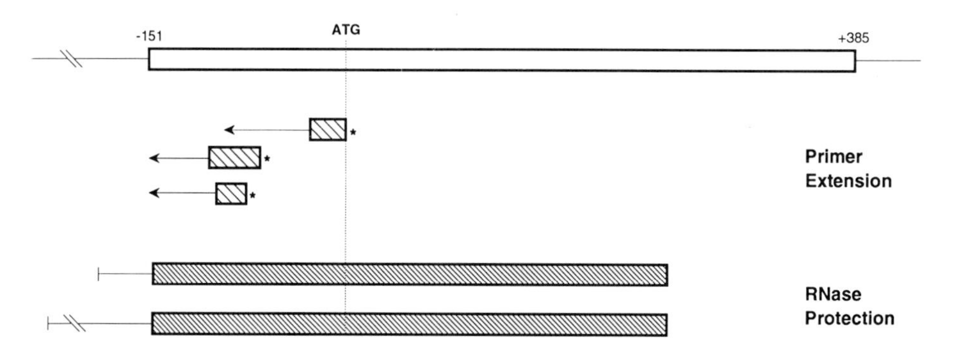

Fig. 5. Mapping the 5' end of the rat IGFBP-2 gene by primer extension and ribonuclease protection. This schematic diagram summarizes the results of primer-extended reverse transcription and ribonuclease protection experiments.[38] Exon 1 (extending from nucleotides -151 to +385 with respect to ATG, +1) is shown (top, *open bar*).

Primer-extended reverse transcription. Oligonucleotides (*hatched bars*) were end-labeled with [γ-^{32}P]ATP, hybridized with total BRL-3A RNA, and extended using reverse transcriptase. The *thin arrows* indicate the size of the extended fragments. Using an oligonucleotide primer that spanned nucleotide +1, the mRNA terminated at nucleotide -88. Using oligonucleotides that spanned nucleotide -88, termination occurred at nucleotide -151.

Ribonuclease protection. Antisense riboprobes extending from nucleotides -189 or -581 to nucleotide +244 were prepared by transcription using SP6 polymerase and [α-^{32}P]UTP. Following hybridization to BRL-3A RNA, unprotected single strand RNA was digested with ribonucleases A and T1. Protected fragments were analyzed by 6% acrylamide gels. The protected region extended from nucleotide +244 to multiple sites between nucleotides -148 and -116. The maximum extent of protection of the two riboprobes was the same, as shown by the *hatched bars.*

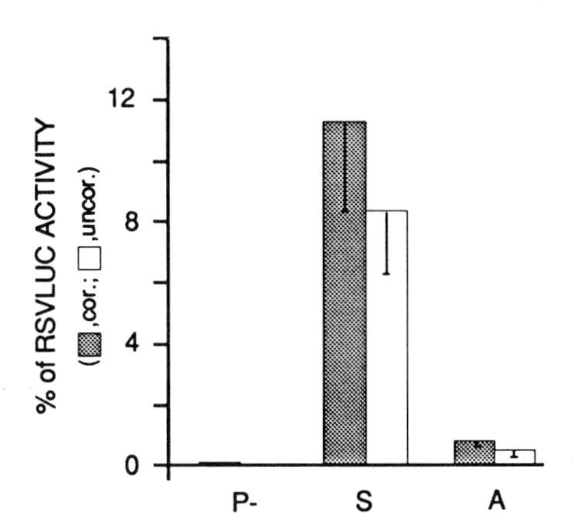

Fig. 6. Promoter activity of the IGFBP-2 5'-flanking region. A restriction fragment corresponding to nucleotides -1295 to -35 of the rat IGFBP-2 gene was inserted upstream of a promoterless luciferase gene (P-) in plasmid pA3LUC (kindly provided by Dr. William M. Wood, University of Colorado) in the sense (S) and antisense (A) orientations. Plasmids were cotransfected with plasmid pTKGH (containing the human growth hormone gene under the control of the thymidine kinase promoter) into BRL-3A cells by electroporation. After 48 h, the cells were lysed and luciferase activity measured in the extracts. Luciferase activity is expressed as the percent of activity in transfections with constructs containing the Rous sarcoma virus promoter (RSVLUC) in the same experiment, and are plotted with *(shaded bars)* or without (*open bars*) correction for the efficiency of growth hormone transfection. (Reprinted from reference 38 with permission).

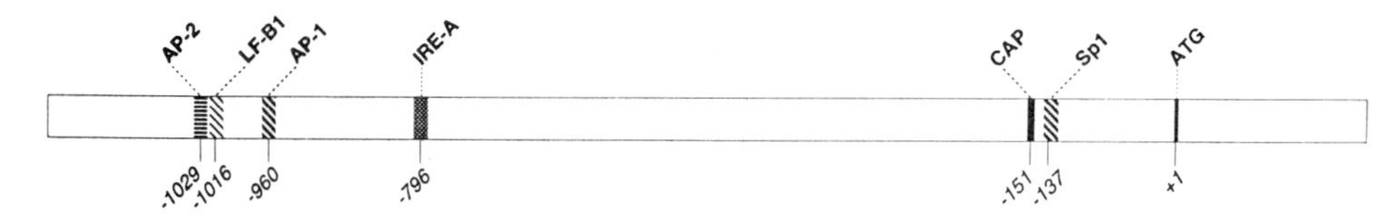

Fig. 7. Schematic diagram showing potential transcription regulatory elements in the 5' flanking region of the rat IGFBP-2 gene. The translation start site (+1, ATG) and the 5' end of IGFBP-2 mRNA (-151, CAP) are shown. Nucleotide positions identify the 5' end of selected potential regulatory elements. LF-B1 is liver-specific transcription factor B1;[45] IRE-A is the minimum consensus sequence of the positive insulin-response element in the glyceraldehyde-3-phosphate dehydrogenase gene.[46]

since transfection of the sense 1144-bp construct into H4-II-E cells that do not express IGFBP-2 did not yield luciferase activity.

The nucleotide sequence of the 5' flanking region was next determined.[38] No TATA box was observed, unlike the human IGFBP-1 and IGFBP-3 genes.[39-41] The region surrounding the transcription initiation site is GC-rich (66% GC) and contains sequences that are potentially recognized by the Sp1 transcription factor.[42] It lacks an initiator sequence.[43] Thus, the IGFBP-2 promoter is similar to GC-rich promoters of housekeeping genes, although the IGFBP-2 gene is more highly regulated than typical housekeeping genes. It has been proposed that in TATA-less promoters of this type, the transcription factor TFIID required for activation of RNA polymerase II is linked via a tethering protein to Sp1 which binds to appropriate sequences in the DNA.[44]

The 5' flanking region also contains several *cis* regulatory motifs that might potentially bind transcription factors that are candidates for mediating the regulation of IGFBP-2 (Fig. 7). It contains an AP-1 site that is recognized by Jun-Fos or Fos-Fos dimers and is frequently regulated by phorbol esters,[42] an AP-2 site that is recognized by transcription factors that are regulated by cyclic AMP or phorbol esters,[42] an insulin response element (IRE-A) similar to the minimum consensus sequence reported in the glyceraldehyde-3-phosphate dehydrogenase gene,[46] and a site for the liver-specific transcription factor LF-B1.[45] The use of these sites and their recognition by appropriately regulated transcription factors remains to be demonstrated.

CONCLUSIONS

The present studies indicate that an important component of IGFBP-1 and IGFBP-2 regulation occurs at the level of mRNA synthesis and abundance. Regulation of IGFBP synthesis would determine the levels of expression of IGFBP-1 and IGFBP-2 in tissues (where they may modulate IGF actions), and contribute to changes in plasma levels of the IGFBPs. We have begun to identify appropriate tools to study the molecular regulation of the IGFBP genes: IGFBP-2 genomic clones, physiologic systems in which IGFBP-1 and IGFBP-2 are regulated, and a cell system in which IGFBP-1 is regulated. These studies should provide insights into the possible tissue-specific functions of the IGFBPs, and help dissect the complex interplay between the IGFs and their receptors. They also will help evaluate whether the IGFBPs have biological actions unrelated to binding IGFs, as suggested by observations showing that partially-purified IGFBP from endothelial cells stimulates glucose uptake and amino acid transport,[47] and that IGFBP mRNA in embryonic rat brain is localized to the floor plate of the spinal cord, a region thought to direct neurite extension, in which neither IGF-I nor IGF-II mRNA is expressed.[26]

REFERENCES

1. M. M. Rechler and S. P. Nissley, Insulin-like growth factors, in: "Peptide growth factors and their receptors I", M. B. Sporn and A. B. Roberts, eds., Springer-Verlag, Heidelberg, Handbook of Experimental Pharmacology vol 95/I,263-367 (1990).
2. S. L. S Drop and R. L. Hintz, "Insulin-like growth factor binding proteins", Excerpta Medica, New York (1989).
3. L. J. Murphy, C. Seneviratne, G. Ballejo, F. Croze, and T. G. Kennedy, Identification and characterization of a rat decidual insulin-like growth factor-binding protein complementary DNA, Mol.Endocrinol. 4:329 (1990).
4. A. L. Brown, L. Chiariotti, C. C. Orlowski, T. Mehlman, W. H. Burgess, E. J. Ackerman, C. B. Bruni, and M. M. Rechler, Nucleotide sequence and expression of a cDNA clone encoding a fetal rat binding protein for insulin-like growth factors, J.Biol.Chem. 264:5148 (1989).
5. J. B. Margot, C. Binkert, J. -L. Mary, J. Landwehr, G. Heinrich, and J. Schwander, A low molecular weight insulin-like growth factor binding protein from rat: cDNA cloning and tissue distribution of its messenger RNA, Mol.Endocrinol. 3:1053 (1989).
6. S. Shimasaki, A. Koba, M. Mercado, M. Shimonaka, and N. Ling, Complementary DNA structure of the high molecular weight rat insulin-like growth factor binding protein (IGF-BP3) and tissue distribution of its mRNA, Biochem.Biophys.Res.Commun. 165:907 (1989).
7. A. L. Albiston and A. C. Herington, Cloning and characterization of the growth hormone-dependent insulin-like growth factor binding protein (IGFBP-3) in the rat, Biochem. Biophys.Res.Commun. 166:892 (1990).
8. S. Shimasaki, F. Uchiyama, M. Shimonaka, and N. Ling, Molecular cloning of the cDNAs encoding a novel insulin-like growth factor-binding protein from rat and human, Mol. Endocrinol. 1451 (1990).
9. M. Roghani, P. Hossenlopp, P. Lepage, A. Balland, and M. Binoux, Isolation from human cerebrospinal fluid of a new insulin-like growth factor-binding protein with a selective affinity for IGF-II, FEBS Lett. 255:253 (1989).
10. J. L. Martin, K. E. Willetts, and R. C. Baxter, Purification and properties of a novel insulin-like growth factor-II binding protein from transformed human fibroblasts, J.Biol.Chem. 265:4124 (1990).
11. B. Forbes, F. J. Ballard, and J. C. Wallace, An insulin-like growth factor-binding protein purified from medium conditioned by a human lung fibroblast cell line (HE[39]L) has a novel N-terminal sequence, J.Endocrinol. 126:497 (1990).
12. J. L. Martin and R. C. Baxter, Production of an insulin-like growth factor (IGF)-inducible IGF-binding protein by human skin fibroblasts, Endocrinology 127:781 (1990).
13. J. Zapf, C. Hauri, M. Waldvogel, and E. R. Froesch, Acute metabolic effects and half-lives of intravenously administered insulinlike growth factors I and II in normal and hypophysectomized rats, J.Clin.Invest. 77:1768 (1986).
14. R. S. Bar, M. Boes, D. R. Clemmons, W. H. Busby, A. Sandra, B. L. Dake, and B. A. Booth, Insulin differentially alters transcapillary movement of intravascular IGFBP-1,

IGFBP-2 and endothelial cell IGF-binding proteins in the rat heart, Endocrinology 127:497 (1990).
15. R. S. Bar, D. R. Clemmons, M. Boes, W. H. Busby, B. A. Booth, B. L. Dake, and A. Sandra, Transcapillary permeability and subendothelial distribution of endothelial and amniotic fluid insulin-like growth factor binding proteins in the rat heart, Endocrinology 127:1078 (1990).
16. W. H. Busby,Jr., D. G. Klapper, and D. R. Clemmons, Purification of a 31,000-dalton insulin-like growth factor binding protein from human amniotic fluid. Isolation of two forms with different biologic actions, J.Biol.Chem. 263:14203 (1988).
17. J. S. M. De Mellow and R. C. Baxter, Growth hormone-dependent insulin-like growth factor (IGF) binding protein both inhibits and potentiates IGF-I-stimulated DNA synthesis in human skin fibroblasts, Biochem.Biophys.Res.Commun. 156:199 (1988).
18. R. C. Baxter and J. L. Martin, Structure of the Mr 140,000 growth hormone-dependent insulin-like growth factor binding protein complex: Determination by reconstitution and affinity-labeling, Proc.Natl.Acad.Sci.USA 86:6898 (1989).
19. M. Binoux and P. Hossenlopp, Insulin-like growth factor (IGF) and IGF-binding proteins: comparison of human serum and lymph, J.Clin.Endocrinol.Metab. 67:509 (1988).
20. A. M. Cotterill, C. T. Cowell, R. C. Baxter, D. McNeil, and M. Silinik, Regulation of the growth hormone-independent growth factor-binding protein in children, J.Clin.Endocrinol.Metab. 67:882 (1988).
21. K. Brismar, M. Gutniak, G. Povoa, S. Werner, and K. Hall, Insulin regulates the 35 kDa IGF binding protein in patients with diabetes mellitus, J.Endocrinol.Invest. 11:599 (1988).
22. Y. W-H. Yang, J-F. Wang, C. C. Orlowski, S. P. Nissley, and M. M. Rechler, Structure, specificity and regulation of the insulin-like growth factor binding proteins in adult rat serum, Endocrinology 125:1540 (1989).
23. G. T. Ooi, manuscript in preparation.
24. G. T. Ooi, C. C. Orlowski, A. L. Brown, R. E. Becker, T. G. Unterman, and M. M. Rechler, Different tissue distribution and hormonal regulation of mRNAs encoding rat insulin-like growth factor binding proteins rIGFBP-1 and rIGFBP-2, Mol.Endocrinol. 4:321 (1990).
25. L. Y-H. Tseng, A. L. Brown, Y. W-H. Yang, J. A. Romanus, C. C. Orlowski, T. Taylor, and M. M. Rechler, The fetal rat binding protein for insulin-like growth factors is expressed in the choroid plexus and cerebrospinal fluid of adult rats, Mol.Endocrinol. 3:1559 (1989).
26. T. L. Wood, A. L. Brown, M. M. Rechler, and J. E. Pintar, The expression pattern of an insulin-like growth factor (IGF)-binding protein gene is distinct from IGF-II in the midgestational rat embryo, Mol.Endocrinol. 4:1257 (1990).
27. J. A. Romanus, L. Y-H. Tseng, Y. W-H. Yang, and M. M. Rechler, The 34 kilodalton insulin-like growth factor binding proteins in human cerebrospinal fluid and the A673 rhabdomyosarcoma cell line are human homologues of the rat BRL-3A binding protein, Biochem.Biophys.Res.Commun. 163:875 (1989).
28. M. Roghani, C. Lassarre, B. Segova, and M. Binoux, Presence in human cerebrospinal fluid (CSF) of two IGF binding proteins (BPs) with a preferential affinity for IGF-II, Program, 72nd Annual Mtg. of the Endocrine Society, Abstract 183 (1990).

29. C. C. Orlowski, A. L. Brown, G. T. Ooi, Y. W-H. Yang, L. Y-H. Tseng, and M. M. Rechler, Tissue, developmental and metabolic regulation of mRNA encoding a rat insulin-like growth factor binding protein (rIGFBP-2), Endocrinology 126:644 (1990).
30. C. Seneviratne, L. Jiangming, and L. J. Murphy, Transcriptional regulation of rat insulin-like growth factor-binding protein-1 expression by growth hormone, Mol.Endocrinol. 4:1199 (1990).
31. M. Boni-Schnetzler, K. Binz, J. -L. Mary, C. Schmid, J. Schwander, and E. R. Froesch, Regulation of hepatic expression of IGF I and fetal IGF binding protein mRNA in streptozotocin-diabetic rats, FEBS Lett. 251:253 (1989).
32. M. Boni-Schnetzler, C. Schmid, J. -L. Mary, B. Zimmerli, P. J. Meier, J. Zapf, J. Schwander, and E. R. Froesch, Insulin regulates the expression of the insulin-like growth factor binding protein 2 mRNA in rat hepatocytes, Mol.Endocrinol. 4:1320 (1990).
33. T. G. Unterman, K. Patel, V. K. Mahathre, G. Rajamohan, D. T. Oehler, and R. E. Becker, Regulation of low molecular weight insulin-like growth factor binding proteins in experimental diabetes mellitus, Endocrinology. 126:2614 (1990).
34. Y. W-H. Yang, A. L. Brown, C. C. Orlowski, D. E. Graham, L. Y-H. Tseng, J. A. Romanus, and M. M. Rechler, Identification of rat cell lines that preferentially express insulin-like growth factor binding proteins rIGFBP-1, 2, or 3, Mol.Endocrinol. 4:29 (1990).
35. C. C. Orlowski, G. T. Ooi, and M. M. Rechler, Dexamethasone stimulates transcription of the insulin-like growth factor binding protein-1 (IGFBP-1) gene in H4-II-E rat hepatoma cells, Mol.Endocrinol. 4:1592 (1990).
36. M. Degerblad, G. Povoa, M. Thoren, I. -L. Wivall, and K. Hall, Lack of diurnal rhythm of low molecular weight insulin-like growth factor binding protein in patients with Cushing's disease, Acta Endocrinol. 120:195 (1989).
37. J. M. Luo, R. E. Reid, and L. J. Murphy, Dexamethasone increases hepatic insulin-like growth factor binding protein-1 (IGFBP-1) mRNA and serum IGFBP-1 concentrations in the rat, Endocrinology 127:1456 (1990).
38. A. L. Brown and M. M. Rechler, Cloning of the rat insulin-like growth factor binding protein-2 gene and identification of a functional promoter lacking a TATA box, Mol.Endocrinol. (1990).(In Press)
39. M. L. Cubbage, A. Suwanichkul, and D. R. Powell, Structure of the human chromosomal gene for the 25 kilodalton insulin-like growth factor binding protein, Mol.Endocrinol. 3:846 (1989).
40. A. Brinkman, C. A. H. Groffen, D. J. Kortleve, and S. L. S. Drop, Organization of the gene encoding the insulin-like growth factor binding protein IBP-1, Biochem. Biophys.Res.Commun. 157:898 (1988).
41. M. L. Cubbage, A. Suwanichkul, and D. R. Powell, Insulin-like growth factor binding protein-3 Organization of the human chromosomal gene and demonstration of promoter activity, J.Biol.Chem. 265:12642 (1990).
42. P. J. Mitchell and R. Tjian, Transcriptional regulation in mammalian cells by sequence-specific DNA binding proteins, Science 245:371 (1989).

43. S. T. Smale and D. Baltimore, The "initiator" as a transcription control element, Cell 57:103 (1989).
44. B. F. Pugh and R. Tjian, Mechanism of transcriptional activation by Sp1: evidence for coactivators, Cell 61:1187 (1990).
45. M. Frain, G. Swart, P. Monaci, A. Nicosia, S. Stampfli, R. Frank, and R. Cortese, The liver-specific transcription factor LF-B1 contains a highly diverged homeobox DNA binding domain, Cell 59:145 (1989).
46. N. Nasrin, L. Ercolani, M. Denaro, X. F. Kong, I. Kang, and M. Alexander, An insulin response element in the glyceraldehyde-3-phosphate dehydrogenase gene binds a nuclear protein induced by insulin in cultured cells and by nutritional manipulation in vivo, Proc.Natl.Acad.Sci.USA 87:5273 (1990).
47. R. S. Bar, B. A. Booth, and B. L. Dake, Insulin-like growth factor-binding proteins from vascular endothelial cells: Purification, characterization, and intrinsic biological activities, Endocrinology 125:1910 (1989).

HORMONAL REGULATION OF INSULIN-LIKE GROWTH FACTOR BINDING PROTEIN-1 EXPRESSION IN THE RAT

Liam J. Murphy, Jiangming Luo, and Charita Seneviratne

Departments of Internal Medicine & Physiology
Faculty of Medicine, University of Manitoba
Winnipeg, Canada, R3E 0W3

INTRODUCTION

The insulin-like growth factors (IGF) are present in the serum, other biological fluids and tissue extracts in association with high affinity binding proteins[1]. When human or adult rat serum is analyzed by gel permeation chromatography two groups of IGF binding proteins are identified. These have apparent molecular mass of 150-200 kDa and 30-45 kDa[1-3]. However, more detailed analyses of sera using SDS polyacrylamide gel electrophoresis and ligand blotting with either ^{125}I-IGF-I or ^{125}I-IGF-II, identifies at least 5 bands. The approximate size of these binding proteins in rat serum when analyzed on reduced gels is 42, 40, 39, 30, 29 and 24 kDa [4-6]. In some case where antibody probes are available it has been possible to identify the various binding proteins detected by ligand blotting however since a number of IGF binding proteins appear to be of similar molecular weight, it is possible that a band identified by ligand blotting may consist of more than one binding protein. Although the exact number of these binding proteins present in serum remains to be determined, evidence for the existence of at least five distinct binding proteins has been presented in the recent literature. Four of these binding proteins have been extensively characterized and cDNAs encoding these binding proteins have been cloned[7-11].

The 42, 40 and 39 kDa binding proteins identified by ligand blotting most probably represent IGFBP-3 which is known to be glycosylated and consequently poorly resolved by SDS-PAGE[7]. This growth hormone dependent binding protein is present in serum from adult rats and human subjects as a complex of approximately 150-200 kDa. This complex is composed of IGF-I or IGF-II, a 100 kDa acid-labile subunit and IGFBP-3[12]. The growth hormone dependence of this binding protein appears to be mediated via IGF-I, since infusion of IGF-I into hypophysectomized rats is able to restore the 150-200 kDa IGF-I binding complex[13]. In our studies of binding proteins in adult rat serum we estimate that greater than 60% of the total available serum ^{125}I-IGF-I binding sites reside on IGFBP-3[14].

In fetal and neonatal rat serum the predominant IGF binding protein is IGFBP-2. This binding protein was originally isolated from buffalo rat liver cell (BRL 3A cells) conditioned medium[9]. IGFBP-2 is not detected in adult rat serum by immunoblotting with specific antisera and expression of this binding protein appears to be developmentally regulated[4,15]. Although low levels of

Molecular Biology and Physiology of Insulin and Insulin-Like Growth Factors
Edited by M.K. Raizada and D. LeRoith, Plenum Press, New York, 1991

IGFBP-2 mRNA are found in multiple tissues in the rat, high levels persist only in the brain stem, hypothalmus and cerebral cortex[16].

The first binding protein to be completely purified was IGFBP-1, a binding protein which is particularly abundant in human amniotic fluid[17]. It is also produced by human hepatoma cells[18]. Under normal circumstances it accounts for considerably less that10% of the total available ^{125}I-IGF-I binding sites in adult rat serum[14], but as will be discussed later in this chapter the serum concentration of this binding protein is increased under a variety of circumstances.

An additional binding protein, IGFBP-4, has been recently purified from rat serum and cDNA encoding rat and human IGFBP-4 have been isolated[11]. This binding protein appears to be equivalent to the 34 kDa binding protein isolated from an SV40 transformed human fibroblast by Martin et al.,[18] and the binding protein found in human bone cell conditioned medium[19]. An IGF binding protein with a unique amino-terminal sequence has been isolated from human cerebrospinal fluid by Roghani et al. [20]. This binding protein may represent another member of this gene family.

Recently, Donovan et al. have provided evidence that the ontogeny of the smallest of the serum IGF binding proteins detected by ligand blotting is different from the other serum IGF binding proteins[4]. This binding protein which has a reported molecular weight of between 17-24 kDa is quite abundant and appears to account for approximately 20% of the total available ^{125}I-IGF-I binding sites in rat serum[14].

THE FUNCTIONAL ROLE OF THE IGF BINDING PROTEINS

The functional role of the binding proteins remains unclear. A simplistic view is that they serve to block the insulin-like activity of the relatively large concentrations of the IGFs present in the circulation[1]. A number of investigators have been able to demonstrate that the IGF binding proteins are able to inhibit the biological actions of the IGFs in certain in vitro bioassays[21-25]. However, several studies suggest that the effects of the binding proteins on IGF-I action may not be simple inhibition. The binding proteins themselves appear to be able to interact with cell surface receptors and may facilitate delivery of the IGFs to target cells[26]. Multiple forms of IGFBP-1 appear to be present in amniotic fluid and a least one form of human IGFBP-1 can actually enhance the action of IGF-I[27]. The multiple forms of IGFBP-1 appear to be the result of post-translational modification including phosphorylation[28].

Another binding protein, IGFBP-3, is also able to inhibit IGF-I action in a variety of assay systems including, DNA synthesis in human fibroblasts[29], lipogenesis in rat adipocytes[30], lipogenesis and glucose oxidation in porcine adipose tissue[25]. However, De Mellow and Baxter observed that under certain conditions, IGFBP-3 can also enhance the effect of IGF-I on DNA synthesis in human fibroblasts[29]. The relevance of these in vitro assays, to the physiological role of the individual binding proteins remains to be determined. Since the ontogeny, hormonal dependence and tissue distribution of the known IGF binding proteins appear to vary, it is highly likely that each of the binding proteins subserve slightly different functions. Clearly, investigating the regulation of these proteins which modulate the actions of the IGFs is important to our overall understanding of the role of the IGFs in the growth process. In this chapter, I will briefly review the published literature on the regulation of

IGFBP-1 in man and report the results of experiments where the regulation of IGFBP-1 expression in the rat has been investigated. Although this binding protein probably constitutes considerably less than 10% of the total IGF binding capacity of adult serum, its abundance in the serum and its expression in various tissues increases in a number of experimental settings, particularly those associated with growth retardation.

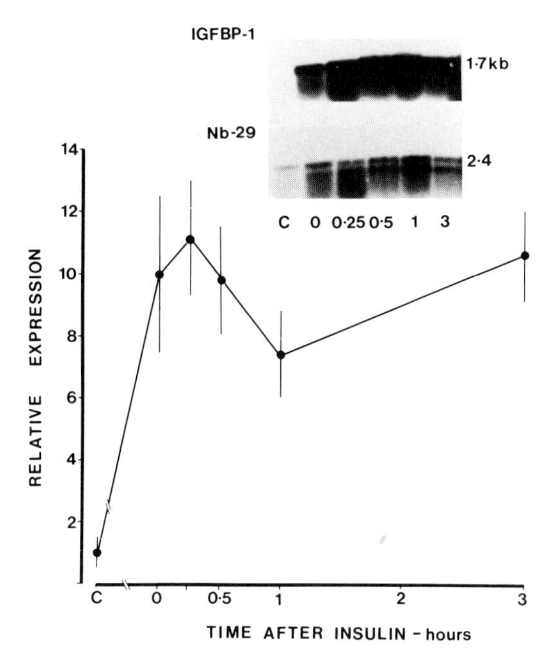

Figure 1. The effect of insulin administration on hepatic IGFBP-1 mRNA abundance in fasted rats. The abundance of IGFBP-1 mRNA was determined by densitometry of autoradiograms and has been expressed relative to the hybridization signal obtained in non-fasted control rats (C). The data represent the mean ± SEM for 5 rats per time-point. A representative Northern blot of RNA from control rats (C), rats fasted for 24 h (time 0) and fasted rats injected with 4U of insulin and killed at various times after insulin injection is shown. The nitrocellulose filter was also hybridized with a rat NB-29 cDNA as a loading control. (Data from Ref. 36).

REGULATION OF IGFBP-1 EXPRESSION BY INSULIN

In human subjects, serum IGFBP-1 concentrations appear to be inversely correlated with serum insulin levels[31,32] and elevated IGFBP-1 concentrations are apparent in normal individuals after an overnight fast[33,34]. In human fetal liver explants, synthesis of IGFBP-1 is suppressed by insulin[35]. In the rat, food-deprivation for a period of 24 hours resulted in a 9.5 fold increase in hepatic IGFBP-1 mRNA abundance[36]. An increase in circulating IGFBP-1 in sera from fasted rats was demonstrated by immunoblotting and an increased abundance of a 30 kDa IGF binding protein in sera from fasted rats was apparent when ^{125}I-IGF-I was used in ligand blotting experiments. This observation was hardly unexpected since it has been shown that in both human subjects and in the pig, fasting is associated with increased serum levels of low molecular weight IGF binding proteins[33,34,37]. Refeeding or glucose infusion in overnight fasted rats resulted in a prompt decline in hepatic IGFBP-1 mRNA[36].

Hepatic IGFBP-1 mRNA could not be detected 1 hour after refeeding and remained suppressed below the level detectable in control non-fasted rats for up to 24 hours. A surprising finding, in view of the discussion above, was the observation that administration of insulin, 0.05 to 4 U i.p. to fasted rats either increased or had no significant effect on hepatic IGFBP-1 mRNA abundance (Fig. 1). The fasted animals, even those that received very small doses of insulin became hypoglycemic[36]. A number of investigators have suggested that the increased serum IGFBP-1 levels in fasted human subjects is a consequence of the low serum insulin levels. This hypothesis is supported by the *in vitro* demonstration that insulin markedly suppresses IGFBP-1 synthesis in human fetal liver explants[35]. However, administration of insulin to fasted rats did not reduce IGFBP-1 mRNA abundance and under some circumstances a paradoxical increase in IGFBP-1 mRNA levels was observed (Fig. 2). Since insulin appears to exert a direct effect on IGFBP-1 expression in rat hepatocytes in organ culture[38], our *in vivo* findings suggest that the hypoglycemia which accompanies insulin administration inhibits the insulin effect on IGFBP-1 expression.

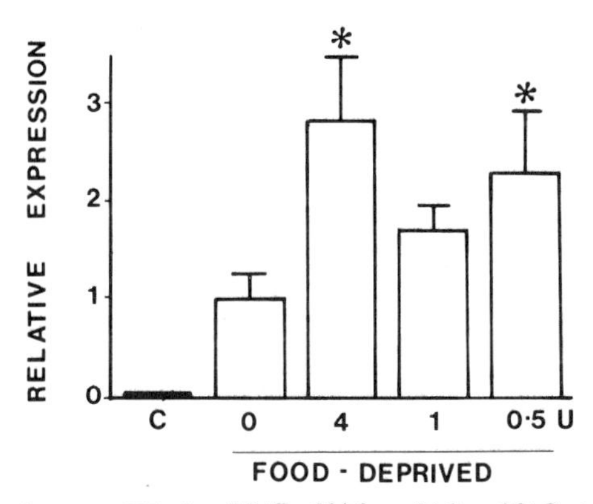

Figure 2. The effect of varying doses of insulin on hepatic IGFBP-1 mRNA abundance in fasted rats. Data represent the mean ± SEM for 4 rats per group. The level of IGFBP-1 mRNA in the insulin-treated rats has been expressed relative to the fasted rats (arbitrarily attributed a value of 1). The serum glucose at the time of death, 1 hour after insulin injection, is shown below. * indicates $p < 0.05$ for the difference between insulin-treated and non-insulin-treated fasted rats. (The data has been reproduced from Ref. 36).

IGFBP-1 EXPRESSION IN DIABETES

Impaired growth is a characteristic feature of uncontrolled diabetes in many species[39,40]. Growth retardation is particularly marked in the diabetic rodent. Administration of insulin, with improved glycemic control, results in an increased growth rate, both in diabetic children and in experimental animals[39,41]. The mechanisms responsible for the growth retardation in diabetic animals are not fully understood. Although circulating growth hormone levels are low in diabetic rats, this alone does not explain the growth retardation since these animals are unresponsive to exogenous growth hormone[42]. Both receptor and post-receptor defects in growth hormone - signal transduction have been reported in diabetic animals[42,43]. Circulating IGF-I levels are low in diabetic rats[42] and administration of exogenous IGF-I is able to stimulate growth in diabetic animals[44], however, it is still not clear whether gowth can be completely normalized in diabetic animals with IGF-I administration alone.

In addition to reduced circulating IGF-I concentrations, plasma from diabetic animals also contains factors which are able to inhibit IGF action in cartilage bioassays[45]. Although the nature and functional role of these inhibitors have not been clarified, the possibility exists that part of this inhibitory activity is due to enhanced IGF binding protein concentrations. An increased concentration of a low molecular weight IGF binding protein in sera from diabetic rats was initially reported by Unterman's group[46]. This binding protein proved to be IGFBP-1. In our studies an increase in IGFBP-1 mRNA was observed in the liver (2-9 fold) and kidney (1.6-2.2 fold) of streptozotocin diabetic rats[47]. IGFBP-1 mRNA abundance increased with increasing duration of diabetes. When data from a large group of diabetic rats were analyzed, a highly significant correlation between hepatic IGFBP-1 mRNA and glucose was observed; $R = 0.75$, $p < 0.001$[47]. Administration of insulin to diabetic rats resulted in a decrease in hepatic and renal IGFBP-1 mRNA levels.

REGULATION OF IGFBP-1 EXPRESSION BY GROWTH HORMONE

Several investigators have demonstrated that serum IGFBP-1 levels are elevated in growth hormone deficient individuals[5,48,49] although this has not been a universal finding[34]. This binding protein, in contrast to IGFBP-3, has been considered by many investigators, to be growth hormone independent. As discussed above, in human subjects, circulating levels of IGFBP-1 correlate inversely with plasma insulin concentrations and *in vitro* experiments with human and rodent hepatic tissue, have demonstrated that secretion of this protein is inhibited by insulin[35,38]. Thus, the enhanced levels of IGFBP-1 in growth hormone deficient individuals may result from the hypoinsulinemia which accompanies growth hormone deficiency. Our studies in the rat suggest that this simplistic explanation may not be correct[6,36].

In the rat, as in human subjects, growth hormone deficiency is associated with elevated hepatic and renal expression of IGFBP-1 and increased circulating IGFBP-1 concentrations[6]. Growth hormone administration to hypophysectomized rats normalizes hepatic IGFBP-1 mRNA levels. The response is, however, followed by a rebound increase in IGFBP-1 mRNA levels. While chronic daily injections of human growth hormone significantly reduced hepatic IGFBP-1 mRNA levels and decreased the 27-30 kDa IGF binding protein in serum and hepatic extracts, these parameters were not completely normalized[6]. Since it is known that the pituitary secretion of growth hormone is episodic in the rat it is not surprising that single daily injections of growth hormone failed to normalize hepatic IGFBP-1 mRNA levels. Furthermore, it is also possible that other factors, deficient in the hypox rats, may also be involved in the regulation of IGFBP-1. It is of interest that IGFBP-2 mRNA is also more abundant in hepatic tissue from hypophysectomized rats. However, it is not clear whether gowth hormone administration can reverse this phenomenon[50].

The decrease in hepatic IGFBP-1 mRNA adundance in hypophysectomized rats following growth hormone administration appears to be the result of a decrease in transcription rate[6]. After a single injection of growth hormone, IGFBP-1 transcription rate was reduced within 30 minutes to the level seen in the sham-operated control rats. These observations established that growth hormone is able to acutely regulate expression of this gene and suggest that at least in the rat, IGFBP-1 may be inversely regulated by growth hormone.

Although the rapidity of the growth hormone effect argues that this is a direct action, the possibility exists that this effect of growth hormone is indirect and mediated via insulin. There are a number of reasons why we believe this is not the case. Firstly, in the fasted rat where pituitary growth hormone secretion is essentially absent[51], hepatic expression of IGFBP-1 is also enhanced. The increase in IGFBP-1 mRNA abundance and transcription is not seen if rats are injected with growth hormone during the period of food-deprivation[36]. Secondly, growth hormone administration to food-deprived rats results in a rapid fall in hepatic IGFBP-1 mRNA abundance (Fig. 3), without any increase in serum insulin concentration[36]. Furthermore, as discussed above, insulin administration to fasted rats results in an increase rather than a decrease in hepatic IGFBP-1 mRNA levels. In addition, insulin administration to hypophysectomized rats did not reduce hepatic IGFBP-l mRNA abundance. Finally, bovine growth hormone is able to suppress IGFBP-1 production by isolated rat hepatocytes[38]. Interestingly, administration of human IGF-I (75 ug/100g) to hypophysectomized rats did lower hepatic IGFBP-1 levels (Fig. 4). Compared to insulin administration, injection of IGF-I had very little effect on serum glucose and thus failure of insulin to reduce IGFBP-1 expression in hypophysectomized rats may be a consequence of the hypoglycemia.

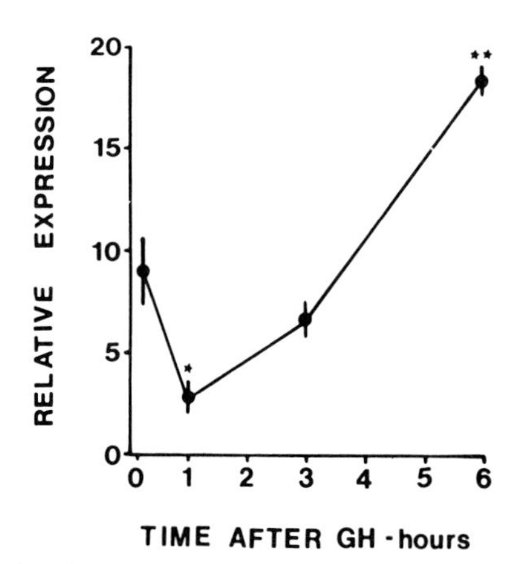

Figure 3. The effect of growth hormone on hepatic IGFBP-1 in fasted rats. Rats were food-deprived for 24 h and then killed at various times after an injection of hGH. IGFBP-1 mRNA abundance has been quantitated in hepatic RNA from 5 individual rats per group by slot-blot hybridization. * and ** represent $p < 0.01$ and 0.001 for the difference between the GH-treated fasted rats and untreated fasted rats (time 0). Data has been reproduced from Ref. 36.

Since IGFBP-1 has been shown to inhibit the mitogenic activity of IGF-I *in vitro*[23,24] it is possible that this inhibitory action of GH on IGFBP-1 expression may be intimately related to the intrinsic growth promoting activity of growth hormone. It will be important to determine what effects administration of IGFBP-1 have on growth in the rat.

As discussed above IGFBP-1 expression is also enhanced during fasting. In both rodents and man, growth hormone release from the pituitary is episodic.

In the rat these secretory episodes occur at approximately 90 minute intervals whereas in the human secretory activity is most marked during sleep. In the rat, unlike man, food-deprivation results in a rapid and profound inhibition of pituitary growth hormone secretory episodes[51]. Since anti-somatostatin antiserum reverses the inhibition, this phenomenon appears to be due to the high levels of somatostatin in the hypothalamic-pituitary portal circulation in the fasted rat[51]. In contrast, in man during fasting pituitary growth hormone secretory activity is enhanced and elevated levels are present in the circulation. We speculated that the enhanced IGFBP-1 expression in the fasted rat may result, at least in part, from the functional growth hormone deficiency that occurs in the food-deprived rodent. We were able to demonstrate that administration of growth hormone to the fasted rat rapidly lowers IGFBP-1 expression and that repeated injections of growth hormone during the period of food-deprivation results in a significant attentuation of the food deprivation-induced, up-regulation of IGFBP-1 expression[36]. This growth hormone effect was not due to secondary hyperinsulinism since there was no significant increase in insulin-levels during the time-course of the experiment. Furthermore as discussed above, insulin administration to fasted animals did not reduce hepatic IGFBP-1 mRNA levels.

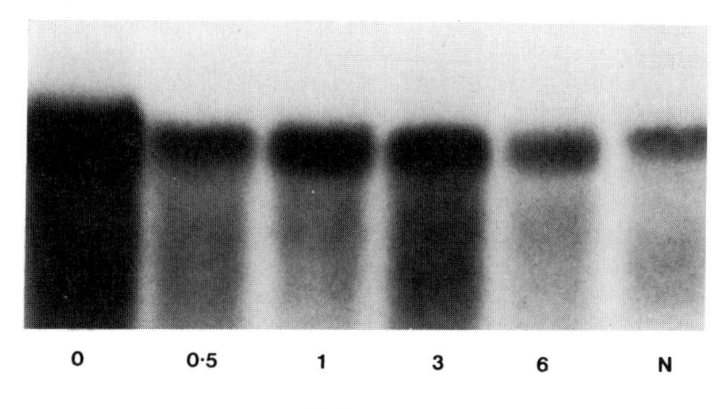

Figure 4. The effect of IGF-I on hepatic IGFBP-1 in hypophysectomized rats. Rats were killed at various times after an injection of IGF-1. Hepatic RNA, 50 ug/lane, was analyzed by the Northern blotting technique. RNA from a pituitary intact rat (N) is included for comparison.

The marked diurnal variation in IGFBP-1 concentrations which are observed in man have been attributed to the hypoinsulinemia which occurs in the early morning hours. However, the rise in serum levels of IGFBP-1 concentrations occurs some 2-4 hours after the time of maximal growth hormone secretory activity. It is tempting to speculate that this phenomenon may be in some way related to the rebound increase in IGFBP-1 expression we observe following the initial suppression after administeration of growth hormone to hypophysectomized or fasted rats[6]. It is of interest that in children the levels of IGFBP-1 are highest pre-pubertally and gradually decline as puberty progresses and growth hormone secretory activity increases[52]. In the latter report an increase in mean serum insulin concentration was also seen and a strong negative correlation between IGFBP-1 and insulin levels could be demonstrated. Suikkari et al. have demonstrated that serum IGFBP-1 concentration declines following an oral glucose load in normal subjects, a situation where insulin is stimulated and growth hormone is likely to be suppressed[31,32]. Since in many situations there appears to be an inverse correlation between serum insulin and growth hormone concentrations, it is difficult to determine the relative role of each of these hormones in regulating IGFBP-1 expression. The data derived from both in vivo and in vitro studies suggest that both hormones may have some role in the regulation of this binding protein in the rat.

REGULATION OF IGFBP-1 EXPRESSION BY GLUCOCORTICOIDS

Glucocorticoid excess, whether iatrogenic or endogenous, is associated with growth retardation. Several different mechanisms have been proposed to explain the growth retarding effect of glucocorticoid excess. Clearly glucocorticoids exert effects at the level of the hypothalamus, pituitary, liver and skeletal tissues. Circulating somatomedin activity and IGF-I concentrations have been reported to be low, normal or high in individuals with endogenous or exogenous glucocorticoid excess. While these differences may in part, result from the different assay techniques used, an additional confounding factor is the possibility that glucocorticoid excess may be associated with excess somatomedin inhibitory activity. An early report from Unterman and Phillips suggested that the steroid-induced fall in net somatomedin bioactivity resulted from an increase in circulating somatomedin inhibitory activity[54]. They suggested this steroid-induced inhibitory activity may be important in the impaired growth observed in children with hypercorticoidism. By analogy with the diabetic state, the steroid inhibitory activity may be due to increased IGF binding protein concentrations.

In our initial attempts to investigate the effects of glucocorticoids on serum IGF binding proteins we used a charcoal binding assay and ligand blotting to quantitate binding proteins in sera from dexamethasone treated rats. These assays demonstrated that glucocorticoid excess is associated with increased serum IGF binding capacity[14]. A significant increase in serum IGF binding capacity was seen with as little as 0.1 ug/100 g body weight. Using specific molecular probes we were able to demonstrate that hepatic expression of both IGFBP-1 and 3 were enhanced by dexamethasone administration. A significant increase in hepatic IGFBP-3 mRNA and IGFBP-1 mRNA levels were apparent as early as 1 hour following dexamethasone administration. IGFBP-1 mRNA levels reached a peak at 1 hour, while IGFBP-3 mRNA levels reached a peak 3 hours after a single injection of dexamethasone[14,53]. The abundance of both transcripts then declined to control levels 6-12 hours after dexamethasone injection. This response was dose-dependent, with as little as 1 ug/100 g body weight, resulting in a significant increase in mRNA levels. An increase in serum IGFBP-1 after dexamethasone was demonstrated by immuno-blotting while the increase in serum IGFBP-3 was quantitated by ligand blotting. A more dramatic increase in IGFBP-1 expression was seen in experiments where rats were treated chronically with varying amounts of dexamethasone. We were unable to detect an effect of dexamethasone on IGFBP-1 transcription rate in hepatic nuclei from rats which had been pretreated for 1 hour with 6 ug/100 g body weight dexamethasone[53].

The increase in serum IGF binding proteins may represent yet another potential mechanism whereby glucocorticoid excess may exert a growth retarding effect. Our observations using specific probes are consistent with the report that glucocorticoids increase IGF-binding proteins released by liver explants[55]. Since an ^{125}I-IGF-I binding assay was used in that study it is unclear as to which of the binding proteins expressed in the liver was increased by glucocorticoids. Our studies are also supported by a more recent report where the effect of dexamethasone on IGFBP-1 expression in H4-II-E rat hepatoma cells was investigated[56]. In the hepatoma cells the authors were able to demonstrated that dexamethasone enhanced IGFBP-1 transcription. In our in vivo experiments, we have only examined the effect of dexamethasone on hepatic IGFBP-1 transcription 1 hour after dexamethasone administration, we can not exclude the possibility that an increase in transcription may have occurred at an earlier time point. In contrast to these studies using rodent tissue, Lewitt & Baxter reported that dexamethasone inhibited production of IGFBP-1 in human fetal liver explants[35]. Furthermore, Gourmelen et al. found that serum IGF binding capacity was reduced in patients with Cushing's syndrome[57] and the nocturnal rise in serum IGFBP-1 concentration was absent in patients with Cushing's disease[58]. Thus, there appears to be a species difference in response of the binding proteins to glucocorticoid excess.

CONCLUSIONS AND FUTURE DIRECTIONS

Our studies in the rat suggest that multiple hormonal factors are involved in the regulation of IGFBP-1 expression and that there may be important species differences between the rat and man in this regulation.

Although IGFBP-1 appears to be a relatively minor component of the total serum IGF binding capacity, expression of this binding protein is up-regulated in a variety of conditions which are associated with growth retardation. These include poorly controlled diabetes, starvation, growth hormone deficiency and glucocorticoid excess. Since IGFBP-1 can inhibit IGF-I action, at least under in vitro assay conditions, these observations provide circumstantial evidence that IGFBP-1 is intimately involved in the growth process. To provide definitive proof for this hypothesis it will be necessary to examine the effects of exogenous IGFBP-1 on weight gain and skeletal growth. An alternative approach would be to over-express IGFBP-1 in transgenic mice.

REFERENCES

1. J. Zapf, M. Waldvogel and E.R. Froesch, Binding of nonsuppressible insulinlike activity to human serum. Arch. Biochem. Biophys. 168:638 (1975).
2. J.L. Martin and R.C. Baxter, Insulin-like growth factor-binding proteins from human plasma. J. Biol. Chem. 261:8754 (1986).
3. J.A. Romanus, J.E. Terrell Y.W.H. Yang S.P. Nissley and M.M. Rechler, Insulin-like growth factor carrier proteins in neonatal and adult rat serum are immunologically different: demonstration using a new radioimmunoassay for the carrier protein from BRL-3A rat liver cells. Endocrinology 118:1743 (1986).
4. S.M. Donovan, Y. Oh, H. Pham and R.G. Rosenfeld, Ontogeny of serum insulin-like growth factor binding proteins in the rat. Endocrinology 125:2621 (1989).
5. P. Hossenlopp, D. Seurin, B. Segovia-Quinson, S. Hardouin and M. Binoux, Analysis of serum insulin-like growth factor binding proteins using western blotting: use of the method for titration of the binding proteins and competitive binding sites. Anal. Biochem .154:138 (1986).
6. C. Seneviratne, J-M. Luo and L.J. Murphy, Transcriptional regulation of rat insulin-like growth factor binding protein-1 by growth hormone. Mol. Endocrinol. 4:1199 (1990).
7. W.I. Wood, G. Cachianes, W.J. Henzel, G.A. Winslow, S.A. Spencer, R. Hellmiss, J.L. Martin, and R.C. Baxter, Cloning and expression of the growth hormone-dependent insulin-like growth factor-binding protein. Mole. Endocrinol. 2:1176 (1988).
8. A. Brinkman, C. Groffen, D.J. Kortleve, A. Geurts van Kessel, S.L.S. Drop, Isolation and characterization of a cDNA encoding the low molecular weight insulin-like growth factor binding protein (IBP-1) EMBO J. 7:2417 (1988).
9. A.L. Brown, L. Chiariotti, C.C. Orlowski, T. Mehlem, W.H. Burgess,

E.J. Ackerman, C.B. Bruni, and M.M. Rechler, Nucleotide sequence and expression of a cDNA clone encoding a fetal rat binding protein for insulin-like growth factors. J. Biol. Chem. 264:5148 (1989)

10. L.J. Murphy, C. Seneviratne, G. Ballejo, F. Croze, T.G. Kennedy, Identification and characterization of a rat decidual insulin-like growth factor binding protein cDNA. Mole. Endocrinol. 4:329 (1990).
11. S. Shimasaki, F. Uchiyama, M. Shimonaka and N. Ling, Molecular cloning of the cDNAs encoding a novel insulin-like growth factor-binding protein from rat and human. Mole. Endocrinol. 4:1451 (1990).
12. R.C. Baxter, J.L. Martin and V.A. Berniac, High molecular weight insulin-like growth factor binding protein complex: Purification and properties of the acid-labile subunit from human serum. J. Biol. Chem. 264:11843 (1989).
13. J. Zapf, C. Hauri, M. Waldvogel, E. Futo, H. Hasler, K. Benz, H.P. Guler, C. Schmid and E.R. Froesch, Recombinant human insulin-like growth factor-I induces its own specific carrier protein in hypophysectomized and diabetic rats. Proc. Natl. Acad. Sci. U.S.A. 86:3813 (1989).
14. J-M. Luo and L.J. Murphy, Regulation of insulin-like growth factor binding protein-3 expression by dexamethasone. J. Mole. Cell. Endocrinol. in press, (1990).
15. J.A. Romanus, J.E. Terrell, Y.W.H. Yang, S.P. Nissley and M.M. Rechler, Insulin-like growth factor carrier proteins in neonatal and adult rat serum are immunologically different: demonstration using a new radioimmunoassay for the carrier protein from BRL-3A rat liver cells. Endocrinology 118:1743 (1986).
16. L.Y-H. Tseng, A.L. Brown, Y. W-H. Yang, J.A. Romanus, C.C. Orlowski, T. Taylor and M.M. Rechler, The fetal rat binding protein for insulin-like growth factors is expressed in the choid plexus and cerebrospinal fluid of adult rats. Mole. Endocrinol. 3:1559 (1989).
17. S.L.S. Drop, G. Valiquette, H.J. Guyda, M.T. Corvol, B.I. Posner, Partial purification and characterization of a binding protein for insulin-like activity (ILAs) in human amniotic fluid: a possible inhibitor of insulin-like activity. Acta Endocrinol. (Copenh.) 90:505 (1979).
18. J.L. Martin, K.E. Willetts, and R.C. Baxter, Purification and properties of a novel insulin-like growth factor-II binding protein from transformed human fibroblasts. J. Biol. Chem. 265:4124 (1990).
19. S. Mohan, C.M. Bautista, J. Wergedal, D.J. Baylink, Isolation of an inhibitory insulin-like growth factor (IGF) binding protein from bone cell conditioned medium: a potential local regulator of IGF action. Proc. Natl. Acad. Sci. U.S.A. 86:8338 (1989).
20. M. Roghani, P. Hossenlopp, P. Lepage, A. Balland and M. Binoux, Isolation from human cerbrospinal fluid of a new insulin-like growth factor-binding protein with a selective affinity for IGF-II. FEBS Letters 255:253 (1989).
21. C. Meuli, J. Zapf and E.R. Froesch, NSILA-carrier protein abolishes the action of nonsuppressible insulin-like activity (NSILA-S) on perfused rat heart. Diabetologia 14:255 (1978).
22. D.J. Knauer and G.L. Smith, Inhibition of biological activity of multiplication-stimulating activity by binding to its carrier protein. Proc. Natl. Acad. Sci. U.S.A. 77:7252 (1980).
23. W.M. Burch, J. Correa, J.E. Shively and D.R. Powell, The 25 kilodalton insulin-like growth factor (IGF)- binding protein inhibits both basal and IGF-I mediated growth of chick embryo pelvic cartilage in vitro. J. Clin. Endocrinol. Metab. 70:173 (1990).
24. O. Ritvos, T. Ranta, J. Jalkanen, A-M. Suikkari, R. Voutilainen, H. Bohn and E-M. Rutanen, Insulin-like growth factor (IGF) binding protein from human decidua inhibits the binding and biological action of IGF-I in cultured choriocarcinoma cells. Endocrinology 122:2150 (1988).
25. P.E. Walton, R. Gopinath and T.D. Etherton, Porcine insulin-like

growth (IGF) binding protein blocks IGF-I action on porcine adipose tissue. Proc. Soc. Exp. Biol. 190:315 (1989).

26. M.A. DeVroede, L.Y. Tseng, P.G. Katsoyannis, S.P. Nissley, and M.M. Rechler, Modulation of insulinlike growth factor-I binding to human fibroblasts monolayer cultures by insulin like growth factor carrier proteins released into the incubation medium. J. Clin. Invest. 77:602 (1986).
27. R.G. Elgin, W.H. Busby, D.R. Clemmons, An insulin-like growth factor (IGF) binding protein enhances the biological response to IGF-I. Proc. Natl. Acad. Sci. U.S.A. 84:3254 (1987).
28. R.A. Frost and L. Tseng, Insulin like growth factor binding protein from human endometrial stromal cells is a mixture of phosphovariants of a single binding protein. Proc. 72th Annual Meeting of the Endocrine Society Abstract 191, (1990).
29. J.S.M. De Mellow and R.C. Baxter, Growth hormone dependent insulin-like growth factor binding protein both inhibits and potentiates IGF-I stimulated DNA synthesis in human skin fibroblasts. Biochem. Biophys. Res. Commun. 156:199 (1988).
30. G.T. Ooi and A.C. Herington, The biological and structural characterization of specific binding proteins for insulin-like growth factors. J. Endocrinol. 118:7 (1988).
31. A-M. Suikkari, V.A. Koivisto, R. Koistinen, M. Seppala and H. Yki-Jarvivnen, Dose-response characteristics for suppression of low molecular weight plasma insulin-like growth factor-binding protein by insulin. J. Clin. Endocrinol. Metab. 68:135 (1989).
32. A-M. Suikkari, V.A. Koivisto, E-M. Rutanen, H. Yki-Jarvinen, S-L. Karonen and M. Seppala, Insulin regulates the serum levels of low molecular weight insulin-like growth factor-binding protein. J. Clin. Endocrinol. Metab. 66:266 (1988).
33. W.H. Busby, D.K. Snyder and D.R. Clemmons, Radioimmunoassay of a 26,000-dalton plasma insulin-like growth factor-binding protein: control by nutritional variables. J. Clin. Endocrinol. Metab. 67:1225 (1988).
34. R.C. Baxter and C.T. Cowell, Diurnal rhythm of growth hormone-independent binding protein for insulin-like growth factors in human plasma. J. Clin. Endocrinol. Metab. 65:432 (1987).
35. M.S. Lewitt and R.C. Baxter, Regulation of growth hormone-independent insulin-like growth factor-binding protein (BP-28) in cultured human fetal liver explants. J. Clin. Endocrinol. Metab. 69:246 (1989).
36. L.J. Murphy, C. Seneviratne, P. Moreira and R. Reid, Enhanced expression of insulin-like growth factor binding protein-I in the fasted rat: The effects of insulin and growth hormone administration. Endocrinology 128: in press (1991).
37. R.H. McCusker, D.R. Campion, W.K. Jones and D.R. Clemmons, The insulin-like growth factor-binding proteins of porcine serum: endocrine and nutritional regulation. Endocrinology 125:501 (1989).
38. Z. Kacchra, C. Yannopoulos, I. Barash, H.J. Guyda, L.J. Murphy and B.I. Posner, The differential regulation by glucagon and growth hormone of IGF-I and IGF binding proteins (IGF-BPs) in cultured rat hepatocytes. Proc. 72th Annual Meeting of the Endocrine Society, Abstract 1133, 1990.
39. L.S. Phillips and A.T. Orawski, Nutrition and somatomedins III. Diabetic control, somatomedin and growth in rats. Diabetes 26:864 (1977).
40. J.A. Birbeck, Growth in juvenile diabetes mellitus. Diabetologia 8:1 (1972).
41. W.V. Tamborlane, R.L. Hintz, M. Bergman, M. Genel, P. Felig, R.S. Sherwin, Insulin-infusion pump treatment of diabetes: influence of improved metabolic control on plasma somatomedin levels. N. Engl. J. Med. 305:303 (1981).
42. M. Maes, L.E. Underwood and J.M. Ketelslegers, Low seum somatomedin-C in insulin-dependent diabetes: evidence for a

postreceptor mechanism. Endocrinology 118:377 (1986).
43. R.C. Baxter, J.M. Bryson and J.R. Turtle, Somatic receptors of rat liver: Regulation by insulin. Endocrinology 107:1176 (1980)
44. E. Scheiwiller, H.P. Guler, J. Merryweather, C. Scandella, W. Maerki, J. Zapf and E.R. Froesch, Growth restoration of insulin-deficient diabetic rats by recombinant human insulin-like growth factor I. Nature 323:169 (1986).
45. L.S. Phillips, D.C. Belosky, H.S. Young and L.A. Reichard, Nutrition and somatomedins VI. Somatomedin activity and somatomedin inhibitory activity in the sera from normal and diabetic rats. Endocrinology 104:1519 (1979).
46. T.G. Unterman, D.T. Oehler, R.E. Becker, Identification of a type-I insulin-like growth factor binding proetin (IGFBP) in serum from rats with diabetes mellitus. Biochem. Biophys. Res. Commun. 163:882 (1989).
47. J-M. Luo and L.J. Murphy, Differential expression of insulin-like growth factor-I and insulin-like growth factor binding protein-1 in the diabetic rat. Mole. Cell. Biochem. in press (1991).
48. S.L.S. Drop, D.J. Kortleve, H.J. Guyda and B.I. Posner, Immunoassay of a somatomedin-binding protein from human amniotic fluid: levels in fetal, neonatal and adult sera. J. Clin. Endocrinol. Metab. 59:908 (1984).
49. G. Povoa, A. Roovete and K. Hall, Cross-reaction of serum somatomedin-binding protein in a radioimmunoassay developed for somatomedin-binding protein isolated from amniotic fluid. Acta Endocrinol. (Copenh.) 107:56 (1984).
50. J.B. Margot, C. Binkert, J.L. Mary, J. Landwehr. G. Heinrich and J. Schwander, A low molecular weight insulin-like growth factor binding protein from rat: cDNA cloning and tissue distribution of its messenger RNA. Mole. Endocrinol. 3:1053 (1989).
51. G.S. Tannenbaum, J. Epelbaum, E. Colle, P. Brazeau and J.B. Martin, Antiserum to somatostatin reverses starvation-induced inhibition of growth hormone but not insulin secretion. Endocrinology 102:1909 (1978).
52. J.M.P. Holly, C.P. Smith, D.B. Dunger, J.A. Edge, R.A. Biddlecombe, A.J.K. Williams, R. Howell, T. Chard, M.O. Savage, L.H. Rees and J.A.H. Wass, Levels of the small insulin-like growth factor-binding protein are strongly related to those of insulin in prepubertal and pubertal children but only weakly so after puberty. J. Endocrinol. 121:383 (1989).
53. J-M. Luo, R.E. Reid and L.J. Murphy, Dexamethasone increases hepatic insulin-like growth factor binding protein-1 (IGFBP-1) mRNA and serum IGFBP-1 concentrations in the rat. Endocrinology 127:1456 (1990).
54. T.G. Unterman and L.S. Phillips LS, Glucocorticoid effects on somatomedins and somatomedin inhibitors. J. Clin. Endocrinol. Metab. 61: 618 (1985).
55. M. Binoux, C. Lassarre and N. Hardouin, Somatomedin production by rat liver in organ culture. III. Studies on the release of insulin-like growth and its carrier protein measured by radioligand assays. Effect of growth hormone, insulin and cortisol. Acta Endocrinol. (Copenh.) 99:422 (1982).
56. C.C. Orlowski, G.T. Ooi and M.M. Rechler, Dexamethasone stimulates transcription of the insulin-like growth factor-binding protein-1 gene in H4-II-E rat hepatoma cells. Mole. Endocrinol. 4:1592 (1990).
57. M. Gourmelen, F. Girard and M. Binoux, Serum somatomedin/insulin-like growth factor (IGF) and IGF carrier levels in patients with Cushing's syndrome or receiving glucocorticoid therapy. J. Clin. Endocrinol. Metab. 54:885 (1982).
58. M. Degerblad, G. Povoa, M. Thoren, I-L Wivall and K. Hall, Lack of diurnal rhythm of low molecular weight insulin-like growth factor binding protein in patients with Cushing's disease. Acta Endocrinol. (Copenh.) 120:195 (1989).

CEREBROSPINAL IGF BINDING PROTEINS : ISOLATION AND CHARACTERIZATION

Michel BINOUX*, Monireh ROGHANI*, Paul HOSSENLOPP*, and Odile WHITECHURCH**

* INSERM U. 142, Hôpital Saint-Antoine, 75012 Paris and
** Transgène S.A., 67082 Strasbourg, France.

INTRODUCTION

In biological fluids, IGFs (insulin-like growth factors I and II) are associated with specific, high-affinity (≈ 10^{10} M^{-1}) binding proteins (BPs) which control both their bioavailability and their action at target cell level (Review in 1). Research done in our laboratory has shown that the BPs are molecularly and functionally heterogenous. Using Western ligand blotting, we have identified five molecular forms in man, of 41.5, 38.5, 34, 30 and 24 kDa. The proportions of these forms vary in different biological fluids and culture media and their regulation and affinities for IGF-I and IGF-II are different (2-6).

Several groups have concentrated on purifying human BPs over the past few years. The major BP in serum, which is GH-dependent, has been purified (7), its cDNA has been cloned (8) and it is now known as IGFBP-3 (9). The 41.5 and 38.5 kDa forms represent two different states of glycosylation of this protein. The BP corresponding to the 34 kDa species has been cloned from a liver cDNA library and is called IGFBP-2 (10). Is has only very recently been purified from serum (11). The BP corresponding to the 30 kDa form has been isolated from amniotic fluid (12, 13) and Hep G2 human hepatoma culture medium (14), its cDNA has been cloned (15-18) and it is known as IGFBP-1 (9). A further BP, recently purified from medium conditioned by TE89 human osteosarcoma cells (19), most probably corresponds to the form which we, in ligand blotting, identify as the 24 kDa BP.

In an earlier study, we reported that BPs extracted from cerebrospinal fluid (CSF) have a preferential affinity for IGF-II (2), which may be physiologically significant, since IGF-II accounts for at least 90% of the IGFs in CSF (20). At the time, we thought that this affinity of the CSF BPs might be attributable to the 34 kDa form which predominates in CSF (3).

Later, we set out to purify the BPs in CSF and our first BP proved to be a form with an N-terminal sequence which differed from that of any other BPs previously characterized (21). We have now isolated other molecular forms (22 and manuscript in preparation) and in this report we summarize our findings.

Molecular Biology and Physiology of Insulin and Insulin-Like Growth Factors
Edited by M.K. Raizada and D. LeRoith, Plenum Press, New York, 1991

MATERIALS AND METHODS

1. CSF samples were collected during neuroradiological examination in adults and during treatment of hydrocephalous children.

Initially, two 700 ml pools of CSF, one from children and one from adults, were used for the purification schedule wich comprised 4 major steps (gel filtration at pH 7.4, chromatofocusing, hydrophobic-interaction chromatography and reverse phase chromatography) and which yielded a BP of ≈ 32 kDa, whose N-terminal sequence we have reported elsewhere (21).

Subsequently, a modified protocol using a 3000 ml pool of child CSF was instituted (60% ammonium sulphate precipitation; gel filtration on 2.5 x 70 cm Ultrogel AcA54 columns in 0.5 M acetic acid, 0.15 M NaCl; affinity chromatography on a Sepharose IGF-I column; and reverse phase chromatography on 4.6 x 220 mm C8 Aquapore RP 300 in a 10 - 60% acetonitrile gradient with 0.1% TFA).

2. An IGFBP-1 preparation purified from human amniotic fluid (13) was generously provided by G. Povoa (Stockholm).

3. Recombinant human (rh) IGF preparations were used : rh IGF-I was a gift from Ciba-Geigy (Basel, Switzerland), rhIGF-II was a gift from Lilly Research Laboratories (Indianapolis) and rh des (1-3) IGF-I (Kabigen AB, Stockholm, Sweden), was kindly provided by V. Sara (Stockholm).

4. Polyclonal antibodies raised against IGFBP-1, IGFBP-3 and IGFBP-2 and -3 (α HEC-1 antibody (23) were generously provided by G. Povoa (Stockholm), R. Baxter (Camperdown, Australia) and G. Lamson (Stanford), respectively.

5. Characterization and assays

The protein assays (Bradford), electrophoretic analyses (SDS-polyacrylamide gel electrophoresis (SDS-PAGE) followed by silver staining, Western ligand or immunoblotting) and measurements of binding activity and competitive binding were the same as those reported previously (21, 22). Glycosylation analyses (N-glycanase, endoglycanase-F, neuraminidase and O-glycanase) were done according to the manufacturer's instructions (Genzyme, Boston). N-terminal amino acid sequences were determined as described previously (21) by a method lacking the ability to detect cysteine residues.

RESULTS AND DISCUSSION

1. Electrophoretic analysis using Western ligand blotting, which specifically detects the IGFBPs, revealed very different profiles for the relative proportions of the various

molecular forms, depending on the media studied (Figure 1). In the serum of normal adults and children over the age of 5, the most abundant BPs are the 41.5 and 38.5 kDa forms which correspond to IGF BP-3 (4, 6). These are barely detectable in CSF at the concentrations studied. It is usually the 34 kDa form which predominates, and is sometimes the only form visible, in CSF (3). It appears as a single band in adults and as a doublet in children. The amounts of 32-30 kDa BP vary widely from one CSF sample to another, appearing sometimes as a poorly defined broad band and sometimes as two or three bands. The 24 kDa form is more often clearly detectable in children.

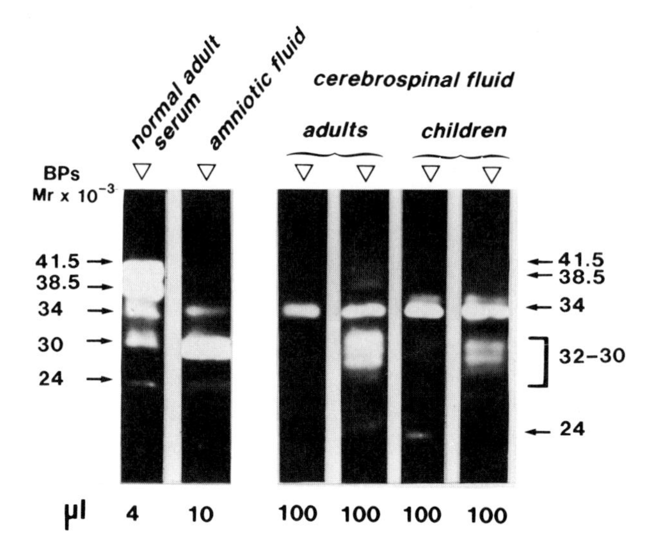

Figure 1 (from Ref. 22) . **Western ligand blot analysis of serum, amniotic fluid and CSF BPs.** After SDS-PAGE (11% homogenous gel) in the absence of reducing agent, the BPs were transferred onto nitrocellulose, incubated with ^{125}I-IGF-II and identified by autoradiography.

Immunoblots done with antibodies directed against IGFBP-3 and IGFBP-1 gave no reaction with CSF BPs (not shown), but the α-HEC-1 antibody yielded a band at the level of the 34 kDa BP (22). From these findings it could be concluded, first, that the three major CSF BPs are immunologically unrelated; secondly, that the 34 kDa form probably corresponds to IGFBP-2; and, thirdly, that the 32-30 kDa material in CSF is not the same as the 30 kDa material in amniotic fluid, which is specifically recognized by the anti-IGFBP-1 antibody (6, 22).

Table 1. **N-terminal amino acid sequence of the 32-30 kDa CSF BP (21) compared with those of the BPs purified from medium conditioned by transformed human lung fibroblasts (24) and from serum (11).**

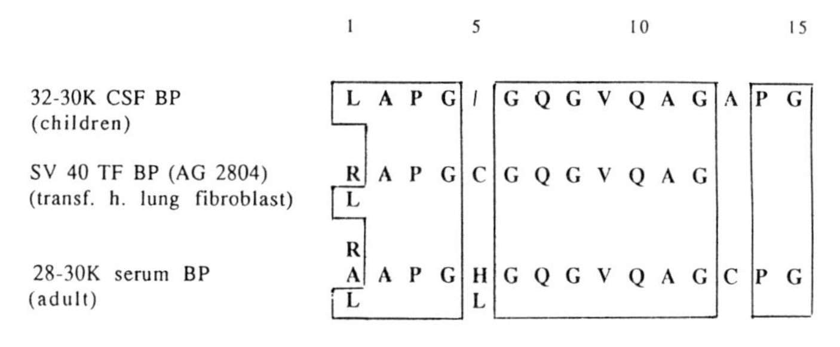

	1	2	3	4	5	6	7	8	9	10	11	12	13	14	15
32-30K CSF BP (children)	L	A	P	G	/	G	Q	G	V	Q	A	G	A	P	G
SV 40 TF BP (AG 2804) (transf. h. lung fibroblast)	R L	A	P	G	C	G	Q	G	V	Q	A	G			
28-30K serum BP (adult)	R A L	A	P	G	H L	G	Q	G	V	Q	A	G	C	P	G

2. The first BP that we purified to homogeneity was the one yielding the 32-30 kDa bands (21). Apart from Residues 12 and 13, the N-terminal sequences over the first 15 amino acid residues in child and adult CSF were identical (-Leu-Leu- in adults, -Gly-Ala- in children). However, they bore no analogy with the N-terminal sequences of any other BP identified at the time.

The identity of this new form of BP has since been confirmed by other authors who have purified it from serum (11) and medium conditionned by lung fibroblast cell lines (J. Ballard et al., personal communication, and 24). The sequences reported are identical to those of our child CSF BP (Table 1).

3. With our second purification schedule, we obtained a new preparation of this 32-30 kDa BP from child CSF and, in addition, three other molecular forms of BP (M. Roghani et al., manuscript inpreparation).

The first was the 34 kDa BP (which is the most hydrophobic in reverse phase chromatography) whose sequence proved to be identical to that deduced for the IGFBP-2 cDNA, except for the first 3 residues which were not found in the CSF BP preparation (Table 2).

Table 2. **N-terminal amino acid sequence of the 34 kDa CSF BP compared with that deduced from the cDNA sequence of h IGFBP-2 (10).**

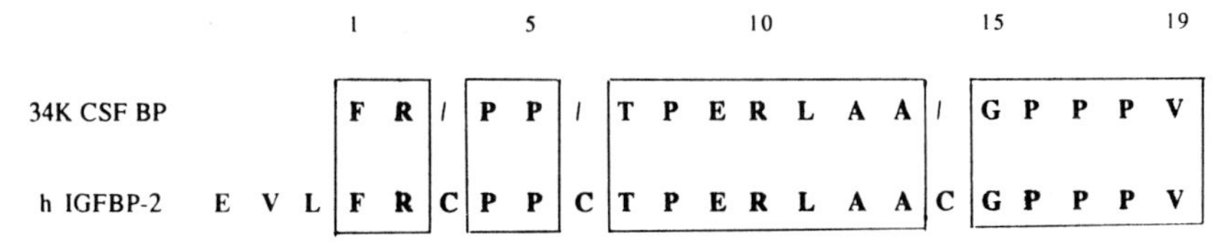

				1	2	3	4	5	6	7	8	9	10	11	12	13	14	15	16	17	18	19
34K CSF BP				F	R	/	P	P	/	T	P	E	R	L	A	A	/	G	P	P	P	V
h IGFBP-2	E	V	L	F	R	C	P	P	C	T	P	E	R	L	A	A	C	G	P	P	P	V

Table 3. **N-terminal amino acid sequence of the 30 kDa form of CSFBP compared with the sequence of h IGFBP-3 purified from serum (GH-dependent BP) (28) and with that deduced from its cDNA (8).**

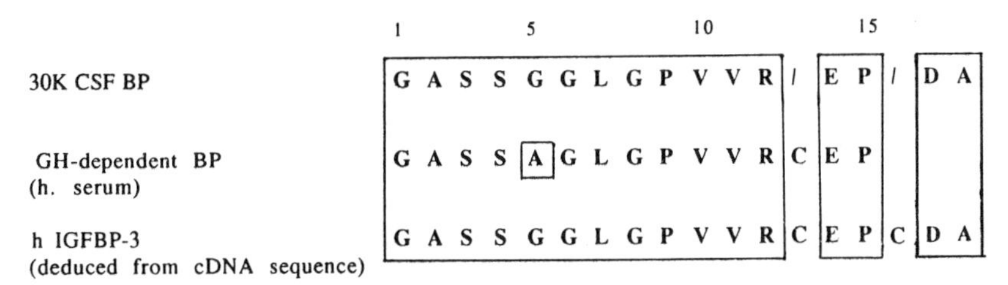

	1				5					10					15		
30K CSF BP	G	A	S	S	G	G	L	G	P	V	V	R	/	E	P	/	D	A
GH-dependent BP (h. serum)	G	A	S	S	A	G	L	G	P	V	V	R	C	E	P			
h IGFBP-3 (deduced from cDNA sequence)	G	A	S	S	G	G	L	G	P	V	V	R	C	E	P	C	D	A

The second was a BP migrating at approximately 30 kDa in SDS-PAGE, whose N-terminal sequence was identical to that of IGFBP-3 (Table 3). The 41.5 and the 38.5 kDa forms eluted later in the acetonitrile gradient. This 30 kDa form may correspond to a form of IGFBP-3 truncated at its C-terminal end. The IGFBP-3 N-terminus has in fact been found in 30 kDa material isolated from rat (25) and human serum (11). Similarly, we have found a 30 kDa band recognizable by anti-IGFBP-3 antibody in acromegalic serum (6) and seen evidence of enzymatic degradation of IGFBP-3 giving rise to 30 kDa material in the serum of pregnant women (26). It is therefore quite possible that this degradation also occurs in CSF.

Table 4. **N-terminal amino acid sequence of the 22 kDa CSF BP compared with those deduced from the cDNAs of h IGFBP-3 (8), h IGFBP-1 (15-18), h IGFBP-2 (10) and with the sequences of the BP purifed from human osteosarcoma cell-conditioned medium (19) and from rat serum (27).**

						1				5					10					15						20
22K CSF BP						/	/	D	S	F	V	P	/	E	P	S	D	E	K	A	L	S	/	/	-	P
h IGFBP-3	G	A	S	S	G	G	L	G	P	V	V	R	C	E	P	C	D	A	R	A	L	A	Q	C	A	P
h IGFBP-1									A	P	W	Q	C	A	P	C	S	A	E	K	L	A	L	C	-	P
h IGFBP-2								E	V	L	F	R	C	P	P	C	T	P	E	R	L	A	A	C	G	P
Inhibit. IGFBP ("25K") (h. osteosarcoma cells TE 89)								D	E	A	/	H	C	P	P	E	S	E	A	K	L	A				
rat serum IGFBP ("36-32K")								D	E	A	/	H	/	P	P	/	S	E	E	K	L	A	R	/	R	P

A third form of CSF BP, of 22 kDa, was also discovered in very small quantities relative to the other molecular forms and has been purified. Its N-terminal sequence bears some analogies to the other BPs (Table 4) : out of the 15 residues identified, 7 are homologous with IGFBP-3, 3, with IGFBP-1 and IGFBP-2, 4, with the BP produced by human osteosarcoma cells (19) and 5, with the murine form of this last BP (27). It seems, therefore, that this 22 kDa form may be a new BP, but further work will be necessary to characterize it.

4. The purified 32-30 kDa and 34 kDa CSF BPs were analysed using different glycanases (Figure 2). With N-glycanase and endoglycanase-F, the electrophoretic profiles of both BPs remained unchanged. After successive incubations with neuraminidase and O-glycanase, however, the upper band of the 34 kDa BP disappeared and the 32-30 kDa BP migrated more rapidly. These results suggest that both BPs are O-glycosylated.

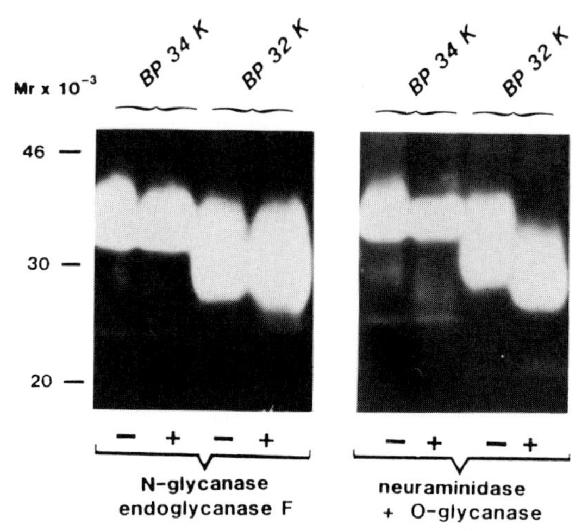

Figure 2 : **Western ligand blot analysis of the purified 34 kDa and 32-30 kDa CSFBPs after incubation with different glycanases.**

5. Competitive binding studies were done in order to determine the affinities of the purified 32-30 kDa and 34 kDa BPs for IGF-I and IGF-II. These were done in comparison with a preparation of IGFBP-1 purified from amniotic fluid. The findings have been reported in detail elsewhere (22) and may be summarized as fellows :

When radio-iodinated IGF-II was used as tracer, IGF-II's competitive potency was 40 times IGF-I's for the 32-30 kDa BP, but only 17 times it for the 34 kDa BP. For IGFBP-1, similar displacement curves were obtained with IGF-I and IGF-II. The des (1-3) IGF-I was ineffective in the cases of the 32-30 kDa and 34 kDa BPs, but with IGFBP-1 it was capable of competing with the IGF-II tracer, although with a 40-fold weaker efficiency than IGF-I.

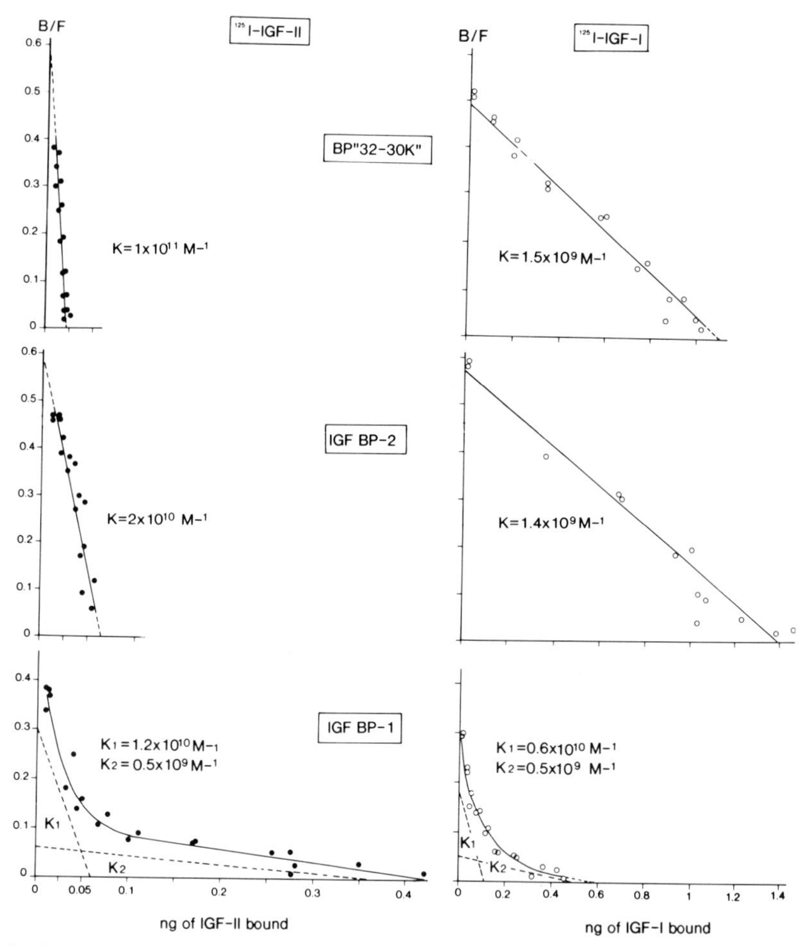

Figure 3 (from Ref. 22) . **Scatchard analysis of competitive binding experiments done with the 32-30 kDa BP and IGFBP-2 (34 kDa) purified from CSF and IGFBP-1 purified from amniotic fluid.**
The plots represent data obtained with cold and labelled rh IGF-II (on the left) and rh IGF-I (on the right).

When IGF-I was used as tracer, the two displacement curves obtained for IGFBP-1 were almost identical to those obtained in experiments using IGF-II as tracer. For the other two BPs, IGF-II's competitive potency was strongly reduced (although it was still much greater than IGF-I's in the case of the 32-30 kDa BP) whereas the IGF-I displacement curves were similar whichever tracer was used. These findings would suggest a high-affinity site for IGF-II and a lesser-affinity site for IGF-I.

Figure 3 shows a Scatchard representation of the results. The affinity constant of the 32-30 kDa BP for IGF-II was ≈ 10^{11} M^{-1} which was similar to that found for the homologue of this BP produced by transformed human lung fibroblasts (24) and is the strongest affinity known among the BPs to date. The 34 kDa BP's affinity for IGF-II was 5 times weaker, and IGFBP-1's, 10 times. The curvilinear shape of the plot for the latter suggests that there are two sets of high- and low-affinity binding sites. IGFBP-1's affinity for IGF-1 was similar to that for IGF-II. The 34 kDa BP's, by contrast, was 15 times weaker. In the case the 32-30 kDa BP, however, its affinity for IGF-I was 70 times weaker than its affinity for IGF-II. This means that the 32-30 kDa BP truly has a selective affinity for IGF-II.

The predominance of the 34 kDa BP (IGFBP-2) and the 32-30 kDa BP over the other molecular forms of BP in CSF and their preferential affinities for IGF-II, the major IGF in this fluid, suggests that these two BPs have some specific role in the central nervous system.

ACKNOWLEDGEMENTS

We are grateful to the staff ot the Hôpital de la Salpêtrière and the Hôpital des Enfants Malades (Paris) who provided us with CSF. We thank B. Segovia, C. Lassarre and B. de Gallé for their technical assistance. This work was supported by the Institut National de la Santé et de la Recherche Médicale. M. Roghani is a recipient of a Nordisk grant for the study of growth. P. Hossenlopp is a Chargé de Recherche at the Centre National de Recherche Scientifique.

REFERENCES

1. Baxter RC, Martin JL 1989 Binding proteins for the insulin-like growth factors : structure, regulation and function. Progress in Growth Factor Research 1:49-68.

2. Binoux M, Hardouin S, Lassarre C, Hossenlopp P 1982 Evidence for production by the liver of two IGF binding proteins with similar molecular weights but different affinities for IGF I and IGF II. Their relations with serum and cerebrospinal fluid IGF binding proteins. J Clin Endocrinol Metab 55:600-602.

3. Hossenlopp P. Seurin D, Segovia-Quinson B, Binoux M 1986 Identification of an insulin-like growth factor binding protein in human cerebrospinal fluid with a selective affinity for IGF II. FEBS Lett 208:439-444.

4. Hardouin S, Hossenlopp P, Segovia B, Seurin D, Portolan G, Lassarre C, Binoux M 1987 Heterogeneity of insulin-like growth factor binding proteins and relationships between structure and affinity. 1. Circulating forms in man. Eur J Biochem 170 : 121-132.

5. Hossenlopp P, Seurin D, Segovia B, Portolan G, Binoux M 1987 Heterogeneity of insulin-like growth factor binding proteins and relationships between structure and affinity. 2. Forms released by human and rat liver in culture. Eur J Biochem 170 : 133-142.

6. Hardouin S, Gourmelen M, Noguiez P, Seurin D, Roghani M, Le Bouc Y, Povoa G, Merimee TJ, Hossenlopp P, Binoux M 1989 Molecular forms of serum insulin-like growth factor (IGF) binding proteins in man : relationships with growth hormone and IGFs and physiological significance. J Clin Endocrinol Metab 69:1291-1301.

7. Martin JL, Baxter RC 1986 Insulin-like growth factor-binding protein from human plasma. Purification and characterization. J Biol Chem 261:8754-8760.

8. Wood WI, Cachianes G, Henzel WJ, Winslow GA, Spencer SA, Hellmiss R, Martin JL, Baxter RC 1988 Cloning and expression of the growth hormone-dependent insulin-like growth factor-binding protein. Molecular Endocrinology 2:1176-1185.

9. Ballard J, Baxter R, Binoux M, Clemmons D, Drop S, Hall K, Hintz R, Rechler M, Rutanen E, Schwander J 1989 On the nomenclature of the IGF binding proteins. Acta Endocrinol (Copenh) 121:751-752.

10. Binkert C, Landwehr J, Mary JL, Schwander J, Heinrich G 1989 Cloning, sequence analysis and expression of a cDNA encoding a novel insulin-like growth factor binding protein (IGF BP-2). EMBO J 8:2497-2502.

11. Zapf J, Kiefer M, Merryweather J, Masiarz F, Bauer D, Born W, Fischer JA, Froesch ER 1990 Isolation from adult human serum of four insulin-like growth factor (IGF) binding proteins and molecular cloning of one of them that is increased by IGF I administration and in extrapancreatic tumor hypoglycemia. J Biol Chem (in press).

12. Drop SLS, Kortleve DJ, Guyda HJ 1984 Isolation of a somatomedin-binding protein from preterm amniotic fluid. Development of a radioimmunoassay. J Clin Endocrinol Metab 59:899-907.

13. Povoa G, Enberg G, Jornvall H, Hall K 1984 Isolation and characterization of a somatomedin-binding protein from mid-term human amniotic fluid. Eur J Biochem 144:199-204.

14. Povoa G, Isaksson M, Jörnvall H, Hall K 1985 The somatomedin-binding protein isolated from a human hepatoma cell line is identical to the human amniotic fluid somatomedin-binding protein. Biochem Biophys Res Commun 128:1071-1078.

15. Brewer MT, Stetler GL, Squires CH, Thompson RC, Busby WH, Clemmons DR 1988 Cloning, characterization, and expression of a human insulin-like growth factor binding protein. Biochem Biophys Res Comm 152:1289-1297.

16. Brinkman A, Groffen C, Kortleve DJ, Geurts van Kessel A, Drop SLS 1988 Isolation and characterization of a cDNA encoding the low molecular weight insulin-like growth factor binding protein (IBP-1). EMBO J 7:2417-2423.

17. Julkunen M, Koistinen R, Aalto-Setälä K, Seppälä M, Jänne OA, Kontula K 1988 Primary structure of human insulin-like growth factor-binding protein/placental protein 12 and tissue-specific expression of its mRNA. FEBS Lett 236:295-302.

18. Lee YL, Hintz RL, James PM, Lee PDK, Shively JE, Powell DR 1988 Insulin-like growth factor (IGF) binding protein complementary deoxyribonucleic acid from human Hep G2 hepatoma cells : predicted protein sequence suggests an IGF binding domain different from those of the IGF-I and IGF-II receptors. Mol Endocrinol 2:404-411.

19. Mohan S, Bautista CM, Wergedal J, Baylink DJ 1989 Isolation of an inhibitory insulin-like growth factor (IGF) binding protein from bone cell-conditioned medium : a potential local regulator of IGF action. Proc Natl Acad Sci USA 86:8338-8342.

20. Haselbacher G, Humbel R 1982 Evidence for two species of insulin-like growth factor II (IGF II and "big" IGF II) in human spinal fluid. Endocrinology 110:1822-1824.

21. Roghani M, Hossenlopp P, Lepage P, Balland A, Binoux M 1989 Isolation from human cerebrospinal fluid of a new insulin-like growth factor binding protein with a selective affinity for IGF II. FEBS Lett 255:253-258.

22. Roghani M, Lassarre C, Zapf J, Povoa G, Binoux M Two insulin-like growth factor binding proteins are responsible for the selective affinity for IGF-II of cerebrospinal fluid binding proteins. J Clin Endocrinol Metab (in press).

23. Lamson G, Pham H, Oh Y, Ocrant I, Schwander J, Rosenfeld RG 1989 Expression of the BRL-3A Insulin-like growth factor binding protein (rBP-30) in the rat central nervous system. Endocrinology 123:1100-1102.

24. Martin JL, Willetts KE, Baxter RC 1990 Purification and properties of a novel insulin-like growth factor-II binding protein from transformed human fibroblasts. J Biol Chem 265:4124-4130.

25. Zapf J, Born W, Chang JY, James P, Froesch ER, Fischer JA 1988 Isolation and NH2-terminal amino acid sequences of rat serum carrier proteins for insulin-like growth factors. Biochem Biophys Res Commun 156:1187-1194.

26. Hossenlopp P, Segovia B, Lassarre C, Roghani M, Bredon M, Binoux M Evidence of enzymatic degradation of insulin-like growth factor binding proteins in the "150 K" complex during pregnancy. J Clin Endocrinol Metab (in press).

27. Shimonaka M, Schroeder R, Shimasaki S, Ling N 1989 Identification of a novel binding protein for insulin-like growth factors in adult rat serum. Biochem Biophys Res Commun 165:189-195.

28. Baxter RC, Martin JL 1987 Binding proteins for insulin-like growth factors in adult rat serum. Comparison with other human and rat binding proteins. Biochem Biophys Res Commun 147:408-415.

THE EFFECT OF QUANTITY AND NUTRITIONAL QUALITY OF DIETARY PROTEINS ON PLASMA CONCENTRATION OF INSULIN-LIKE GROWTH FACTOR BINDING PROTEINS (IGFBP) AND THE SATURABILITY OF IGFBP WITH ENDOGENOUS IGF-I

Hisanori Kato, Tsutomu Umezawa, Yutaka Miura and
Tadashi Noguchi

Department of Agricultural Chemistry, Faculty of Agriculture
The University of Tokyo, Bunkyo-ku, Tokyo 113, Japan

INTRODUCTION

Plasma concentrations of insulin-like growth factor-I (IGF-I) have been shown to be controlled by growth hormone, insulin and nutritional status of animals.[1-3] In previous papers, the present authors demonstrated that quantity and nutritional quality of dietary proteins significantly affect the plasma concentration of IGF-I.[4] The concentration correlated well with the growth rate or the rate of whole body protein synthesis.[5] We also showed that liver content of IGF-I mRNA was also strongly affected by the status of protein nutrition.[6,7] It decreased greatly under protein deprivation and it was highly correlated with the plasma IGF-I concentration. The content of the mRNA species of larger molecular weight was most severely affected. These results strongly suggest that the rate of body protein synthesis or growth rate is regulated by the plasma concentration of IGF-I through liver IGF-I mRNA content (or the rate of synthesis and secretion of IGF-I by liver).

Plasma IGF-I is found as complexes with several binding proteins. Yang et al.[8] showed that rat plasma IGF-I is rich in the 150 kDa IGFBP-3 complex and in the complex with IGFBP-1 forming 40 kDa complex. The complex of IGF-I with IGFBP-2 has been reported to be a minor component in adult rat plasma.

Although the physiological role of these IGF-I and IGFBP complexes has not been elucidated, the concentration of these complexes seemed be regulated also by the nutritional status of animals. Employing a gel filtration technique, the present authors have shown that the plasma concentration of each complex and the saturability of each binding protein with endogenous IGF-I is also regulated by the nutritional status of animals.[4]

The present studies extended the previous work and showed that the concentration and the saturability of IGFBPs are strictly regulated by the

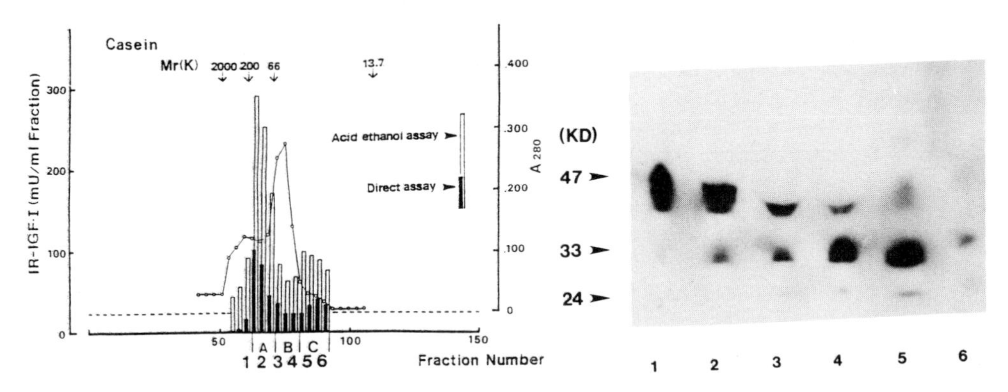

Fig. 1. LEFT: Typical gel filtration profiles of plasma of rats fed on the casein diet. Acid extracted immunoreactive IGF-I (□), unextracted immunoreactive IGF-I (■) and absorption at 280nm(O-O) are shown. RIGHT: Detection of IGFBPs in pooled fractions of the gel filtration by ligand blotting.

nutritional status of animals. The results strongly suggest that the plasma concentration and saturability of IGFBPs are simple and practical indices for the assessment of the status of protein nutrition in the children of undernutrition or the patients suffering from nutritional deficiency.

I. GEL FILTRATION STUDIES

Rats were fed a 12 % casein diet, a 12 % wheat gluten diet, or a protein-free diet for 1 week.[9] After 1 week, their plasma was gel-filtrated on Sephacryl S-200 and the fractions were assayed for immunoreactive IGF-I. Figure 1 shows a typical result of the casein-fed rats. Two large peaks were observed one at 150 kDa and the other at about 40 kDa. The peak at 150 kDa was larger than that at 40 kDa in this case. The peak at 40 kDa was more prominent in the rats fed the gluten or protein-free diet. These results show that IGF-I forms preferentially the 150 kDa complex under good nutritional conditions.

II. LIGAND BLOTTING STUDIES[9]

Figure 2 shows the relationship between the proteins in the fractions of gel filtration and those found in ligand blotting. In this experiment, sera were subjected to sodium dodecylsulfate gel electrophoresis (SDS-PAGE), the separated proteins transferred to a nylon membrane and probed with ^{125}I-IGF-I.[10, 11] Peak A in the gel filtration (Fig. 1) contained two bands in ligand blotting with ^{125}I-IGF-I at the molecular mass approximately 40 kDa. Peak C in Fig. 1 contained two bands at the molecular mass about 30 kDa. The two bands in the peak A and those in the peak C were sometimes difficult to differentiate by the ligand blotting technique. Another minor peak was found at about 22 kDa. These results were compared with those of Yang et al.[8], summerized and depicted in Fig. 2 (lane 2 and 3). Although this picture is hypothetical, the following results will be explained by referring this figure.

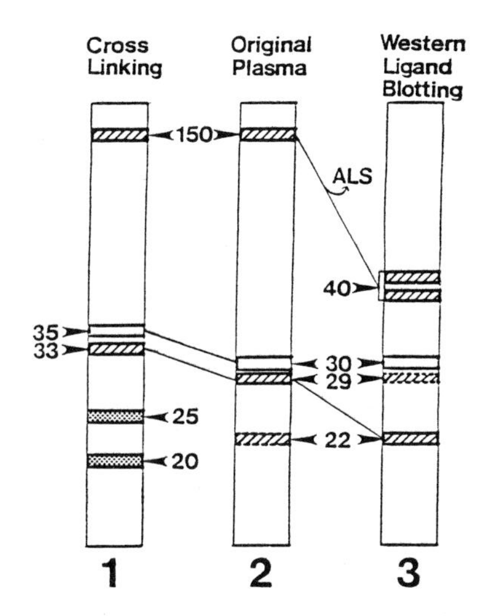

Fig. 2. Schematic representation of the relationship among the IGFBPs detected by ligand blotting (Fig. 1, 3), detected by ligand cross linking (Fig. 4) and in original rat plasma (8,9). Hatched bands; IGFBP-3. The band at 29 kDa in the lane 3 is the N-terminal 29 kDa protein of 40 kDa IGFBP-3 (12). ALS; acid labile subunit of 150 kDa complex. Open bands: IGFBP-1. Shaded bands; unidentified IGFBPs. Numbers mean molecular mass, kDa.

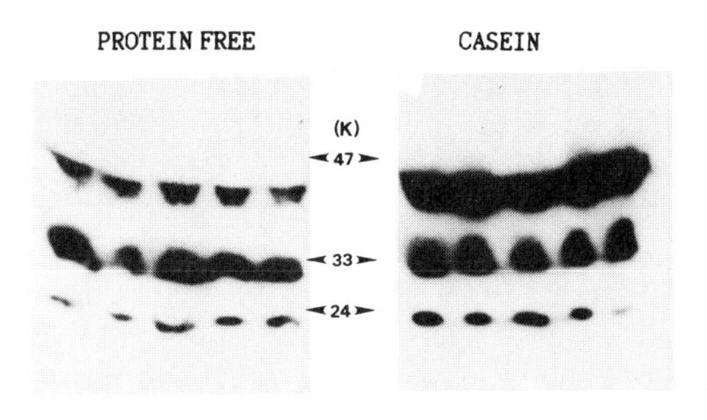

Fig. 3. Ligand blot analysis (10,11) of plasma of rats fed on the 12% casein diet or the protein free diet (9). Each lane represents plasma from one rat.

Figure 3 shows the results of ligand blotting of the plasma obtained from the rats fed the casein or protein-free diet. The band at 40 kDa (IGFBP-3) decreased significantly under protein deprivation. The band at about 30 kDa was not affected extensively by protein deprivation. This band is supposed to be composed of two bands, i.e., one of them is IGFBP-1 and the other is a N-terminal fragment of IGFBP-3[12] (gp 29,[8] see Fig. 2). This means that the sum of the amount of IGFBP-1 and gp 29 was not affected significantly by dietary protein deficiency.

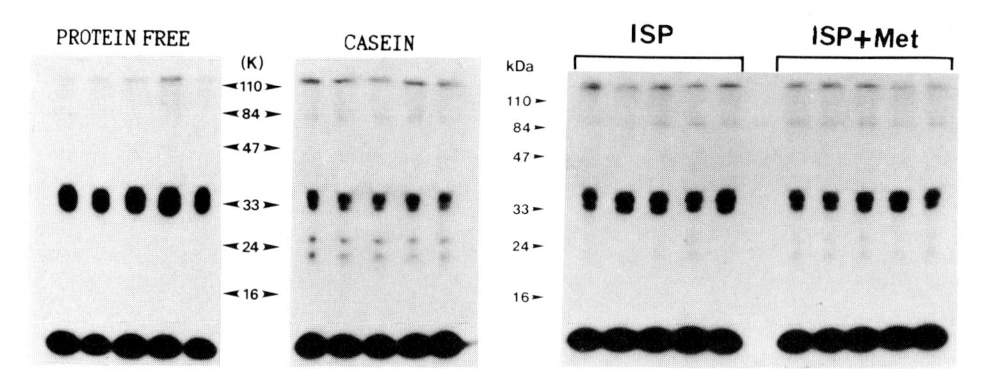

Fig. 4. SDS-PAGE of the rat plasma covalently bound with ^{125}I-IGF-I (ligand cross linking (11)). LEFT: Rats were fed on the casein diet or the protein free diet. RIGHT: Rats were fed on the isolated soybean protein diet (ISP) or ISP supplemented with methionine. Each lane represents plasma from one rat.

III. LIGAND CROSS LINKING STUDIES[9]

The rat plasma was mixed with ^{125}I-labeled IGF-I, then treated with disuccinimidyl suberate (DSS). By this method, IGFBPs unoccupied by endogenous (unlabeled) IGF-I was covalently conjugated with labeled IGF-I thereby making it possible to estimate the amount of IGFBPs unoccupied with endogenous IGF-I.[11] Figure 2 (lane 1,2) also shows the relationship of the bands detected by this technique and the IGF-I-IGFBP complex in the original plasma.[8] Figure 4 (left) shows that the band at 150 kDa decreased under protein deprivation. As shown in Fig. 2, the band at 150 kDa is composed of IGF-I, IGFBP-3 and the acid labile subunit reported by Baxer et al.[13] Therefore, the results in Fig. 4 show that the unoccupied IGFBP-3 (or IGFBP-3 unsaturated with endogenous IGF-I) decreases under protein deprivation. On the contrary, unoccupied IGFBP-1 greatly increases in the plasma of the rats fed the protein-free diet. Although the origin of the two bands at 25 and 22 kDa is not explained clearly at present, these might be the degradation products of IGFBP-3 which is relatively sensitive to the endogenous protease(s) in plasma and is degraded into fragments.[14] Some of the fragments have been demonstrated to retain the activity to form complexes with IGF-I in the case of human IGFBP-3.[13 15]

The present results show that (i) plasma concentration of IGFBP-3 decreases under protein deprivation, (ii) unsaturated IGFBP-3 also decreases concomitantly, (iii) unoccupied IGFBP-1 increases under protein deficiency, (iv) and although it is not conclusive, the total concentration of IGFBP-1 is supposed to increase under protein deficiency, judging from the following observations. As described above, the band of IGFBP-1 is found at the position very close to the N-terminal fragment of IGFBP-3. Therefore, if N-terminal fragment of IGFBP-3 decreases as IGFBP-3 itself, the band at 30 kDa must decrease corresponding to the decrease in IGFBP-3 at 40 kDa. However, the band at 30 kDa did not change significantly under protein deprivation.

This suggests that IGFBP-1 increased in protein deprivation. Besides, plasma IGFBP-1 is known to increase in fasting[16-18] and in diabetes,[19-21] and to decrease under insulin administration.[22] Furthermore, IGFBP-1 mRNA also increases in liver in food deprivation or in diabetic animals.[23,24] This suggests that animals respond to adverse physiological conditions by increasing the plasma concentration of IGFBP-1 or its mRNA content in liver. These observations favor the above assumption that plasma IGFBP-1 concentration increases under protein deprivation. Our studies also showed that liver or kidney IGFBP-1 mRNA increases greatly under protein deprivation (A. Takenaka et al. to be published).

IV. CHANGES IN UNOCCUPIED IGFBP CONCENTRATIONS DURING NUTRITIONAL IMPROVEMENT

Figure 4 (right) shows the effect of methionine supplementation of an isolated soybean protein (ISP) diet on the concentration of plasma unoccupied IGFBPs. In this case, the original ISP diet was supplemented with lysine and threonine up to the level of the recommendations by National Research Council.[25] Therefore, the original ISP diet is only deficient in methionine. As it is evident from Fig. 4, supplementation of methionine to ISP diet, a decrease in 30 kDa band were observed. In other words, the autoradiogram of the plasma obtained from the ISP-methionine diet-fed rats showed a similar pattern to that from the casein diet-fed rats, showing the shift to the status of good nutrition. These autoradiograms after cross-linking with ^{125}I-IGF-I reflect with great sensitivity the effect of quantity and nutritional quality of dietary proteins.

CONCLUSIONS

Protein deprivation or feeding a low quality protein cause a decrease in plasma IGFBP-3 concentration and an increase in IGFBP-1 concentration. This is consistent with the observations that plasma IGFBP-1 and liver IGFBP-1 mRNA increase in fasting or in diabetic animals. As reported previously, IGFBP-1 inhibits[26,27] or potentiates[28,29] the activity of IGF-I. However, the physiological significance or adaptive advantages under conditions of poor nutrition remains to be clarified. In contrast to the relationship between nutritional status and IGFBP-1 levels, plasma IGFBP-3 concentration increased under good nutritional conditions. This suggests that IGFBP-3 might potentiate the activity of IGF-I by binding it or through inhibiting the binding of IGF-I with IGFBP-1 or through some other unknown mechanisms.[4,30,31] Regardless of their physiological importance, these results suggest that studies of ligand blotting and ligand cross linking of IGFBPs with ^{125}I-labeled IGF-I are useful and simple indices of the assessment of the status of protein nutrition in patients suffering from various nutritional deficiencies.

ACKNOWLEDGMENT

The authors greatly appreciate Dr. J. J. Van Wyk, University of North Carolina, for his helpful comments and advices.

REFERENCES

1. J. J. Van Wyk, The somatomedins: Biological actions and physiologic control mechanisms. in:"Hormonal Proteins and Peptides. vol. 12," pp. 8 (1984).

2. E. R. Froesch, C. Schmid, J. Schwander, and J. Zapf, Actions of insulin-like growth factors. Ann. Rev. Physiol., 47:443 (1985).

3. D. LeRoith, and M. K. Raizada, (eds.) "Molecular and Cellular Biology of Insulin-like Growth Factors and Their Receptors", Plenum Press, New York and London (1989).

4. S. Takahashi, M. Kajikawa, T. Umezawa, S.-I. Takahashi, H. Kato, Y. Miura, T. J. Nam, T. Noguchi, and H. Naito, Effect of dietary proteins on the plasma immunoreactive insulin-like growth factor-1/somatomedin C concentration in the rat. Brit. J. Nutr. 63:521 (1990).

5. T. J. Nam, T. Noguchi, R. Funabiki, H. Kato, Y. Miura, and H. Naito, Correlation between the urinary excretion of acid-soluble peptides, fractional synthesis rate of whole body proteins, and plasma immunoreactive insulin-like growth factor-1/somatomedin C concentration in the rat. Brit. J. Nutr. 63:515 (1990).

6. H. Kato, Y. Miura, A. Okoshi, T. Umezawa, S.-I. Takahashi, and T. Noguchi, Dietary and hormonal factors affecting the mRNA level of IGF-I in rat liver in vivo and in primary cultures of rat hepatocytes. in:"Molecular and Cellular Biology of Insulin-like Growth Factors and Theier Receptors," D. LeRoith, and M. K. Raizada, (eds.) Plenum Press, New York and London. pp.125, (1989).

7. D. S. Straus, and C. D. Takemoto, Effect of protein deprivation on insulin-like growth factor(IGF)-I and -II, IGF binding protein-2, and serum albumin gene expression in rat. Endocrinology 127:1849 (1990).

8. Y. W.-H. Yang, J.-F. Wang, C. C. Orlowski, S. P. Nissley, and M. M. Rechler, Structure, specificity, and regulation of the insulin-like growth factor-binding proteins in adult rat serum. Endocrinology 125:1540 (1989).

9. T. Umezawa, Y. Ohsawa, Y. Miura, H. Kato, and T. Noguchi, Effect of protein deprivation on insulin-like growth factor binding proteins in rats. Brit. J. Nutr. in press.

10. P. Hossenlopp, D. Seurin, B. Segovia-Quinson, S. Hardouin, and M. Binoux, Analysis of serum insulin-like growth factor binding proteins using western blotting: Use of the method for titration of the binding proteins and competitive binding studies. Anal. Biochem. 154:138 (1986).

11. S. Hardouin, P. Hossenlopp, B. Segovia, D. Seurin, G. Portolan, C. Lassarre, and M. Binoux, Heterogeneity of insulin-like growth factor binding proteins and relationships between structure and affinity. (1) Circulating forms in man. Eur. J. Biochem. 170:121 (1987).

12. J. Zapf, W. Born, J.-Y. Chang, P. James, E. R. Froesch, and J. A. Fischer, Isolation and NH_2-terminal amino acid sequences of rat serum carrier proteins for insulin-like growth factors. Biochem. Biophys. Res. Commun. 156:1187 (1988).

13. R. C. Baxter, and J. L. Martin, Structure of the Mr 140,000 growth hormone-dependent insulin-like growth factor binding protein complex: Determination by reconstitution and affinity-labeling. Proc. Natl. Acad. Sci. USA. 86:6898 (1989).

14. J. L. Martin, and R. C. Baxter, Insulin-like growth factor-binding protein from human plasma: Purification and characterization. J. Biol. Chem. 261:8754 (1986).

15. G. T. Ooi, and A. C. Herington, Recognition of insulin-like growth factor (IGF) serum binding proteins by an antibody raised against a specific IGF-inhibitor. Biochem. Biophys. Res. Commun. 156:783 (1988).

16. W. H. Busby, D. K. Snyder, and D. R. Clemmons, Radioimmunoassay of a 26,000-dalton plasma insulin-like growth factor-binding protein: Control by nutritional variables. J. Clin. Endocrinol. Metab. 67:1225 (1988).

17. S. I. Yeoh, and R. C. Baxter, Metabolic regulation of the growth hormone independent insulin-like growth factor binding protein in human plasma. Acta Endocrinol. 119:465 (1988).

18. R. H. McCusker, D. R. Campion, W. K. Jones, and D. R. Clemmons, The insulin-like growth factor-binding proteins of porcine serum: Endocrine and nutritional regulation. Endocrinology 125:501 (1989).

19. K. Brismar, M. Gutniak, G. Povoa, S. Werner, and K. Hall, Insulin regulates the 35 kDa IGF binding protein in patients with diabetes mellitus. J. Endocrinol. Invest. 11:599 (1988).

20. A.-M. Suikkari, V. A. Koivisto, E.-M. Rutanen, H. Yki-Jarvinen, S.-L. Karonen, and M. Seppala, Insulin regulates the serum levels of low molecular weight insulin-like growth factor-binding protein. J. Clin. Endocrinol. Metab. 66:266 (1988).

21. T. G. Unterman, D. T. Oehler, and R. E. Becker, Identification of a type 1 insulin-like growth factor binding protein (IGF BP) in serum from rats with diabetes mellitus. Biochem. Biophys. Res. Commun. 163:882 (1989).

22. A.-M. Suikkari, V. A. Koivisto, R. Koistinen, M. Seppala, and H. Yki-Jarvinen, Dose-response characteristics for suppression of low molecular weight plasma insulin-like growth factor-binding protein by insulin. J. Clin. Endocrinol. Metab. 68:135 (1989).

23. G. T. Ooi, C. C. Orlowski, A. L. Brown, R. E. Becker, T. G. Unterman, and M. M. Rechler, Different tissue distribution and hormonal regulation of messenger RNAs encoding rat insulin-like growth factor-binding proteins-1 and -2. Mol. Endocrinol. 4:321 (1990).

24. L. J. Murphy, C. Seneviratne, G. Ballejot, F. Croze, and T. G. Kennedy, Identification and characterization of a rat decidual insulin-like growth factor-binding protein complementary DNA. Mol. Endocrinol. 4:329 (1990).

25. National Research Council. Nutrient requirements of the laboratory rat, in:"Nutrient Requirements of Laboratory Animals, 3rd ed", p.23. National Academy of Science, Washington, DC (1978).

26. D. J. Knauer and G. L. Smith, Inhibition of biological activity of multiplication-stimulating activity by binding to its carrier protein. Proc. Natl. Acad. Sci. USA. 77:7252 (1980).

27. O. Ritvos, T. Ranta, J. Jalkanen, A.-M. Suikkari, R. Voutilainen, H. Bohn, and E.-M. Rutanen, Insulin-like growth factor (IGF) binding protein from human decidua inhibits the binding and biological action of IGF-I in cultured chondriocarcinoma cells. Endocrinology 122:2150 (1988).

28. R. G. Elgin, W. H. Busby, Jr. and D. R. Clemmons, An insulin-like growth factor (IGF) binding protein enhances the biologic response to IGF-I. Proc. Natl. Acad. Sci. USA. 84:3254 (1987).

29. W. H. Busby, P. Hossenlopp, M. Binoux, and D. R. Clemmons, Purified preparations of the amniotic fluid-derived insulin-like growth factor-binding protein contain multimeric forms that are biologically active. Endocrinology 125:773 (1989).

30. W. F. Blum, E. W. Jenne, F. Reppin, K, Kietzmann, M. B. Ranke, and J. R. Bierich, Insulin-like growth factor I (IGF-I)-binding protein complex is a better mitogen than free IGF-I. Endocrinology 125:766 (1989).

31. J. S. M. DeMellow, and R. C. Baxter, Growth hormone-dependent insulin-like growth factor (IGF) binding protein both inhibits and potentiates IGF-I stimulated DNA synthesis in human skin fibroblasts. Biochem. Biophys. Res. Commun. 156:199 (1988).

PRESENCE OF INSULIN-LIKE GROWTH FACTORS AND THEIR BINDING PROTEINS IN RAT MILK

Anthony F. Philipps, Jean M. Wilson, Radhakrishna Rao, David M. McCracken, and Otakar Koldovsky

Department of Pediatrics and Anatomy
AHSC, The Children's Research Center
1501 N. Campbell Avenue
Tucson, Arizona 85724

Introduction

The insulin-like growth factors (IGF's) are potent mitogens in a variety of mammalian cell types *in vitro* and *in vivo* (1-3). Because of the rapid growth rate during the perinatal period, recent interest has focussed on the IGF's as possible modulators of fetal and neonatal somatic and organ growth. Significant concentrations of IGF-1 and IGF-2 have been observed in fetal blood in a variety of mammalian species (4-6). In addition, recent demonstration of active IGF synthesis using probes for IGF mRNA in fetal and neonatal tissues has led credence to this hypothesis. Active synthesis and expression of Type 1 and 2 IGF receptors has been documented for a wide variety of cell types in several animal species in fetal life as well (7-8). Interestingly, although the production of IGF receptors in developing gastrointestinal (GI) tissues is high (9), expression of IGF specific mRNA is low in these tissues in comparison to that found in other organs such as liver and brain (7, 17) Since, in the neonate, GI tissues represent a rapidly dividing and differentiating cell mass, the possibility exists that exogenously derived IGF's might play a role in postnatal GI growth. IGF's are also found in most biological fluids tightly bound to at least several molecular weight species of binding proteins (BP's). Whether these BP's act to carry the IGF's to distant sites, fulfilling an endocrine function; act as a reservoir of IGF; or act to protect IGF from proteolytic degradation is unclear at present.

For these reasons, several investigators have examined milks from various species, including human, for their concentrations of IGF's and their binding proteins. Concentrations of IGF-1 are appreciable in several species studied, with bovine colostrum containing levels approaching 1 µg/ml (10). IGF-2 levels in milk have been less well studied, but appeared to be 2-5 fold higher than concentrations of IGF-1 in the few reports identified (11,12). Commercial infant formulas have virtually no IGF's present (unpublished data), but do contain some IGF-binding proteins.

Since the suckling rat represents a rapidly growing neonatal mammal that feeds exclusively on milk for the first three weeks of postnatal life, and in which developmental changes in IGF synthesis have been delineated, we sought to use this species to determine whether or not breast milk IGF's might be important stimulators of gut development. Thus, the characterization of milk derived IGF's and their binding proteins was an appropriate first step. We also sought to characterize relative amounts and developmental changes in IGF-BP's to determine whether or not these BP's could potentially serve a carrier or protective function in milk.

METHODS

Milking

Sprague-Dawley rats of timed gestation were observed for parturition (day "0"). Eight dams were selected with post delivery times from 2 through 18 days. Each dam selected was injected with 2 I.U. of oxytocin (Sigma Chemicals, St. Louis, MO) 5 minutes prior to milking. Light sedation with ketamine was also given approximately 15 minutes before milking. Milk was expressed by hand using a gentle squeezing motion of thumb and forefinger. Milk from each animal was placed in polyethylene tubes and stored whole at -20°C until time of assay. Milk was collected from each rat at only one milking because of the concerns regarding potential trauma on milk production and IGF concentration.

Biochemical Assays

Prior to study, the volume of each sample was determined before and after fat removal. Each sample was centrifuged at 3000 rpm, 15 minutes, 4°C and the water soluble portion ("cytosol") removed using a pipette. Protein concentration was determined with standard methods (Sigma Kit, St. Louis, MO).

IGF's were assayed from frozen rat milk samples of cytosol. Radio labelled IGF-1, when mixed with whole rat milk, was recovered from cytosol almost completely (95% recovery) after centrifugation. Because of the presence on G50 Sephadex chromatography of IGF-BP's, it was necessary to perform acid-gel chromatography of all specimens prior to IGF assay in order to separate BP's from the IGF's (13). Reconstituted fractions were then assayed for IGF-1 and IGF-2 activity. IGF-1 and 2 assays utilized human recombinant IGF's (M. Lake, M.D., Kabi Gen, Stockholm, Sweden) for both standard and radio ligand. The IGF-1 radio immunoassay (RIA-IGF-1) also utilized rabbit antihuman IGF-1 immunoglobulin (kindly supplied by S. Raiti, National Pituitary and Hormone Program, Baltimore, MD) as first antibody.

IGF-2 concentrations were measured by radio-receptor assay (RRA-IGF-2), as previously reported (13). In both assays, serial dilutions of a standard rat serum and milk sample yielded curves parallel to IGF standards.

Unextracted milk cytosol specimens were examined for binding protein activity using G-150 Sephadex chromatography after incubation with ^{125}I-IGF-1 . Subsequent specimens (10 ml) from milk of postnatal days 2-18 were subjected to SDS-PAGE (sodium dodecyl sulfate polyacrylamide gel electrophoresis-non reducing conditions) in order to separate IGF associated binding proteins by molecular weight. After Western blot transfer to nylon membranes, overnight incubation of identical preparations in 2 x 10^5 cpm/ml of ^{125}I-IGF-1 or ^{125}I-IGF-2 (700 - 900 Ci/mmole specific (5ml) activity) was performed (ligand blot method (14-15). Membranes were subjected to autoradiography and density of individual bands assessed by counting in a gamma counter. Results of SDS-PAGE studies are expressed as % activity found in binding of ^{125}I-IGF-1 or ^{125}I-IGF-2 to SDS-PAGE of a standard adult rat serum (1 μl)(in order to counter variation in labelling activity and sample volume). IGF concentrations are presented as mean±SEM.

Results

Protein concentrations of individual specimens were similar and ranged between 45 and 52 gm/L. No changes in protein concentration were observed related to postnatal age. Rat milk samples exposed to acid gel chromatography exhibited significant peaks of activity for RIA-IGF-1 and RRA-IGF-2 in the 7-8K molecular weight range. These peaks were also in a position identical to human r-^{125}I-IGF-1 or -2 when placed over the same columns. A small larger molecular weight IGF-2 peak was also apparent.

TABLE 1

IGF Concentrations in Rat Milk and Other Species

	RIA-IGF-1 ng/ml	RRA-IGF-2 ng/ml	RIA-IGF-2 ng/ml
Rat Milk			
d_{2-18}	16.0±6.9	105.7±30.0	
Bovine Milk	3.5±0.7 (ref 11)	-	106.6±9 (ref 11)
Bovine Milk			
Colostrum	300 (ref 10)	-	-
d_{1-5}	75	-	-
$>d_5$	30 - 60	-	-
Human Milk			
d_{1-9}	17.6 (ref 22)	-	-
$>d_9$	6 - 8	-	-

Rat milk of postnatal age 2-18 days contained concentrations of RIA-IGF-1 of 16.0±6.9 ng/ml (Table) with a range from 3.0 to 56.0 ng/ml. Fractions also assayed for IGF-2 activity detected concentrations in these same specimens of RRA-IGF-2 of 105.7±30.0 ng/ml with a range of 36.4 to 163.0 ng/ml. There were tendencies for RIA-IGF-1 concentrations to fall with advancing postnatal age of milk specimens and of RRA-IGF-2 concentrations to rise postnatally. However, no statistically significant changes were noted when compared to postnatal age of specimen.

Daily IGF intake could be extrapolated from known milk intake at 10 and 18 days postnatally (6 and 8 ml/day, respectively (16)) and mean IGF concentration. Thus at 10 and 18 days of age, suckling rat IGF-1 ingestion would approximate 100 and 150 ng/day, respectively. Similar calculated values for IGF-2 equalled approximately 500 and 1000 ng/day at 10 and 18 days, respectively.

Figure 1 shows a typical autoradiograph of rat milk cytosols separated by SDS-PAGE and stained against ^{125}I-IGF-1 (A) or ^{125}I-IGF-2 (B). Vertical numbers represent molecular weight standards (x 10^3). Adult rat serum (on far right, "s") exhibited at least 3 families of bands consistent with IGF-binding proteins previously identified. (16). These correspond in molecular weight to BP3 (39-45K), BP2 (30-34K) and BP1 (24K). However, rat milk contained only 2 major bands consistent with BP2 and BP3 at all postnatal ages studied. The ability of BP's to bind to ^{125}I-IGF's in milk was approximately 10% of that in a similar volume of adult rat serum. Overall, approximately twice as much ^{125}I-IGF-2 bound to rat milk than did ^{125}I-IGF-1 when corrected for specific activity of ligand and sample protein concentration.

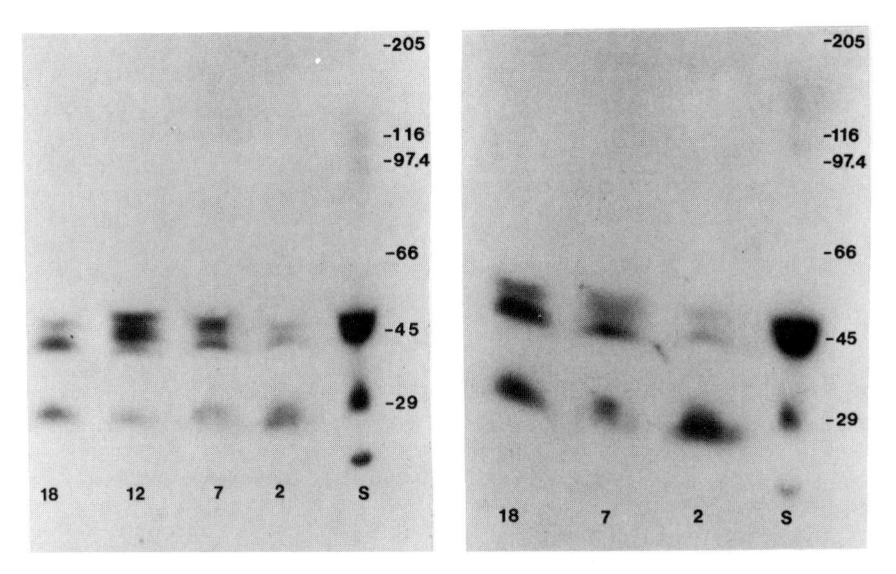

Figure 1 (A) (B)
SDS - Page of rat milk of varying postnatal age (d 2-18) and adult rat serum ("S") stained by ligand blot against (A) ^{125}I-IGF-1 and (B) ^{125} I-IGF-2.

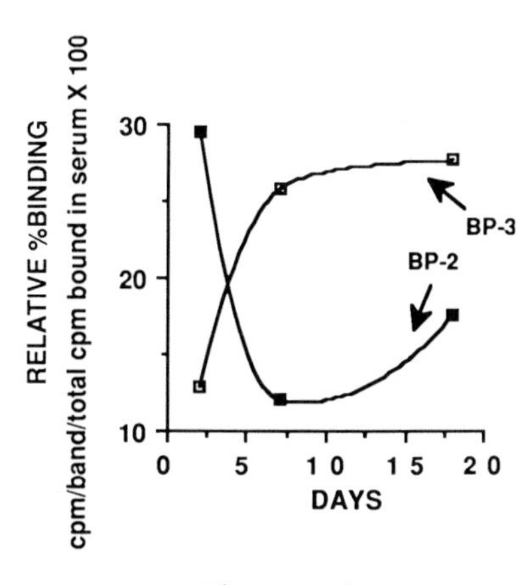

Figure 2

Postnatal changes in IGF - binding protein activity. When expressed as percent of binding of ^{125} I-IGF-2 to a standard serum.

$$\left(\frac{\text{CPM per band in milk specimen}}{\text{total cpm in sera standard}} \times 100\right)$$

When the activities of bound ^{125}I-IGF-2 in the BP-2 and BP-3 peaks of each milk specimen were calculated as a percentage of cpm bound to a standard rat serum (Figure 2), a decrement in BP-2 activity but a rise in BP-3 activity were observed. Similar, but less prominent changes were observed for activity after ligand-blots of milk with ^{125}I-IGF-1.

Maternal sera of similar postnatal ages was also subjected to SDS-PAGE and ligand-blot with ^{125}I-IGF-1 and 2. Bands found were similar to those noted in nonpregnant adults with little change in activity of specific bands between days 2 and 18.

Discussion

The insulin-like growth factors (IGF's) may be of importance in regulating growth in the mammalian fetus and neonate. Both cellular receptors for IGF's as well as IGF specific mRNA's have been detected in at least several species during the perinatal period (7, 9, 17). In addition, administration of artificial IGF to growing neonatal rats increased somatic and skeletal growth (18). Certain organs, such as neonatal stomach and intestine possess large concentrations of IGF receptors, but appear to have relatively less IGF synthetic activity when compared to other organs (7, 9, 17). Thus, the possibility exists that exogenously derived IGF's, such as might be present in breast milk, might play specific roles in neonatal development.

We have shown that significant concentrations of IGF-1 and IGF-2 exist in rat milk. These concentrations are similar to those found in milks of other species (Table) and suggest the possible usefulness of the suckling rat as a model for studying the fate and effects of enterally derived IGF's in this species.

IGF-related binding proteins constitute a major reservoir of IGF in mammalian serum (19). Although activity in milk appears less than in serum, significant quantities of at least two major BP's were present in rat milk. Despite no changes in serum IGF BP's, we found an age related increase in activity of BP-3. Several studies have documented significant quantities of BP-3 in milk of other species (10,20,21) with definite increased specificity for IGF-2 over IGF-1. However, age related changes in milk-borne IGF-BP's have not been addressed in other species as yet. Whether or not these carrier proteins function to stabilize IGF's in milk or in the neonatal intestine is not clear. However, the documentation of daily injestion of significant quantities of IGF-1 and IGF-2 by the suckling rat suggest that this intake may be of biological importance prior to weaning.

References

1. Sara, V.R. and Hall, K. Insulin - Like Growth Factors and Their Binding Proteins. Physiol. Rev.. 70:591 - 614, 1990

2. D'Ercole, A.J. Somatomedins/Insulin-Like Growth Factors and Fetal Growth. J.Develop. Physiol. 9: 481-495, 1987.

3. Hill,J.J. Crace,C.J., Strain, A.J. and Milner, R.D.G. Regulation of Amino Acid Uptake and Deoxyribonucleic Acid Synthesis in Isolated Human Fetal Fibroblasts and Myoblasts: Effect of Human Placental Lactogen, Somatomedin - C, Multiplication - Stimulating Activity, and Insulin. J. Clin. Endocrin. Metab.. 62: 753 - 760, 1986.

4. D'Ercole, A.J., Applewhite, G.T., and Underwood, L.E. Evidence That Somatomedin is Synthesized by Muliple Tissues in the Fetus. Develop. Biol, 75: 315-328, 1980.

5. Lund, P.K., Moats-Staats, B.M., Hynes, M.A. Simmons, J.G., Jansen, M., D'Ercole, A.J. and VanWyk, J.J., Somatomedin - c/Insulin-Like Growth Factor - 1 and Insulin-Like Growth Factor - 2 MRNAS in Rat Fetal and Adult Tissues. J. Biol. Chem. 261: 14539-14544, 1986.

6. D'Ercole, A.J. and Underwood, L.E. Ontogeny of Somatomedin During Development in the Mouse. Develop. Biol. 79: 33-45, 1980

7. Han, V.K.M., Lund, P.K., Lee, D.C., and D'Ercole, A.J. Expression of Somatomedin/Insulin-Like Growth Factor Messenger Ribonucleic Acids in the Human Fetus: Identification, Characterization, and Tissue Distribution. J.Clin. Endocrinol Metab. 66: 422-429, 1988

8. Young, G.P., Taranto, T.M., Jonas, H.A., Cox, A.J., Hogg, A. and Werther,G.A. Insulin-Like Growth Factors and the Developing and Mature Rat Small Intestine: Receptors and Biological Actions. Digestion 46 (suppl.): 240 - 252, 1990.

9. Werner, H., Woloschak, M., Adams M., Shen-Orr,Z, Roberts, C.T., Jr., and LeRoith, D. Developmental Regulation of the Rat Insulin-Like Growth Factor Receptor Gene.l Proc. Natl. Acad. Sci. 86:7451-5455, 1989

10. Campbell, P.G. and Baumrucher, C.R. Insulin-Like Growth Factor -1 and its Association With Binding Proteins in Bovine Milk. J. Endocrinology 120: 21-29, 1989

11. Francis, G.L., Upton, F.M., Ballard, F.J., McNeil, K.A., and Wallace, J.C. Insulin-Like Growth Factors 1 and 2 in Bovine Colostrum. Biochemical Journal 251:95-103, 1988.

12. Juskevich,J.C. and Guyer, C.G., Bovine Growth Hormomne: Human Food Safety Evaluation. Science 249: 875-885, 1990

13. Philipps, A., Drakenberg, K., Persson, B., Sjogren, B., Eklof, A.C., Hall, K and Sara, V. The Effects of Altered Nutritional Status Upon Insulin-Like Growth Factors and Their Binding Proteins in Neonatal Rats. Pediatric Res. 26: 128-134, 1989

14. Hossenlopp, P., Seurin, D., Segovia-Quinson, B. Hardouin, S., Binous, M. Analysis of Serum Insulin-Like Growth Factor Binding Proteins and Competitive Binding Studies. Anal. Biochem. 154:138-143, 1986.

15. Clemmons, D.R., Thissen, J.P., Maes, M., Ketelslegers, J.M., and Underwood,L.E., Insulin-Like Growth Factor -1 Infusion Into Hypophysectomized or Protein Deprived Rats Induces Specific IGF-Binding Proteins in Serum. Endocrinology 125: 2967-2972, 1989

16. Hahn, P. and Koldovsky, O. Utilization of Nutrients During Postnatal Development. Pergamon Press, Oxford, London, 1967.

17. Schober, D.A., Simmen, F.A., Hadsell, D.L., and

Baumrucher, C.R. Perinatal Expression of Type I IGF Receptors in Porcine Small Intestine. Endocrinology 126: 112-1132, 1990

18. Philipps, A.F., Persson, B., Hall,K., Lake, M., Skottner, A., Sanengen, T., and Sara, V.R., The Effects of Biosynthetic Insulin-Like Growth Factor - 1 Supplementation on Somatic Growth, Maturation, and Erythropoiesis in the Neonatal Rat. Pediatr. Res. 23: 298-305, 1988.

19. Ballard, J., Baxter, R., Binoux, M., Clemmons, D., Drop, S., Hall, K., Hintz, R., Rechler, M., Rutanen, E., and Schwander, J., On the Nomenclature of the IGF Binding Proteins. Acta Endocrinol. (Copenh.) 121: 751-752, 1989.

20. Baxter, R.C., Zatlsman, Z., Turtle, J.R., Immunoreactive Somatomedin - C/Insulin-Like Growth Factor I and its Binding Protein in Human Milk. J. Endocrinol Metab 58:955 - 959, 1984

21. Hodgkinson, S.C., Moore, L., Napier, J.R., Davis, S.R., Bass, J.J. and Gluckman, P.D., Characterization of Insulin-Like Growth Factor Binding Proteins in Ovine Tissue Fluids. J. Endocrinol 120: 429-438, 1989.

MOLECULAR HETEROGENEITY OF INSULIN RECEPTORS IN RAT TISSUES

Barry J. Goldstein, and Alana L. Dudley

Research Division, Joslin Diabetes Center, and
Department of Medicine, Brigham and Women's Hospital
Harvard Medical School, Boston, MA 02215

INTRODUCTION

Heterogeneity in the structure and function of the insulin receptor in various tissues and animal species has recently been reported by several laboratories. Our laboratory has been interested in determining the molecular basis for this variation in the insulin receptor, which may be involved in the regulation of certain actions of insulin in target cells. This brief presentation will summarize our recent work in this area which has used a variety of molecular approaches to study the structure of the coding region of insulin receptor mRNA transcripts in several rat tissues.

STRUCTURE AND BIOSYNTHESIS OF THE INSULIN RECEPTOR

The insulin receptor is a plasma membrane glycoprotein that mediates the pleiotropic effects of insulin in target cells (1). The mature insulin receptor has a native heterotetrameric structure consisting of two α-subunits which bind insulin and two β-subunits which have tyrosine kinase activity toward the receptor itself as well as exogenous substrates (2). Insulin induces a rapid autophosphorylation of the receptor β-subunit which activates that receptor kinase and causes conformational changes in the receptor structure. However, the mechanism of signal transduction by the insulin receptor into divergent metabolic pathways is not understood (1-3).

The human insulin receptor is encoded on chromosome 19 by a single gene that spans more than 130 kb and consists of 22 exon segments (4,5). Transcription of the insulin receptor gene in rodent and human tissues produces multiple large mRNA species which are 1.5 to 5.4 kb larger than the 4.2 kb cDNA coding region, primarily due to extensive tracts of 3' untranslated RNA (6-10). The proreceptor is synthesized in cells as a single polypeptide chain of approximately 155 kDa which contains co-linear α and β subunit sequences (6,7,11). This receptor precursor undergoes glycosylation and a complex series of post-translational processing events which include cleavage into the component subunits prior to its assembly as the mature α_2/β_2 holoreceptor and its insertion into the plasma membrane (12).

Molecular Biology and Physiology of Insulin and Insulin-Like Growth Factors
Edited by M.K. Raizada and D. LeRoith, Plenum Press, New York, 1991

EVIDENCE FOR HETEROGENEITY OF THE INSULIN RECEPTOR

Despite the presence of a single allelic form of the insulin receptor gene, naturally-occurring receptor subtypes with different functional properties and altered structural features have recently been identified. Since insulin affects several metabolic pathways in its target tissues, it has been postulated that subtypes of the insulin receptor may play a role in the regulation or transmission of particular actions of insulin in specific cell types.

Functional Studies of the Insulin Receptor Kinase

The kinetic properties of the tyrosine kinase activity of insulin receptors from various sources have been directly compared in a few studies. For example, the autophosphorylation activity of the human muscle insulin receptor was significantly greater than the activity observed with human adipose tissue or liver insulin receptors (13). The rat liver insulin receptor has been shown to have an insulin-stimulated phosphotransferase activity that was 2.8-fold more rapid than the insulin receptor from human placenta, apparently due to increased V_{max} of the intrinsic kinase activity (14). Also, the ATP K_m for autophosphorylation has been shown to be 5-fold greater for insulin receptors from rat compared to insulin receptors from human cells (15).

Heterogeneity in the Receptor Structure

Rat liver and muscle insulin receptor α-subunits had respective M_r values of 135,000 and 131,000 in one report, and glycosidase treatment did not affect the apparent size difference, suggesting that differences might exist in the polypeptide chain of the receptor in these tissues (16). In contrast, another study reported that the size difference in the insulin receptor α-subunit in human liver and muscle was abolished by neuraminidase treatment (13). Both subunits of the rat brain insulin receptor have been shown to be smaller than those of the receptor in rat adipocytes and liver by approximately 7 to 10 kDa (17,18). Although this difference may be in part due to alterations in the extent of N-linked-glycosylation (19), antigenic differences were retained between the human brain and placental insulin receptors after neuraminidase treatment, suggesting that tissue differences might exist in the receptor polypeptide sequence (20). The insensitivity of most protein electrophoretic techniques in detecting small differences in molecular mass has made it difficult to conclude whether subtle alterations in receptor primary structure exist that may have important influences on receptor function in various tissues.

Alternative Splicing of Insulin Receptor mRNAs

Direct evidence for structural variation in the insulin receptor polypeptide chain was found in the two initial cDNA sequences for the human placental insulin receptor, which differed by 36 bases in the distal α-subunit sequence near the subunit cleavage site (6,7). This mRNA heterogeneity was shown to arise by tissue-specific alternative splicing of Exon 11 of the human insulin receptor gene that apparently alters the mature receptor structure by 12 amino acids (21,22). A recent report indicated that the receptor isoform lacking Exon 11 has approximately a 2-fold higher affinity for insulin than the receptor which contains Exon 11 sequences when expressed in heterologous fibroblast cells (23). However, in preliminary studies from another laboratory, the small difference in insulin binding between the two types of insulin receptors was not statistically significant (24). A direct comparison of the activation of the receptor tyrosine kinase by insulin for these two receptor isoforms has not yet been reported.

MOLECULAR STUDIES OF INSULIN RECEPTOR mRNA IN RAT TISSUES

Translation *in vitro*

In studies of the initial biosynthesis of the insulin receptor precursor, we provided further evidence for structural insulin receptor heterogeneity by *in vitro* translation of rat liver poly(A)$^+$ RNA which produces two immunoprecipitable insulin receptor precursor species in the range of 160-164 kDa (11). Further refinement of these results was

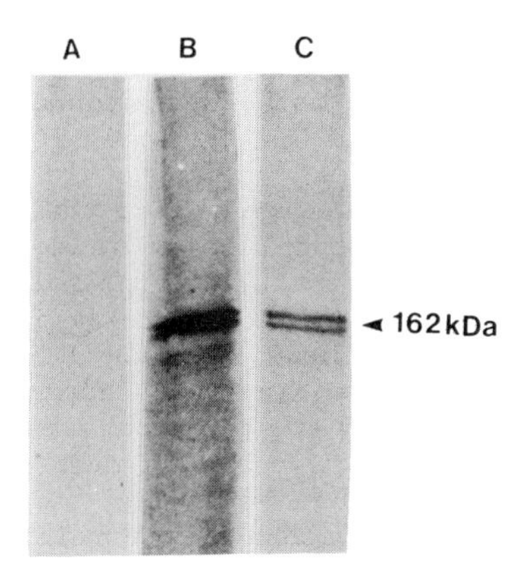

Figure 1. Cell-free translation of rat liver insulin receptor precursors. Rat liver poly(A)+ RNA was translated in a reticulocyte lysate system and the products were immunoprecipitated with rat insulin receptor antiserum and analyzed by gel electrophoresis and fluorography as described (11). Lane A, no RNA control; lane B, unfractionated poly(A)+ RNA; lane C, prior to translation, rat liver poly(A)+ RNA was fractionated in a methylmercury-agarose gel (25).

recently achieved by fractionation of rat liver poly(A)$^+$ RNA on a 1% low melting temperature agarose gel containing 10 mM methylmercuric hydroxide prior to cell-free translation (25). After electrophoresis, gel fractions containing insulin receptor mRNA were identified by visualization of RNA size markers run in parallel lanes. Aliquots of melted gel slices were then analyzed by *in vitro* translation in a reticulocyte lysate with labeled cysteine, followed by immunoprecipitation with a rat liver insulin receptor antiserum, SDS-polyacrylamide gel electrophoresis and fluorography. This denaturation and partial purification of insulin receptor mRNA apparently enhanced the translation and resolution of the two immunoprecipitable proreceptor forms (Figure 1).

To insure that the multiple rat liver proreceptors did not occur artifactually in the cell-free system by a mechanism such as limited proteolysis, synthetic insulin receptor mRNA templates were used in control translation assays. These uniform insulin receptor mRNAs were prepared by a transcription reaction with SP6 RNA polymerase (26) from a pGEM-SP72 plasmid construct containing the intact human insulin receptor cDNA (6) in sense orientation (pSP72.hIR) and translated directly in the cell-free system. Only a single receptor precursor of approximately 160 kDa was produced by the synthetic insulin receptor mRNAs that was further identified by its immunoprecipitation with a monoclonal antibody to the human insulin receptor (Figure 2). These results confirmed that the rat

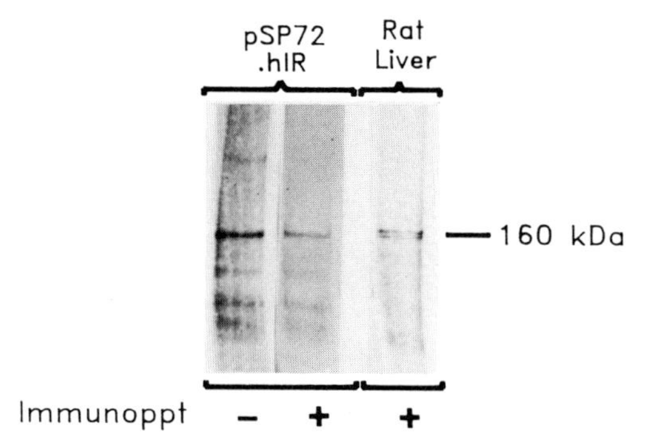

Figure 2. Cell-free translation of human insulin receptor mRNAs prepared by *in vitro* transcription with SP6 RNA polymerase. The intact coding region of the human insulin receptor (6) was subcloned into a pGEM-SP72 vector in sense orientation (pSP72.hIR). This plasmid was linearized and translated directly in the cell-free system (26). Translation products were analyzed directly as well as after immunoprecipitation with insulin receptor antibodies as indicated. Products of cell-free translation of rat liver poly(A)+ RNA run on the same gel are also shown for comparison.

liver insulin receptor mRNAs contained heterogeneous sequence information in their coding regions to produce structurally variant insulin proreceptor forms that do not arise simply as an artifact of the cell-free translation system.

Studies with hepatoma cells treated with tunicamycin to block core glycosylation and further processing of the proreceptor provided evidence that both proreceptor forms also occurred *in vivo* (11). The studies in metabolically-labeled cells and the results of the cell-free translation experiments suggested that subtle variations in the receptor protein sequence were detectable by gel electrophoresis of the unmodified polypeptide chains of the proreceptor. These proreceptor isoforms most likely arose from two types of insulin receptor mRNA with heterogeneous sequences in their coding regions.

Molecular Cloning of the Rat Liver Insulin Receptor cDNA

To investigate further the occurrence of structural variation in insulin receptor mRNA sequences in a species and tissue type that had not yet been characterized, we obtained the complete primary sequence of a rat liver insulin proreceptor by molecular cloning (27). The cDNA consisted of 5397 bp, including 347 bp of 5′ untranslated domain and an open reading frame of 4149 bp.

In comparison with the known human insulin receptor sequences, scattered amino acid changes occur starting with position 139, with a notable difference being the inclusion of a threonine and a proline residue in the rat insulin receptor at positions 547 and 548, respectively (27). These two additional amino acids also occur in the mouse insulin receptor sequence (28). A cluster of amino acid alterations is also found in the N-terminal portion of the rat insulin receptor β-subunit sequence, immediately following the subunit cleavage site. Compared to the human receptor, in this region 12 differences are present in 28 residues from position 740 to 767. Interestingly, the transmembrane segment, from residues 932 to 954, has full sequence identity with the human and mouse sequences. This conservation might suggest a potential role for the transmembrane segment itself in signalling by the insulin receptor.

Conservation of Exon 11 Domain Alternative Splicing: The amino acid sequence of the rat insulin receptor α-subunit near the subunit cleavage site contained sufficient residues to align with the human sequences which contain the 12 amino acids encoded by the alternatively-spliced domain of Exon 11. Using the polymerase chain reaction (PCR) to amplify a cDNA segment corresponding to this mRNA region in several rat tissues, 2 products that differed by 36 nucleotides were observed with a tissue-specific distribution (27). Interestingly, the relative abundance of rat insulin receptor mRNA transcripts that contained the additional residues in brain 17 $\pm$ 5%, kidney 66 $\pm$ 3%, liver 85 $\pm$ 4%, placenta 44 $\pm$ 2% and spleen 6 $\pm$ 3% (mean $\pm$ SEM, n=4) closely matched data reported for human tissues (21,22).

The structure of the two PCR products was further evaluated by direct sequence analysis of the amplified cDNA. These data revealed that the mRNA splice sites flanked the segment encoding amino acids 719-730, and interrupt arginine codons in precisely identical positions to those found for Exon 11 splicing in the human receptor (Figure 3).

```
     Rat IR                 *                   *
Pro Arg Lys Thr Ser Ser Gly Asn Gly Ala Glu Asp Thr Arg Pro Ser Arg Lys
CCC AGA AAA ACC TCT TCA GGC AAT GGT GCT GAG GAC ACT AGG CCA TCC CGA AAG

     Human IR Exon 11       *                   *
Pro Arg Lys Thr Ser Ser Gly Thr Gly Ala Glu Asp Pro Arg Pro Ser Arg Lys
CCC AGA AAA ACC TCT TCA GGC ACT GGT GCC GAG GAC CCT AGG CCA TCT CGG AAA
```

Figure 3: Conservation of mRNA splice sites in Rat and Human insulin receptors in the region of Human Exon 11. The alternatively-spliced segment is underlined. Residues that are not identical are highlighted with an asterisk.

Further comparison with the reported amino acid sequence for a Drosophila insulin receptor homolog (29) revealed additional residues in this domain, near the subunit cleavage site, that were similar in organization to the larger form of the vertebrate insulin proreceptor (Figure 4).

The strict quantitative and structural conservation of this alternative mRNA splicing event through the divergent evolution of these species provides further evidence that these two forms of the insulin receptor may have particular functional or regulatory roles in different tissues. The extracellular localization of the domain encoded by Exon 11 has lead to studies that are closely examining the kinetics of insulin binding and the interaction of insulin with the extracellular domain of the receptor (23,24). Perhaps the presence of these additional amino acids in the receptor at the cell surface affects the aggregation of oligomeric forms of the receptor or the association between the insulin receptor and other plasma membrane macromolecules involved in regulation of the insulin receptor or its transmembrane signalling activity.

Use of RNA Heteroduplex Mapping to Detect Receptor mRNA Splicing Events

The available data on insulin receptor structural heterogeneity between tissues and the occurrence of a well-characterized alternative mRNA splicing event suggest that additional variants of the insulin receptor mRNA structure might occur. Cloning and sequence analysis of the rat liver insulin receptor has enabled us to examine the entire coding region of the insulin receptor mRNA for tissue-specific sequence variation. We are currently using RNA heteroduplex mapping to detect variations of the insulin receptor mRNA structure (30). This method uses highly labeled antisense RNA probes from plasmids containing segments of the rat insulin receptor cDNA which are hybridized to mRNA from various rat tissues. The heteroduplexes are then digested with a mixture of RNAse A and RNAse T1 and the labeled probe fragments are analyzed by denaturing polyacrylamide gel electrophoresis. Perfectly matched mRNA:RNA probe hybrids are resistant to cleavage by the ribonucleases. However, regions of probe mismatch, such as exon insertions or deletions in the tissue poly(A)$^+$ RNA, will be cut by the nucleases and the size of the resulting labeled fragments will indicate where the duplex inhomogeneity occurred.

In initial studies, we have validated this technique using a 686 base probe from the rat liver cDNA that contains the additional Exon 11 sequences (Figure 5). Cleavage products close to the expected size of approximately 443 and 207 bases were detected using mRNA from rat brain and muscle where the alternative form of the receptor RNA is known to occur. These RNA probe fragments are not observed with mRNA from rat liver since our previous studies have shown that 85% of the insulin

```
                       Human Exon 11
HIR-11: FEDYLHNVVFVP..R..............................PSRKRR..SLG....DV.GNV
HIR+11: FEDYLHNVVFVP..RKT...SSGT..GAE....DP.........RPSRKRR..SLG....DV.GNV
   RIR: FEDYLHNVVFVP..RKT...SSGN..GAE....DT.........RPSRKRR..SLE....EV.GNV
   DIR: FENALQNFIFVPNIRKSKNGSSDKSDGAEGAALDSNAIPNGGATNPSRRRRDVALEPELDDVEGSV
```

Figure 4: Optimal alignment of human (HIR), rat liver (RIR) and Drosophila (DIR) insulin receptor sequences in the region of the distal α-subunit, flanking the tetrabasic subunit cleavage site.

receptor transcripts in this tissue contain the additional 36 nucleotides. Thus, almost all of the liver mRNAs would exactly match the RNA probe sequence and be protected from cleavage.

CONCLUSIONS

Detailed recent studies of the insulin receptor have provided interesting data on subtle alterations in its structure and function that occur across species and in a tissue-specific fashion. As we begin to understand the potential role of certain domains of the insulin receptor in signalling to specific metabolic pathways (3), it will be important to fully characterize molecular alterations that occur in the insulin receptor mRNA as well as post-translational processing events that may contribute to this receptor heterogeneity. The use of sensitive techniques such as RNA heteroduplex mapping with the cloned rat liver insulin receptor cDNA will allow us to detect whether additional alterations in receptor primary structure exist among various tissues. Further characterization of these receptor isoforms will enable us to determine how they might influence receptor signalling in various type of target cells for insulin.

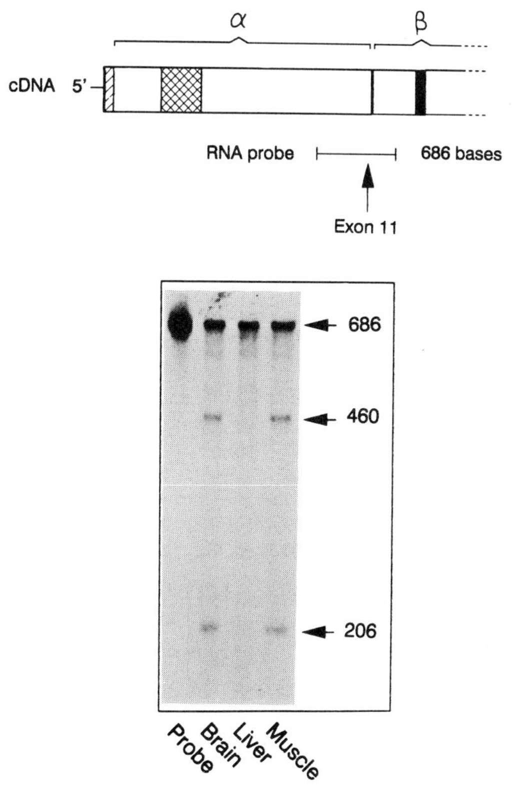

Figure 5. Ribonuclease heteroduplex mapping of insulin receptor mRNA structure in the region of Exon 11 using mRNA from rat brain, liver and muscle (30). The ^{32}P-labeled antisense RNA probe was prepared from a plasmid vector carrying a segment of the rat insulin receptor cDNA that spanned the subunit cleavage region as shown. The intact 686 base RNA probe, prior to hybridization and ribonuclease treatment, was included as a control along with samples generated by heteroduplex mapping with RNA from the rat tissues.

ACKNOWLEDGEMENTS

This work was supported by a Pilot and Feasibility Award from the Joslin Diabetes Endocrinology Research Center (DERC) Grant DK-36836, a research award from The Medical Foundation of Boston, and NIH Grant DK 31036.

REFERENCES

1. O.M. Rosen, After insulin binds, Science 237:1452 (1987).
2. C.R. Kahn, and M.F. White, The insulin receptor and the molecular mechanism of insulin action, J. Clin. Invest. 82:1151 (1988).
3. J.M. Olefsky, The insulin receptor: a multifunctional protein, Diabetes 39:1009 (1990).
4. T.L. Yang Feng, U. Francke, and A. Ullrich, Gene for human insulin receptor: localization to site on chromosome 19 involved in pre-B-cell leukemia, Science 228:728 (1985).
5. S. Seino, M. Seino, S. Nishi, and G.I. Bell, Structure of the human insulin receptor gene and characterization of its promoter, Proc. Natl. Acad. Sci. USA 86:114 (1989).
6. A. Ullrich, J.R. Bell, E.Y. Chen, R. Herrera, L.M. Petruzzelli, T.J. Dull, A. Gray, L. Coussens, Y-C. Liao, M. Tsubokawa, A. Mason, P.H. Seeburg, C. Grunfeld, O.M. Rosen, and J. Ramachandran, Human insulin receptor and its relationship to the tyrosine kinase family of oncogenes, Nature 313:756 (1985).
7. Y. Ebina, L. Ellis, K. Jarnagin, M. Edery, L. Graf, E. Clauser, J-H. Ou, F. Masiarz, Y.W. Kan, I.D. Goldfine, R.A. Roth, and W.J. Rutter, Human insulin receptor cDNA: the structural basis for hormone activated transmembrane signalling, Cell 40:747 (1985).
8. B.J. Goldstein, D. Muller-Wieland, and C.R. Kahn, Variation in insulin receptor mRNA expression in human and rodent tissues, Molecular Endocrinology 1:759 (1987).
9. B.J. Goldstein, and C.R. Kahn, Analysis of mRNA heterogeneity by ribonuclease H mapping: Application to the insulin receptor, Biochem. Biophys. Res. Commun. 159:664 (1989).
10. D.S. Tewari, D.M. Cook, and R.A. Taub, Characterization of the promoter region and 3' end of the human insulin receptor gene, J. Biol. Chem. 264:16238 (1989).
11. B.J. Goldstein, and C.R. Kahn, Initial processing of the insulin receptor precursor *in vivo* and *in vitro*, J. Biol. Chem. 263:12809 (1988).
12. P. Gorden, R. Arakaki, E. Collier, and J-L. Carpentier, Biosynthesis and regulation of the insulin receptor, Yale J. Biol. Med. 62:521 (1989).
13. J.F. Caro, S.M. Raju, M.K. Sinha, I.D. Goldfine, and G.L. Dohm, Heterogeneity of human liver, muscle, and adipose tissue insulin receptor, Biochem. Biophys. Res. Commun. 151:123 (1988).
14. T. O'Hare, and P.F. Pilch, Intrinsic kinase activity of the insulin receptor: the intact (α_2 β_2) insulin receptor from rat liver contains a kinase domain with greater intrinsic activity than the intact insulin receptor from human placenta, J. Biol. Chem. 264:602 (1989).
15. D.J. Brillon, R.R. Henry, H.H. Klein, J.M. Olefsky, and G.R. Freidenberg, Functional and structural differences in human and rat-derived insulin receptors: characterization of the ß-subunit kinase activity, Endocrinology 123:1837 (1988).
16. C.F. Burant, M.K. Treutelaar, N.E. Block, and M.G. Buse, Structural differences between liver- and muscle-derived insulin receptors in rats, J. Biol. Chem. 261:14361 (1986).

17. K.A. Heidenreich, N.R. Zahniser, P. Berhanu, D. Brandenburg, J.M. Olefsky, Structural differences between insulin receptors in the brain and peripheral target tissues, J. Biol. Chem. 258:8527 (1983).
18. W.L. Lowe, Jr., F.T. Boyd, D.W. Clarke, M.K. Raizada, C. Hart, and D. LeRoith, Development of brain insulin receptors: Structural and functional studies of insulin receptors from whole brain and primary cell cultures, Endocrinology 119:25 (1986).
19. K A. Heidenreich, and D. Brandenburg, Oligosaccharide heterogeneity of insulin receptors. Comparison of N-linked glycosylation of insulin receptors in adipocytes and brain, Endocrinology 118:1835 (1986).
20. R.A. Roth, D.O. Morgan, J. Beaudoin, and V. Sara, Purification and characterization of the human brain insulin receptor, J. Biol. Chem. 261:3753 (1986).
21. S. Seino, and G.I. Bell, Alternative splicing of human insulin receptor messenger RNA, Biochem. Biophys. Res. Commun. 159:312(1989).
22. D.E. Moller, A. Yokota, J.F. Caro, and J.S. Flier, Tissue-specific expression of two alternatively spliced insulin receptor mRNAs in man. Molecular Endocrinology 3:1263 (1989).
23. L. Mosthaf, K. Grako, T.J. Dull, L. Coussens, A. Ullrich, and D.A. McClain, Functionally distinct insulin receptors generated by tissue-specific alternative splicing, EMBO J. 9:2409 (1990).
24. J.L. Gu, R.M. Shymko, B. Wallach, J. Whittaker, and P. De Meyts, Receptor binding properties of the two natural variants of the insulin receptor, Diabetes 39:79A (1990).
25. T.L. Brandt, and P.B. Hackett, Characterization of messenger RNA by direct translation from agarose gels, Anal. Biochem. 135:401 (1983).
26. P.A. Krieg, and D.A. Melton, *In vitro* RNA synthesis with SP6 RNA polymerase, Methods Enzymol. 155:397 (1987).
27. B.J. Goldstein, and A.L. Dudley, The rat insulin receptor: primary structure and conservation of tissue-specific alternative messenger RNA splicing, Molecular Endocrinology 4:235 (1990).
28. J.R. Flores-Riveros, E. Sibley, T. Kastelic, and M.D. Lane, Substrate phosphorylation catalyzed by the insulin receptor tyrosine kinase: kinetic correlation to autophosphorylation of specific sites in the β subunit, J. Biol. Chem. 264:21557 (1989).
29. Y. Nishida, M. Hata, Y. Nishizuka, W.J. Rutter, and Y. Ebina, Cloning of a Drosophila cDNA encoding a polypeptide similar to the human insulin receptor precursor, Biochem. Biophys. Res. Commun. 141:474 (1986).
30. B.J. Goldstein, and C.R. Kahn, Insulin receptor messenger ribonucleic acid sequence alterations detected by ribonuclease cleavage in patients with syndromes of insulin resistance, J. Clin. Endocrinol. Metab. 69:15 (1989).

MUTATIONS IN THE INSULIN RECEPTOR GENE IN PATIENTS WITH GENETIC SYNDROMES OF INSULIN RESISTANCE

Simeon I. Taylor, Domenico Accili, Alessandro Cama, Hiroko Kadowaki, Takashi Kadowaki, Eiichi Imano, and Maria de la Luz Sierra

Diabetes Branch
National Institute of Diabetes, Digestive, and Kidney Diseases
National Institutes of Health
Bethesda, MD

INTRODUCTION

Insulin resistance contributes importantly to the pathogenesis of noninsulin-depedendent diabetes mellitus (NIDDM) (1, 2). Furthermore, in longitudinal studies, insulin resistance has been shown to be among the best predictors of future development of NIDDM (3,4). Therefore, we have focused upon genetic causes of insulin resistance in the hope that this will help us to identify genetic factors that predispose to the development of NIDDM. Because of the central role of the insulin receptor in mediating the first step in insulin action (5), we have begun by examining the insulin receptor gene in insulin resistant patients. We have selected patients who manifest an extreme degree of insulin resistance in the hope that the severe insulin resistance would be associated with major biochemical defects, thereby simplifying the task of identifying the molecular defect. In this review, we will summarize the mutations that have been identified in the insulin receptor genes of patients with extreme insulin resistance. Interestingly, multiple different types of mutations have been identified (5). In addition to elucidating the molecular genetics of human insulin resistance, these studies have begun to give new insights into structure-function relationships of the insulin receptor protein.

Genetic Syndromes of Insulin Resistance

Patients with inborn errors in the pathways of insulin action represent intriguing "experiments of nature" (5). The severity of the insulin resistance facilitates investigation of the biochemical and molecular mechanisms of insulin resistance. Although these patients manifest an extreme degree of resistance to the biological actions of insulin, many of the patients are not diabetic. In some patients, the levels of insulin rise to the point at which they are sufficiently high (often 10- to 100-fold above the normal range) to maintain

Molecular Biology and Physiology of Insulin and Insulin-Like Growth Factors
Edited by M.K. Raizada and D. LeRoith, Plenum Press, New York, 1991

relatively normal glucose tolerance. Nevertheless, some patients—especially the patients with the most severe degree of insulin resistance—develop fasting hyperglycemia and overt diabetes. Two mechanisms contribute to the development of hyperinsulinemia in insulin resistant patients. First, as reflected by the increase in the levels of C-peptide in plasma, the beta cell increases the rate of insulin secretion. Second, because receptor-mediated endocytosis is the principal route by which insulin is cleared from plasma, a decrease in the number of insulin receptors on the cell surface decreases the insulin clearance rate (6).

Two clinical features are commonly observed in patients with all of the syndromes of extreme insulin resistance, irrespective of the biochemical mechanism that causes the insulin resistance:

1. *Acanthosis Nigricans.* This is a hyperkeratotic, hyperpigmented skin lesion located primarily in skin folds such as the axillae and back of the neck. It tends to correlate with hyperinsulinemia. In patients with insulin resistance caused by anti-receptor autoantibodies, acanthosis nigricans waxes and wanes in association with the appearance and disappearance of the insulin resistance. This has led to the hypothesis that acanthosis nigricans may be caused by a "toxic" effect of hyperinsulinemia upon the skin (7).

2. *Hyperandrogenism.* Levels of plasma testosterone are commonly elevated in premenopausal women with extreme insulin resistance (8-10). The elevated levels of testosterone result from overproduction of testosterone by the ovaries. As with acanthosis nigricans, the elevated levels of testosterone correlate with hyperinsulinemia. Clinically, the elevated levels of testosterone are manifested as a syndrome of polycystic ovaries, oligomenorrhea, and hirsutism.

While all of the syndromes of extreme insulin resistance share some features in common, multiple distinct syndromes can be defined based upon the presence or absence of specific clinical features (5). For example, *type A extreme insulin resistance* is defined by the triad of insulin resistance, acanthosis nigricans, and hyperandrogenism in the absence of obesity or lipoatrophy (10). In *lipoatrophic diabetes*, there is atrophy of subcutaneous fat, hypertriglyceridemia, and fatty metamorphosis of liver (11). Patients with *leprechaunism* have multiple abnormal features, including intrauterine growth retardation and fasting hypoglycemia (12). The *Rabson-Mendenhall syndrome* is associated with abnormalities of teeth and nails and, reportedly, pineal hyperplasia (13, 14).

Subunit Structure of the Insulin Receptor

The binding of insulin to its receptor on the surface of the target cell is the first step in insulin action. The receptor is an oligomeric glycoprotein that consists of two α- and two β-subunits that are assembled into a heterotetrameric structure that is held together by both noncovalent interactions as well as disulfide bonds (Fig. 1). Despite the fact that the receptor consists of two different types of subunits, it is encoded by a single gene located on the short arm of chromosome 19 (15). Thus, the α- and β-subunits are produced by proteolytic cleavage of a single high molecular

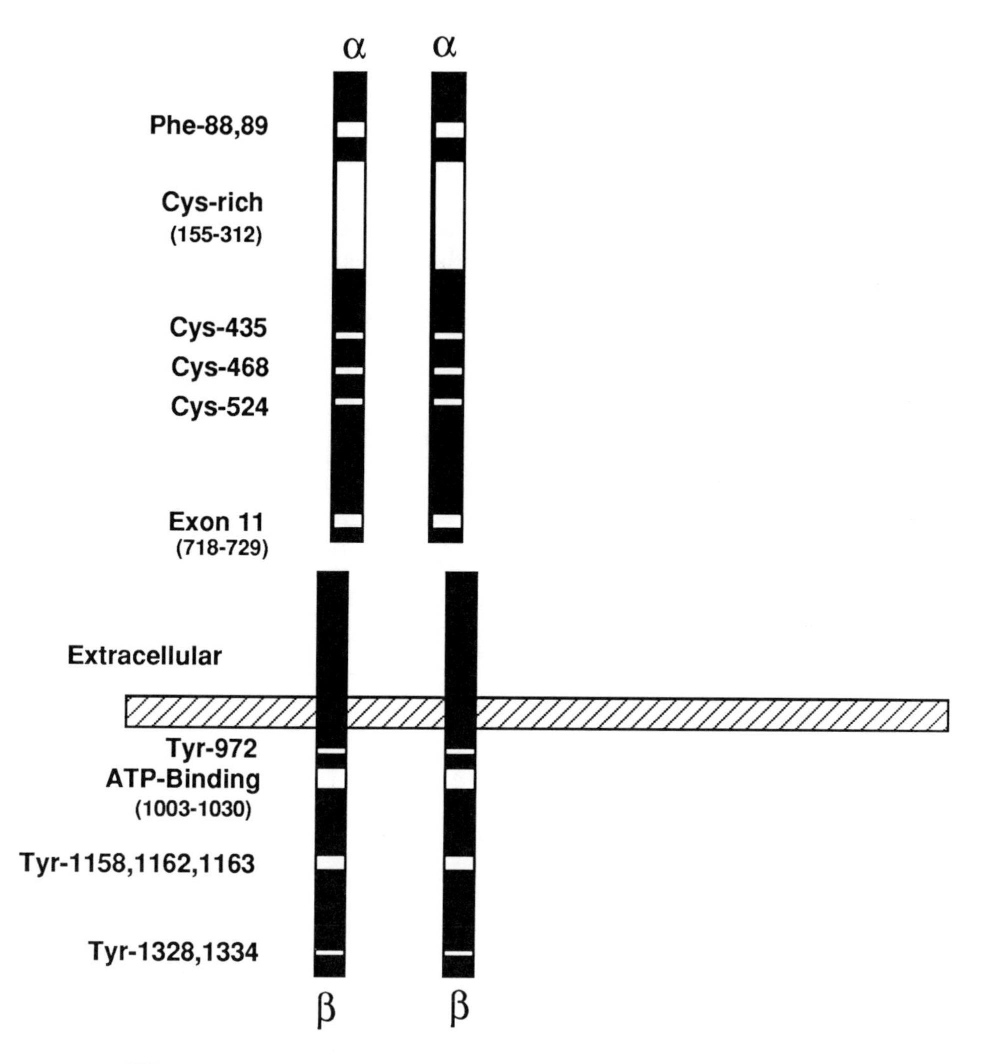

Figure 1. Structural map of the human insulin receptor.

Key structural landmarks are identified at the left of the drawing of the receptor. Phe[88], Phe[89] (16), and the cysteine-rich domain (17, 18) have all been implicated as playing a role in the insulin binding domain. Cys[435], Cys[468], and Cys[524] are candidates to contribute sulfhydryl groups for formation of the disulfide bonds between adjacent α-subunits (19). Exon 11 is an exon that has been described to undergo variable splicing (20-22). As described in the text, the five (23, 24) or six (25) tyrosine residues that are sites of autophosphorylation are indicated. The consensus sequence for an ATP binding domain is located between amino acid residues 1003-1030 (26).

weight precursor molecule ($M_r \approx 190,000$). The α--subunit ($M_r \approx 135,000$) is entirely extracellular and provides the insulin binding site (17, 18, 27, 28). The β-subunit ($M_r \approx 95,000$) is a transmembrane subunit that contains a single transmembrane domain (17, 18, 28). The β-subunit anchors the α-subunit to the plasma membrane. In addition, the intracellular portion of the β-subunit contains the tyrosine kinase domain of the receptor. When insulin binds to the extracellular domain of the receptor, this activates autophosphorylation of 5-6 tyrosine residues in the β-subunit by the receptor tyrosine kinase (23-25, 29, 30). Receptor autophosphorylation activates the receptor to phosphorylate other intracellular proteins. There is abundant evidence suggesting that activation of the receptor tyrosine kinase is necessary for the receptor to mediate insulin action (29-34).

Insulin Receptor Biosynthesis

The first step in insulin receptor biosynthesis is transcription of the gene. The transcript undergoes splicing prior to transport to the cytosol where the mRNA is translated by ribosomes on the rough endoplasmic reticulum. Subsequently, the receptor precursor undergoes multiple post-translational processing steps within the endoplasmic reticulum and Golgi apparatus. These post-translational processing steps include proteolytic cleavage of the receptor precursor into α- and β-subunits (17, 18, 35-37), N-linked glycosylation (35, 38), O-linked glycosylation (39, 40), and acylation (41). The mature receptor is transported to the cell surface where it is inserted into the plasma membrane. Insulin binds to the receptor on the cell surface, thereby activating the receptor tyrosine kinase. In addition, subsequent to insulin binding, the insulin receptor complex undergoes receptor-mediated endocytosis. Once within the cell, the receptor partitions between two alternate fates: recycling back to the cell surface for reutilization or intracellular degradation (possibly, within lysosomes). Clearly, this is an extremely complicated pathway. Defects in any one of these steps can impair insulin receptor function and cause insulin resistance. In the remainder of the chapter, we will review the various types of mutations that have been identified in the insulin receptor gene, and classify them according to the molecular mechanisms by which they impair the function of the insulin receptor.

MUTATIONS IN THE INSULIN RECEPTOR GENE

It is possible to classify mutations in the insulin receptor gene into at least five classes (Table I) (5). This classification, similar to that proposed by Brown and Goldstein for mutations in the low density lipoprotein receptor gene (42), is based upon the molecular mechanism whereby the mutation impairs the function of the receptor. (Note that some mutations may cause multiple defects in receptor function so that some mutations are classified in more than one class.)

Class 1. Decreased Rate of Receptor Biosynthesis

Premature Chain Termination Mutations. Several mutations have been identified that impair receptor biosynthesis. Four nonsense mutations have been identified: at codons 133, 672, 897, and 1000 (43-45). In addition, two deletion mutations have been identified that result in premature chain

Table I. Classification of mutations in the insulin receptor gene.

Mutations in the insulin receptor gene are classified according to the mechanism whereby they cause insulin resistance. The classification is based upon a modification of the classification proposed by Brown and Goldstein for mutations in the low density lipoprotein receptor gene (42). Some mutations (Lys^{15} and Val^{382}) are classified in two classes because they impair insulin receptor by more than one mechanism. Several mutations are listed as unclassified because sufficient data are not available to classify them.

Class 1. ***Decreased receptor biosynthesis.***

—Nonsense mutations [codons 133, 672, 897, and 1000 (43-45)]
—Unidentified cis-acting mutations that decrease mRNA levels (44)

Class 2. ***Impaired transport of receptors to plasma membrane.***

—Lys^{15}, Arg^{209}, and Val^{382} (45-47)

Class 3. ***Decreased affinity to bind insulin.***

—Lys^{15} and Ser^{735} (45, 47, 48-51)

Class 4. ***Defect in insulin-stimulated tyrosine kinase activity.***

—Val^{382}, Val^{1008}, Thr^{1134}, Glu^{1135}, Ile^{1153}, and Ser^{1200} (46, 52-58)

Class 5. ***Impaired ability of acid pH to dissociate insulin from its receptor.***

—Glu^{460} and Ser^{462} (43, 45, 60)

Unclassified.

—Pro^{233} (59)
—Deletions: exon 14 or exons 17(3')-22 (61, 62)

termination mutations: (i) deletion of exon 14 that causes a frame shift and the introduction of an in-frame chain termination codon (61); and (ii) deletion of the entire gene downstream from codon 1012 in exon 17 resulting in a fusion protein with a premature chain termination (62). Some of these mutations—the deletion of exon 14 and also the nonsense mutations at codons 133, 672, and 897—truncate the receptor upstream from the transmembrane domain. Accordingly, these truncated receptors appear not to be expressed on the cell surface. Furthermore, three of the nonsense mutations—at codons 133, 897, and 1000—have been demonstrated to cause an 80-90% decrease in the level of insulin receptor mRNA (44, 45). This reduction in the level of mRNA is predicted to decrease the rate of receptor biosynthesis. Thus, mutations that result in premature chain termination of translation impair receptor function by one or more of the following mechanisms: (i) reduction of the number of insulin receptors by reducing the level of mRNA (44, 45, 63); (ii) truncation of the receptor upstream from the transmembrane domain so that the receptor is not expressed on the cell surface (43, 45, 61); or (iii) deletion of important functional domains in the intracellular portion of the receptor (45, 62).

Other Cis-Acting Mutations that Decrease Levels of mRNA. Many different types of mutations are known to decrease levels of mRNA without altering the coding sequence of the gene. For example, there might be mutations in regulatory regions of the gene that decrease the rate at which the gene is transcribed. In addition, there might be mutations in introns that impair the splicing of the transcript. Finally, there might be mutations in untranslated portions of the mRNA that would either decrease the translatability of the mRNA or decrease the stability of the mRNA. Although none of these mutations have been explicitly identified in the insulin receptor gene, there is strong evidence suggesting the existence of this type of mutation. We have investigated a patient with leprechaunism (leprechaun/Minn-1) whose cells have a marked decrease in the level of insulin receptor mRNA (44, 63, 64). In the allele inherited from the father, there is a nonsense mutation at codon 897 that acts in a cis-dominant fashion to decrease mRNA. However, the allele inherited from the mother is also underexpressed even though the coding sequences of all 22 exons of the gene are normal. Nevertheless, studies of the expression of this allele in cells from the patient's mother provided compelling evidence that there is a mutation in this allele. The existence of a silent polymorphism at codon 234 (GAC^{Asp} vs. GAT^{Asp}) enabled us to quantitate separately the transcripts derived from each of the two alleles of the insulin receptor gene. In the mother's cells, approximately 90% of the mRNA was derived from the GAC^{Asp}-allele. Because both alleles were exposed to the same array of trans-acting factors in the mother's cells, it is possible to conclude that there is an unidentified cis-acting mutation in the GAT^{Asp}-allele (44). Investigation of two other patients has suggested that they may have similar types of cis-acting mutations that decrease insulin receptor mRNA by mechanisms that have not yet been elucidated (65).

Class 2, Impaired Transport of Receptors to the Cell Surface

Patients A-5 and A-8 are two sisters who are members of a consanguineous kindred in which the parents are first cousins (46, 66). Both parents are heterozygous carriers of a mutant allele that was inherited from one of the great-grandparents. In addition, the patients have four unaffected siblings, all of whom are heterozygous carriers of the mutation. However, the two sisters with type A extreme insulin resistance are both homozygous for a mutation substituting valine for phenylalanine at position 382 in the α-subunit of the wild type insulin receptor (46). To evaluate the significance of the substitution of valine for phenylalanine at position 382, the mutant form of the insulin receptor cDNA was expressed by transfection in cultured cells. These studies demonstrated that the Val^{382}-mutation impairs transport of the receptor through the endoplasmic reticulum and Golgi (46, 63, 67). As a consequence of the defect in intracellular transport of the receptor, there is a decrease in the number of receptors transported to the plasma membrane, and there is a decrease in the number of receptors expressed on the cell surface (46). We have identified two additional mutations that impair intracellular transport of receptors through the endoplasmic reticulum and Golgi to the cell surface: substitution of arginine for His^{209} (45) and substitution of lysine for Asn^{15} (45, 47). Associated with the defect in transport of the receptor through the endoplasmic reticulum and Golgi, there is an impairment in the post-translational processing of the receptor. For example, in most of the receptor molecules, the high mannose form of N-linked oligosaccharide does not undergo terminal processing to the complex form of oligosaccharide that contains sialic acid (46, 47). It seems likely that the missense mutation results in a receptor that does not fold properly to assume its normal conformation. Most of the abnormal receptors are somehow targeted for degradation rather than transport to the cell surface. Nevertheless, because the block is incomplete, a small percentage of the mutant receptors proceed through the normal processing pathway and are eventually inserted in the plasma membrane.

Apparently, even the small percentage of receptors that "leak" through the block do not necessarily function normally. For example, both the Val^{382}- and the Lys^{15}-mutations impair the function of the receptor. The Lys^{15}-mutation caused a fivefold reduction in the affinity of the receptor to bind insulin (47). Although the Val^{382}-mutation does not decrease the affinity with which the receptor binds insulin, it inhibits the ability of insulin to activate receptor tyrosine kinase activity (52). Thus, the patients' insulin resistance results from two defects, both of which are caused by these two mutations. First, there is a decrease in the number of receptors on the cell surface as a result of impaired transport of receptors to the cell surface (46, 47, 67). Second, the small number of receptors on the cell surface are impaired in their function: a defect in insulin binding in the case of the Lys^{15}-mutation (47), and a defect in the ability of insulin to activate the receptor tyrosine kinase in the case of the Val^{382}-mutation (52). In contrast, in the case of the Arg^{209}-mutation which also impairs intracellular transport and post-translational processing of the receptor, we did not detect any defect in either insulin binding or tyrosine kinase activation (68).

Class 3. Decreased Affinity of Insulin Binding

The fact that the Lys^{15}-mutation decreases the affinity of insulin binding has already been discussed (see above in discussion of Class 2 mutations) (47). In addition, another mutation (i.e., substitution of serine for Arg^{735}) has been reported to decrease the affinity of insulin binding. Arg^{735} is the fourth amino acid in an Arg-Lys-Arg-Arg sequence that separates the α-subunit from the β-subunit in the insulin receptor precursor. The Ser^{735}-mutation inhibits the cleavage of the precursor into two separate subunits (48-51). In addition, this mutation has been reported to decrease the affinity with which the receptor binds insulin.

Class 4. Defects in Receptor Tyrosine Kinase Activity

At least five missense mutations have been identified in the intracellular domain of β-subunit of the insulin receptor: $Gly^{1008} \rightarrow Val$ (53), $Met^{1153} \rightarrow Ile$ (57), $Ala^{1134} \rightarrow Thr$ (56), $Ala^{1135} \rightarrow Glu$ (58), $Trp^{1200} \rightarrow Ser$ (54, 55). The Val^{1008}-mutation was identified in a young man with insulin resistance and acanthosis nigricans (53); the remainder of the mutations were identified in young women with the syndrome of hyperandrogenism, insulin resistance, and acanthosis nigricans.

Gly^{1008} is the third glycine residue in the highly conserved Gly^{1003}-X-Gly^{1005}-X-X-**Gly^{1008}**...........Lys^{1030} motif (26) that provides part of the binding site for ATP, the phosphate donor in the tyrosine kinase reaction. Thus it is likely that the Val^{1008}-mutation impairs tyrosine kinase activity because it distorts the ATP binding site in the tyrosine kinase domain (53). The other four mutations alter amino acid residues (Met^{1153}, Ala^{1135}, Ala^{1134}, and Trp^{1200}) that are highly conserved in the amino acid sequences of tyrosine kinases in the insulin receptor family although their precise roles in enzyme structure and function are not understood.

What is the mechanism whereby these mutations cause the phenotype of insulin resistance in a dominant fashion? Although this question has not been answered with certainty, the leading hypothesis relates to the oligomeric structure of the receptor. If mutant insulin receptor heterodimers ($\alpha\beta_m$) and wild type insulin receptor heterodimers ($\alpha\beta_{wt}$) associate with one another randomly and with equal affinity, then three different heterotetramers would form in a ratio of 1:2:1—$\alpha_2(\beta_{wt})_2$, $\alpha_2\beta_m\beta_{wt}$, and $\alpha_2(\beta_m)_2$. If the hybrid heterotetramer ($\alpha_2\beta_m\beta_{wt}$) were impaired in its tyrosine kinase activity, then a mutation in a single allele might lead to a 75% reduction in insulin receptor tyrosine kinase activity.

Class 5. Accelerated Degradation of the Insulin Receptor

A mutation substituting glutamic acid for Lys460 was identified in a patient with leprechaunism (leprechaun/Ark-1) (43). This mutation, inherited from the patient's mother, is recessive in that the mother has normal glucose tolerance and does not appear to be insulin resistant. Prior to the identification of the mutation by cDNA cloning, we had identified multiple abnormalities in insulin binding to receptors on the surface of the patient's EBV-transformed lymphoblasts including decreased sensitivity to changes in temperature and pH (69, 70). Interestingly, the patient's receptor has a fivefold increase in binding affinity at physiological temperature (37 °C) and pH (pH 7.4). Furthermore, insulin stimulates the receptor-associated tyrosine kinase normally in receptors from the patient's EBV-transformed lymphoblasts.

How does the Glu^{460} mutation impair receptor function to cause insulin resistance *in vivo* ? An analogy to a site-directed mutant of the LDL receptor suggests an answer to the question (71). When the portion of the extracellular domain of the LDL receptor homologous to the epidermal growth factor (EGF) precursor molecule is deleted, this causes LDL binding to become insensitive to changes in pH. After LDL receptors bind LDL, the ligand-receptor complex is internalized into endocytic vesicles. These endocytic vesicles develop an acidic pH within their lumens. This acidic pH plays a crucial role in dissociating the ligand from its receptor. Subsequent to internalization, at least two distinct pathways are available to the receptor: *recycling* to the cell surface for reutilization or *degradation* within the lysosomes. In the case of the deletion mutant of the LDL receptor, desensitization to changes in pH was associated with inhibition of the recycling pathway. When ligand bound to the receptor, it was not dissociated from the receptor in the endocytic vesicle. The receptor was targeted preferentially for degradation within the lysosome.

With normal insulin receptors (Lys^{460}), decreasing the pH from 7.8 to 6.0 causes a tenfold acceleration in the rate at which [^{125}I]insulin dissociates from its receptor. The effect of acid pH to accelerate [^{125}I]insulin dissociation is markedly blunted with the Glu^{460}-mutant receptor. In analogy to the deletion mutant of the LDL receptor (see above), we have obtained evidence that the Glu^{460}-mutation causes insulin resistance by accelerating the rate of receptor degradation (60). According to this hypothesis, the cause of insulin resistance is a decrease in the number of insulin receptors on the surface of target cells that results from an accelerated rate of receptor degradation.

A similar defect in pH sensitivity of insulin binding was observed with another mutant receptor in which serine was substituted for Asn^{462} (45). It is intriguing that the Glu^{460}- and Ser^{462}-mutations map within two amino acid residues of one another.

GENETICS OF INSULIN RESISTANCE

Many of the patients with genetic forms of extreme insulin resistance have mutations in the insulin receptor gene (5, 43-62). Nevertheless, there may be genetic heterogeneity within these syndromes. The mechanism of insulin action is complex, and normal insulin action requires the participation of the products of many genes. Thus, it is possible that, in some patients, insulin resistance may be caused by mutations at other genetic loci.

The degree of insulin resistance is extremely variable among these patients. For example, some patients are so severely insulin resistant that they have fasting hyperglycemia and overt diabetes despite having fasting plasma insulin levels >100 μU/ml (normal, ≤20 μU/ml). This type of patient may require several thousand units of insulin per day to achieve acceptable control of the level of glucose in the plasma. There are relatively few patients with such severe insulin resistance. We have investigated several patients in this category, all of whom have had two mutant alleles of the insulin receptor gene (Table II). For example, two sisters (patients A-5 and A-8) were members of a consanguineous kindred and were homozygous for a missense mutation in the insulin receptor gene (46, 66). Another patient (patient A-1) was a compound heterozygote, having inherited two different mutations, one from her father and one from her mother (45). However, it is more common for patients to have less severe insulin resistance. In fact, in many patients with type A insulin resistance, the fasting insulin levels are elevated in the range 20-100 μU/ml. This is sufficient to prevent fasting hyperglycemia in this group of patients although most of these patients have impaired glucose

Table II. **Patients with mutations in the insulin receptor gene.**

Listed below are the mutations in the insulin receptor gene that have been identified in patients with genetic forms of insulin resistance. Missense mutations are identified by the amino acid substitution encoded by the mutation. WT refers to a wild type allele that encodes a receptor with a normal amino acid sequence. A question mark refers to an allele that is thought to be normal, but where the nucleotide sequence has been reported for only a portion of the protein coding domain. Amber and opal refer to nonsense mutations corresponding to codons UAG and UGA, respectively. Deletion mutations are abbreviated as $\Delta^{\text{Exon 17(3')-22}}$ (a deletion beginning after codon 1012 in the 3' part of exon 17 and extending downstream through exons 18-22) and $\Delta^{\text{Exon 14}}$ (a deletion of exon 14).

Patient	Genotype	References
Type A extreme insulin resistance		
A-1	Amber^{133} / Ser^{462}	45
A-5 and A-8	Val^{382} (homozygous)	46
BI-1	Thr^{1134} / WT	56
A-3	Glu^{1135} / ?	58
A-6	Met^{1153} / ?	57
BI-2	Ser^{1200} / WT	54, 55
Chiba-1	$\Delta^{\text{Exon 17(3')-22}}$ / ?	62
Chiba-2	$\Delta^{\text{Exon 14}}$ / ?	61
Kyushu-1 and -2	Ser^{735} (homozygous)	48, 49
Sapporo-1	Val^{1008} / ?	45
Rabson-Mendenhall syndrome		
RM-1	Lys^{15} / Opal^{1000}	45
Leprechaunism		
Ark-1	Glu^{460} / Amber^{672}	43
Geldermalsen	Pro^{233} (homozygous)	59
Minn-1	Opal^{897} / Unidentified mutation	44
Winnipeg	Arg^{209} (homozygous)	45

tolerance. Several patients in this category have been investigated. They appear to be heterozygous for a single mutation that causes insulin resistance in a dominant (or co-dominant) fashion (Table II) (45, 53-62, 65). Note, however, that it is difficult to completely eliminate the possibility that there is a mutation in a gene—particularly in a regulatory domain of the gene. Because the regulatory domains in the insulin receptor gene have not been completely identified, it is not possible to determine the sequence of all the regulatory domains. Moreover, it is likely that there are regulatory elements that play a role in tissue specific regulation in insulin target cells (e.g., liver or muscle). It might not be possible to detect mutations in these tissue-specific regulatory elements if the studies are carried out in EBV-transformed lymphoblasts or cultured skin fibroblasts.

CORRELATION OF CLINICAL SYNDROME WITH GENOTYPE

Mutations have been identified in the insulin receptor gene in patients with several distinct syndromes: leprechaunism, the Rabson-Mendenhall syndrome, and type A insulin resistance. What determines which syndrome a patient will develop? Because the clinical syndromes do not correlate with the type of mutation, it seems most likely that it is the severity of insulin resistance that determines the clinical manifestations. For example, patients with leprechaunism appear to have the most extreme degree of insulin resistance. All of the patients with leprechaunism have had two mutant alleles of the insulin receptor gene; two were compound heterozygotes (43, 44) and two were homozygotes (45, 59). Some patients with type A extreme insulin resistance have also been reported to have two mutant alleles: patient A-1 who is a compound heterozygote as well as two sisters in each of two consanguineous pedigrees, all of whom were homozygous for mutations in the insulin receptor gene (46, 48-51). However, some patients with type A extreme insulin resistance have been reported to be heterozygous for a single mutant allele of the insulin receptor gene (53-62). As exemplified by the comparison between leprechaun/Ark-1 and her father (43, 70), the degree of insulin resistance observed in heterozygotes is less severe than the insulin resistance in homozygotes and compound heterozygotes. The father of leprechaun/Ark-1 is heterozygous for a nonsense mutation in the insulin receptor gene while his other allele encodes a normal receptor. The presence of a second mutant allele in the daughter is the cause not only of her more severe degree of insulin resistance but also the multiple phenotypic abnormalities associated with the syndrome of leprechaunism. Similarly, the mother of leprechaun/Winnipeg—an obligate heterozygote for the Arg^{209}-mutation—is insulin resistant and hyperinsulinemic (12). Thus, although the phenotype of leprechaunism is recessive, the phenotype of insulin resistance caused by the Arg^{209}-mutation is inherited in a co-dominant fashion.

NONINSULIN-DEPENDENT DIABETES MELLITUS

In light of the central role of insulin resistance in predisposing to development of noninsulin-dependent diabetes mellitus (NIDDM) (1, 2), it is reasonable to inquire whether patients with NIDDM may have mutations in the insulin receptor gene. In favor of this hypothesis, some patients with extreme insulin resistance have had relatives in whom heterozygosity for mutations in the insulin receptor gene is associated with a moderate degree of insulin resistance. For example, as described above, the father of leprechaun/Ark-1, who is heterozygous for a nonsense mutation in the insulin receptor gene, has a moderate degree of insulin resistance comparable to what is observed in patients with NIDDM (72, 73). Furthermore, the mutations detected thus far in the insulin receptor gene have all caused major defects in the function of the insulin receptor. It is possible that patients with a milder degree of insulin resistance might have mutations that cause less severe disruption of the function of the insulin receptor.

Analysis of restriction fragment length polymorphisms (RFLP's) have yielded equivocal results (5). However, there are many reasons that this approach might fail to detect association of the disease with the insulin receptor gene. For example, there may be more than one mutation causing insulin resistance in the diabetic population, and these mutations may not all be in linkage disequilibrium with the same RFLP. If this were true, population studies would not detect strong association of NIDDM with a particular RFLP. Furthermore, it may be difficult to demonstrate linkage of NIDDM to a particular gene even in studies of the inheritance of RFLP's in families. For example, if development of NIDDM requires simultaneous mutations at more than one locus, linkage may be difficult to detect unless all of the relevant loci are analyzed simultaneously.

Accordingly, several laboratories have embarked upon a more direct approach—to determine the nucleotide sequence of the insulin receptor gene in diabetic patients. Thus far, at least three insulin-resistant Pima Indians have been studied (72, 73). No mutations were detected in the protein coding domain of the insulin receptor gene in these three patients. Of course, even if there are no mutations in the protein coding domain, it remains possible that there may be a mutation in a regulatory domain of the gene. However, as described above, the regulatory domains of the gene have not yet been clearly delineated so that it is more difficult to address this question at the present time.

The mechanism by which insulin elicits its multiple biological responses in target cells is extremely complex. Insulin has multiple effects upon multiple different target cells. For target cells to respond to insulin, this requires the function of many proteins encoded by many genes. At least in theory, each of these genes is a candidate to be the locus of a mutation causing insulin resistance in NIDDM. As progress is made in the identification and cloning of the many genes that allow for the normal response to insulin, it seems likely that it will be possible to identify the genes that are targets of mutations that cause insulin resistance in NIDDM.

REFERENCES

1. Reaven GM: Role of insulin resistance in human disease. *Diabetes* 37:1595-1607, 1988.
2. DeFronzo RA: Lilly lecture 1987. The triumvirate: beta-cell, muscle, liver. A collusion responsible for NIDDM. *Diabetes* 37:667-687, 1988.
3. Saad MF, Knowler WC, Pettitt DJ, Nelson RG,Mott DM, Bennet PH. The natural history of impaired glucose tolerance in the Pima Indians. *N Engl J Med.* 319: 1500-1506, 1988.

4 Lillioja S, Mott DM, Howard BV, *et al.* Impaired glucose tolerance as a disorder of insulin action. Longitudinal and cross-sectional studies in Pima Indians. *N Engl J Med.* 318:1217-1225, 1988.
5. Taylor SI, Kadowaki T, Kadowaki H, Accili D, Cama A, McKeon C: Mutations in insulin-receptor gene in insulin-resistant patients. *Diabetes Care* 13:257-279, 1990.
6. Flier JS, Eastman RC, Minaker KL, Matteson D, Rowe JW: Acanthosis nigricans in obese women with hyperandrogenism. Characterization of an insulin-resistant state distinct from the type A and B syndromes. *Diabetes* 34:101-107, 1985.
7. Fradkin JE, Eastman RC, Lesniak MA, Roth, J: Specificity spillover at the hormone receptor--exploring its role in human disease. *N. Engl. J. Med.* 320:640-645, 1989.
8. Taylor SI, Dons RF, Hernandez E, Roth J, Gorden P: Insulin resistance associated with androgen excess in women with autoantibodies to the insulin receptor. *Ann Intern Med* 97:851-855, 1982.
9. Barbieri RL, Ryan KJ: Hyperandrogenism, insulin resistance, and acanthosis nigricans syndrome: a common endocrinopathy with distinct pathophysiologic features. *Am J Obstet Gynecol* 147:90-101, 1983.
10. Kahn CR, Flier JS, Bar RS, Archer JA, Gorden P, Martin MM, Roth J: The syndromes of insulin resistance and acanthosis nigricans. Insulin-receptor disorders in man. *N Engl J Med* 294:739-745, 1976.
11. Rossini AA, Cahill GF, Jr.: Lipoatrophic diabetes in *Endocrinology* , De Groot LJ, Cahill GF, Jr., Martini L et al. (Eds.) Volume 2, pp. 1093-1097. New York; Grune & Stratton.
12. Rosenberg AM, Haworth JC, Degroot GW, et al: A case of leprechaunism with severe hyperinsulinemia. *Am J Dis Child* 134:170-175, 1980.
13. Rabson SM, Mendenhall EN: Familial hypertrophy of pineal body, hyperplasia of adrenal cortex and diabetes mellitus. *Amer J. Clin Path*, 26:283- ,1956.
14. West RJ, Leonard JV: Familiar insulin resistance with pineal hyperplasia: metabolic studies and effects of hypophysectomy. *Arch Dis Child* 55:619-621, 1980.
15. Yang-Feng TL, Francke U, Ullrich A: Gene for human insulin receptor: localization to site on chromosome 19 involved in pre-B-cell leukemia. *Science* 228:728-731, 1985.
16. DeMeyts, P., J.-L. Gu, R.M. Shymko, B.E. Kaplan, G.I. Bell, J. Whittaker: Identification of a ligand binding region of the human insulin receptor encoded by second exon of the gene. *Molecular Endocrinology*, 4:409-416, 1990.
17. Ullrich A, Bell JR, Chen EY, Herrera R, Petruzelli LM, Dull TJ, Gray A, Coussens L, Liao YC, Tsubokawa M, Mason A, Seeburg PH, Grunfeld C, Rosen OM, Ramachandran J: Human insulin receptor and its relationship to the tyrosine kinase family of oncogenes. *Nature* 313:756-761, 1985.
18. Ebina, Y., Ellis, L., Jarnagin, K., Edery, M., Graf, L., Clauser, E., Ou, J.H., Masiarz F., Kan, Y.W., Goldfine, I.D., Roth, R.A., and Rutter, W.J. : The human insulin receptor cDNA: the structural basis for hormone-activated transmembrane signalling. *Cell* 40:747-758, 1985.
19. Frias I, Waugh SM: Probing the a-a subunit interface region in the insulin receptor, location of interhalf disulfide(s). *Diabetes* 38 (suppl. 2) : 60A (Abstract #238), 1989.

20. Seino S, Bell GI: Alternative splicing of human insulin receptor messenger RNA. *Biochem Biophys Res Commun* 159:312-316, 1989.
21. Moller DE, Yokota A, Caro JF, Flier JS: Tissue-specific expression of two alternatively spliced insulin receptor mRNAs in man. *Mol. Endocrinol.* 3:1263-1269, 1989.
22. Mosthaf L, Grako K, Dull TJ, Coussens L, Ullrich A, McClain DA: Functionally distinct insulin receptors generated by tissue-specific alternative splicing. *EMBO J* 9:2409-2413, 1990.
23. White MF, Shoelson SE, Keutmann H, Kahn CR: A cascade of tyrosine autophosphorylation in the beta-subunit activates the phosphotransferase of the insulin receptor. *J Biol Chem* 263:2969-2980, 1988.
24. White MF, Livingston JN, Backer JM, Lauris V, Dull TJ, Ullrich A, Kahn CR: Mutation of the insulin receptor at tyrosine 960 inhibits signal transmission but does not affect its tyrosine kinase activity. *Cell* 54:641-649, 1988.
25. Tornqvist HE, Gunsalus JR, Nemenoff RA, Frackelton HR, Pierce MW, Avruch J: Identification of the insulin receptor tyrosine residues undergoing insulin-stimulated phosphorylation in intact rat hepatoma cells. *J Biol Chem* 263:350-359, 1988.
26. Hanks SK, Quinn AM, Hunter T: The protein kinase family: conserved features and deduced phylogeny of the catalytic domains. *Science* 241:42-52, 1988.
27. Pilch PF, Czech MP: Interaction of cross-linking agents with the insulin effector system of isolated fat cells. Covalent linkage of ^{125}I-insulin to a plasma membrane receptor protein of 140,000 daltons. *J Biol Chem* 254:3375-3381, 1979.
28. Grunfeld C, Shigenaga JK, Ramachandran J: Urea treatment allows dithiothreitol to release the binding subunit of the insulin receptor from the cell membrane: implications for the structural organization of the insulin receptor. *Biochem Biophys Res Commun* 133:389-396, 1985.
29. Kasuga M, Karlsson FA, Kahn CR: Insulin stimulates the phosphorylation of the 95,000-dalton subunit of its own receptor. *Science* 215:185-187, 1982.
30. Kasuga M, Zick Y, Blithe DL, Crettaz M, Kahn CR: Insulin stimulates tyrosine phosphorylation of the insulin receptor in a cell-free system. *Nature* 298:667-669, 1982.
32. Chou CK, Dull TJ, Russell DS, Gherzi R, Lebwohl D, Ullrich A: Human insulin receptors mutated at the ATP-binding site lack protein tyrosine kinase activity and fail to mediate postreceptor effects of insulin. *J Biol Chem* 262:1842-18447, 1987.
33. Ebina Y, Araki E, Taira M, Shimada F, Mori M, Craik CS, Siddle K, Pierce SB, Roth RA, Rutter WJ: Replacement of lysine residue 1030 in the putative ATP-binding region of the insulin receptor abolishes insulin-and antibody-stimulated glucose uptake and receptor kinase activity. *Proc Natl Acad Sci USA* 84:704-708, 1987.
34. McClain DA, Maegawa H, Lee J, Dull TJ, Ullrich A, Olefsky JM: A mutant insulin receptor with defective tyrosine kinase displays no biologic activity and does not undergo endocytosis. *J Biol Chem* 262:14663-14671, 1987.
35. Hedo JA, Kahn CR, Hayashi M, Yamada Km, Kasuga M: Biosynthesis and glycosylation of the insulin receptor. Evidence for a single polypeptide precursor of the two major subunits. *J Biol Chem* 258:10020-10026, 1983.

36. Deutsch PJ, Wan CF, Rosen OM, Rubin CS: Latent insulin receptors and possible receptor precursors in 3T3-L1 adipocytes. *Proc Natl Acad Sci USA* 80:133-136, 1983.
37. Jacobs SJ, Kull FCJ, Cuatrecasas P: Monensin blocks the maturation of receptors for insulin and somatomedin C: Identification of receptor precursors. *Proc Natl Acad Sci USA* 80:1228-1231, 1983.
38. Hedo JA, Kasuga M, Van Obberghen E, Roth J, Kahn CR: Direct demonstration of glycosylation of insulin receptor subunits by biosynthetic and external labeling: evidence for heterogeneity. *Proc Natl Acad Sci USA* 78:4791-4795, 1981.
39. Herzberg VL, Grigorescu F, Edge AS, Spiro RG, Kahn CR: Characterization of insulin receptor carbohydrate by comparison of chemical and enzymatic deglycosylation. *Biochem Biophys Res Commun* 129:789-796, 1985.
40. Collier E, Gorden P: The insulin receptor contains O-linked oligosaccharide. *Diabetes* 38 *(suppl. 2)* :178A (Abstract #686), 1989.
41. Hedo JA, Collier E, Watkinson A: Myristyl and palmityl acylation of the insulin receptor. *J Biol Chem* 262:954-957, 1987.
42. Brown MS, Goldstein JL: A receptor-mediated pathway for cholesterol homeostasis. *Science* 232:34-47, 1986.
43. Kadowaki T, Bevins CL, Cama A, Ojamaa K, Marcus-Samuels B, Kadowaki H, Beitz L, McKeon C, Taylor SI: Two mutant alleles of the insulin receptor gene in a patient with extreme insulin resistance. *Science* 240:787-790, 1988.
44. Kadowaki T, Kadowaki H, Taylor SI: A nonsense mutation causing decreased levels of insulin receptor mRNA: Detection by a simplified technique for direct sequencing of genomic DNA amplified by polymerase chain reaction. *Proc Natl Acad Sci USA* 87:658-662, 1990.
45. Kadowaki T, Kadowaki H, Rechler MM, Serrano-Rios M, Roth J, Gorden P, Taylor SI: Five mutant alleles of the insulin receptor gene in patients with genetic forms of insulin resistance. *J Clin Invest* 86:254-264, 1990.
46. Accili D, Frapier C, Mosthaf L, McKeon C, Elbein S, Permutt MA, Ramos E, Lander E, Ullrich A, Taylor SI: A mutation in the insulin receptor gene that impairs transport of the receptor to the plasma membrane and causes insulin resistant diabetes. *EMBO J* 8:2509-2517, 1989.
47. Kadowaki T, Kadowaki H, Accili D, Taylor SI: Substitution of lysine for asparagine-15 in the human insulin receptor impairs intracellular transport of the receptor to the cell surface and decreases the affinity of insulin binding, *J. Biol. Chem.*, 265: 19143-19150, 1990.
48. Yoshimasa Y, Seino S, Whittaker J, Kakehi T, Kosaki A, Kuzuya H, Imura I, Bell GI, Steiner DF: Insulin-resistant diabetes due to a point mutation that prevents insulin proreceptor processing. *Science* 240:784-787, 1988.
49. Kobayashi M, Sasaoka T, Takata Y, Ishibashi O, Sugibayashi M, Shigeta Y, Hisatomi A, Nakamura E, Tamaki M, Teraoka H: Insulin resistance by unprocessed insulin proreceptors point mutation at the cleavage site. *Biochem Biophys Res Commun* 153:657-663, 1988.
50. Kakehi T, Hisatomi A, Kuzuya H, Yoshimasa Y, Okamoto M, Yamada K, Nishimura H, Kosaki A, Nawata H, Umeda F, Ibayashi H, Imura H: Defective processing of insulin receptor precursor in cultured lymphocytes from a patient with extreme insulin resistance. *J Clin Invest* 81:2020-2022, 1988.

51. Kobayashi M, Sasaoka T, Takata Y, Hisatomi A, Shigeta Y: Insulin resistance by uncleaved insulin proreceptor. Emergence of binding site by trypsin. *Diabetes* 37:653, 1988.
52. Accili D, Mosthaf L, Ullrich A, Taylor S.I. (1990) A mutation in the extracellular domain of the insulin receptor impairs the ability of insulin to stimulate receptor autophosphorylation. *J Biol Chem* 266:in press, 1990.
53. Odawara M, Kadowaki T, Yamamoto R, Shibasaki Y, Tobe K, Accili D, Bevins C, Mikami Y, Matsuura N, Akanuma Y, Takaku F, Taylor SI, Kasuga M: Human diabetes associated with a mutation in the tyrosine kinase domain of the insulin receptor. *Science* 245:66-68, 1989.
54. Moller DE, Flier JS: Detection of an alteration in the insulin-receptor gene in a patient with insulin resistance, acanthosis nigricans, and the polycystic ovary syndrome (type A insulin resistance). *N Engl J Med* 319:1526-1529, 1988.
55. Moller DE, Yokota A, Ginsberg-Fellner F, Flier JS: Funtional properties of a naturally occurring $Trp^{1200} \rightarrow Ser^{1200}$ mutation of the insulin receptor. *Mol Endocrinol* 4:1183-1191, 1990.
56. Moller DE, Yokota A, White MF, Pazianos AG, Flier JS: A naturally occurring mutation of insulin receptor Ala^{1134} impairs tyrosine kinase function and is associated with dominantly inherited insulin resistance. *J Biol Chem* 265:14979-14985, 1990.
57. Cama A, Sierra M, Ottini L, Imperato J, Taylor SI: A mutation in the tyrosine kinase domain of the insulin receptor causing insulin resistance in an obese woman. *Diabetes*, in press (abstract), 1991.
58. Cama A, Sierra M, Ottini L, Taylor SI: A mutation at Ala^{1135} in one allele of the insulin receptor gene of a patient with type A extreme insulin resistance. *Endocrinology*, in press (abstract), 1991.
59. Kadowaki H, Kadowaki T, Cama A, Marcus-Samuels B, Rovira A, Bevins C, Taylor SI: Mutagenesis of lysine-460 in the human insulin receptor: effects upon receptor recycling and site-site interactions among binding sites, *J. Biol. Chem.*, 265: in press, 1990.
60. Klinkhamer M, Groen NA, van der Zon GCM, Lindhout D, Sandkuyl LA, Krans HM, Möller W, Maassen JA: A leucine-to-proline mutation in the insulin receptor in a family with insulin resistance. *EMBO J* 8:2503-2507, 1989.
61. Shimada F, Taira M, Suzuki Y, Hashimoto N, Nozaki O, Taira M, Tatibana M, Ebina Y, Tawata M, Onaya T, Makino H, Yoshida S: Insulin-resistant diabetes associated with partial deletion of insulin-receptor gene. *Lancet* 335:1179-1181, 1990.
62. Taira M, Taira M, Hashimoto N, Shimada F, Suzuki Y, Kanatsuka A, Nakamura F, Ebina Y, Tatibana M, Makino H, Yoshida S: Human diabetes associated with a deletion of the tyrosine kinase domain of the insulin receptor. *Science* 245:63-66, 1989.
63. Ojamaa K, Hedo JA, Roberts CT Jr, Moncada VY, Gorden P, Ullrich A, Taylor SI: Defects in human insulin receptor gene expression. *Mol Endocrinol* 2:242-247, 1988.
64. Taylor SI, Samuels B, Roth J, Kasuga M, Hedo JA, Gorden P, Brasel DE, Pokora T, Engel RR: Decreased insulin binding in cultured lymphocytes from two patients with extreme insulin resistance. *J Clin Endocrinol Metab* 54:919-930, 1982.
65. Imano E, Kadowaki H, Kadowaki T, Iwama N, Watarai T, Kawamori R, Kamada T, Taylor SI: Two patients with insulin resistance due to

decreased levels of insulin receptor mRNA. *Diabetes*, in press (abstract), 1991.
66. Barnes ND, Palumbo PJ, Hayles AB, Folgar H: Insulin resistance, skin changes, and virilization: a recessively inherited syndrome possibly due to pineal gland dysfunction.. *Diabetologia* 10:285-289, 1974.
67. Hedo JA, Moncada VY, Taylor SI: Insulin receptor biosynthesis in cultured lymphocytes from insulin-resistant patients. *J Clin Invest* 76:2355-2361, 1985.
68. Kadowaki H, Kadowaki T, Taylor SI: A mutation in the cysteine-rich domain of the insulin receptor that impairs post-translational processing and transport of receptors to the plasma membrane. *Diabetes*, in press (abstract), 1991.
69. Taylor SI, Roth J, Blizzard RM, Elders MJ: Qualitative abnormalities in insulin binding in a patient with extreme insulin resistance: Decreased sensitivity to alterations in temperature and pH. *Proc Natl Acad Sci USA* 78:7157-7161, 1981.
70. Taylor SI, Marcus-Samuels B, Ryan-Young J, Leventhal S, Elders MJ: Genetics of the insulin receptor defect in a patient with extreme insulin resistance. *J Clin Endocrinol Metab* 62:1130-1135, 1986.
71. Davis CG, Goldstein JL, Sudhof TC, Anderson RG. Russell DW, Brown MS: Acid-dependent ligand dissociation and recycling of LDL receptor mediated by growth factor homology region. *Nature* 326:760-765, 1987.
72. Moller DE, Yokota A, Flier JS: Normal insulin-receptor cDNA sequence in Pima Indians with NIDDM. *Diabetes* 38:1496-1500, 1989.
73. Cama A, Patterson A, Kadowaki T, Siegel G, D'Ambrosio D, Lillioja S, Roth J, Taylor SI: Cloning of insulin receptor cDNA from an insulin resistant Pima Indian. *J Clin Endocrinol Metab* 70:1155-1161, 1990.

ROLE OF RECEPTOR INTERNALIZATION IN INSULIN SIGNALLING

Susan C. Frost and Robert Risch

The Department of Biochemistry and Molecular Biology
University of Florida 32610
Gainesville, FL 32610

INTRODUCTION

Many hormones and nutrient molecules are internalized into their target cells by the common mechanism of receptor-mediated endocytosis. This process involves the clustering of receptors into invaginated regions of the plasma membrane coated on their intracellular surface by the protein clathrin which forms a basket-like casing. This pitted region internalizes to form an independent vesicle which eventually loses its clathrin coat. Movement through the vesicular pathway allows intracellular sorting. The uptake of insulin appears to follow this path. In a variety of tissues, insulin stimulates the internalization of its own receptor upon binding (Hedo and Simpson, 1984; Krupp and Lane, 1982; Knutson et al., 1983). At some point along the endocytotic path, the hormone dissociates from the receptor and is degraded (Krupp and Lane, 1982). The receptor recycles to the plasma membrane where it can bind another insulin molecule to reinitiate the cycle. Besides stimulating receptor internalization, insulin stimulates rapid receptor autophosphorylation (Kohanski et al., 1986; White et al., 1985) activating the tyrosine kinase. The receptor is internalized in this state (Backer et al., 1989). However, prior to recycling, the receptor is dephosphorylated (Backer et al., 1989) which inactivates the catalytic activity (Klein et al., 1987). How this kinase activity and endocytotic cycle is involved in insulin signalling is not well defined.

Critical to the internalization of the receptor is the formation of the coated pit. This structure may provide molecular determinants to cause clustering or may direct the formation of coated vesicles. Larkin et al. (1983, 1986) have shown that depletion of intracellular potassium inhibits coated pit formation which blocks internalization of the LDL receptor. We have used this later approach to study insulin receptor internalization in 3T3-L1 adipocytes. We show here that potassium depletion in 3T3-L1 adipocytes leads to an increase in cell surface insulin receptor number, a decrease in receptor-specific insulin degradation, but litte change in insulin-sensitive glucose transport. These results suggest that receptor internalization is not required for insulin enhanced glucose transport.

Molecular Biology and Physiology of Insulin and Insulin-Like Growth Factors
Edited by M.K. Raizada and D. LeRoith, Plenum Press, New York, 1991

EFFECT OF HYPOTONIC SHOCK ON POTASSIUM LEVELS IN 3T3-L1 ADIPOCYTES

To examine the effect of hypotonic exposure on intracellular K^+, control adipocytes were compared to adipocytes that were treated for 10min in K^+-free hypotonic buffer followed by incubation for various times in K^+-free isotonic buffer. Within one hour of hypotonic treatment, intracellular K^+ levels fell to 50%, and within 4h to 25% of the initial value (Figure 1). By comparison, the control cells, which

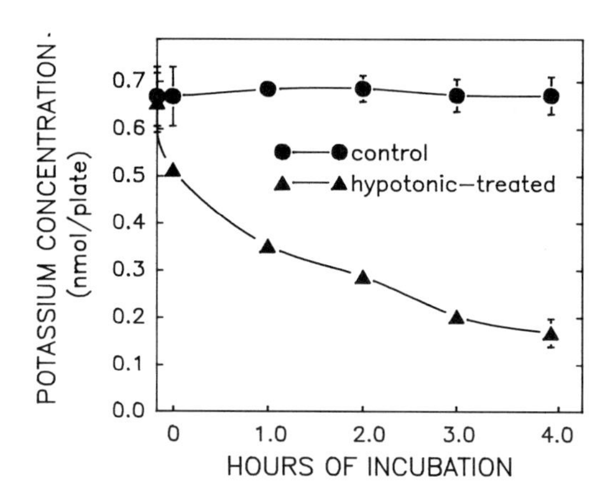

Figure 1. Effect of Hypotonic Treatment on Potassium Levels in 3T3-L1 Adipocytes. Potassium depletion was accomplished by a modification of the method of Larkin et al. (1983). 3T3-L1 adipocytes in 35mm plates were treated with a hypotonic buffer (buffer A:water 1:1) for 10min. Following this, plates were washed 3 times with 3.0ml of buffer A (25mM Hepes, 125mM NaCl, 1mM CaCl2), and incubated for up to 4h in buffer A containing 5mM glucose, which was replaced hourly. At the appropriate time, cells were washed 3 X 3.0ml with MgCl2 and then air dried. After lysing in 500μl of 0.1N NaOH, the extracts were centrifuged at 6500g for 15min. An aliquot (25μl) was diluted in 1.975ml dH20 and analyzed for potassium content by flame photometry. Controls were treated as above except that KRP (128mM NaCl, 5mM $NaPO4^{--}$ pH7.4, 4.7mM KCl, 1.25mM CaCl2, 1.25mM MgSO4) was used instead of buffer A. The data shown represent the average of duplicates with standard error in a single experiment. The experiment was performed 3 times with nearly identical results.

were incubated in KRP for the same period of time, showed a constant intracellular level of K^+ over the incubation period. Based on the intracellular water volume of 3T3-L1 adipocytes (1.8μl/10^6 cells; Frost and Lane, 1985) and the number of cells on a 35mm plate (2.1 X 10^6), the concentration of K^+ in control 3T3-L1 adipocytes was about 170mM. Thus hypotonic treatment combined with incubation in K^+-free buffer reduced intracellular K^+ to 42mM. At 4h, both control and hypotonically-treated adipocytes excluded trypan blue, indicating the integrity of the plasma membranes.

INSULIN BINDING IN POTASSIUM-DEPLETED 3T3-L1 ADIPOCYTES

To determine if cell surface receptor number changes in response to K^+ depletion, insulin binding was measured over the time of hypotonic treatment. As shown in Figure 2, insulin binding increased after one

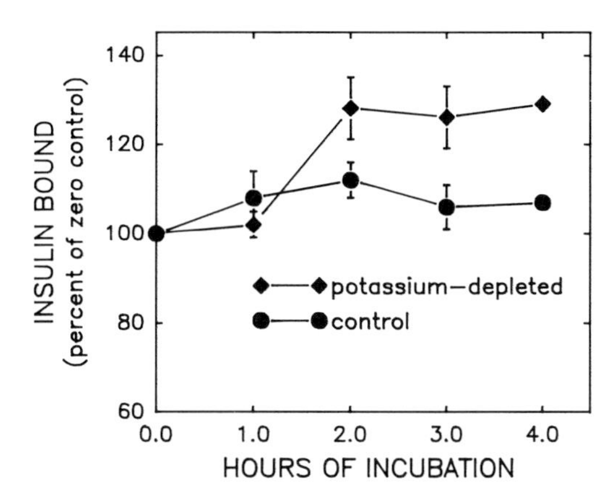

Figure 2. Effect of Hypotonic Treatment on Insulin Binding. Cells were hypotonically treated and then incubated over 4h in K^+-free buffer as described in Figure 1. Control cells were maintained in KRP. At each time point, cells were cooled to 4°C and insulin binding was assayed at 1nM ^{125}I-insulin in the presence or absence of 1μM unlabeled insulin. After equilibration overnight, plates were washed 5 times with 3.0 ml of cold PBS over 20 sec. Cells were extracted in 1.0 ml of 0.1% SDS and counted for radioactivity. The data represent the average of specific binding (total minus non-specific) ± standard error of duplicate assays.

hour of K^+ depletion. The difference in binding between control and K^+-depleted cells was stable from 2 to 4h. It should be noted that non-specific binding did not vary more than 7% over the treatment period. To further characterize this difference, equilibrium binding experiments were performed. A comparison between control and 4 hour hypotonically-treated cells revealed that K^+-depleted cells showed an increase in binding capacity relative to controls (Figure 3A). The Scatchard plot of these data (Figure 3B) demonstrates that K^+ depletion did not alter the apparent Kd of the receptor but increased the number of binding sites by 22% (from 237,000 to 301,000 receptors/cell). It has been estimated previously that 20% of the total receptor population in 3T3-L1 adipocytes resides within the cell (Reed et al., 1984). Thus, these data suggest that K^+ depletion inhibits receptor internalization while permitting recycling of the internal pool to the plasma membrane.

INSULIN DEGRADATION IN POTASSIUM-DEPLETED CELLS

If insulin degradation occurs via a receptor-mediated process, then K^+ depletion should decrease the amount of insulin degraded. It has been previously shown that insulin is degraded in isolated adipocytes by both receptor-mediated and receptor-independent pathways (Gliemann and Sonne, 1978). The receptor-mediated pathway represents only about 30% of total degradation in 3T3-L1 adipocytes (data not shown), which is consistent with earlier work in isolated adipocytes (Gliemann and Sonne,

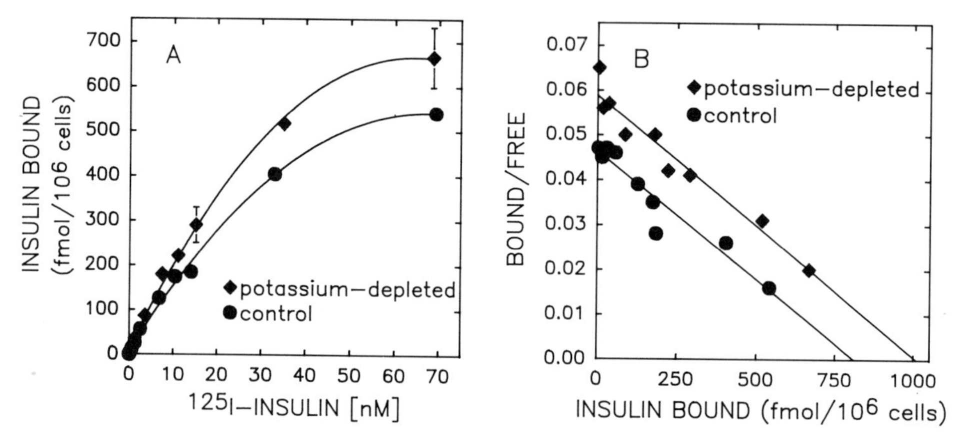

Figure 3. Effect of Potassium Depletion on Insulin Binding. A. Cells were hypotonically-treated and then incubated for 4h in K^+-free buffer. Control cells were maintained in KRP for an equal time. All cells were cooled to 4°C and insulin binding performed as described above at the indicated concentrations with or without 1μM unlabeled insulin. The data represent the average of duplicate points ± standard error. The experiment was repeated 10 times demonstrating an identical trend. The 2nd-order equation which defines the fit for the hypotonic data is $y = 10.8 + 20.2x - 0.15\ x^2$ (R =.999). The equation which defines the control data is $y = 7.4 + 15.9x - 0.12x^2$ (R = .998). B. The data from A were replotted according to Scatchard (1949).

1978). Very little of this degradation occurs within lysosomes since chloroquine, which neutralizes the lysosomal compartment, had little effect on receptor-mediated degradation. However, K^+-depleted cells exhibited a 70% reduction in total receptor-mediated degradation (Figure 4A). The amount of receptor-mediated degradation associated with the cell (Figure 4B) showed a similar trend but radioactivity accumulated at a very slow rate compared to total degradation. In fact, the cell-associated degradation represented 5% or less of total receptor-mediated degradation over the time examined.

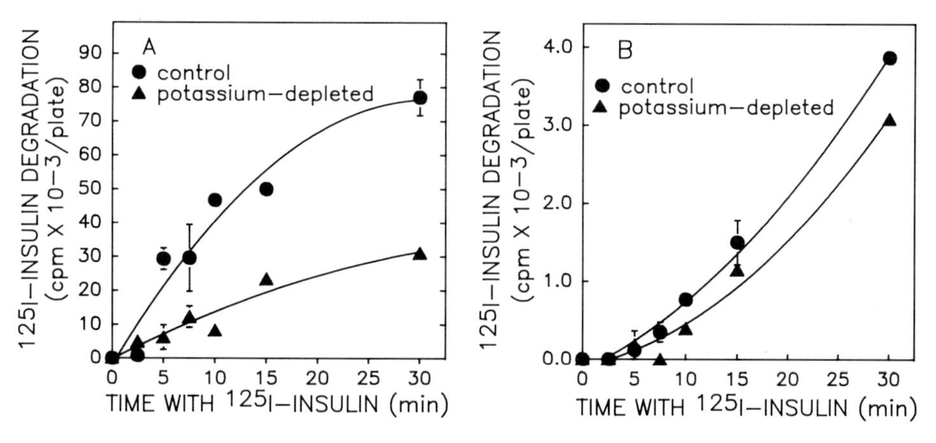

Figure 4. Effect of Potassium Depletion on Insulin Degradation. Cells were treated hypotonically and then incubated with K^+-free buffer for 4h. Control cells were incubated in KRP. All cells were then incubated at 37°C in buffer containing 0.1% BSA with 1nM ^{125}I-insulin with or without 1μM unlabeled insulin for the indicated times. An aliquot of the buffer overlay was removed and protein precipitated with 10% TCA. The cells were then washed 4 times with 3.0ml ice-cold PBS and precipitated with 1.0ml of 10% TCA at 4°C for 30 min. Subsequently, plates were scraped and washed with 1.0 ml of 10% TCA. Samples were then centrifuged at 4000rpm at 4°C for 55min. An aliquot of the supernatants (TCA-soluble radioactivity) and pellets (TCA-precipitable radioactivity) from the buffer and cells were counted separately to determine total insulin degradation. Released (Panel A) and cell-associated (Panel B) degraded insulin is shown. The data represent the mean ± S.E. of duplicate plates. The experiment was repeated 3 times with nearly identical results. The 2nd order equation which defines the fit for the hypotonic data in Panel A is $y = -.17 + 1.57x - 0.17x^2$ (R = .964). The equation which defines the fit for the control data is $y = -0.11 + 4.97x - .08x^2$ (R = .979).

EFFECT OF POTASSIUM DEPLETION ON INSULIN-STIMULATED GLUCOSE TRANSPORT

It is well documented that insulin stimulates the rate of glucose transport in 3T3-L1 adipocytes (Frost and Lane, 1985; Resh, 1982). The control cells demonstrated a consistently high rate of insulin-stimulated transport over the time frame of the experiment: a 10-fold rate enhancement was achieved over basal values (Figure 5A). During the time course of K^+ depletion, the basal values in treated cells increased as did the insulin-stimulated rate. At 4h, the insulin-stimulated rate was only 2.7 times that of basal. At face value, it appeared that insulin stimulation was reduced. However, if the basal values for control and K^+-depleted cells were subtracted from their respective insulin-stimulated values, it could be seen that there was little difference between control and K^+-depleted cells (Figure 5B). At 4h, the value for insulin-stimulated, K^+-depleted cells was nearly identical to that of insulin-stimulated, control cells (1.61nmol/10^6 cells/min vs 1.56nmol/10^6 cells/min, respectively). Thus K^+ depletion

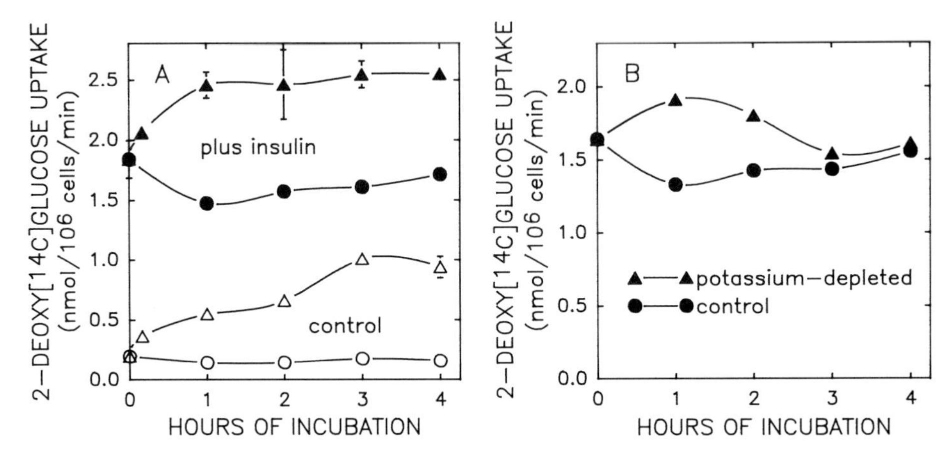

Figure 5. Effect of Potassium Depletion on Glucose Transport. A. After 4h of pretreatment as described in Figure 1, all cells were incubated in buffer A for 10 min with or without 1μM insulin followed by 0.2 mM [^{14}C]deoxyglucose for 10 min. Cells were then washed 3 times with 3.0 ml PBS and lysed with 1.0 ml of 0.1% SDS. An aliquot was taken for radioactivity. The values represent the average of quadruplicate points with a standard deviation of less than 5%. B. The average basal values from panel A, for both control and K^+-depleted cells, were subtracted from their respective insulin-stimulated values.

enhanced basal uptake but had little effect on maximal insulin-stimulated transport. Interestingly, sensitivity of the transport system increased with K^+ depletion as shown in Figure 6. The concentration of insulin which gave half-maximal stimulation in control cells was 5nM while the K^+-depleted cells showed a shift to the left to about 0.5nM. It is important to note that cytochalasin B which inhibits facilitated glucose transport (Jung and Rampal, 1977) blocked greater than 95% of deoxyglucose uptake in both control and K^+-depleted cells (data not shown). This indicates that non-specific uptake was minimal in any of the transport assays. It also verifies that the plasma membranes were intact.

DISCUSSION

We show for the first time the effect of hypotonic treatment, which depletes intracellular K^+, on glucose transport in an insulin-sensitive cell line. The data are consistent with inhibition of internalization of insulin receptors in that the cell surface receptor number increases while receptor-mediated insulin degradation decreases when intracellular K^+ concentration is reduced from 170mM to 40mM. In addition, maximal rates of insulin-sensitive transport do not change with K^+ depletion suggesting that the signal(s) generated to activate (or translocate) the insulin-sensitive glucose transporter (also referred to as GLUT 4) do not require the internalization of the receptor.

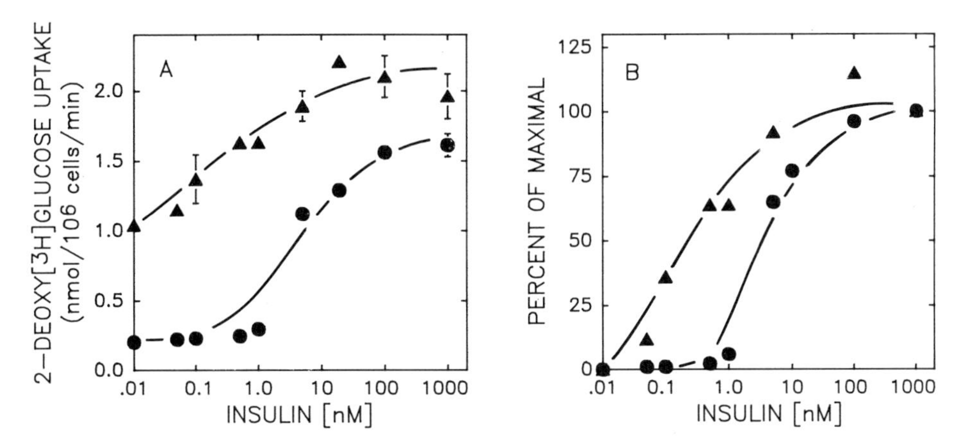

Figure 6. Effect of Potassium Depletion on Dose-dependent, Insulin-sensitive Glucose Transport. A. Control and K^+-depleted cells at 4h were incubated with varying concentrations of insulin for 10 min. Transport was determined in a terminal 10min uptake assay with 0.2 mM [^{3}H]deoxyglucose. The values represent quadruplicate points with a standard deviation of less than 8%. B. The data from panel A were replotted according to percent of maximal stimulation.

The role of K^+ in insulin action has been under consideration since the 1950's when it was shown in vivo that insulin stimulates K^+ uptake into muscle which correlated with an increase in glucose metabolism (Zierler, 1957). The insulin-stimulated K^+ uptake has been verified with isolated cells including muscle (Zierler and Rogus, 1981), isolated rat adipocytes (Resh et al., 1980) and 3T3-L1 adipocytes (Resh, 1982). In addition, insulin causes significant hyperpolarization of both muscle cells (Zierler and Rogus, 1981) and adipocytes (Cheng et al., 1981) within just a few seconds of insulin addition, potentially representing the first metabolic effect of insulin. It has been suggested that K^+ uptake is involved in or causes hyperpolarization. Two mechanisms for the increase in K^+ uptake have been suggested. One possibility is by activation of the Na^+/K^+ ATPase. This enzyme causes the efflux of 3 moles of Na^+ per mole of ATP hydrolyzed but less than 3 moles of K^+ taken up, thus the hyperpolarization. However, ouabain, which inhibits the Na^+/K^+ ATPase, has no effect on insulin-stimulated glucose transport (Resh, 1982). In fact, hyperpolarization occurs in the presence of ouabain (Lantx et al., 1980). Thus the alternative explanation for increased K^+ uptake, that insulin increases membrane permeability to K^+, seems favored.

Another insulin-sensitive pathway which appears to have K^+-dependency

Another insulin-sensitive pathway which appears to have K^+-dependency is protein synthesis. Insulin stimulates protein synthesis within minutes of addition to 3T3-L1 adipocytes (Pettengell and Frost, 1989). It has been shown that K^+ depletion inhibits protein synthesis in human fibroblasts (Ledbetter and Lubin, 1977). We have noted a similar effect in the 3T3-L1 adipocytes (Frost and Risch, unpublished data). However, we have also shown that insulin-sensitive transport occurs in the absence of new protein synthesis in 3T3-L1 adipocytes (Pettengell and Frost, 1989) as have others in isolated rat adipocytes (Kono et al., 1981).

It has been documented in cultured fibroblasts that K^+-depletion inhibits the internalization of the LDL receptor (Larkin et al., 1983). Further investigation revealed that K^+ was required for the formation of the polygonal clathrin basket associated with the vesicle (Larkin et al., 1986). One of the first steps in receptor-mediated endocytosis is the aggregation of receptors into coated pits. The coated pit matures to form a coated vesicle as the structural proteins surrounding the vesicle dissociate from the plasma membrane. Thus when basket formation was prevented by K^+ depletion, the pit does not mature into a vesicle. However, internal LDL receptors can still recycle, i.e., return to the cell surface, since this process is K^+-independent (Larkin et al., 1983). To reduce intracellular K^+ to the same extent as reported by Larkin et al. (1983), we found that we had to extend not only the hypotonic treatment from 5 to 10 min but also the K^+-free, isotonic incubation from 1 to 4h. Despite this, our results appear similar to those analyzing LDL receptor internalization based on the increase in the number of cell surface insulin receptors and reduced insulin degradation which suggests that internalization, but not recycling, is blocked. Based on other data, we have postulated that recycling occurs by a coated-vesicle independent pathway (Frost et al., 1989) which may underlie the K^+-independent nature of the process (Larkin et al., 1983).

Our conclusion that insulin receptor internalization is not required for insulin-stimulated glucose transport is based on the fact that insulin receptors in 3T3-L1 adipocytes traverse the coated vesicle pathway when internalized. In a recent study, it was demonstrated that insulin stimulated the association of clathrin with the plasma membrane (Corvera, 1990) which provides indirect but supportive data. Two other studies have examined this issue at the ultrastructural level. The first study by Fan et al. (Fan et al., 1983) showed that during the first two minutes of incubation, ^{125}I-insulin became associated with microvilli and coated pits. Over 60 min, clear endocytotic vesicles and lysosomal-like structures became labeled. Another study revealed that within 1 min of insulin addition, 50% of the cell surface receptors migrated to both coated and non-coated regions of the plasma membrane where 70% of these of these receptors were aggregated (Smith et al., 1985). Although occupied receptors were found at the same level in both the non-coated and coated regions, the non-coated regions represented about 90% of the total aggregation areas at steady-state. Since these micrographic studies were of fixed cells, there was no way to determine if the non-coated regions eventually became coated pits while maintaining the steady-state distribution. Migration to non-coated areas may actually represent the first step in coated pit maturation and thus receptor internalization.

Many studies have addressed the role of receptor internalization. Our data in addition to other studies show that the receptor mediates insulin degradation. This process removes the activating ligand from the receptor reducing the stimulus. Interestingly, this process in 3T3-

L1 adipocytes is chloroquine-insensitive which differs from data in isolated adipocytes where chloroquine completely blocked insulin degradation (Olefsky and Sackow, 1982; Smith and Jarett, 1982). Thus initially (over 30min), lysosomes play little role in insulin degradation. On the other hand, the degradation process in 3T3-L1 adipocytes is completely inhibited by bacitracin, an antibiotic which inhibits non-lysosomal proteases. The location of these proteases are unknown. Based on the rate of insulin degradation and the amount released from the cells compared to the amount associated with the cells, it seems likely that insulin degradation occurs close to the cell surface and near the site of receptor/ligand dissociation.

Potassium depletion blocked 70% of receptor-mediated degradation. As mentioned, based on the increase in cell surface binding, we interpret this to mean that insulin is not transported to its site of degradation. However, these data could also mean that degradation, like protein synthesis, is specifically K^+-dependent. Although we have been unable to find any studies addressing this type of regulation, arguing against this is the fact that K^+ depletion did not affect receptor-independent degradation.

Few studies have been able to dissociate the process of receptor internalization with that of insulin action. Within one minute of insulin addition, insulin receptors in 3T3-L1 adipocytes become maximally phosphorylated (Kohanski et al., 1986), a response which is mimicked by Concanavalin A in hepatocytes and CHO cells (Shiba et al., 1990). Thus receptor aggregation and autophosphorylation occur within the same time frame. Receptor internalization (Knutson et al., 1983) and glucose transport activation (Frost and Lane, 1985) clearly occur after this step. Thus one of the difficulties in addressing internalization versus activated transport has been the similar time frame over which the processes occur. In addition, the internalized receptor has at least partial activity with respect to activation of glucose transport (Ueda et al., 1985). In previous studies, we were able to show that phenylarsine oxide, a trivalent arsenical which binds vicinal sulphydryls, blocked insulin-stimulated glucose transport (Frost and Lane, 1985) without affecting insulin association (Frost and Lane, 1985) or receptor autophosphorylation (Frost et al., 1987). These data suggested that the signalling cascade was interrupted. However, phenylarsine oxide also blocks fluid-phase (Frost et al., 1989) as well as receptor-mediated endocytosis (Knutson et al., 1983). Although this supports the notion that receptor internalization is important for signal transmission, the present study shows that receptor internalization can be uncoupled from transport activation. In fact, we have recently demonstrated that radiolabeled phenylarsine oxide binds to specific cellular proteins potentially involved in signal transmission (Frost and Schwalbe, 1990).

In conclusion, we have shown that potassium depletion appears to block internalization of the insulin receptor supporting processing by a coated vesicle pathway. However, internalization of the receptor is not a prerequisite for insulin-stimulated glucose transport suggesting that signals generated at the surface can be transmitted to the transport system independent of the vesicular sorting pathway.

REFERENCES

Backer, J. M., Kahn, C. R., and White, M. F. (1989) Proc. Natl. Acad. Sci. USA 86, 3209-3213.

Bishayee, B., and Das, M. (1982) Arch. Biochem. Biophy. 214, 425-430.
Cheng, K., Groarke, J., Osotimehin, B., Haspel, H. C., and Sonenberg, M. (1981) J. Biol. Chem. 256, 649-655.
Corvera, S. (1990) J. Biol. Chem. 265. 2413-2416.
Ellis, L., Clauser, E., Morgan, D. O., Edery, M., Roth, R. A., Rutter, W. J. (1986) Cell 45, 721-732.
Fan, J. Y., Carpentier, J. L., VanObberghen, E., Blackett, N. M., Grunfeld, C., Gorden, P., and Orci, L. (1983) J. Histochem. Cytochem. 31, 859-870.
Frost, S. C., and Lane, M. D. (1985) J. Biol. Chem. 260, 2646-2652.
Frost, S. C., Kohanski, R. A., and Lane, M. D. (1987) J. Biol. Chem. 262, 9872-9876.
Frost, S. C., Lane, M. D., and Gibbs, E. M. (1989) J. Cell. Physiol. 141, 467-474.
Frost, S. C., and Schwalbe, M. S. (1990) Biochem. J. 269, 589-595.
Gliemann, J., and Sonne, O.. (1978) J. Biol. Chem. 253, 7857-7863.
Hedo, J. A. and Simpson, I. A. (1984) J. Biol. Chem. 259, 11083-11089.
Jung, C. Y., and Rampal, A. L. (1977) J. Biol. Chem. 252, 5456-5463.
Kasuga, M., White, M. F., and Kahn, C. R. (1985) in Methods of Enzymology 109, 609-621.
Klein, H. H., Freidenberg, G. R., Matthaei, S., and Olefsky, J. M. (1987) J. Biol. Chem. 262, 10557-10564.
Knutson, V. P., Ronnett, G. V., and Lane, M. D. (1983) J. Biol. Chem. 258, 12139-12142.
Kohanski, R. A., Frost, S. C., and Lane, M. D. (1986) J. Biol. Chem. 261, 12272-12281.
Kono, T., Suzuki, K., Dansey, L. E., Robinson, F. W., and Blevins, T. L. (1981) J. Biol. Chem. 256, 6400-6407.
Krupp, M. N., and Lane, M. D. (1982) J. Biol. Chem. 257, 1372-1377.
Lantx, R. C., Elsas, L. J., and R. L DeHaan (1980) Proc. Natl. Acad. Sci. USA 77, 3062-3066.
Larkin, J. M., Brown, M. S., Goldstein, J. L., and Anderson, R. G. W. (1983) Cell 33, 273-285.
Larkin, J. M., Donzell, W. C., and Anderson, R. G. W. (1986) J. Cell Biol. 103, 2619-2627.
Ledbetter, M. L. S., and Lubin, M. (1977) Experimental Cell Res. 105, 223-236.
Makinen, K. K. (1972) Int. J. Protein. Res. 4, 21-28.
Morgan, D. O., and Roth, R. A. (1987) Proc. Natl. Acad. Sci. USA 84, 41-45.
Olefsky, J. M., and Saekow, M., (1982) Molec. and Cell. Biochem. 47, 23-29.
Pettengell, K., and Frost, S. C. (1989) Biochem. Biophys. Res. Comm. 161, 633-639.
Poole, D. T., Butler, T. C., and Williams, M. E. (1972) Biochim. Biophys. Acta 266, 463-470.
Reed, B. C., Glasted, K., and Miller, B. (1984) J. Biol. Chem. 259, 8134-8143.
Resh, M. D., Nemenoff, R. A., and Guidotti, G. (1980) J. Biol. Chem. 255, 10938-10945.
Resh, M. C. (1982) J. Biol. Chem. 257, 6978-6986.
Scatchard, G. (1949) Ann. N. Y. Acad. Sci. 51, 660-672.
Shiba, T., Tobe, K., Koshio, O., Yamamoto, R., Shibasaki, Y., Matsumoto, N., Toyoshima, S., Osawa, T., Akanuma, Y., Takaku, F., Kasuga, M. (1990) Biochem. J. 267, 787-794.
Smith R. M., and Jarett, L. (1982) Proc. Natl. Acad. Sci. USA 79, 7302-7306.
Smith, R. M., Cobb, M. H., Rosen, O. M., and Jarett, L. (1985) J.Cell. Physiol. 123, 167-179.

Thies, R. S., Webster, N. J., and McClain, D. A. (1990) J. Biol. Chem. 265, 10132-10137.
Ueda, M., Robinson, F. W., Smith, M. M., and Kono, T. (1985) J. Biol. Chem. 260, 3941-3946.
White, M. F., Takayama, S., and Kahn, C. R. (1985) J. Biol. Chem. 260, 9470-9478.
Wilden, P. A., Backer, J. M., Kahn, C. R., Cahill, D. A., Shroeder, G. J., and White, M. F. (1990) Proc. Natl. Acad. Sci. USA 87, 3358-3362.
Zierler, K. L. (1957) Science 126, 1067-1068.
Zierler, K. L., and Rogus, E. M., (1981) Biochim. Biophys. Acta 640, 687-692.

This work was supported by research grants from the National Institutes of Health.

ENDOGENOUS SUBSTRATES OF THE INSULIN RECEPTOR: STUDIES WITH CELLS EXPRESSING WILD-TYPE AND MUTANT RECEPTORS

Kazuyoshi Yonezawa, Sarah Pierce, Cynthia Stover, Martine Aggerbeck*, William J. Rutter* and Richard A. Roth

Department of Pharmacology,
Stanford University School of Medicine
Stanford, CA 94305
and
*Hormone Research Institute and
Department of Biochemistry and Biophysics
University of California, San Francisco
San Francisco, CA 94143

INTRODUCTION

The insulin receptor has an intrinsic tyrosine kinase activity which appears to be required for insulin to elicit its various biological responses (1). Identification of endogenous substrates of the insulin receptor kinase and other tyrosine kinases has been greatly advanced in the last few years by the development of high affinity polyclonal and monoclonal antibodies to phosphotyrosine (2-4). These reagents can be used to identify substrates of tyrosine kinases by immunoblotting and immunoprecipitation as well as to purify substrates by utilizing affinity columns composed of anti-phosphotyrosine antibodies. Via these and other techniques, numerous proteins have been identified as becoming phosphorylated on tyrosine residues. These include a number of proteins with unknown functions. One such protein, M_r ~160,000 to 180,000, has been reported to be tyrosine phosphorylated in response to both insulin and insulin-like growth factor I in a variety of cell types (5-6). Other proteins that have been identified as substrates of tyrosine kinases include various enzymes as well as other defined molecules. One such identified molecule is another tyrosine kinase, called $pp60^{C-SRC}$. Activation of the platelet derived growth factor (PDGF) receptor tyrosine kinase was shown to increase the extent of tyrosine phosphorylation of the SRC protein and increase its enzymatic activity (7, 8). More recently, a phospholipase C (9-11), the type I phosphatidylinositol kinase (12-13), the GTPase activating protein of Ras (called GAP) (14, 15)and several serine/threonine kinases (the MAP2 kinase, the proto-oncogene Raf kinase and a cell cycle dependent kinase, CDC-2) (16-18) have all been shown to be tyrosine phosphorylated. The role of the tyrosine phosphorylation of these different molecules is not always clear. Only the physiological role of the tyrosine phosphorylation of the yeast cell cycle dependent kinase has been directly demonstrated (19). However, the tyrosine kinase which phosphorylates this enzyme has not yet been identified. Tyrosine phosphorylation of phospholipase Cγ by the epidermal growth factor (EGF) receptor tyrosine kinase has been extensively documented in vitro and in vivo (9-11). However, this phosphorylation does not appear to affect the activity of the isolated enzyme and overexpression of this enzyme did not affect various biological responses to EGF (11, 20). In contrast, tyrosine phosphorylation of MAP2 and Raf kinases appears to stimulate their enzymatic activities (21-22).

To identify endogenous substrates of the insulin receptor kinase, we have been testing cells which overexpress insulin receptors for insulin-stimulated tyrosine phosphorylation of various proteins. We and others have been unable to detect insulin-stimulated tyrosine phosphorylation of phospholipase C, the GAP protein, CDC-2 and the Raf kinase (23-25). We and others have been able to demonstrate that insulin and insulin-like growth factor I (IGF-I) stimulate tyrosine phosphorylation of the type I phosphatidylinositol kinase (26-29). In addition, we have been able to detect numerous proteins of molecular weight ranging from 15,000 to 250,000 daltons which become tyrosine phosphorylated in response to insulin and IGF-I (30). To attempt to further elucidate the role of these proteins in insulin action, studies were now performed on cells expressing various mutated receptors as well as on various cell types overexpressing the insulin receptor.

IDENTIFICATION OF SUBSTRATES BY IMMUNOBLOTTING WITH ANTIPHOSPHOTYROSINE ANTIBODIES

Chinese hamster ovary (CHO) cells overexpressing either wild-type or mutant insulin receptors (31) or rat hepatoma (32) or 3T3 cells (33) overexpressing the wild-type receptor were treated with or without insulin, lysed and the lysates were analyzed by immunoblotting. In addition, cells were also treated with or without vanadate, an inhibitor of phosphotyrosine phosphatases (34). The blots were probed with antibodies directed against either phosphotyrosine (Fig. 1), the insulin receptor β subunit (Fig. 2), or control immunoglobulin (Fig. 3). The bound antibodies were detected with an alkaline-phosphatase conjugated second antibody and a histochemical substrate. In CHO cells overexpressing wild-type receptor, insulin alone stimulated the tyrosine phosphorylation of the β subunit of the receptor as well as several proteins of M_r of 35,000 to 55,000 (Fig. 1). As previously observed (30), the inclusion of vanadate in the media greatly increased the extent of phosphorylation of the receptor as well as the number of proteins that were observed to be tyrosine phosphorylated in response to insulin (Fig. 1). These tyrosine phosphorylated proteins did not appear to be proteolytic fragments of the receptor since they did not react with antibodies to the receptor (Fig. 2). This effect of vanadate required the presence of vanadate in the media of the cells for several hours prior to the addition of insulin and could not be duplicated by the addition of vanadate in the lysis buffer. The potentiating effect of vanadate was also observed in a line of rat hepatoma cells overexpressing the insulin receptor (Fig. 1). Indeed, in these cells it was difficult to observe an effect of insulin on tyrosine phosphorylation in the absence of vanadate. A line of 3T3 cells overexpressing the insulin receptor (33) exhibited a huge increase in tyrosine phosphorylation of numerous proteins when treated with vanadate (Fig. 1). This was not due to a nonspecific reaction since blots with control immunoglobulin did not exhibit any such reaction (Fig. 3). The effect of vanadate could be mediated through the insulin receptor kinase since vanadate treatment was observed to result in increased tyrosine phosphorylation (Fig. 1) and activation of the receptor kinase (data not shown).

CHO cells expressing various mutated receptors were also examined. Receptors lacking 30 or 14 amino acids in the carboxy-tail of the β subunit were examined since this region contains identified tyrosine (35) and threonine phosphorylation sites (36). In addition, a receptor with a carboxy-tail deletion of 43 amino acids has been reported to have increased kinase activity (37). In contrast, the receptors lacking 30 or 14 amino acids in their carboxy-tail appeared to have the same kinase activity as the wild-type receptor and to phosphorylate the same spectrum of substrates (Fig. 1). Another mutant receptor tested (YF-3) has had the two critical autophosphorylation sites (tyrosines 1162 and 1163) changed to phenylalanine (31). This mutant receptor was previously found to exhibit little insulin-stimulated receptor kinase activity in vitro. However, in vivo insulin was found to be capable of stimulating receptor autophosphorylation to a partial extent in this mutant receptor (31). In agreement with the latter findings, this receptor was found capable of mediating an insulin-stimulated increase in tyrosine phosphorylation of most of the same proteins as observed with the wild-type receptor although the extent of phosphorylation was less than the wild-type receptor (Fig. 1). These results indicate that this mutant receptor still retains an insulin-stimulated kinase activity in the intact cell. Since the in vitro studies could not detect such an effect of insulin, these results would suggest that in vitro assessments of the kinase activities of mutant receptors may not always reflect their in vivo activities.

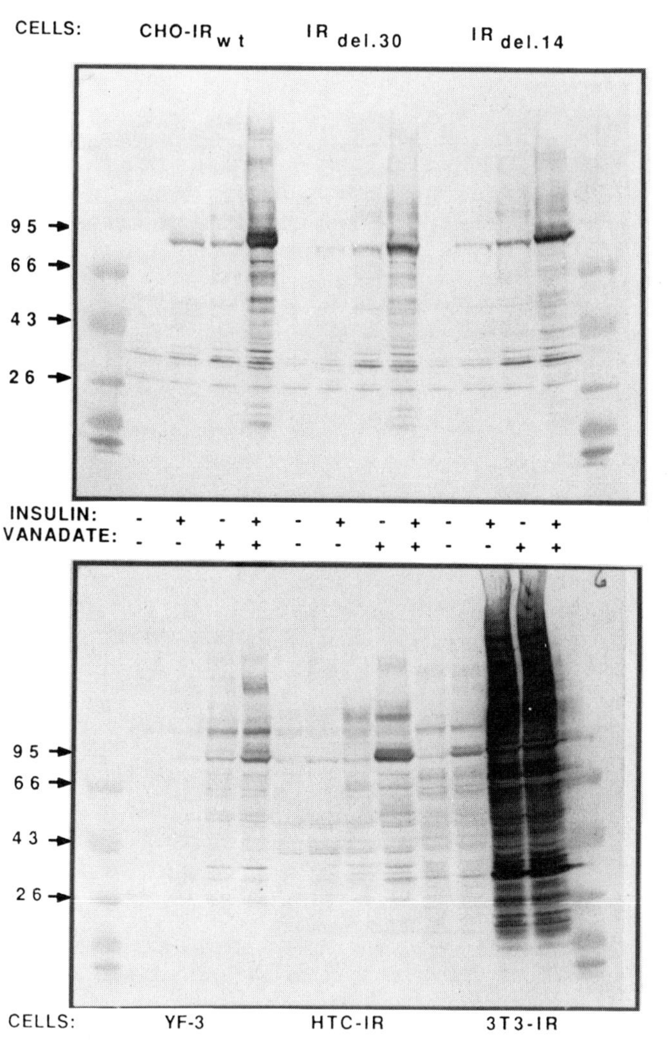

Figure 1. Western with anti-phosphotyrosine antibodies. The indicated cells were treated with or without vanadate (0.5 mM, 4 hrs) and then insulin (100 nM, 10 min) as indicated. Lysates were then prepared, run on a SDS-PAGE, transferred to nitrocellulose and reacted with polyclonal antiphosphotyrosine antibodies.

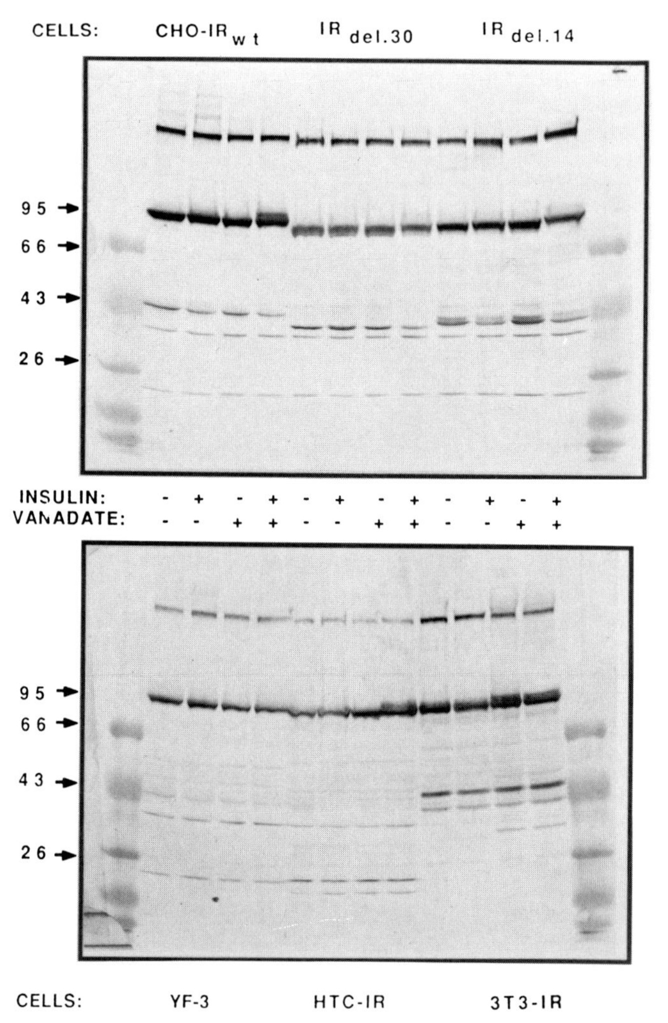

Figure 2. Western with anti-receptor antibodies. Equivalent amounts of the extracts as in Fig. 1 were run on SDS-PAGE, transferred to nitrocellulose and reacted with a polyclonal antibody to the cytoplasmic domain of the insulin receptor.

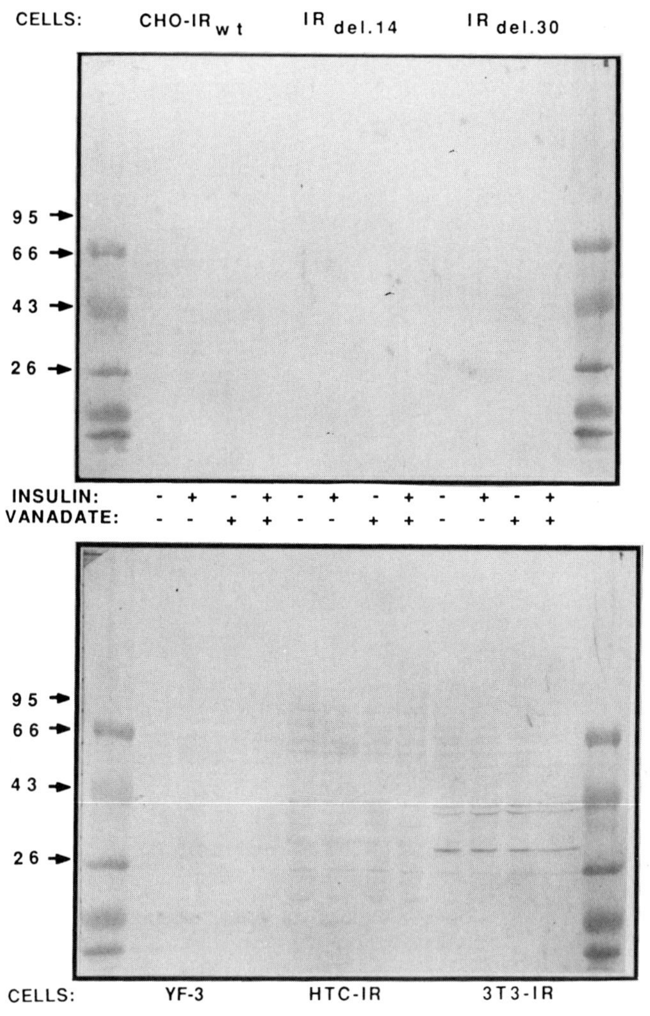

Figure 3. Western with normal immunoglobulin. Equivalent amounts of the extracts as in Fig. 1 were run on SDS-PAGE, transferred to nitrocellulose and reacted with control rabbit immunoglobulin.

STUDIES ON THE TYROSINE PHOSPHORYLATION OF THE TYPE I PHOSPHATIDYLINOSITOL KINASE IN VARIOUS CELLS

Insulin and IGF-I have been previously shown to stimulate the tyrosine phosphorylation of the type I phosphatidylinositol (PtdIns) kinase in CHO cells overexpressing their respective receptors (26-28). In addition, insulin and IGF-I have been shown to stimulate the formation of PtdIns-3, 4-P_2 in the transfected CHO cells and in a clonal strain of cultured Leydig cells (27, 29). We have now studied the tyrosine phosphorylation of PtdIns kinase in the hepatoma and 3T3 cells expressing wild-type insulin receptors and CHO cells expressing mutant receptors. Insulin was found to stimulate tyrosine phosphorylation of the PtdIns kinase in both the hepatoma and 3T3 cells expressing wild-type receptor (Fig. 4). In addition, insulin was found to stimulate the tyrosine phosphorylation of this enzyme in primary rat adipocytes (Fig. 4). These results indicate that the PtdIns kinase is a substrate for the insulin receptor kinase in a variety of cells and therefore can be utilized as a sensitive monitor of the activation of the insulin receptor kinase in vivo.

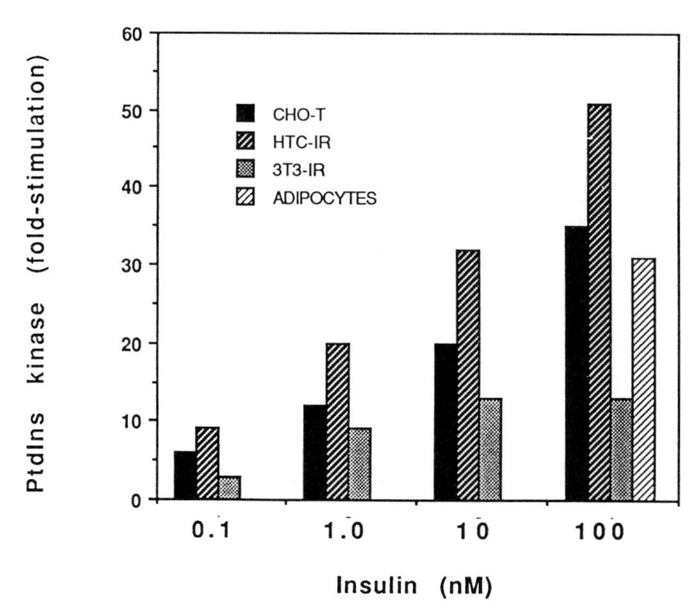

Figure 4. Insulin stimulation of tyrosine phosphorylation of PtdIns kinase in different cell types. The different cell types were incubated with the indicated concentrations of insulin, lysed and the lysates were adsorbed with a monoclonal antiphosphotyrosine (py20) antibody (4) bound to Protein G-sepharose. The beads were then washed and assayed for PtdIns kinase activity as described (26).

The CHO cells expressing the mutant IR were then examined. In cells expressing the receptor with the 30-amino acid carboxy-tail deletion, insulin was found capable of stimulating the tyrosine phosphorylation of the PtdIns 3-kinase with about the same sensitivity and to about the same level as the wild-type receptor (Fig. 5). In contrast, cells expressing mutant receptors with the critical tyrosine autophosphorylation sites 1162/1163 changed to phenylalanine exhibited ~1/3 the ability of the wild-type receptor in stimulating tyrosine phosphorylation of the PtdIns kinase (38). These results are in good agreement with the studies utilizing immunoblotting to assess their in vivo kinase activities.

STUDIES ON THE CELLULAR LOCALIZATION AND RECEPTOR CROSS-LINKING OF THE PtdIns KINASE

The rapid tyrosine phosphorylation of the type I PtdIns kinase in response to insulin is consistent with this enzyme being a direct substrate of the insulin receptor tyrosine kinase (26, 27). To further test this hypothesis, we attempted to crosslink the PtdIns kinase to the insulin receptor. Intact CHO cells overexpressing the wild-type insulin receptor were treated with various concentrations of two different bifunctional cross-linkers, lysed and the lysates were adsorbed with either monoclonal antibodies to phosphotyrosine or to the insulin receptor. Approximately 50% of the PtdIns kinase was found to be cross-linked to the receptor with either cross-linker (Fig. 6). However, both cross-linkers caused a decrease in the amount of PtdIns kinase activity recovered with the anti-phosphotyrosine antibodies. Lower concentrations of cross-linker did not cause as much of a loss in PtdIns activity but also did not increase the yield of cross-linked PtdIns kinase activity.

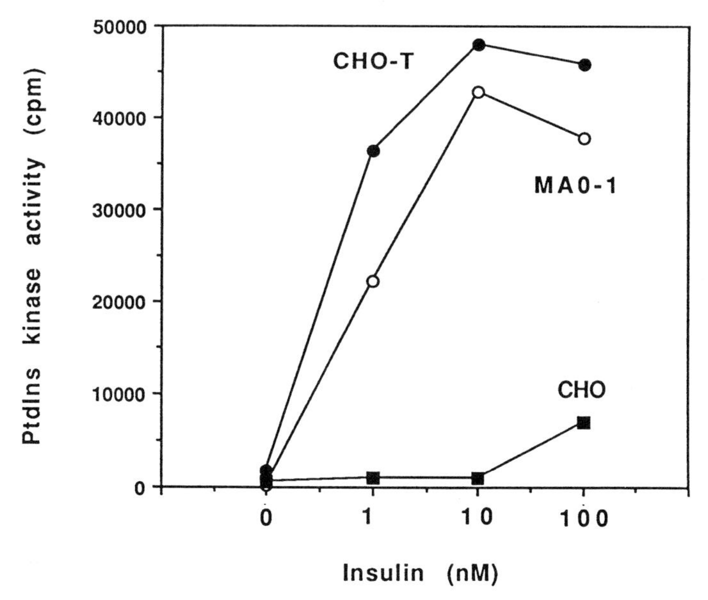

Figure 5. PtdIns kinase activity in cells expressing a truncated receptor. CHO cells overexpressing the wild-type receptor (CHO-T), a receptor lacking 30 amino acids at its carboxy-tail (MAO-1) or the parental cells (CHO) were treated with the indicated concentrations of insulin, lysed and the lysates were analyzed for tyrosine phosphorylated PtdIns kinase activity as described (26).

To further study the interaction of the PtdIns kinase and the insulin receptor, lysates of insulin treated cells were fractionated into membrane and cytosolic preparations and the amount of tyrosine phosphorylated and receptor associated PtdIns kinase in each fraction was determined (Fig. 7). Approximately 4% of the tyrosine phosphorylated PtdIns kinase was found to be in the membrane fraction. Of the membrane associated fraction, ~25% was immunoprecipitated with a monoclonal anti-receptor antibody. Thus, overall ~1% of the tyrosine PtdIns kinase was associated with the receptor. This agrees with our finding that ~1 to 3% (average of seven experiments was 2.1%) of the PtdIns kinase activity associates with receptor in immunoprecipitates from total cell lysates. This may be an underestimate of the fraction which is associated with the receptor in the cell since some of the PtdIns kinase could dissociate from the receptor during the wash procedure. Attempts to increase the receptor associated fraction by utilizing different wash conditions, milder detergents for lysis (digitonin, CHAPS, octylglucoside), or varying the time and temperature of insulin treatment were unsuccessful.

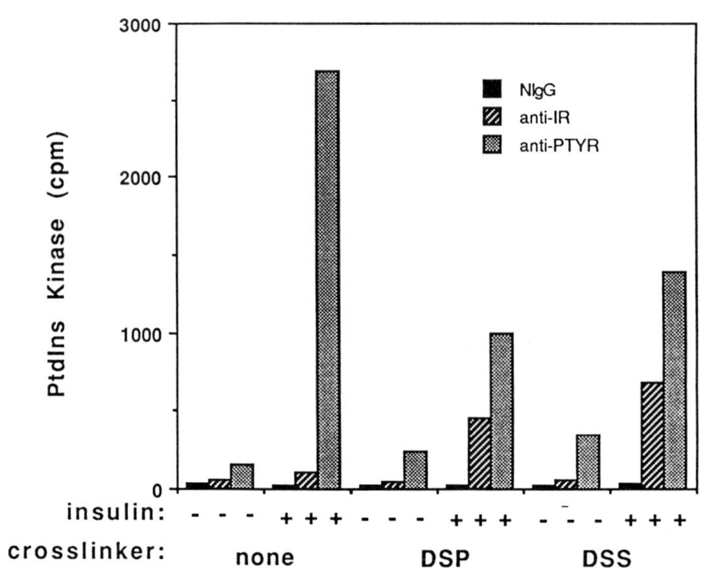

Figure 6. Cross-linking PtdIns kinase to the insulin receptor. Intact CHO-T cells were treated with 100 nM insulin for 5 min at 37°C then either buffer, 150 μg/ml of dithiobis (succinimidyl propionate) (DSP) or 150 μg/ml disuccinimidyl suberate (DSS) was added for 30 min at 4°C. Cells were then lysed and the lysates were adsorbed with either control IgG (NIgG), monoclonal anti-insulin receptor antibody 5D9 (anti-IR) or monoclonal anti-phosphotyrosine antibody py20 (anti-PTYR). Shown are the amounts of PtdIns kinase activities present in the different precipitates.

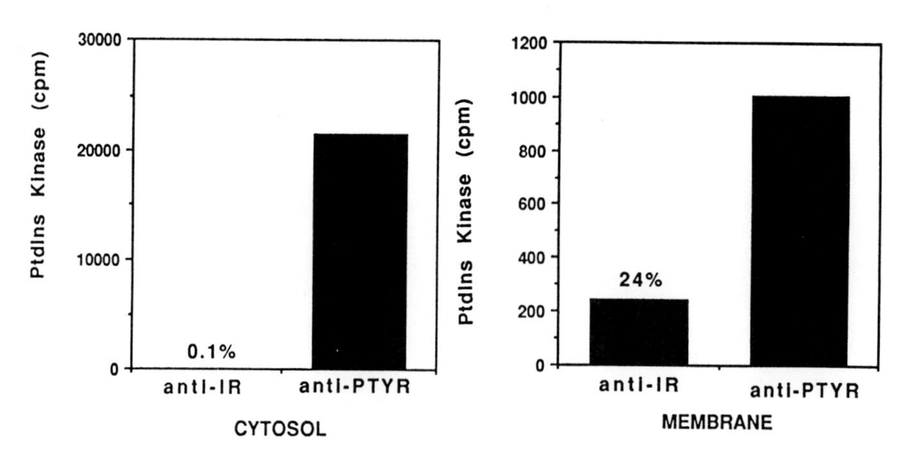

Figure 7. Association of PtdIns kinase with the insulin receptor. CHO-T cells were treated with 100 nM insulin for 5 min at 37°C and then the cells were lysed and membrane and cytosol fractions were prepared (38). These fractions (after solubilization of the membranes) were adsorbed with the monoclonal anti-insulin receptor (anti-IR) or anti-phosphotyrosine (anti-PTYR) and the precipitates assayed for PtdIns kinase activity.

STUDIES ON INSULIN STIMULATION OF THYMIDINE INCORPORATION

Insulin stimulation of thymidine incorporation into DNA was examined in CHO cells overexpressing either the wild-type receptor, the receptor lacking the two critical tyrosine autophosphorylation residues 1162 and 1163, the parental CHO cells, or the receptor lacking 30 amino acids at the carboxy-tail. As previously observed, overexpression of the wild-type receptor resulted in an increased responsiveness to low concentrations of insulin in comparison to the parental cells (Fig. 8). The cells overexpressing the mutant receptor lacking the twin tyrosines also exhibited a small increase in sensitivity to insulin in comparison to the parental cells. However, they did not exhibit as much of a shift in sensitivity as cells overexpressing the wild-type receptor (Fig. 8). In contrast, cells expressing the receptor with the carboxy-tail deletion exhibited the same increase in sensitivity to insulin as the cells overexpressing the wild-type receptor (Fig. 9).

DISCUSSION

In the present studies we have analyzed substrates of the insulin receptor kinase in various cell lines overexpressing the wild-type and mutant insulin receptors. In agreement with prior studies (30), a large number of proteins can be observed by immunoblotting to be tyrosine phosphorylated in response to insulin when cells are incubated with insulin in the presence of vanadate (Fig. 1). This effect of vanadate may be due to its ability to inhibit tyrosine phosphatases (34) or to some other property of this molecule. Phenylarsine oxide has also been observed to allow one to increase the number of proteins that appear to be tyrosine phosphorylated in response to insulin (30). The presence of either phenylarsine oxide or vanadate allows one to increase the extent of tyrosine phosphorylation of both the insulin receptor (Fig. 1) and the PtdIns kinase (unpublished studies) over that observed with insulin alone. Thus, it is possible that these compounds increase the extent of tyrosine phosphorylation of various proteins which are normally phosphorylated to a low level in the presence of insulin alone. However, it is also possible that the presence of these compounds allows the tyrosine phosphorylation of proteins which normally do not get tyrosine phosphorylated.

The various proteins which became tyrosine phosphorylated in response to insulin appeared to be the same in CHO cells expressing either wild-type receptors or mutant receptors lacking 30 or 14 amino acids from the carboxy-tail or a mutant receptor lacking the two critical tyrosine autophosphorylation residues 1162 and 1163 (Fig. 1). The carboxy-tail deleted mutants appeared to have approximately the same kinase activity as the wild-type receptor in the immunoblotting assay. In addition, the mutant receptor lacking the 30 amino acids also appeared to have the same kinase activity as the wild-type receptor when its activity was assessed by its ability to mediate the tyrosine phosphorylation of PtdIns kinase (Fig. 5) and the same ability as the wild-type receptor to stimulate thymidine incorporation (Fig. 9). These results would suggest that the two tyrosine autophosphorylation sites (residues 1328 and 1334) and the threonine phosphorylation site (residue 1348) lacking in this mutant do not normally play a role in regulating the receptor kinase. In contrast, a mutant receptor lacking 43 amino acids in its carboxy-tail has been reported to have increased tyrosine kinase activity and ability to stimulate thymidine incorporation in comparison to the wild-type receptor (37). It is possible that the difference between these results is due to the removal of a regulatory residue(s) in this longer deletion mutant but still present in the shorter mutant.

The mutant insulin receptor with tyrosine residues #1162 and 1163 changed to phenylalanine was observed to have a decreased tyrosine kinase activity in comparison to the wild-type receptor when this activity was assessed via either immunoblotting (Fig. 1) or the tyrosine phosphorylation of PtdIns kinase (38). However, the activity observed with this mutant receptor was still more than had previously been observed when its kinase activity was assessed in vitro (31). These results would suggest that in vitro assays of the insulin receptor kinase activity may not always be a good reflection of its activity in the cell.

This same mutant receptor was also previously observed to have a decreased ability to mediate a biological response, insulin stimulation of glucose uptake (31). In addition, insulin stimulation of receptor internalization was also observed to be decreased in this mutant receptor (39). However, another study suggested that this mutant receptor was equal to the wild-type receptor in its ability to mediate a long term response, the stimulation of thymidine incorporation into DNA (40). However, in the present studies we have found that this mutant receptor exhibits a decreased ability to mediate this response in comparison to the wild-type receptor (Fig. 8). The decrease in this response is consistent with the observed decrease in the tyrosine kinase activity of this receptor.

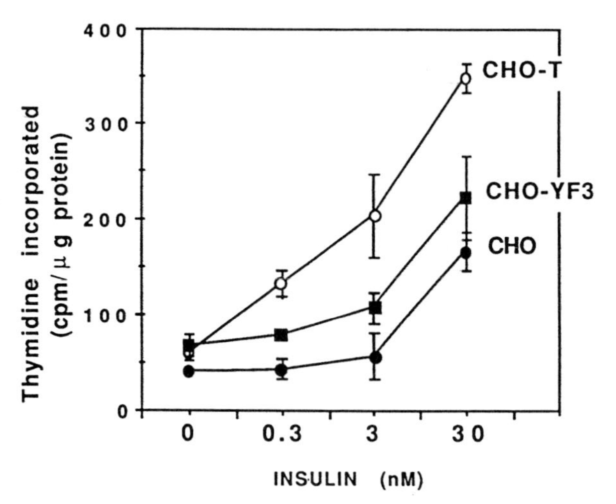

Figure 8. Insulin stimulation of thymidine incorporation into cells expressing mutant receptors lacking two autophorylation sites. CHO cells overexpressing wild-type receptors (CHO-T), mutant receptors with tyrosines 1162 and 1163 changed to phenylalanine (CHO-YF3) or the parental cells (CHO) were treated with the indicated concentrations of insulin and [^{3}H]-thymidine incorporation into DNA was measured as described (28). Results have been normalized for the amount of protein present in the lysates and are means ± SD of triplicate determinations.

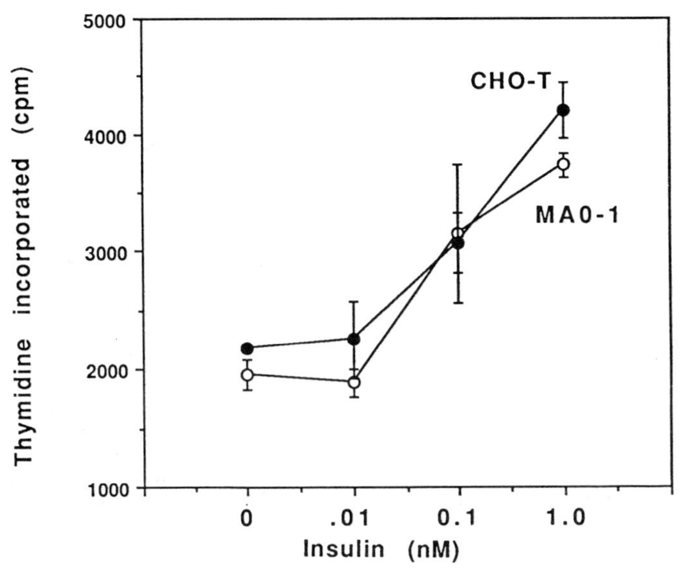

Figure 9. Insulin stimulation of thymidine incorporation into cells expressing mutant receptors lacking the carboxy-tail. CHO cells overexpressing wild-type receptors (CHO-T) or mutant receptors lacking 30 amino acids from the carboxy-tail (MAO-1) were treated with the indicated concentrations of insulin and [^{3}H]-thymidine incorporation into DNA was measured as described (28).

The twin tyrosine mutant receptor's decreased ability to mediate biological responses also paralleled its decreased ability to cause the tyrosine phosphorylation of the PtdIns kinase (38). These results are consistent with a role for this enzyme in mediating a biological response to insulin. In addition, the finding that the PtdIns kinase can be cross-linked to the receptor is consistent with this enzyme being a direct substrate of the receptor kinase (Fig. 6). The finding that a small percentage of the PtdIns kinase is associated with the receptor (~1% to 2% in our experiments) even in the absence of cross-linker is in agreement with the studies of Ruderman et al. (27) and is also consistent with the enzyme directly interacting with the receptor. The actual percent which is found associated with the receptor may vary depending on the lysis and wash conditions. The percentage of PtdIns kinase associated with the receptor may actually be much higher in the cell. Additional studies on the association of this enzyme with the receptor as well as its role in insulin action must await its purification and generation of antibodies to the molecule.

ACKNOWLEDGMENT

The authors are grateful to Drs. John Glenney for the py20 antibody, Ira Goldfine for the HTC-IR cells, Jonathan Whittaker for 3T3-IR cells, Jan Stagsted and Lennart Olsson for rat adipocytes and Suvan Gerlach for preparation of the manuscript. Work in the authors laboratories were supported by NIH grants DK 34926, DK 41765 and DK 37661.

REFERENCES

1. O.M. Rosen, After insulin binds, Science 237:1452 (1987).
2. H.A. Ross, D. Baltimore and H.N. Eisen, Phosphotyrosine-containing proteins isolated by affinity chromatography with antibodies to a synthetic hapten, Nature 294:654 (1981).
3. A.R. Frackelton, A.H. Ross and H.N. Eisen, Characterization and use of monoclonal antibodies for isolation of phosphotyrosyl proteins from retrovirus-transformed cells and growth factor-stimulated cells, Mol. Cell. Biol. 3:1343 (1983).
4. J.R. Glenney, L. Zokas and M.J. Kamps, Monoclonal antibodies to phosphotyrosine, J. Immunol. Meth. 109:277 (1988).
5. M.F. White, R. Maron and C.R. Kahn, Insulin rapidly stimulates tyrosine phosphorylation of a M_r-185,000 protein in intact cells, Nature 318:183 (1985).
6. M. Kasuga, T. Izumi, K. Tobe, T. Shiba, K. Momomura, Y. Tashiro-Hashimoto and T. Kadowaki, Substrates for insulin-receptor kinase, Diabetes Care 13:317 (1990).
7. R. Ralston and J.M. Bishop, The product of the protooncogene c-src is modified during the cellular response to platelet-derived growth factor, Biochemistry 82:7845 (1985).
8. K.L. Gould and T. Hunter, Platelet-derived growth factor induces multisite phosphorylation of $pp60^{c-src}$ and increases its protein-tyrosine kinase activity, Mol. Cell. Biol. 8:3345 (1988).
9. M.I. Wahl, T.O., Daniel and G. Carpenter, Antiphosphotyrosine recovery of phospholipase C activity after EGF treatment of A-431 cells, Science 241:968 (1988).
10. S. Nishibe, M.I. Wahl, S.G. Rhee and G. Carpenter, Tyrosine phosphorylation of phospholipase C-II in vitro by the epidermal growth factor receptor, J. Biol. Chem. 264:10335 (1989).
11. J. Meisenhelder, P.G. Suh, S.G. Rhee and T. Hunter, Phospholipase C-gamma is a substrate for the PDGF and EGF receptor protein-tyrosine kinases in vivo and in vitro, Cell 57:1109 (1989).
12. S.A. Courtneidge and A. Heber, An 81 kd protein complexed with middle T antigen and pp 60^{c-src}: A possible phosphatidylinositol kinase, Cell 50:1031 (1987).
13. D.R. Kaplan, M. Whitman, B. Schaffhausen, D.C. Pallas, M. White, L. Cantley and T.M. Roberts, Common elements in growth factor stimulation and oncogenic transformation: 85 kd phosphoprotein and phosphatidylinositol kinase activity, Cell 50:1021 (1987).
14. C.J. Molloy, D.P. Bottaro, T.P. Fleming, M.S. Marshall, J.B. Gibbs and S.A. Aaronson, PDGF induction of tyrosine phosphorylation of GTPase activating protein, Nature 342:711 (1989).
15. C. Ellis, M. Moran, F. McCormick and T. Pawson, Phosphorylation of GAP and GAP-associated proteins by transforming and mitogenic tyrosine kinases, Nature 343:377 (1990).
16. A.J. Rossomando, D.M. Payne, M.J. Weber and T.W. Sturgill, Evidence that pp42, a major tyrosine kinase target protein, is a mitogen-activated serine/threonine protein kinase, Proc. Natl. Acad. Sci. USA, 86:6940 (1989).
17. D.K. Morrison, D.R. Kaplan, U. Rapp and T.M. Roberts, Signal transduction from membrane to cytoplasm: Growth factors and membrane-bound oncogene products increase Raf-1 phosphorylation and associated protein kinase activity, Proc. Natl. Acad. Sci. USA, 85:8855 (1988).

18. A.O. Morla, G. Draetta, D. Beach and J.Y.J. Wang, Reversible tyrosine phosphorylation of cdc2: dephosphorylation accompanies activation during entry into mitosis, Cell 58:193 (1989).
19. K.L. Gould and P. Nurse, Tyrosine phosphorylation of the fission yeast cdc2$^+$ protein kinase regulates entry into mitosis, Nature, 342:39 (1989).
20. B. Margolis, A. Zilberstein, C. Franks, S. Felder, S. Kremer, A. Ullrich, S.G. Rhee, K. Skorecki, J. Schlessinger, Effect of phospholipase C-γ overexpression on PDGF-induced second messengers and mitogenesis, Science 248:607 (1990).
21. D.K. Morrison, D.R. Kaplan, J.A. Escobedo, U.R. Rapp, T.M. Roberts and L.T. Williams, Direct activation of the serine/threonine kinase activity of Raf-1 through tyrosine phosphorylation by the PDGF β-receptor, Cell 58:649 (1989).
22. N.G. Anderson, J.L. Maller, N.K. Tonks and T.W. Sturgill, Requirement for integration of signals from two distinct phosphorylation pathways for activation of MAP kinase, Nature 343:651 (1990).
23. K. Yonezawa, G. Endemann, K.S. Kovacina, J.E. Chin, C. Stover and R.A. Roth, Substrates of the insulin receptor kinase, in: "The Biology and Medicine of Signal Transduction," Y. Nishizuka et al., Publisher, New York (1990).
24. P.J. Blackshear, D. McNeill Haupt, H. App and U.R. Rapp, Insulin activates the Raf-1 protein kinase, J. Biol. Chem., 265:12131 (1990).
25. K.S. Kovacina, K. Yonezawa, D.L. Brautigan, N.K. Tonks, U.R. Rapp and R.A. Roth, Insulin activates the kinase activity of the Raf-1 proto-oncogene by increasing its serine phosphorylation, J. Biol. Chem., 265:12115 (1990).
26. G. Endemann, K. Yonezawa and R.A. Roth, Phosphatidylinositol kinase or an associated protein is a substrate for the insulin receptor tyrosine kinase, J. Biol. Chem., 265:396 (1990).
27. N.B. Ruderman, R. Kapeller, M.F. White and L.C. Cantley, Activation of phosphatidylinositol 3-kinase by insulin, Proc. Natl. Acad. Sci. USA, 87:1411 (1990).
28. G. Steele-Perkins and R.A. Roth, Monoclonal antibody αIR-3 inhibits the ability of insulin-like growth factor II to stimulate a signal from the type I receptor without inhibiting its binding, Biochem. Biophys. Res. Comm., in press.
29. O.P. Pignataro and M. Ascoli, Mol. Endocrinol, 4:758 (1990).
30. R.A. Roth, G. Steele-Perkins, J. Hari, C. Stover, S. Pierce, J. Turner, J.C. Edman and W.J. Rutter, Insulin and insulin-like growth factor receptors and responses, Cold Spring Harbor Symp., LIII:537 (1988).
31. L. Ellis, E. Clauser, D.O. Morgan, M. Edery, R.A. Roth and W.J. Rutter, Replacement of insulin receptor tyrosine residues 1162 and 1163 compromises insulin-stimulated kinase activity and uptake of 2-deoxyglucose, Cell, 45:721 (1986).
32. D.M. Hawley, B.A. Maddux, R.G. Patel, K-Y. Wong, P.W. Mamula, G.L. Firestone, A.Brunetti, E. Verspohl and I.D. Goldfine, Insulin receptor monoclonal antibodies that mimic insulin action without activating tyrosine kinase, J. Biol. Chem., 264:2438 (1989).
33. J. Whittaker, A.K. Okamoto, R. Thys, G.I. Bell, D.F. Steiner and C.A. Hofmann, High-level expression of human insulin receptor cDNA in mouse NIH 3T3 cells, Proc. Natl. Acad. Sci. USA 84:5237 (1987).
34. G. Swarup, S. Cohen, D.L. Garbers, Inhibition of membrane phosphotyrosyl protein phosphatase activity by vanadate, Biochem. Biophys. Res. Commun., 107:1104 (1982).
35. J. Avruch, H.E. Tornqvist, J.R. Gunsalus, E.J. Yurkow, Insulin regulation of protein phosphorylation, in: "Insulin," P. Cuatrecasas and S. Jacobs, ed., Springer-Verlag Berlin Heidelberg (1990).
36. R.E. Lewis, L. Cao, D. Perregaux and M.P. Czech, Threonine 1336 of the human insulin receptor is a major target for phosphorylation by protein kinase C, Biochemistry, 29:1807 (1990).
37. R.S. Thies, A. Ullrich and D.A. McClain, Augmented mitogenesis and impaired metabolic signaling mediated by a truncated insulin receptor, J. Biol. Chem., 264:12820 (1989).
38. K. Yonezawa and R.A. Roth, Assessment of the in situ tyrosine kinase activity of mutant insulin receptors lacking tyrosine autophosphorylation sites 1162 and 1163, Mol. Endocrinol., Submitted (1990).
39. C. Reynet, M. Caron, J. Magré, G. Cherqui, E. Clauser, J. Picard and J. Capeau, Mutation of tyrosine residues 1162 and 1163 of the insulin receptor affects hormone and receptor internalization, Mol. Endocrinol., 304 (1990).
40. A. Debant, E. Clauser, G. Ponzio, C. Filloux, C. Auzan, J.O. Contreres, B. Rossi, Replacement of insulin receptor tyrosine residues 1162 and 1163 does not alter the mitogenic effect of the hormone, Proc. Natl. Acad. Sci. USA, 85:8032 (1988).

MOLECULAR MECHANISMS INVOLVED IN THE ANTILIPOLYTIC ACTION OF INSULIN: PHOSPHORYLATION AND ACTIVATION OF A PARTICULATE ADIPOCYTE cAMP PHOSPHODIESTERASE

Vincent C. Manganiello*, Carolyn J. Smith*, Eva Degerman[++], Valeria Vasta**, Hans Tornqvist[++] and Per Belfrage[++]

*Laboratory of Cellular Metabolism, National Heart, Lung, and Blood Institute, National Institutes of Health, Building 10, Room 5N-307, Bethesda, MD 20892; **Department of Biochemistry, University of Florence, Florence, Italy; [++]Department of Medical and Physiological Chemistry, University of Lund, Lund, Sweden

INTRODUCTION

cAMP is an important intracellular second messenger in hormonal regulation of many physiological processes, including lipolysis, glycogenolysis, platelet aggregation, myocardial contractility, and smooth muscle relaxation (1-5). The isolated rat adipocyte has served as a useful model system in which to study hormonal and cAMP-mediated regulation of lipolysis. As outlined in Fig. 1, in rat adipocytes lipolytic hormones (e.g., catecholamines, glucagon, ACTH) and certain antilipolytic effectors (such as adenosine and prostaglandin E_1) interact with specific cell surface receptors and transmit stimulatory or inhibitory signals to the catalytic unit of adenylate cyclase via stimulatory or inhibitory guanyl nucleotide binding proteins, respectively. cAMP activates cAMP-dependent protein kinase (cAMP-PrK) which phosphorylates, on serine-563 (6,7), and activates the hormone-sensitive lipase, leading to hydrolysis of stored triglyceride with release of glycerol and free fatty acids. Steady state concentrations of cAMP are also regulated by cyclic nucleotide phosphodiesterases, enzymes that catalyze hydrolysis of cAMP to 5'AMP. From the scheme presented in Fig. 1, it is obvious that lipolysis can be regulated at several loci, i.e., at the level of cAMP formation or destruction, cAMP-PrK, protein phosphatase(s), etc.

Insulin is a physiologically important inhibitor of lipolysis (1,8). Although the precise mechanism(s) of the antilipolytic action of insulin is not completely understood and may very well involve integration of multiple insulin-regulated pathways, the antilipolytic effect can be correlated with an insulin-induced reduction in hormone-activated cAMP-PrK (9), which presumably reflects a decrease in a physiologically 4 relevant adipocyte cAMP pool (10,11).

Molecular Biology and Physiology of Insulin and Insulin-Like Growth Factors
Edited by M.K. Raizada and D. LeRoith, Plenum Press, New York, 1991

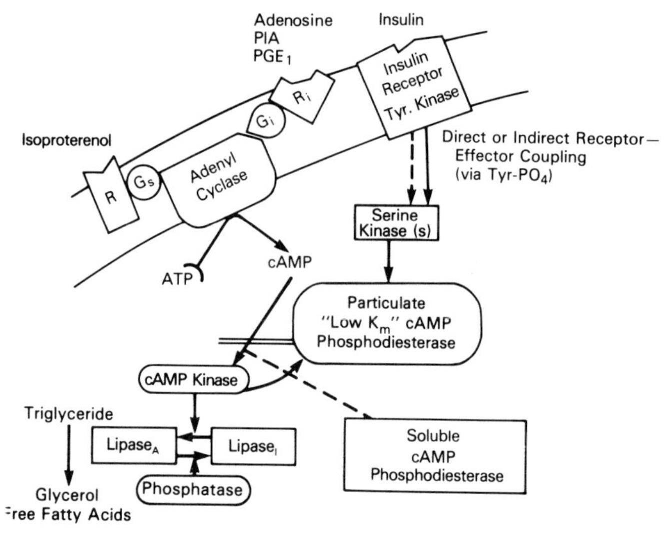

Fig. 1 Scheme for regulation of lipolysis.

Insulin could reduce adipocyte cAMP by inhibiting adenylate cyclase, activating cAMP PDE, or both. This review will focus on recent studies from our laboratories and others, which strongly suggest that, in intact dipocytes, insulin-induced activation of an intracellular serine kinase results in phosphorylation and activation of a specific, membrane-associated (i.e., particulate) cAMP PDE, a member of the so-called cGMP-inhibited "low K_m" cAMP PDE (cGI PDE) family (12-14). Insulin-induced activation of the particulate cGI PDE leads to an increase in cAMP hydrolysis which is important in (if not responsible for) insulin-induced reduction of hormone-stimulated cAMP and cAMP-PrK, and thus in insulin-induced reduction of hormone-activated triglyceride lipase and lipolysis (12-14).

Characteristics of cGMP-inhibited cAMP PDEs

Multiple types of cyclic nucleotide phosphodiesterases are found in different amounts and proportions in most mammalian cells. Members of five major PDE families have been identified, purified and characterized (Table 1). These PDE families, which seem to be products of distinct but related genes, differ in substrate (i.e., cAMP and cGMP) affinities, kinetic characteristics, physicochemical properties, responsiveness to specific effectors and inhibitors, regulatory properties and immunological epitopes (15,16). While the precise nature of the molecular and evolutionary relationships between the major PDE families remains to be

Table 1. Cyclic Nucleotide Phosphodiesterases

PDE Families	Isoenzyme Sub-Families (18)
1) Ca^{++}-, calmodulin-sensitive	7
2) cGMP-stimulated	2
3) cGMP-inhibited (Milrinone, Enoximone, Cilostamide, etc.)	2
4) RO 20-1724-inhibited (Rolipram)	8
5) cGMP-specific	
a) Photoreceptor	3
b) cGMP-binding	1

determined, each major PDE family seems to include a number of isoenzyme subfamilies identified by biochemical characterization and molecular genetics (Table 1) (15,16).

The important characteristics of the cGI PDE family are summarized in Table 2 (for review see ref. 12). This family exhibits high affinities for both cAMP and cGMP, with V_{max} 5-10 times greater for cAMP than cGMP, and is specifically and potently inhibited by cGMP and several recently developed inotropic/vasodilatory drugs designed to improve myocardial contractility in failing human hearts and to inhibit platelet aggregation (Table 2) (17,18). Other PDE families are not sensitive to inhibition by these drugs (17,18).

Table 2. Characteristics of the cGI PDE Family

- K_m cAMP = cGMP (< 1μM); V_{max} cAMP > cGMP
- Specifically and potently inhibited by cGMP, cilostamide, and certain inotropic/vasodilatory drugs including Milrinone, Enoximone, OPC 3911, Imazodan, LY195115, Anagrelide
- Related enzymes present in adipocytes, ventricular myocardium, platelets, vascular smooth muscle
- Associated with intracellular membranes during differentiation of 3T3-L1 adipocytes and perhaps cardiac sarcoplasmic reticulum
- Rapidly regulated, activated by insulin in adipocytes and by effectors that increase cAMP in adipocytes and platelets
- Involved in regulation of a cAMP pool important in control of lipolysis, platelet aggregation, myocardial contractility, smooth muscle relaxation

Table 3. Catalytic Properties and Inhibitor Specificities of cGI PDEs

Catalytic Properties	Rat Adipose	Bovine Aorta Smooth Muscle	Bovine Platelets	Bovine Heart
K_m (μM)				
cAMP	0.4	0.2	0.2	0.2
cGMP	0.3	0.1	0.02	0.1
V_{max} (μmol/min/mg)				
cAMP	3.5	3.1	3.0	6.0
cGMP	2.0	0.3	0.3	0.6
Inhibitors				
IC_{50} (μM)				
OPC-3911 or Cilostamide	0.04	0.05	0.04	.01
Milrinone	0.6	0.4	0.5	0.3
CI 930	0.4	0.3	-	-
RO 20-1724	190	>30	220	62
cGMP	0.2	0.3	0.1	0.1
References	19	21	25,27	26

Because of their responsiveness to these specific inhibitors, cGI PDEs from rat (19) and bovine adipose tissue (20), bovine aorta (21), and human platelets (unpublished results) were purified via chromatography employing the N-(2-isothiocyanato)ethyl derivative of cilostamide (CIT) coupled to aminoethyl-agarose. cGI PDEs seem to be associated with intracellular membranes in rat adipocytes (22), with dog (23) and human (24) sarcoplasmic reticulum, and with cytosol in platelets (25) and bovine aorta (21). As seen in Table 3, very similar kinetic properties and inhibitor specificities have been described for cGI PDEs purified by chromatography on CIT-agarose (19-21) or by other techniques (26-29). As summarized in Table 4, at least some of the cGI PDEs display similar regulatory properties and share immunological epitopes. Immunoprecipitating antibodies have been raised against cGI PDEs purified from bovine adipose tissue (20) and heart (27,30) and human platelets (unpublished observations). Using these immunoprecipitating antibodies, it has been demonstrated that agents that increase cAMP increase phosphorylation (presumably via activation of cAMP-PrK) and activity of cGI PDE in intact rat adipocytes (13,14), and human platelets (30,31). In rat adipocytes, insulin-activation of an intracellular serine kinase also results in phosphorylation (on a site(s)) distinct from that phosphorylated by cAMP-PrK) and activation of particulate cGI PDE (14,32). The catalytic unit of cAMP-dependent protein kinase also apparently directly phosphorylates cGI PDE _in vitro_ in adipocyte particulate fractions and human cardiac sarcoplasmic reticulum (unpublished observations) as well as human platelet extracts (30).

Table 4. Properties of cGMP-inhibited PDEs

	Rat Adipose	Human SR	Human Platelets
Subunit Size[a]	135 kDa	130 kDa	110-120
Phosphorylation intact cells	insulin	n.d.	n.d.
	cAMP	n.d.	cAMP
in vitro	cAMP-PrK	cAMP-PrK	cAMP-PrK
Crossreact with			
Antibovine adipose	+	n.d.	0
Antihuman platelet	+	+	+
Antibovine heart (monoclonal)	0	n.d.	+

[a]based on SDS-PAGE/autoradiography of immunoprecipitates
n.d., not determined.

Table 5. Biological Responses Mediated by cAMP and Altered by Specific cGI PDE Inhibitors in Cells/Tissues Apparently Enriched in cGI PDE

Cell/Tissue	Biological Responses	cAMP Change	Specific Inhibitor Effect
Adipocytes	Lipolysis	increase	Yes (12)
Platelets	Aggregation	decrease	Yes (35)
Heart	Contractility	increase	Yes (4,33)
Smooth Muscle	Relaxation	increase	Yes (34)

With respect to cellular functions regulated by cGI PDEs, selective inhibition of cGI PDEs by specific inhibitors can alter cAMP-mediated biological responses in cells/tissues apparently enriched in cGI PDEs (Table 5). Thus, this PDE family may be quite important in regulation of cAMP pools involved in control of lipolysis (12), myocardial contractility (4,33), smooth muscle relaxation (34), and platelet aggregation (35) (Table 5).

Hormonal Regulation of Adipocyte Particulate cGI PDE and Its Role in the Antilipolytic Action of Insulin

Incubation of intact adipocytes with lipolytic hormones, such as catecholamines, which activate adenylate cyclase and increase cAMP formation, or with cAMP analogs, which directly activate cAMP-PrK, results in activation of both particulate cGI PDE and triglyceride lipase (13). For example, incubation of intact adipocytes with isoproterenol induced rapid activation (maximal response in < 2 min) of both cAMP-PrK (7-8 fold) and cGI PDE (~2-fold); the concentration dependency for isoproterenol activation of hormone-sensitive lipase (lipolysis) was similar to that for activation of particulate cGI PDE (13). Removal of the inhibitory ligand adenosine by incubation of adipocytes with adenosine deaminase removed inhibitory constraints on adenylate cyclase (Fig 1) and resulted in the rapid activation of cAMP-PrK, cGI PDE, and lipolysis (32).

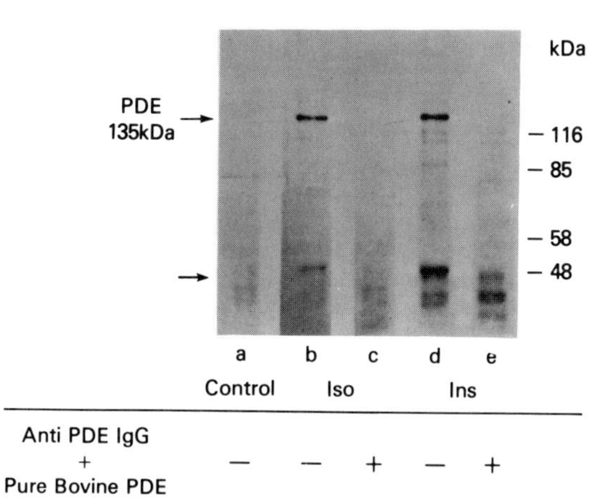

Fig. 2. Insulin- and isoproterenol-induced ^{32}P-phosphorylation of cGI PDE. Rat fat cells, labelled with ^{32}P, were incubated at 37°C with vehicle (contr), insulin (ins)(1 nM, 12 min) or isoproterenol (iso) (300 nM, 3 min). Particulate fractions were solubilized and immunoprecipitated with anti-cGI PDE (-) or with anti-cGI PDE blocked by preincubation with pure rat cGI PDE (+), and subjected to SDS-PAGE and autoradiography of the dried slab gels. The data from the insulin- and isoproterenol-treated cells are not directly comparable; approximately 40% less material from insulin-treated cells was applied to SDS-PAGE in this particular experiment. (Reproduced with permission from Proc. Natl. Acad. Sci. USA [14]).

In ^{32}P-labelled adipocytes incubated with isoproterenol, there was rapid phosphorylation of a 135 kDa protein specifically immunoprecipitated from detergent solubilized extracts of adipocyte particulate fractions and identified by several criteria as the subunit of native particulate cGI PDE (Fig. 2) (14). The time course and concentration dependency for isoproterenol-induced phosphorylation correlated with activation of cGI PDE (32). The ß-adrenergic blocker propanalol inhibited isoproterenol-induced phosphorylation and activation (32). Similarly, effects of adenosine deaminase on phosphorylation/activation of cGI PDE were prevented or reversed by the non-hydrolyzable adenosine analog phenylisopropyladenosine (PIA) (32). These experiments, which demonstrated that inhibition of adenylate cyclase by PIA can reverse phosphorylation and activation of cGI PDE, suggest that dephosphorylation of cGI PDE (presumably by protein phosphatase(s)) can lead to inactivation of cGI PDE. Taken together these experiments were consistent with the notion that the activation state of cGI PDE is regulated by phosphorylation/dephosphorylation (32).

This work and that of others (for review, see ref. 12) strongly suggest that in adipocytes cAMP analogs and lipolytic effectors, which activate adenylate cyclase directly (e.g., isoproterenol) or by removal of endogenous inhibitory ligands (e.g., adenosine deaminase), rapidly activate cAMP-PrK which in turn phosphorylates and activates both particulate cGI PDE and hormone-sensitive lipase. The purified adipocyte cGI PDE or cGI PDE in particulate fractions can be readily phosphorylated in vitro by the catalytic subunit of cAMP-PrK. Although we have not been able to demonstrate consistent changes in activation with phosphorylation, other workers have reported 50% increases in particulate cGI PDE activity in adipocyte particulate fractions incubated with cAMP-PrK and ATP (36). Human platelet cGI PDE, immunoprecipitated by a monoclonal antibody against cardiac cGI PDE, was activated by ~40% when phosphorylated by cAMP-PrK and ATP (30).

Since lipolytic effectors activate with the same time course and concentration dependency both synthesis (adenylate cyclase) and degradation (particulate cGI PDE) of cAMP, we suggest that in rat adipocytes a physiologically important, close functional coupling exists between activation of adenylate cyclase, particulate cGI PDE and cAMP-PrK. cAMP-induced phosphorylation/activation of particulate cGI PDE may be important in "feedback" regulation of cAMP "turnover," cAMP content, and the activation state of cAMP-PrK, and thus indirectly that of hormone-sensitive lipase. In rat adipocytes activation of cGI PDE, with increased hydrolysis of cAMP and production of 5'AMP, might also play a role in hormone-induced production of adenosine, a potential "feedback" inhibitory ligand of adenylate cyclase.

Incubation of ^{32}P-labelled rat adipocytes with insulin also results in serine phosphorylation/activation of particulate cGI PDE, with little or no effect on supernatant PDE activity (13,14). Insulin did not increase cAMP-PrK activity (13,14). The time course of phosphorylation and activation of cGI PDE induced by insulin is somewhat slower than that for isoproterenol (32). The concentration dependency for insulin-induced phosphorylation is similar to that for activation (32). Although both insulin and isoproterenol increased cGI PDE phosphorylation to the same extent (6-25 fold), activation of cGI PDE with isoproterenol was usually about twice that with insulin. Although insulin induced phosphorylation/activation of cGI PDE in the absence of lipolytic effectors, the functional significance of this response to insulin is unknown.

In adipocytes the same particulate cGI PDE can apparently be phosphorylated and activated by physiologically opposing effectors,

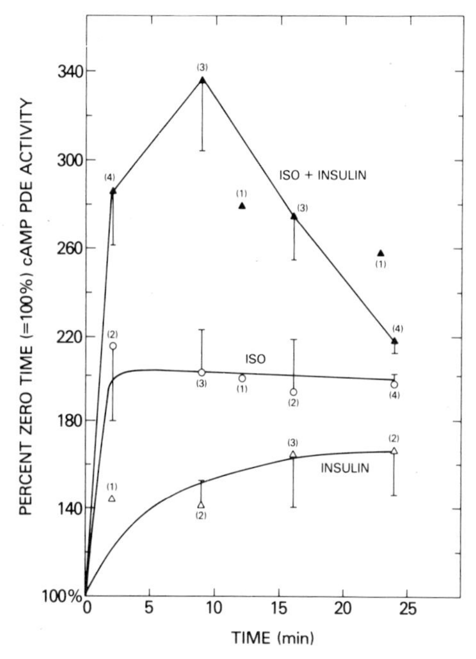

Fig. 3. Time-course of insulin activation of cGI PDE in the presence and absence of isoproterenol. Adipocytes were incubated in duplicate with 3 nM PIA plus 1 unit ADA/ml and 100 nM isoproterenol (ISO) (O), 0.1 nM insulin (△), or isoproterenol plus insulin (▲). Particulate cGI PDE activity was normalized relative to time-zero specific activities. (Reproduced with permission from Mol. Pharmacol. [13]).

insulin, an antilipolytic agent, and the lipolytic hormone, isoproterenol. Our data, however, suggest that this dual regulation by insulin and lipolytic effectors in a "physiologic setting" is an important component in the antilipolytic action of insulin since there seems to be an interaction between cAMP and insulin in regulation of phosphorylation and activation of the cGI PDE.

Incubation of rat adipocytes with both insulin and isoproterenol was associated with a reduction in isoproterenol-activated cAMP-PrK and lipolysis (13). With insulin and isoproterenol, there was a rapid and synergistic activation of cGI PDE which was maximal within 10 min and which temporally correlated with insulin-induced reduction in cAMP-PRK (13) (Fig. 3). In ^{32}P-labelled adipocytes there was a rapid, synergistic phosphorylation of cGI PDE which correlated temporally with enhanced activation of the enzyme (32).

Thus, in adipocytes exposed to both insulin and catecholamines synergistic activation of particulate cGI PDE may result in increased hydrolysis of cAMP PDE which is important in (if not responsible for) insulin-induced reduction of hormone-stimulated cAMP-PrK activity and lipolysis. This phenomenon could perhaps also explain the failure of insulin to inhibit lipolysis induced by very high concentrations of lipolytic hormones, when insulin-activation of cGI-PDE does not effectively reduce elevated cAMP content to a level at which cAMP-PrK activation and hormone-sensitive lipase phosphorylation/activation are also reduced.

These synergistic effects on phosphorylation/activation, the different time courses for insulin and isoproterenol-induced phosphorylation/activation, and the greater activation induced by isoproterenol despite the equivalent effects of both insulin and isoproterenol on phosphorylation, strongly implicate cAMP-PrK and an unidentified insulin-sensitive serine kinase in phosphorylation of discrete sites in particulate cGI PDE.

In addition to these kinetic studies, other experiments also strongly suggest that insulin-induced phosphorylation/activation of the particulate cGI PDE is an important component of the antilipolytic action of insulin. The concentration dependency of phosphorylation/activation of cGI PDE correlates with that for insulin-inhibition of hormone-stimulated lipolysis (13,32). In cultured 3T3-L1 cultured adipocytes and in rat adipocytes the antilipolytic action of insulin was blocked by cilostamide and OPC-3911 (a derivative of cilostamide), specific inhibitors of cGI PDE (12). Beebe et al. (37) reported that of a number of cAMP analogs, all of which increased lipolysis and activated cAMP-PrK in adipocytes, only those analogs that are substrates for cGI PDE were effectively inhibited by insulin (37). Insulin did not inhibit lipolysis stimulated by analogs that are not substrates for cGI PDE and which, therefore, would not be hydrolyzed by insulin-activated cGI PDE (37). These results further indicated that insulin-inhibition of adenylate cyclase was not required for inhibition of lipolysis. In another type of study with hypothyroid rats, Goswami and Rosenberg reported that in adipocytes from hypothyroid rats, particulate cGI PDE activity was markedly increased and catecholamine-stimulated lipolysis was reduced (38). Incubation of adipocytes from hypothyroid rats with cGMP (which would presumably inhibit particulate cGI PDE activity) restored catecholamine-stimulated lipolysis to control values (38). These different types of studies indicate an important role for the particulate cGI PDE in regulation of a cAMP pool involved in control of lipolysis, as well as in the antilipolytic action of insulin.

Other mechanisms have also been proposed to contribute to the antilipolytic action of insulin, i.e., inhibition of adenylate cyclase (29,39), activation of protein phosphatase(s) which act on hormone-sensitive lipase (9), reduction in sensitivity of cAMP-PrK to cAMP, as well as production of insulin "mediators", including phosphatidylinositolglycans (40). The relative importance of these effects and the integration of these and other insulin-regulated pathways in the overall antilipolytic action of insulin remain to be established.

Based on these and other studies, we suggest that dual regulation of the adipocyte particulate cGI PDE is important in hormonal regulation of lipolysis in rat adipocytes. Agents (Fig 1) that increase cAMP activate cAMP-PrK which catalyzes phosphorylation of serine site(s) on cGI PDE. The phosphorylation state determined by the balancing of kinase and phosphatase activities determines the state of activation of cGI PDE, which in turn participates in processes that regulate cellular cAMP content and hence the activation state of cAMP-PrK. There seems to be a close functional coupling between hormonal activation of adenylate cyclase, cGI PDE, and cAMP PrK. The cGI PDE is also phosphorylated at different serine site(s) by an unidentified intracellular insulin-sensitive serine kinase, which in the presence of agents that increase cAMP, produces synergistic phosphorylation/activation of the cGI PDE. This synergistic phosphorylation/activation of cGI PDE may be an important component of the antilipolytic action of insulin. Whether this insulin-sensitive serine kinase is activated by tyrosine phosphorylation catalyzed by the activated insulin-receptor tyrosine kinase is not known.

The cGI PDE may present a useful model system in which to investigate mechanisms for regulation of insulin-signalling as well as for the coupling of an insulin-signal to intracellular effector systems, i.e., to cAMP-mediated pathways such as regulation of triglyceride lipase in adipose tissue. It is also important to note that cGI PDEs seem to be involved in regulation of platelet aggregation, myocardial contractility and smooth muscle relaxation. We hope that understanding of structure/ function relationships, regulation of gene expression, etc. of cGI PDEs will advance understanding of the pathophysiology of and the design of rational therapy for certain diabetic and obese states, and perhaps certain cardiovascular diseases, especially those complications related to diabetes.

REFERENCES

1. Steinberg, D., Mayer, S.E., Khoo, J.C., Miller, E.A., Friedholm, B., and Eichner, R. (1975) Adv. Cyclic Nucleotide Res., 5, 549-568.
2. Sutherland, E.W. (1971) Cyclic AMP, Academic Press, New York, pp. 5-17.
3. Salzman, E.W., and Weisinberger, H. (1972) Adv. Cyclic Nucleotide Res. 1, 231-248.
4. Colucci, W.S., Wright, R.F., and Braunwald, E. (1986) New Engl. J. Med., 314, 290-299, 349-358.
5. Kramer, G.L., and Hardman, J.G. (1980) Handbook of Physiology: The Cardiovascular System II, American Physiological Society, Bethesda, MD pp. 179-199.
6. Stralfors, P., Bjorgell, P., and Belfrage, P. (1984) Proc. Natl. Acad. Sci. U.S.A., 81, 3317-3321.
7. Holm, C., Kirchgessner, T.G., Svenson, K.L., Fredrickson, G., Nelsson, S., Miller, C.G., Shively, J.E., Heinzmann, C., Sparkes, R.S., Mohandas, T., Lusis, A.J., Belfrage, P., and Schotz, M.C. (1988) Science, 241, 1503-1506.
8. Belfrage, P., Fredrickson, G., Olsson, H., and Stralfors, P. (1983) The Adipocyte and Obesity: Cellular and Molecular Mechanisms, Raven Press, New York, pp. 217-224.
9. Londos, C., Honnor, R.C., and Dhillon, G.S. (1985) J. Biol. Chem. 260, 15139-15145.
10. Butcher, R.W., Sneyd, J., Parks, C.R., and Sutherland, E.W. (1966) J. Biol. Chem., 241, 1651-1653.
11. Soderling, T.R., Corbin, J.D., and Park, C.R. (1973) J. Biol. Chem., 248, 1822-1829.
12. Manganiello, V.C., Smith, C.J., Degerman, E., and Belfrage, P. (1990) Cyclic Nucleotide Phosphodiesterases: Structure, Regulation, and Drug Action, John Wiley and Sons Ltd., Chichester, pp. 87-116.
13. Smith, C.J., and Manganiello, V.C. (1989) Mol. Pharm., 35, 381-386.
14. Degerman, E., Smith, C.J., Tornqvist, H., Vasta, V., Belfrage, P., and Manganiello, V.C. (1990) Proc. Natl. Acad. Sci. U.S.A., 87, 533-537.
15. Beavo, J. (1990) Cyclic Nucleotide Phosphodiesterases: Structure, Regulation, and Drug Action, John Wiley and Sons Ltd., Chichester, pp. 3-19.
16. Beavo, J. (1988) Adv. Second Messenger Phosphoprotein Res., 22, 1-38.
17. Weishaar, R.E., Carn, M.H., and Bristol, J.A. (1985) J. Med. Chem., 28, 537-545.
18. Beavo, J., and Reifsnyder, D.H. (1990) Trends in Pharmacological Sciences, 11, 150-156.
19. Degerman, E., Belfrage, P., Newman, A.H., Rice, K.C., and Manganiello, V.C. (1987) J. Biol. Chem., 262, 5797-5807.

20. Degerman, E., Manganiello, V.C., Newman, A.H., Rice, K.C., and Belfrage, P. (1988) Adv. Second Messengers and Phosphoproteins, 12, 171-182.
21. Rascon, A., Belfrage, P., Lindgren, S., Andersson, K.E., Newman, A.H., Manganiello, V.C., and Degerman, E. (1990) Purine Nucleosides and Nucleotides in Cell Signalling: Targets for New Drugs, Springer-Verlag, New York, pp. 353-358.
22. Kono, T., Robinson, F.W., and Sarver, J.A. (1975) J. Biol. Chem., 250, 7826-7835.
23. Kauffman, R.F., Crowe, V.G., Utterback, B.G., and Robertson, D.W. (1987) Mol. Pharm., 30, 609-616.
24. Masuoka, H., Ito, M., Nakano, T., Naka, M., and Tanaka, T. (1990) J. Cardiovasc. Pharm., 15, 302-307.
25. Grant, P.W., and Colman, R.W. (1984) Biochemistry, 23, 1801-1807.
26. Harrison, S.A., Reifsnyder, D.H., Gallis, B., Cadd, C.G., and Beavo, J.A. (1986) Mol. Pharm., 25, 506-514.
27. Macphee, C.H., Harrison, S.A., and Beavo, J.A. (1986) Proc. Natl. Acad. Sci. U.S.A., 83, 6660-6663.
28. Boyes, S., and Loten, E.G. (1988) Eur. J. Biochem., 174, 303-309.
29. Houslay, M.D., and Kilgour, E. (1990) Cyclic Nucleotide Phosphodiesterases: Structure, Regulation, and Drug Action, John Wiley and Sons, Ltd., Chichester, pp. 185-226.
30. Macphee, C.H., Reifsnyder, D.H., Moore, T.A., Levea, K.M., and Beavo, J.A. (1988) J. Biol. Chem., 263, 10353-10358.
31. Grant, P.H., Marinaro, A.F., and Colman, R.W. (1988) Proc. Natl. Acad. Sci. U.S.A., 85, 9071-9075.
32. Smith, C.J., Vasta V., Degerman, E., Tornqvist, H., and Manganiello, V.C., ms. submitted.
33. Reeves, M.L., and England, P.J. (1990) Cyclic Nucleotide Phosphodiesterases: Structure, Regulation, and Drug Action, John Wiley and Sons Ltd., Chichester, pp. 299-316.
34. Lindgren, S., Rascon, A., Andersson, K.-E., Manganiello, V.C., and Degerman, E. (1990) ms. submitted.
35. Alvarez, R., Banerjie, G.L., Brunz, J.J., Jones, G.L., Littachwager, K., Strossberg, A., and Venuti, M.C. (1986) Mol. Pharm., 29, 554-560.
36. Gettys, T.W., Vine, A.J., Simonds, M.F., and Corbin, J.D. (1988) J. Biol. Chem., 263, 10359-10363.
37. Beebe, S.J., Redman, J.B., Blackmore, P.W., and Corbin, J.D. (1985) J. Biol. Chem., 260, 15781-15788.
38. Goswami, A., and Rosenberg, I. (1985) J. Biol. Chem., 260, 82-85.
39. Hepp, K.O., and Renner, R. (1972) FEBS Lett., 20, 191-194.
40. Low, M.G., and Saltiel, A.R. (1988) Science, 239, 268-275.

REGULATION OF THE GLUCOSE TRANSPORTER IN ANIMAL MODELS OF DIABETES

Jeffrey E. Pessin, Jeanne M. Richardson, and William I. Sivitz

Departments of Physiology & Biophysics and Internal Medicine, The University of Iowa College of Medicine Iowa City, IA

Introduction

Glucose oxidation, a major source of metabolic energy for mammalian cells, depends upon the transport of glucose across the cell surface membrane by specific carrier proteins. Two general classes of glucose transporters are found in mammalian cells. The first category of glucose transporters are the Na^+/dependent glucose co-transporters which are found in the brush border membrane of epithelial cells in the small intestine and proximal tubule of the kidney.[1] These carriers actively transport glucose from the lumen into the epithelial cell against its concentration gradient by coupling the uphill movement of glucose with the downhill movement of Na^+ across its concentration gradient. The necessary Na^+ gradient for this transport process is maintained by the Na^+/K^+ATPase found on the basolateral membrane of these epithelial cells.

The second category of mammalian glucose carriers are the facilitative glucose transporters which mediate the uptake of glucose in a stereospecific, energy independent manner. Currently, five members of this mammalian multi-gene family have been identified with related structural and functional properties.[2] Recently, a sixth glucose transporter-like sequence was isolated but since it contains multiple stop codons in the open reading frame it does not produce a functional protein and therefore represents a pseudo-gene.[3]

Facilitative glucose transporter gene-family

The generally accepted nomenclature for the facilitative glucose transporter gene family is based upon the order in which the various cDNA clones were isolated. GLUT1 was the first member to be cloned from both human HepG2 and rat brain cDNA libraries.[4,5] Based upon both immunological and partial amino acid sequence analysis, this transporter is equivalent to the extensively studied human erythrocyte glucose transporter protein.[4] The human erythrocyte glucose transporter is heterogeneously glycosylated and has a Km for glucose of 1-5mM when assayed by zero-trans kinetics and 20mM under equilibrium exchange conditions.[6,7] This transporter also displays asymmetric transport in that substrate loading of the inward facing carrier increases the uptake of glucose.[8] Hydropathy analysis of the 492 GLUT1 amino acid sequence

Molecular Biology and Physiology of Insulin and Insulin-Like Growth Factors
Edited by M.K. Raizada and D. LeRoith, Plenum Press, New York, 1991

has predicted that this protein spans the cell membrane 12 times with both the amino and carboxy termini oriented towards the cytoplasm.[4] A single asparagine-linked glycosylation consensus site is present in the extracellular loop between membrane spanning segments 1 and 2 and a large intracellular hydrophilic domain is situated between transmembrane segments 6 and 7 (Fig. 1). Several features of this predicted structural orientation have been confirmed by the use of site-specific antibodies, proteolytic digestion and mutagenesis of the GLUT1 cDNA.[4,8-11]

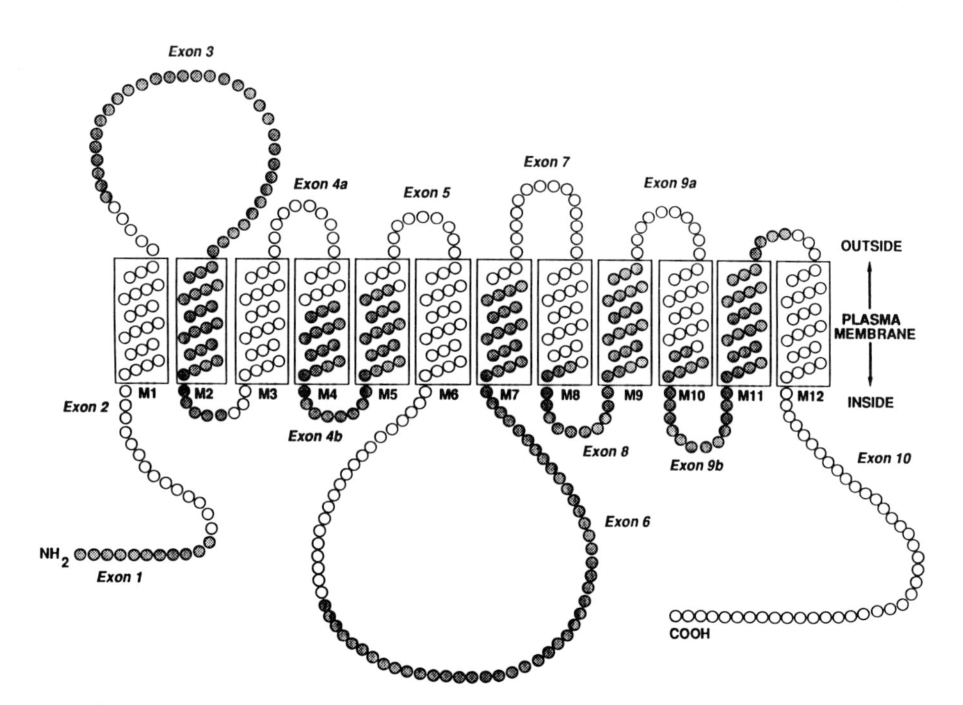

Figure 1. General model for the orientation of the facilitative glucose transporters in the plasma membrane. The putative membrane spanning α-helices are numbered M1-M12. The consensus site for asparagine-linked glycosylation is found in the extracellular loop between M1 and M2. The exon-intron organization of the putative ancestral gene which gave rise to the different members of the mammalian glucose transporter family is indicated by the filled and open circles. Reprinted from Bell *et al.* Diabetes Care 13:198 (1990).

Prior to the cloning of the GLUT1 isoform, it was recognized that hepatocytes expressed a facilitative glucose transporter with different characteristics than the human erythrocyte transporter, including a significantly higher Km for glucose (~15-20mM by zero-trans analysis) and a ten-fold weaker Kd for cytochalasin B binding.[12-14] Subsequent to the cloning of the GLUT1 isoform it was rapidly recognized that the liver did not express GLUT1 mRNA which is consistent with the presence of a second glucose transporter gene.[15] Low stringency screening of both rat and human liver cDNA libraries with the GLUT1 cDNA resulted in the isolation of a related glucose transporter-like sequence which possessed a 55%

amino acid identity with GLUT1.[14,16] This glucose transporter was subsequently called GLUT2. The rat (522 amino acids) and human (524 amino acids) were predicted to have an identical transmembrane topology. Interestingly, the length of the extracellular loop between the transmembrane segments 1 and 2 are markedly different, being 64 amino acids in GLUT2 but only 32 amino acids in GLUT1. Although there is little amino acid sequence conservation in this region, GLUT2 also contains a potential site for N-linked glycosylation. Expression of the GLUT2 cDNA in both a glucose transport defective E. coli mutant and Xenopus oocytes demonstrated that this clone encodes a functional glucose transport protein[15,17] with an estimated Km of 9 mM when expressed in Xenopus oocytes.[17]

Screening of a human fetal skeletal muscle cDNA library under low stringency conditions identified a third member of this gene family, termed GLUT3. This cDNA encodes a protein with 496 amino acids having 64% and 52% identity with GLUT1 and GLUT2, respectively.[18] Functional properties of this glucose transporter isoform have not been extensively investigated but it appears to have a Km similar to that of GLUT1 (G. Bell, personal communication).

Surprisingly, even though this cDNA was isolated from a fetal skeletal muscle library, GLUT3 mRNA was not detectable in adult skeletal muscle.[18] In addition, previous immunological evidence had indicated that the glucose transporter in insulin-responsive tissues (skeletal muscle and adipose tissue) was a unique isoform in the glucose transporter gene family.[19] Several groups isolated this apparent insulin-responsive glucose transporter cDNA (GLUT4) from human, mouse and rat libraries.[20-24] The human GLUT4 cDNA predicted a protein of 509 amino acids with 65%, 54% and 58% identity with the human GLUT1, GLUT2 and GLUT3 proteins, respectively. Again, the predicted membrane topology of GLUT4 is identical to that of the other glucose transporter isoforms. Expression of the GLUT4 isoform in Xenopus oocytes demonstrated a Km of 1.8 mM under equilibrium exchange conditions.[7]

The most recent member of the facilitative glucose transporter gene-family to be identified is GLUT5 which encodes a protein of 532 amino acids.[3] This protein is also predicted to span the membrane 12 times with an identical secondary structure. The human GLUT5 protein has 42%, 40%, 39% and 42% amino acid identity with human GLUT1, GLUT2, GLUT3 and GLUT4, respectively. Analogous proteins in other species as well as the expression and kinetic properties of this glucose transporter isoform have not been examined to date.

Tissue distribution

The tissue distribution of the five currently identified glucose transporter mRNAs reveals a distinct but overlapping pattern. Northern blot analysis indicates that the GLUT1 transcript is usually the predominant or only glucose transporter isoform expressed in cultured cells. *In vivo* the highest levels are found in fetal tissues, brain microvessels and placenta.[14,25,26] It is also present but significantly less abundant in kidney, intestines, adipose and cardiac tissues. The distribution of GLUT2 mRNA is primarily limited to the liver and the beta cells of the pancreatic islet although lower levels have been found in the small intestines and kidney.[14,16] Histochemical studies indicate that islet beta cells are polarized with respect to GLUT2 which is localized to microvilli that exclusively face the other islet endocrine cells.[27] On the other hand, GLUT3 mRNA is expressed in all adult human tissues examined including placenta, brain, colon, stomach, jejunum, kidney, cerebrum, gallbladder and to a smaller extent adipocytes.[18] However, in adult rodents this transcript is primarily found in brain tissue and is apparently not generally expressed. The fourth glucose transporter, GLUT4, is predominantly localized to insulin-responsive tissues such as adipose, skeletal and cardiac muscle[20-24] whereas the distribution of GLUT5 appears to be primarily limited to the small intestines and to a lesser extent in kidney, adipose and cardiac tissue.[3] A summary of the tissue distribution of the glucose transporter gene-family is shown in Table 1.

Table 1

Tissue distribution of the human facilitative glucose transporter mRNAs.

Name	Size (amino acids)	Major locations of mRNA
GLUT1	492	Placenta, brain, kidney & colon
GLUT2	524	Liver, β cells, kidney & ileum
GLUT3	496	Many tissues including brain, placenta & kidney
GLUT4	509	Adipose, cardiac & skeletal muscle
GLUT5	501	Ileum & kidney

Insulin stimulation of adipocyte glucose transport activity

Prior to the identification and cloning of the various glucose transporter gene-family members, it was well established that insulin regulated glucose transport activity in adipocytes in a rapid and protein synthesis-independent fashion.[28] It has been generally accepted that acute insulin stimulation of glucose transport activity primarily results from a redistribution or recruitment of preformed intracellular glucose transporters to the cell surface.[29,30] Recent evidence has indicated that the GLUT1 isoform mainly resides on the cell surface of primary isolated rat adipocytes whereas the GLUT4 isoform is predominantly stored in a subpopulation of intracellular vesicles.[31] In contrast, insulin activation results in the translocation primarily of the intracellular GLUT4 protein to the cell surface with only a marginal redistribution of the GLUT1 isoform. These data in conjunction with the broad tissue type expression of the GLUT1 isoform suggest that this species functions as the basal transporter.

However, stable transfectants expressing the GLUT1 isoform in both 3T3-L1 and CHO cell lines were capable of displaying insulin-dependent activation of both glucose transport activity and translocation.[32-34] Several studies have also demonstrated that pre-differentiated 3T3-L1 cells uniquely express the GLUT1 protein and mRNA but during the process of differentiation into adipocytes there is an increase in GLUT4 expression.[21,24,35,36] Recently, quantization of the glucose transporter isoforms in differentiated 3T3-L1 cells demonstrated 3.4-fold greater levels of GLUT1 compared to GLUT4.[37] This is in contrast to primary isolated rat adipocytes in which the relative amount of GLUT4 protein is approximately 10-fold greater than GLUT1.[31] Further, although insulin stimulated the translocation primarily of GLUT4 in the differentiated 3T3-L1 cells, both the GLUT4 and GLUT1 proteins were found to be co-localized in the same intravesicular compartment.[37] In contrast, unique intracellular pools of GLUT1 and GLUT4 have been reported in isolated primary rat adipocytes.[31] Although a molecular basis for the experimental differences between the differentiating 3T3-L1 adipocyte cell line and primary isolated rat adipocytes have not been determined, it is nevertheless clear that the major acute insulin-dependent regulation of adipocyte glucose transport activity results from GLUT4 translocation.

Adipose-specific expression of the GLUT4 glucose transporter

Insulin has also been observed to be a potent regulator of GLUT4 protein and mRNA levels in adipose tissue *in vivo*.[38-41] Streptozotocin-induced diabetes (as a model of Type I diabetes or IDDM), results in a marked reduction (~10-fold) of adipose GLUT4 protein[38-40] and mRNA[39-41] levels (Fig. 2) with no significant effect on the levels of GLUT1. Similarly, fasting which is also associated with reduced insulin levels, results in a marked decrease in adipocyte GLUT4 mRNA and protein.[38,41] These observations suggest that the insulin resistance associated with type I diabetes may be due to a down-regulation of GLUT4 mRNA resulting in a decreased intracellular pool of GLUT4 protein available for insulin-dependent recruitment.

Interestingly, insulin therapy resulted in the superinduction of both GLUT4 mRNA and protein to levels approximately 3-4 fold above those of control animal (Fig. 3B). This occurred in a transient fashion with GLUT4 mRNA maximally induced 2-3 days subsequent to the initiation of insulin therapy whereas maximal GLUT4 protein induction required approximately 7 days. Previous studies have demonstrated that the recovery of insulin-resistant glucose transport activity in streptozotocin diabetic rats also superinduces with maximal stimulation occurring at approximately 8 days.[42] A comparison of GLUT4 protein recovery and insulin-stimulated glucose transport activity demonstrates a direct temporal relationship between these two phenomenon (Fig. 3A). Similar results were also observed upon refeeding of fasted animals (data not shown).

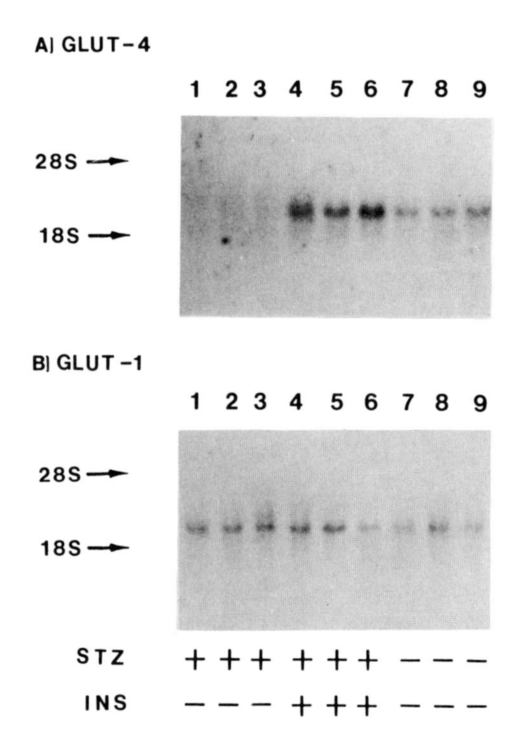

Figure 2. Expression of the GLUT4 (A) and GLUT1 (B) glucose transporter mRNAs in adipose tissue from control (lanes 7-9), STZ-diabetic (lanes 1-3) and insulin-treated STZ-diabetic (lanes 4-6) rats. Rats were untreated or made insulin deficient by a single intraperitoneal injection of STZ (125mg/kg). After three days, half the diabetic animals (blood glucose greater than 300mg/100ml) were treated with insulin (3 units regular plus 4 units of intermediate acting insulin) for 18h before sacrifice. Changes in RNA levels were examined by Northern blot analysis using the human GLUT4 or rat GLUT1 cDNA.

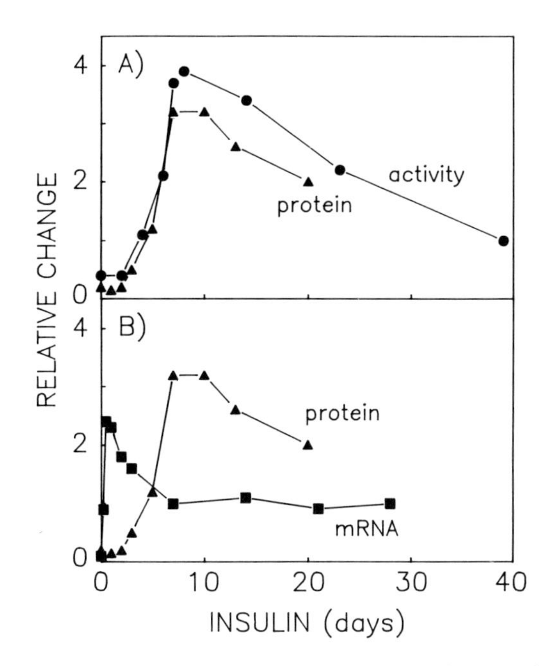

Figure 3. Comparison of adipocyte insulin-stimulated glucose transport activity, GLUT4 protein and mRNA levels in insulin-treated STZ-diabetic rats. STZ-diabetic rats were treated with insulin as described in Figure 2. The relative level of GLUT4 mRNA was determined by Northern blotting and GLUT4 protein by Western blotting using the monoclonal antibody 1F8. The relative insulin stimulation of glucose transport activity was replotted from the data of Kahn and Cushman _J. Biol. Chem._ 262:5118 (1987). A) Time course of GLUT4 protein levels and insulin-stimulated glucose transport activity subsequent to the initiation of insulin therapy. B) Comparison between GLUT4 protein and mRNA levels.

The above data demonstrate that streptozotocin-induced diabetes (hypoinsulinemia with hyperglycemia) as well as fasting (hypoinsulinemia with hypoglycemia) result in a specific decrease in expression of GLUT4 in adipose tissue. Since fasting and streptozotocin-induced diabetes have opposite effects on the glycemic state but are similar with respect to a reduction in insulin levels, this suggests that insulin, not glucose, is the central regulator of adipose GLUT4 expression. In order to test this hypothesis[43], we examined the effect of euglycemia on adipose GLUT4 expression in the presence of hypoinsulinemia (Fig. 4). This was accomplished by the use of phlorizin which inhibits the renal reabsorption of glucose via the Na+/dependent-glucose transporter. Under these conditions, the streptozotocin-diabetic rats were euglycemic but remained insulin-deficient. Northern blot analysis clearly demonstrated that normalization of circulating glucose levels under hypoinsulinemic conditions is not sufficient to induce GLUT4 mRNA expression in adipose tissue. In order to rule out any potential non-specific cytotoxic effect of phlorizin, a group of streptozotocin-diabetic animals were treated with a combination of phlorizin and insulin which resulted in the complete recovery of adipose GLUT4 mRNA levels (Fig. 5).

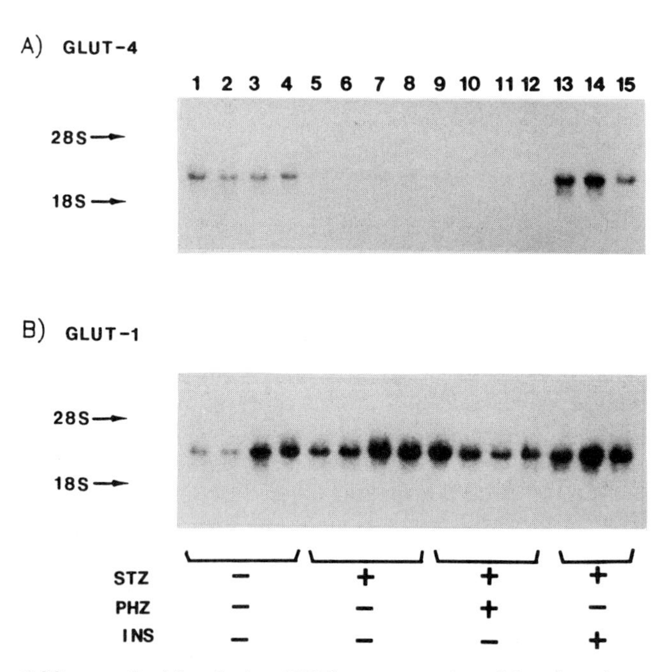

Figure 4. Effect of phlorizin (PHZ) versus insulin (INS)-induced glucose normalization on expression of the GLUT4 and GLUT1 mRNAs in STZ-diabetic rats. Northern blot analysis of GLUT4 (A) and GLUT1 (B) mRNA in rat adipose tissue from control (lanes 1-4), untreated STZ-diabetic (lanes 5-8), phlorizin-treated STZ-diabetic (lanes 9-12) and insulin-treated STZ-diabetic (lanes 13-15) rats was performed as described in Sivitz *et al*. Mol. Endocrinol. 4:583 (1990). The STZ-diabetic and insulin-treated STZ-diabetic rats were prepared as described in Figure 2. Phlorizin treatment (133 mg/kg every 8h for 24h) was initiated 3 days after the induction of diabetes. Using this protocol, the degree of glycemic change induced by phlorizin was similar to that induced by insulin.

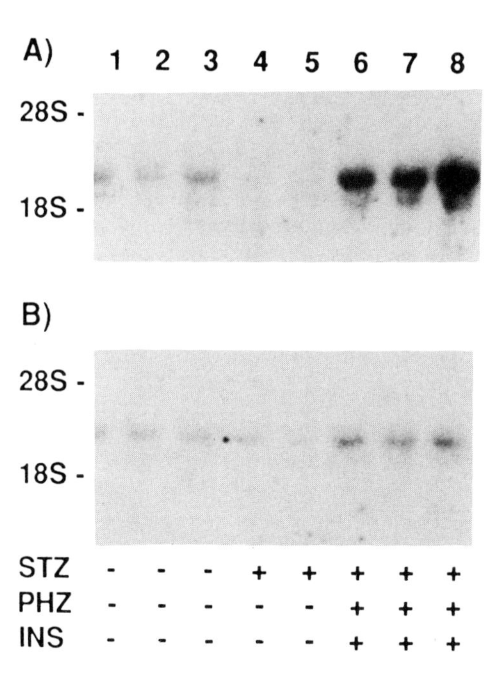

Figure 5. Expression of GLUT4 (A) and GLUT1 (B) mRNA in STZ-diabetic rats treated with phlorizin (PHZ) and insulin (INS). Rat adipose tissue RNA was isolated from control (lanes 1-3), untreated STZ-diabetic (lanes 4 and 5) and STZ-diabetic rats treated with both phlorizin and insulin (lanes 6-8) as described in Figure 4.

Muscle-specific expression of the GLUT4 glucose transporter

In contrast to adipose tissue, skeletal muscle is the primary site for postprandial absorption of glucose and under most circumstances glucose transport into skeletal muscle is the rate limiting step for glucose metabolism.[44-46] Several studies have demonstrated that slow contracting oxidative red skeletal muscle fibers are significantly more insulin-sensitive than fast contracting glycolytic white muscle fibers.[47,48] This is probably due in part to greater insulin receptor levels and kinase activity in red skeletal muscle fibers[49] in addition to 2 to 5-fold greater levels of GLUT4 protein and mRNA.[50] Although it is significantly more difficult to isolate subcellular membrane fractions from skeletal muscle, recent studies have demonstrated that acute insulin treatment induces a redistribution of GLUT4 protein to the cell surface in an analogous fashion to that previously established for adipocytes.[51,52]

Streptozotocin-induced diabetes was also observed to decrease expression of rat hindquarter muscle GLUT4 protein and mRNA with no significant effect on GLUT1 levels.[39] Insulin therapy restored the relative levels of the GLUT4 glucose transporter toward control values but did not result in a superinduction as was observed for adipocytes treated in an identical manner. Surprisingly, although fasting decreases GLUT4 expression in adipocytes it has been recently reported that fasting increased the expression of both GLUT4 protein and mRNA in hindquarter skeletal muscle.[53] The basis and physiological consequences of this apparent discordant relationship between adipocyte and skeletal muscle GLUT4 expression is unclear at the present time.

GLUT4 expression in models of insulin resistance

Based upon the identification of the major insulin-responsive glucose transporter and its marked regulation both in fasting/refeeding and streptozotocin-induced diabetes, it has been speculated that alterations in GLUT4 expression may play a central role in insulin resistant states associated with obesity and non-insulin dependent diabetes mellitus (NIDDM). Studies in five week old hyperinsulinemic and hyperglycemic obese db/db mice, a situation resembling human NIDDM, demonstrated a decrease in GLUT4 mRNA in quadriceps skeletal muscle without any significant change in adipose, diaphragm and cardiac muscle tissues. However, GLUT4 protein levels in all these tissues were unaffected.[54] GLUT4 protein levels in skeletal muscle of obese Zucker rats, which are markedly insulin resistant but are not overtly diabetic, were not significantly different from control lean littermates[55]. However, strenuous exercise, which increases insulin sensitive glucose uptake into skeletal muscle, resulted in a 3-fold increase in quadriceps muscle GLUT4 protein levels in the Zucker rats.[56]

GLUT4 expression in NIDDM patients is also apparently quite complex which may reflect the heterogeneity of this disease process. For example, GLUT4 protein and mRNA levels in skeletal muscle biopsy samples from obese and NIDDM patients were reported not to be significantly different than control values.[57] In contrast, GLUT4 levels in the rectus muscle of NIDDM patients have recently been found to be significantly reduced in comparison to obese non-diabetic and control samples (J. Caro, personal communication). Similarly, a decrease in GLUT4 protein and mRNA levels in adipocytes obtained from obese and NIDDM patients has also been observed (T. Garvey, personal communication). Thus, although a decreased expression of the GLUT4 glucose transporter plays a central role in insulin resistance associated with animal models of IDDM, its contribution to NIDDM in patients has not been generally established.

Summary

The identification and cloning of the five mammalian facilitative glucose transporters has given us the necessary tools to begin to dissect the intricate properties of transporter function and expression. Although several of these general issues have been addressed, numerous questions remain unresolved. A uniform model for the molecular basis of facilitative glucose transport function in mammalian cells must account for the specific roles that each member of this gene-family play in the control of glucose homeostasis. In particular, the diverse distribution and kinetic properties of the glucose transporter isoforms suggest that specific functional properties of the glucose transporter are required in a given tissues. Yet many tissues and cell types can express more than one glucose transporter isoform. Even though reasonable predictions have been proposed for the tissue-specific functions of GLUT1, GLUT2 and GLUT4, currently a role for GLUT3 and GLUT5 as well as other potential facilitative glucose transporters has not yet been addressed.

Another key issue which is currently under investigation is the identification of structural determinants that define intracellular targeting and insulin-dependent translocation. Since several studies have demonstrated that insulin is capable of recruiting both GLUT1 and GLUT4, albeit to different extents depending upon the cell type examined, it will be necessary to establish the relationship between sequence-specific and cell type-mediated targeting of the glucose transporter isoforms. Moreover, since insulin deficiency has marked effects on both adipose and skeletal muscle expression of GLUT4, future studies directed at determining the molecular basis for transcriptional and post-transcriptional regulation of this gene may provide important information directly related to the insulin resistance associated with IDDM.

References

1. D.L. Baly and R. Horuk, The biology and biochemistry of the glucose transporter, Biochim. Biophys. Acta. 947:571 (1988).
2. G.I. Bell, T. Kayano, J.B. Buse, C.F. Burant, J. Takeda, D. Lin, H. Fukumoto and S. Seino, Molecular biology of mammalian glucose transporters, Diabetes Care 13:198 (1990).
3. T. Kayano, C.F. Burant, H. Fukumoto, G.W. Gould, Y. Fan, R.L. Eddy, M.G. Byers, T.B. Shows, S. Seino and G.I. Bell, Human facilitated glucose transporters, J. Biol. Chem. 265:13276 (1990).
4. M. Mueckler, C. Caruso, S.A. Baldwin, M. Panico, M. Blench, H.R. Morris, W.J. Allard, G.E. Lienhard and H.F. Lodish, Sequence and structure of a human glucose transporter, Science 299:941 (1985).
5. M.J. Birnbaum, H.C. Haspel and O.M. Rosen, Cloning and characterization of a cDNA encoding the rat brain glucose transporter protein, Proc. Natl. Acad. Sci. USA 83:5784 (1986).
6. T.J. Wheeler and P.C. Hinkle, The glucose transporter of mammalian cells, Ann. Rev. Physiol. 47:503 (1985).
7. K. Keller, M. Strube and M. Mueckler, Functional expression of the human hepG2 and rat adipocyte glucose transporters in Xenopus oocytes, J. Biol. Chem. 264:18884 (1989).
8. H.C. Haspel, M.G. Rosenfeld and O.M. Rosen, Characterization of antiserum to a synthetic carboxyl-terminal peptide of the glucose transporter protein, J. Biol. Chem. 263:398 (1988).
9. A. Davies, K. Meeran, M.T. Cairns and S.A. Baldwin, Peptide-specific antibodies as probes of the orientation of the glucose transporter inn the human erythrocyte membrane, J. Biol. Chem. 262:9347 (1987).

10. M.T. Cairns, D.A. Elliot, P.R. Scudder and S.A. Baldwin, Proteolytic and chemical digestion of the human erythrocyte glucose transporter, Biochem. J. 221:179 (1984).
11. M. Mueckler and H.F. Lodish, The human glucose transporter can insert posttranslationally into microsomes Cell 44:629 (1986).
12. J.D. Axelrod and P.F. Pilch, Unique cytochalasin B binding characteristics of the hepatic glucose carrier, Biochemistry 22:2222 (1983).
13. M. Mueckler, Family of glucose-transporter genes: Implications for glucose homeostasis and diabetes, Diabetes 39:6 (1990).
14. B. Thorens, H.K. Sarkar, H.R. Kaback and H.F. Lodish, Cloning and functional expression in bacteria of a novel glucose transporter present in liver, intestine, kidney, and β-pancreatic islet cells, Cell 55:281 (1988).
15. J.S. Flier, M. Mueckler, A.L. McCall and H.F. Lodish, Distribution of glucose transporter messenger RNA transcripts in tissues of rat and man, J. Clin. Invest. 79:657 (1987).
16. H. Fukumoto, S. Seino, H. Imura, Y. Seino, R.L. Eddy, Y. Fukushima, M. Byers, T.B. Shows and G.I. Bell, Sequence, tissue distribution and chromosomal localization of mRNA encoding a human glucose transporter-like protein, Proc. Natl. Acad. Aci. USA 85:5434 (1988).
17. M.A. Permutt, L. Koranyi, K. Keller, P.E. Lacy, D.W. Scharp and M. Mueckler, Cloning and functional expression of a human pancreatic islet glucose-transporter cDNA, Proc. Natl. Acad. Sci. USA 86:8688 (1989).
18. T. Kayano, H. Fukumoto, R.L. Eddy, Y-S. Fan, M.G. Byers, T.G. Showa and G.I. Bell, Evidence for a family of human glucose transporter-like proteins: sequence and gene localization of a protein expressed in fetal skeletal muscle and other tissues, J. Biol. Chem. 263:15245 (1988).
19. D.E. James, R. Brown, J. Navarro and P.F. Pilch, Insulin-regulatable tissues express a unique insulin-sensitive glucose transport protein, Nature 333:183 (1988).
20. D.E. James, M. Strube and M. Mueckler, Molecular cloning and characterization of an insulin-regulatable glucose transporter, Nature 338:83 (1989).
21. M.J. Birnbaum, Identification of a novel gene encoding an insulin-responsive glucose transporter protein, Cell 57:305 (1989).
22. M.J. Charron, F.C. Brosius, S.L. Alper and H.F. Lodish, A glucose transport protein expressed predominantly in insulin-responsive tissues, Proc. Natl. Acad. Sci. USA 86:2535 (1989).
23. H. Fukumoto, T. Kayano, J.B. Buse, Y. Edwards, P.F. Pilch, G.I. Bell and S. Seino, Cloning and characterization of the major insulin-responsive glucose transporter expressed in human skeletal muscle and other insulin-responsive tissues, J. Biol. Chem. 264:7776 (1989).
24. K.H. Kaestner, R.J. Christy, J.C. McLenithan, L.T. Braiterman, P. Cornelius, P.H. Pekala and M.D. Lane, Sequence, tissue distribution and differential expression of mRNA for a putative insulin-responsive glucose transporter in mouse 3T3-L1 adipocytes, Proc. Natl. Acad. Sci. USA 86:3150 (1989).
25. H. Fukumoto, S. Seino, H. Imura, Y. Seino and G.I. Bell, Characterization and expression of human HepG2/erythrocyte glucose-transporter gene, Diabetes 37:657 (1988).
26. W.M. Pardridge, R.J. Boado and C.R. Farrell, Brain-type glucose transporter (GLUT-1) is selectively localized to the blood-brain barrier, J. Biol. Chem. 265:18035 (1990).

27. L. Orci, B. Thorens, M. Ravazzola and H.F. Lodish, Localization of the pancreatic beta cell glucose transporter to specific plasma membrane domains, Science 245:295 (1989).

28. J.E. Pessin and M.P. Czech, In: The Enzymes of Biological Membranes (A.N. Martonosi, Ed.) Vol. 3, pp. 497, Plenum Press, NY (1985).

29. S.W. Cushman and L.J. Wardzala, Potential mechanism of insulin action on glucose transport in the isolated rat adipose cell, J. Biol. Chem. 255:4758 (1980).

30. Y. Suzuki and T. Kono, Evidence that insulin causes translocation of glucose transport activity to the plasma membrane from an intracellular storage site, Proc. Natl. Acad. Sci. USA 77:2542 (1980).

31. A. Zorzano, W. Wilkinson, N. Kotliar, G. Thoidis, B.E. Wadzinkski, A. Ruoho and P. Pilch, Insulin-regulated glucose uptake in rat adipocytes is mediated by two transporter isoforms present in at least two vesicle populations, J. Biol. Chem. 264:12358 (1989).

32. G.W. Gould, V. Derechins, D.E. James, K. Tordjman, S. Ahern, E.M. Gibbs, G.E. Lienhard and M. Mueckler, Insulin-stimulated translocation of the HepG2/erythrocyte-type glucose transporter expressed in 3T3-L1 adipocytes, J. Biol. Chem. 264:2180 (1989).

33. T. Asano, Y. Shibasaki, S. Ohno, H. Taira, J.-L. Lin, M. Kasuga, Y. Kanazawa, Y. Akanuma, F. Takaku and Y. Oka, Rabbit brain glucose transporter responds to insulin when expressed in insulin-sensitive Chinese hamster ovary cells, J. Biol. Chem. 264: 3416 (1989).

34. S.A. Harrison, J.M. Buxton, A.L. Helgerson, R.G. MacDonald, F.J. Chlapowski, A. Carruthers and M.P. Czech, Insulin action on activity and cell surface disposition of human HepG2 glucose transporters expressed in Chinese hamster ovary cells, J. Biol. Chem. 265:5793 (1990).

35. A.G. De Herreros and M.J. Birnbaum, The acquisition of increased insulin-responsive hexose transport inn 3T3-L1 adipocytes correlates with expression of a novel transporter gene, J. Biol. Chem. 264:19994 (1989).

36. K.M. Tordjman, K.A. Leingang, D.E. James and M.M. Mueckler, Differential regulation of two distinct glucose transporter species expressed in 3T3-L1 adipocytes: Effect of chronic insulin and tolbutamide treatment, Proc. Natl. Acad. Sci. USA 86:7761 (1989).

37. D.M. Calderhead, K. Kitagawa, L.I. Tanner, G.D. Holman and G.E. Lienhard Insulin regulation of the two glucose transporters in 3T3-L1 adipocytes, J. Biol. Chem. 265:13800 (1990).

38. J. Berger, C. Biswas, P. Vicario, H.V. Strout, R. Saperstein and P. Pilch, Decreased expression of the insulin-responsive glucose transporter in diabetes and fasting, Nature 340:70 (1989).

39. W.T. Garvey, T.P. Hueckstead and M.J. Birnbaum, Pretranslational suppression of an insulin-responsive glucose transporter in rats with diabetes mellitus, Science 245:60 (1989).

40. B.B. Kahn, M.J. Charron, H.F. Lodish, S.W. Cushman and J.S. Flier, Differential regulation of two glucose transporters in adipose cells from diabetic and insulin-treated diabetic rats, J. Clin. Invest. 84:404 (1989).

41. W.I. Sivitz, S.L. DeSautel, T. Kayano, G.I. Bell and J.E. Pessin, Regulation of glucose transporter messenger RNA in insulin-deficient states, Nature 340:72 (1989).

42. B.B. Kahn and S.W. Cushman, Mechanism for markedly hyperresponsive insulin-stimulated glucose transport activity in adipose cells from insulin-treated streptozotocin diabetic rats, J. Biol. Chem. 262:5118 (1987).
43. W.I. Sivitz, S.L. DeSautel, T. Kayano, G.I. Bell and J.E. Pessin, Regulation of glucose transporter messenger RNA levels in rat adipose tissue by insulin, Mol. Endocrinol. 4:583 (1990).
44. A.D. Baron, G. Brechtel, P. Wallace and S.V. Edelman, Rates and tissue sites of non-insulin- and insulin-mediated glucose uptake in humans, Am. J. Physiol. 255:E769 (1988).
45. J. Elbrinck and I. Bihler, Membrane transport: it's relation to cellular metabolic rates: glucose transport into animal cells is adapted to their metabolic rate and often controls rate of glucose use, Science 183:1177 (1975).
46. A. Katz, B. Nyomba and C. Bogardus, No accumulation of glucose in human skeletal muscle during euglycemic hyperinsulinemia, Am. J. Physiol. 255:E942 (1988).
47. D.E. James, A.B. Jenkins and E.W. Kraegen, Heterogeneity of insulin action in individual muscles in vivo: euglycemic clamp studies in rats, Am. J. Physiol. 248:E567 (1985).
48. A. Bonen, M.H. Tan and W.M. Watson-Wright, Insulin binding and glucose uptake differences in rodent skeletal muscles, Diabetes 30:702 (1981).
49. D.E. James, A. Zorzano, M. Boni-Schnetzler, R.A. Nemenoff, A. Powers, P.F. Pilch and N.B. Ruderman, Intrinsic differences of insulin receptor kinase activity in red and white muscle, J. Biol. Chem. 261:14939 (1986).
50. M. Kern, J.A. Wells, J.M. Stephens, C.W. Elton, J.E. Friedman, E.B. Tapscott, P.H. Pekala and G.L. Dohm, Insulin responsiveness in skeletal muscle is determined by glucose transporter (Glut-4) protein level, Biochem. J. 270:397 (1990).
51. M.F. Hirshman, L.J. Goodyear, L.J. Wardzala, E.D. Horton and E.S. Horton, Identification fo an intracellular pool of glucose transporters from basal and insulin-stimulated rat skeletal muscle, J. Biol. Chem. 265:987 (1990).
52. A. Klip, T. Ramlal, D.A. Young and J.O. Holloszy, Insulin-induced translocation of glucose transporters in rat hindlimb muscles, FEBS Lett. 224:224 (1987).
53. M.J. Charron and B.B. Kahn, Divergent molecular mechanisms for insulin-resistant glucose transport in muscle and adipose cells *in vivo*, J. Biol. Chem. 265:7994 (1990).
54. L. Koranyi, D. James, M. Mueckler and A. Permutt, Glucose transporter levels in spontaneously obese (db/db) insulin-resistant mice, J. Clin. Invest. 85:962 (1990).
55. J.E. Friedman, W.M. Sherman, M.J. Reed, C.W. Elton and G.L. Dohm, Exercise training increases glucose transporter protein GLUT-4 in skeletal muscle of obese Zucker (fa/fa) rats, FEBS Lett 268:13 (1990).
56. K.J. Rodnick, J.O. Holloszy, C.E. Mondon and D.E. James, Effects of exercise training on insulin-regulatable glucose transporter protein levels in rat skeletal muscle, Diabetes 39:1425 (1990).
57. O. Pedersen, J.F. Bak, P.H. Andersen, S. Lund, D.E. Moller, J.S. Flier and B.B. Kahn, Evidence against altered expression of GLUT1 or GLUT4 in skeletal muscle of patients with obesity or NIDDM, Diabetes 39:865 (1990).

REGULATION OF INSULIN-LIKE GROWTH FACTOR I RECEPTOR GENE EXPRESSION IN NORMAL AND PATHOLOGICAL STATES

Haim Werner, Bethel Stannard, Mark A. Bach, Charles T. Roberts, Jr., and Derek LeRoith

Section of Molecular and Cellular Physiology
Diabetes Branch, NIDDK, National Institutes of Health, Bethesda, Maryland

INTRODUCTION

Most of the biological actions of insulin-like growth factor I (IGF-I)/somatomedin C are initiated by its binding to the IGF-I receptor, a heterotetrameric glycoprotein structurally related to the insulin receptor[1]. The presence of IGF-I receptors in most body tissues suggests that it mediates many different effects. Indeed, IGF-I has been shown to be involved not only in endocrine functions, such as the mediation of growth hormone's effect on longitudinal growth, but it is also involved in many autocrine/paracrine systems at the local tissue level[2,3]. These biological actions include both short-term, metabolic effects (similar in nature to those stimulated by insulin) as well as long term, growth promoting actions. It is not surprising then, that the IGF- I receptor should be able to respond to various tissue - and development-specific stimuli. To study the expression of the IGF-I receptor gene in both physiological and pathological conditions in a convenient animal model, we undertook the cloning of rat IGF-I receptor cDNAs.

CLONING OF IGF-I RECEPTOR cDNAs

Rat IGF-I receptor cDNAs were isolated from an SV40-transformed rat granulosa cell cDNA library which was screened with a human IGF-I receptor cDNA probe[4]. This cell line had been previously shown to express high levels of IGF-I receptor mRNA and abundant high affinity IGF-I receptors[5]. Upon screening 2 x 10^6 recombinant phage, five independent clones were isolated[6]. The cDNA inserts, ranging in size from 0.6 kb to 1.6 kb, were recovered by digestion of phage DNA with EcoR1, and subsequently cloned into pGEM vectors (Fig. 1). Sequence analyses revealed that these were overlapping clones that spanned a region including ~1 kb of 5' untranslated region and sequences encoding the signal peptide and the first 333 amino acids of the α-subunit, including most of the cysteine-rich region. The similarity between the rat sequence

Molecular Biology and Physiology of Insulin and Insulin-Like Growth Factors
Edited by M.K. Raizada and D. LeRoith, Plenum Press, New York, 1991

and the corresponding human sequence is 94% at the nucleotide level and 97% at the amino acid level. In addition, all cysteine residues are conserved, as well as all five potential N-glycosylation sites present in this region. This high degree of homology may reflect the evolutionary conservation of IGF-I receptor function.

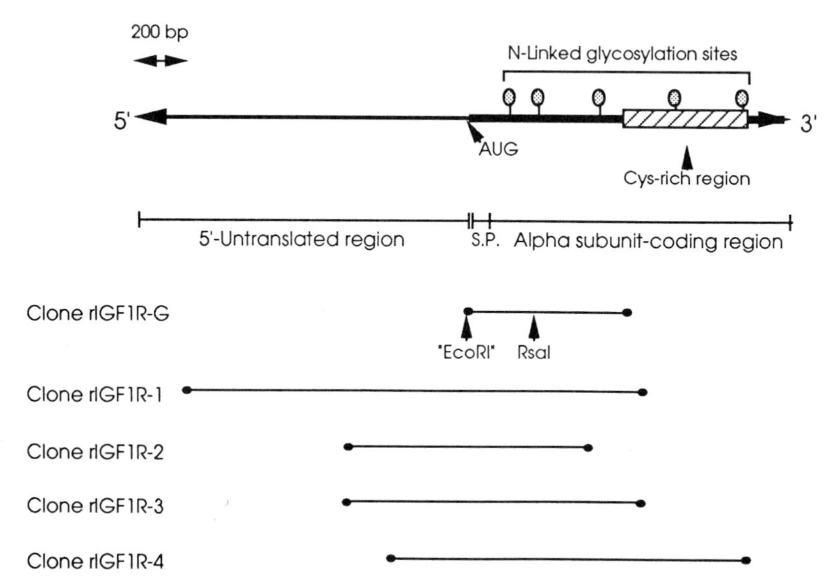

Fig. 1. Schematic representation of rat IGF-I receptor cDNA clones. Five partial rat IGF-I receptor cDNAs were isolated from an SV40-transformed rat granulosa cell cDNA library which was screened with a ^{32}P-labeled human IGF-I receptor cDNA probe. The cysteine-rich region, as well as potential N-linked glycosylation sites, are indicated. S.P.: signal peptide. Restriction sites for enzymes Rsa I and EcoRI (linker) are shown.

DEVELOPMENTAL REGULATION OF IGF-I RECEPTOR GENE EXPRESSION

To study the regulation of IGF-I receptor mRNA in rat tissues during ontogeny, an antisense IGF-I receptor RNA probe was generated from a 265-bp EcoR1-Rsa1 cDNA fragment encoding 15 bases of 5' untranslated sequence, the signal peptide, and the first 53 amino acids of the α-subunit (Fig. 1). This riboprobe was used in solution hybridization/RNase protection assays with total RNA obtained at various developmental stages[6].

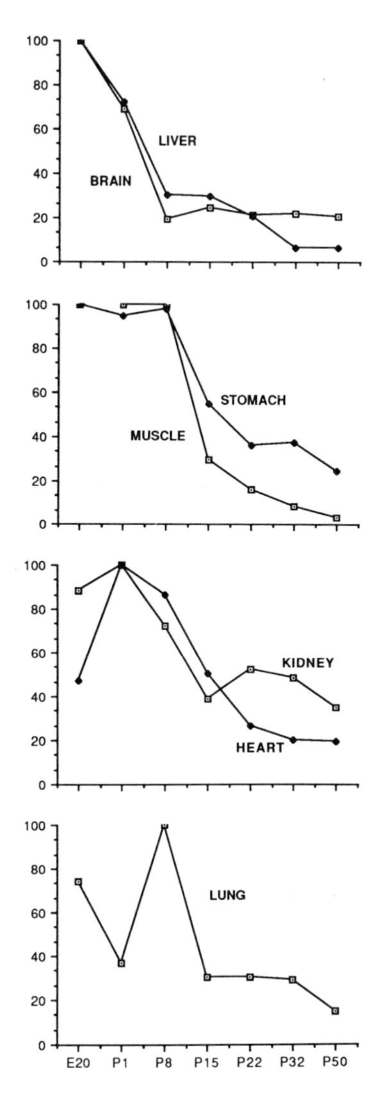

Fig. 2. Developmental regulation of IGF-I receptor gene expression in rat tissues. The steady-state levels of IGF-I receptor mRNA were measured in different rat tissues at various developmental stages by solution hybridization/-RNase protection assays using a ^{32}P-labeled IGF-I receptor antisense RNA probe. The autoradiograms obtained were quantitated by scanning densitometry and, for each tissue, a value of 100% was assigned to the stage showing the highest levels of the message. E, embryonic; P, postnatal.

Maximal levels of expression of IGF-I receptor mRNA were generally detected at perinatal stages, decreasing subsequently in a tissue-specific manner. Thus, whereas in liver the levels at postnatal day 50 (P50) were ≈6% of those at embryonic day 20 (E20), in stomach and brain the levels of the message at P50 were ≈20% of those seen at E20. In muscle, the levels of mRNA at P50 were ≈3% of those at P1, whereas in heart and kidney the levels at P50 were ≈17% and ≈30%,

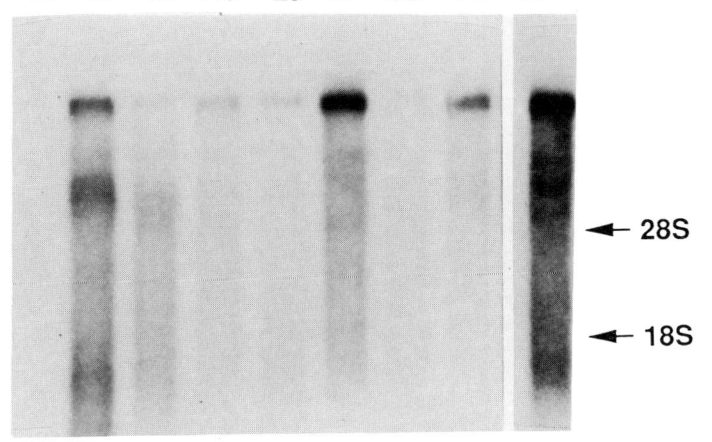

Fig. 3. Northern blot analysis of IGF-I receptor mRNA in rat tissues. Four micrograms of poly (A)+mRNA from different adult rat tissues were electrophoresed on a 1.5% agarose/formaldehyde gel and, following electro-blotting, were hybridized to an homologous ^{32}P-labeled IGF-I receptor cDNA probe, as described. Li, liver; ki, kidney; st, stomach; he, heart; lu, lung; br, brain; mu, muscle; te, testis; gr, SV40-transformed granulosa cell line.

respectively, of those observed at P1 (Fig. 2). The results obtained are, therefore, in general accordance with previous studies performed by other investigators who demonstrated that the general trend in IGF-I binding to tissues was a decrease from very high levels at fetal and neonatal stages to lower levels at adult stages[7-9]. Thus, this decrease in binding could be a consequence of altered gene expression of the IGF-I receptor.

The expression of the IGF-I receptor gene was also studied by Northern blot analysis using poly (A)+ mRNA. In most of the tissues, a single major hybridizing band of ≈11 kb was observed (Fig. 3). However, this method was not sensitive enough to detect any expression of IGF-I receptor mRNA in adult liver, consistent with previous observations which showed that IGF-I binding by this tissue is extremely low.

IGF-I RECEPTOR GENE EXPRESSION IN DIABETES AND IN ALTERED NUTRITIONAL STATES

One of the major complications of both insulin-dependent and non-insulin-dependent diabetes is nephropathy, the loss of glomerular function and subsequent kidney failure due to structural lesions in the glomeruli. Among different mitogenic factors studied, IGF-I was shown to be especially potent in stimulating mesangial cell proliferation[10]. Moreover, compensatory hypertrophy subsequent to unilateral nephrectomy led to an increase in IGF-I in collecting duct cells, suggesting a causative role for this growth factor in hypertrophy of this organ[11].

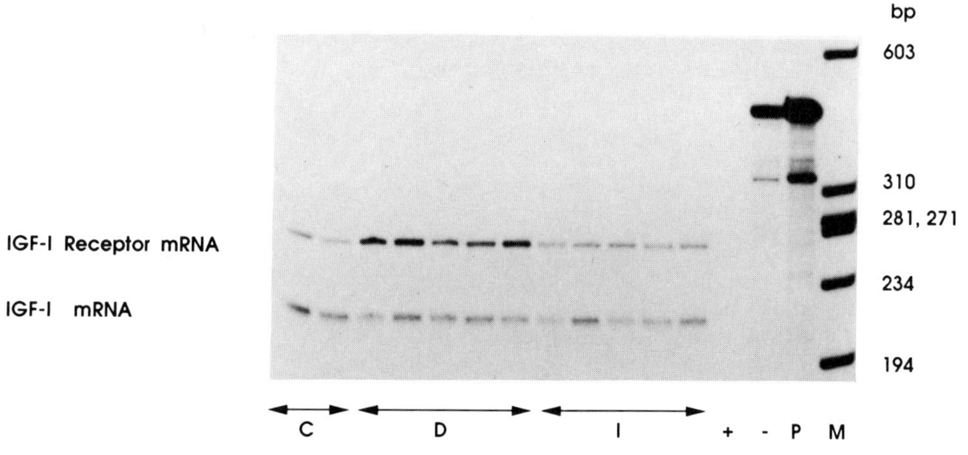

Fig. 4. Expression of IGF-I and IGF-I receptor genes in control, diabetic, and insulin-treated kidneys. The levels of IGF-I and IGF-I receptor mRNAs were measured by solution hybridization/RNase protection assays using twenty ug of kidney RNA and ^{32}P-labeled IGF-I and IGF-I receptor antisense RNA probes. Each lane represents an individually processed organ. C, control; D, diabetic; I, insulin-treated diabetic rat; +, probe alone with RNase; -, probe alone without RNase; P, native probe; M, marker phi-x174 DNA digested with HaeIII.

Table 1. Binding of ^{125}I-IGF-1 and ^{125}I-IGF-II to kidney membranes from control, diabetic, and insulin-treated rats.

	Specific Binding (% of Total Binding)	
	^{125}I-IGF-I	^{125}I-IGF-II
Controls	4.72±0.18	2.82±0.60
Diabetics	10.77±2.16*	9.52±0.81*
Diabetics + Insulin	5.21±0.39	4.44±0.85

Values represent the mean ± S.E.M. of the percentage of total binding by each specific ligand per 60 ug protein. The values of five different animals were averaged in each group. *Significantly different from respective controls ($p<0.005$).

Because most IGF-I effects are mediated via the IGF-I receptor, it was important to study the possible involvement of IGF-I and its receptor in the development of diabetic nephropathy. To this end, the levels of IGF-I and IGF-I receptor mRNAs, as well as IGF-I binding, were measured in kidney and in other tissues of the streptozocin-diabetic rat[12]. A ≈2.5- fold increase in the steady-state levels of IGF-I receptor mRNA was observed in the diabetic kidney (Fig. 4); this increase was accompanied by a ≈2.3-fold increase in IGF-I binding (Table 1). Both receptor mRNA levels and binding returned to control levels upon insulin treatment. No effect of the diabetic state in this model was seen on the levels of kidney IGF-I mRNA levels. Because the IGF-I receptor is able to transduce mitogenic signals upon activation of its tyrosine kinase domain following IGF-I (or IGF-II) binding, we hypothesize that, among other factors, high levels of receptor in the untreated diabetic kidney may contribute to the development of early renal changes and perhaps the later development of diabetic nephropathy.

Interestingly, the levels of IGF-II/mannose-6-phosphate receptor mRNA and of IGF-II binding (Table 1) were also increased in the diabetic kidney. It is conceivable that higher concentrations of this receptor and the concommitant intracellular transport and packaging of lysosomal enzymes may contribute to tissue remodelling processes.

Similarly to the kidney, the levels of IGF-I receptor mRNA were also elevated in the diabetic heart and skeletal muscle, however, no change was seen in brain or testes, which are organs separated from the general circulation by an endothelial barrier.

Altered nutritional status was also shown to affect both IGF-I binding and the expression of the IGF-I receptor gene[13]. Thus, fasting in rats was correlated with a significant increase in ^{125}I-IGF-I binding in lung, testes, stomach, kidney and heart. These elevated receptor concentrations, as determined by Scatchard analysis, were associated with a ≈1.6-2.5 fold increase in the steady-state levels of IGF-I receptor mRNA.

CLONING AND CHARACTERIZATION OF THE RAT IGF-I RECEPTOR GENE PROMOTER

To study the molecular mechanisms responsible for the regulation of expression of the rat IGF-I receptor gene, we have cloned the putative promoter region of this gene[14]. Using primer extension assays with a synthetic oligonucleotide complementary to the 5' end of the rat IGF-I receptor cDNA clone 1 (Fig. 5), we found a single major transcription start site. This site was located only 9 bases upstream of the 5' end of the cDNA clone. The presence of a unique initiation site was corroborated by RNase protection assays using antisense RNA probes generated from genomic clones extending up to 415 bases upstream of this start site, and probes derived either from genomic or cDNA clones which are complementary to the first 520 bases of 5' untranslated region.

The 5' flanking region of the IGF-I receptor gene is devoid of TATA or CAAT boxes which are generally required for transcription initiation at a unique site (Fig. 6). However, the sequence surrounding the transcription start site is homologous to an "initiator" element recently described in the terminal deoxynucleotidyl transferase and in the Adenovirus middle late promoters, and which directs transcription initiation from a single site in the absence of TATA or CAAT boxes[15]. Both the 5' flanking and 5' untranslated regions are very GC-rich (≈70%) and several potential SP1 binding sites are located in these regions[16]. In addition, the 5' flanking region contains potential binding sites for the transcription factors ETF, at position -369 to -357, and AP2, at position

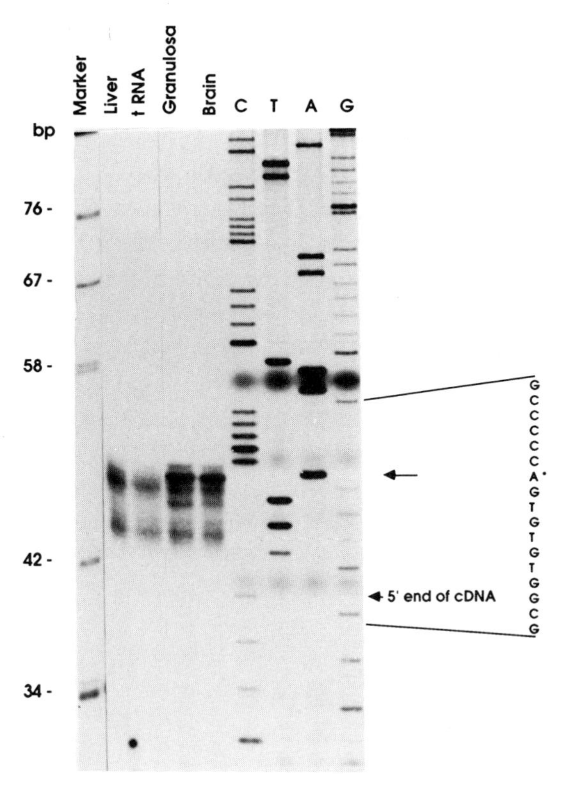

Fig. 5. Primer extension mapping of the 5' end of the rat IGF-I receptor mRNA. Fifteen ug of poly(A)+ mRNA from SV40 transformed-rat granulosa cells, neonate brain and adult liver, and tRNA were annealed to a ^{32}P-labeled oligonucleotide complementary to the 5' end of the rat IGF-I receptor cDNA clone 1 (rIGFI-R-1, Fig. 1). Following incubation with AMV-reverse transcriptase, the extended products were resolved on an 8% polyacrylamide/8M urea denaturing gel. The granulosa cells and neonatal brain were previously shown to contain high levels of IGF-I receptor mRNA while there is very little mRNA in adult liver. The sequence ladder shown on the four right lanes was obtained by using the same oligonucleotide as a sequencing primer of a genomic fragment covering this region. The sequence shown at the right is similar to the "initiator" sequence described in reference 15. The transcription start site is indicated by an arrow and an asterisk.

-166 to -159. ETF is able to stimulate transcription of genes which lack TATA and CAAT boxes, whereas the AP2 element may mediate cAMP or phorbol ester control of transcription[17,18]. Preliminary analyses of human genomic clones containing the putative promoter of the human IGF-I receptor gene indicate that the initiator sequence and the flanking sequences are highly conserved. We are curently mapping the specific DNA elements which interact with nuclear trans-acting factors by means of gel shift, DNA footprinting and transient expression assays.

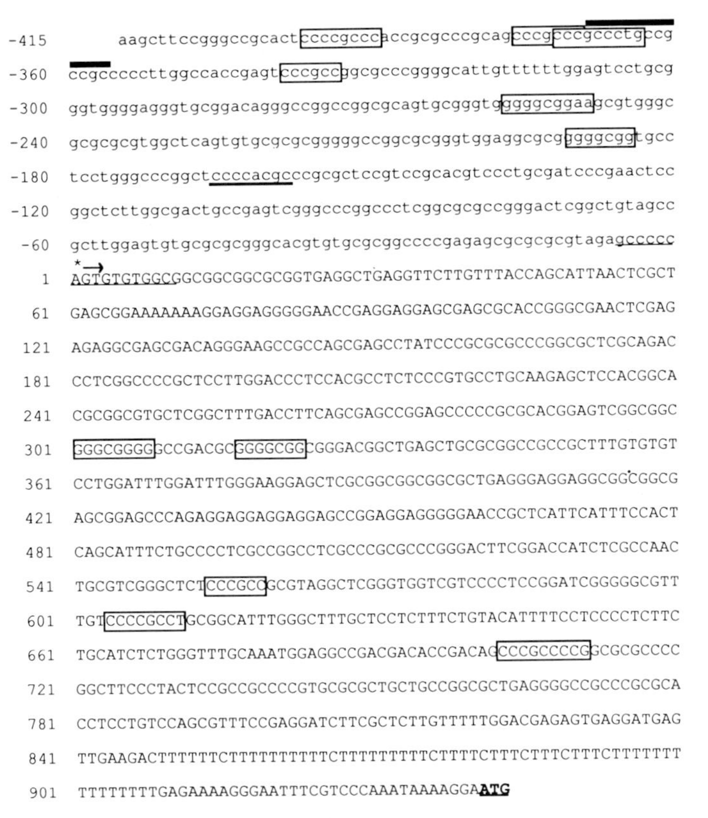

Fig. 6. DNA sequence of the 5' flanking and 5' untranslated regions of the rat IGF-I receptor gene. Nucleotide 1 corresponds to the transcription initiation site. Lower case letters are used for the 5' flanking sequences, whereas upper case letters are used for the 5' untranslated region. The ATG codon for translation initiation is shown at the end of the sequence. Potential SP1 sites are boxed, a putative ETF-binding element is overlined by a thick bar (-369 to -357), and a potential AP-2 binding site is underlined by a thick line (-165 to -158). Also shown is the "initiator" element which encompasses the transcription start site.

A COMPARISON BETWEEN IGF-I AND INSULIN RECEPTOR PROMOTERS

The IGF-I and insulin receptors, though structurally and phylogenetically related, seem to differ in the mechanisms which regulate their expression. Both gene promoters lack TATA or CAAT boxes and are very GC-rich. However, whereas the insulin receptor contains several transcription initiation sites, the rat IGF-I receptor appears to contain only one major start site[19,20]. In addition, both gene promoters contain several Sp1 binding sites. In the case of the insulin receptor promoter, no other binding sites for known transcription factors have thus far been identified. On the other hand, binding elements for the trans-acting factors ETF and AP2 were described in the IGF-I receptor. Thus, the insulin receptor gene promoter appears to conform with the definition of a "housekeeping" gene, namely a gene which is expressed at a relatively constant level, whereas the IGF-I receptor gene is highly regulated during differentiation and development as well as in certain disease states.

REFERENCES

1. Rechler, M.M., and Nissley, S.P., The nature and regulation of the receptors for insulin-like growth factors, Ann. Rev. Physiol. 47: 425 (1985).

2. Daughaday, W.H., and Rotwein, P., Insulin-like growth factors I and II. Peptide, messenger ribonucleic acid and gene structures, serum, and tissue concentrations, Endocrine Rev. 10: 68 (1989).

3. D'Ercole, A.J., Stiles, A.D., and Underwood, LE., Tissue concentrations of somatomedin C: further evidence for multiple sites of synthesis and paracrine or autocrine mechanism of action, Proc. Nat. Acad. Sci. USA 81:935 (1984).

4. Ullrich, A., Gray, A., Tam, A.W., Yang-Feng, T., Tsubokawa, M., Collins, C., Henzel, W., LeBon, T., Kathuria, S., Chen, E., Jacobs, S., Francke, U., Ramachandran, J., and Fujita-Yamaguchi, Y., Insulin-like growth factor I receptor primary structure: comparison with insulin receptor suggests structural determinants that define functional specificity, EMBO J. 5: 2503 (1986).

5. Zilberstein, J., Chou, J.Y., Lowe, Jr., W.L., Shen-Orr, Z., Roberts, Jr., C.T., LeRoith, D., and Catt, K.J., Expression of insulin-like growth factor-I and its receptor by SV40-transformed rat granulosa cells, Mol. Endocrinol. 3: 1488 (1989).

6. Werner, H., Woloschak, M., Adamo, M., Shen-Orr, Z., Roberts, Jr., C.T., and LeRoith, D, Developmental regulation of the rat insulin-like growth factor I receptor gene, Proc. Natl. Acad. Sci. USA 86: 7451 (1989).

7. Bassas, L., DePablo, F., Lesniak, M.A., and Roth, J., Ontogeny of receptors for insulin-like peptides in chick embryo tissues: early dominance of insulin-like growth factor over insulin receptors in brain, Endocrinol. 117: 2321 (1985).

8. Pomerance, M., Gavaret, J-M., Jacquemin, C., Matricon, C., Toru-Delbauffe, D., and Pierre, M., Insulin and insulin-like growth factor I receptors during postnatal development of rat brain, Develop. Brain Res. 42:77 (1988).

9. Alexandrides, T., Moses, A.C., and Smith, R.J., Developmental expression of receptors for insulin, insulin-like growth factor I (IGF-I), and IGF-II in rat skeletal muscle, Endocrinol. 124: 1064 (1989).

10. Conti, F.G., Striker, L.J., Elliot, S.J., Andreani, D., and Striker, G.E., Synthesis and release of insulin-like growth factor I by mesangial cells in culture, Am. J. Physiol. 255:F1214 (1988).

11. Lajara, R., Rotwein, P., Bortz, J.D., Hansen, V.A., Sadow, J.L., Betts, C.R., Rogers, S.A., and Hammerman, M.R., Dual regulation of insulin-like growth factor I expression during renal hypertrophy, Am. J. Physiol. 257:F252 (1989).

12. Werner, H., Shen-Orr, Z., Stannard, B., Burguera, B., Roberts, Jr., C.T., and LeRoith, D., Experimental diabetes increases insulin-like growth factor I and II receptor concentration and gene expression in the kidney, Diabetes, in press (1990).

13. Lowe, Jr., W.L., Adamo, M., Werner, H., Roberts, Jr., C.T., and LeRoith, D., Regulation by fasting of rat insulin-like growth factor I and its receptor, J. Clin. Invest. 84: 619 (1989).

14. Werner, H., Stannard, B., Bach, M.A., LeRoith, D., and Roberts, Jr., C.T., Cloning and characterization of the proximal promoter region of the rat insulin-like growth factor I (IGF-I) receptor gene, Biochem. Biophys. Res. Comm. 169: 1021 (1990).

15. Smale, S.T., and Baltimore, D., The "initiator" as a transcription control element, Cell 57:103 (1989).

16. Kadonaga, J.T., and Tjian, R., Affinity purification of sequence-specific DNA binding proteins, Proc. Natl. Acad. Sci. USA 83: 5889 (1986).

17. Kageyama, R., Merlino, G.T., and Pastan, I., Nuclear factor ETF specifically stimulates transcription from promoters without a TATA box, J. Biol. Chem. 264: 15508 (1989).

18. Roesler, W.J., Vandenbark, G.R., and Hanson, R.W., Cyclic AMP and the induction of eukaryotic gene transcription, J. Biol. Chem. 263: 9063 (1988).

19. Araki, E., Shimada, F., Uzawa, H., Mori, M., and Ebina, Y., Characterization of the promoter region of the human insulin receptor gene, J. Biol. Chem. 262: 16186 (1987).

20. Seino, S., Seino, M., Nishi, S., and Bell, G.I., Structure of the human insulin receptor gene and characterization of its promoter, Proc. Natl. Acad. Sci. USA 86:114 (1989).

IGF-I MEDIATED RECRUITMENT OF GLUCOSE TRANSPORTERS FROM INTRACELLULAR MEMBRANES TO PLASMA MEMBRANES IN L6 MUSCLE CELLS

Philip J. Bilan, Toolsie Ramlal, and Amira Klip

Division of Cell Biology, The Hospital for Sick Children

555 University Avenue, Toronto, Ontario M5G 1X8, Canada

INTRODUCTION

IGF-I has long been known to have mitogenic effects on isolated skeletal tissue (Salmon and Daughaday, 1957) and insulin-like metabolic effects on isolated adipose tissue and muscle tissue (Froesch et al., 1966; Poggi et al., 1979). Stimulation of glucose utilization by IGF-I has been observed *in vivo* in rats and humans (Zapf et al., 1986; Guler et al., 1987). Moreover, Giacca et al. (1990) have demonstrated that skeletal muscle is the preferred site for the stimulation of glucose utilization by IGF-I in completely insulin-deficient diabetic dogs. Simultaneously, Moxely III et al. (1990) demonstrated that *in vivo*, IGF-I stimulates hexose uptake directly into rat muscles. These specific effects of IGF-I on skeletal muscle are consistent with the observation that skeletal muscle expresses abundant amounts of IGF-I receptors (Livingston et al., 1988, Dohm et al., 1990), whereas adipose tissue has very low numbers of IGF-I receptors (Rechler and Nissley, 1985; Sinha et al., 1990). Skeletal muscle is also the primary site of action for the stimulation of glucose utilization by insulin *in vivo* (Defronzo et al., 1981). Although IGF-I and insulin have common biological actions, IGF-I and insulin receptors may function independently. For example, rat 1 fibroblasts expressing a mutant insulin receptor with an inactive tyrosine kinase are unresponsive to insulin stimulation of glucose uptake but can still respond to IGF-I through their endogenous IGF-I receptors (McClain et al., 1990). Secondly, Lammers et al. (1989) demonstrated that chimeric receptors consisting of the extracellular insulin receptor domain and the intracellular IGF-I receptor domain were ten times more responsive to insulin for stimulation of DNA synthesis than was the native insulin receptor. This suggested that the intracellular kinase of the IGF-I receptor is more active than the insulin receptor kinase and is thus inherently different from its insulin receptor counterpart. Hence, IGF-I and insulin and their receptors share common responses but can trigger them independently.

The rat L6 cell line of muscle origin shows many characteristics of skeletal muscle including differentiation into myotubes (Yaffe, 1968); development of action potentials and contractile activity (Kidoboro, 1975); and production of muscle specific proteins (Shainberg et al., 1971). During differentiation of L6 myoblasts into myotubes, insulin receptors increase in number by 3-fold, whether expressed per unit protein or per unit DNA (Klip et al., 1983 and Beguinot et al., 1986). As L6 cells differentiate, their basal glucose transport rate decreases but the degree of glucose transport stimulation by acute exposure to insulin (Klip et al., 1984) and chronic exposure to insulin (Beguinot et al., 1986) increases. L6 cells also express IGF-I and IGF-II receptors (Beguinot et al., 1985) but the

Molecular Biology and Physiology of Insulin and Insulin-Like Growth Factors
Edited by M.K. Raizada and D. LeRoith, Plenum Press, New York, 1991

regulation of glucose transport by these hormones has not been fully explored.

Glucose transport into muscle and fat occurs by facilitated diffusion (Widdas, 1988). These tissues express the GLUT 1 and GLUT 4 isoforms of facilitative glucose transporters (James et al., 1989, Birnbaum, 1989). The transmembrane segment of the GLUT 1 transporter predicted to be photolabelled in a glucose competitive manner by the potent glucose transport inhibitor, cytochalasin B (Holman and Rees, 1987), has 90% amino acid similarity with its GLUT 4 counterpart (James et al.; 1989; Birnbaum, 1989). This explains why the GLUT 4 transporter can also be photolabelled by cytochalasin B (James et al. 1988). Since the GLUT 1 transporter (Baldwin et al., 1982) and presumably the GLUT 4 transporter bind cytochalasin B in a one to one molar ratio, cytochalasin B is a useful probe for determining the number of GLUT 1 and GLUT 4 transporters. In contrast, the intracellular domains of the GLUT 1 and GLUT 4 transporters are distinct. This difference has allowed investigators to raise antisera against synthetic peptides corresponding to the last 12-15 amino acids of the C-terminal end of each transporter. With these probes the particular glucose transporter isoform can be identified.

In recent years it has been observed that insulin induces the rapid recruitment of GLUT 4 transporters from internal membranes to the plasma membranes of muscle (Douen et al., 1990) and fat (James et al., 1988) cells. In fat cells, but not in muscle, the hormone also causes the recruitment of the GLUT 1 transporter (Douen et al., 1990; Zorzano et al., 1989). We have recently shown that insulin induces the acute recruitment of glucose transporters in L6 cells, as measured by cytochalasin B binding (Ramlal et al., 1988). L6 cells express the GLUT 1 and GLUT 4 transporter isoforms, the former in greater abundance (Walker et al., 1989; Koivisto et al., 1990). However the specific transporter isoform(s) recruited by insulin in L6 muscle have not been identified.

Although stimulation of glucose transport by IGF-I in muscle and L6 cells has been documented, the mechanism of this stimulation is not known. In particular, it is not known whether IGF-I can induce the recruitment of glucose transporters. L6 cells offer a good model system in which to compare the effects of insulin and IGF-I on muscle cells. The objectives of the present study were to compare the actions of IGF-I and insulin on glucose transport in L6 cells and to investigate the subcellular distribution of the glucose transporter isoforms in response to acute exposure to IGF-I.

MATERIALS AND METHODS

Materials

Tissue culture medium, serum and reagents were obtained from GIBCO. 2-deoxy-D-[^{3}H]glucose and [^{125}I]protein A were purchased from ICN. [^{3}H]Cytochalasin B was from Amersham. Porcine insulin, D- and L-glucose, 2-deoxy-D-glucose, cytochalasins B and E, PMSF, E-64, leupeptin, pepstatin A, EGTA, Hepes and BSA were obtained from Sigma. Nitrocellulose membrane was from Schleicher & Schuell. Electrophoresis grade chemicals were from BioRad. Gradient grade sucrose was from Merck. Recombinant human IGF-I was a kind gift from Dr. M. Vranic. The polyclonal antiserum, R820 that recognizes the C-terminus of GLUT 4 was kindly provided by Dr. D. James. The anti-GLUT 1 antibody was raised against purified Band 4.5 (glucose transporter) from human erythrocytes as reported earlier (Bilan and Klip, 1990), and was affinity purified on an affi-gel BioRad column crosslinked with a 12-residue polypeptide containing the human GLUT 1 C-terminal sequence (Bilan and Paquet, unpublished).

Cell cultures

Spontaneously fusing L6 rat skeletal muscle cells were maintained in monolayer culture in α-Minimal Essential Medium containing 2% (v/v) fetal bovine serum and 1% (v/v) antibiotic/antimycotic solution (10 000 units/ml penicillin, 10 mg/ml streptomycin, 25 μg/ml amphotericin B) in 10 cm diameter dishes for membrane fractionation experiments or 6-well plates for hexose transport experiments, in an atmosphere of 5% CO_2 at 37°C as previously described (Klip et al., 1984). Cell monolayers in the plates and flasks were allowed to reach confluence, align and fuse into myotubes prior to experimental determinations. Myogenesis was monitored by phase contrast microscopy. L6 cells were used for experiments when the majority became multinucleated myotubes (4 or 5 days after alignment). The cells were fed fresh medium every 48 h; prior to experimental manipulations the cells were pre-incubated in serum-free α-MEM containing 25 mM D-glucose for 16 h.

Hormone incubations and hexose transport measurements

Cells pre-incubated for 16 h in serum-free medium were incubated for the indicated periods of time with the indicated concentrations of insulin or IGF-I, followed by removal of the medium and assay of hexose transport as follows. The cells were rinsed twice with glucose-free Hepes-buffered saline solution (140 mM NaCl, 1 mM $CaCl_2$, 5 mM KCl, 2.5 mM $MgSO_4$, 20 mM Hepes, pH 7.4) and subsequently incubated for 5 minutes with 10 μM 2-deoxy-D-[^{3}H]glucose (0.83 μCi/well) in the same solution. Uptake was ended by washing the cells 3 times with 2 ml of ice-cold 0.9% NaCl. Under the conditions of this assay, hexose transport is rate limiting (Klip et al., 1984). Carrier-independent transport was determined in parallel wells containing 10 μM cytochalasin B. The protein content was determined by the method of Bradford (1976) and the specific uptake was expressed in pmol/mg/min.

Membrane fractionation

The procedure was essentially that described earlier (Ramlal et al., 1988) with some modification. Briefly, cells in 20 culture dishes were incubated with 20 nM IGF-I, 200 nM insulin or no additions in α-MEM containing 25 mM D-glucose for 45 min at 37°C, then scraped with a rubber policeman and pelleted at 760 x g for 5 min 4°C. In each case, the cell pellet was resuspended in 12 ml of buffer 1 (250 mM sucrose, 5 mM NaN_3, 2 mM EGTA, 200 μM PMSF, 1 μM leupetin, 1 μM pepstatin A, 10 μM E-64 and 20 mM Hepes pH 7.4) and homogenized by 15 strokes with a 30 ml Wheaton glass homogenizer using a Eberbach drive motor. The homogenate was centrifuged at 760 x g for 3 min at 4°C. The supernatant was separated from the pellet and put on ice. The pellet was rehomogenized in 8 ml of buffer 1 (15 strokes), followed by another 3 min centrifugation at 760 x g, 4°C. The pellet was discarded and the resultant supernatant was pooled with the first supernatant and centrifuged at 31,000 x g for 20 min at 4°C. The supernatant obtained was used for isolation of the internal membranes and the pellet (P-1) for separation of the plasma membranes. The supernatant was centrifuged at 177,000 x g for 60 min to yield the light microsome pellet. The P-1 pellet was resuspended to 6 ml in buffer 1 and layered on top of 3 ml each of 30% and 40% w/w sucrose in buffer 2 (2 mM EGTA, 5 mM NaN_3, 20 mM Hepes pH 7.4) and centrifuged at 210,000 x g_{max} for 150 min. Membranes were recovered from each sucrose layer, washed by a 10-fold dilution in buffer 2, recovered by centrifugation at 177,000 x g for 60 min and resuspended in buffer 1. All samples were stored at -20°C. Protein concentrations were determined according to Lowry et al. (1951). Cytochalasin B binding was assayed within 24 h and enzyme markers were assayed within 48 h.

Cytochalasin B binding

75-100 μg membrane protein (75 μl) were incubated with 0.2 μM [^{3}H]cytochalasin B, 200 mM D- or L-glucose and 5.5 μM cytochalasin E in a total volume of 100 μl at room temperature for 15 min. The samples were filtered through Whatman GF/B filters (2.4 cm diameter) on a Millipore filtration unit under vacuum and quickly (2 seconds) washed 3 times with 1.5 ml of 0.9% NaCl (4°C). Filters were removed and placed into scintillation vials along with 15 ml scintillation cocktail. Glucose transporter-specific cytochalasin B binding was calculated by subtracting the D-glucose values (pmol/mg) from the L-glucose values.

γ-Glutamyl transferase activity

The activity was measured by the method of Yasumoto et al. (1983). The enzyme is a useful plasma membrane marker in skeletal muscle (Fushiki et al., 1989). Briefly, 50 μg membrane protein (50 μl) was incubated with solution A (3.5 mM γ-glutamyl-*p*-nitroanilide, 20 mM glycylglycine, 100 mM Tris-HCl pH 8.5, total vol = 1 ml) for 20 min at 37°C. The reaction was stopped with 3 ml of 1.7 N acetic acid and the samples were quickly cooled to room temperature. Blanks for each type of membrane fraction were prepared by mixing 50 μg protein, solution A and 3 ml of 1.7 N acetic acid (total vol = 4 ml), followed by a 20 min incubation at 37°C. Sample absorbance was measured at 410 nm and was converted to μmol/mg/min using a standard curve of known concentrations of *p*-nitroaniline.

Immunoblot analysis

Fifty μg samples of the plasma membrane and intracellular membrane fractions were solubilized in electrophoresis sample buffer (10% glycerol, 2% SDS, 5% ß-mercaptoethanol, 50 mM dithiothreitol, 0.02% bromophenol blue, 0.1 M Tris-HCl, pH 6.8) and separated by SDS-polyacrylamide electrophoresis according to Laemmli (1970), in Bio-Rad Mini-protean® II dual slab cells using 5% stacking and 10% separating polyacrylamide gels. The samples were subsequently transferred electrophoretically to nitrocellulose membranes. Nitrocellulose membranes were briefly stained with Ponceau S to mark the position of molecular weight standards and to assess equal transfer of proteins. The membranes were then blocked for 1 hour with 3% (w/v) BSA in Tris-buffered saline (TBS) containing 0.04% NP-40 and incubated overnight at 4°C with a 1:500 dilution of R820 antibody or a 1:25 dilution of affinity purified α-GLUT 1 antibody. After 3 times for 15 min each time washes in Tris-saline NP-40 solution, the filters were incubated at room temperature with either 1 μCi/10 ml ^{125}I-labelled protein A for 1 h to detect antibody binding. Subsequently the nitrocellulose membranes were washed 3 times for 15 min each time with NP-40-TBS buffer, air-dried and autoradiographed at -120°C with Kodak XAR-5 film.

RESULTS

Short term effects of IGF-I and insulin on hexose uptake.

Incubation of L6 myotubes with 20 nM IGF-I and 200 nM insulin for varying times up to 120 minutes, resulted in equivalent stimulation of 2-deoxy-D-glucose transport by both hormones (fig 1A). After only 15 minutes with either hormone, hexose transport stimulation had reached a common plateau and remained constant for up to 90 minutes. This observation led us to examine the dose-response relationships for the short stimulation of hexose transport by each hormone (fig 2A). The maximal response to insulin or IGF-I was similar, i.e. approximately 90 to 100% above basal uptake (table 1). However, the cells were more sensitive to IGF-I than to insulin (fig 2A): the average hormone concentration required for 50% stimulation (ED_{50} value) was about 10-fold lower for IGF-I than for insulin (table 1). The stimulation of hexose transport during short term incubations (45 min)

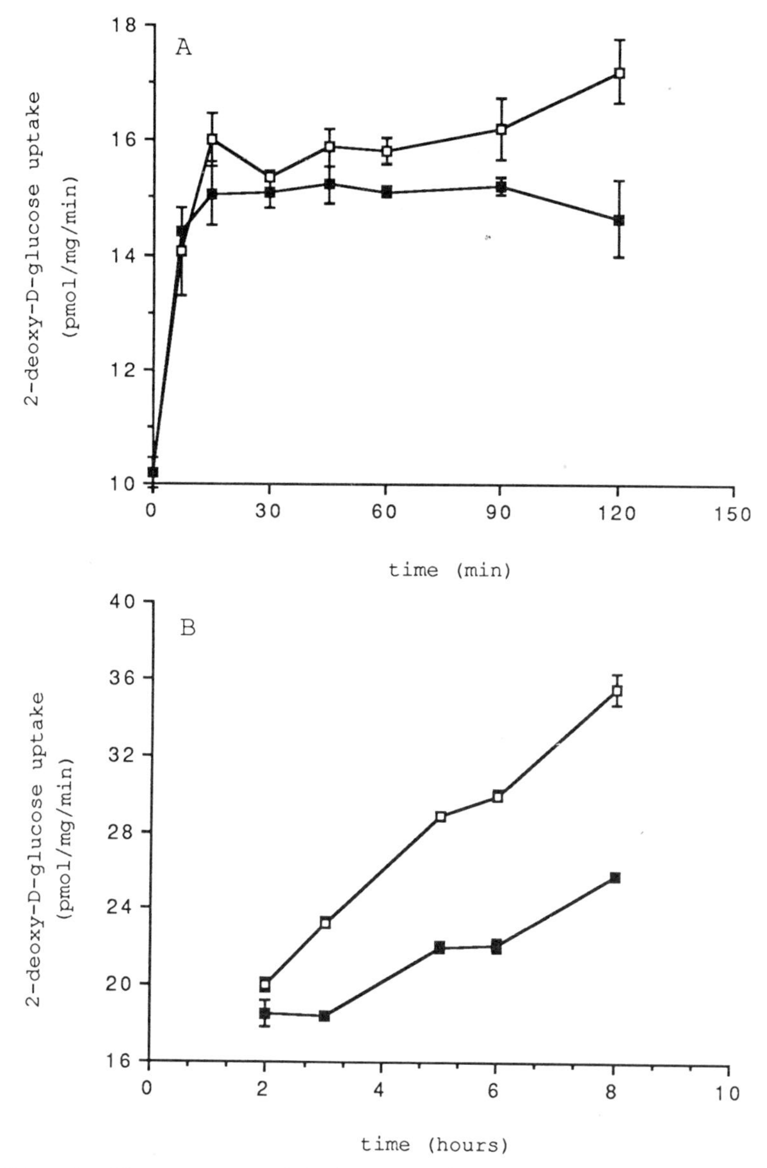

Fig. 1. Time course of 2-deoxy-D-glucose uptake stimulation by 20 nM IGF-I (open squares) and 200 nM insulin (closed squares) in L6 myotubes. (A) Short term (up to 120 min); (B) long term (up to 8 h). Data points are averages of three determinations ± standard error of (A) one experiment or (B) two experiments. Where no S.E. bars are given, these were smaller than the symbol.

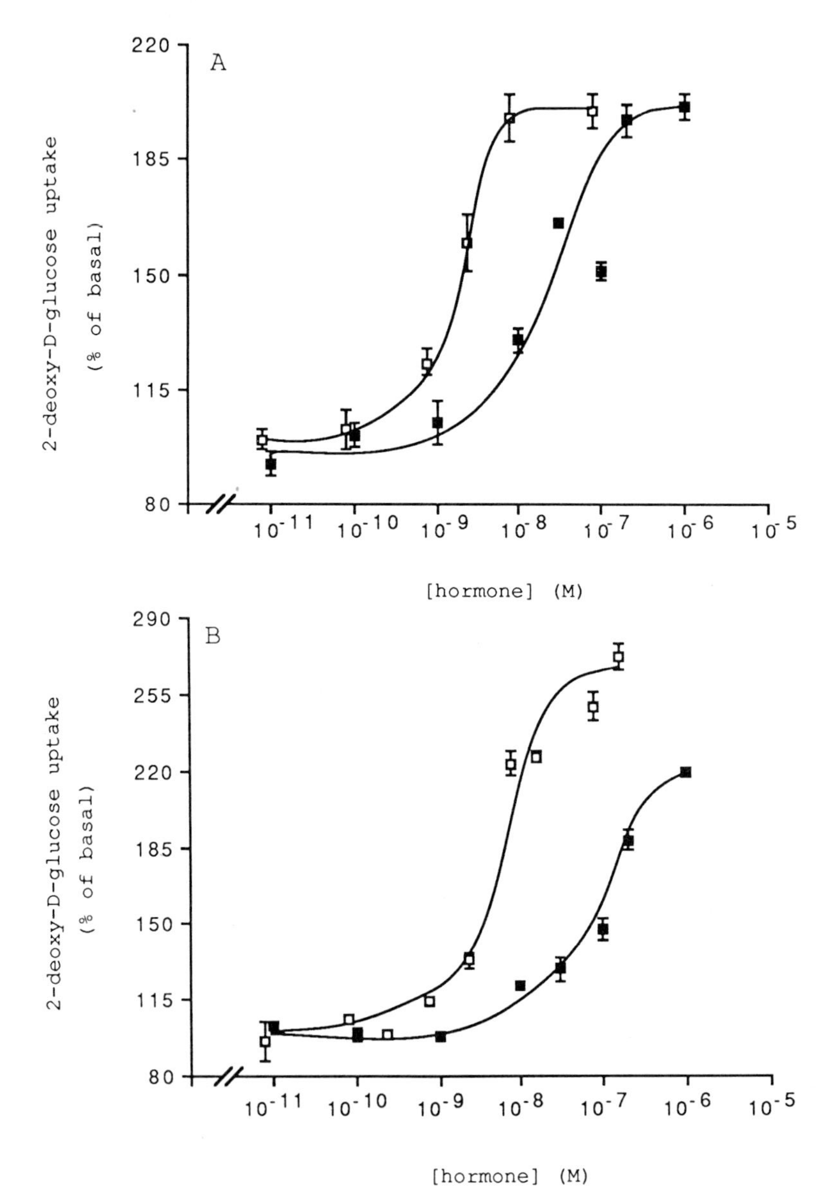

Fig. 2. Dose-response curves of IGF-I (open squares) and insulin (closed squares) stimulated 2-deoxy-D-glucose in L6 myotubes. (A) Short term incubation (45 min); (B) long term incubation (8 h). Data points are averages of three determinations ± standard error of one experiment representative of (A) five or (B) three. Where no S.E. bars are given, these were smaller than the symbols.

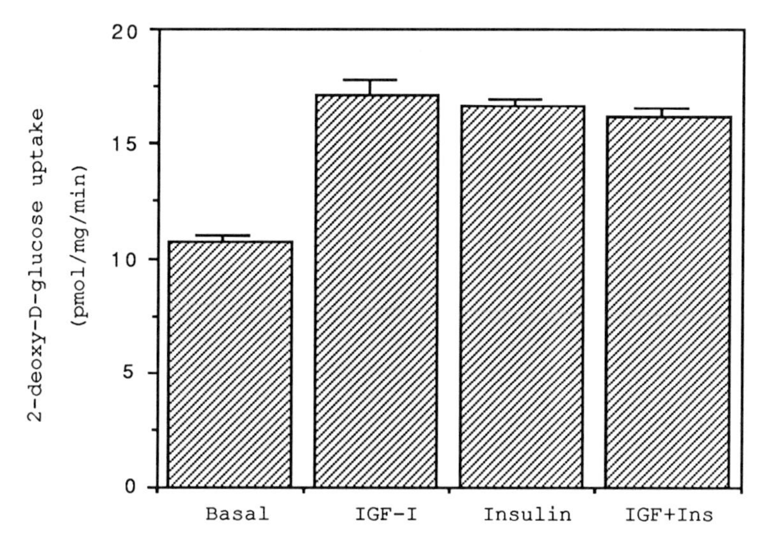

Fig. 3. Combined effects of IGF-I and insulin on 2-deoxy-D-glucose uptake. L6 myotubes were incubated for 45 min with IGF-I (100 nM); insulin (1 μM); or IGF-I+insulin (100 nM+1 μM). Data are averages of 3 determinations ± S.E. This is representative of 3 similar experiments.

with IGF-I and insulin at maximally effective concentrations was not additive (fig 3).

Long-term effects of IGF-I and insulin on hexose uptake.

Interestingly, after incubating L6 cells with IGF-I for 2 hours a second phase of hexose transport stimulation was noted, of approximately 10% above the acute phase plateau. This trend continued in a linear fashion as

Table 1. Sensitivity (ED_{50}) and responsiveness (% maximal stimulation) of hexose uptake in L6 myotubes stimulated with IGF-I or insulin.

Hormone	ED_{50} (nM)		% Stimulation at maximal [hormone]	
	45 min	8 hr	45 min	8 hr
IGF-I	2.2 ± 0.22	7.7 ± 0.6	192 ± 14	319 ± 29
Insulin	27.0 ± 4.7	84.7 ± 65.0	194 ± 20	239 ± 21

L6 myotubes were incubated with varying concentrations of the hormones for short (45 min) or long (8 hours) pre-incubation periods. Hexose uptake was subsequently determined for 5 min. Data are averages of 5 experiments (45 min) and 3 experiments (8 hours) ± standard error. ED_{50} = Effective dose which produced 50% stimulation.

the exposure time of L6 myotubes to IGF-I was increased (fig 1B). Although insulin also further increased hexose transport with longer incubation times, its effect lagged behind that of IGF-I and the rate of the insulin-dependent increase, which began at 3 hours, was slower (fig 1B). Consequently, the stimulation of hexose transport by insulin at 8 hours was significantly lower than that by IGF-I. The second phase of hexose transport stimulation by insulin into L6 myotubes has been previously reported and found to be protein synthesis-dependent (Walker et al., 1990). Further investigation is warranted to determine if a protein synthesis-dependent mechanism is involved in the second phase stimulation by IGF-I.

The dose-response curves for each hormone at 8 hours of incubation (fig 2B) indicated that the concentrations of IGF-I and insulin used in the time course experiments illustrated in fig 1B were near maximally effective doses. At maximally effective doses the percent stimulation of hexose transport at 8 hours was 319 ± 29% for IGF-I and 239 ± 21% for insulin, and the ED_{50} value was about 10-fold lower for IGF-I than for insulin (table 1). Interestingly, prolonged exposure to either hormone resulted in a 3-fold decrease in the hormone sensitivities compared to those measured at short exposure times (table 1).

Effect of short-term exposure to IGF-I on the subcellular distribution of glucose transporters.

In order to investigate the effect of IGF-I on the subcellular distribution of glucose transporters, we isolated membrane fractions from L6 myotubes incubated for 45 minutes without or with 20 nM IGF-I as described in Materials and Methods. In the 30% sucrose membrane fraction we found a 2-fold increase in the activity of the plasma membrane marker γ-glutamyl transferase relative to P-1 (table 2). Thus this fraction was designated the plasma membrane fraction. The light microsomal fraction, by virtue of its lower density and relatively low activity of the plasma membrane marker enzyme , was designated the internal membrane fraction. The 40% sucrose membrane fraction was slightly depleted in the plasma membrane marker enzyme . Further characterization of these membranes was reported earlier (Ramlal et al., 1988).

The number of cytochalasin B binding sites (glucose transporters) was significantly increased in the plasma membrane fractions from L6 myotubes treated with IGF-I, relative to controls. Furthermore, there was a concomitant decrease in the number of cytochalasin B binding sites in the internal membranes from cells treated with IGF-I (fig 4). Although the 40% sucrose membrane fraction (not shown) contains significant amounts of cytochalasin B binding sites (about 80% of the amount found in the plasma

Table 2. Activity of the plasma membrane marker γ-glutamyl transferase in membrane fractions derived from L6 myotubes.

Condition	γ-Glutamyl transferase activity			
	(µmol/mg protein/min)			
	P-1	30%	40%	LM
Control	7.7 ± 1.3 (4)	15.6 ± 2.5 (6)	5.9 ± 1.4 (6)	7.8 ± 1.0 (6)
IGF-I	8.2 ± 1.3 (3)	15.3 ± 2.5 (6)	6.4 ± 1.6 (6)	9.1 ± 1.1 (6)

L6 myotubes were incubated for 45 min with no additions (Control) or with 20 nM IGF-I. Membrane fractions were prepared as described under Materials and Methods. The number of different fractions tested is shown in brackets, each assayed in duplicate. P-1, crude plasma membranes; 30%, 30% sucrose fraction; 40%, 40% sucrose fraction; LM, light microsomes.

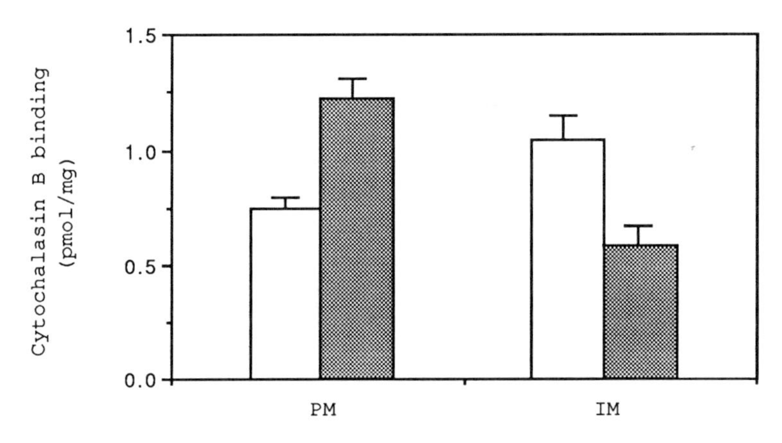

Fig. 4. Cytochalasin B binding to plasma membrane (PM) and internal membrane (IM) fractions from L6 myotubes incubated for 45 min with no additions (open bars) or 20 nM IGF-I (shaded bars). Data are pmoles of D-glucose inhibitable cytochalasin B binding per mg protein. The data are the mean of 6 independent experiments ± standard error. Within each experiment both fractions were assayed in quintuplicate.

membrane), this fraction did not exhibit a net change in the number of cytochalasin B binding sites in response to insulin (0.51 ± 0.12 pmol/mg *versus* a control value of 0.48 ± 0.11 pmol/mg) or in response to IGF-I (0.58 ± 0.15 pmol/mg *versus* a control of 0.69 ± 0.18 pmol/mg).

When corrected for protein yields, the increment in glucose transporters in the plasma membrane fractions of the IGF-I stimulated cells relative to controls was almost identical to the decrement in transporters in the internal membrane fractions of IGF-I stimulated cells relative to controls (table 3).

Table 3. Recovery of cytochalasin B binding sites in isolated membranes after cellular stimulation with IGF-I.

Fraction	pmol/20 dishes (Control)	pmol/20 dishes (IGF-I)	Δ pmol/ 20 dishes
PM	2.23 ± 0.23 (6)	4.09 ± 0.53 (6)	(+) 1.86 ± 0.41
LM	3.66 ± 0.78 (6)	1.97 ± 0.52 (6)	(-) 1.69 ± 0.31

Results are expressed in pmoles of D-glucose inhibitable cytochalasin B binding per 20 culture dishes. Data are averages of 6 experiments ± standard error. Within each experiment at least 5 determinations of cytochalasin B binding per fraction were performed. Delta (Δ) values are the mean of the differences calculated from six experiments. They represent the pmoles of cytochalasin B binding sites/20 dishes gained (+) or lost (-) relative to fractions from control cells.

Table 4. Percent increases in cytochalasin B binding in the plasma membrane, and in glucose transport activity, by IGF-I or insulin in L6 myotubes.

Hormone	% increase in CB binding in the PM fraction	% increase in 2-deoxyglucose uptake
IGF-I	73 ± 12 (6)	78 ± 23 (5)
Insulin	83 ± 32 (4)	66 ± 17 (5)

For a direct comparison of the increase in hexose transport with the increase of glucose transporter number in the plasma membrane, 2-deoxy-D-glucose transport was measured directly in the 10 cm culture dishes in which L6 myotubes were grown for membrane preparations. Three sets of 25 culture dishes each were incubated with insulin (200 nM), IGF-I (20 nM) or no hormone for 45 min at 37°C. Five dishes from each set were used to measure hexose uptake while the rest were used for membrane fractionation. Data are presented as the average percent increase above basal ± standard error. The number of independent experiments are indicated in brackets. CB = cytochalasin B.

The increase of cytochalasin B binding sites in the plasma membrane fractions was compared with the increase in hexose uptake in L6 myotubes incubated for 45 min with IGF-I. On average, IGF-I caused a 73% increase in the number of cytochalasin B binding sites in the plasma membrane relative to controls. This was in excellent agreement with the 78% increase of IGF-I-stimulated 2-deoxy-D-glucose in the same L6 myotubes used for membrane fractionation (table 4). A similar observation was made for insulin, confirming our earlier observations (Ramlal et al., 1988).

Immunoblot analysis with glucose transporter isoform-specific antibodies demonstrated that the GLUT 1 glucose transporter was evenly distributed (per μg protein) in the internal membranes and the plasma membranes from unstimulated cells. In contrast, the GLUT 4 glucose transporter was predominantly localized to the internal membranes, at a 2- to 3-fold higher density per μg protein than in the plasma membrane. Figure 5 illustrates representative immunoblots of L6 membrane fractions probed with C-terminal specific antibodies to either GLUT 4 or GLUT 1 transporters. IGF-I treatment of L6 myotubes caused a decrease in the GLUT 4 isoform in the internal membranes with a concomitant increase in the plasma membranes (fig 5A). These observations are consistent with recruitment of GLUT 4 glucose transporters. The average increase in GLUT 4 transporters in the plasma membrane was +0.33 ± 0.11 above controls and the average decrease in the internal membrane was -0.07 ± 0.08 below controls (5 membrane preparations, 7 immunoblots). Although the average decrease appears insignificant, a decrease was clearly observed in 3 out of 5 membrane preparations. In contrast, IGF-I did not appear to elicit recruitment of GLUT 1 transporters in these membrane preparations (fig 5B). On average the relative change of GLUT 1 transporter density/μg protein in the plasma membrane was +0.10 ± 0.06 above controls and in the internal membrane it was +0.08 ± 0.07 above controls (5 membrane preparations, 9 immunoblots).

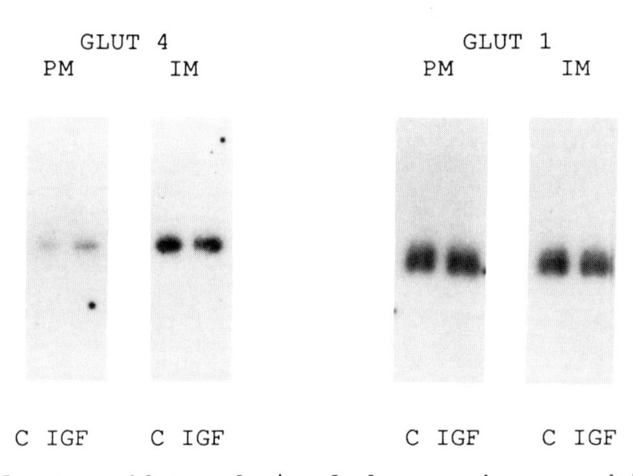

Fig. 5. Immunoblot analysis of plasma membranes and internal membranes prepared from L6 myotubes incubated 45 min with 20 nM IGF-I (IGF) or no hormone (C). Nitrocellulose blots were probed with anti-GLUT 1 or -GLUT4 isoform-specific antibodies. Antibody binding was detected by [^{125}I]-Protein A. The average molecular weight of GLUT 1 was 53 000 and of GLUT 4 was 57 000.

DISCUSSION

The present study demonstrates the effect of IGF-I as compared to insulin on glucose transport in L6 myotubes. Acute activation of hexose transport was demonstrated for IGF-I and insulin (92% and 98% above basal, respectively). The initial time course of stimulation by each hormone was identical, and at 15 minutes of incubation the stimulation had reached a plateau. This plateau was maintained for up to 90 min in the presence of IGF-I and 3 hours in the presence of insulin. Since L6 myotubes were 10-fold more sensitve to IGF-I than to insulin, this could suggest that both hormones perhaps stimulate hexose transport through the L6 myotube IGF-I receptors. It is often difficult to assign a hormone-stimulated biological response to a specific receptor when the hormone can bind to more than one type of receptor. It is known that L6 myotubes have IGF-I receptors that bind IGF-I 50 to 100 times more effectively than insulin. They also have high affinity insulin receptors; however, these receptors apparently bind IGF-I with greater affinity since IGF-I can displace 50% of bound ^{125}I-insulin at concentrations 10-fold lower than required for the unlabelled insulin molecule (Burant et al., 1987). The presence of these unusual receptors may explain the 10-fold higher sensitivity of L6 myotubes to IGF-I compared to insulin. Taking into account these observations, insulin may signal through the insulin receptor while IGF-I conceivably could signal through both the insulin and the IGF-I receptors. Since the effects of both hormones on hexose transport activation were not additive (fig 3), it can be suggested that the short term activation by the two hormones is through a common early signalling mechanism. Stimulation of hexose transport could in principle involve translocation of glucose transporters from internal membranes to plasma membranes, intrinsic activation of the turnover rate of glucose transporters present in the plasma membrane and/or an increase of glucose transporter biosynthesis. In the present report, we studied the involvement of glucose transporter recruitment in L6 myotubes activated by short incubations with IGF-I.

The recruitment of glucose transporters was assesed by cytochalasin B binding and immunoblot analysis of enriched plasma membrane and internal membrane fractions from L6 myotubes. Initial characterization of the membrane fractions was reported earlier (Ramlal et al., 1988). In those studies Mg^{2+}-ATPase and adenylate cyclase activities were enriched 2- to 3-

fold in the 30% sucrose fraction relative to the P-1 fraction. In the present study we used γ-glutamyl transferase, another plasma membrane marker enzyme, to demonstrate the 30% sucrose fraction was enriched 2-fold in plasma membranes (relative to the P-1 fraction). Cytochalasin B binding measurements in these fractions demonstrated that short incubations with a maximal concentration of IGF-I caused a decrease in binding sites in the internal membrane fraction and an increase in sites in the plasma membrane fraction. When corrected for protein yields, the increment in glucose transporters in the plasma membrane fraction of the IGF-I stimulated cells relative to controls was almost identical to the decrement in transporters in the internal membrane fraction of IGF-I stimulated cells relative to controls (table 3). These observations are compatible with the recruitment hypothesis for the acute stimulation of hexose transport in fat and muscle described by Simpson and Cushman (1986). Moreover, the fold increase of cytochalasin B binding sites in the plasma membrane stimulated by IGF-I was equivalent to the fold increase in hexose transport in intact L6 cells. This suggests that the sole mechanism for the acute stimulation of hexose transport by IGF-I (for the first 90 min) is the recruitment of glucose transporters. A similar observation was made for insulin in our previous study (Ramlal et al., 1988)

Using the immunoblot method and amino acid sequence-specific antibodies one can investigate which particular glucose transporter isoform is recruited in response to a stimulus. L6 myotubes express GLUT 1 and GLUT 4 glucose transporter isoforms (Koivisto et al., 1990) in contrast to other muscle cell lines (BC3H-1, C2 and G8) that do not express the GLUT 4 isoform (Calderhead et al., 1990). In the present study, GLUT 1 transporters were found to be equally distributed between the plasma membrane and internal membrane fractions, whereas GLUT 4 transporters were found predominantly in the internal membranes (2- to 3-fold higher based on autoradiographic density per μg protein). In response to IGF-I, GLUT 4 transporters increased in the plasma membrane fraction and decreased slightly in the internal membrane fraction. The magnitude of the change was small and did not account for the total change observed in cytochalasin B binding. However, they qualitatively support the notion that GLUT 4 transporters are recruited to the plasma membrane by IGF-I. The changes in GLUT 1 transporters were smaller, and although they could potentially contribute to the change in cytochalasin B binding, it is presently ambiguous whether IGF-I induces translocation of GLUT 1 transporters. The combined increase of GLUT 1 and GLUT 4 transporters in the plasma membrane was only 44 ± 12% above controls, whereas the increase in cytochalasin B binding sites was an average of 73 ± 12%. This discrepancy may be due to methodological unequivalences, or it may also suggest that a third type of glucose transporter could be expressed in these cells that was not detected immunologically but could contribute to cytochalasin B binding. This possibility remains to be tested.

During longer times of hormone incubation (2-8 h) other mechanisms are activated which will contribute to the increase in hexose transport. For example in L6 cells, long incubations with insulin are known to stimulate glucose transporter biosynthesis (Walker et al., 1990). In the present study we observed that longer periods of incubation with IGF-I, like insulin, lead to further activation of hexose transport above the initial plateau of the short term activation. However, IGF-I was more effective than insulin in activating hexose transport during this second phase of stimulation. The interpretation of experiments involving hormone incubation for long periods of time are complicated by the observation that L6 cells secrete IGF binding proteins that could potentially affect the response to IGF-I. However, it has been shown that the binding proteins in the medium inhibit the effectiveness of IGF-I on L6 cells (Ross et al., 1989) and hence these proteins could not be responsible for the increased response to IGF-I during the second phase stimulation of hexose transport (IGF binding proteins would presumably play only a minor role in experiments using short time incubations, since the hormones are added to the cells after a medium change). Other complications envisaged during prolonged incubations are the lability of IGF-I and insulin through degradation by the cell cultures (Smith R.J., personal communication). Assuming that the degredation rate of insulin and IGF-I are similar, the different responsiveness of L6 myotubes to each hormone is inconsistent with the suggestion that IGF-I and insulin only signal through the IGF-I receptor. Therefore if IGF-I and insulin signal through different receptors, the acute activation of hexose transport must involve a common post-receptor pathway, whereas chronic activation could involve diverse post-receptor pathways.

In conclusion IGF-I has both acute and chronic effects on hexose transport in L6 myotubes. In the short term, stimulation of hexose transport is mainly due to the recruitment of glucose transporters to the plasma membrane. GLUT 1 and GLUT 4 isoforms are expressed in L6 myotubes and IGF-I appears to preferentially translocate the GLUT 4 isoform from the internal membranes to the plasma membranes. At longer periods of time, L6 myotubes were more sensitive and responsive to IGF-I than to insulin. Whether this involves signalling through different receptors and/or post-receptor pathways, remains to be determined experimentally.

ACKNOWLEDGEMENTS

We thank Dr. Nava Bashan for careful reading of this manuscript and helpful suggestions, Dr. Mladen Vranic for the gift of IGF-I and Dr. David James for the gift of the R820 antibody. We also thank Ms. Loretta Lam for help with the immunoblot work. P.J.B. is the recipient of a M.R.C. Studentship, and A.K. is the recipient of a M.R.C. Scientist Award. This work was supported by a grant from the Muscular Dystrophy Association of Canada.

REFERENCES

Baldwin, S.A., Baldwin, J.M., and Lienhard, G.E., 1982, Monosaccharide transporter of the human erythrocyte. Characterization of an improved preparation, Biochemistry, 21: 3836-3842.

Beguinot, F., Kahn, C.R., Moses, A.C., and Smith, R.J., 1985, Distinct biologically active receptors for insulin, insulin-like growth factor I and insulin-like gorwth factor II in cultured skeletal muscle cells, J. Biol. Chem., 260: 15892-15898.

Beguinot, F., Kahn, C.R., Moses, A.C., and Smith R.J., 1986, The development of insulin receptors and responsiveness is an early marker of differentiation in the muscle cell line L6, Endocrinology, 118: 446-455.

Bilan, P.J., and Klip, A., 1990, Glycation of the human erythrocyte glucose transporter *in vtiro* and its functional consequences, Biochem. J., 268: 661-667.

Birnbaum, M.J., 1989, Identification of a novel gene encoding an insulin-responsive glucose transporter protein, Cell, 57: 305-315.

Bradford, M.M., 1976, A rapid and sensitive method for the quantitation of microgram quantities of protein utilizing the principle of protein-dye binding, Anal. Biochem., 72: 248-254.

Burant, C.F., Treutelar, M.K., Allen, K.D., Sens, D.A., and Buse, M.G., 1987, Comparison of insulin and insulin-like growth factor I receptors from rat skeletal muscle and L6 myocytes, Biochem. Biophys. Res. Commun., 147: 100-107.

Calderhead, D.M., Kitagawa, K., Lienhard, G.E., and Gould G.W., 1990, Translocation of the brain-type glucose transporter largely accounts for insulin stimulation of glucose transport in BC3H-1 myocytes, Biochem. J., 269: 597-601.

Defronzo, R.A., Ferranninni, E., Sato, Y., Felig, P., and Wahren, J., 1981, Synergistic interaction between exercise and insulin on peripheral glucose uptake, J. Clin. Invest., 68: 1468-1474.

Dohm, C.L., Elton, C.W., Raju, M.S., Mooney, N.D., DiMarchi, R., Pories, W.J., Flickinger, E.G., Atkinson, S.M. Jr., and Caro, J.F., 1990, IGF-I-Stimulated glucose transport in human skeletal muscle and IGF-I resistance in obesity and NIDDM, Diabetes, 39: 1028-1032.

Douen, A.G., Ramlal, T., Rastogi, S., Bilan, P.J., Cartee, G.D., Vranic, M., Holloszy, J.O., and Klip, A., 1990, Exercise induces recruitment of the insulin-responsive glucose transporter, J. Biol. Chem., 265:13427.

Froech, E.R., Muller, W.A., Burgi, H., Waldvogel, M., and Labhart, A., 1966, Nonsuppressable insulin-like activity of human serum. I. Physicochemical properties, extraction and partial purification, Biochim. Biophys. Acta, 121: 360-374.

Fushiki, T., Wells, J.A., Tapscott, E.B., and Dohm, G.L., 1989, Changes in glucose transporters in muscle in reponse to exercise, Am.J. Physiol., 256: E580-E587.

Giacca, A., Gupta, R., Efendic, S., Hall, K., Skottner, A., Lickley, L., and Vranic, M., 1990, Differential effects of IGF-I and insulin of glucoregulation and fat metabolism in depancreatized dogs, Diabetes, 39: 340-347.

Guler, H.P., Zapf, J., and Froesch, E.R., 1987, Short-term metabolic effects of recombinant human insulin-like growth factor I in healthy adults, N. Engl. J. Med., 317: 137-140.

Holman, G.D., and Rees, W.D., 1987, Photolabelling of the hexose transporter at external and internal sites: Fragmentation patterns and evidence for a conformational change, Biochim. Biophys. Acta, 897: 395-405.

James, D.E., Brown, R. Navarro, J., and Pilch, P.F., 1988, Insulin-regulatable tissues express a unique insulin-sensitive glucose transport protein, Nature, 333: 183-185.

James, D.E., Strube, M., and Mueckler, M., 1989, Molecular cloning and characterization of an insulin-regulatable glucose transporter, Nature, 338: 83-87.

Kidokoro, Y., 1975, Developmental Changes of membrane electrical properties in a rat skeletal muscle cell line, J. Physiol.(London), 244: 129-143.

Klip, A., Li, G., and Walker, D., 1983, Insulin binding to differentiating muscle cells in culture, Can. J. Biochem. Cell Biol., 61: 644-649.

Klip, A., Li, G., and Logan, W.J., 1984, Induction of sugar uptake response to insulin by serum depetion in fusing L6 myoblasts, Am. J. Physiol., 247: E291-E296.

Koivisto, U.-M., Martinez-Valdez, H., Bilan, P.J., Burdett, E., Ramlal, T., and Klip, A., Differential regulation of GLUT-1 and GLUT-4 transport systems by glucose and insulin in L6 muscle cells in culture, J. Biol. Chem. (in press).

Laemmli, U.K., 1970, Cleavage of structural proteins during the assembly of the head of bacteriophage T4, Nature, 227, 680-685.

Lammers, R., Gray, A., Schlessinger, J., and Ullrich, A., 1989, Differential signalling potential of insulin and IGF-I receptor cytoplasmic domains, EMBO J., 8: 1369-1375.

Livingston, N., Pollare, T., Lithell, H., and Arner, P., 1988, Characterization of insulin-like growth factor I receptor in skeletal muscles of normal and insulin resistant subjects, Diabetologia, 31: 871-877.

Lowry, O.H., Rosebrough, N.J., Farr, A.L., Randall, R.J., 1951, Protein measurement with the folin phenol reagent, J. Biol. Chem., 193: 265-275.

McClain, D.A., Maegawa, H., Thies, R.S., and Olefsky, J.M., 1990, Dissection of the growth *versus* metabolic effects of insulin and insulin-like growth factor-I in transfected cells expressing kinase-defective human insulin receptors, J. Biol. Chem., 265: 1678-1682.

Moxley III, R.T., Arner, P., Moss, A., Skottner, A., Fox, M., James, D., Livingston, J.N., 1990, Acute effects of insulin-like growth factor I and insulin on glucose metabolism in vivo, Am. J. Physiol., 259: E561-E567.

Poggi, C., Le Marchand-Brustel, Y., Zapf, J., Froesch, E.R., and Freychet, P., 1979, Effects and binding on insulin-like growth factor I (IGF I) in the isolated soleus muscle of lean and obese mice: Comparison with insulin, Endocrinology, 105: 723-730.

Ramlal, T., Sarabia, V., Bilan, P.J., and Klip, A., 1988, Insulin-mediated translocation of glucose transporters from intracellular membranes to plasma membranes: Sole mechanism of stimulation of glucose transport in L6 muscle cells, Biochem. Biophys. Res. Commun., 157: 1329-1335.

Rechler, M.M., and Nissley, S.P., 1985, The nature and regulation of the receptors for insulin-like growth factors, Ann. Rev. Physiol., 47: 425-442.

Ross, M., Francis, G.L., Szabo, L., Wallace, J.C., and Ballard, F.J., 1989, Insulin-like growth factor (IGF)-binding proteins inhibit the biological activities of IGF-1 and IGF-2 but not des-(1-3)-IGF-1, Biochem. J., 258: 267-272.

Salmon, W.D. Jr., and Daughaday, W.H., 1957, A hormonally controlled serum factor which stimulates sulfate incorporation by cartilage *in vivo*, J. Lab. Clin. Med., 49: 825-836.

Shainberg, A., Yagil, G., and Yaffe, D., 1971, Alteration of enzymatic activities during muscle differentiation in vitro, Dev. Biol., 25: 1-29.

Simpson, I.A., and Cushman, S.W., 1988, Hormonal regulation of mammalian glucose transport, Annu. Rev. Biochem., 55: 1059-1089.

Sinha, M.K., Buchanan, C. Leggett, N., Martin, L., Khazanie, P.G., DiMarchi, R., Pories, W.J., and Caro, J.F., 1989, Mechanism of IGF-I-stimulated glucose transport in human adipocytes: Demontration of specific IGF-I receptors not involved in stimulation of glucose transport, Diabetes, 38: 1217-1225.

Walker, P.S., Ramlal, T., Sarabia, V., Koivisto, U.-M., Bilan, P.J., Pessin, J.E., and Klip, A., 1990, Glucose transport activity in L6 muscle cells is regulated by the coordinate control of subcellular glucose transproter distribution, biosynthesis and mRNA transcription, J. Biol. Chem., 265: 1516-1523.

Widdas, W.F., 1988, Old and new concepts of the membrane transport for glucose in cells, Biochim. Biophys. Acta, 947: 385-404.

Yaffe, D., 1968, Retention of differentiation potentialities during prolonged cultivation of myogenic cells, Proc. Natl. Acad. Sci. U.S.A., 61: 477-483.

Yasumoto, K., Iwami, K., Fushiki, T., and Mitsuda, H., 1978, Purification and enzymatic properties of γ-glutamyl transferase from bovine colostrum, J. Biochem., 84: 1227-1236.

Zapf, J., Hauri, C. Waldvogel, M., and Froesch, E.R., 1986, Acute metabolic effects and half-lives of intravenously administered insulin-like growth factors I and II in normal and hypophysectomized rats, J. Clin. Invest., 77: 1768-1775.

Zorzano, A., Wilkinson, W., Kotliar, N., Thoidis, G., Wadzinski, B.E., Ruoho, A.E., and Pilch, P.F., 1989, Insulin-regulated glucose uptake in rat adipocytes mediated by two transporter isoforms present in at least two vesicle populations, J. Biol. Chem., 264: 1235-1263.

INSULIN-LIKE GROWTH FACTORS AND THEIR RECEPTORS IN MUSCLE DEVELOPMENT

Gyorgyi Szebenyi and Peter Rotwein

Washington University School of Medicine
Departments of Medicine and Genetics
Division of Biology and Biomedical Science
St. Louis, MO. 63110

INTRODUCTION

Insulin-like growth factors (IGF) I and II exert pleiotropic actions on target tissues. In muscle both IGF-I and II exhibit a range of physiological effects, including a concentration-dependent stimulation of metabolic functions such as glucose and amino acid uptake, acceleration of the rate of DNA synthesis, and enhancement of myoblast differentiation (1-4). The mitogenic and metabolic effects of the IGFs can be distinguished from their differentiation-promoting properties: the stimulation of differentiation persists in the presence of inhibitors of cell replication (5,6), and antisense myogenin oligonucleotides, which block IGF-I-induced differentiation, do not influence IGF-I stimulated metabolite uptake or proliferation (7). These observations imply that distinct signal-transduction pathways may be involved in each process.

In muscle cells IGFs bind to several cell-surface receptors, including the IGF-I receptor, the IGF-II/cation independent mannose 6-phosphate receptor (IGF-II/CIMPR), and the insulin receptor, as evidenced by both competition-binding studies and cross-linking assays (1,4). The relative contribution of each receptor toward mediating the biological actions of each IGF in muscle is under active investigation. The IGF-I receptor appears to transduce signals for both IGF-I and IGF-II, as suggested by several lines of evidence. Skeletal muscle cells prepared from the body walls of embryonic chickens can be stimulated to differentiate in serum-free medium in response to nanomolar concentrations of IGF-I or IGF-II (3), while the chicken CIMPR does not bind IGF-II (8,9). In the rat L6 myoblast cell line a lower concentration of IGF-I than IGF-II stimulates both metabolic and mitogenic actions (4). In addition, an antibody that blocked the binding of IGF-II to the IGF-II/CIMPR did not decrease the metabolic actions of IGF-II in L6 myoblasts, but did inhibit the disappearance of the growth factor from the culture medium (10). These results and others have led to the suggestion that the IGF-I receptor is the signaling receptor for both IGFs, and that the IGF-II/CIMPR participates in a degradative pathway for IGF-II (11).

Molecular Biology and Physiology of Insulin and Insulin-Like Growth Factors
Edited by M.K. Raizada and D. LeRoith, Plenum Press, New York, 1991

Other observations, however, are consistent with the IGF-II/CIMPR playing a role in signal transduction in mammalian species. Recently this receptor has been linked to second messenger systems involving guanine nucleotide binding proteins (12), inositol phosphate turnover (13), and calcium influx (14), with all responses being stimulated by IGF-II. IGF-II also may have biological effects through the IGF-II/CIMPR by interfering with targeting or endocytosis of lysosomal enzymes (15,16). In addition, in rat muscle satellite cells in primary culture IGF-II enhances cellular proliferation even in the presence of concentrations of insulin that saturate the IGF-I receptor (17). Similar observations were made with regard to metabolic actions of IGF-II in human myoblasts (2). These results suggest that another receptor, possibly the IGF-II/CIMPR, mediates IGF-II-specific responses in these systems.

Our laboratory has been interested in the expression and actions of the IGFs and their receptors in muscle growth and development. The experiments presented in this chapter indicate that mouse C2 myoblasts synthesize IGF-I and II, the IGF-I receptor, the IGF-II/CIMPR, and an IGF binding protein during its terminal myogenic differentiation. This cell line thus will be useful in characterizing the mechanisms involved in the regulation of these factors in muscle development and their autocrine actions in the differentiating myoblast.

EXPERIMENTAL METHODS AND RESULTS

The C2 cell line, originally established from crushed mouse skeletal muscle (18), was used in all of our experiments. C2 cells were grown in DME supplemented with 20% fetal calf serum (growth medium) until they reached 70-80% of confluent density, and then were washed and refed with 2% horse serum in DME (differentiation medium). C2 cells exposed to differentiation medium undergo terminal myogenic differentiation: cells align within 24 hours, and progressive fusion is observed from 48 to 96 hours. Creatine kinase mRNA and protein are first detectable at 48 hours of differentiation; also by 48 hours mRNA for the muscle-specific form of actin, α-actin, appears (19,20). For our experiments, undifferentiated cells, cells in different stages of terminal myogenic development, and conditioned culture media were analyzed for IGF-I and IGF-II, IGF receptors, and IGF binding protein expression. In addition, cell extracts were prepared by freeze-thaw lysis for measurement of lysosomal enzyme activities (21), and total cellular RNA was isolated (22) for assessment of IGF, lysosomal enzyme and receptor mRNAs.

During C2 cell differentiation there is marked accumulation of IGF-II in conditioned culture media. As determined by radioimmunoassay IGF-II increased progressively from 3 ng/ml 4 hours after the switch to differentiation medium to 110 ng/ml by 120 hours. Accumulation of IGF-I was much less extensive than IGF-II, with values rising from 1 to 4 ng/ml over the same developmental period (20). The cell surface expression of the IGF-I receptor and the IGF-II/CIMPR also changed during C2 cell differentiation. As determined by competitive binding studies IGF-II/CIMPR number substantially increased during differentiation, with a 7-fold elevation by 24 hours and an 11-fold rise by 72 hours with no change in affinity for IGF-II. The rise in binding activity was accompanied by a comparable increase in total receptor number, indicating that the elevation in cell-surface binding resulted from enhanced receptor synthesis. By competitive cross-linking assay followed by SDS-polyacrylamide gel electrophoresis and autoradiography, it was shown that IGF-II bound principally to a cell-surface protein of apparent Mr 220,000, a protein consistent in size with the IGF-II/CIMPR (19). By contrast, cell-surface expression of the IGF-I receptor rose by only 2-fold during the first 48 hours of differentiation and then declined over the subsequent 48 hours to the lower values found in proliferating myoblasts. By competitive cross-linking analysis IGF-I was found to bind to a cell-surface protein of Mr 135,000, consistent in mass with the a subunit of the IGF-I receptor (20). Since C2 cells express low levels of the insulin receptor (our unpublished observations), and since binding of ^{125}I-IGF-I was minimally displaced by an excess of unlabeled insulin, it is likely that this protein represents the authentic IGF-I receptor a subunit.

In addition to ligands and receptors, potential third components of IGF signaling are IGF binding proteins (IGF BP). We assessed secretion of IGF BPs during C2 cell development using both a competitive binding assay and a modified Western blot, the so-called ligand blot (20). An IGF BP of apparent Mr of 29,000 was found in conditioned C2 differentiation medium. Its presence was detected within 4 hours of the onset of differentiation, with a progressive accumulation of nearly 100-fold over the subsequent 116 hours (20).

In order to gain insight into the mechanisms involved in the regulation of the components of IGF signaling during C2 myoblast differentiation IGF-I, IGF-II, and IGF-II/CIMPR mRNAs were measured by solution-hybridization nuclease protection assays, using total cellular RNA (23). All the probes used were derived from mouse genomic DNA and contained exons which code for translated portions of the corresponding proteins (15,20,24). In addition we sought to characterize the nature of the single IGF BP secreted by C2 cells. Low levels of IGF-I, IGF-II, and IGF-II/CIMPR mRNAs were detected in proliferating, undifferentiated myoblasts. IGF-I mRNA abundance increased transiently by nearly 10-fold by 48-72 hours of differentiation, and then declined. IGF-II mRNA levels rose progressively by nearly 25-fold over 96 hours after exposing cells to differentiation medium, with a 4-fold increase being noted in the initial 16 hours. Steady-state values for IGF-II/CIMPR mRNA rose nearly 13-fold by 48 hours with kinetics of accumulation that paralleled the increase in IGF-II mRNA (24). The relative concentrations of IGF-I and IGF-II mRNAs in differentiating myoblasts were assessed by simultaneous hybridization in solution to the same RNA samples. Based on these experiments IGF-II mRNA levels were found to be least 15 times higher than those of IGF-I in developing myotubes (20). These results were consistent with the presence of 20-25 times more IGF-II than IGF-I in conditioned differentiation medium, as noted above.

The IGF-II/CIMPR is a multifunctional protein with binding sites for both mannose 6-phosphate-containing ligands and IGF-II (25). Although the rise in receptor expression paralleled the increase in IGF-II during C2 cell development, it seemed possible that comparable changes would be found in other ligands of this receptor or in other components of the lysosomal targeting pathway, since, during the tissue remodeling that accompanies muscle development there would seem to be a need for increased lysosomal activity. To address this question we measured transcript levels for the cation-dependent mannose 6-phosphate receptor (CDMPR) (25) and selected lysosomal enzymes as a function of differentiation, and assessed the enzymatic activity of two lysosomal glycosidases, β-glucuronidase and β-hexosaminidase, in cell extracts. Complementary DNA probes were obtained, using polymerase chain reaction amplification techniques (26), for the mouse CDMPR (27,28), for two lysosomal proteases, cathepsins B and L (29,30), and for a lysosomal glycosidase, β-hexosaminidase (31). Transcript abundance of these lysosomal enzymes and of the CDMPR was measured during C2 cell differentiation by either Northern blotting or solution hybridization methods. No change was detected in CDMPR mRNA content and only a transient 2-4-fold increase was detected at a single time point, at 24 hours after the initiation of differentiation, in mRNA levels of the three lysosomal enzymes. Measurements of enzymatic activities of β-hexosaminidase and β-glucuronidase in cell extracts showed little increase. Thus, minimal changes occurred in other components of lysosomal enzyme signaling during C2 cell differentiation.

In order to characterize the IGF BP that was secreted by C2 cells during their terminal myogenic differentiation, we first attempted to determine whether this BP might be the mouse homologue of either IGF BP1 or BP2, since these proteins have similar mobilities on SDS polyacrylamide gels (32). Northern blots were performed under conditions of reduced stringency using cDNA probes for human IGF BP1 (33) or rat IGF BP2 (34). No signal was detected with either cDNA, indicating that the mouse C2 IGF BP was not similar to the known IGF BPs. We have now purified the C2 IGF BP to apparent homogeneity and have obtained partial amino-terminal amino acid sequence. Our preliminary results indicate that the C2 BP is a novel member of the IGF BP family. Further sequence analysis and cDNA cloning are in progress.

SUMMARY AND CONCLUSIONS

Several proteins involved in IGF action are expressed in C2 cells and their abundance was found to vary as a function of development. IGF-I and II mRNA levels rose 10 and 25-fold, respectively, during differentiation, and were accompanied by an increase in growth factor secretion. The accumulation of IGF-II in conditioned culture medium was much greater than that of IGF-I. There was also an increase in the number of IGF-I receptors and IGF-II/CIMPR on the cell surface during differentiation. The sustained rise in IGF-II/CIMPR expression appeared to be a consequence of a similar increase in its mRNA abundance. The mechanisms responsible for the transient increment in IGF-I receptor number were not assessed, although it is likely that the decline in IGF-I receptor content after 72 hours in differentiation medium was a consequence of down-regulation by the IGF-II that accumulated in the medium (35). In contrast to the 13-fold rise in IGF-II/CIMPR mRNA levels, transcript levels for the CDMPR remained constant during C2 cell development, enzymatic activities of two lysosomal enzymes did not change, and only a small increment was detected at a single time point in the expression of several lysosomal enzyme mRNAs. In addition, during C2 muscle differentiation, a novel IGF binding protein was induced. These results demonstrate modulation of several components of IGF signaling pathways in differentiating myoblasts, and argue for a local role for IGFs in muscle development.

ACKNOWLEDGEMENT

The work presented in this chapter was supported by NIH Grant DK37449 and by Basic Research Grant #1-1223 from the March of Dimes Birth Defects Foundation. We thank Janet Seavitte for assistance in preparation of the manuscript.

REFERENCES

1. Beguinot, F., Kahn, C.R., Moses, A.C., and Smith, R.J. Distinct biologically active receptors for insulin, insulin-like growth factor I, and insulin-like growth factor II in cultured skeletal muscle cells. J. Biol. Chem. 260: 15892-15898, 1985.

2. Shimizu, M., Webster, C., Morgan, D.O., Blau, H.M., and Roth, R.A. Insulin and insulin-like growth factor receptors and responses in cultured human muscle cells. Am. J. Physiol. 251: E611-E615, 1986.

3. Schmid, C., Steiner, T., and Froesch, E.R. Preferential enhancement of myoblast differentiation by insulin-like growth factors (IGF I and IGF II) in primary cultures of chicken embryonic cells. FEBS Letters 161: 117-121, 1983.

4. Ewton, D.Z., Falen, S.L., and Florini, J.R. The type II insulin-like growth factor (IGF) receptor has low affinity for IGF-I analogs: Pleiotypic actions of IGFs on myoblasts are apparently mediated by the type I receptor. Endocrinol. 120: 115-123, 1987.

5. Turo, K.A., and Florini, J.R. Hormonal stimulation of myoblast differentiation in the absence of DNA synthesis. Am. J. Phyiol. 243: C278-284, 1982.

6. Florini, J.R., Ewton, D.Z., Falen, S.L., and Van Wyk, J.J. Biphasic concentration dependency of stimulation of myoblast differentiation by somatomedins. Am. J. Physiol. 250: C771-C778, 1986.

7. Florini, J.R., and Ewton, D.Z. Highly specific inhibition of IGF-I-stimulated differentiation by an antisense oligodeoxyribonucleo-tide to myogenin mRNA. J. Biol. Chem. 265: 13435-13437, 1990.

8. Clairmont, K.B., and Czech, M.P. Chicken and Xenopus mannose 6-phosphate receptors fail to bind insulin-like growth factor II. J. Biol. Chem. 264: 16390-16392, 1989.

9. Canfield, W.M., and Kornfeld, S. The chicken liver cation-independent mannose 6-phosphate receptor lacks the high affinity binding site for insulin-like growth factor II. J. Biol. Chem. 264: 7100-7103, 1989.

10. Kiess, W., Haskell, J.F., Lee, L., Greenstein, L.A., Miller, B.E., Aarons, A.L., Rechler, M.M., and Nissley, S.P. An antibody that blocks insulin-like growth factor (IGF) binding to the type II IGF receptor is neither an agonist nor an inhibitor of IGF-stimulated biologic responses in L6 myoblasts. J. Biol. Chem. 262: 12745-12754, 1987.

11. Czech, M.P. Signal transmission by the insulin-like growth factors. Cell 59: 235-238, 1989.

12. Okamoto, T., Katada, T., Murayama, Y., Ui, M., Ogata, E., and Nishimoto, I. A simple structure encodes G protein-activating function of the IGF-II/mannose 6-phosphate receptor. Cell 62: 709-717, 1990.

13. Rogers, S.A., and Hammerman, M.R. Mannose 6-phosphate potentiates insulin-like growth factor II-stimulated inositol triphosphate production by proximal tubular basolateral membranes. J. Biol. Chem. 264: 4273-4276, 1989.

14. Nishimoto, I., Hata, Y., Ogata, E., and Kojima, I. Insulin-like growth factor II stimulates calcium influx in competent BALB/c 3T3 cells primed with epidermal growth factor. J. Biol. Chem. 262: 12120-12126, 1987.

15. Kiess, W., Thomas, C.L., Greenstein, L.A., Lee, L., Sklar, M.M., Rechler, M.M., Sahagian, G.G., and Nissley, S.P. Insulin-like growth factor-II (IGF-II) inhibits both the cellular uptake of β-galactosidase and the binding of β-galactosidase to purified IGF-II/mannose 6-phosphate receptor. J. Biol. Chem. 264: 4710-4714, 1989.

16. Braulke, T., Tippmer, S., Chao, H-J., and von Figura, K. Insulin-like growth factors I and II stimulate endocytosis but do not affect sorting of lysosomal enzymes in human fibroblasts. J. Biol. Chem. 265: 6650-6655, 1990.

17. Dodson, M.V., Allen, R.E., and Hossner, K.L. Ovine somatomedin, multiplication-stimulating activity, and insulin promote skeletal muscle satellite cell proliferation *in vitro*. Endocrinol. 117: 2357-2363, 1985.

18. Yaffe, D., and Saxel, O. Serial passaging and differentiation of myogenic cells isolated from dystrophic mouse muscle. Nature 270: 725-727, 1977.

19. Tollefsen, S.E., Sadow, J.L., and Rotwein, P. Coordinate expression of insulin-like growth factor II and its receptor during muscle differentiation. Proc. Natl. Acad. Sci. USA 86: 1543-1547, 1989.

20. Tollefsen, S.E., Lajara, R., McCusker, R.H., Clemmons, D.R., and Rotwein, P. Insulin-like growth factors (IGF) in muscle development. J. Biol. Chem. 264: 13810-13817, 1989.

21. Warren, L. Stimulated secretion of lysosomal enzymes by cells in culture. J. Biol. Chem. 264: 8835-8842, 1989.

22. Chirgwin, J.M., Przybyla, A.E., MacDonald, R.J., and Rutter, W.J. Isolation of biologically active ribonucleic acid from sources enriched in ribonuclease. Biochem. 24: 5294-5299, 1979.

23. Rotwein, P., Burgess, S.K., Milbrandt, J.D., andd Krause, J.E. Differential expression of insulin-like growth factor genes in rat central nervous system. Proc. Natl. Acad. Sci. USA 85: 265-269, 1988.

24. Szebenyi, G., and Rotwein, P. submitted.

25. Kornfeld, S. and Mellman, I. The biogenesis of lysosomes. Annu. Rev. Cell Biol. 5: 483-525, 1989.

26. Newman, P.J., Gorski, J., White, G.C., Gidwitz, S., Cretney, C. J., and Aster, R.H. Enzymatic amplification of platelet-specific messenger RNA using the polymerase chain reaction. J. Clin. Invest. 82: 739-743, 1988.

27. Pohlmann, R., Nagbi, G., Schmidt, B., Stein, M., Lorkowski, G., Krentler, C., Cully, J., Meyer, H.E., Grzeschik, K-H., Mersmann, G., Hasilik, A., and von Figura, K. Cloning of a cDNA encoding the human cation-dependent mannose 6-phosphate-specific receptor. Proc. Natl. Acad. Sci. USA 84:5575-5579, 1987.

28. Dahms, N.M., Lobel, P., Breitmeyer, J., Chirgwin, J.M., and Kornfeld, S. 46 kd mannose 6-phosphate receptor: Cloning, expression, and homology to the 215 kd mannose 6-phosphate receptor. Cell 50: 181-192, 1987.

29. Chan, S.J., Segundo, B.S., McCormick, M.B., and Steiner, D.F. Nucleotide and predicted amino acid sequences of cloned human and mouse preprocathepsin B cDNAs. Proc. Natl. Acad. Sci. USA. 83: 7721-7725, 1986.

30. Joseph, L.J., Chang, L.C., Stamenkovich, D., and Sukhatme, V.P. Complete nucleotide and deduced amino acid sequences of human and murine preprocathepsin L. J. Clin. Invest. 81: 1621-1629, 1988.

31. Bapat, B., Ethier, M., Neote, K., Mahuran, D., and Gravel, R.A. Cloning and sequence analysis of a cDNA encoding the β-subunit of mouse β-hexosaminidase. FEBS Letters 237: 191-195, 1988.

32. Baxter, R.C., and Martin, J.L. Binding proteins for the insulin-like growth factors: Structure, regulation and function. Progress in Growth Factor Research, 1: 49-68, 1989.

33. Brewer, M.T., Stetler, G.L., Squires, C.H., Thompson, R.C., Busby, W.H., and Clemmons, D.R. Cloning, characterization, and expression of a human insulin-like growth factor binding protein. Biochem. Biophys. Res. Comm. 152: 1289-1297, 1988.

34. Brown, A.L., Chiariotti, L., Orlowski, C.C., Mehlman, T., Burgess, W.H., Ackerman, E.J., Bruni, C.B., and Rechler, M.M. Nucleotide sequence and expression of a cDNA clone encoding a fetal rat binding protein for insulin-like growth factors. J. Biol. Chem. 264: 5148-5154, 1989.

35. De Vroede, M.A., Romanus, J.A., Standaert, M.L., Pollet, R.J., Nissley, S.P., and Rechler, M.M. Interaction of insulin-like growth factors with a nonfusing mouse muscle cell line: Binding, action, and receptor down-regulation. Endocrinol. 114: 1917-1929, 1984.

INSULIN-LIKE GROWTH FACTOR RECEPTORS IN TESTICULAR VASCULAR TISSUE FROM NORMAL AND DIABETIC RATS

Joyce F. Haskell and Russell B. Myers

Department of Obstetrics and
Gynecology, University of Alabama
at Birmingham, Birmingham, Alabama

INTRODUCTION

The complex endocrine regulation of the testicular vascular system is ill defined. Many blood vessels contain growth factor receptors, including those for insulin, IGF-I, and IGF-II. These growth factors may interact to modulate growth and/or repair of damaged vessels, and possibly regulate blood vessel permeability and responsiveness to other regulatory substances. Diabetic vascular complications have many common features whether large or small blood vessels are involved. Whether growth hormone or other growth factors are involved in the altered vascular cell proliferation, thickening of the basement membrane and altered vascular permeability is not yet known. Retinopathy, peripheral vascular disease and glomerular nephropathy remain frequent diabetic abnormalities. Male reproductive functions are also commonly compromised in diabetics. Impotency and decreased androgen levels may result from vascular complications, in addition to diabetic neuropathy. In the present study, we examined insulin-like growth factor receptor characteristics from normal and diabetic testicular vascular tissue using immunohistochemical and affinity crosslinking techniques.

Production of Testicular Insulin-like Growth Factors

Levels of IGFs in human serum and tissues vary with age (1). IGF-I is primarily produced in the liver and is regulated by growth hormone. IGF-I levels, which are low in children, gradually increase around puberty, then decrease post-pubertally. Until recently, IGF-II was thought to be primarily a fetal growth factor since serum IGF-II levels are highest during the third trimester of gestation, fall at birth, and decline continuously throughout life. Both IGF-I and IGF-II are mitogenic growth factors which have many effects on cultured cells, interacting with two types of receptor molecules (2). In contrast to insulin, both IGF-I and IGF-II circulate in the serum bound to serum binding proteins. The IGF-I receptor resembles the insulin receptor, in that both

Molecular Biology and Physiology of Insulin and Insulin-Like Growth Factors
Edited by M.K. Raizada and D. LeRoith, Plenum Press, New York, 1991

are tetrameric glycoproteins with intrinsic tyrosine protein kinase activity. The IGF-II/mannose-6-phosphate receptor is a single chain glycoprotein which is mainly extracellular and binds IGF II and M6P to different sites. Most cells have both types of IGF receptors as well as insulin receptors (2).

Tissue distribution of IGF-I and IGF-II have been investigated by several groups (3-8). Although in fetal tissues IGF-I is primarily synthesized by the liver under growth hormone regulation, serum IGF-II levels are greater than that of IGF-I. Casella et al.(3) initially isolated IGF-I cDNA from rat testes. It has been clearly established that the testis contains mRNA for both IGF-I and IGF-II (3). Steroid producing organs are particularly interesting since IGFs have been shown to influence reproductive function (7,9).

Vascular IGFs, binding proteins and diabetes

Within the circulatory system, IGFs are bound to two types of large molecular weight proteins, approximately M_r=150,000 and 30,000. Various forms of binding proteins continue to be identified in rat and human tissues, making this a very complex field to study (10-16). Physiological status, such as time of day, fasting, hyperglycemia and various diets, etc. have been shown to affect circulating levels of IGF-I and the various forms of IGF binding proteins (13-16). Insulin may interact with IGFs by varying the circulating levels of IGFs and/or their binding proteins (14) as well as interacting with the IGF receptors. However, recent studies show that endothelial cells produce binding proteins, which then contribute to circulating levels of IGF binding proteins. Recently, IGF binding proteins have been implicated to interact with cellular IGF receptors (16,17), suggesting that IGF binding proteins may be involved in the regulation of the physiological actions of IGFs, in addition to increasing the biological half-life of the IGFs.

Diabetic vascular complications have been studied for numerous years, yet the etiology remains obscure (18). Whether deficiencies or excesses in vascular insulin or glucose levels contribute to the vascular defects by altering IGF function has not yet been elucidated. Diabetic complications may result from alterations in receptors, tissue production of IGFs or their binding proteins (13-18). Although vascular complications have many similarities regardless of the tissue involved, the interaction of IGFs with vascular tissue from the testis is particularly interesting. Many diabetics also experience impotency, which is thought to involve both vascular and neural etiology (18). Testicular diabetic complications are extremely common (18, 19) and may be influenced by the lowering of the levels of testosterone produced during diabetes. Leydig cells contain insulin, IGF I and IGF II/M6P receptors. These peptides have been shown to regulate testosterone production by purified cells in vitro (20,21). Additionally, in vivo injections with hCG or LH increase IGF I binding in normal and hypophysectomized rats (21,22).

Streptozotocin-treated Sprague Dawley rats have been utilized in many research laboratories to examine the effect of diabetes on cellular function. Recently it has been noted that the kidneys, gastrointestinal organs, and bladder are enlarged in streptozotocin-induced diabetic rats; other tissues such as the epididymal fat pad, spleen, heart, liver and testes decrease dramatically in size (23,24).

The rat testicular vessels which we are currently studying consist of a mixture of peritubular and interstitial vessels. Peritubular vessels are the vessels which are situated alongside and embedded among the seminiferous tubules (25,26). The blood-testes barrier, however, is apparently formed by the tight junctions of the Sertoli cells, not tight junctions between adjacent endothelial cells which are responsible for the blood-tissue barriers in the brain and retina (26).

Gonadotropins and other growth factors have been implicated in vascular functions (27-35). Endothelial cells and blood vessels from many different tissues have been shown to bind insulin and IGFs (28-36). Vascular smooth muscle cell proliferation is increased dramatically by insulin and IGF-I (29). The retinal and cerebral microvessels have high affinity receptors for insulin and IGF I similar to those seen in peripheral tissues (36,37). However, brain and retinal nonvascular tissues have slightly lower molecular weight forms of insulin and IGF I receptors compared with those seen in the microvessels and peripheral tissue (36,37). These differences in size have been attributed to differences in glycosylation (38). Insulin and IGF I receptors in isolated rat Leydig cells and testicular vascular tissue appear to be similar to the peripheral receptors in size and binding characteristics (37). While the role of growth factors in vascular tissue is still poorly understood, peptide growth factor binding, processing and degradation occurs in both endothelial cells and whole blood vessels (30-39).

The IGF-II receptor has been identified as the cation-independent mannose-6-phosphate (M6P) receptor (40-44). This has important ramifications in vascular tissue. Transport of glycoproteins across blood vessels, for example, could be affected via this receptor. The IGF II/M6P receptor has also been shown to circulate in fetal and adult serum from several different species (45). Although IGF I and IGF II are both mitogenic effectors, the majority of biological actions stimulated by IGF II may occur via crossreactivity with the IGF I receptor rather than acting through the IGF II/M6P receptor (46). Murayama et al. (47) have shown that a G protein binds and possibly conveys a transmembrane signalling function for the IGF-II/M6P receptor.

RESULTS

Localization of IGF-II/M6P receptors in rat testicular vessels

Immunohistochemical localization of IGF-II/ M6P receptors in Sprague Dawley rats was examined using a goat anti-rabbit antiserum coupled to Texas red to visualize specific binding of an anti-IGF-II receptor polyclonal antibody. We examined deparaffinized thin sections of rat testicular arteries from normal (FIG.1 A) and three week diabetic (FIG. 1 B) Sprague Dawley rats. Testicular blood vessels clearly bound the affinity purified IgG (Ab3637) while nonvascular tissue was not as intensely labelled. Diabetic vessels were not as uniformly labelled as normal testicular arteries. An increase in non-membrane attached florescence was apparent, leading us to speculate that the IGF-II/M6P receptor may not be as

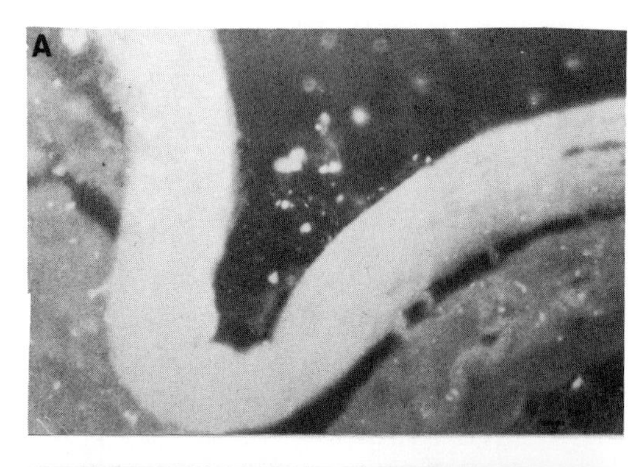

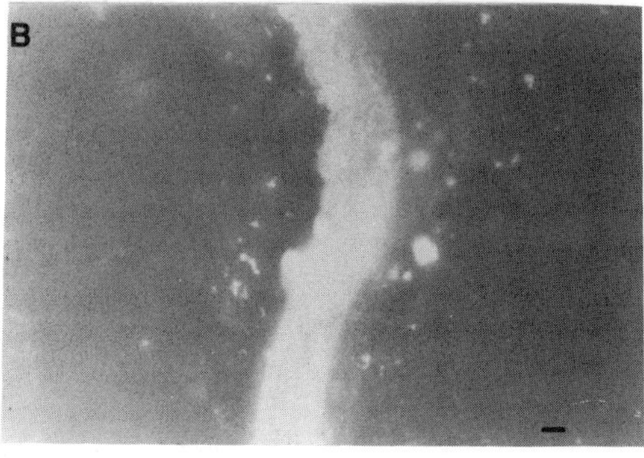

FIG.1 Immunohistochemical localization of IGF II/M6P receptors in rat testicular thin sections. Tissue sections were prepared from normal (panel A) and diabetic (panel B) rat testes fixed with Bouins and stained for the presence of the IGF II/M6P receptor using purified IgG (Ab3637) and goat antirabbit second antibody conjugated with Texas Red.

tightly associated with the vascular tissue in diabetic as compared to nondiabetic rats. We cannot rule out a methodological problem, since fluorescence artifacts are quite common. It is also possible that the presence of Fc or other factors within the vessels may influence the quality of the antibody binding. Normal rabbit IgG was incubated with normal and diabetic tissue; no fluorescence was apparent in the control sections.

IGF-I and IGF-II Receptor in Testicular Microvessels

Rat testicular microvascular tissue was obtained from normal and diabetic rats by a modification of the method of Meezan et al.(48) for the preparation of renal glomeruli. Rats were anesthetized with sodium pentobarbital and perfused via the renal artery with Krebs Ringer phosphate buffer, pH 7.3 for five minutes. The kidneys and testes were then perfused with magnetic ferric oxide until the tissues were gray. Testes were removed and decapsulated. The testicular artery was manually dissected upon removal of the testicular capsule. The remaining tissue was then gently homogenized in ice cold phosphate buffered saline. Microvessels were obtained by filtering the testicular homogenate over a section of nylon sieving material with 220 μm openings while gently stirring with a teflon pestle to isolate the microvascular component. The seminiferous tubules and interstitial cells are easily fragmented and are separated by filtration. However, testicular blood vessels are more durable and remain on top of the 220 μm pore size nylon sieve. Testicular blood vessels and whole decapsulated testes were detergent solubilized with 1% Triton-X-100 in 50 mM Hepes buffer, pH= 7.6, for 45 min at 4 C. We have examined testicular vessels using affinity crosslinking techniques.

The effect of M6P on the binding of IGFs to adult rat vascular tissue is shown in Figs. 2 and 3. The effect of M6P (1 mM) on binding in detergent solubilized vascular tissue was examined using covalent crosslinking techniques and SDS-polyacrylamide electrophoresis, followed by autoradiography of the dried gel. M6P had a dramatic effect on the IGF-I binding to the testicular vascular Type II IGF/M6P receptor. Using detergent solubilized rat tissue preparations, M6P has been seen to amplify ^{125}I-IGF-I and ^{125}I-IGF-II binding to the IGF-II/M6P receptor band (37).

Also in Figs. 2 and 3, we have examined the effect of diabetes on detergent solubilized testicular microvascular IGF receptors. ^{125}I-IGF-I crosslinking to the lower (M_r=135,000-145,000) band from detergent solubilized testicular vascular tissue is clearly enhanced in tissue from diabetic rats. In the presence of M6P, binding to the upper (M_r=260,000) is present and blocked by excess unlabeled IGFs in either normal or diabetic detergent solubilized microvascular tissue. Binding to the alpha subunit of the IGF I receptor is enhanced in the diabetic tissue from animals which were hyperglycemic for four weeks (Fig.2, lanes 3,4) or for only two weeks (Fig.3, lanes 3,4). Results of physiological findings in these two week diabetic animals is presented in Table 1. Androgen, testis and body weights, as well as IGF I levels are decreased in these diabetic animals. Similar results were recently shown by Bach and Jerums (24).

Crosslinking results of two week diabetic animals are depicted in Fig. 3. IGF-I binding to the IGF I alpha subunit band in animals four weeks after streptozotin was visibly greater, as shown in Fig.2 compared with Fig.3. IGF I binding in one week streptozotocin-diabetic animals did not seem to be affected. Excess IGF-I totally displaced the intensely labeled lower band observed in diabetic vascular tissue (data not shown). Moreover, M6P (1-5 mM) enhanced radiolabeled IGF-I or the IGF-II binding to the higher molecular weight band (Mr=260,000) in testicular vascular tissue from normal and diabetic animals. Incubated in the presence of M6P, a specific anti-IGF-II receptor antibody (Ab 3637) displaces this band, indicating that it represents the IGF-II/M6P receptor and not merely binding to a dimer of the IGF-I receptor (data not shown). Testicular vascular IGF-I and IGF-II receptors from streptozotocin-diabetic rats are of a slightly lower apparent molecular weight than control, suggesting that hyperglycemia may cause an alteration in glycosylation. We propose that this smaller receptor may be involved with the vascular pathophysiology observed in patients with diabetes mellitus.

TABLE 1. Effect of diabetes on testes weight, total body weight, plasma testosterone, and plasma insulin-like growth factor levels

GROUP	glucose mg/dl	testes wt, g	body wt, g	T ng/ml	IGF-I ng/ml
control (n=5)	124±12	1.68±.15	370±33	3.5±1.0	206±4
diabetic (n=5)	323±19	1.53±.23	323±19	1.67±.6	125±13

The effects of diabetes and M6P on relative IGF I and IGF II binding to rat testicular vessels is summarized in Table 2. The intensity of the IGF-I labeling to diabetic rat testicular microvessels (Fig. 3) was distinctly greater than normal rats, leading us to conclude that IGF-I binding in diabetic tissues is different from normal tissue. M6P lowers IGF-I binding to the IGF-I receptor while enhancing both IGF-I and IGF-II binding to the IGF II/M6P receptor band. Since these results were conducted using detergent solubilized tissue, we hesitate to propose that this augmentation has any physiological significance. Nonetheless, IGF I binding in diabetic testicular vascular tissue is altered compared to that observed in normal tissue.

Table 2. Relative autoradiographic intensities of ^{125}I-IGF-I and ^{125}I-IGF-II crosslinked to detergent solubilized rat testicular microvessels

EXPERIMENTAL GROUP	M6P (1 mM)	^{125}I-IGF-I M_r=130k band	^{125}I-IGF-I M_r=260k band	^{125}I-IGF-II M_r=260k band
control	-	20.2	5.0	19.0
control	+	24.3	36.6	65.0
diabetic	-	40.2	7.9	24.5
diabetic	+	6.1	29.2	41.7

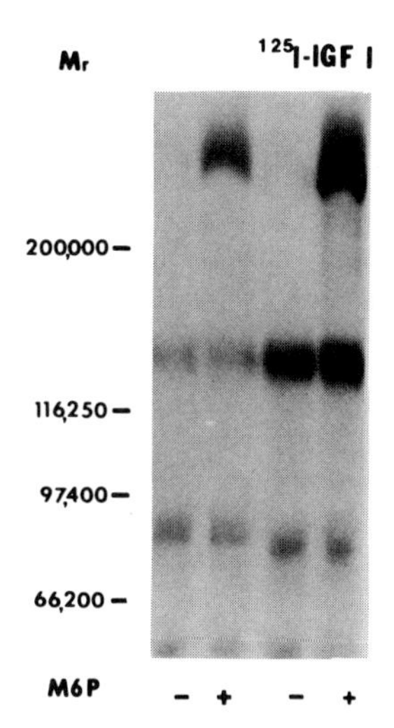

FIG.2 Effect of M6P on IGF I crosslinked to detergent solubilized testicular blood vessels from normal and diabetic rats. M6P (5 mM) was added to detergent solubilized testicular vessels from normal (lanes 1,2) and 4 week diabetic (lanes 3,4) rats.

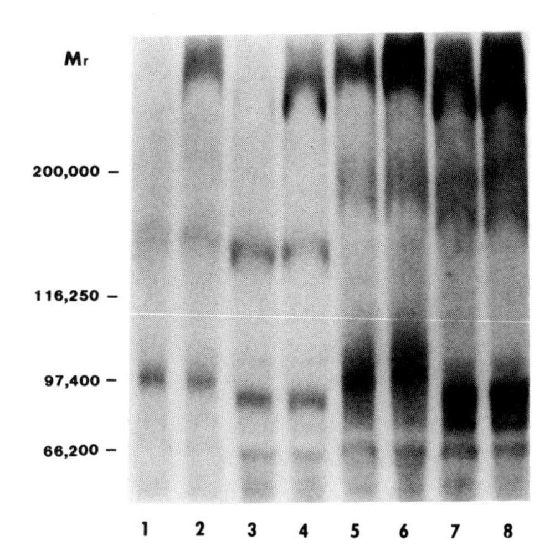

FIG.3 Effect of M6P on IGF I and IGF II crosslinked to detergent solubilized testicular blood vessels from normal and diabetic rats. M6P (5 mM) was added to detergent solubilized rat testicular vessels (even numbered lanes) incubated with ^{125}I- IGF-I (lanes 1-4) or ^{125}I-IGF-II (lanes 5-8). Testicular vessels were obtained from normal (lanes 1,2,5,6) or two week diabetic (lanes 3,4,7,8) adult Spraque Dawley rats.

DISCUSSION

Generally rat vascular tissues bind more IGF-II than IGF-I. We have also reported that M6P enhanced IGF-I and IGF-II binding to rat vascular and nonvascular tissue (37). Others have demonstrated that the IGF-II/M6P receptor circulates in maternal and fetal sheep serum and monkey serum (46). The physiological importance of both of these findings has not yet been established. However, it appears that the circulating form of the IGF II/M6P receptor may be a truncated form of the membrane receptor (46). Recent analysis of this receptor supports the hypothesis that most of the IGF-II/M6P receptor is hydrophilic and therefore probably extracellular. It is tempting to speculate that in rapidly growing tissues or tissues undergoing regenerative growth that the IGF-II/M6P receptor may actually be removed from the cell membrane following IGF-II binding. This might serve as a mechanism for terminating growth once final size is achieved. This dissociation model is not unique since many enzymes, such as protein kinase A, may be composed of regulatory and catalytic subunits. The hCG/lutropin receptor, according to Bhalla et al.(51), has been suggested to have a similar mechanism. More recently the v-erb oncogene has been shown to be homologous to a truncated form of the EGF or TGF-α receptor (52). In this example, growth proceeds in an unregulated fashion. These findings along with the appearance of the IGF II/M6P receptor in animal serum, suggest that further studies is this area are warranted.

The IGF II/M6P receptor may be involved with cellular growth and differentiation. This hypothesis is supported by recent studies in which regenerating liver has been shown to contain more IGF II/M6P receptors than seen in normal tissue (49,50). We have recently demonstrated that the IGF II/M6P receptor is enhanced in regenerating testes following treatment with ethane dimethylsulfonate (EDS). However, no increase was noted for the isolated testicular arteries from these animals (Endocrine Society Abstract 606, 1990).

SUMMARY

Testicular blood vessels contain IGF-I and IGF-II/M6P receptors. Binding to these receptors was altered following treatment with streptozotocin to induce diabetes. Intensity of labelling and size of receptors were examined using SDS-gel electrophoresis and autoradiography. The IGF-I and IGF-II/M6P receptor of the diabetic rat testicular microvessels appear to have a lower molecular weight as compared to controls. Macro- and microvascular tissues from diabetic rats apparently contain more IGF-I receptors than normal Sprague-Dawley rats. Using immunohistochemical techniques, the IGF-II/M6P receptor appears to dissociate easier from diabetic rat testicular arteries than from control animal blood vessels. M6P appears to increase both IGF-I and IGF-II binding to the rat IGF-II/M6P receptor, at least as visualized using affinity crosslinking analysis. Whether these differences in the IGF receptors are involved in the development of diabetic vascular disease is not yet known.

ACKNOWLEDGEMENTS

Our gratitude to Drs. S.P. Nissley and W. Kiess, for providing purified IgG from Ab 3637 and Dr. M. Smith (Eli Lilly, Indianapolis for supplying the human recombinant IGF-II which we iodinated for this study. We also appreciate the technical assistance of L.E. Haws, D.S.Long, and S.D. Rose, and the immunohistochemical assistance of Dr. T. Borg and the Dept of Pathology, University of South Carolina School of Medicine, Columbia, SC.

The authors are grateful for support from the National Institute of Health (HL 38442) and a VA Research Service merit award.

REFERENCES

1. S. Salardi, E. Cacciari, D. Ballardini, F. Righetti, M. Capello, A. Cicognani, S. Zucchini, G. Natali, D. Tassinari: Relationships between growth factors (somatomedin-C and growth hormone) and body development, metabolic control, and retinal changes in children and adolescents with IDDM. Diabetes 35:832 (1986).

2. S. P. Nissley, J. F. Haskell, N. Sasaki, M. DeVroede, M. M. Rechler: Insulin-like growth factors. J Cell Sci Suppl 3:39 (1985).

3. S. J. Casella, E. P. Smith, J. J. Van Wyk, D. R. Joseph, M. A. Hynes, E. C. Hoyt, P. K. Lund: Isolation of rat testis cDNA encoding an insulin-like growth factor I precursor. DNA 6:325 (1987)

4. L. J. Murphy, G. I. Bell, H. G. Friesen: Tissue distribution of insulin-like growth factor I and II messenger ribonucleic acid in the adult rat. Endocrinology 120:1279 (1987).

5. P. K. Lund, B. M. Moats-Saats, M. A. Hynes, J. G. Simmons, M. Jansen, J. D. D'Ercole, J. J. Van Wyk: Somatomedin-C/insulin-like growth factor-II mRNAs in rat fetal and adult tissues. J Biol Chem 261:14539 (1986).

6. R. Voutilainen, W. L. Miller: Developmental and hormonal regulation of mRNAs for insulin-like growth factor II and steroidogenic enzymes in human fetal adrenals and gonads. DNA 7:9 (1988).

7. D. J. Handelsman, J. A. Spaliviero, C. D. Scott, R. C. Baxter: Identification of insulin-like growth factor-I and its receptors in the rat testis. Acta Endocrinology 109:543 (1985).

8. C. T. Roberts, S. R. Lasky, W. L. Lowe, W. T. Seaman, D. LeRoith: Molecular cloning of rat insulin-like growth factor complementary deoxyribonucleic processing and regulation by growth hormone in extrahepatic tissues. Mol Endocrinol 1:2431 (1987).

9. T. Lin, J. F. Haskell, N. Vinson, L. Terracio: Characterization of insulin and IGF-I receptors of purified Leydig cells and their role on steroidogenesis in primary culture: A comparative study. Endocrinology 119:1641 (1987).

10. C. C. Orlowski, A. L. Brown, G. T. Ooi, Y. W. H. Yang, L. Y. H. Tseng, M. M. Rechler: Tissue, developmental, and regulation of messenger ribonucleic acid encoding a rat insulin-like growth factor-binding protein. Endocrinology 126:644 (1990).

11. R. S. Bar, B. A. Booth, M. Boes, B. L. Dake: Insulin-like growth factor-binding proteins from vascular endothelial cells: purification, characterization, and intrinsic biological activities. Endocrinology 125:1910 (1989).

12. R. G. Elgin, W. H. Busby Jr, D. R. Clemmons: An insulin-like growth factor (IGF) binding protein enhances the biologic response to IGF-I. Proc Natl Acad Sci USA 84:3254 (1987).

13. J. Zapf, C. Hauri, M. Waldvogel, E. Futo, H. Hasler, K. Binz, H. P. Guler, C. Schmid, E. R. Froesch: Recombinant human insulin-like growth factor I induces its own specific carrier protein in hypophysectomized and diabetic rats. Proc Natl Acad Sci USA 86:3813 (1989).

14. K. Hall, B. L. Johansson, G. Povoa, B. Thalme: Serum levels of insulin-like growth factor (IGF) I, II and IGF binding protein in diabetic adolescents treated with continuous subcutaneous insulin infusion. J Int Med 225:273 (1989).

15. A. M. Suikkari, V. A. Koivisto, E. M. Rutanen, H. Yki-Jarvinen, S. L. Karonen, M. Seppala: Insulin regulates the serum levels of low molecular weight insulin-like growth factor-binding protein. J Clin Endo Metab 66:266 (1988).

16. C. D. Scott, R. C. Baxter: Production of insulin-like growth factor I and its binding protein in rat hepatocytes cultured from diabetic and insulin-treated diabetic rats. Endocrinology 119:2346 (1986).

17. D. R. Clemmons, J. P. Thissen, M. Maes, J. M. Ketekskegers, L. E. Underwood: Insulin-like growth factor I (IGF-I) infusion into hypophysectomized or protein-deprived rats induces specific IGF-binding proteins in serum. Endocrinology 125: 2967 (1989).

18. K. F. Hanssen, K. Dahl-Jorgensen, T. Lauritzen, B. Feldt-Rasmussen, O. Brinchmann-Hansen, T. Deckert: Diabetic control and microvascular complications: the near-normoglycaemic experience. Diabetologia 29:677 (1986).

19. L. Rodreguez-Rigau: Diabetes and male reproductive function. J. Andrology 1:105 (1980).

20. P. Chatelain, A. Penhoat, M. H. Perrard-Sapori, C. Jaillar, D. Naville, J. Saez: Maturation of steroidogenic cells: a target for IGF-I, Acta Paediatr Scand (Suppl) 347: 104 (1988).

21. T. Lin, J. Blaisdell, J. F. Haskell: Type I IGF receptors of Leydig cells are up-regulated by human chorionic gonadotropin. Biochem Biophys Res Commun 149:852 (1987).

22. T. Lin, J. Blaisdell, J. F. Haskell: Hormonal regulation of Type I IGF receptors of Leydig cells in hypophysectomized rats. Endocrinology 123:134 (1988).

23. D. J. Pillion, R. L. Jenkins, J. A. Atchison, C. R. Stockard, R. S. Clements, W. E. Grizzle: Paradoxical organ-specific adaptations to streptozotocin diabetes mellitus in adult rats. Amer J Physiol 254:E749 (1988).

24. L. A. Bach, G. Jerums: Effect of puberty on initial kidney growth and rise in kidney IGF-I in diabetic rats. Diabetes 39:557 (1990).

25. M. Dym: The fine structure of the monkey (macaca) Sertoli cell and its role in maintaining the blood-testis barrier. Anat Rec 175:639 (1973).

26. H. Takayama and T. Tomoyoshi: Microvascular architecture of rat and human testes. Invest. Urol.18:341 (1981).

27. B. P. Setchell, R. M. Sharpe: Effect of injected human chorionic gonadotropin on capillary permeability, extracellular fluid volume and the flow of lymph and blood in the testes of rats. J. Endocr. 91:245, (1981).

28. G. L. King: Cell biology as a approach to the study of the vascular complications of diabetes. Metabolism 34:17 (1985).

29. B. Pfeifle, H. Hamann, R. Fussganger, H. Ditschuneit: Insulin as a growth regulator of arterial smooth muscle cells: effect of insulin on IGF-I. Diabet Metab 13: 326 (1987).

30. M. Grant, J. Jerdan, T.J. Merimee: Insulin-like growth factor-I modulates endothelial cell chemotaxis. J Clin Endocrinol Metab 65: 370 (1987).

31. H. A. Hansson, E. Jennische, A. Skottner: IGF-I expression in blood vessels varies with vascular load. Acta Physiol Scand 129: 165 (1987).

32. H. L. Hachiya, J. L. Carpentier, G. L. King: Comparative studies on insulin-like growth factor-II and insulin processing by vascular endothelial cells. Diabetes 35:1065 (1986).

33. K. E. Bornfeldt, H. J. Arnquist, H. H. Dahlkvist, A. Skottner, J. E. S. Wikberg: Receptors for insulin-like growth factor-I in plasma membranes isolated from bovine mesenteric arteries. Acta Endocrinol (Copenh) 117:428 (1988).

34. R. S. Bar, M. Boes, M. Yorek: Processing of insulin-like growth factors I and II by capillary and large vessel endothelial cells. Endocrinology 118:1072 (1986).

35. P. Pekala, M. Marlow, D. Heuvelman, D. Connolly: Regulation of hexose transport in aortic endothelial cells by vascular permeability factor and tumor necrosis factor-α, but not insulin. J Biol Chem 265:18051 (1990).

36. J. F. Haskell, E. Meezan, D. J. Pillion: Identification of the insulin receptor of cerebral microvessels. Amer J Physiol 248:E115 (1985).

37. J. F. Haskell, L. E. Haws, A. Davis, R. Hunt: Comparison of insulin-like growth factor receptors in human retinal cells. in: Molecular and Cellular Biology of insulin-like growth factors and their receptors, D. LeRoith and M. K. Raizada, eds., Plenum Publishing Co, New York, p.297 (1989).

38. K. A. Heidenreich, P. R. Gilmore: Structural and functional characteristics of insulin receptors in rat neuroblastoma cells. Journal of Neurochemistry 45:1642 (1985).

39. J. F. Haskell, D. J. Pillion, E. Meezan: Specific high-affinity receptors for insulin-like growth factor II (IGF-II) in rat kidney glomerulus. Endocrinology 123:774 (1987).

40. R. S. Bar, W. L. Lowe, R. G. Spanheimer: Interactions of insulin and IGFS with cellular components of the arterial wall: Potential impact on atherosclerosis.in: Molecular and Cellular Biology of insulin-like growth factors and their receptors, D. LeRoith and M. K. Raizada, eds., Plenum Publishing Co, New York, p.473 (1989).

41. D. O. Morgan, J. C. Edman, D. N. Standring, V. A. Fried, M. C. Smith, R. A. Roth, W. J. Rutter: Insulin-like growth factor II receptor as a multifunctional binding protein. Nature 329:301 (1987).

42. T. Braulke, C. Causin, A. Waheed, U. Junghans, A. Hasilik, P. Maly, R. E. Humbel, K. von Figura: Mannose-6-phosphate/insulin-like growth factor II receptor: Distinct binding sites for mannose 6-phosphate and insulin-like growth factor II. Biochem Biophys Res Commun 150:1287 (1988).

43. R. A. Roth, C. Stover, J. Hari, D. O. Morgan, M. C. Smith, V. Sara, V. A. Fried: Interaction of the receptor for insulin-like growth factor II with mannose-6-phosphate and antibodies to the mannose-6-phosphate receptor. Biochem Biophys Res Commun 149:600 (1987).

44. P. Y. Tong, S. E. Tollefsen, S. Kornfeld: The cation-independent mannose-6-phosphate receptor binds insulin-growth factor II. J Biol Chem 263:2585 (1988).

45. W. Kiess, J. F. Haskell, L. Lee, L. A. Greenstein, B. E. Miller, A. L. Aarons, M. M. Rechler, S. P. Nissley: An antibody that blocks insulin-like growth factor (IGF) binding to the type II IGF receptor is neither an agonist nor an inhibitor of IGF-stimulated biologic responses in L6 myoblasts. J Biol Chem 262:12745 (1987).

46. M. C. Gelato, C. Rutherford, R. I. Stark, S. S. Daniel: The insulin-like growth factor II/mannose-6-phosphate receptor is present in fetal and maternal sheep serum. Endocrinology 124: 7100 (1989).

47. Y. Murayama, Okamoto, E. Pgata, T. Asano, T. Katada, M. Ui, J. Grubb, W. S. Sly, I Nishimoto: Distinctive regulation of the human cation-independent mannose 6-phosphate receptor and GTP-binding proteins by insulin-like growth factor-II and mannose-6-phosphate. J Biol Chem 265:17456 (1990).

48. E. Meezan, K. Brendel, J. Ulreich, E. C. Carlson: Properties of a pure metabolically active glomerular preparation from rat kidneys I. Isolation. J. Pharmacol Exp Ther 187:332 (1973).

49. C. D. Scott, M. Ballesteros, R. C. Baxter: Increased expression of insulin-like growth factor-II/mannose-6-phosphate receptor in regenerating rat liver. Endocrinology 127:2210 (1990).

50. B. Buruera, H. Werner, M. Sklar, Z. Shen-Orr, B. Stannard, C. T. Roberts,Jr, S. P. Nissley, S. J. Vore, J.F. Caro, D. LeRoith: Liver regeneration is associated with increased expression of the insulin-like growth factor II/mannose-6-phosphate receptor. Mol Endocrinol 4:1539 (1990).

51. V. B. Bhalla, E. S. Browne, G. S. Sohal: Demonstration of hCG binding sites and hCG stimulated steroidogenesis in different populations of interstitial cells. Advances in Experimental Medicine and Biology 219:489 (1987).

52. S. J. Decker: Epidermal growth factor-induced truncation of the epidermal growth factor receptor. J Biol Chem 264: 17641 (1989).

RECIPROCAL MODULATION OF BINDING OF LYSOSOMAL ENZYMES AND INSULIN-LIKE GROWTH FACTOR-II (IGF-II) TO THE MANNOSE 6-PHOSPHATE/IGF-II RECEPTOR

Peter Nissley[1] and Wieland Kiess[2]

[1]Metabolism Branch, National Cancer Institute, National Institutes of Health, Bethesda, MD 20892
[2]Children's Hospital, University of Munich, Munich, Germany

INTRODUCTION

The discovery by Morgan et al.[1] that the human IGF-II receptor is 80% homologous to the bovine cation-independent mannose 6-phosphate (Man-6-P) receptor led them to propose that the receptor is bifunctional, binding both a large number of lysosomal enzymes through the Man-6-P recognition site, and the growth factor, IGF-II. Later, Kornfeld's and Czech's laboratories reported that the avian and amphibian Man-6-P receptors did not bind IGF-II, suggesting that the bifunctional property of the receptor may be confined to mammals.[2,3] Did the mammalian receptor gene simply pick up a nucleotide sequence encoding an IGF-II binding site, enabling the receptor to provide a degradative pathway for IGF-II, or are there important interactions of the two disparate classes of ligands for binding to the receptor? These interactions could result in reciprocal modulation of the targeting of lysosomal enzymes by IGF-II, on the one hand, and the modulation of IGF-II degradation and IGF-II stimulated biologic responses by lysosomal enzymes, on the other. We will briefly summarize our experimental results which provide evidence for reciprocal inhibition of binding of the two classes of ligands for the mammalian Man-6-P/IGF-II receptor.

INHIBITION OF IGF-II BINDING BY LYSOSOMAL ENZYMES

Inhibition of ^{125}I-IGF-II Binding to the Man-6-P/IGF-II Receptor by β-Galactosidase

For a particular lysosomal enzyme, the population of molecules is heterogeneous with respect to the number of mannose 6-phosphate residues and whether or not they are in phosphodiester linkage. We purified β-galactosidase from bovine testis and then used DEAE-Sephacel ion-exchange chromatography to isolate a subpopulation of molecules which exhibit high cellular uptake.[4] This preparation of β-galactosidase inhibited the binding of ^{125}I-IGF-II to pure receptor from rat placenta by 80% (Fig. 1). The concentration of β-galactosidase which exhibited half-maximal inhibition of ^{125}I-IGF-II binding was 25 nM. This value agrees with published data on the concentration of β-galactosidase required for half-maximal binding to cells and purified receptor.[5-7] Coincubation of Man-6-P with the receptor, ^{125}I-IGF-II, and

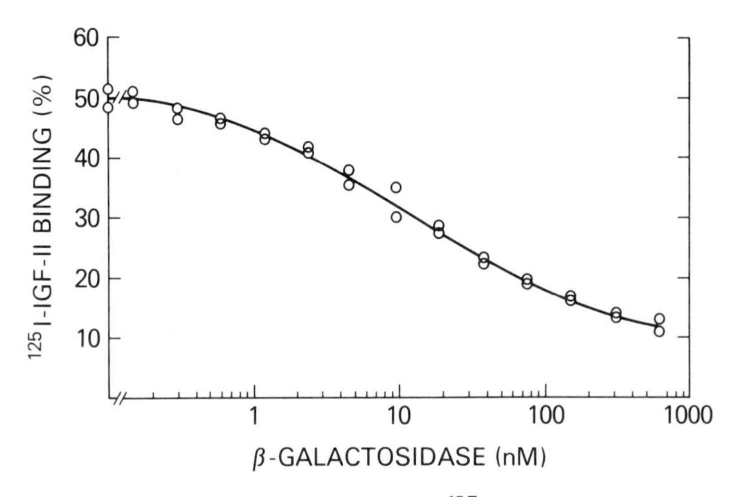

*Fig. 1. **Dose-response curve for inhibition of ^{125}I-IGF-II binding to the Man-6-P/IGF-II receptor by β-galactosidase.** β-galactosidase was purified from bovine testis by affinity chromatography on concanavalin-A-Sepharose 4B and PATG-agarose columns and high uptake forms of the enzyme were isolated on a DEAE-Sephacel column. The Man-6-P/IGF-II receptor was purified to homogeneity from rat placenta. ^{125}I-IGF-II and the indicated concentrations of β-galactosidase were incubated with the receptor for 3 h at 22°C. Bound radioligand was separated from free using activated charcoal. Binding of ^{125}I-IGF-II is expressed as a percentage of input radioactivity (24,580 cpm). Reprinted with permission from reference 4.*

β-galactosidase, showed that the phosphorylated sugar could completely block the inhibitory activity of β-galactosidase.[4] Thus the ability of β-galactosidase to inhibit the binding of ^{125}I-IGF-II depended upon binding to the Man-6-P recognition site. These results extend earlier affinity crosslinking experiments in which a single concentration of β-galactosidase inhibited the binding of ^{125}I-IGF-II to the bovine liver Man-6-P receptor and this inhibition was prevented by coincubation with Man-6-P.[8] In addition, β-glucuronidase and cathepsin D have been shown by others to inhibit the binding of ^{125}I-IGF-II to the Man-6-P/IGF-II receptor.[9,10]

Scatchard analysis of IGF-II binding to the receptor in the presence and absence of β-galactosidase showed that the lysosomal enzyme decreased the binding affinity of the receptor for IGF-II (Fig. 2).[4] This result is consistent with several different models. One possibility would be that β-galactosidase and IGF-II share a common binding site. This is unlikely since Man-6-P itself did

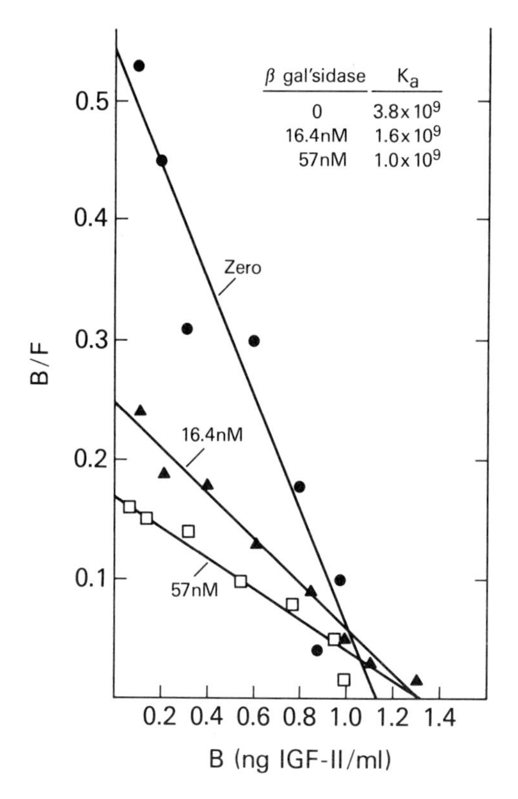

Fig. 2. ***Scatchard analysis of IGF-II binding to the Man-6-P/IGF-II receptor in the presence and absence of β-galactosidase.*** *Competitive binding dose-response curves for IGF-II were generated in the absence or presence of β-galactosidase (16.4 nM and 57 nM) and serial dilutions of unlabeled IGF-II. At the conclusion of the 3 h incubation at 22°C, bound radioligand was separated from free with activated charcoal. The IGF-II binding data were analyzed according to the method of Scatchard. Values for the binding affinity constants (K_a, M^{-1}) are given. Reprinted with permission from reference 4.*

not inhibit ^{125}I-IGF-II binding. Also, receptor antibodies have been shown to preferentially inhibit either the binding of IGF-II or pentamannosyl-6-O-phosphate-substituted BSA.[11] In a second model, the lysosomal enzyme could bind nearby to the IGF-II binding site and sterically inhibit the binding of the growth factor. A third possibility would be that β-galactosidase binds at a more distant site and changes the conformation of the receptor. The data of Tong et al.[7]

are consistent with only two of the 15 repeating domains of the extracellular portion of the receptor being utilized for high affinity binding of Man-6-P and only one molecule of β-galactosidase was found to bind to each receptor molecule. Thus, whichever of the above models is correct, the binding of one molecule of the lysosomal enzyme appears to be sufficient to block the binding of IGF-II.

Stimulation of IGF-II Binding to the Man-6-P/IGF-II Receptor by Man-6-P

At first glance, earlier reports of the ability of Man-6-P to stimulate ^{125}I-IGF-II binding to the receptor[12,13] would seem to be in direct conflict with the finding that lysosomal enzymes inhibit the binding of ^{125}I-IGF-II. We observed that the stimulation of ^{125}I-IGF-II binding by Man-6-P depended upon the method of receptor preparation.[4] Crude receptor preparations or receptor that had been purified by IGF-II affinity chromatography exhibited the positive effect of Man-6-P whereas receptor purified on a lysosomal enzyme affinity column did not show the positive effect. A possible explanation for these results would be

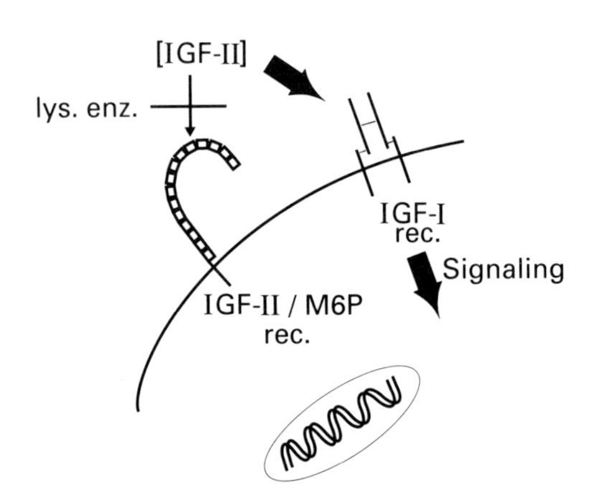

Fig. 3. Possible consequences of the inhibition of IGF-II binding to the Man-6-P/IGF-II receptor by extracellular lysosomal enzymes.

that Man-6-P is simply stripping endogenous lysosomal enzymes from the receptor. Since these endogenous lysosomal enzymes inhibit IGF-II binding, their removal results in enhanced IGF-II binding. Receptors purified on a lysomal enzyme affinity column would have had to exchange endogenous acid hydrolases for the lysomal enzyme coupled to the affinity column matrix. Support for this explanation comes from the experiments of Polychronakos et al.[14] who showed that extensive washing of membranes with Man-6-P yielded membrane preparations which no longer exhibited the positive effect of Man-6-P on IGF-II binding. In addition, these investigators further purified by sucrose gradient centrifugation a receptor preparation that had been initially purified on an IGF-II affinity column and thereby eliminated the enhancing effect of Man-6-P on ^{125}I-IGF-II binding to the receptor.

Possible Consequences of the Inhibition of IGF-II Binding by Lysosomal Enzymes

The possible consequences of the ability of lysosomal enzymes to inhibit the binding of IGF-II to the Man-6-P/IGF-II receptor are illustrated in Fig. 3. Although the majority of newly synthesized acid hydrolases are targeted directly to lysosomes, some escape this intracellular pathway and are secreted.[15] If present in sufficient concentration, these extracellular lysosomal enzymes could block the binding of IGF-II to cell surface receptors. Extracellular IGF-II is known to be degraded by Man-6-P/IGF-II receptor mediated internalization,[16,17] degradation presumably occurring in lysosomes. Inhibition of binding of IGF-II to the receptor would result in higher extracellular concentrations of IGF-II, making it more likely that IGF-II could stimulate biologic responses by acting through the IGF-I receptor. The IGF-I receptor is a member of group of growth factor receptors that contain a tyrosine kinase domain in the cytoplasmic portion of the receptor.[18] Although IGF-I binds with higher affinity to the IGF-I receptor

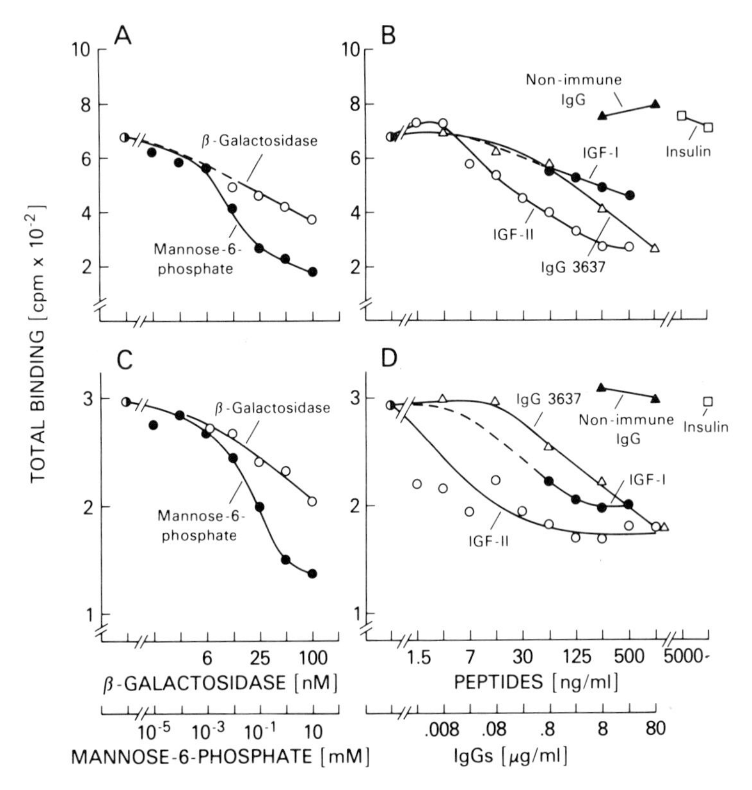

Fig. 4. Uptake of ^{125}I-β-galactosidase in BRL 3A2 rat liver cells (A and B) and C6 rat glial cells (C and D). IgGs were purified from nonimmune serum and anti-Man-6-P/IGF-II receptor serum 3637. Monolayer cultures were incubated with ^{125}I-β-galactosidase and unlabeled peptides, IgGs, mannose 6-phosphate and β-galactosidase for 2.5 h at 22°C. Cell-associated radioactivity is plotted versus the concentration of unlabeled ligands and IgGs that were present during the incubation. Reprinted with permission from reference 27.

than does IGF-II, the IGF-I receptor can be activated by IGF-II. In human skin fibroblasts, for example, IGF-II has been shown to stimulate growth responses by acting through the IGF-I receptor.[19,20] Alternatively, there are an increasing number of reports which suggest that IGF-II stimulates certain cellular events by acting through the Man-6-P/IGF-II receptor.[21-26] Consequently, inhibition of IGF-II binding to the Man-6-P/IGF-II receptor by extracellular lysosomal enzymes would block these biologic responses.

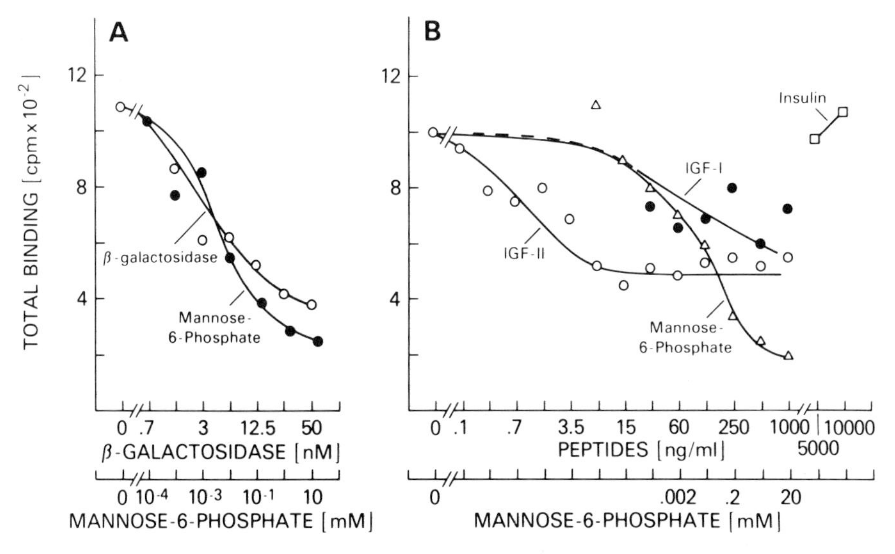

Fig. 5. ***Binding of*** ^{125}***I-β-galactosidase to pure Man-6-P/IGF-II receptor from rat placenta.*** *^{125}I-β-galactosidase and indicated concentrations of unlabeled β-galactosidase, mannose 6-phosphate, and peptides were incubated for 2.5 h at 22°C. Bound ^{125}I-β-galactosidase was separated from free using receptor antiserum 3637 and staphylococcal protein A. Binding of ^{125}I-β-galactosidase is plotted versus the concentrations of unlabeled ligands that were present during the incubation. Reprinted with permission from reference 27.*

INHIBITION OF THE BINDING OF β-GALACTOSIDASE TO THE MAN-6-P/IGF-II RECEPTOR BY IGF-II

Inhibition of Cellular Uptake of ^{125}I-β-Galactosidase by IGF-II

We found that IGF-II inhibited the cellular uptake of ^{125}I-β-galactosidase by C6 rat glial cells and BRL3A2 rat liver cells (Fig. 4).[27] IGF-I was considerably less potent than IGF-II in inhibiting uptake and insulin was without effect. This relative potency pattern corresponds to the binding profile of these ligands for the Man-6-P/IGF-II receptor. In addition, antiserum 3637 which blocks the binding of ^{125}I-IGF-II to the receptor also blocked the uptake of

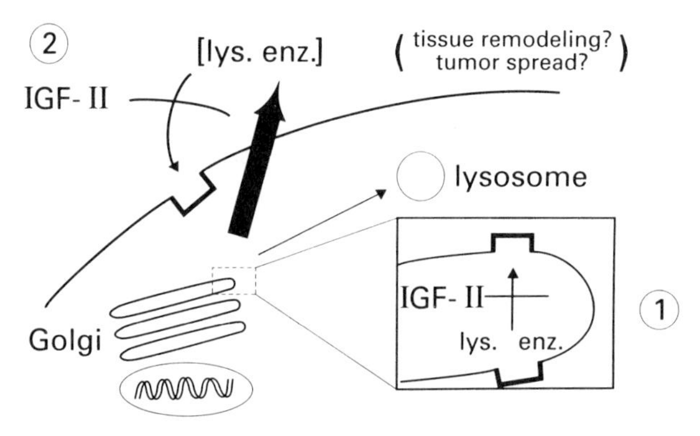

*Fig. 6. **Possible consequences of the inhibition of lysosomal enzyme binding to the Man-6-P/IGF-II receptor by IGF-II.** Potentially, this inhibition could occur intracellularly (1) or extracellularly (2).*

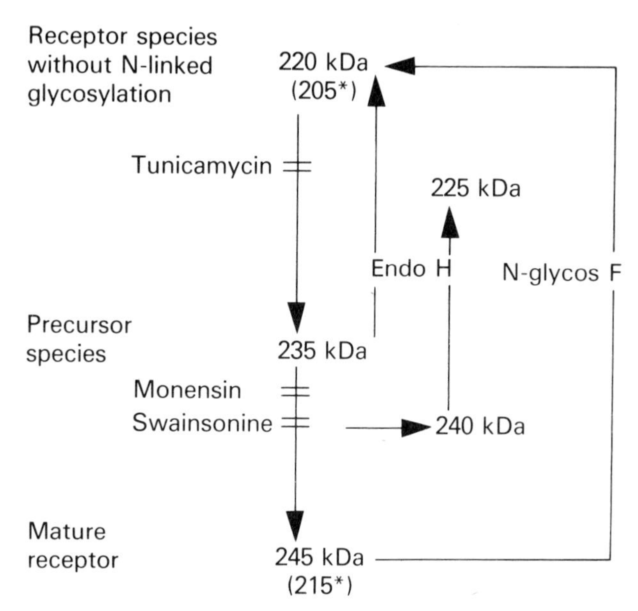

*Fig. 7. **Biosynthetic precursors of the Man-6-P/IGF-II receptor in C6 rat glial cells.** These species were identified in metabolic labeling experiments in which tunicamycin, monensin, and swainsonine were used to block specific biosynthetic steps. Endoglycosidase H and N-glycosidase F were also employed to remove N-linked carbohydrate of receptor precursors and mature receptor, respectively. The molecular mass values given in parentheses were obtained by SDS acrylamide gel electrophoresis under nonreducing conditions whereas the values not enclosed in parentheses were obtained under reducing conditions. The data are from reference 29.*

^{125}I-β-galactosidase. Since IGF-II does not bind to the 47 kDa cation dependent Man-6-P receptor[8,28] and the 47 kDa receptor has never been shown to mediate the cellular uptake of lysosomal enzymes, the above results strongly suggest that the cellular uptake of ^{125}I-β-galactosidase was via the Man-6-P/IGF-II receptor.

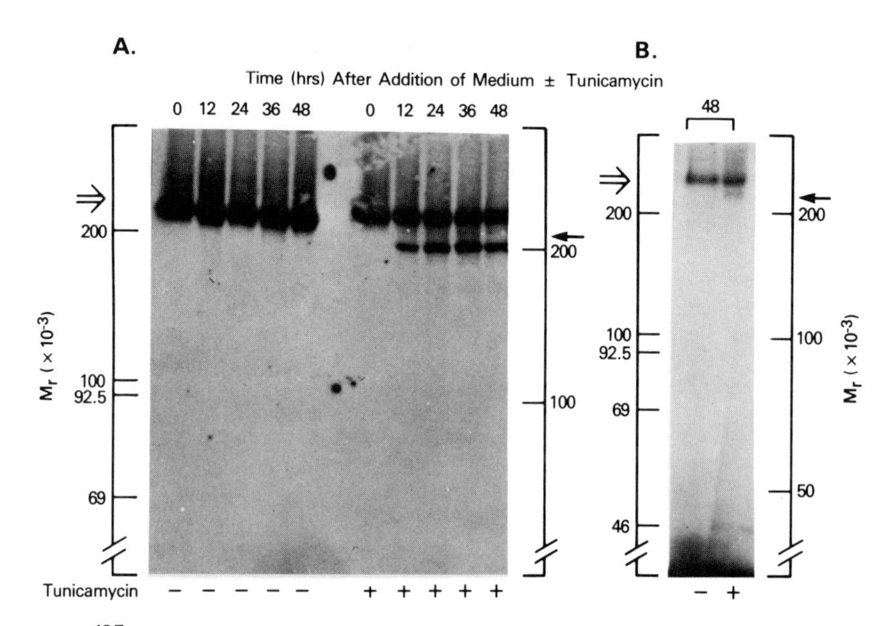

*Fig. 8. **^{125}I-IGF-II binding to the Man-6-P/IGF-II receptor precursor which accumulates in tunicamycin treated cells.** **Panel A.** C6 glial cells were cultured in the presence or absence of tunicamycin (250 ng/ml) for the indicated time periods. The cells were then lysed and immunoblotted using receptor antiserum 3637 and ^{125}I-Protein A. SDS-polyacrylamide gel electrophoresis was performed without reduction of disulfide bonds. **Panel B.** Confluent C6 glial cells were cultured for 48 h in the presence or absence of tunicamycin. The cell monolayers were than incubated for 2.5 h at 22°C with ^{125}I-IGF-II and radioligand-receptor complexes were chemically crosslinked with disuccinimidyl suberate. The cells were lysed and subjected to SDS acrylamide gel electrophoresis under reducing conditions. Reprinted with permission from reference 29.*

The most likely explanation for the cellular uptake results is that IGF-II inhibited the binding of ^{125}I-β-galactosidase to the receptor. In the experiment in Fig. 5 we showed that IGF-II inhibited the binding of ^{125}I-β-galactosidase to pure receptor. The inhibition was 60% compared to the 70-80% inhibition observed in the cellular uptake experiments, suggesting that inhibition of binding to the receptor could account for most of the inhibition of cellular uptake.

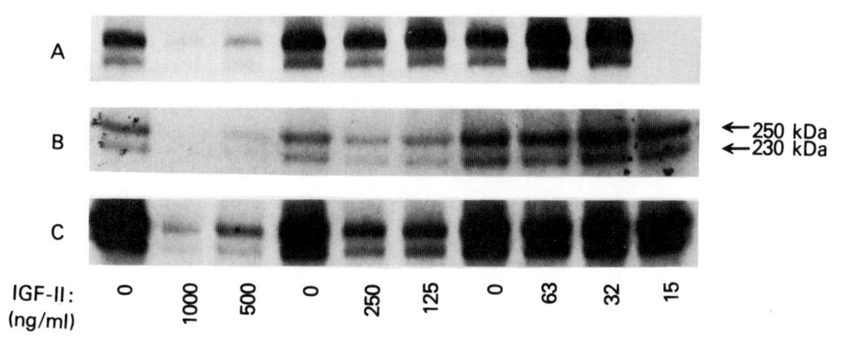

Fig. 9. ***Affinity crosslinking of*** ^{125}***I-IGF-II to tunicamycin treated C6 glial cells; competitive binding with unlabeled IGF-II.*** *In three separate experiments, monolayer cultures of C6 glial cells were first treated with tunicamycin for 48 h. The cultures were then incubated with* ^{125}*I-IGF-II (300,000 cpm, 4 ng/ml) in the presence of the indicated concentrations of unlabeled IGF-II. Affinity crosslinking was performed with disuccinimidyl suberate (0.1 mM) and the monolayers were solubilized with 2% Triton X-100. The extracts were analysed by SDS-acrylamide electrophoresis and autoradiography. Reprinted with permission from ref. 29.*

Possible Consequences of the Inhibition of Lysosomal Enzyme Binding to the Man-6-P/IGF-II Receptor by IGF-II

Inhibition of the cellular uptake of lysosomal enzymes by IGF-II would result in a higher extracellular concentration of these acid hydrolases and perhaps would result in more effective digestion of extracellular matrix in processes such as tissue remodeling or tumor spread (Fig. 6). We also speculate that intracellularly, newly synthesized IGF-II could inhibit the binding of lysosomal enzymes to intracellular receptor which would result in decreased targeting to lysosomes and an increase in the secretory pathway.

Binding of IGF-II to Biosynthetic Precursors of the Man-6-P/IGF-II Receptor

The prediction in Fig. 6 that IGF-II could modulate the targeting of lysosomal enzymes to lysosomes intracellularly may require that IGF-II bind with high affinity to precursor forms of the receptor. The various sized intermediates in the biosynthetic pathway for the receptor in C6 glial cells is shown in Fig. 7. These intermediates have been identified in biosynthetic labeling experiments carried out in the presence of inhibitors.[29] Tunicamycin inhibits N-linked glycosylation so the biosynthetic precursor which accumulates in the presence of this inhibitor is unglycosylated. In the experiment shown in panel A of Fig. 8, the accumulation of this unglycosylated precursor was documentated by Western blotting. Affinity crosslinking of ^{125}I-IGF-II was performed in the experiment shown in panel B of Fig. 8. ^{125}I-IGF-II was able to bind to the

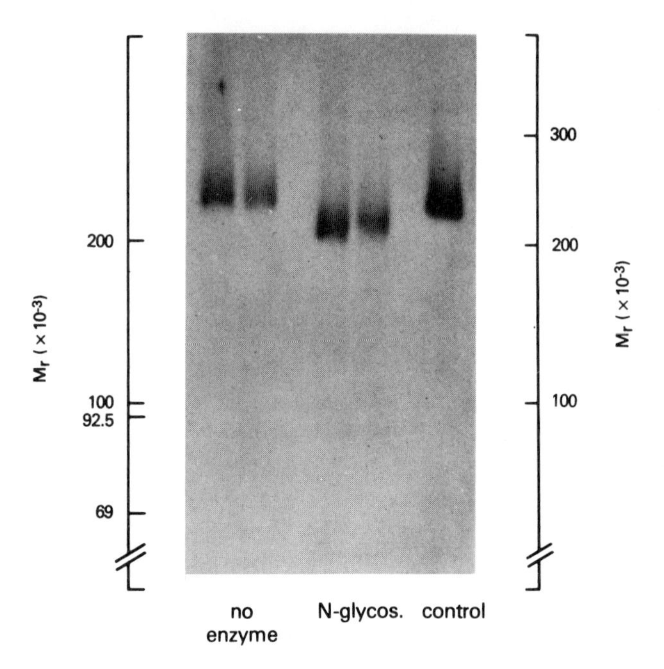

Fig. 10. N-glycosidase F digestion of the Man-6-P/IGF-II receptor. Pure Man-6-P/IGF-II receptor was digested with 80 units of N-glycosidase F at 37°C for 18 h. Analysis of digested and undigested receptor was performed by immunoblotting with receptor antiserum 3637 and ^{125}I-Protein A. The lanes labeled 'no enzyme' contain two different concentrations of receptor that were incubated without N-glycosidase F. The next two lanes contain the same concentrations of receptor that were incubated with enzyme and the last lane contains receptor that was not incubated at 37°C.

unglycosylated precursor which accumulated in the presence of tunicamycin. In order to compare the binding affinity of IGF-II for this precursor and the native receptor, affinity crosslinking was performed in the presence of increasing concentrations of unlabeled IGF-II (Fig. 9). Densitometric analysis of these data showed that unlabeled IGF-II competed equally well for binding of ^{125}I-IGF-II to the unglycosylated precursor and the native receptor, providing evidence that the binding affinities were indistinguishable. As a second approach to examine the binding affinity of IGF-II for unglycosylated receptor we purified the receptor from rat placenta and used N-glycosidase F to remove N-linked carbohydrate (Fig. 10). Scatchard analysis of IGF-II binding data for the deglycosylated receptor and the native receptor indicated that the binding affinities were indistinguishable (Fig. 11). We conclude that IGF-II can bind to precursor forms of the receptor. It had been previously demonstrated that unglycosylated receptor could bind to phosphomannan.[30]

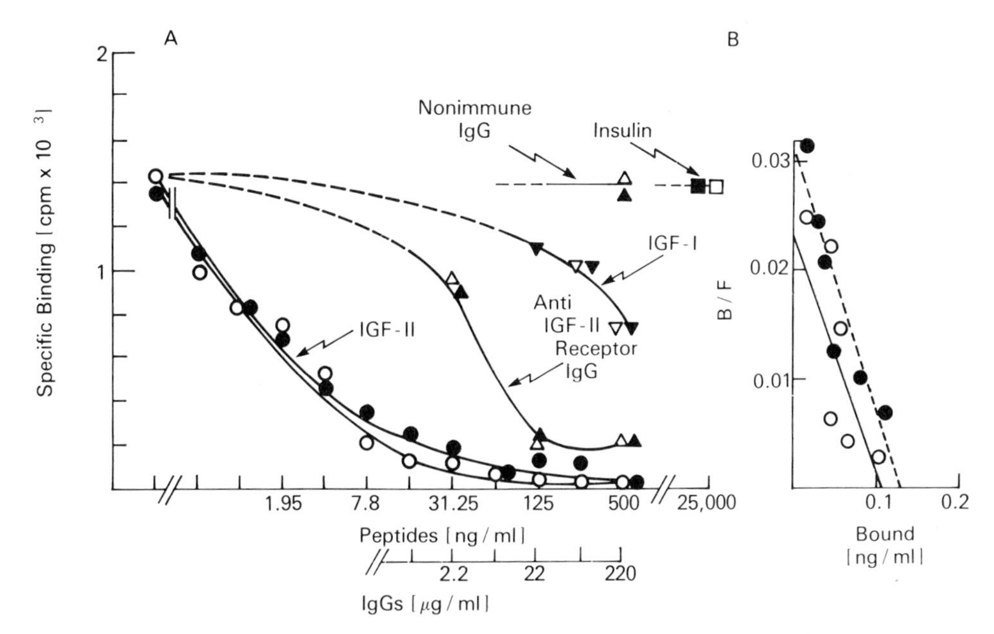

Fig. 11. ***The effect of N-glycosidase digestion of the Man-6-P/IGF-II receptor on IGF-II binding to the receptor.*** *Purified receptor was digested with N-glycosidase F or incubated without enzyme as described in Fig. 10.* ***Panel A.*** *^{125}I-IGF-II was incubated with digested (solid circle) and undigested (open circle) receptor in the presence of the indicated concentrations of IGF-II, IGF-I, insulin, anti-receptor IgG 3637 or control IgG for 2.5 h at 22°C.* ***Panel B.*** *The IGF-II binding data in Panel A was graphed according to the method of Scatchard. Bound/free (B/F) is plotted versus bound (ng/ml). Reprinted with permission from reference 29.*

CONCLUSION

We have summarized our data which support reciprocal modulation of binding of the two classes of ligands for the Man-6-P receptor: lysosomal enzymes and the growth factor, IGF-II. We emphasize that the possible functional consequences of these interactions outlined in Figures 3 and 6 remain speculative. Obviously, additional experiments need to be performed to demonstrate that these predictions are correct.

REFERENCES

1. D. O. Morgan, J. C. Edman, D. N. Standring, V. A. Fried, M. C. Smith, R. A. Roth, and W. J. Rutter. Insulin-like growth factor II receptor as a multifunctional binding protein. *Nature* 329: 301 (1987).

2. W. M. Canfield and S. Kornfeld. The chicken liver cation-independent mannose 6-phosphate receptor lacks the high affinity binding site for insulin-like growth factor II. *J. Biol. Chem.* 264: 7100 (1989).
3. K. B. Clairmont and M. P. Czech. Chicken and *Xenopus* mannose 6-phosphate receptors fail to bind insulin-like growth factor II. *J. Biol. Chem.* 264: 16390 (1989).
4. W. Kiess, C. L. Thomas, M. M. Sklar, and S. P. Nissley. Beta-galactosidase decreases the binding affinity of the insulin-like growth factor-II/mannose 6-phosphate receptor for the insulin-like growth factor-II. *Eur. J. Biochem.* 190: 71 (1990).
5. A. R. Robbins, R. Myerowitz, R. J. Youle, G. J. Murray, and D. M. Neville. The mannose 6-phosphate receptor of Chinese hamster ovary cells. Isolation of mutants with altered receptors. *J. Biol. Chem.* 256: 10618 (1981).
6. G. G. Sahagian, J. Distler, and G. W. Jourdian. Characterization of a membrane-associated receptor from bovine liver that binds phosphomannosyl residues of bovine testicular β-galactosidase. *Proc. Natl. Acad. Sci. USA.* 78: 4289 (1981).
7. P. Y. Tong, W. Gregory, and S. Kornfeld. Ligand interactions of the cation-independent mannose 6-phosphate receptor. The stoichiometry of mannose 6-phosphate binding. *J. Biol. Chem.* 264: 7962 (1989).
8. W. Kiess, G. D. Blickenstaff, M. M. Sklar, C. L. Thomas, S. P. Nissley, and G. G. Sahagian. Biochemical evidence that the type II insulin-like growth factor receptor is identical to the cation-independent mannose 6-phosphate receptor. *J. Biol. Chem.* 263: 9339 (1988).
9. M. Mathieu, H. Rochefort, B. Barenton, C. Prebois, and F. Vignon. Interactions of cathepsin-D and insulin-like growth factor-II (IGF-II) on the IGF-II/mannose-6-phosphate receptor in human breast cancer cells and possible consequences on mitogenic activity of IGF-II. *Mol. Endocrinol.* 4: 1327 (1990).
10. C. M. Nolan, J. W. Kyle, H. Watanabe, and W. S. Sly. Binding of insulin-like growth factor II (IGF-II) by human cation-independent mannose 6-phosphate receptor/IGF-II receptor expressed in receptor-deficient mouse L cells. *Cell Regulation* 1: 197 (1990).
11. T. Braulke, C. Causin, A. Waheed, U. Junghans, A. Hasilik, P. Maly, R. E. Humbel, and K. von Figura. Mannose 6-phosphate/insulin-like growth factor II receptor: distinct binding sites for mannose 6-phosphate and insulin-like growth factor II. *Biochem. Biophys. Res. Comm.* 150: 1287 (1988).
12. R. A. Roth, C. Stover, J. Hari, D. O. Morgan, M. C. Smith, V. Sara, and V. A. Fried. Interactions of the receptor for insulin-like growth factor II with mannose 6-phosphate and antibodies to the mannose 6-phosphate receptor. *Biochem. Biophys. Res. Comm.* 149: 600 (1987).
13. R. G. MacDonald, S. R. Pfeffer, L. Coussens, M. A. Tepper, C. M. Brocklebank, J. E. Mole, J. K. Anderson, E. Chen, M. P. Czech, and A. Ullrich. A single receptor binds both insulin-like growth factor II and mannose 6-phosphate. *Science* 239: 1134 (1988).

14. C. Polychronakos, H. J. Guyda, and B. I. Posner. Mannnose 6-phosphate increases the affinity of its cation-independent receptor for insulin-like growth factor II by displacing inhibitory endogenous ligands. *Biochem. Biophys. Res. Commun.* 157: 632 (1988).
15. N. M. Dahms, P. Lobel, and S. Kornfeld. Mannose 6-phosphate receptors and lysosomal enzyme targeting. *J. Biol. Chem.* 264: 12115 (1989).
16. Y. Oka, L. M. Rozek, and M. P. Czech. Direct demonstration of rapid insulin-like growth factor II receptor internalization and recycling in rat adipocytes. Insulin stimulates ^{125}I-insulin-like growth factor II degradation by modulating the IGF-II receptor recycling process. *J. Biol. Chem.* 260: 9435 (1985).
17. W. Kiess, J. F. Haskell, L. Lee, L. A. Greenstein, B. E. Miller, A. L. Aarons, M. M. Rechler, and S. P. Nissley. An antibody that blocks insulin-like growth factor (IGF) binding to the type II IGF receptor is neither an agonist nor an inhibitor of IGF-stimulated biologic responses in L6 myoblasts. *J. Biol. Chem.* 262: 12745 (1987).
18. Y. Yarden, and A. Ullrich. Growth factor receptor tyrosine kinases. *Ann. Rev. Biochem.* 57: 443 (1988).
19. R. W. Furlanetto, J. N. DiCarlo, and C. Wisehart. The type II insulin-like growth factor receptor does not mediate deoxyribonucleic acid synthesis in human fibroblasts. *J. Clin. Endocrinol. Metab.* 64: 1142 (1987).
20. C. A. Conover, P. Misra, R. L. Hintz, and R. G. Rosenfeld. Effect of an anti-insulin-like growth factor I receptor antibody on insulin-like growth factor II stimulation of DNA synthesis in human fibroblasts. *Biochem. Biophys. Res. Commun.* 139: 501 (1986).
21. S. A. Rogers and M. R. Hammerman. Insulin-like growth factor II stimulates production of inositol trisphosphate in proximal tubular basolateral membranes from canine kidney. *Proc. Natl. Acad. Sci. USA.* 85: 4037 (1988).
22. M. Tally, C. H. Li, and K. Hall. IGF-2 stimulated growth mediated by the somatomedin type 2 receptor. *Biochem. Biophys. Res. Commun.* 148: 811 (1987).
23. B. Bhaumick and R. M. Bala. Parallel effects of insulin-like growth factor-II and insulin on glucose metabolism of developing mouse embryonic limb buds in culture. *Biochem. Biophys. Res. Commun.* 152: 359 (1988).
24. J. Hari, S. B. Pierce, D. O. Morgan, V. Sara, M. C. Smith, and R. A. Roth. The receptor for insulin-like growth factor II mediates an insulin-like response. *EMBO J.* 6: 3367 (1987).
25. I. Kojima, I. Nishimoto, T. Iiri, E. Ogata, and R. Rosenfeld. Evidence that type II insulin-like growth factor receptor is coupled to calcium gating system. *Biochem. Biophys. Res. Commun.* 154: 9 (1988).
26. I. Nishimoto, Y. Murayama, Y. Katada, M. Ui, and E. Ogata. Possible direct linkage of insulin-like growth factor-II receptor with guanine nucleotide-binding proteins. *J. Biol. Chem.* 264: 14029 (1989).
27. W. Kiess, C. L. Thomas, L. A. Greenstein, L. Lee, M. M. Sklar, M. M. Rechler, G. G. Sahagian, and S. P. Nissley. Insulin-like growth factor-II (IGF-II) inhibits both the cellular uptake of β-galactosidase and the

binding of β-galactosidase to purified IGF-II/mannose 6-phosphate receptor. *J. Biol. Chem.* 264: 4710 (1989).

28. P. Y. Tong, S. E. Tollefsen, and S. Kornfeld. The cation-independent mannose 6-phosphate receptor binds insulin-like growth factor II. *J. Biol. Chem.* 263: 2585 (1988).
29. W. Kiess, L. A. Greenstein, L. Lee, C. Thomas, and S. P. Nissley. Biosynthesis of the insulin-like growth factor-II (IGF-II)/mannose 6-phosphate receptor in rat C6 glial cells. The role of N-linked glycosylation in binding of IGF-II to the receptor. *Mol. Endocrinol.*, in press (1991).
30. G. G. Sahagian and E. F. Neufeld. Biosynthesis and turnover of the mannose 6-phosphate receptor in cultured Chinese hamster ovary cells. *J. Biol. Chem.* 258: 7121 (1983).

EXPRESSION OF IGF-II, THE IGF-II/MANNOSE-6-PHOSPHATE RECEPTOR AND IGFBP-2 DURING RAT EMBRYOGENESIS

John E. Pintar, Teresa L. Wood, Randal D. Streck, Leif Havton, Leslie Rogler, and Ming-Sing Hsu

Department of Anatomy and Cell Biology
Centers for Reproductive Science and Neurobiology and Behavior
Columbia University College of Physicians and Surgeons
New York, N.Y. 10032

INTRODUCTION

It is becoming increasingly apparent that the IGF system (including both IGF-I and IGF-II, at least two IGF receptors, and a family of at least six IGF binding proteins; 1-3) has a fundamental role in the normal progression of prenatal development. The initial observation that IGF-II peptide levels were high in the fetus but decreased post-natally (4) was the first to suggest that this peptide might have functional importance during ontogeny. *In situ* hybridization studies (5-7) extended Northern analysis and confirmed that the developmental regulation of IGF-II expression extends to the level of RNA regulation and demonstrated that the IGF-II gene expression pattern during ontogeny is precisely regulated both temporally and spatially. These results suggested that if the autocrine-paracrine mode of action classically proposed for the IGFs (1,2) was in fact correct, then interference with normal IGF-II synthesis would be expected to produce regional rather than systemic deficits.

Two distinct IGF receptors have been identified (8,9). At least one of these receptors, the IGF-II/mannose-6-phosphate receptor (IGF-II/M-6-P), is also developmentally regulated (10,11) and is present at high levels in many fetal tissues that express the IGF-II gene (5,7). Although these observations also were consistent with a local action of IGF-II, they were limited by the fact that the precise subcellular distribution of the receptor was unknown. In tissue culture cells, where receptor trafficking has been most completely studied (12), it is known that the IGF-II/M-6-P receptor is distributed among different intracellular compartments (Golgi, cell surface, and lysosomes). It is still unclear whether this receptor, which has distinct binding sites for IGF-II and mannose-6-phosphate ligands (9,13), can mediate responses to extracellular ligands. Recently, however, it has been shown to be capable of binding G proteins following activation (14). The discovery of this putative signal transduction mechanism raises the possibility that this receptor acually does function in signal transduction *in vivo*. Moreover, recent studies have determined that an existing sex-linked mutation (*Tme*), that is lethal when maternally inherited, consists of a deletion in a chromosomal region that includes the IGF-II/M-6-P gene (15). Since these observations suggest a potential role for this receptor in development, they emphasize the necessity to determine the state of IGF-II/M-6-P receptor cycling in the fetus.

Finally, six insulin-like growth factor binding proteins that likely mediate IGF actions have thus far been cloned and sequenced (3,16-25). It has been established during the past two years that the gene for at least one of these binding proteins, the insulin-like growth factor binding protein 2 (IGFBP-2) gene, also is expressed in high abundance prenatally and decreases significantly after birth (17,26,27). Initial studies of fetal IGFBP-2 expression using *in situ* hybridization demonstrated a pattern of expression distinct from IGF-II during mid-gestational development (28). High levels of IGFBP-2 expression were observed in specific regions of the nervous system (such as the floor plate) that appear to organize surrounding tissues during development and in certain ectoderm derivatives that are characterized by complex morphogenetic movements (28).

To provide a complete understanding of the role of the IGF system in embryogenesis, it is important to understand both the relationship between the synthetic sites of IGFs and mediators of the IGF response, as well as the consequences of *in vivo* interference with individual components of the system. Here we reevaluate the expression pattern of IGF-II in light of recent gene targeting experiments in which *in vivo* IGF-II levels have been altered, report that the IGF-II/M-6-P receptor is present on the surfaces of fetal cells that express it, and show that IGFBP-2 mRNA is expressed in numerous cell populations of early post-implantation embryos that participate in a variety of growth and morphogenetic processes.

RESULTS AND DISCUSSION

IGF-II Expression During Development

IGF-II mRNA is present in most, if not all, extraembryonic tissue types during early post-implantation stages of rodent development (7). These tissues include all trophectoderm derivatives that are precursors to the chorioallantoic placenta. In contrast, progenitors of the embryo proper that reside in the embryonic ectoderm or epiblast do not express levels of IGF-II gene activity detectable by *in situ* hybridization until mesoderm formation begins during gastrulation (7). Derivatives of embryonic mesoderm express the highest levels of IGF-II during subsequent stages of mid-and late gestation development (5), although selected ectodermal and endodermal tissue types also express IGF-II. Based on the transient expression of IGF-II in the early stage of pituitary development (5,28,29), it was suggested that IGF-II expression might be shared by all fetal placodes (5). It has now become clear that the IGF-II gene in fact is expressed not only in a variety of ectodermal placodes (including the newly-forming nasal placode (Figure 1B)), but also at lower levels in restricted regions of the surface ectoderm , including the ectoderm surrounding the developing mandibular arch (Figure 1B).

Disruption of IGF-II expression using a gene targeting strategy has been achieved by DeChiara, Efstratiadis and Robertson and has clearly established that IGF-II is essential for normal prenatal growth (30). Mouse embryonic stem (ES) cells were produced in which a mutated IGF-II allele replaced one of the normal IGF-II alleles. These cells were used to produce chimeric mice that were able to transmit the mutated IGF-II allele through the germ line. Heterozygote offspring that inherited the mutated IGF-II allele from the paternal genome exhibited a growth deficiency (to about 60% normal size) that is apparent in both the embryo proper and in the placenta as early as examined in detail (embryonic day 16; 30). Further, the effects of IGF-II gene disruption on development depends on the parental origin of the mutated allele; heterozygous mice that have inherited the altered IGF-II allele from the maternal, rather than paternal, genome are normal in size (31). Finally, mice homozygous for the IGF-II deficiency are essentially indistinguishable from growth deficient heterozygous mice (31).

Therefore, not only is the IGF-II locus imprinted (i.e. allelic expression in the fetus is modified by passage through the germ line; 31) but expression of the paternal allele is also required for normal growth. In contrast to predictions based on changes in IGF-II expression during development of mesoderm derivatives, IGF-II does not appear to have an essential role in morphogenesis or cell differentiation. It is possible that the loss of IGF-II may be compensated for by increases in the activities of genes that have not yet been identified or that are normally expressed and provide natural redundancy. However, no apparent changes in IGF-I, type 1 IGF receptor, or IGFBP-2 expression have been noted in preliminary examinations of mid-gestational embryos by solution or *in situ* hybridization (DeChiara, Wood, Streck, Efstratiadis, and Pintar, unpublished observations).

Hypotheses about the mechanisms that produce the distinct maternal and paternal phenotypes can be formulated and await testing. On one hand, the very similar decreases in the

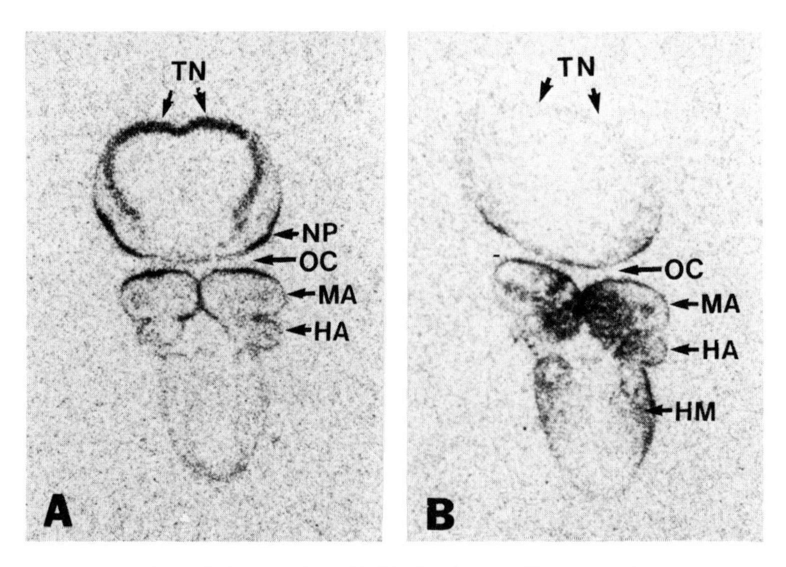

Figure 1. Expression of IGFBP-2 and IGF-II. Autoradiograms show transverse sections from an embryonic day 11 embryo hybridized with ^{35}S-labeled RNA probes to either rat IGFBP-2 (Panel A) or rat IGF-II (Panel B). The RNA probes and *in situ* hybridization conditions were described in detail previously (28). TN, telencephalon; NP, nasal placode; OC, oral cavity; MA, mandibular arch; HA, hyoid arch; HM, head mesenchyme.

sizes of the growth-deficient placentas and embryos have led to the suggestion that deficits in IGF-II action have similar and independent effects in both cell populations (31). In this case, IGF-II most likely acts as a modulator of growth at a systemic rather than local level; this in turn would suggest that the precise spatial pattern of IGF-II expression in the embyro proper (5,7) has no obvious function. Alternatively, it is possible that the general deficit in embryonic growth associated with inheritance of an altered paternal IGF-II allele results primarily from retarded placental growth. In this model, impaired placental growth is a direct consequence of the IGF-II deficit, with secondary effects on growth of the fetus proper. This hypothesis is consistent with the observations that: 1) IGF-II expression is detected in progenitors of the extraembryonic cell lineages prior to its expression in the fetus and throughout development in the placenta; and that 2) the paternal genome appears to contribute more information required for normal differentiation of extraembryonic structures than the maternal genome (32).

Fetal Expression of the IGF-II/M-6-P Receptor

In order to determine the pattern of IGF-II receptor appearance *in vivo*, selected embryonic and fetal sections were immunostained with antiserum to the IGF-II/M-6-P receptor. Receptor immunoreactivity was present in high abundance in many of the cell types that express IGF-II including the heart (7; Pintar, in preparation). In order to determine whether this receptor can be detected at the surface of fetal cell populations, two approaches were used. In the first, fetal heart sections were prepared for EM immunocytochemistry using a post-embedding staining protocol. Immunostaining of ultrathin sections of the fetal heart demonstrated that immunoreactivity was present not only intracellularly in a perinuclear pattern near the Golgi apparatus, but also on the cell surface (Figure 2). A second set of experiments showed that

FIGURE 2. Localization of the IGF-II receptor on the surface of fetal rat heart cells. Electron micrograph shows immunogold labeling of the IGF-II/M-6-P receptor in the fetal heart. Arrows correspond to gold particles associated with the cell surface. Whole rat embryonic hearts (e12) were immersion-fixed for 4 hours in a fixative containing 0.01M sodium periodate, 0.075M L-lysine monohydrochloride and 4% paraformaldehyde in 0.1M sodium phosphate buffer, pH 7.4, washed in 0.1M sodium phosphate buffer, pH 7.4, dehydrated in methanol and embedded in LRWhite. Ultrathin sections were incubated overnight at 4°C with the IgG fraction of rabbit anti-IGF-II receptor immune serum (46) diluted 1:100 with a solution containing 4% horse serum in 0.1M TBS, pH 7.6, washed, and incubated for 2-3 hours at room temperature with goat anti-rabbit antibodies labelled with 15nm gold particles. The tissue was then fixed for 10 minutes at room temperature in 2.5% glutaraldehyde in 0.1M sodium phosphate buffer, washed and stained with OsO_4, uranyl acetate and lead citrate. In control experiments, the tissue was incubated overnight at 4°C with non-immune rabbit serum or with the IgG fraction of the rabbit anti-IGF-II receptor immune serum preabsorbed with a 250 kilodalton IGF-II receptor antigen. No labeling was detected following incubation with these control sera.

iodinated mannose-6-phosphate ligands can associate with the surfaces of fetal heart cells and branchial arch cells during short-term culture *in vitro* and that this association can be blocked with 5 mM M-6-P (Pintar, in preparation). Therefore, both biochemical and ultrastructural approaches have demonstrated that the IGF-II/M-6-P receptor cycles to the surface of fetal cells *in vivo*, apparently via pathways similar to those elucidated in lines of tissue culture cells (12). These results suggest that prospective ligands for this receptor (including precursor forms of lysosomal enzymes, IGF-II, proliferin, and the precursor forms of TGF-beta 1) may associate with it not only during biosynthesis, but also following their release. It has been suggested that the IGF-II receptor may be the key gene in the sex-linked *Tme* mutant, since it is the only gene thus far known to be imprinted in the chromosomal region deleted in this mutant (15). If this turns out to be correct, it would raise the question of which of these IGF-II/M-6-P receptor ligands normally provide functions that, in the absence of the receptor, result in lethality.

Fetal Expression of IGFBP-2

In addition to the IGF receptors, IGF responses also can be modulated by IGF binding proteins (3,16,33). The most detailed studies in cultured cells have shown that one RGD-containing IGFBP (IGFBP-1) potentiates the mitogenic response to IGF-I (33). A distinct IGFBP, IGFBP-2, is developmentally regulated and significantly decreases as postnatal development proceeds (17,26,27). In the fetus, IGFBP-2 exhibits a pattern of expression distinct from that of IGF-II (28). Specifically, the IGFBP-2 gene is not expressed at detectable levels in most mesodermal derivatives that express IGF-II (e.g. heart, cartilage) during mid-gestational development. Instead, its presence is restricted to specific regions of surface

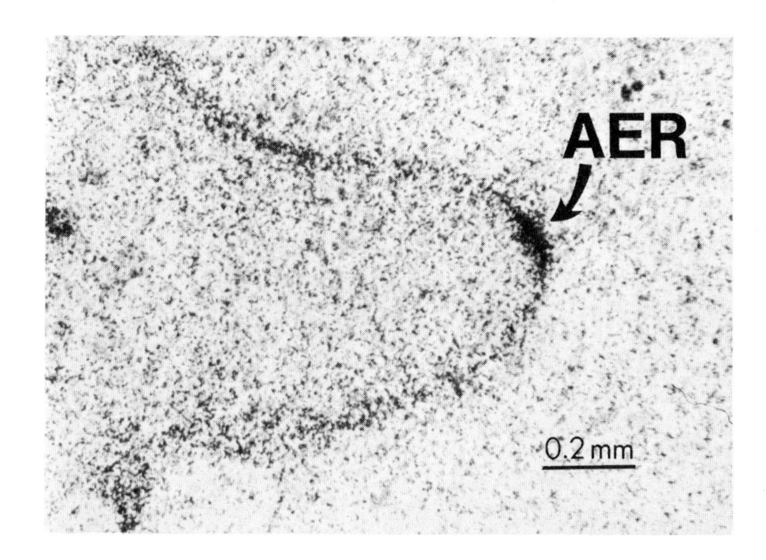

FIGURE 3. IGFBP-2 expression in the apical ectodermal ridge of a developing limb bud. Autoradiogram showing a transverse section through a hindlimb bud of an embryonic day 11.5 rat embryo hybridized with an ^{35}S-labeled RNA probe to rat IGFBP-2. The position of the strongly labeled apical ectodermal ridge (AER) is indicated. Note that there is little labeling of the ectoderm covering the dorsal and ventral surfaces of the limb bud, and no detectable labeling internal to the limb bud.

ectoderm, the central nervous system and the endoderm lining of the gut (28). We have examined earlier ages in the rat and have demonstrated that IGFBP-2 mRNA is present in early postimplantation ages in cells of the embryonic epiblast and that, following gastrulation, IGFBP-2 expression is retained in many ectodermal derivatives of this population (34). In general, the highest levels of IGFBP-2 expression during development characterize regions of surface ectoderm that either undergo complex morphogenetic processes or that direct such processes.

The results shown in Figure 1 document the differences in expression patterns between IGFBP-2 and IGF-II. High levels of IGFBP-2 expression are apparent in the ectoderm of the mandibular and hyoid arches (Figure 1A). During this stage (embryonic day 11), these regions of the presumptive face are being invaded by mesenchymal cells derived from both the neural crest and mesoderm. In addition, high levels of IGFBP-2 hybridization are observed in the ectoderm that will form the nasal placode, and IGFBP-2 expression in fact is characteristic of all other placodes (28,34). This expression pattern is consistent with the possibility that IGFBP-2 may have a role in the extensive folding of these epithelial sheets that accompanies their development. As mentioned above, lower levels of IGF-II expression also are detectable in the surface ectoderm surrounding the arches and lining the oral cavity along the ventral surface of the head, but IGF-II mRNA is most prominent in the mesoderm of the mandibular and hyoid arches, particularly in the medial regions (Figure 1B). Also apparent in Figure 1 is that IGFBP-2, but not IGF-II, is expressed in the neuroepithelium of the telencephalon. This CNS expression is more restricted spatially than at earlier ages (34) when IGFBP-2 expression extends throughout the rostral-caudal extent of the neural tube.

The pattern of IGFBP-2 expression in the surface ectoderm of the developing limb suggests a putative role for this protein in directing a different morphogenetic process, the patterned outgrowth of cells during development. A high level of IGFBP-2 expression characterizes the apical ectodermal ridge (AER) which covers the distal edge of the developing limb bud (Figure

3; 35). This group of ectodermal cells assumes a characteristic morphology soon after limb bud outgrowth begins and is essential for continued outgrowth of the limb during subsequent development (36,37). The mesodermal cells directly underlying the AER divide more rapidly than more proximal mesodermal limb cells (38,39), and *in vitro* experiments have confirmed that the AER has a mitogenic effect on limb mesoderm (40). IGFBP-2 mRNA is expressed in the surface ectoderm covering the distal limb bud even before the AER becomes multilayered, and a high level of expression persists in the AER throughout its existence as a distinct structure (35). IGFBP-2 mRNA disappears from the distal limb ectoderm as the AER regresses (35). Thus, the pattern of IGFBP-2 expression in limb surface ectoderm is consistent with a role in influencing the directional outgrowth of limb mesenchyme through a localized stimulation of cell division within the adjacent limb mesoderm.

The preceding results suggest two models for the function of IGFBP-2 that are not mutually exclusive. The first is based on the finding that, in several cases, IGFBP-2 is expressed either within (embryonic epiblast, anterior nervous system) or adjacent to (limb buds) cell populations that proliferate more rapidly than their neighbors. Since these sites of expression are distinct from the synthetic sites of both IGF-I and IGF-II, it is conceivable that IGFBP-2 synthesized by these cell types is deposited near its sites of synthesis where it could potentiate the effects of the IGFs synthesized by distinct cell types. This *in vivo* model is consistent with *in vitro* results with the related binding protein IGFBP-1 which, when added concurrently with exogenous IGF-I, can increase the mitogenic response to IGF-I (33,41). Thus, IGFBP-2 may be an important modulator (and in this model a stimulator) of the mitogenic actions of the IGFs in the developing embryo.

A second model for the function of IGFBP-2, also consistent with the fetal expression pattern, is that IGFBP-2 functions during morphogenesis of epithelial structures as a constituent of the extracellular matrix. In this model, IGFBP-2, synthesized at high levels by epithelia during periods of tissue remodelling, directs or participates in this remodelling by acting independently of IGF action. Consistent with this proposal is the presence of an RGD sequence within the carboxyl region of the IGFBP-2 protein (17,18) that could direct its association with the extracellular matrix.

At least two distinct mutant mouse strains are characterized by defects in the development of many of the tissues that normally express high levels of IGFBP-2 during embryogenesis. In the first of these, the insertional mutant *legless*, the mutant phenotype is characterized by abnormalities in the limb, anterior brain, and gut (42,43). In the second, the mutant *dominant hemomelia*, deficits are observed both in the limb and in the differentiation of endodermal organs, especially those associated with the anterior splanchnic mesodermal plate (44), a restricted region of mesoderm that expresses IGFBP-2, but not IGF-II (34).

Although linkage analysis and immunocytochemical localization of the IGFBP-2 protein product may give some evidence of its function, a direct disruption of IGFBP-2 expression *in vivo* will be necessary to determine whether any morphogenetic defects result from its absence. To this end, the production of mice deficient in IGFBP-2 using the homologous recombination/embryonic stem cell approach is in progress. As an early step in such an analysis, we have determined that the IGFBP-2 gene is expressed in embryonic stem cells (CCE cells), but not in the fibroblast feeder layer (STO cells) used to maintain these cells in an undifferentiated state (Figure 4). Since this gene is expressed in ES cells, the selection of ES cells in which most of the IGFBP-2 locus is replaced by an IGFBP-2/neo cassette should be readily accomplished. It was shown recently that both alleles of genes expressed in ES cells can be disrupted by gene targetting *in vitro* (45). If IGFBP-2 expression has a necessary function in this population (or in the embryonic epiblast *in vivo* at a slightly later developmental stage), ablation of IGFBP-2 expression may affect survival, growth, and/or the differentiation of ES cells *in vitro*. Finally, the production of chimeric animals with ES cells containing a disrupted IGFBP-2 allele(s) should allow the *in vivo* consequences of IGFBP-2 loss to be determined directly.

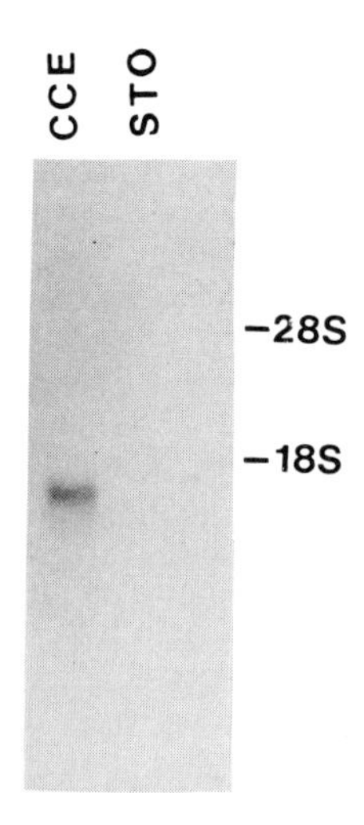

FIGURE 4. IGFBP-2 expression in embryonic stem cells. RNA blot analysis of total RNA isolated from CCE embryonic stem cells (initially obtained from E. Robertson) (10ug) and from STO fibroblast feeder cells (20ug) hybridized with a ^{32}P-labeled random-primed probe to rat IGFBP-2. Total RNA was isolated by lithium chloride precipitation (47), fractionated by electrophoresis on a 1.2% agarose gel containing formaldehyde and transferred onto Genescreen (Dupont) nylon membrane. The bound RNA was hybridized at 42°C for 12 hours to a ^{32}P-labeled IGFBP-2 probe labeled by random priming. Following hybridization, the blot was washed to 0.1xSSC, 1%SDS at 50°C and exposed to Kodak X-Omat XAR film at -70°C with an intensifying screen.

REFERENCES

1. M.M. Rechler and S.P. Nissley, Insulin-like growth factors, in: "Peptide Growth Factors and Their Receptors, Handbook of Experimental Pharmacology, Vol. 95/1," M.B. Sporn and A.B. Roberts, eds., Berlin: Springer–Verlag (1990).
2. W.H. Daughaday and P. Rotwein, Insulin-like growth factors I and II. Peptide, messenger ribonucleic acid and gene structures, serum and tissue concentrations, Endocr. Rev. 10:68 (1989).
3. D.R. Clemmons, Insulinlike growth factor binding proteins, Trends Endocrinol. Metab. 1:412 (1990).
4. A.C. Moses, S.P. Nissley, P.A. Short, M.M. Rechler, R.M. White, A.B. Knight and O.Z. Higa, Increased levels of multiplication-stimulating activity, an insulin-like growth factor, in fetal rat serum, Proc. Natl. Acad. Sci. USA 77:3649 (1980).
5. F. Stylianopoulou, A. Efstratiadis, J. Herbert and J. Pintar, Pattern of the insulin-like growth factor II gene expression during rat embryogenesis, Development 103:497 (1988).
6. F. Stylianopoulou, J. Herbert, M.B. Soares and A. Efstratiadis, Expression of the insulin-like growth factor-II gene in the choroid plexus and the leptomeninges of the adult rat central nervous system, Proc. Natl. Acad. Sci. USA 85:141 (1988).
7. J.E. Lee, J.E. Pintar and A. Efstratiadis, Pattern of insulin-like growth factor II gene expression during early mouse embryogenesis, Development 110:151 (1990).

8. A. Ullrich, A. Gray, A.W. Tam, T. Yeng-Feng, M. Tsubokawa, C. Collins, W. Henzel, T. LeBon, S. Kathuria, E. Chen, S. Jacobs, U. Franke, J. Ramachandran and Y. Fulita-Yamaguchi, Insulin-like growth factor I receptor primary structure: comparison with insulin receptor suggests structural determinants that define functional specificity, EMBO J. 5:2503 (1986).
9. D.O. Morgan, J.C. Edman, D.N. Standring, V.A. Fried, M.C. Smith, R.A. Roth and W.J. Rutter, Insulin-like growth factor II receptor as a multifunctional binding protein, Nature 329:301 (1987).
10. M.M. Sklar, W. Kiess, C.L. Thomas and S.P. Nissley, Developmental expression of the tissue insulin-like growth factor II/mannose 6-phosphate receptor in the rat. Measurement by quantitative immunoblotting, J. Biol. Chem. 264:16733 (1989).
11. P.V. Senior, S. Byrne, W.J. Brammar and F. Beck, Expression of the IGF-II/mannose-6-phosphate receptor mRNA and protein in the developing rat, Development 109:67 (1990).
12. J.R. Duncan and S. Kornfeld, Intracellular movement of two mannose 6-phosphate receptors: return to the Golgi apparatus, J. Cell Biol. 106:617 (1988).
13. R.G. MacDonald, S.R. Pfeffer, L. Coussens, M.A. Tepper, C.M. Brocklebank, J.E. Mole, J.K. Anderson, E. Chen, M.P. Czech and A. Ullrich, A single receptor binds both insulin-like growth factor II and mannose-6-phosphate, Science 239:1134 (1988).
14. T. Okamoto, T. Katada, Y. Murayama, M. Ui, E. Ogata and I. Nishimoto, A simple structure encodes G protein-activating function of the IGF-II/mannose 6-phosphate receptor, Cell 62:709 (1990).
15. D.P. Barlow, R. Stoger, B.G. Herrmann, K. Saito and N. Schweifer, The mouse insulin-like growth factor type-2 receptor is imprinted and closely linked to the *Tme* locus, Nature 349:84 (1991).
16. R.C. Baxter and J.L. Martin, Binding proteins for the insulin-like growth factors: structure, regulation, and function, Prog. Growth Factor Res. 1:49 (1989).
17. A.L. Brown, L. Chiariotti, C. Orlowski, T. Mehlman, W.H. Burgess, E.J. Ackerman, C.B. Bruni and M.M. Rechler, Nucleotide sequence and expression of a cDNA clone encoding a fetal rat binding protein for insulin-like growth factors, J. Biol. Chem. 264:5148 (1989).
18. J.B. Margot, C. Binkert, J–L. Mary, J. Landwehr, G. Heinrich and J. Schwander, A low molecular weight insulin-like growth factor binding protein from rat: cDNA cloning and tissue distribution of its messenger RNA, Mol. Endocrinol. 3:1053 (1989).
19. M. Julkunen, R. Koistinen, K. Aalto–Setala, M. Seppala, O.A. Janne and K. Kontula, Primary structure of human insulin-like growth factor binding protein placental 12 and tissue specific expression of its mRNA, FEBS Lett. 236:295 (1988).
20. M.T. Brewer, G.L. Stetler, C.H. Squires, R.C. Thompson, W.H. Busby and D.R. Clemmons, Cloning, characterization, and expression of a human insulin-like growth factor–binding protein, Biochem. Biophys. Res. Commun. 152:1289 (1988).
21. W.I. Wood, G. Cachianes, W.J. Henzel, G.A. Winslow, S.A. Spencer, R. Hellmiss, J.L. Martin and R.C. Baxter, Cloning and expression of the growth hormone-dependent insulin-like growth factor–binding protein, Mol. Endocrinol. 2:1176 (1988).
22. A. Brinkman, C. Groffen, D.J. Kortleve, A. Geurts van Kessel and S.L.S. Drop, Isolation and characterization of a cDNA encoding the low molecular weight insulin-like growth factor binding protein (IBP-1), EMBO J. 7:2417 (1988).
23. L.J. Murphy, C. Seneviratne, G. Ballejo, F. Croze and T.G. Kennedy, Identification and characterization of a rat decidual insulin-like growth factor-binding protein complementary DNA, Mol. Endocrinol. 4:329 (1990).
24. S. Shimasaki, F. Uchiyama, M. Shimonaka and N. Ling, Molecular cloning of the cDNAs encoding a novel insulin-like growth factor-binding protein from rat and human, Mol. Endocrinol. 4:1451 (1990).
25. S. Shimasaki, M. Shimonaka, T. Bicsak and N. Ling, Isolation and molecular characterization of three novel IGFBPs, 2nd International Symposium on Insulin-like Growth Factors/Somatomedins, Abstr. #C54 (1991).
26. L.Y-H. Tseng, A.L. Brown, Y.W-H. Yang, J.A. Romanus, C.C. Orlowski and M.M. Rechler, The fetal rat binding protein for insulin-like growth factors is expressed in the choroid plexus and cerebrospinal fluid of adult rats, Mol. Endocrinol. 3:1559 (1989).
27. C.C. Orlowski, A.L. Brown, G.T. Ooi, Y.W-H. Yang, L.Y-H Tseng and M.M. Rechler, Tissue, developmental and metabolic regulation of mRNA encoding a rat insulin-like growth factor binding protein (rIGFBP–2), Endocrinology 126:644 (1990).

28. T.L. Wood, A.L. Brown, M.M. Rechler and J.E. Pintar, The expression pattern of an insulin-like growth factor (IGF)-binding protein gene is distinct from IGF-II in the midgestational rat embryo, Mol. Endocrinol. 4:1257 (1990).
29. C.A. Bondy, H. Werner, C.T. Roberts, Jr. and D. LeRoith, Cellular pattern of insulin-like growth factor-I (IGF-I) and type I IGF receptor gene expression in early organogenesis: comparison with IGF-II gene expression, Mol. Endocrinol. 4:1386 (1990).
30. T.M. DeChiara, A. Efstratiadis and E.J. Robertson, A growth–deficiency phenotype in heterozygous mice carrying an insulin-like growth factor II gene disrupted by targeting, Nature 345:78 (1990).
31. T.M. DeChiara, E.J. Robertson and A. Efstratiadis, Parental imprinting of the mouse insulin-like growth factor II gene, Cell 64, in press (1991).
32. M.A.H. Surani, W. Reik, M.L. Norris and S.C. Barton, Influence of germline modifications of homologous chromosomes on mouse development. J. Embryol. exp. Morph. 97 Supplement:123 (1986).
33. G.R. Elgin, W.H. Busby and D.R. Clemmons, An insulin-like growth factor (IGF) binding protein enhances the biologic response to IGF-I, Proc. Natl. Acad. Sci. USA. 84:3254 (1987).
34. T.L. Wood and J.E. Pintar, Expression pattern of the IGFBP–2 gene in postimplantation rat embryos, submitted (1991).
35. R. D. Streck, T. L. Wood, M.-S. Hsu and J. E. Pintar, The transcript for insulin-like growth factor binding protein-2 is extremely abundant in the apical ectodermal ridge of rat embryonic limbs, submitted (1991).
36. J. W. Saunders, Jr., The proximo-distal sequence of origin of the parts of the chick wing and the role of the ectoderm, J. Exp. Zool. 108:363 (1948).
37. E. Zwilling, The role of epithelial components in the developmental origin of the "Wingless" syndrome of chick embryos, J. Exp. Zool. 111:175 (1949).
38. A. Hornbruch and L. Wolpert, Cell division in the early growth and morphogenesis of the chick limb, Nature 226:764 (1970).
39. D. Summerbell and L. Wolpert, Cell density and cell division in the early morphogenesis of the chick wing, Nature New Biol. 239:24 (1972).
40. R. S. Reiter and M. Solursh, Mitogenic property of the apical ectodermal ridge, Dev. Biol. 93:28 (1982).
41. D.R. Clemmons and L.I. Gardner, A factor contained in plasma is required for IGF binding protein-1 to potentiate the effect of IGF-I on smooth muscle cell DNA synthesis, J. Cell. Physiol. 145:129 (1990).
42. J. D. McNeish, W. J. Scott and S.S. Potter, *Legless*, a novel mutation found in PHT1-1 transgenic mice, Science 241:837 (1988).
43. J. D. McNeish, J. Thayer, K. Walling, K. K. Sulik, S. S. Potter and W. J. Scott, Phenotypic characterization of the transgenic mouse insertional mutation, *Legless*, J. Exp. Zool. 253:151 (1990).
44. M. C. Green, A defect of the splanchnic mesoderm caused by the mutant gene *dominant hemimelia* in the mouse, Dev. Biol. 15:62 (1967).
45. H. Riele, E.R. Maandag, A. Clarke, M. Hooper and A. Berns, Consecutive inactivation of both alleles of the *pim-1* proto-oncogene by homologous recombination in embryonic stem cells, Nature 348:649 (1990).
46. D.E. Goldberg, C.A. Gabel and S. Kornfeld, Studies of the biosynthesis of the mannose 6-phosphate receptor in receptor-positive and receptor-deficient cell-lines, J. Cell Biol. 97:1700 (1983).
47. G. Cathala, J.F. Savouret, B. Mendz, B.L. West, M. Karin, J.A. Martial and J.D. Baxter, A method for isolation of intact, translationally active ribonucleic-acid, DNA 2:329 (1983).

EARLY EVENTS IN THE HORMONAL REGULATION OF GLIAL GENE EXPRESSION: EARLY RESPONSE GENES

Alaric Arenander, Janet Cheng, and Jean de Vellis

Department of Anatomy and Cell Biology, Mental Retardation Research Center, NPI, UCLA School of Medicine, Los Angeles, CA

INTRODUCTION

Glial cells play important roles in the development and the functioning of the brain (Arenander and de Vellis, 1983, Arenander and de Vellis, 1989). Glial cell development and functions are regulated by growth factors, hormones, neurotransmitters and other autocrine and paracrine factors and cell-cell contact. How are these factors mediating their effects? The diversity of cell types and complexity of cellular interaction necessitated developing *in vitro* methods to study mechanisms of glial growth and function. Cell culture techniques allow experimental control of developmental events and molecular markers allow precise identification of cell types. As a result, glial research has greatly advanced our understanding of the molecular and cellular mechanisms regulating neural development.

Cultured glia were orIginally maintained in the presence of fetal calf serum to provide the necessary "factors" for survival and growth. In developing chemically defined media to replace serum, a number of growth factors, hormones and metabolic substrates were found to be essential for glial cells *in vitro*. Glial cells under serum-free conditions require insulin (Morrison and de Vellis, 1981; Fischer , 1984; Michler-Stuke et al., 1984; Weibel et al., 1984). The concentration of insulin (INS) needed *in vitro* is higher than physiological levels suggesting INS may be working through an insulin-like growth factor-I (IGF-I) receptor. The ability of INS and/or IGF-I to influence glial growth and development suggests these growth factors are capable of influencing gene expression necessary for survival and phenotypic response.

An important question is how brief exposures of a cell to a growth factor like insulin can lead to long term structural and functional changes. What mechanisms are available for a cell to link the transitory receptor-mediated alterations in cell physiology to long term phenotypic adaptations? We discuss some of the early events in the hormonal regulation of glial gene expression and present data suggesting INS and IGF-I can alter glial physiology by inducing early response genes as part of an

Molecular Biology and Physiology of Insulin and Insulin-Like Growth Factors
Edited by M.K. Raizada and D. LeRoith, Plenum Press, New York, 1991

intracellular cascade mediating extracellular regulation of gene expression and phenotypic change.

EARLY RESPONSE GENES

Until about 8 years ago our understanding of the intracellular cascade that coupled extracellular signals with gene expression was predominately composed of a set of second messages. Following ligand-receptor interaction a number of molecules were produced or activated encoding the external information into specific intracellular pathways. These transduction pathways then regulated gene activity. The best understood second messenger pathways included elevation of cAMP leading to activation of protein kinase A (PKA) and elevation of Ca^{++} and diacylglycerol leading to activation of protein kinase C (PKC). Direct activation of membrane receptors, such as the insulin receptor that possess intrinsic kinase activity and steriod hormone-receptor complex formation and translocation represent additional parallel signal transduction pathways. It appears much of gene regulation is the consequence of kinase-mediated phosphorylation of protein substrates that function as nuclear regulatory factors. These activated factors binding to or influencing promoters sequences of different genes lead to selective suppression or enhancement of transcription and, hence, differential gene expression.

Several years ago it became apparent that a third-messenger system was functioning in the ligand-stimulated cascade leading to gene regulation (Fig.1). This third system is composed of proteins encoded by a family of genes, whose key induction characteristic is rapid transcriptional activation independent of protein synthesis. In addition, treatment with the protein synthesis inhibitor, cycloheximide, usually leads to a characteristic superinduction of message levels. Within this family of early response genes (ERGs), transcriptional control and half-lives of mRNAs delineate two main categories of family members. ERGs can be separated based on whether their expression is transient or prolonged: nuclear regulatory factors, or proteins functioning as either secretory proteins, membrane proteins and receptors or cytoplasmic structural and enzymatic components, respectively (see Almendral et al., 1989; Curran, 1988; Herschman, 1989, 1991; Arenander and Herschman, 1991).

The majority of the work on ERGs has been the characterization of the genes encoding nuclear factors. Historically, c-fos was discovered to act as a nuclear proto-oncogene, rapidly and transiently inducible in many cell types by serum (Curran, 1988). Subsequently, an entire family of fos-related antigens (FRAs) have been described and shown to play a central role in promoter activation or inhibition. A number of genes have been identified as targets of Fos (for review, see Arenander and Herschman, 1991). In neurons and glia, Fos is considered to help control the transcription of nerve growth factor (Mocchetti et al., 1989;), tyrosine hydroxylase (Gizang-Ginsberg and Ziff, 1990), transin (Machida et al., 1989) and proenkephlin (Sonnenberg et al., 1989). Since the expression of these ERG target genes follows the expression of ERGs and is sensitive to blockers of protein synthesis, they can be referred to as late response genes (LRGs), whose transcription is dependent upon the prior expression and translation of one or more ERG mRNAs. The ERG superfamily is estimated to reach about 100-200 members, many of whom share nucleotide sequence homology and are considered to be members of distinct families, such as the fos, jun, krox and TIS11 families (see Table 1).

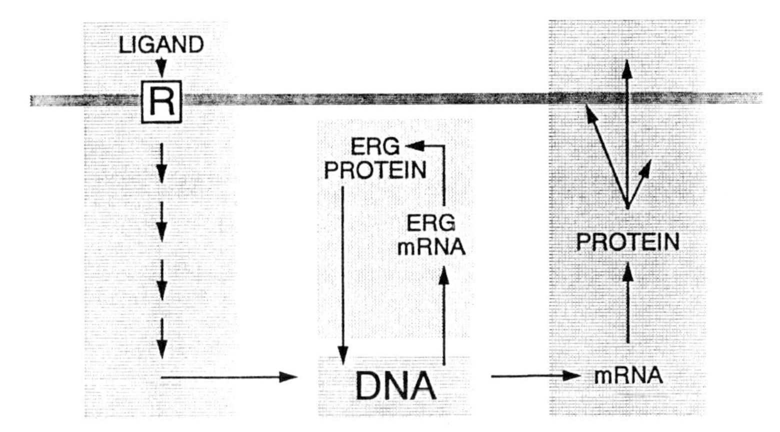

Fig. 1. Early Response Genes as Mediators of Phenotypic Change. This figure illustrates the central role that ERGs can play in determining the genomic response of a cell to environmental ligands. The rapid and transient expression of ERG mRNA and protein form a self-referral loop of genetic control, whereby a third-messenger system in the cascade of signal transduction relies on the expression of nuclear factors which return to the nucleus to coordinate, through a combinatorial process, cell type-specific gene expression.

Table 1. Classification of ERGs. ERGs can be organized by family and known or putative function (see Herschman, 1989,1991). This list of sequenced ERGs and additional isolates, include ten ERGs that encode for nuclear proteins involved in transcriptional regulation.

GENE		FUNCTION
c-myc		Transcription modulator
c-fos	TIS28	
fra-1		
fos-B		
c-jun		Transcription factor
junB		
junD		
egr-1	krox24/zif268 TIS8/NGF1A/PC1	
krox20	egr-2	
NGF1B	TIS1/nur77/N10	
JE		Cytokine
TCA3		
N51		Unknown
TIS7	PC4	
TIS10		
TIS11		
TIS21		

Specificity of ERG Expression

The first few years of research on ERGs were perplexing. While it was expected the ERGs may function as specific determinants of ligand-mediated phenotypic change, research showed that c-fos and most of the ERGs could be induced by a diverse range of cellular ligands in a wide variety of cell types. The apparent lack of expression specificity was puzzling because of the assumption that each ERG could alone function as a determinant of phenotypic responsiveness. Current data now strongly suggests, while some of the ERGs may be crucial switches for phenotypic alteration, specificity is most likely to be found in the combinatorial control exerted by many ERGs interacting to coordinate the readout of genomic programs.

The specificity of encoding environmental information may be a function of the combinatorial interaction among specific subsets of ERGs, some of which may also exhibit developmentally-restricted expression. One can envision the existence of both quantitative and qualitative parameters. Quantitative parameters of a specific ligand-induced biological response are defined as the summation of levels and kinetics of induction of ERGs. Thus, variation in the onset, duration, offset and the extent of message transcription, translation and post-translational modification among ERGs could yield a very complex time course of expression of numerous nuclear factors capable of interacting among themselves and at LRG promoters. Differential patterns of expression of many ERG mRNAs have been shown for both glial and neurons (Kujubu et al., 1987; Arenander et al., 1989a,b,c, 1991a,b,c; Bartel et al., 1989). Furthermore, since the extracellular environment is a complex and variable mixture of many ligands, simultaneous stimulation of several intracellular pathways may be more the rule than the exception. Published reports support the notion that simultaneous activation of cells by multiple ligands can quantitatively alter ERG expression patterns (Arenander et al., 1989a,c). Both additive and synergistic interactions are observed (see Arenander and Herschman, 1991).

Combinatorial mechanisms of intracellular signal transduction may also depend upon qualitative parameters of ERG expression. Qualitative changes can arise from either ligand-specific or cell type-specific restriction of ERG expression. Selective ligand-restricted expression is documented in the lack of inducibility of c-jun by elevated KCl in PC12 cells, even though elevated KCl leads to cell depolarization and induction of other ERGs and c-jun is inducible in PC12 by other ligands (Bartel et al., 1989). Cell-type restriction of ERG expression is present when the ability to induce one or more ERGs is completely extinguished. Reports are available demonstrating that a particular ERG may not be inducible in a given cell type by any ligand or cell perturbation. For example, TIS10 is not inducible in PC12 cells (Kujubu et al., 1987; Arenander et al., 1989b). Using *in situ* hybridization we have preliminary evidence of restriction of TIS10 in some immune cells (unpublished observations with H. R. Herschman and P. Koeffler). NGF1B/TIS1 is also extinguished in several types of immune cells (Varnum et al., 1989; unpublished observations). In addition, preliminary work with early primary glial precursor populations indicates cell-type specific restriction of all the ERGs studied so far (Arenander et al., 1991c). These restrictions suggests each cell type may exhibit a developmentally-specified set of ERGs that participates in determining cell phenotype and represent a unique configuration of third-messenger systems or 'filter' through which cells display differential sensitivity to environmental signals.

In vivo work is consistent with cell-type specific restriction observed in cell culture. Developmental data shows differential ERG expression in the body and CNS during normal development (Dony and Gruss, 1987; Chavrier et al., 1988; Tippetts et al., 1988; Caubet, 1989; Wilkinson et al., 1989a,b; Arenander and de Vellis, unpublished observations). In addition, chemical and mechanical stimulation of brain leads to selective anatomical and cell-type specific patterns of ERG mRNA or protein expression in adult animals (see Herschman, 1991; Arenander and Herschman, 1991). These studies suggest that subsets of ERG proteins may be differentially expressed in a cell-type specific manner during development and in the adult animal. The development and maintenance of such patterns of ERG expression may serve to coordinate the selective transcription of LRGs and thus act as determinants of cell fate and functional responsiveness to environmental cues.

ERG Expression in Glial Cells

In rat neocortical astrocytes and oligodendrocytes, a wide range of ligands are able to induce ERG mRNA expression (Arenander et al., 1988, 1989a,b,c, 1991a,b,c; Arenander and Herschman, 1991). The ligands include growth factors, such as fibroblast growth factor or insulin, tumor promoters capable of activating PKC, such as tetradecanoyl phorbol acetate (TPA), and agents that increase PKA activity such as norepinephrine (Fig. 2). Our findings support two important conclusions. First, the patterns of expression differ in several quantitative parameters. Examination of nine ERGs shows that both the level and temporal pattern of expression of ERGs can vary considerably. For example, for each ligand a characteristic dose response relationship is observed for each ERG determining the maximum level of induction through the respective intracellular pathway activated. Also, for a given ERG, the onset of appearance of mRNA can vary markedly depending upon the ligand-receptor pathway stimulated (see Arenander et al., 1989a). Second, our studies have shown that the induction of ERG mRNAs occurs via several independent and interacting intracellular pathways. Co-treatment of TPA and FGF or epidermal growth factor (Arenander et al., 1989c) gives, at least, additive elevation of ERGs without altering the temporal course of induction.

INSULIN AND INSULIN-LIKE GROWTH FACTORS

Insulin and the Control of Gene Expression

Insulin is a major determinant of cellular metabolism. Classically, insulin exerts opposite effects on gluconeogenesis (reduces) and glycolysis (enhances) in the liver. The mechanisms by which it exerts its broad control over cell metabolism and phenotypic change depend in large part on its ability to induce a coordinated pattern of gene expression (for review, see Rosen, 1987; Standaert and Pollet, 1988; Adamo et al., 1989). The binding of insulin to the insulin receptor leads to activation of the receptor kinase. The substrates of the INS receptor kinase activity are rare and not well characterized. Equally complex is the nature and consequence of the various types of receptor phosphorylation including the autophosphorylation-mediated kinase activation. Mutant receptor studies indicate the actions of INS on the cell depend upon a functional kinase domain of the receptor. Many of the effects of INS on gene expression, in particular, on c-fos (see below), glucose transporter (Mudd et al., 1990) and PEPCK (see O'Brien et al., 1990), are mimicked by treatment with TPA.

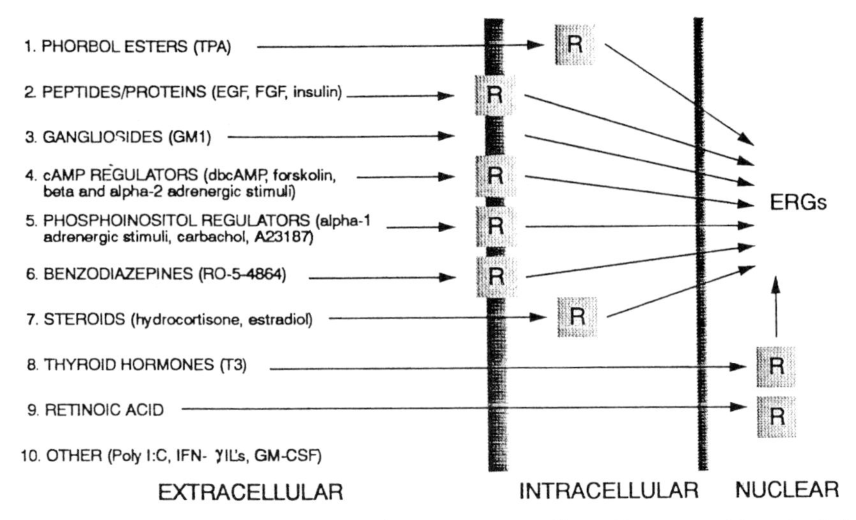

Fig. 2. Multiple Pathways Regulating ERG Induction in Glia. Ten types of ligands, defined by common intracellular components, represent separate, and possibly, interacting, pathways that contribute to the combinatorial mode of encoding extracellular information. The activated pathways converge on the nucleus as a multifactorial signal controlling ERG expression in glial cells. Data is derived from cultures of neocortical astrocytes. Similar, but less extensive, findings come from C6 glioma cells, oligodendrocytes and glial progenitor cell cultures (see text for details).

Although INS increases the levels of diacylglycerols in cells, INS effects on gene expression are considered to be mainly PKC-independent.

The effects of insulin are spread across a large time domain. Some events occur within minutes, such as stimulation of sugar transport, while others take hours, such as increase glucose transporter mRNA levels (Werner et al., 1989). Many of the early events of insulin action do not require protein synthesis. Several of these early events involve gene regulation (see Table 2). The INS-mediated inhibition of PEPCK expression is the best understood example of INS control on gene regulation. Transcription of PEPCK is stimulated by cAMP and glucocorticoids activating pathways leading to protein-DNA interaction at specific PEPCK promoter sequences. The promoter sequences mediating INS inhibition have recently been described. Two copies of a 15 base pair INS response sequence (IRS) in the PEPCK promoter (O'Brien et al., 1990) are both required for maximal inhibition of ligand-induced PEPCK expression. Gel retardation assays reveal a protein that can bind to this promoter "silencing" element. The binding of the unknown protein is considered to be modulated by phosphorylation, possibly by the INS receptor kinase. Recent work suggests PEPCK is an ERG (Laz et al., 1991).

The most rapid effect of INS on gene transcription is, however, the induction of c-fos (Stumpo and Blackshear, 1986, 1988; Ong et al., 1987; Taub et al., 1987; Messina, 1990), suggesting part of INS influence on expression of many genes may be mediated by expression of ERGs. INS and IGF-I can also induce c-myc in different cell types (Banskota et al., 1989; Gai et al., 1989). INS induction of c-fos requires a functional receptor kinase domain (Stumpo et al., 1988) and cycloheximide

Table 2. Control of Insulin-responsive Genes

GENE	TISSUE	EFFECT ON mRNA			
		INS	GC	cAMP	T3
NEGATIVE EFFECTS:					
1. PEPCK	Liver	----	++	++	++
2. Growth hormone	Pituitary	----	++	++	++
3. Adipsin	Preadipocytes	----	+/-		+?
	Adipocytes	----	++		+?
4. Glut. Synthetase	Astrocytes(m/r)	--/+	++	++	++
	Adipocytes	----	++		
POSITIVE EFFECTS:					
5. GPDH	Preadipocytes	++	----	---?	+?
	Adipocytes	++	----	---?	+?
	Glia	-?	++	++	
6. Glucokinase	Liver	++	++	----	++
7. Pyruvate kinase	Liver	++		----	
8. α-Amylase	Pancreas	++	++		
9. Casein*	Mammary	++	++		
10. GAPDH	Numerous	++			
11. c-fos	Numerous	++	++	++	++

Insulin is known to alter the expression of a large number of genes, most of which are involved in glucose metabolism. Genes are listed depending upon whether their expression is inhibited or stimulated by INS. Most of the work has been carried out in liver and fat cells with some examples from glial and other cells. Note that inhibitory effects of INS are generally opposed by cAMP and glucocorticoids (GC). Also note the opposite transcriptional control of GPDH by ligands in mouse adipocytes and rat glia cells. See text for references. * = protein determination; ? = not fully defined.

treatment leads to superinduction of c-fos by INS (Stumpo and Blackshear, 1986). The influence of INS on ERG expression is considered to be independent of short term changes in intracellular Ca^{++} or pH (Cook et al., 1988) and independent of activation of PKC (Stumpo and Blackshear, 1987) and PKA (Blackshear et al., 1985). Transient transfection of heterologous fos promoter constructs demonstrate that the serum response element (SRE) is sufficient for INS activation of the fos promoter. The simplest scenario of INS-mediated fos induction starts with INS/receptor interaction leading to autophosphorylation (Shemer et al., 1989). Activation of receptor tyrosine kinase, in turn, activates one or more nuclear factors directly, or indirectly by phosphorylating another kinase. The activated serum response factor(s) then bind to the SRE sequence and stimulate transcription.

The influence of INS and TPA occur via separate, independent and additive intracellular pathways converging on common kinase substrates or at the SRE of the fos promoter. The observed antagonism between glucocorticoids and INS (or TPA?) in the control of gene expression (see Table 2) is, in part, probably a result of direct interaction between the glucocorticoid receptor and a member of the AP-1 transcription complex, in particular, Jun or Fos. This direct association between members of two distinct transcription factor families results in mutual antagonism of normal transcriptional control exerted by either factor (Diamond et al., 1990; Jonat et al., 1990; Schule et al., 1990; Yang-Yen et al., 1990).

IGF-I, a peptide homologous to INS is considered to participate in the development and normal metabolism of the brain (see Adamo et al., 1989). Like INS, IGF-I can rapidly induce c-fos (Ong et al., 1987; Damante et al., 1988; Merriman et al., 1990) and the glucose transporter gene (GT1; Werner et al., 1989). Transfection experiments carried out with fos promoter constructs demonstrate IGF-I can activate the fos promoter, although the sequence responsible has not been identified. Steroid hormones can not only antagonize INS effects on gene expression coordinating glucose metabolism, but can also alter the expression of IGF receptor levels. For example, dexamethasone has been shown to partially suppress IGF-I mRNA levels (Adamo et al., 1988). Because of the cross-reactivity of INS at high concentrations with the IGF-I receptor, many previous studies using high INS levels may have been detecting IGF-I-mediated ERG induction. For example, PC12 cells do (Greenberg, et al., 1985) and do not (Kruijer et al., 1985) show an INS-mediated increase in fos mRNA. The former positive result (Greenberg et al., 1985) was obtained using high INS concentrations suggesting IGF, not INS, receptor stimulation was the source of c-fos induction in these studies.

Insulin and IGF-I in Glial Cells

INS and IGF-I are considered to influence the development and function of glial cells (for review, see Adamo et al., 1989; Kiess et al., 1989). INS and IGF-I receptors are present in the brain on both glial and neuronal cells. In addition, evidence clearly show endogenous synthesis of INS and IGF-I by neuronal cells. Since INS selectively stimulates glucose uptake in astrocytes, but not neuronal cells (Clark et al., 1984), INS is another example of functional coupling between neuronal and glial cells (Arenander and de Vellis, 1983, 1989). IGF-I, on the other hand, appears to be made by both cell types. In addition, glial, but not neuronal, cultures have been reported to express IGF-II mRNA. Thus, each cell type appears to produced a distinct set of growth factors. INS has been shown to increase glucose uptake in glial cells and to alter neurotransmitter uptake processes. The optimal level of INS in cultured astrocytes for growth (5-10 μg/ml) is higher than physiological levels suggesting cross-stimulation of IGF-I receptors. IGF-I has been reported to be a potent inducer of oligodendrocyte development *in vitro* promoting both proliferation and commitment of precursor cells (McMorris et al., 1986; van der Pal et al., 1988; McMorris and Dubois-Dalcq, 1988) and may play a role in glioma tumor cell growth *in vivo* (Merrill and Edwards, 1990). Insulin also has an influence on oligodendrocyte growth and differentiation (see also, Saneto et al., 1988; Fressinaud et al., 1989).

Interaction among INS, T3 and HC

The determinants of glial cell phenotype are not known. Analogous to the specific pattern of gene expression in differentiating adipocytes, a number of extracellular ligands may play significant parts in the orchestration of glial lineage development. Adipocyte share some characteristics of glial cells. These similarities may extend to the expression of transcription factors responsible for differentiation-dependent gene expression such as PKC-mediated increases in c-fos and the nuclear tri-iodothyronine and steriod receptors (see Gaskin et al., 1989). For example, the promoters of the adipocyte aP2 gene, the homolog of the myelin P2 gene in Schwann cells, and of the GPDH gene, whose expression is restricted to oligodendrocytes in the CNS, contain many of the same key elements (Cook et al., 1988). T3 receptor binding sites homologous to those found in the growth hormone promoter are present several times in the promoters of GPDH, aP2 and adipsin (Hunt et al., 1986). These T3 responsive elements are found in close proximity to the Fos binding site on the aP2 promoter. This proximity may represent a site of interaction between Fos and T3 receptor and may explain how Fos suppresses, while T3 enhances aP2 expression. The promoter of c-fos itself has a T3-like response element (Haynes et al., 1987).

Steriod hormones are also induce differentiation of both adipocytes (Chapman et al., 1984) and glia (Aizeman and de Vellis, 1981; Kumar et al., 1986; Kumar et al., 1989). It is therefore possible that mechanisms of transcriptional activation leading to cell differentiation in these two cell types are partially conserved. For example, hydrocortisone or dexamethasone stimulate rat glial GPDH transcription presumably through a glucocorticoid response element (GRE). GREs are present in the mouse adipocyte GPDH promoter region (Phillips et al., 1986; Cook et al., 1988). cAMP also enhances glial GPDH transcription (Kumar et al., 1986). Thus, evidence suggests that a number of ligands, in particular, HC, T3 and INS may interact not only in adipocyte differentiation but, by analogy, in glial cell differentiation as well. Since GPDH is a LRG, depending upon ligand-mediated ERG expression, ERGs appear to participate in the intracellular activation of both peptide and steriod hormone pathways. There are a number of probable sites of interaction between HC, T3 and INS pathways. Since T3 binds to its nuclear receptor and HC binds to and translocates with its receptor to the nucleus, interaction may occur between the ligand-activated nuclear receptor and INS-induced ERGs (Jun or Fos) either directly or indirectly at the ERG promoter. Interaction between these ligands would thus suggest that ligand-induced or activated nuclear transcription factors are capable of altering each others ability to bind to their cognate promoter response elements and to regulate gene transcription. Studies carried out in glial cells treated with a variety of ligands are described below. Our findings demonstrate cross-coupling between pathways activated by INS, T3 and HC controlling ERG expression in glial cells.

LIGAND INTERACTION AND ERG EXPRESSION IN GLIAL CELLS

Pathway Interaction in Astrocytes

Previous studies showed insulin is a good inducer of several ERGs (Arenander et al., 1988) and levels of ERG mRNA accumulation can vary due to interactions or cross-talk between intracellular signalling pathways (Arenander et al., 1989a,c). We have expanded our investigation and examined whether HC, T3 and INS or IGF-I can induce ERGs in glial cells and whether their corresponding signal transduction pathways interact or converge on the process of ERG induction (Arenander et al., 1991a,b). Secondary cultures of rat neocortical astrocytes grown in 10% fetal calf serum (FCS) were treated with insulin (INS, 5 μg/ml), tri-iodothyronine (T3, 50 nM) and/or hydrocortisone (HC, 1 μM) for up to 90 mins. Full time courses were carried out in the presence of various combinations of the three ligands. In most experiments, a high level of INS was used similar to concentrations in most tissue culture media known to stimulate both insulin and IGF-I receptors. In other experiments (see below), low serum cultures were studied and low INS levels (20 ng/ml) were tested to selectively activate INS receptors. Northern blot analysis was carried out on total cellular RNA using ^{32}P-labelled cDNAs (see Arenander et al., 1989b). To aid comparison, each experiment was carried out 2 or 3 times and, for each experiment, all ligand treatments were carried out on a single set of sister cultures and run on the same agarose gel (60 samples/gel). We probed the filters for several ERGs: NGF1B/TIS1, a member of the steriod hormone superfamily of ligand-inducible transcription factors (Milbrandt, 1988; Lim et al., 1987), egr1/TIS8, a zinc-finger containing transcription factor (Sukhatme et al., 1987; Lim et al., 1987), c-fos (Curran, 1988), and TIS11, a rapidly inducible ERG with no known function (Herschman, 1991). Data on other ERGs is not reported here.

In general, the data show 4 ERGs respond in a similar fashion. Although each of the ligands was capable of inducing the ERGs (Fig. 3), the effectiveness of ERG induction by each ligand differed. T3 was a weak inducer compared to INS (5 μg/ml). The induction of ERGs by either INS or T3 can be superinduced by co-treatment with cycloheximide. Combinations of ligands yielded interesting results. Unlike other previously described interacting ligand-activated pathways (Arenander et al., 1989c), co-treatment of cells with T3+INS had no apparent effect on ERG mRNA levels. HC by itself stimulated the rapid expression of ERG mRNAs but the kinetics appeared to vary for each ERG mRNA, in contrast to results with INS. Adding HC to cultures with T3 had no effect above that observed for HC alone. In contrast, HC appears to be additive with INS. This interaction suggests these two different signalling pathways are coupled at one or more levels.

Remarkably, the combined stimulation of T3+INS+HC gave a further increase in the level of ERG mRNA expression without altering the induction pattern. This suggests, under conditions of high serum, the intracellular pathways employed by T3 and INS do not normally interact, whereas the presence of HC appears to alter the transduction process.

The effects of serum on pathway interaction in astrocytes

Research suggests high levels of serum can increase the activity of the cAMP second-messenger pathway. For example, previous work in our lab showed that serum enhanced levels of both the number and responsiveness of β-adrenergic receptors (Wu et al., 1985). Since cAMP acts as a strong antagonist of INS action, it was of interest to compare the previous findings carried out in 10% serum conditions to experiments conducted in low levels of serum (2% FCS). Cultures were grown in 10% serum and shifted to 2% two days before treatment with T3, INS and/or HC. We expected, if low serum conditions lead to lowering of the activity of the cAMP system, astrocytes would exhibit enhance responsiveness to INS, either alone or in combination with T3 and/or HC. Fig. 4 shows the results of such an experiment. Note that all three ligands alone are capable of inducing the ERGs examined. And as before, T3 and INS showed (1) little or no interaction in the absence of HC and (2) showed an additive response in the presence of HC.

The main difference between high and low serum conditions appeared to be the interaction of INS and HC. In the presence of HC, INS induction of ERG mRNA appear to be synergistic. Previous work in our lab has shown that HC synergistically enhances INS-mediated induction of glutamine synthetase activity in astrocytes (Aizeman and de Vellis, 1981). Table 3 summarizes the findings regarding low serum culture conditions. It seems serum has a marked influence on astrocyte responsiveness to extracellular signals. The ability of HC to augment the effects of INS appears to be abolished by high concentrations of serum.

TABLE 3. HORMONAL INTERACTION: LEVELS OF ERG EXPRESSION

LIGAND:	T3	INS	T3+INS	---
- HC	+	++	++	_
+ HC	++	+++ +++	+++	++

Data from Fig. 4 is tabulated to describe the interactions of three different ligands on the expression of ERGs in astrocytes under low serum conditions. See text for details.

Pathway interaction in C6 glioma cells

Different results were obtained for C6 cells under low serum conditions (Arenander et al., 1991a,b; data not shown). INS (5μg/ml), T3 and HC all rapidly and transiently induce c-fos and egr1 mRNA. In addition, the influence of T3 is additive with INS and HC is additive with INS+T3. Two experiments suggest the effects of INS, however, are due to stimulation of IGF-I, and not INS, receptors. First, treatment of C6 cultures with low levels of INS (20 ng/ml) does not lead to ERG induction, even under superinducing (CHX) conditions. Second, IGF-I at 100 ng/ml induces rapid and transient induction of these ERG mRNAs. There appears to be no interaction between IGF-I and either T3 or HC in the induction of c-fos or egr1 mRNA. Dose course experiments and previous culture work (Rechler and Nissley, 1985) indicate this amount of IGF-I gives maximum responses.

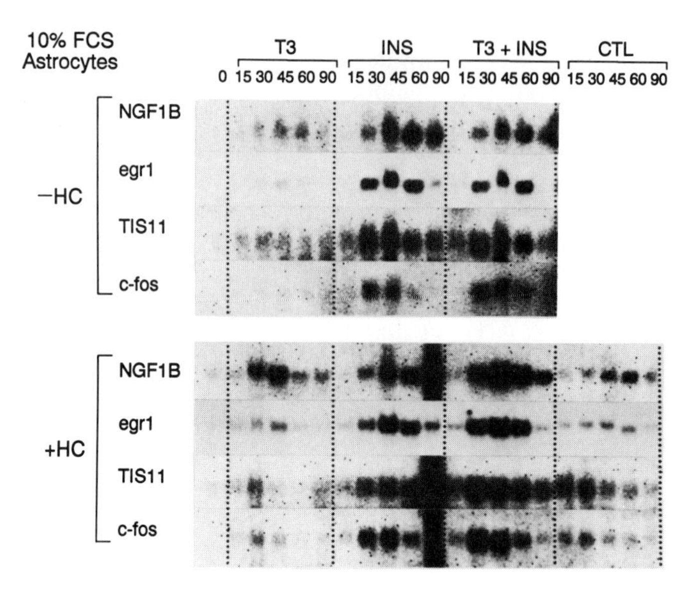

Fig. 3. ERG induction by INS, T3 and HC in 10% serum. Secondary cultures of rat neocortical astrocytes were grown and treated in the presence of 10% FCS. Ligands were added individually or in combination for the times shown. Total cellular RNA was harvested and run on agarose/formaldehyde gels and probed with ^{32}P-labeled cDNAs. The pattern of ribosomal RNA and constitutive mRNA indicated that RNA loading was even (Arenander et al., 1989b; data not shown). The smearing of the 90 min INS sample of RNA is an artifact (from Arenander et al., 1991a).

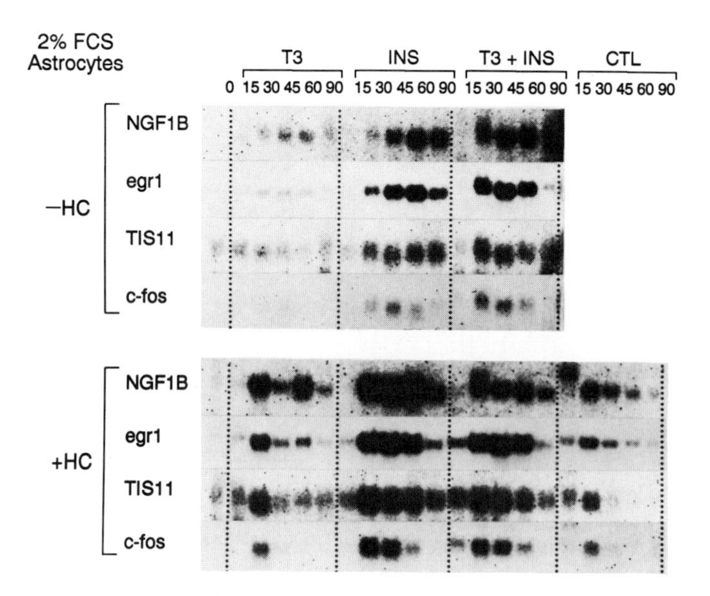

Fig. 4. ERG induction by INS, T3 and HC in 2% serum. Secondary cultures of rat neocortical astrocytes were grown and treated in the presence of 2% FCS. The experiment was identical in design and processing to that in Fig. 3 (from Arenander et al., 1991a).

CONCLUSIONS

The effects of insulin on metabolism are well documented. Less certain are the mechanisms by which INS directs the coordinate shift in gene expression necessary to alter a cell's function. Changes in INS levels alter the expression of many genes -- some induced, some suppressed. Recent evidence suggests that INS binding to its receptor triggers a complex series of intracellular events that lead to both direct and indirect effects of INS on transcriptional control. Inhibition of PEPCK, the key enzyme involved in gluconeogenesis, by INS occurs via activation of a protein capable of binding to a specific promoter sequence, blocking the ability of cAMP and/or dexamethasone to stimulate transcription of PEPCK. The induction c-fos and other ERGs encoding nuclear factors by INS in many cell types suggests another, indirect, pathway involved in the expression of many INS-regulated genes. Thus, the INS-mediated expression of FOS and other nuclear ERG proteins represent an additional layer of cellular control over gene expression and cell phenotype.

In the adult CNS, INS may serve as counterbalancing agent. INS produced by neurons may exert an important role in balancing the influence of cAMP-elevating neurotransmitters and steriod compounds. Such interaction may involve altered patterns of expression of ERGs encoding nuclear transcription factors. During CNS development, environmental factors necessary for proper growth and development of glia may interact. The combined impact of external signals on gene regulation may also involve, in part, the integrated expression of various ERGs encoding nuclear proteins. It seems likely future experiments will find ERGs participating in INS control of glial gene expression. Such control is envisioned to be primarily a matter of activation of multiple intracellular pathways and ERG combinatorial interaction. Since INS and IGF-I receptors are developmentally regulated, INS and IGF-I may participate in developmental decisions regulating growth and differentiation of glial and neuronal cells by influencing the cell-type specific and stage-dependent pattern of ERG expression in the brain.

REFERENCES

Adamo, M., Werner, H., Farnsworth, W., Roberts, C.T., Raizada, M.K. and LeRoith, D. (1988) Dexamethasone reduces steady state insulin-like growth factor-I messenger ribonucleic acid levels in rat neuronal and glial cells in primary culture. **Endoc.** 123: 2526.

Adamo, M., Raizada, M.K. and LeRoith, D. (1989) Insulin and insulin-like growth factor receptors in the nervous system. **Mol. Neurobiol.** 3: 71.

Aizenman, Y. and de Vellis, J. (1987) Synergistic action of thyroid hormone, insulin and hydrocortisone on astrocyte differentiation. **Brain Res.** 414: 301.

Almendral, J.M., Sommer, D., MacDonald-Bravo, H., Burckhardt, J., Perera, J. and Bravo, R. (1988) Complexity of the early genetic response to growth factors in mouse fibroblasts. **Mol. Cell. Biol.** 8: 2140.

Arenander, A.T. and de Vellis, J. (1983) Frontiers of glial physiology. In: R. Rosenberg (ed.), **The Clinical Neurosciences., Section V., Neurobiology** (W.D. Willis, Assoc. ed.) Chapter IV, New York: Churchill Livingstone, p. 53. Arenander, A.T and de Vellis, J. (1989) Development, In: G. Siegel, R. W. Albers, B. W. Agranoff, P. Molinoff (Eds.), **Neurochemistry: Molecular, Cellular and Medical Aspects** 4th Edition, New York: Raven Press, p 479.

Arenander, A.T., de Vellis, J. and Herschman, H.R. (1989a) Induction of c-fos and TIS genes in cultured rat astrocytes by neurotransmitters. **J. Neurosci. Res.** 24: 107.

Arenander, A.T., Lim, R.W., Varnum, B.C., Cole, R., de Vellis, J. and Herschman, H.R. (1989b) TIS gene expression in cultured rat astrocytes: Induction by mitogens and stellation agents. **J. Neurosci. Res.** 23: 247.

Arenander, A.T., Lim, R.W., Varnum, B.C., Cole, R. and de Vellis, J. (1989c) TIS gene expression in cultured rat astrocytes: Multiple pathways of induction by mitogens. **J. Neurosci. Res.** 23: 257.

Arenander, A.T. and Herschman, H.R. (1991) Primary response gene expression in the nervous system, In: J.H. Fallon and S.E. Loughlin (Eds.), **Neurotrophic Factors**, New York: Academic Press, In press.

Arenander, A.T.. Cheng, J. and de Vellis, J. (1991a) The induction of c-fos and other early response genes in glial cultures by insulin, IGF-I, tri-iodothyronine and hydrocortisone: Evidence for pathway interaction. submitted.

Arenander, A.T.. Cheng, J. and de Vellis, J. (1991b) The pattern of induction of c-fos and other early response genes in C6 glioma cells treated with mitogens, phorbol ester and hormones. Manuscript in preparation.

Arenander, A.T., Herschman, H.R. and de Vellis, J. (1991c) Cell-type restricted patterns of expression of c-fos and other primary response genes encoding nuclear factors in developing cultures of rat neocortical glial progenitor cells using *In situ* hybridization techniques. Manuscript in preparation.

Banskota, N.K., Taub, R., Zellner, K., Olsen, P. and King, G.L. (1989) Characterization of induction of protooncogene c-myc and cellular growth in human vascular smooth muscle cells by insulin and IGF-I. **Diab.** 38: 123.

Bartel, D, Sheng,M., Lau,L. and Greenberg, L. (1989) Growth factors and membrane depolarization activate distinct programs of early response gene expression: Dissociation of fos and jun induction. **Genes Dev.** 3: 304.

Blackshear, P.J., Witters, L.A, Girard, P.R., Kuo, J.F. and Quamo, S.N. (1985) Growth factor-stimulated protein phosphorylation in 3T3-L1 cells. **J. Biol. Chem.** 260: 13304.

Caubet, JF. (1989) c-fos proto-oncogene expression in the nervous system during mouse development. **J. Cell. Biol.** 9: 2269.

Chapman, A.B., Knight, D.M., Dieckmann, B.S. and Ringold, G.M. (1984) Analysis of gene expression during differentiation of adipogenic cells in culture and hormonal control of the developmental program. **J. Biol. Chem.** 259: 15548.

Chavrier, P., Lemaire, P., Revelant, O., Bravo, R. and Charnay, P. (1988) Characterization of a mouse multigene family that encodes zinc finger structures. **Mol. Cell. Biol.** 8: 1319.

Clarke, D.W., Boyd, F.T., Kappy, M.S. and Raizada, M.K. (1984) Insulin binds to specific receptors and stimulates 2-deoxy, D-glucose uptake in cultured glial cells from rat brain. **J. Biol. Chem.** 259: 11672. Cook, K.S., Hunt, C.R. and Spiegelman, B.M. (1985) Developmentally regulated mRNAs in 3T3 adipocytes: Analysis of transcriptional control. **J. Cell Biol.** 100: 514.

Cook, P.W., Weintraub, W.H., Swanson, K.T., Machen, T.E. and Firestone, G.L. (1988) Glucocorticoids confer normal serum/growth factor-dependent growth regulation to Fu5 rat hepatoma cells in vitro. **J. Biol. Chem.** 263: 19296.

Curran, T. (1988) The fos oncogene. In: E.P.Reddy, A.M.Skalka, T.Curran (Eds.) **The Oncogene Handbook**, Amsterdam: Elsevier, p.307.

Damante, G., Cox, F. and Rapoport, B. (1988) IGF-I increases c-fos expression in FRTS5 rat thyroid cells by activating the c-fos promoter. **Biochem. Biophy. Res. Comm.** 151: 1194.

Diamond, M.I., Miner, J.N., Yoshinaga, S.K. and Yamamoto, K.R. (1990) Transcription Factor Interactions: Selectors of Positive or Negative Regulation from a Single DNA Element. **Science** 249: 1266.

Dony, C. and Gruss, P. (1987) Proto-oncogene c-fos expression in growth regions of fetal bone and mesodermal web tissue. **Nature** 328: 711.

Fischer, G. (1984) Growth requirements of immature astrocytes in serum-free hormonally defined media. **J. Neurosci. Res.** 12: 543.

Fressinaud, C., Sarlieve, L.L. and Labourdette, G. (1989) Regulation of cerebroside sulfotransferase activity in cultured oligodendrocytes: Effect of growth factors and insulin. **J. Cell. Physiol.** 141: 667.

Gai, X., Rizzo, M-G, Valpreda, S. and Baserga, R. (1989) Regulation of c-myc mRNA levels by insulin or platelet-poor plasma. **Oncog. Res.** 5: 111.

Gaskins, H.R., Hausman, G.J. and Martin, R.J. (1989) Regulation of gene expression during adipocyte differentiation: A review. **J. Anim. Sci.** 67: 2263.

Gizang-Ginsberg, E. and Ziff, E.B. (1990) Nerve growth factor regulates tyrosine hydroxylase gene transcription through a nucleoprotein complex that contain c-Fos. **Genes Dev.** 4: 477.

Greenberg, M., Green, L. and Ziff, E. (1985) Nerve growth factor and epidermal growth factor induce rapid transient changes in proto-oncogene transcription in PC-12 cells. **J. Biol. Chem.** 260: 14101.

Haynes, T.E., Kitchen, A.M. and Cochran, B.H. (1987) Inducible binding of a factor to the c-fos regulatory region. **Proc. Natl. Acad. Sci. USA** 84: 1272.

Herschman, H.R. (1991) Primary Response genes induced by growth factors and tumor promoters. **Ann. Rev. Biochem.** 60: in press.

Hunt, C.R., Ro, J.H., Dobson, D.E., Min, H.Y. and Spiegelman, B.M. (1986) Adipocyte P2 gene: Developmental expression and homology of 5'flanking sequences among fat cell-specific genes. **Proc. Natl. Acad. Sci. USA** 83: 3786.

Jonat, C., Rahmsdorf, H.J., Park, K-K., Cato, A.C.B., Gebel, S., Ponta, H. and Herrlich, P. (1990) Antitumor promotion and antiinflammation: Down-modulation of AP-2 (Fos/Jun) activity by glucocoricoid hormone. **Cell** 62: 1189.

Kiess, W., Lee, L., Graham, D.E., Greenstein, L., Tseng, L., Rechler, M. and Nissley, S.P. (1989) Rat C6 glial cells synthesize insulin-like growth factor-I (IGF-I) and express IGF-I receptors and IGF-II/mannose 6-phosphate receptors. **Endoc.** 124: 1727.

Kruijer, W., Schubert, D., Verma, I.M. and (1985) Induction of the proto-oncogene fos by nerve growth factor. **Proc. Nat. Acad. Sci. USA** 82: 7330.

Kujubu, D.A., Lim, R.W., Varnum, B.C. and Herschman, H.R. (1987) Induction of transiently expressed genes in PC-12 pheochromocytoma cells. **Oncogene** 1: 257.

Kumar, S., Holmes, E., Scully, S., Birren, B.W., Wilson, R.H. and de Vellis, J. (1986) The hormonal regulation of gene expression of glial markers: Glutamine synthetase and glycerol phosphate dehydrogenase in primary cultures of rat brain and in C6 cell line. **J. Neurosci. Res.** 16: 251.

Kumar, S., Cole, R., Chiappelli, F. and de Vellis, J. (1989) Differential regulation of oligodendrocyte markers by glucocorticoids: Post-transcriptional regulation of both proteolipid protein and myelin basic protein and transcriptional regulation of glycerol phosphate dehydrogenase. **Proc. Natl. Acad. Sci USA** 86: 6807.

Laz, T.M., Mohn, K.L., Melby, A.E., Hsu, J-C. and Taub, R. (1991) The identification of 41 novel immediate-early genes in regenerating liver and insulin stimulated H35 cells. **J. Cell. Biochem.** 15B(Suppl): 194.

Lim, R., Varnum, B.C. and Herschman, H., (1987) Cloning of sequences induced as a primary response following mitogen treatment of density arrested Swiss 3T3 cells. **Oncogene** 1:263.

McMorris, F.A. and Dubois-Dalcq, M. (1988) Insulin-like growth factor-I promotes cell proliferation and oligodendroglial commitment in rat glial progenitor cells developing in vitro. **J. Neurosci Res.** 21: 199.

McMorris, F.A., Smith, T.M., DeSalvo, S. and Furlanetto, R.W. (1986) Insulin-like growth factor-I/somatomedin C: A potent inducer of oligodendrocyte development. **Proc. Natl. Acad. Sci. USA** 83: 822.

Machida, C.M., Rodland, K., Matrisian, L., Magun, B.E. and Ciment, G. (1989) NGF induction of the gene encoding the protease transin accompanies neuronal di fferentiation in PC12 cells. **Neuron** 2: 1587.

Merrill, M.J. and Edwards, N.A. (1990) Insulin-like growth factor-I receptors in human glial tumors. **J. Clin. Endoc. Metab.** 71: 199.

Merriman, H.L., La Tour, D., Linkhart, T.A., Mohan, S., Baylink, D.J. and Strong, D.D. (1990) Insulin-like growth factor-I and insulin-like growth factor-II induce c-fos in mouse osteoblastic cells. **Calif. Tiss. Intl.** 46: 258.

Messina, J.L. (1990) Insulin's regulation of c-fos gene transcription in hepatoma cells. **J. Biol. Chem.** 265: 11700.

Michler-Stuke, A., Wolff, J.R. and Bottenstein, J.E. (1984) Factors influencing astrocyte growth and development in defined media. **Int. J. Dev. Neurosci.** 2: 575

Milbrandt, J. (1988) Nerve growth factor induces a gene homologous to the glucocorticoid receptor gene. **Neuron** 1: 183.

Mochetti, I., De Bernardi, M., Szekely, A., Alho, H., Brooker, G. and Costa, E. (1989) Regulation of nerve growth factor synthesis by beta-adrenergic receptor activation in astrocytoma cells: a potential role of c-fos protein. **Proc. Natl. Acad. Sci. USA** 86: 3891.

Morrison, R.S. and de Vellis, J. (1981) Growth of purified astrocytes in a chemically defined medium. **Proc. Natl. Acad. Sci. USA** 78: 7205.

Mudd, L.M., Werner, H., Shen-Orr, Z., Roberts, C.T., LeRoith, D., Haspel, H.C. and Raizada, M.K. (1990) Regulation of rat brain/HepG2 glucose transporter gene expression by phorbol esters in primary cultures of neuronal and glial cells. **Endoc.** 126: 545.

O'Brien, R.M., Lucas, P.C., Forest, C.D., Magnuson, M.A. and Granner, D.K. (1990) Identification of a sequence in the PEPCK gene that mediates a negative effect of insulin on transcription. **Science** 249: 533.

Ong, J., Yamashita, S. and Melmed, S. (1987) Insulin-like growth factor-I induces c-fos messenger ribonucleic acid in L6 rat skeletal muscle cells. **Endoc.** 120: 353.

Phillips, M.P., Djian, P. and Green, H. (1986) The nucleotide sequence of three genes participating in the adipose differentiation of 3T3 cells. **J. Biol. Chem.** 261: 10821.

Rechler, M.M. and Nissley, S.P. (1985) The nature and regulation of the receptors for insulin-like growth factors. **Ann. Rev. Physiol.** 47:425.

Rosen, O.M. (1987) After insulin binds. **Science** 237: 1452.

Saneto, R.P., Low, K.G., Melner, M.H. and de Vellis, J. (1988) Insulin/insulin-like growth factor-I and other epigenetic modulators of myelin basic protein expression in isolated oligodendrocyte progenitor cells. **J. Neurosci. Res.** 21: 210.

Schule, R., Rangarajan, P., Kliewer, S., Ransone, L.J., Bolado, J., Yang, N., Verma, I.M. and Evans, R.M. (1990) Functional antagonism between oncoprotein c-Jun and the glucocorticoid receptor. **Cell** 62: 1217.

Shemer, J., Adamo, M., Raizada, M.K., Heffez, D., Zick, Y. and LeRoith, D. (1989) Insulin and IGF-I stimulate phosphorylation of their respective receptors in intact neuronal and glial cells in primary culture. **J. Mol. Neurosci.** 1: 3.

Sonnenberg, J.L., Rauscher, F.J., Morgan, J.I. and Curran, T. (1989) Regulation of proenkephalin by proto-oncogenes fos and jun. **Science** 246: 1622.

Standaert, M.L. and Pollet, R.J. (1988) Insulin-glycerolipid mediators and gene expression. **FASEB J.** 2: 2453.

Stumpo, D. and Blackshear, P. (1986) Insulin and growth factor effects on c-fos expression on normal and protein kinase C-deficient 3T3-L1 fibroblast and adipocytes. **Proc. Natl. Acad. Sci. USA** 83: 9453.

Stumpo, D.J., Stewart, T.N., Gilman, M.Z. and Blackshear, P.J. (1988) Identification of c-fos sequences involved in induction by insulin and phorbol esters. **J. Biol. Chem.** 263: 1611.

Sukhatme, V.P., Kartha, S., Toback, F.G., Taub, R., Hoover, R.G. and Tsai-Morris, C. (1987) A novel early growth response gene rapidly induced by fibroblast, epithelial cell and lymphocyte mitogens. **Oncogene Res.** 1: 343.

Taub et al., 1987, Roy, A., Dieter, R. and Koontz, J. (1987) Insulin as a growth factor in rat hepatoma cells. **J. Biol. Chem.** 262: 10893.

Tippetts, M.T., Varnum, B.C., Lim, R.W. and Herschman, H.R. (1988) Tumor promoter-inducible genes are differentially expressed in the developing mouse. **Mol. Cell. Biol.** 8: 4570.

van der Pal, R.H.M., Koper, J.W., van Golde, L.M. and Lopes-Cardozo, M. (1988) Effects of insulin and insulin-like growth factor (IGF-I) on oligodendrocyte-enriched glial cultures. **J. Neurosci. Res.** 19: 483.

Varnum, B.C., Lim, R.W., Kaufman, S.E., Gasson, J.C., Greenberger, J.S. and Herschman, H.R. (1989) Granulocyte-macrophage colony-stimulating factor induces a unique pattern of primary response TIS genes in both proliferating and terminally differentiated myeloid cells. **Mol. Cell. Biol.** 9: 3580.

Weibel, M., Pettmann, B., Duane, G., Labourdette, G. and Sensenbrenner, M. (1984) Chemically defined medium for rat astroglial cells in primary culture. **Int. J. Dev. Neurosci.** 2: 355.

Werner, H., Raizada, M.K., Mudd, L.M., Foyt, H.L., Simpson, I.A., Roberts, C.T. and LeRoith, D. (1989) Regulation of rat brain/HepG2 glucose transporter gene expression by insulin and insulin-like growth factor-I in primary cultures of neuronal and glial cells. **Endoc.** 125: 314.

Wilkinson, D., Bhatt, S., Chavrier, P., Bravo, R. and Charnay, P. (1989) segment-specific expression of a zinc-finger gene in the developing nervous system of the mouse. **Nature** 337: 461.

Wilkinson, D., Bhatt, S., Ryseck, R-P. and Bravo, R. (1989) Tissue-specific expression of c-jun and junB during organogenesis in the mouse. **Dev.** 106: 465.

Wu, D.K., Morrison, R.S. and de Vellis, J. (1985) Modulation of beta-adrenergic response in rat brain astrocytes by serum and hormones. **J. Cell. Physiol.** 122: 73.

Yang-Yen, J-F., Chambard, J-C., Sun, Y-L., Smeal, T., Schmidt, T.J., Drouin, J. and Karin, M. (1990) Transcriptional interference between c-Jun and the glucocorticoid receptor: Mutual inhibition of DNA binding due to direct protein-protein interaction. **Cell** 62: 1205.

TROPHIC ACTIONS OF IGF-I, IGF-II AND INSULIN ON CHOLINERGIC AND DOPAMINERGIC BRAIN NEURONS

Beat Knusel and Franz Hefti

Ethel Percy Andrus Gerontology Center
University of Southern California
Los Angeles, CA 90089-0191

INTRODUCTION

Insulin, IGF-I and IGF-II have only recently been proposed to possess physiological functions distinctive for the central nervous system (review: Baskin et al., 1988). Receptors for insulin and the IGFs are found in rat brain and seem to be heterogeneously distributed (Hill et al., 1986; Mendelson, 1987; Bohannon et al., 1988). While evidence that insulin occurs in the brain is still equivocal (Baskin et al., 1988), mRNAs for IGF-I and IGF-II were detected in many brain areas and are differentially regulated during development (Rotwein et al., 1988). With cell culture methods it has been shown that insulin, IGF-I and IGF-II can have effects on neurons which are generally described as "neurotrophic". Thus, in cultures of brain cells these hormones promote neuron survival, neurite extension, and expression of neuron-specific genes and, in astrocytes but possibly also in neurons, DNA synthesis (Bhat, 1983; Lenoir and Honegger, 1983; Mill et al., 1985; Recio-Pinto et al., 1986; Aizenman and de Vellis, 1987; Kyriakis et al., 1987; Avola et al., 1988; DiCicco-Bloom and Black, 1988). These effects are reminiscent of those produced by NGF in cell cultures. NGF and other polypeptide growth factors are believed to play decisive roles in the mammalian nervous system, particularly during early development but also in the adult organism (reviews: Thoenen et al., 1987; Snider and Johnson, 1989; Barde, 1989). Best established are the roles of NGF and brain-derived neurotrophic factor (BDNF) in the control of neuronal survival during development of certain peripheral neuron populations (Thoenen et al., 1987; Barde, 1989) and there is initial evidence for a similar action of neurotrophin-3, a polypeptide related to NGF and BDNF (Hohn et al., 1989; Maisonpierre et al., 1990). However, BDNF has been purified from brain (Barde et al., 1987) and NGF and its receptors have been known for several years to exist in the central nervous system where NGF is believed to play a role in the development of cholinergic neurons of the basal forebrain and the corpus striatum (Gnahn et al., 1983; Hartikka and Hefti, 1988a; Hatanaka and Tsukui, 1986; Hatanaka et al., 1987; Hefti et al., 1985; Honegger and Lenoir, 1982; Johnston et al., 1987; Martinez et al., 1987; Mobley et al., 1985; Mobley et al., 1986; Koh and Loy, 1989). NGF and its receptor are also found in adult tissue (review: Hefti et al., 1989) and it is assumed that NGF and, possibly, other neurotrophic factors play significant roles during the entire lifespan of the mammalian organism.

The cholinergic neurons of the basal forebrain are among the most extensively studied CNS neurons responsive to trophic factors. In cell culture these neurons respond to NGF (see above), to basic fibroblast growth factor (Grothe et al., 1989; Knusel et al., 1990a), to insulin, to the insulin-like growth factors I and II (Knusel et al., 1990a) and to BDNF (Knusel et al., 1990b) by increasing the activity of the cholinergic marker enzyme choline acetyltransferase (acetyl-CoA: choline O-acetyltransferase; EC 2.3.1.6; ChAT). This susceptibility to the trophic factors insulin and the IGFs, NGF, BDNF and bFGF suggests that different polypeptide growth factors may control specific mechanisms necessary for normal developmental growth and maintenance of function of these cells. The presence of at least some of these factors in the adult hippocampus, a principal target area of basal forebrain cholinergic neurons, is particularly intriguing since this group of cells seems to be critically involved in processes of learning and memory (e.g. Flicker et al., 1983; Murray and Fibiger, 1985; Tilson et al., 1988; Cassel and Kelche, 1989). Another group of intensively studied brain neurons are the dopaminergic cells of the substantia nigra which are also believed to be trophically influenced by their target area (e.g. Prochiantz et al., 1979; Shalaby et al., 1983; Doucet et al., 1990) but for which no growth factor has been molecularly identified and characterized yet. The analysis of the role of trophic factors in development and adult function of forebrain cholinergic and mesencephalic dopaminergic neurons has been an important goal of our laboratory during the past years. Findings summarized above, suggesting a neurotrophic role of insulin and the IGFs, prompted us to evaluate their actions on developing cholinergic and dopaminergic cells, using our well established cell culture systems.

CELL CULTURES

Cell culture methods used have been described in detail elsewhere (Knusel et al., 1990a). Briefly, basal forebrain, pontine and ventral mesencephalic areas were dissected from fetal rat (E16). Cells were dissociated and grown in modified Leibovitz' L-15 medium containing 5% horse serum and 0.5% fetal bovine serum. Trophic factors, unless otherwise specified in the text, were present in the medium during the entire culture time. Fresh factors were added after each medium change. The cultures were taken for biochemical determination of ChAT activity or of dopamine uptake rate usually after 1 week *in vitro*. NGF was purified from mouse salivary glands. Bovine insulin was purchased from Sigma (St. Louis, MO). Human recombinant bFGF was provided by Synergen (Boulder, CO). This bFGF stimulated proliferation of various cells at concentrations of 1-10 ng/ml. Human recombinant IGF-I and IGF-II were obtained from Eli Lilly Laboratories (Indianapolis, IN). Frozen aliquots of bFGF at 2 mg/ml in 10 mM sodium phosphate buffer, pH 7.0, containing 0.3 M glycerol and 0.3 mg/ml heparin, of IGF-I and IGF-II at 100 μg/ml in 5 mM HCl and of NGF at 0.2 mg/ml in 0.2% acetic acid were stored at -70°C. Aliquots were thawed and diluted with medium immediately before application. Insulin was dissolved freshly for each experiment.

TROPHIC ACTIONS ON BASAL FOREBRAIN CHOLINERGIC NEURONS

Addition of insulin, IGF-I or IGF-II to the growth medium of basal forebrain cultures increased the activity of ChAT in a dose dependent manner (fig. 1A). The maximal effects on ChAT activity were very similar with each of the three factors. The rank order of potency was IGF-I > IGF-II > insulin and with IGF-I half-maximal response was achieved at 3.2×10^{-9} M. The rank order of potency is in accordance with the relative affinities of the three ligands for a recently characterized neuronal

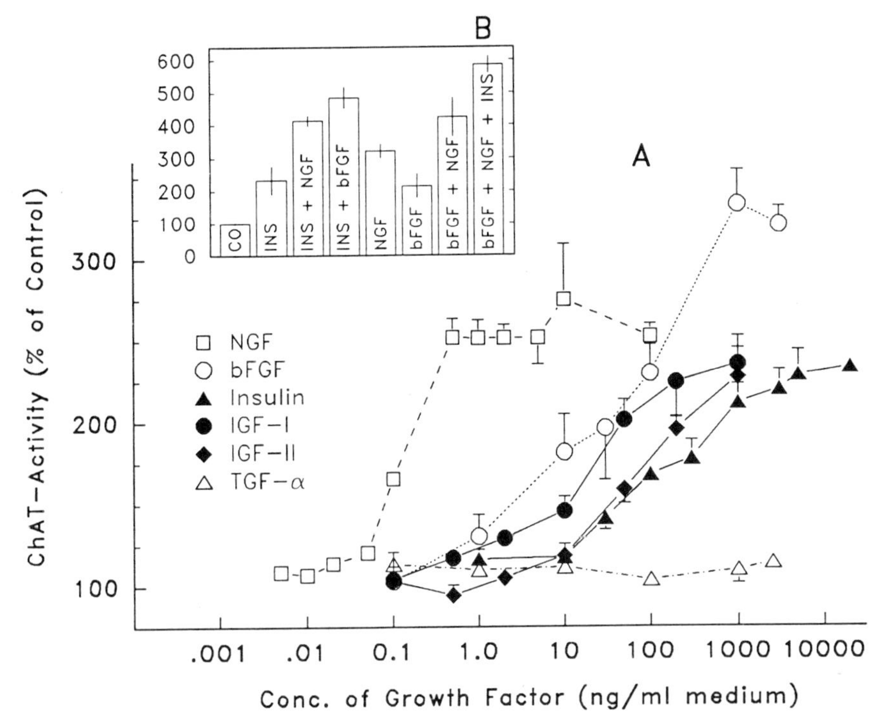

Fig. 1. ChAT - activity in basal forebrain cultures treated with insulin, IGF-I, IGF-II, NGF, bFGF or TGF-α. A: Dose-response curves in cultures grown for 6 days. B: Additivity of effects in cultures grown for 8 days. Growth factors were present during the entire culture period. Error bars represent standard errors and, in fig. A, were omitted where they would have appeared smaller than the symbol.

IGF-I receptor (Burgess et al., 1987). For comparison, the EC_{50} for NGF in identical cultures was 6 x 10^{-12} M and for bFGF 5.4 x 10^{-9} M. Transforming growth factor-α (TGF-α) did not increase ChAT activity in our cultures. Similarly, substance P, epidermal growth factor (EGF) and platelet derived growth factor (PDGF) were ineffective (data not shown). It should be noted that insulin, the IGFs, bFGF, EGF and TGF-α, but not NGF, substance P or PDGF increased cell density and protein content of the cultures (data not shown). For each factor affecting both, ChAT activity and protein content, the concentration requirements were very similar for both effects. When combinations of insulin, bFGF and NGF at maximally active concentrations were simultaneously added to the cultures (fig. 1B), the effects of each factor on ChAT activity and protein content was additive to the effect(s) of the other(s). However, combining maximally effective concentrations of insulin and the IGFs did not elevate ChAT activity (or protein content) above the level which was attained by either of the three factors alone (table 1), suggesting that these factors might act via the same receptor(s), most likely an IGF-I receptor. Together, these findings are evidence that insulin, IGF-I and IGF-II, stimulate different cellular mechanisms than NGF or bFGF. Using antibodies and antisera against bFGF and NGF we further

Table 1. Effects of insulin, IGF-I and IGF-II on ChAT - activity in basal forebrain cultures. Equal effectiveness and lack of additivity.

Growth Factors	ChAT/WELL (pmol/min)
CONT	82.6 ± 2.9
IGF-I	167.6 ± 4.1
IGF-II	153.0 ± 9.1
Insulin	149.0 ± 3.0
IGF-I + IGF-II	163.3 ± 6.4
IGF-I + Insulin	153.1 ± 7.0
IGF-II + Insulin	159.3 ± 8.3

Cultures were prepared and grown as those for fig. 1; concentrations: IGF-I and IGF-II 1 ug/ml; insulin 30 ug/ml. Values are mean ± S.E.M.; n = 4 to 8. All differences to control are statistically significant ($P < 0.001$; Student's t test).

established that neither the effects of insulin, nor of bFGF, are indirectly mediated via stimulation of endogenous release of NGF in our cell cultures (data not shown).

To establish whether cholinergic neurons exhibit different time windows of responsiveness to various growth factors cultures were treated with insulin, bFGF or NGF at different times during their development *in vitro*. In an experiment employing only three days application of the growth factors after varying time delays between preparation and treatment of the cultures we observed that insulin was equally effective when applied immediately after plating or when applied to cultures which had been grown for several days under control conditions (fig. 2). In contrast, bFGF early *in vitro* was particularly effective, increasing ChAT activity to 327% of control after 3 days treatment, but had no significant effect if applied late, while with NGF ChAT activity increase was only 184% of control in young, but 266% in more developed cultures. The relatively minor response to insulin and to bFGF after 8 days *in vitro* did not reflect a principal inability of the cultures to respond. If the cultures were grown in presence of growth factors for 6 days instead of 3 after a delay of 8 days, bFGF increased ChAT activity to 166 %, insulin to 165 % and NGF to 574 % of control (data not shown). The large effect of bFGF on ChAT activity in young cultures is not due to enhanced survival of freshly dissociated neurons during the first days in culture as observed with cells from various brain areas in serum-free medium (Morrison et al., 1986; Walicke et al., 1986; Walicke 1988) or to increased mitosis of neuronal precursors (Gensburger et al., 1987). The total number of neurons identified with neurofilament immunocytochemistry after two days in vitro was not significantly increased with any growth factor treatment (data not shown). These findings provide further evidence that insulin and the IGFs, bFGF or NGF stimulate different mechanisms, possibly at different times, during cholinergic neuronal development.

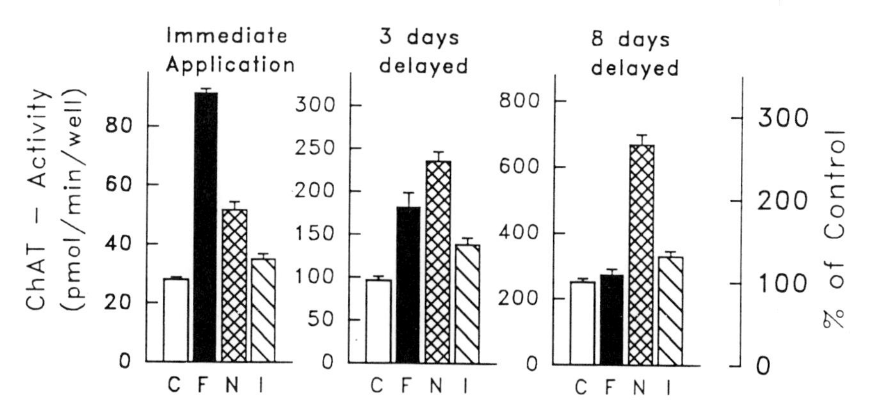

Fig. 2. Differences in responsiveness of basal forebrain cholinergic neurons to insulin, bFGF and NGF during development in culture. Growth factors were added to the cultures after varying time intervals after plating of the cells. Each set of cultures was treated for 3 days and then taken for ChAT assay. While the response to insulin did not differ with time in culture, the response to bFGF was particularly pronounced early *in vitro*, to NGF late *in vitro*. Shown are means and standard errors.

Recently, two protein kinase inhibitors of microbial origin, K-252a and staurosporine, were shown to selectively inhibit the actions of NGF on the clonal PC12 cell line and on peripheral neurons while they did not interfere with the binding of NGF to its receptor (Koizumi et al., 1988; Matsuda and Fukuda, 1988; Hashimoto and Hagino, 1989). We found with basal forebrain cultures that K-252a inhibits the NGF effect also on brain cholinergic neurons but that the structurally related compound K-252b is a much less toxic and similarly potent, and therefore particularly useful, inhibitor of the trophic action of NGF on neurons (Knusel and Hefti, 1990). K-252b, as K-252a or staurosporine abolished the trophic actions of NGF, but not of insulin or bFGF in our cultures (fig. 3). This finding further demonstrates that both, insulin and most likely the IGFs, as well as bFGF, after they bind to the respective receptors, stimulate different cellular mechanisms in our cultures than NGF and that neither insulin, nor bFGF stimulation of ChAT are the result of increased endogenous release of NGF. Since staurosporine has been shown to be a potent inhibitor of the tyrosine-specific protein kinase of the insulin receptor, but, incidently, not of the IGF-I receptor (Fujita-Yamaguchi and Kathuria, 1988) our results with these inhibitors support the conclusion stated above that stimulatory effects of insulin on cholinergic neurons are mediated by IGF-I receptors.

TROPHIC ACTIONS ON PONTINE CHOLINERGIC NEURONS AND MESENCEPHALIC DOPAMINERGIC NEURONS

To assess the selectivity of trophic actions of insulin and the IGFs we treated cultures of pons, containing the cholinergic neurons of the pedunculopontine and dorsolateral tegmental nuclei with these factors. Pontine cholinergic neurons share some of the morphological characteristics of basal forebrain cholinergic neurons, having medium to large sized cell bodies and long centrally ascending axons (Woolf

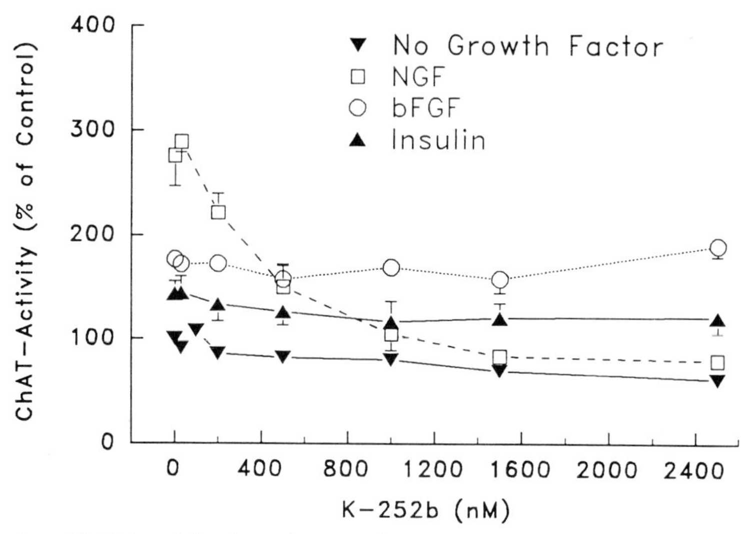

Fig. 3. ChAT activity in cultures of rat basal forebrain treated with insulin, NGF or bFGF and the protein kinase inhibitor K-252b. ChAT stimulation mediated by NGF, but not by bFGF or insulin, is completely abolished by K-252b. The cultures were grown for 7 days and treatment was for 5 days from day 2 of culture. Error bars as in fig. 1.

and Butcher, 1986). However, this cell group has been shown *in vitro* not to respond to NGF (Knusel and Hefti, 1988). Insulin, IGF-I and IGF-II markedly increased ChAT activity in pontine cultures to between 200% and 300% of control (table 2). Dose requirements for these effects were indistinguishable from the ones for the effects in basal forebrain cultures. bFGF or NGF, in contrast to insulin and the IGFs, did not enhance ChAT activity in pontine cultures.

As a third group of neurons we tested the dopaminergic cells of the substantia nigra in cultures of ventral mesencephalon for possible trophic factor effects. Insulin and the IGFs did increase transmitter-specific development of these cultures as evident in enhanced dopamine uptake in cultures which were treated for 1 week (table 2). The dose requirements were very similar to the ones for the effects on the cholinergic neurons of the basal forebrain or the pons. bFGF, but not NGF, also increased dopamine uptake in these cultures. The increase, however, was less then obtained with the members of the insulin family of growth factors (table 2).

Our results suggest a widespread action of insulin and the IGFs on neural development and differentiation. Their pronounced regulation in the rat brain during early development (Pomerance et al., 1988; Rotwein et al., 1988; Bassas et al., 1985) could indicate that their time- and site-specific presence is required for proper differentiation and growth of developing nerve cells and their function during this time might be different from later, more general, stimulating influences for many biochemical parameters. The obvious lack of selectivity for specific populations of neurons makes insulin and the IGFs seem unlikely candidates for the control of survival and development of neurons in a specific, target-dependent way as shown for NGF and also likely for the other neurotrophins (Thoenen et al., 1987; Barde, 1989; Hohn et al., 1990; Maisonpierre et al., 1990).

Table 2. Effects of growth factors on pontine cholinergic and mesencephalic dopaminergic neurons.

Growth Factor	Conc.	PONTINE CULTURES ChAT - Activity (pmol ACh/min/well)	MESENCEPHALIC CULTURES DA - Uptake (fmol DA/min/well)
Control		17.2 ± 0.2	55.3 ± 0.3
Insulin	10 μg/ml	43.2 ± 5.5*	120.8 ± 6.2*
IGF-I	1 μg/ml	42.8 ± 6.5*	136.7 ± 1.7*
IGF-II	1 μg/ml	42.7 ± 5.4*	120.5 ± 13.3*
bFGF	1 μg/ml	15.4 ± 1.6	93.3 ± 4.9*
NGF	100 ng/ml	16.9 ± 1.2	53.7 ± 7.1

Insulin and the IGFs increase transmitter-specific development in both cultures, bFGF only affects dopaminergic development, NGF has no effect. Cultures were grown for 7 days. Shown are means ± standard errors; n = 4 to 6; Differences to control: *P < 0.01 (Student's t test).

CONCLUSIONS

Insulin, IGF-I and IGF-II, their mRNA and receptors have been reported to occur and to be heterogeneously distributed in the adult mammalian brain (Dorn et al., 1982; Rotwein et al., 1988; Sara et al., 1982; Hill et al., 1986; Mendelson, 1987; Bohannon et al., 1988). None of the areas used for our cell cultures or the target regions of the neuronal populations studied have been recognized as being rich in insulin or IGF receptors in these studies. Thus, it seems surprising that insulin, IGF-I and IGF-II act similarly and profoundly on cholinergic neurons of the basal forebrain and pontine areas as well as on dopaminergic neurons of the substantia nigra in culture. However, it has been shown that the expression of insulin and IGFs and of their receptors is developmentally regulated in a complex way with generally high levels of mRNA and receptors for the IGFs (Pomerance et al., 1988; Rotwein et al., 1988; Bassas et al., 1985). The regional distributions of receptors for insulin and IGFs or mRNAs coding for receptor proteins in the embryonic brain are not known. It is conceivable that insulin and, in particular, the IGFs play a different and more widespread role during early neuronal development than in the adult brain. This concept is supported by findings that insulin and the IGFs can enhance neuronal survival and induce neurite formation and the synthesis of neural proteins in developing neurons (Bothwell, 1982; Bhat, 1983; Recio-Pinto et al., 1986; Kyriakis et al., 1987; Aizenman and de Vellis, 1987; Puro and Agardh, 1984; DiCicco-Bloom and Black, 1988). Our results, showing similar promotion of transmitter-specific development of different cholinergic and dopaminergic CNS neurons by insulin, IGF-I and IGF-II, suggest that these effects are not limited to specific neuronal populations.

Different roles in neural development of insulin and the IGFs, of bFGF and of NGF are suggested by our experiments which demonstrate that, (i) the responsiveness of basal forebrain cholinergic neurons to these factors distinctly differs with time during development of the neurons in culture; (ii) the actions of insulin (and IGF-I), NGF and bFGF are additive to each other; (iii) clear variations are observed in the selectivity of these factors for different sets of cells; and (iv) susceptibility of the insulin and bFGF trophic actions to inhibition by protein kinase inhibitors is different from that of NGF action. Many recent studies on neurotrophic actions have been based on the concept of target derived survival factors as it is suggested by the well known biology of nerve growth factor (for reviews see Barde, 1989; Thoenen et al., 1987; Thoenen and Edgar, 1985; Purves, 1986). This concept does not exclude that a multitude of different factors are required at specific ontogenetic stages for a neuron to develop properly and to maintain its structural integrity and function during adult life. Since the stimulatory actions of insulin and the IGFs on neural development and differentiation most likely are very widespread and, at least in vitro, not limited to a specific time in development, this family of growth factors seems unlikely to be involved in target controlled neuronal survival. Nevertheless, their time- and site-specific presence might be required during development and their function during this time could be different from later, more general, stimulating influences on many biochemical parameters.

REFERENCES

Avola, R., Condorelli, D.F., Surrentino, S., Turpeenoja, L., Costa, A., and Giuffrida Stella, A.M. 1988, Effect of epidermal growth factor, and insulin on DNA, RNA, and cytoskeletal protein labeling in primary rat astroglial cell cultures. J. Neurosci. Res., 19:230-238.

Barde, Y.A. 1989, Trophic factors and neuronal survival. Neuron, 2:1525-1534.

Barde, Y.A., Davies, A.M., Johnson, J.E., Lindsay, R.M. and Thoenen, H. 1987, Brain-derived neurotrophic factor. Prog. Brain Res., 71:185-189.

Baskin, D.G., Wilcox, B.J., Figlewicz, D.P., and Dorsa, D.M. 1988, Insulin and insulin-like growth factors in the CNS. Trends in Neurosci., 11:107-111.

Bassas, L., De Pablo, F., Lesniak, M.A., and Roth, J. 1985, Ontogeny of receptors for insulin-like peptide in chick embryo tissues: Early dominance of insulin-like growth factor over insulin receptors in brain. Endocrinology, 117:2321-2329.

Bhat, N. 1983, Insulin dependent neurite outgrowth in cultured embryonic mouse brain cells. Dev. Brain Res., 11:315-318.

Bohannon, N.J., Corp, E.S., Wilcox, B.J., Figlewicz, D.P., Dorsa, D.M., and Baskin, D.G. 1988, Localization of binding sites for insulin-like growth factor I (IGF-I) in the rat brain by quantitative autoradiography. Brain Res., 444:205-213.

Bothwell, M. 1982, Insulin and somatomedin MSA promote nerve growth factor-independent neurite formation by cultured chick dorsal root ganglionic sensory neurons. J. Neurosci. Res., 8:225-231.

Burgess, S.K., Jacobs, S., Cuatrecasas, P., and Sahyoun, N. 1987, Characterization of a neuronal subtype of insulin-like growth factor I receptor. J. Biol. Chem., 262:1618-1622.

Cassel, J.C. and Kelche, C. 1989, Scopolamine treatment and fimbria-fornix lesions: Mimetic effects on radial maze performance. Physiol. Behav., 46:347-353.

DiCicco-Bloom, E., and Black, I. 1988, Insulin growth factors regulate the mitotic cycle in cultured rat sympathetic neuroblasts. Proc. Natl. Acad. Sci. USA, 85:4066-4070.

Dorn, A., Bernstein, H.G., Rinne, A., Hahn, H.J., and Ziegler, M. 1982, Insulin-like immunoreactivity in the human brain. Histochemistry, 74:293-300.

Doucet, G., Brundin, P., Descarries, L. and Bjorklund, A. 1990, Effect of prior dopamine denervation on survival and fiber outgrowth from intrastriatal fetal mesencephalic grafts. Eur. J. Neurosci., 2:279-290.

Flicker, C., Dean, R.L., Watkins, D.L., Fisher, S.K. and Bartus, R.T. 1983, Behavioral and neurochemical effects following neurotoxic lesions of a major cholinergic input to the cerebral cortex in the rat. Pharmacol. Biochem. Behav., 18:973-981.

Fujita-Yamaguchi, Y. and Kathuria, S. 1988, Characterization of receptor tyrosine-specific protein kinases by the use of inhibitors. Staurosporine is a 100-times more potent inhibitor of insulin receptor than IGF-I receptor. Biochem. Biophys. Res. Commun., 157:955-962.

Gensburger, C., Labourdette, G., and Sensenbrenner, M. 1987, Brain basic fibroblast growth factor stimulates the proliferation of rat neuronal precursor cells in vitro. FEBS Lett., 217:1-5.

Gnahn, H., Hefti, F., Heumann, R., Schwab, M. and Thoenen, H. 1983, NGF-mediated increase of choline acetyltransferase (ChAT) in the neonatal forebrain: Evidence for a physiological role of NGF in the brain? Dev. Brain Res., 9:45-52.

Grothe, C., Otto, D. and Unsicker, K. 1989, Basic fibroblast growth factor promotes in vitro survival and cholinergic development of rat septal neurons: Comparison with the effects of nerve growth factor. Neuroscience, 31:649-661.

Hartikka, J. and Hefti, F. 1988a, Development of septal cholinergic neurons in culture: plating density and glial cells modulate effects of NGF on survival, fiber growth, and expression of transmitter-specific enzymes. J. Neurosci., 8:2967-2985.

Hartikka, J., and Hefti, F. 1988b, Comparison of nerve growth factor's effects on development of septum, striatum, and nucleus basalis cholinergic neurons in vitro. J. Neurosci. Res., 21:352-364.

Hashimoto, S. and Hagino, A. 1989, Blockage of nerve growth factor action in PC12h Cells by staurosporine, a potent protein kinase inhibitor. J. Neurochem., 53:1675-1685.

Hatanaka, H., and Tsukui, H. 1986, Differential effects of nerve growth factor and glioma-conditioned medium on neurons cultured from various regions of fetal rat central nervous system. Dev. Brain Res., 30:47-56.

Hatanaka, H., Tsukui, H. and Nihonmatsu, I. 1987, Septal cholinergic neurons from postnatal rat can survive in the dissociate culture conditions in the presence of nerve growth factor. Neurosci. Lett., 79:85-90.

Hefti, F., Hartikka, J. and Knusel, B. 1989, Function of neurotrophic factors in the adult and aging brain and their possible use in the treatment of neurodegenerative diseases. Neurobiol. Aging, 10:515-533.

Hefti, F., Hartikka, J., Eckenstein, F., Gnahn, H., Heumann, R. and Schwab, M. 1985, Nerve growth factor (NGF) increases choline acetyltransferase but not survival or fiber outgrowth of cultured fetal septal cholinergic neurons. Neuroscience, 14:55-68.

Hill, J.M., Lesniak, M.A., Pert, C.B., and Roth, J. 1986, Autoradiographic localization of insulin receptors in rat brain: prominence in olfactory and limbic areas. Neuroscience, 17:1127-1138.

Hohn, A., Leibrock, J., Bailey, K. and Barde, Y.A. 1990, Identification and characterization of a novel member of the nerve growth factor/brain-derived neurotrophic factor family. Nature 344:339-341.

Honegger, P., Lenoir, D. 1982, Nerve growth factor (NGF) stimulation of cholinergic telencephalic neurons in aggregating cell cultures. Dev. Brain Res., 3:229-239.

Johnston, M.V., Ruthkowski, J.L., Wainer, B.H., Long, J.B. and Mobely, W.C. 1987, NGF effects on developing forebrain cholinergic neurons are regionally specific. Neurochem. Res., 12:985-994.

Knusel, B., and Hefti, F. 1988, Nerve growth factor promotes development of rat forebrain but not pedunculopontine cholinergic neurons in vitro; lack of effect of ciliary neuronotrophic factor and retinoic acid. J. Neurosci. Res., 21:365-375.

Knusel, B., Michel, P.P., Schwaber, J.S. and Hefti, F. 1990a, Selective and nonselective stimulation of central cholinergic and dopaminergic development in vitro by nerve growth factor, basic fibroblast growth factor, epidermal growth factor, insulin and the insulin-like growth factors I and II. J. Neuroscience, 10:558-570.

Knusel, B. and Hefti, F. 1990, NGF stimulation, but not bFGF or insulin stimulation, of brain neuron development *in vitro*, is blocked by protein kinase inhibitors K-252a, K-252b, and staurosporine. (Submitted).

Knusel, B., Winslow, J.W., Rosenthal, A., Burton, L.E., Seid, D.P., Nikolics, K. and Hefti, F. 1990b, Promotion of central cholinergic and dopaminergic neuron differentiation by brain-derived neurotrophic factor but not neurotrophin-3. (Submitted).

Koh, S. and Loy, R. 1989, Localization and development of nerve growth factor-sensitive rat basal forebrain neurons and their afferent projections to hippocampus and neocortex. J. Neurosci., 9:2999-3018.

Koizumi, S., Contreras, M.L., Matsuda, Y., Hama, T., Lazarovici, P. and Guroff, G. 1988, K-252a: a specific inhibitor of the action of nerve growth factor on PC12 cells. J. Neurosci., 8:715-721.

Kyriakis, J.M., Hausman, R.E., and Peterson, S.W. 1987, Insulin stimulates choline acetyltransferase activity in cultured embryonic chicken retina neurons. Proc. Natl. Acad. Sci. USA, 84:7463-7467.

Lenoir, D., and Honegger, P. 1983, Insulin-like growth factor I (IGF-I) stimulates DNA synthesis in fetal rat brain cell cultures. Dev. Brain Res., 7:205-213.

Maisonpierre, P.C., Belluscio, L., Squinto, S., Ip, N.Y., Furth, M.E., Linsay, R.M. and Yancopoulos, G.D. 1990, Neurotrophin-3: A neurotrophic factor related to NGF and BDNF. Science, 247:1446-1451.

Martinez, H.J., Dreyfus, C.F., Jonakait, G.M. and Black, I.B. 1987, Nerve growth factor selectively increases cholinergic markers but not neuropeptides in rat basal forebrain in culture. Brain Res., 412:295-301.

Matsuda, Y. and Fukuda, J. 1988, Inhibition by K-252a, a new inhibitor of protein kinase, of nerve growth factor-induced neurite outgrowth of chick embryo dorsal root ganglion cells. Neurosci. Lett., 87:11-17.

Mendelsohn, L.G. 1987, Visualization of IGF-II receptors in rat brain, in: "Insulin, Insulin-like Growth Factors, and Their Receptors in the Central Nervous System," M.K. Raizada, M.I. Phillips, D. LeRoith, eds. Plenum Press, New York, pp. 269-275.

Mill, J.F., Chao, M.V., and Ishii, D.N. 1985, Insulin, insulin-like growth factor II, and nerve growth factor effects on tubulin mRNA levels and neurite formation. Proc. Natl. Acad. Sci. USA, 82:7126-7130.

Mobley, W.C., Rutkowski, J.L., Tennekoon, G.I., Buchanan, K.and Johnston, M.V. 1985, Choline acetyltransferase activity in striatum of neonatal rats increased by nerve growth factor. Science, 229:284-287.

Mobley, W.C., Rutkowski, J.L., Tennekoon, G.I., Gemski, J. Buchanan, K.and Johnston, M.V. 1986, Nerve Growth Factor increases choline acetyltransferase activity in developing basal forebrain neurons. Mol. Brain Res., 1:53-62.

Morrison, R.S., Sharma, A., DeVellis, J. and Bradshaw, R.A. 1986, Basic fibroblast growth factor supports the survival of cerebral cortical neurons in primary culture. Proc. Natl. Acad. Sci. USA, 83:7537-7541.

Pomerance, M., Gavaret, J.-M., Jacquemin, C., Matricon, C., Toru- Delbauffe, D., and Pierre, M. 1988, Insulin and insulin-like growth factor 1 receptors during postnatal development of rat brain. Dev. Brain Res., 42:77-83.

Prochiantz, A.M., Di Porzio, U., Kato, A., Berger, B. and Glowinsky, J. 1979, *In vitro* maturation of mesencephalic dopaminergic neurons from mouse embryos is enhanced in presence of their striatal target cells. Proc. Natl. Acad. Sci. USA, 76:5387-5391.

Puro, D.G., Agardh, E. 1984, Insulin-mediated regulation of neuronal maturation. Science 225,1170-1172.

Purves, D. 1986, The trophic theory of neural connections. Trends Neurosci., 9:486-489.

Recio-Pinto, E., Rechter, M.M., and Ishii, D.N. 1986, Effects of insulin, insulin-like growth factor-II and nerve growth factor on neurite formation and survival in cultured sympathetic and sensory neurons. J. Neurosci., 6:1211-1219.

Rotwein, P., Burgess, S.K., Milbrandt, J.D., and Krause, J.E. 1988, Differential expression of insulin-like growth factor genes in rat central nervous system. Proc. Natl. Acad. Sci. USA, 85:265-269.

Sara, V.R., Hall, K., von Holtz, H., Humber, R., Sjogren, B., and Wetterberg, L. 1982, Evidence for the presence of specific receptors for insulin-like growth factors 1 (IGF-1) and 2 (IGF-2) and insulin throughout the adult human brain. Neurosci. Lett., 34:39-44.

Shalaby, I.A., Kotake, C., Hoffmann, P.C. and Hellelr, A. 1983, A release of dopamine from coaggregate cultures of mesencephalic tegmentum and corpus striatum. J. Neurosci., 3:1565-1571.

Snider, W.D. and Johnson, E.M. Jr. 1989, Neurotrophic molecules. Annals Neurol., 26:489-506.

Thoenen, H., and Edgar, D. 1985, Neurotrophic factors. Science, 229:238-242.

Thoenen, H., Bandtlow, C., and Heumann, R. 1987, The physiological function of nerve growth factor in the central nervous system: comparison with the periphery. Rev. Physiol. Biochem. Pharmacol., 109:145-178.

Tilson, H.A., McLamb, R.L., Shaw, S., Rogers, B.C., Pediaditakis, P. and Cook, L. 1988, Radial-arm maze deficits produced by colchicine administered into the area of the nucleus basalis are ameliorated by cholinergic agents. Brain Res., 438:83-94.

Walicke, P. Cowan, W.M., Ueno, N., Baird, A. and Guillemin, R. 1986, Fibroblast growth factor promotes survival of dissociated hippocampal neurons and enhances neurite extension. Proc. Natl. Acad. Sci. USA, 83:3012-3016.

Walicke, P.A. 1988, Basic and acidic fibroblast growth factors have trophic effects on neurons from multiple CNS regions. J. Neurosci., 8:2618-2627.

Woolf, N.J., and Butcher, L.L., 1986, Cholinergic systems in the rat brain: III. Projections from the pontomesencephalic tegmentum to the thalamus, tectum, basal ganglia, and basal forebrain. Brain Res. Bull,. 16:603-637.

SECOND MESSENGERS MEDIATING GENE EXPRESSION ESSENTIAL TO NEURITE FORMATION DIRECTED BY INSULIN AND INSULIN-LIKE GROWTH FACTORS

Douglas N. Ishii, Chiang Wang, and Yi Li

Physiology and Biochemistry Departments
Colorado State University
Fort Collins, Colorado 80523

INTRODUCTION

The nervous system confers an enormous survival advantage to higher animals by permitting behavioral adaptation to both external and internal environmental cues. Insulin and insulin-like growth factors (IGFs) are among regulatory signals implicated in the development of the neural circuitry (reviewed in Ishii et al., 1985; Recio-Pinto and Ishii, 1988a). Although the structure of the nervous system is profoundly complex, it is hoped that a reductionist approach, aimed at first understanding the cellular and molecular mechanisms regulating neurite (axon and dendrite) formation and extension, will eventually provide a foundation for understanding higher orders of organization. With this in mind, the present article contains discussion on our strategy and recent findings directed at elucidating the biochemical pathway through which insulin homologs and nerve growth factor (NGF) regulate the long term, stable extension of neurites.

NEUROPHYSIOLOGY

The neurobiological, behavioral, and electrical effects of insulin and IGFs have been reviewed, as well as the CNS distribution of these factors, their transcripts, and their receptors in the brain of vertebrates (Recio-Pinto and Ishii, 1988a). It is revealing that IGF-II mRNA content is highest in brain and spinal cord of adult rats (Soares et al., 1985; 1986). A close correlation has been found between expression of the IGF-II gene in skeletal muscle with development and regeneration of synapses (Ishii, 1989; Ishii, 1990). The prediction from hypothesis (Ishii, 1989) that IGF administration would increase motor nerve terminal sprouting has been tested and validated in rodents (Caroni and Grandes, 1990). Insulin, IGF-I and/or IGF-II increase neurite outgrowth in cultured sensory, sympathetic (Recio-Pinto et al., 1986), and motor (Caroni and Grandes, 1990) neurons, and human neuroblastoma cells (Recio-Pinto and Ishii, 1984). The insulin homologs additionally support survival of peripheral and central neurons (Recio-Pinto et al., 1986; Aizenman and de Vellis, 1987).

MECHANISM OF NEURITE OUTGROWTH

Brain Insulin and IGF Receptors

In order to understand more clearly how axons and dendrites grow in response to a neuritogenic polypeptide, elucidation of the biochemical pathway is essential. One approach is to identify the receptor, which is the initial site of interaction between ligand and cell, then study the sequence of events triggered by receptor occupancy.

In this discussion we shall refer to the type I and type II IGF binding sites. The type I IGF site has a higher affinity for IGF-I than IGF-II, and a low affinity for insulin (Kasuga et al., 1982; Rechler et al., 1983). It is similar in structure to the insulin receptor (Massague et al., 1980).

The type II site has a high affinity for IGF-II and does not bind insulin. Early studies reported that the type II site binds IGF-I with affinity lower than IGF-II. However, it is recently found that highly purified IGF-I from recombinant sources does not bind to type II sites (Rosenfeld et al., 1987). The type II site is structurally dissimilar to insulin receptors, and is simultaneously the mannose-6-phosphate receptor (Morgan et al., 1987; MacDonald et al., 1988), a lysosomal enzyme transport protein. The function of type II sites in relation to neurite outgrowth is still unclear. All actions of IGF-II on neurite outgrowth seems readily explained at the moment by its binding to type I sites.

Multiple Receptors Regulate Neurite Outgrowth on SH-SY5Y Cells

Insulin and IGF Receptors on SH-SY5Y Cells. The heterogeneous nature of binding to brain tissues has made it very difficult to assign binding to specific neuronal functions, and the role of brain insulin and IGF receptors have remained largely speculative. The cloned SH-SY5Y cell provides an attractive alternative model (Recio-Pinto and Ishii, 1988b). The K_d of 3 nM for reversible binding to high affinity insulin sites compares favorably with the half-maximally effective concentration of 4 nM for insulin-enhanced neurite outgrowth in sensory and sympathetic, as well as SH-SY5Y cells. Type I IGF sites are also present on SH-SY5Y cells.

At low physiological concentrations, insulin enhances neurite outgrowth through insulin receptors, whereas IGF-I and IGF-II enhances neurite outgrowth through type I IGF receptors. Because IGF-I from recombinant sources does not bind to type II sites (Rosenfeld et al., 1987), the effects of low IGF-I concentration must be due to occupancy of type I receptors. Cross occupancy of insulin and type I sites at supra-physiological concentrations of insulin homologs is likely to explain their broad dose-response curves for neurite outgrowth.

Nerve Growth Factor Receptors on SH-SY5Y Cells. High (slow) and low (fast) affinity nerve growth factor (NGF) receptors have been identified on human neuroblastoma cell lines, and the neurite outgrowth response is correlated with the high affinity sites (reviewed in Sonnenfeld and Ishii, 1985). Only high affinity NGF receptors are detected on SH-SY5Y cells.

PHOSPHORYLATION AND TRANSMEMBRANE SIGNALING

It is now well known that the insulin and type I IGF receptors contain a tyrosine kinase on beta-subunits. The potential role of phosphorylation as the transmembrane signal for NGF as well as insulin homologs was previously reviewed (Ishii and Mill, 1987). A number of phosphorylated substrates have been identified, and an overlapping pattern of phosphorylation is produced by insulin and NGF in PC12 cells (Halegoua and Patrick, 1980). Insulin and IGF-I likewise produce an overlapping pattern of phosphorylation in neuroblastoma (Shemer et al., 1987) and other cell types (Kadowaki et al., 1987). Although it seems likely that a substrate molecule is phosphorylated in common by NGF, insulin and IGF and sits near the beginning of a shared biochemical pathway for neurite outgrowth, a difficult challenge is faced in determining and identifying the cellular functions of phosphorylated substrates.

Another difficulty is that a hormone such as insulin triggers a large number of responses in cells, such as increased DNA, RNA, protein, and lipid synthesis, and glucose uptake. How does one, then, assign a particular phosphorylated substrate to a specific pathway, in this case neurite outgrowth?

Paradigm to Elucidate Pathway for Neurite Outgrowth

The biochemical pathway regulating neurite outgrowth would be difficult to elucidate in primary neurons which require NGF or insulin homologs for survival. An apparent increase in content of a metabolite in treated cells might instead be due to prevention of decline in that metabolite relative to slowly dying, untreated cells. Primary neuronal cultures may also contain non-neurons or several types of neurons which can confound the analysis. This problem is circumvented in clonal SH-SY5Y cells in which serum or other exogenous factors are not required for survival and cells can survive without loss of numbers for at least a week (Sonnenfeld and Ishii, 1982).

In order to overcome the various difficulties discussed above, a general paradigm was proposed (Ishii et al., 1985). One might be able to identify a biochemical pathway for neurite outgrowth under shared regulation by multiple receptors for insulin, IGFs, and NGF in the cloned human neuroblastoma SH-SY5Y cell. It was hoped to identify metabolites regulated by all three of the neuritogenic factors. This would improve the probability that such metabolites play a role in neurite outgrowth. Moreover, certain key metabolites known to be major components in the neurite growth pathway could be selected for study (see below). Variant cell lines unresponsive by neurite outgrowth to NGF, insulin, and tumor promoting phorbol esters have been identified, and their receptors studied (Sonnenfeld and Ishii, 1982; Recio-Pinto and Ishii, 1984; Spinelli et al., 1982).

The paradigm has proven to be a powerful tool for analysis, and segments of a common pathway have been identified (Ishii and Mill, 1987). Advances and refinements to the model are discussed below.

SHARED UP REGULATION OF α- AND β-TUBULIN GENE EXPRESSION

The long-term growth of axons over substantial distances, such as occurs during development or regeneration, requires synthesis of those proteins which constitute the cytoskeleton of neurites. The three major classes of filamentous proteins in neurites are microtubules, neurofilaments, and microfilaments.

Alpha- and beta-tubulin heterodimers polymerize to form microtubules, which serve as components of the cytoskeleton and, also, as "tracks" for axonal transport. It seemed plausible that neuritogenic polypeptides might selectively increase mRNAs encoding these filamentous proteins, and this possibility was tested.

Dose response studies show that concentrations of insulin which occupy insulin receptors (Recio-Pinto and Ishii, 1988b) and promote neurite outgrowth (Recio-Pinto and Ishii, 1984) also selectively increase alpha- and beta-tubulin mRNA content in SH-SY5Y cells (Mill et al., 1985). The relative synthesis of tubulins is increased (Fernyhough et al., 1989). IGF-I (Fernyhough et al., 1989) and IGF-II (Mill et al., 1985), likewise, elevate tubulin mRNAs. Interestingly, insulin by itself is unable to induce neurites (Recio-Pinto et al., 1984) or increase tubulin mRNAs (Fernyhough and Ishii, 1987) in PC12 cells. However, both responses are synergistically potentiated in combination with NGF.

Comparison studies with NGF are highly revealing. NGF has a number of effects on microtubules. It can increase the cellular content of tubulin mRNAs in PC12 (Fernyhough and Ishii, 1987), sensory (Ishii and Mill, 1987), and sympathetic (unpublished) cultures. Tubulin proteins (Drubin et al., 1985), and microtubules (Angeletti et al., 1971) are increased. Moreover, microtubules are stabilized against depolymerization (Black and Greene, 1982). Tubulin mRNA content is increased during nerve regeneration (Neumann et al., 1983), although the signal for this event has not yet been identified. Thus, there is a strong correlation between the capacity of insulin, IGFs, and NGF to increase tubulin mRNA content and enhance neurite outgrowth.

SHARED UP REGULATION OF 68 kDa AND 170 kDa NEUROFILAMENT GENES

Neurofilaments (NF) belong to another abundant class of fibrillar elements of neurites, and NF genes are expressed only in neural tissues (Julien et al., 1985; Myers et al., 1987). The 68 kDa and 170 kDa NF proteins form the central core of neurofilaments, whereas the 200 kDa NF proteins forms cross-bridges (Metuzals et al., 1981; Hirokawa et al., 1984; Trojanowski et al., 1985). NF protein content increases during neurite outgrowth (Angeletti et al., 1971; Black et al., 1986; Lindenbaum et al., 1987), and the caliber of neurites is suggested to be regulated by NF gene expression (Hoffman et al., 1987).

NF protein content and expression of the 68 kDa and 170 kDa NF genes is increased during neurite outgrowth directed by NGF in PC12 cells (Lindenbaum et al., 1987; 1988). Moreover, 68 kDa and 170 kDa NF gene expression is increased by insulin and IGFs in SH-SY5Y cells (Wang et al., 1988; Ishii et al., 1989), showing that these are other key metabolites of the neurite outgrowth pathway regulated in common by all of these neuritogenic polyeptides.

The elevation of tubulin and NF mRNAs is selective. Increases can be shown relative to total RNA, poly(A)$^+$ RNA, histone mRNA, and actin mRNA. With respect to the latter, there is actually a small increase in response to NGF (Lindenbaum et al., 1988). We also have sometimes observed a small increase in actin mRNA in response to insulin. However, due to difficulty in characterizing small increases, we have considered actin mRNA as relatively invariant in our analyses. Actin polymerizes to form microfilaments.

"LOOSE COORDINATE" REGULATION OF TUBULIN AND NEUROFILAMENT GENE EXPRESSION

Some form of temporal regulation would be needed to ensure that the various proteins which comprise the axonal cytoskeleton were produced together at the proper time. It therefore became of interest to determine whether there was tight temporal regulation of tubulin and NF gene expression.

Alpha- and beta-tubulin genes are dispersed across different chromosomes (Havercroft and Cleveland, 1984). Equal number of alpha- and beta-tubulins would be needed to produce the heterodimers that assemble to form microtubules. It was intriguing to consider whether there was coordinate gene expression. The time courses for expression of alpha- and beta-tubulin mRNAs are not precisely the same in response to insulin (Mill et al., 1985) or NGF (Fernyhough and Ishii, 1987), indicating independent regulation. Yet the expression of heterogeneous tubulin transcripts is sufficiently close in time and space as to be considered approximately coordinate, and the term "loose coordinate" regulation will be used in reference to this phenomenon.

Would approximate "loose coordinate" expression also be found for NFs, which belong to a different class of structural genes? Time course studies show that 68 kDa NF and 170 kDa NF genes indeed are elevated independently, but in a temporal manner that is "loose coordinate".

It is interesting that α-tubulin, β-tubulin, 68 kDa, and 170 kDa mRNAs are all independently regulated yet can accumulate at about the same times. The composite data for SH-SY5Y cells is shown in Fig. 1. It is curious that these mRNAs all peak then decline to a lower level (but above baseline), although neurite outgrowth remains elevated (discussed in Ishii and Mill, 1987). A similar time-course is seen in response to NGF in PC12 cells (Fernyhough and Ishii, 1987) and in sensory neurons (unpublished).

Thus, there is "loose coordinate" regulation in expression not only of genes which contribute different subunits to a given class of filaments such as tubulin, but also of genes which contribute subunits to a separate class of filaments such as neurofilaments.

These data suggest that "loose coordinate" regulation may permit greater flexibility in individual gene expression needed

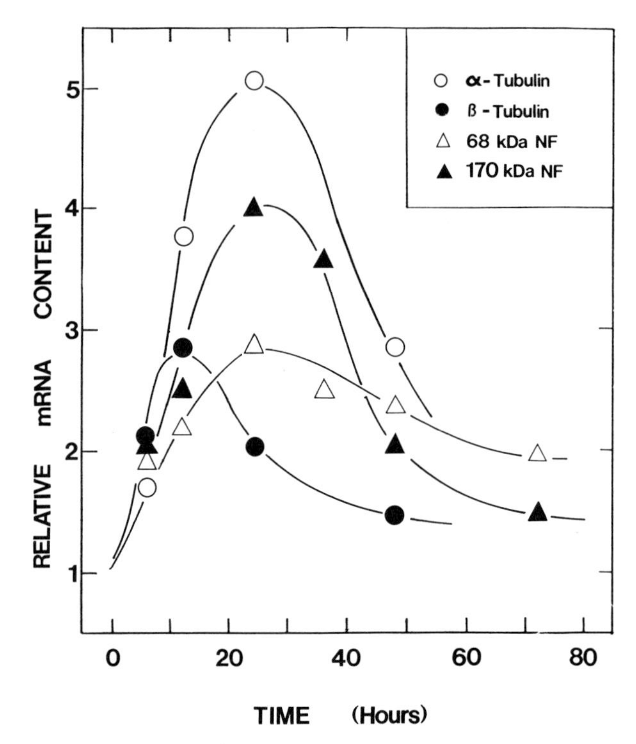

Fig. 1. Time-dependent "loose coordinate" change in tubulin and neurofilament mRNA content in human neuroblastoma SH-SY5Y cells exposed to insulin. Cells were cultured without or with 0.1 μM insulin. The change in specific mRNA content of insulin treated cells is shown relative to values in untreated cultures, which remains essentially invariant hence are not shown. Tubulin and NF genes are dispersed across different chromosomes and are independently regulated. Nevertheless, a "loose coordinate" regulation in time and space of α-tubulin, β-tubulin, 68 kDa NF, and 170 kDa NF mRNAs is observed.

to meet various circumstances. For example, it is reported that during nerve regeneration NF gene expression is temporarily reduced while tubulin gene expression is elevated (Hoffman et al., 1987). One reason this may occur is that the caliber of axons is reduced during regeneration, and NF gene expression is correspondingly down regulated. It is thought that cross-linking by NFs might be inhibitory to the early axon extension process.

On the other hand one would not expect down regulation of NF gene expression in SH-SY5Y or PC12 cells. In this situation neuritogenic factors induce more and longer neurites, and it is not unexpected that NF gene expression up regulates.

TUBULIN AND NF GENES ARE REGULATED BY DIFFERENT MECHANISMS

Normal coordinate regulation might be achieved by a single mechanism. However, a more complex scheme is not unexpected for "loose coordinate" regulation, and has been uncovered.

Both α- and β-tubulin mRNAs are stabilized against degradation by NGF in PC12 cells (Fernyhough and Ishii, 1987) and insulin and IGFs in SH-SY5Y cells (Fernyhough et al., 1989). Histone and actin mRNAs are neither increased nor stabilized. Nuclear run-off experiments show that insulin has no significant effect on transcription rates (Fernyhough et al., 1989), and it appears that tubulin mRNA increases are regulated predominantly by transcript stabilization.

In contrast, the 68 kDa and 170 kDa NF transcripts appear to be only very weakly stabilized by NGF in PC12 cells (Lindenbaum et al., 1988) and insulin in SH-SY5Y cells (unpublished). However, there is a substantial increase in nuclear run-off rates in response to NGF (Lindenbaum et al., 1988) and insulin (unpublished). The effects of IGFs on these processes is still under examination. Nevertheless, it appear from limited data that the tubulin mRNAs and the NF mRNAs are up regulated by fundamentally different strategies.

This line of study has revealed that there is a common biochemical pathway regulated by insulin, IGFs, and NGF. The predominant regulation of tubulin mRNAs is by transcript stabilization, whereas regulation of NF mRNAs is by increased transcription (Fig 2). With these results the problem of identification of the transmembrane event has been enormously simplified. Instead of trying to correlate changes in putative second messengers with the amorphous endpoint of neurite outgrowth, it now seems possible to correlate changes in second messenger content with increased stabilization or transcription of specific mRNAs.

These data also permit the inference that there must be a common transmembrane signal(s). This is consistent with the observation that the insulin receptor and type I IGF receptors are tyrosine kinases, and that binding to high affinity NGF receptors results in phosphorylation. Preliminary studies reveal a close correspondence between occupancy of insulin receptors and phosphorylation of a 34 kDa protein in SH-SY5Y cells (Ishii and Mill, 1987).

IS TUBULIN AND NEUROFILAMENT GENE EXPRESSION REGULATED BY cAMP?

The hypothesis that cAMP might be a signal that mediates the increases in tubulin and NF mRNAs was tested in SH-SY5Y cells (Wang et al., 1990). Dibutyryl cAMP increased α-tubulin, β-tubulin, 68 kDa NF, and 170 kDa NF mRNA content in SH-SY5Y

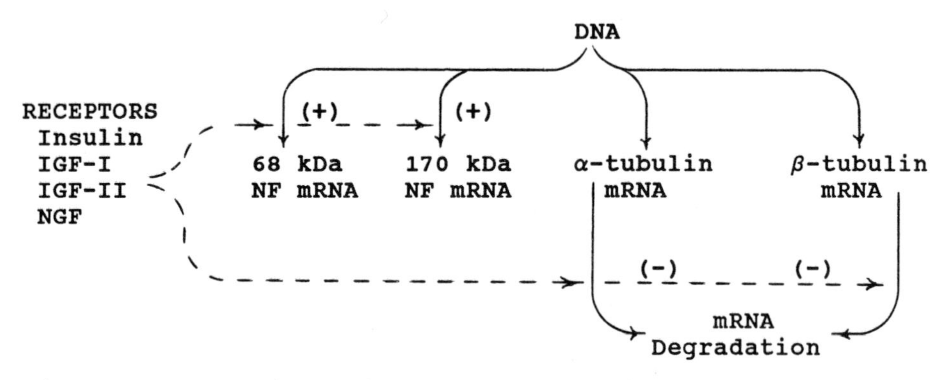

Fig. 2. Common biochemical pathway utilized by insulin, IGFs, and NGF to selectively elevate content of mRNAs encoding structural proteins of neurites in SH-SY5Y cells. Dashed lines show predominant regulation: (+) stimulation, (-) inhibition.

cells. Moreover, there was a synergistic potentiation when cells were treated with the combination of dibutyryl cAMP and insulin. In contrast, there was no change in histone mRNA content.

Forskolin and cholera toxin, which respectively are direct and indirect activators of adenylate cyclase, increased the content of tubulin and NF mRNAs. Increases in tubulin and NF mRNAs were also registered upon treatment of cells with caffeine, theophylline, and IBMX, which are all inhibitors of cAMP phosphodiesterase. These effects were not restricted to SH-SY5Y cells, and 8-bromo-cAMP increased tubulin and NF mRNAs in PC12 cells as well.

If cAMP were to mediate the effects of insulin and IGFs, one might predict that increased cAMP would increase neurite outgrowth. This prediction was tested, and SH-SY5Y cells responded with neurite outgrowth to dibutyryl cAMP and to forskokin. These results agree with the observation that the analogs 8-(4-chlorophenylthio)-cAMP and 8-bromo-cAMP can increase neurite outgrowth, but do not mimic all of the effects of NGF in sympathetic and sensory neurons (Rydel and Greene, 1988).

It is intriguing that cAMP replicates the capacity of NGF, insulin, and IGFs to increase α-tubulin mRNA, β-tubulin mRNA, 68 kDa NF mRNA, and 170 kDa NF mRNA as well as neurite outgrowth. On the other hand, cAMP would not be expected to replicate all actions of these factors if other second messengers were generated by the transmembrane event.

Nevertheless, caution is necessary in the interpretation of these data. For example, IBMX may not be specific for cAMP phosphodiesterase, and conceivably may inhibit at other sites. Although forskolin has been shown to increase cAMP content in SH-SY5Y cells (Mei et al, 1989), it must be kept in mind that other proteins may be inhibited (Laurenza et al., 1989).

Moreover, NGF increases adenylate cyclase activity, but only transiently in PC12 cells (Schubert and Whitlock, 1977; Golubeva et al., 1989). It remains to be studied whether insulin and IGFs actually increase cAMP content in SH-SY5Y cells.

PUTATIVE ROLE OF G PROTEINS

IGF-I binding to its receptor can modify activity of G-proteins in 3T3 cells (Nishimoto et al., 1987). The transient activation of adenylcyclase is temporally linked to G-proteins in PC12 cells treated with NGF (Golubeva et al., 1989). Moreover, microinjection of ras p21 protein, which has properties similar to G-proteins, causes neurite outgrowth in PC12 cells (Satoh et al., 1987), whereas injection of antibodies to v-ras p21 inhibits NGF-dependent neurite outgrowth (Hagag et al., 1986). These studies suggest the potential involvement of G-proteins in neurite outgrowth.

Our observation that cholera toxin can increase NF mRNA content suggests G-proteins might be signal transducers in SH-SY5Y cells as well. Cholera toxin is known to ADP-ribosylate an arginine residue and persistently activate $G_{s\alpha}$, resulting in stimulation of adenylate cyclase (reviewed in Freissmuth et al., 1989). These results together support the following provisional model: Insulin, IGFs, and NGF may activate G proteins which subsequently increase adenylate cyclase activity. This, in turn, may increase the content of tubulin and NF mRNAs. If this were correct it would be an exciting development. However, additional studies are needed before this model can have credence.

PROTEIN KINASE C MEDIATES ACTIONS OF NGF, INSULIN AND IGFs

PKC Agonists Induce Neurite Outgrowth

Tumor promoting phorbol esters, such as 12-O-tetradecanoyl-phorbol-13-acetate (TPA), are activators of protein kinase C (PKC). They can induce neurite outgrowth (Spinelli et al., 1982) in a manner closely correlated with binding to PKC in SH-SY5Y cells (Spinelli and Ishii, 1983; Ishii et al., 1985). NGF induces PKC activity (Chan et al., 1989; Contreras and Guroff, 1987), and TPA then becomes able to potentiate the response to NGF (Hall et al., 1988) in PC12 cells. Insulin stimulates PKC in adipocytes (Draznin et al., 1988) and in mammary tumor cells (Gomez et al., 1988).

PKC Antagonists Inhibit Neurite Outgrowth

Inhibitors of PKC, such as sphingosine (Hall et al., 1988), K252a (Koizumi et al., 1988) and staurosporine (Hashimoto and Hagino, 1989) inhibit NGF-dependent neurite outgrowth and phosphorylation. Sphingosine is a reversible inhibitor of neurite outgrowth. It competitively inhibits TPA binding and activation of PKC (Hannun et al., 1986; Merrill et al., 1986). It is particularly interesting that sphingosine can inhibit with the same sensitivity neurite outgrowth directed by NGF, insulin, and IGF in cultured sympathetic neurons (Ishii et al., 1989). This suggests that a common mechanism is involved, and that PKC is in a shared segment of the biochemical pathway regulating neurite outgrowth by these factors.

It is well known that these inhibitors of PKC can be nonspecific and act at other sites. However, NGF is continuously required for neurite outgrowth in PC12 cells, and inhibition of NGF action by sphingosine can be reversed by TPA (Hall et al., 1988). This is quite important, because if sphingosine were inhibiting at any other site the inhibition would not be reversed by TPA. Moreover, Linweaver-Burke type double reciprical plots show that sphingosine competitively inhibits neurite outgrowth stimulated by tumor promoter agonists in SH-SY5Y cells, and sphingosine inhibition can be antagonized by various phorbol ester analogs in a manner which clearly identifies PKC as the site of sphingosine inhibition (P. Fernyhough, F.L. Hall, and D.N. Ishii, unpublished). These results strongly support the interpretation that PKC is a mediator of insulin, IGF, and NGF-directed neurite outgrowth.

While PKC inhibitors do block neurite outgrowth, they do not prevent the increase in tubulin mRNAs caused by NGF in PC12 cells (Hall et al., 1988) or insulin and IGF on NF mRNAs in SH-SY5Y cells (unpublished). PKC evidently acts at a distal site in the neurite outgrowth pathway.

Down Regulation and Isozymes of PKC

A source of potential confusion is that NGF is able to induce neurites in PC12 cells despite down regulation of PKC caused by exposure to a high concentration of TPA (Reinhold and Neet, 1989; Damon et al., 1990). However, down regulation is a quite unreliable test of total PKC activity.

Multiple PKC isozymes (7 presently) have been discovered (Jaken and Kiley, 1987; Kikkawa et al., 1987; Nishizuka, 1988), and these isozymes have differential susceptibilities to being down regulated (Huang et al., 1989a). Reinhold and Neet (1989) tested down regulation using histone as a substrate for phosphorylation. However, TPA treatment of cells can lead to down regulation of PKC-dependent histone but not PKC-dependent vinculin phosphorylation (Cochet et al., 1986; Cooper et al., 1987). Since histone is not implicated in the neurite growth pathway, its selection as PKC substrate has no relevance to the problem at hand.

The activity of detergent insoluble species of PKC (Huang et al., 1989b; Moss et al., 1990) was not measured by Reinhold and Neet (1989). This is particularly important in neural tissues because there the majority of PKC is membrane bound (Neary et al., 1988). Therefore the isotype of PKC which is down regulated by TPA in PC12 cells may well be different from the isotype which regulates neurite outgrowth. This interpretation may help explain the observation that TPA does not down regulate binding to PKC in SH-SY5Y cells (Spinelli et al., 1982; Spinelli and Ishii, 1983). Neurite outgrowth remains elevated in cultures grown in the presence of TPA. We and others caution against the use of down regulation unless it can be shown which PKC isotype is involved.

SUMMARY AND DISCUSSION

In order for a second messenger to be considered as a signal for neurite outgrowth in our identification scheme, the signal will have to be increased by insulin and IGFs as well as

NGF. Because neurite outgrowth is reversible on withdrawal of factors, the signal would have to be constantly elevated in the presence of a neuritogenic factor. Moreover, neurite outgrowth directed by insulin, IGF, or NGF should show the same sensitivity to inhibition by compounds which antagonize a messenger located in a common pathway.

Signals that activate PKC are likely to fulfill these criteria. PKC does not mediate the actions of factors on NF and tubulin mRNAs, but does mediate "downstream" events. The responses to NGF have been categorized as transcription-dependent and transcription-independent in PC12 cells (Greene et al., 1980). These results suggest PKC may mediate at least some of the transcription independent events. This is in keeping with its localization in neurites and growth cones (Wood et al., 1986; Girard et al., 1988; Saito et al., 1988), and capacity to phorphorylate GAP43 (reviewed in Gispen et al., 1990), a protein enriched in growth cones and correlated with neuronal growth and regeneration (Skene and Willard, 1981). The localization of PKC at nerve terminals is important support for the hypothesis that IGFs produced in tissues during development and regeneration act on the axon terminals (Ishii, 1989).

Occupancy of insulin, IGF, and NGF receptors results in selective accumulation of mRNAs encoding structural proteins of neurites. These transcription-dependent events are regulated through two different mechanisms. In the case of 68 and 170 kDa NF, the predominant mechanism is increased transcription or nuclear processing of transcripts. On the other hand, for α- and β-tubulin mRNAs the predominant mechanism is transcript stabilization. It is thought that other transcription-dependent events may fall into one of these two classes, and this hypothesis is being investigated with results to be reported elsewhere.

Our findings are consistent with a previous report that NGF actions are mediated in part by cAMP- and Ca^{++}/phospholipid-dependent protein kinases (Cremins et al., 1986). Although as yet no second messenger has been found that fulfills all of our criteria for a signal molecule(s) that regulates tubulin and NF mRNAs, cAMP and G-proteins are among promising candidates inviting further study.

ACKNOWLEDGEMENTS

We thank Diane M. Guertin for assistance in preparing this manuscript. This work was supported by National Institute of Neurological Disorders and Stroke grants RO1 NS24787 and PO1 NS28323.

REFERENCES

Aizenman, Y., and de Vellis, J., 1987, Brain neurons develop in a serum and glial free environment: effects of transferrin, insulin, insulin-like growth factor-I and thyroid hormone on neuronal survival, growth and differentiation, Brain Res., 406:32-42.

Angeletti, P. U., Levi-Montalcini, R., and Caramia, F., 1971, Analysis of the effects of the antiserum to the nerve growth factor in adult mice, Brain Res., 27:343-355.

Black, M. M., and Greene, L. A., 1982, Changes in the colchicine susceptibility of microtubules associated with neurite outgrowth: studies with nerve growth factor-responsive PC12 pheochromocytoma cells, J. Cell Biol., 95:379-386.

Black, M. M., Keyser, P., and Sobel, E., 1986, Interval between the synthesis and assembly of cytoskeletal proteins in cultured neurons, J. Neurosci., 6:1004-1012.

Caroni, P., and Grandes, P., 1990, Nerve sprouting in innervated adult skeletal muscle induced by exposure to elevated levels of insulin-like growth factors, J. Cell Biol., 110:1307-1317.

Chan, B. L., Chao, M. V., and Saltiel, A. R., 1989, Nerve growth factor stimulates the hydrolysis of glycosyl-phosphatidylinositol in PC-12 cells: A mechanism of protein kinase C regulation, Proc. Natl. Acad. Sci. USA, 86:1756-1760.

Cochet, C., Souvignet, C., Keramidas, M., and Chambaz, E. M., 1986, Altered catalytic properties of protein kinase C in phorbol ester treated cells, Biochem. Biophys. Res. Commun., 134:1031-1037.

Contreras, M. L., and Guroff, G., 1987, Calcium-dependent nerve growth factor-stimulated hydrolysis of phosphoinositides in PC12 cells, J. Neurochem., 48:1466-1472.

Cooper, D. R., de Ruiz Galaretta, C. M., Fanjul, L. F., Mojsilovic, L., Standaert, M. L., Pollet, R. J., and Farese, R. V., 1987, Insulin but not phorbol ester treatment increases phosphorylation of vinculin by protein kinase C in BC3H-1 myocytes, FEBS Lett., 214:122-126.

Cremins, J., Wagner, J. A., and Halegoua, S., 1986, Nerve growth factor action is mediated by cyclic AMP- and Ca^{+2}/phospholipid-dependent protein kinases, J. Cell Biol., 103:887-893.

Damon, D. H., D'Amore, P. A., and Wagner, J. A., 1990, Nerve growth factor and fibroblast growth factor regulate neurite outgrowth and gene expression in PC12 cells via both protein kinase C- and cAMP-independent mechanisms, J. Cell Biol., 110:1333-1339.

Draznin, B., Leitner, J., Sussman, K., and Sherman, N., 1988, Insulin and glucose modulate protein kinase C activity in rat adipocytes, Biochem. Biophys. Res. Commun., 156:570-575.

Drubin, D. G., Feinstein, S. C., Shooter, E. M., and Kirschner, M. W., 1985, Nerve growth factor-induced neurite outgrowth in PC12 cells involves the coordinate induction of microtubule assembly and assembly-promoting factors, J. Cell Biol., 101:1799-1807.

Fernyhough, P., and Ishii, D. N., 1987, Nerve growth factor modulates tubulin transcript levels in pheochromocytoma PC12 cells, Neurochem. Res., 12:891-899.

Fernyhough, P., Mill, J. F., Roberts, J. L., and Ishii, D. N., 1989, Stabilization of tubulin mRNAs by insulin and insulin-like growth factor I during neurite formation, Mol. Brain Res., 6:109-120.

Freissmuth, M., Casey, P. J., and Gilman, A. G., 1989, G proteins control diverse pathways of transmembrane signaling, FASEB J., 3:2125-2131.

Girard, P., Wood, J., Frenschi, J., and Kuo, J., 1988, Immunocytochemical localization of protein kinase C in developing brain tissue and in primary neuronal cultures, Dev. Biol., 126:98-107.

Gispen, W. H., Boonstra, J., De Graan, P. N. E., Jennekens, F. G. I., Oestreicher, A. B., Schotman, P., Schrama, L. H., Verhaagen, J., and Margolis, F. L., 1990, B-50/GAP-43 in neuronal development and repair, Restor. Neurol. Neurosci., 1:237-244.

Golubeva, E. E., Posypanova, G. A., Kondratyev, A. D., Melnik, E. I., and Severin, E. S., 1989, The influence of nerve growth factor on the activities of adenylate cyclase and high affinity GTPase in pheochromocytoma PC12 cells, FEBS Lett., 247:232-234.

Gomez, M., Medrano, E., Cafferatta, E., and Tellez-Inon, M., 1988, Protein kinase C is differentially regulated by thrombin, insulin, and epidermal growth factor in human mammary tumor cells, Exp. Cell Res., 175:74-80.

Greene, L. A., Burstein, D. E., and Black, M. M., 1980, The priming model for the mechanism of action of nerve growth factor: evidence derived from clonal PC12 pheochromocytoma cells, in: "Tissue Culture in Neurobiology," E. Giacobini, and A. Vernadakis, eds., Raven Press, New York, pp. 313-319.

Hagag, N., Halegoua, S., and Viola, M., 1986, Inhibition of growth factor-induced differentiation of PC12 cells by microinjection of antibody to ras p21, Nature, 319:680-682.

Halegoua, S., and Patrick, J., 1980, Nerve growth factor mediates phosphorylation of specific proteins, Cell, 22:571-581.

Hall, F. L., Fernyhough, P., Ishii, D. N., and Vulliet, P. R., 1988, Suppression of nerve growth factor-directed neurite outgrowth in PC12 cells by sphingosine, an inhibitor of protein kinase C, J. Biol. Chem., 263:4460-4466.

Hannun, Y. A., Loomis, C. R., Merrill, A. H. , Jr., and Bell, R. M., 1986, Sphingosine inhibition of protein kinase C activity and of phorbol dibutyrate binding in vitro and in human platelets, J. Biol. Chem., 261:12604-12609.

Hashimoto, S., and Hagino, A., 1989, Blockage of nerve growth factor action in PC12h cells by staurosporine, a potent protein kinase inhibitor, J. Neurochem., 53:1675-1685.

Havercroft, J. C., and Cleveland, D. W., 1984, Programmed expression of beta-tubulin genes during development and differentiation of the chicken, J. Cell Biol., 99:1927-1935.

Hirokawa, N. M., Glicksman, M. A., and Willard, M. B., 1984, Organization of mammalian neurofilament polypeptides within neuronal cytoskeleton, J. Cell Biol., 98:1523-1536.

Hoffman, P. N., Cleveland, D. W., Griffin, J. W., Landes, P. W., Cowan, N. J., and Price, D. L., 1987, Neurofilament gene expression: a major determinant of axonal caliber, Proc. Natl. Acad. Sci. USA, 84:3472-3476.

Huang, F. L., Yoshida, Y., Cunha-Melo, J. R., Beaven, M. A., and Huang, K. -P., 1989a, Differential down-regulation of protein kinase C isozymes, J. Biol. Chem., 264:4238-4243.

Huang, F. L., Chuang, D. -M., and Huang, K. -P., 1989b, Protein kinase C isozymes in primary cultures of cerebellar granule cells: Glutamate-stimulated membranous association of type II PKC, Soc. Neurosci. Abstr., 15:832.

Ishii, D. N., Recio-Pinto, E., Spinelli, W., Mill, J. F., and Sonnenfeld, K. H., 1985, Neurite formation modulated by nerve growth factor, insulin, and tumor promoter receptors, Int. J. Neurosci., 26:109-127.

Ishii, D. N., Glazner, G. W., Wang, C., and Fernyhough, P., 1989, Neurotrophic effects and mechanism of insulin, insulin-like growth factors, and nerve growth factor in spinal cord and peripheral neurons, in: "Molecular and cellular biology of insulin-like growth factors and their receptors," D. LeRoith, and M. K. Raizada, eds., Plenum Publishing Corporation, New York, pp. 403-425.

Ishii, D. N., 1990, Insulin-like growth factor II gene expression: Relationship to the development and regeneration of neuromuscular synapses, Restor. Neurol. Neurosci., 1:205-210.

Ishii, D. N., 1989, Relationship of insulin-like growth factor II gene expression in muscle to synaptogenesis, Proc. Natl. Acad. Sci. USA, 86:2898-2902.

Ishii, D. N., and Mill, J. F., 1987, Molecular mechanisms of neurite formation stimulated by insulin-like factors and nerve growth factor, Curr. Top. Membr. Transp., 31:31-78.

Jaken, S., and Kiley, S. C., 1987, Purification and characterization of three types of protein kinase C from rabbit brain cytosol, Proc. Natl. Acad. Sci. USA, 84:4418-4422.

Julien, J. -P., Ramachandran, K., and Grosveld, F., 1985, Cloning of a cDNA encoding the smallest neurofilament protein from the rat, Biochim. Biophys. Acta, 825:398-404.

Kadowaki, T., Koyasu, S., Nishida, E., Tobe, K., Izumi, T., Takaku, F., Sakai, H., Yahara, I., and Kasuga, M., 1987, Tyrosine phosphorylation of common and specific sets of cellular proteins rapidly induced by insulin, insulin-like growth factor 1, and epidermal growth factor in intact cells, J. Biol. Chem., 262:7342-7350.

Kasuga, M., Karlsson, F. A., and Kahn, C. R., 1982, Insulin stimulates the phosphorylation of the 95,000-dalton subunit of its own receptor, Science, 215:185-187.

Kikkawa, U., Ono, Y., Ogita, K., Fujii, T., Asaoka, Y., Sekiguchi, K., Kosaka, Y., Igarashi, K., and Nishizuka, Y., 1987, Identification of the structures of multiple subspecies of protein kinase C expressed in rat brain, FEBS Lett., 217:227-231.

Koizumi, S., Contreras, M. L., Matsuda, Y., Hama, T., Lazarovici, P., and Guroff, G., 1988, K-252a: a specific inhibitor of the action of nerve growth factor on PC12 cells, J. Neurosci., 8:715-721.

Laurenza, A., McHugh Sutkowski, E., and Seamon, K. B., 1989, Forskolin: a specific stimulator of adenyl cyclase or a diterpene with multiple sites of action?, Trends Pharmacol. Sci., 10:442-447.

Lindenbaum, M. H., Carbonetto, S., and Mushynski, W. E., 1987, Nerve growth factor enhances the synthesis, phosphorylation, and metabolic stability of neurofilament proteins in PC12 cells, J. Biol. Chem., 262:605-610.

Lindenbaum, M. H., Carbonetto, S., Grosveld, F., Flavell, D., and Mushynski, W. E., 1988, Transcriptional and post-transcriptional effects of nerve growth factor on expression of the three neurofilament subunits in PC-12 cells, J. Biol. Chem., 263:5662-5667.

MacDonald, R. G., Pfeffer, S. R., Coussens, L., Tepper, M. A., Brocklebank, C. M., Mole, J. E., Anderson, J. K., Chen, E., Czech, M. P., and Ullrich, A., 1988, A single receptor binds both insulin-like growth factor II and mannose-6-phosphate, Science, 239:1134-1137.

Massague, J., Pilch, P. F., and Czech, M. P., 1980, Electrophoretic resolution of three major insulin receptor structures with unique subunit stoichiometries, Proc. Natl. Acad. Sci. USA, 77:7137-7141.

Mei, L., Roske, W. R., and Yamamura, H. I., 1989, The coupling of muscarinic receptors to hydrolysis of inositol lipids in human neuroblastoma SH-SY5Y cells, Brain Res., 504:7-14.

Merrill, A. H. , Jr., Sereni, A. M., Stevens, V. L., Hannun, Y. A., Bell, R. M., and Kinkade, J. M. , Jr., 1986, Inhibition of phorbol ester-dependent differentiation of human promyelocytic leukemic (HL-60) cells by sphinganine and other long-chain bases, J. Biol. Chem., 261:12610-12615.

Metuzals, J., Montpetit, V., and Clapin, D. F., 1981, Organization of the neurofilamentous network, Cell Tissue Res., 214:455-482.

Mill, J. F., Chao, M. V., and Ishii, D. N., 1985, Insulin, insulin-like growth factor II, and nerve growth factor effects on tubulin mRNA levels and neurite formation, Proc. Natl. Acad. Sci. USA, 82:7126-7130.

Morgan, D. O., Edman, J. C., Standring, D. N., Fried, V. A., Smith, M. C., Roth, R. A., and Rutter, W. J., 1987, Insulin-like growth factor II receptor as a multifunctional binding protein, Nature, 329:301-307.

Moss, D. J., Fernyhough, P., Chapman, K., Baizer, L., Bray, D., and Allsopp, T., 1990, Chicken growth-associated protein GAP-43 is tightly bound to the actin-rich neuronal membrane skeleton, J. Neurochem., 54:729-736.

Myers, M. W., Lazzarini, R. A., Lee, V. M. -Y., Schlaepfer, W. W., and Nelson, D. L., 1987, The human mid-size neurofilament subunit: a repeated protein sequence and the relationship of its gene to the intermediate filament gene family, EMBO J., 6:1617-1626.

Neary, J., Norenberg, L., and Norenberg, M., 1988, Protein kinase C in primary astrocyte cultures: Cytoplasmic localization and translocation by a phorbol ester, J. Neurochem., 50:1179-1184.

Neumann, D., Scherson, T., Ginzburg, I., Littauer, U. Z., and Schwartz, M., 1983, Regulation of mRNA levels for microtubule proteins during nerve regeneration, FEBS Lett., 162:270-276.

Nishimoto, I., Ogata, E., and Kojima, I., 1987, Pertussis toxin inhibits the action of insulin-like growth factor-I, Biochem. Biophys. Res. Commun., 148:403-411.

Nishizuka, Y., 1988, The molecular heterogeneity of protein kinase C and its implications for cellular recognition, Nature, 334:661-665.

Rechler, M. M., Kasuga, M., Sasaki, N., De Vroede, M. A., Romanus, J. A., and Nissley, S. P., 1983, Properties of insulin-like growth factor receptor subtypes, in: "Insulin-like growth factors/somatomedins: basic chemistry, biology, and clinical importance," E. M. Spencer, ed., W. de Gruyter, New York, pp. 459-490.

Recio-Pinto, E., and Ishii, D. N., 1984, Effects of insulin, insulin-like growth factor-II and nerve growth factor on neurite outgrowth in cultured human neuroblastoma cells, Brain Res., 302:323-334.

Recio-Pinto, E., Rechler, M. M., and Ishii, D. N., 1986, Effects of insulin, insulin-like growth factor-II, and nerve growth factor on neurite formation and survival in cultured sympathetic and sensory neurons, J. Neurosci., 6:1211-1219.

Recio-Pinto, E., and Ishii, D. N., 1988a, Insulin and related growth factors: effects on the nervous system and mechanism for neurite growth and regeneration, Neurochem. Int., 12:397-414.

Recio-Pinto, E., and Ishii, D. N., 1988b, Insulin and insulinlike growth factor receptors regulating neurite formation in cultured human neuroblastoma cells, J. Neurosci. Res., 19:312-320.

Reinhold, D. S., and Neet, K. E., 1989, The lack of a role for protein kinase C in neurite extension and in the induction of ornithine decarboxylase by nerve growth factor in PC12 cells, J. Biol. Chem., 264:3538-3544.

Rosenfeld, R. G., Conover, C. A., Hodges, D., Lee, P. K. K., Misra, P., Hintz, R. L., and Li, C. H., 1987, Heterogeneity of IGF-I affinity for the IGF-II receptor: Comparison of natural, synthetic and recombinant DNA-derived IGF-I, Biochem. Biophys. Res. Commun., 143:199-205.

Rydel, R. E., and Greene, L. A., 1988, cAMP analogs promote survival and neurite outgrowth in cultures of rat sympathetic and sensory neurons independently of nerve growth factor, Proc. Natl. Acad. Sci.USA, 85:1257-1261.

Saito, N., Kikkawa, U., Nishizuka, Y., and Tanaka, C., 1988, Distribution of protein kinase C-like immunoreactive neurons in rat brain, J. Neurosci., 8:369-382.

Satoh, T., Nakamura, S., and Kaziro, Y., 1987, Induction of neurite formation in PC12 cells by microinjection of proto-oncogenic Ha-ras protein preincubated with guanosine-5'-O-(3-thiotriphosphate), Mol. Cell. Biol., 7:4553-4556.

Schubert, D., and Whitlock, C., 1977, The alteration of cellular adhesion by nerve growth factor, Proc. Natl. Acad. Sci. USA, 74:4055-4058.

Shemer, J., Adamo, M., Wilson, G. L., Heffez, D., Zick, Y., and LeRoith, D., 1987, Insulin and insulin-like growth factor 1 stimulate a common endogenous phosphoprotein substrate (pp185) in intact neuroblastoma cells, J. Biol. Chem., 262:15476-15482.

Skene, J. H. P., and Willard, M., 1981, Changes in axonally transported proteins during regeneration in toad retinal ganglion cells, J. Cell Biol., 89:86-95.

Soares, M. B., Ishii, D. N., and Efstratiadis, A., 1985, Developmental and tissue-specific expression of a family of transcripts related to rat insulin-like growth factor II mRNA, Nucleic Acids Res., 13:1119-1134.

Soares, M. B., Turken, A., Ishii, D. N., Mills, L., Episkopou, V., Cotter, S., Zeitlin, S., and Efstratiadis, A., 1986, Rat insulin-like growth factor II gene: a single gene with two promoters expressing a multitranscript family, J. Mol. Biol., 192:737-752.

Sonnenfeld, K. H., and Ishii, D. N., 1982, Nerve growth factor effects and receptors in cultured human neuroblastoma cell lines, J. Neurosci. Res., 8:375-391.

Sonnenfeld, K. H., and Ishii, D. N., 1985, Fast and slow nerve growth factor binding sites in human neuroblastoma and rat pheochromocytoma cell lines: relationship of sites to each other and to neurite formation, J. Neurosci., 5:1717-1728.

Spinelli, W., Sonnenfeld, K. H., and Ishii, D. N., 1982, Effects of phorbol ester tumor promoters and nerve growth factor on neurite outgrowth in cultured human neuroblastoma cells, Cancer Res., 42:5067-5073.

Spinelli, W., and Ishii, D. N., 1983, Tumor promoter receptors regulating neurite formation in cultured human neuroblastoma cells, Cancer Res., 43:4119-4125.

Trojanowski, J. Q., Obracka, M. A., and Lee, V. M., 1985, Distribution of neurofilament subunits in neurons and neuronal processes, J. Histochem. Cytochem., 33:557-563.

Wang, C., Wible, B., Angelides, K., and Ishii, D. N., 1988, Insulin and insulinlike growth factor-I increase neurofilament mRNA and neurite formation, Soc. Neurosci. Abstr., 14:1169.

Wang, C., Li, Y., Wible, B., Angelides, K., and Ishii, D. N., 1990, Mechanism of insulin and insulin-like growth factor directed neurite formation: Role of cyclic AMP in neurofilament and tubulin gene expression, Soc. Neurosci. Abstr., 16:180.

Wood, J., Girard, P., Mazzei, G., and Kuo, J., 1986, Immunocytochemical localization of protein kinase C in identified neuronal compartments of rat brain, J. Neurosci., 6:2571-2577.

REGULATION OF PROTEIN PHOSPHORYLATION BY INSULIN AND INSULIN-LIKE GROWTH FACTORS IN CULTURED FETAL NEURONS

K.A. Heidenreich, S.P. Toledo, and K.A. Kenner

Department of Medicine; M-023E
University of California, San Diego
La Jolla, CA 92093

INTRODUCTION

Abundant evidence indicates that protein phosphorylation and dephosphorylation of tyrosine, serine, and threonine residues are key mechanisms by which insulin regulates cell function (for review see Ref. 1). The β-subunit of the insulin receptor is a tyrosine kinase activated by insulin binding that catalyzes autophosphorylation and phosphorylation of other proteins (2,3). The intrinsic tyrosine kinase activity of the insulin receptor appears to be critical for insulin action since cells transfected with kinase-deficient mutant insulin receptors (4) or cells injected with antibodies that inhibit the receptor kinase (5) become insensitive to insulin. A current working hypothesis is that activation of the insulin receptor kinase catalyzes the phosphorylation of tyrosine residues on several proteins. These proteins, in turn, regulate the more abundant serine kinases that phosphorylate proteins involved in the end-response.

Insulin and insulin receptors are found throughout the central nervous system. Insulin has multiple effects in nervous tissue all outside the realm of glucose homeostasis. These roles include regulation of growth and development, neuromodulation, and regulation of food intake (for review, see Ref. 6). Insulin's role as a neurotrophic factor is the best documented. Many laboratories have shown that insulin in the absence of any other growth factors is capable of supporting the growth and differentiation of a population of neurons in serum-free medium (7,8). Insulin increases neurite outgrowth, stimulates protein synthesis, and increases mRNA levels for tubulin and neurofilament proteins in cultured neurons (8-10). It also appears to regulate the formation of nascent synapses in culture (11). Based on dose-response data, the above actions of insulin appear to be mediated by neuronal insulin receptors rather than the closely related IGF-I receptors, although in some cases, IGF-I appears to have similar effects. Neuronal insulin receptors are structurally distinct from insulin receptors found on other cell types (12). The heterogeneity is based, at least in part, on differences in N-linked glycosylation (13).

Little is known about insulin or IGF signal transduction pathways in neurons. Previous experiments using partially purified insulin receptors from rat brain demonstrated that the brain subtype of insulin receptor undergoes autophosphorylation <u>in vitro</u> in response to insulin binding and is capable of phosphorylating exogenous proteins (14). This article will

Molecular Biology and Physiology of Insulin and Insulin-Like Growth Factors
Edited by M.K. Raizada and D. LeRoith, Plenum Press, New York, 1991

summarize our recent studies examining the *in vivo* regulation of tyrosine phosphorylation by insulin and IGFs in cultured neurons. Studies of 2 serine kinases implicated in neuronal insulin receptor signalling are also presented.

IN VIVO AUTOPHOSPHORYLATION OF NEURONAL INSULIN AND IGF-I RECEPTORS

Neuronal tyrosine phosphorylation in response to insulin, IGF-I, and IGF-II was examined by immunoblotting neuronal cell extracts with affinity purified polyclonal anti-phosphotyrosine antibodies. Under basal conditions approximately 5 major and 9 minor phosphoproteins were detected in neurons (15, manuscript submitted). Three different anti-phosphotyrosine antibodies detected the same phosphorylated proteins, although the intensity of the iodinated bands varied suggesting differences in the affinity of the antibodies for the cellular substrates. Detection of neuronal phosphoproteins was prevented by coincubation of the nitrocellulose with antibodies plus 2 mM phosphotyrosine, but not by coincubation with 2 mM phosphothreonine or phosphoserine indicating that the antibodies are specific for phosphotyrosine.

Exposure of the neurons to insulin or IGF-I increased the tyrosine phosphorylation of an 87 kDa protein. The dose-response of pp87 phosphorylation to increasing concentrations (1-20 nM) of insulin and IGF-I was comparable, however, the maximal response to each peptide differed. Insulin stimulated phosphorylation of pp87 by 2-3-fold, whereas, IGF-I stimulated the phosphorylation of a similar sized protein by about 20-fold. It is likely that pp87 represents the β-subunit of both the insulin and IGF-I receptors. The larger stimulation of pp87 phosphorylation by IGF-I is consistent with an 8-fold higher number of IGF-I receptors in these cultures. IGF-II also stimulated pp87 phosphorylation in the same concentration range as insulin and IGF-I. To determine which receptor type was mediating the effects of IGF-II, the affinity of IGF-II for neuronal insulin and IGF-I receptors was examined using radioligand binding assays. IGF-II was only slightly less (5-10-fold) effective than IGF-I in inhibiting the binding of 125I-IGF-I. Interestingly, IGF-II had almost an equal affinity to insulin in displacing 125I-insulin from neuronal insulin receptors. The apparent high affinity of IGF-II for both insulin and IGF-I receptors indicate that IGF-II is capable of stimulating the autophosphorylation of both receptor types. The affinity of IGF-II for neuronal insulin receptors is much higher than its affinity for peripheral insulin receptors. These data raise the possibility that IGF-II, which is abundant in the brain throughout development, may be an important endogenous ligand for the insulin receptor in brain.

ENDOGENOUS SUBSTRATES FOR THE INSULIN AND IGF-I RECEPTOR KINASES IN NEURONS

Although large increases in insulin and IGF-I receptor autophosphorylation were detected in 5 day cultures, tyrosine phosphorylation of other proteins by the activated receptor kinases was not detected at this time in culture. In constract, neurons maintained in culture less than 24 hr contained a 70 kDa protein (pp70) that was phosphorylated on tyrosine residues in response to insulin, IGF-I, and IGF-II (16). Phosphorylation of pp70 by all 3 growth factors was detected within 30 sec and was maximal by 5 min. Both the time course and dose-response of pp70 phosphorylation correlated very closely with the time course and dose-response of receptor autophosphorylation, suggesting that this protein is a substrate for the receptor kinases in neurons.

Interestingly, growth factor-stimulated phosphorylation of pp70 was transient, reaching a maximum after 2-5 hours in culture and decreasing to nondetectable levels after 24 hr. Insulin and IGF-I receptor autophosphorylation was constant in neurons cultured from 1-24 hr, suggesting that the extent of pp70 phosphorylation was not limited by the number or extent of phosphorylation of these receptors. The highest level of stimulated pp70 phosphorylation occurred when neurite outgrowth in the cultures was maximal. The close temporal relationship between neurite outgrowth and pp70 phosphorylation taken together with previous data showing insulin-stimulated neurite outgrowth suggests that phosphorylation of pp70 may mediate insulin's effect on neuronal process formation.

NEURONAL SERINE KINASES REGULATED BY INSULIN

As an initial attempt to identify early steps that may be involved in the growth responses of neurons to insulin, studies were carried out to examine whether insulin receptor activation increases the phosphorylation of ribosomal protein S6 in cultured fetal neurons and whether activation of a protein kinase is involved in this process. When neurons were incubated for 2 hr with 32Pi, the addition of insulin for the final 30 min increased the incirporation of 32Pi into a 32K microsomal protein (17). The incorporation of 32Pi into the majority of other neuronal proteins was unaltered by the 30-min exposure to insulin. Cytosolic extracts from insulin-treated neurons incubated in the presence of exogenous rat liver 40S ribosomes and [γ-32P] ATP displayed a 3- to 8-fold increase in the phosphorylation of ribosomal protein S6 compared to extracts from untreated cells. Inclusion of cycloheximide during exposure of the neurons to insulin did not inhibit the increased cytosolic kinase activity. Activation of S6 kinase activity by insulin was dose dependent (seen at insulin concentrations as low as 0.1 ng/ml) and reached a maximum after 20 min of incubation. Addition of phosphatidylserine, diolein, and Ca++ to the <u>in vitro</u> kinase reaction had no effect on the phosphorylation of ribosomal protein S6. Likewise, treatment of neurons with (Bu)2cAMP did not alter the phosphorylation of ribosomal protein S6 by neuronal cytosolic extracts. Thus, insulin activates a cytosolic protein kinase that phosphorylates ribosomal S6 in neurons and is distinct from protein kinase C and cAMP dependent protein kinase. Recently, homologs to the Zenopus S6 kinase II (referred to as pp90 rsk proteins) have been identified in human, mouse, and chicken cells (18). These enzymes differ from the 70,000 Mr S6 kinase activities that have been previously purified from a variety of animals cells. pp90 rsk and pp70 S6 kinases share substrates and are activated by growth factors, however, they have distinct molecular size, chromatographic properties, phosphopeptide maps, and kinetics of activation. Further study is required to determine if the neuronal S6 kinase activity represents pp90 rsk, pp70, or both of these S6 kinases.

During the course of the S6 kinase studies, we detected kinase activity in neuronal cytosolic extracts that was distinct from S6 kinase, dependent on the presence of phospholipids, and elevated in insulin-treated cells. Further study showed that insulin stimulates protein kinase C (PKC) in both cytosolic and membrane fractions from cultured neurons (19,20). The increment in PKC activity was apparent in cells that were pretreated with cycloheximide and was evident in partially purified PKC fractions from DEAE columns. The stimulation of PKC activity by insulin was also apparent when neuronal proteins were phosphorylated in vitro by PKC. Addition of phosphatidylserine and diolein to neuronal cytosolic extracts resulted in the phosphorylation of 4 proteins and the phosphorylation of all 4 proteins was 2-fold higher in insulin-treated cells. In neuronal membrane fractions, a 70 kDa protein was phosphorylated by the PKC activators and prior treatment of the cells with insulin increased the phosphorylation of this protein by 200%.

To identify the types of PKC isoforms present in cultured fetal chick neurons, whole cell extracts were analyzed by immunoblot analysis using antibodies against PKC α, β, γ, δ, and ε (20). The antibodies were raised against synthetic peptides corresponding to unique sequences found in PKC α (residues 318-331, ISPSEDRRQPSNNL), PKC β(residues 319-332, EKTTNTISKFDNNG), PKC γ (residues 316-322, VRTGPSSSPIPSPSPT), PKC δ (residues 318-334, KTAVSGNDIPDNNGTYG), and PKC ε (residues 3836-853, NQEEFKGFSYFGEDLMP). Antibodies against PKC α, β, γ, and δ failed to detect proteins in neuronal cell extracts. In contrast, antibodies against PKC ε detected a major protein with an apparent molecular weight of 90 kDa. The molecular weight of this protein corresponds to the reported apparent molecular weight of PKC ε (21). To determine the cellular distribution of PKC ε, neurons were fractionated into membrane and cytosolic fractions and immunoblotted. Results showed that 90% of the enzyme residues in the cytosol, consistent with our calculations for the distribution of PKC activity in the cells. Insulin had no effect on the total amount of distribution of PKCε. To confirm that PKCε was the isoform activated by insulin in neurons, neuronal PKC activity was reassayed using a synthetic peptide substrate similar to the psuedosubstrate binding site of PKCε in the absence of calcium. Calcium was ommitted since recent studies have shown that PKCε lacks the calcium regulatory domain found in PKC α,β, and γ. In the presence of phosphatidylserine and diolein, the phosphorylation of the synthetic peptide was about 50% higher in insulin-treated cells compared to control cells. Thus, the stimulation of PKC activity by insulin was apparent when the assay was modified to specifically measure PKCε.

These studies indicate that insulin stimulates PKCε in neurons by a mechanism that does not involve translocation of the enzyme from the cytosol to the membrane. Interestingly, insulin treatment of neurons resulted in a decrease in the mobility of PKC ε in cytosolic and membrane fractions. Autophosphorylation of PKC ε is known to occur under our assay conditions. Thus, the shift in the mobility of the enzyme after exposure to insulin could indicate increased phosphorylation of the enzyme due to either enhanced autophosphorylation or phosphorylation by another kinase.

SUMMARY

The pathways depicted in Figure 1 summarize the data discussed in this article. In neurons, the binding of insulin and IGF-I to their respective receptors triggers autophosphorylation of the receptor β-subunits. IGF-II binds to both neuronal insulin and IGF-I receptors and can stimulate autophosphorylation of either receptor type. In addition to enhancing insulin and IGF-I receptor autophosphorylation, all 3 peptides stimulate the tyrosine phosphorylation of a 70 kDa protein with a similar time course and dose response to receptor phosphorylation. The identity of pp70 is unknown, although the close temporal relationship between pp70 phosphorylation and neurite outgrowth suggests a potential role for this protein. Subsequent to these very early events, two neuronal serine kinases are activated by insulin. One has S6 kinase activity and may represent either the pp90rsk or pp70 class of S6 kinases. Since S6 kinases are activated by direct phosphorylation rather than by second messengers, it is likely that a neuronal S6 kinase kinase exists. The activation of S6 kinase is likely to mediate insulin's effects on neuronal protein synthesis or other growth-related processes. The second serine kinase that is activated by insulin is PKC ε. This enzyme is largely restricted to the nervous system, so this signalling pathway may be neuronal-specific. The mechanism of activation of PKCε is unknown, although preliminary data suggests that enhanced phosphorylation of the enzyme is involved. Studies are currently underway to investigate the potential role of diacylglycerol, a potential second messenger generated from either phosphotidylinositol or phosphotidylcholine hydrolysis, in the activation of PKCε by insulin.

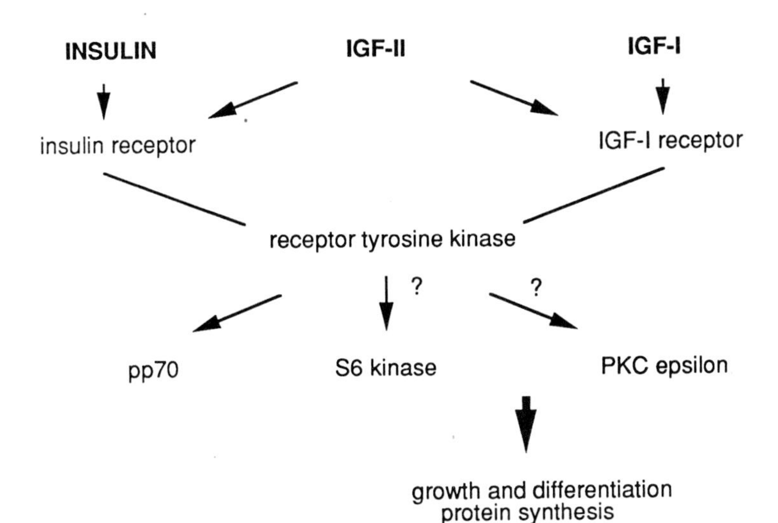

FIG. 1. Potential pathways for insulin and IGF signal transduction in neurons.

ACKNOWLEDGEMENTS

This work was supported by research grants from the National Science Foundation (BNS-8742067) and the Juvenile Diabetes Foundation International. We wish to thank Mrs. Elizabeth Martinez for preparing this manuscript.

REFERENCES

1. O.M. Rosen, After insulin binds, Science 237:1452 (1987)
2. M. Kasuga, Y. Zick, D.L. Blithe, M. Crettaz, and C.R. Kahn, Insulin stimulates tyrosine phosphorylation of the insulin receptor in a cell-free system, Nature 298:667 (1982)
3. M.F. White, R. Maron, and C.R. Kahn, Insulin rapidly stimulates tyrosine phosphorylation of a Mr 185,000 protein in intact cells, Nature 318:183 (1985)
4. D.A. McClain, H. Maegawa, J. Lee, T.J. Dull, A. Ullrich, and J.M. Olefsky, A mutant insulin receptor with defective tyrosine kinase displays no biological activity and does not undergo endocytosis. J. Biol. Chem. 262:14663 (1987)
5. D.O. Morgan and R.A. Roth, Acute insulin action requires insulin receptor kinase activity: Introduction of an inhibitory monoclonal antibody into mammalian cells blocks the rapid effects of insulin, Proc. Natl. Acad. Sci USA 84:41 (1987)
6. D.G. Baskin, D.P. Figlewicz, S.C. Woods, D. Porte, Jr., and D.M. Dorsa, Insulin in the brain, Ann. Rev. Physiol. 49:335 (1987)
7. K.A. Heidenreich, G. de Vellis, and P.R. Gilmore, Functional properties of the subtype of insulin receptor found on neurons, J. Neurochem. 51:878 (1988)
8. E. Recio-Pinto, M.M. Rechler, and D.N. Ishii, Effects of insulin, insulin-like growth factor II, nerve growth factor on neurite formation and survival in cultured sympathetic and sensory neurons, J. Neurosci. 6:1211 (1986)
9. K.A. Heidenreich and S.P. Toledo, Insulin receptors mediate growth effects in cultured fetal neurons. I. Rapid stimulation of protein synthesis, Endocrinology 125:1451 (1989)

10. J.F. Mill, M.V. Chao, and D.N. Ishii, Insulin, insulin-like growth factor II, and nerve growth factor effects on tubulin mRNA levels and neurite formation, Proc. Natl. Sci. USA 82:7126 (1985)
11. D.G. Puro and E. Agardh, Insulin-mediated regulation of neuronal maturation, Science 225:1170 (1984)
12. K.A. Heidenreich, N.R. Zahniser, P. Berhanu, D. Brandenburg, and J.M. Olefsky, Structural differences between insulin receptors in the brain and peripheral target tissues, J. Biol. Chem. 285:8527 (1983)
13. K.A. Hendenreich and D. Brandenberg, Oligosaccharide heterogeneity of insulin receptors. Comparison of N-linked glycosylation of insulin receptors in adipocytes and brain, Endocrinology 118:1835 (1986)
14. R.W. Rees-Jones, S.A. Hendricks, M. Quarum, and J. Roth, The insulin receptors of rat brain are coupled to tyrosine kinase activity, J. Biol. Chem. 259:3470 (1984)
15. K.A. Kenner and K.A. Heidenreich, Regulation of tyrosine phosphorylation by insulin and IGF-I in cultured fetal neurons, Soc. Neurosci. 15:708 (1989)
16. K.A. Kenner and K.A. Heidenreich, Tyrosine phosphorylation of a 70kDa protein by insulin and insulin-like growth factors in cultured fetal neurons, Soc. Neurosci. 16: (1990)
17. K.A. Heidenreich, and S.P. Toledo, Insulin receptors mediate growth effects in cultured fetal neurons. II. Activation of a protein kinase that phosphorylates ribosomal protein S6, Endocrinology 125:1458 (1989)
18. R.H. Chen and J. Blenis, Identification of Xenopus S6 protein kinase homologs (pp90rsk) in somatic cells: Phosphorylation and activation during initial cell proliferation, Mol. Cell. Biol. 10:3204 (1990)
19. K.A. Heidenreich and S.P. Toledo, Insulin increases protein kinase C activity in both cytosolic and membrane fraction of cultured fetal neurons, Diabetes (Suppl. 1):65A (1989)
20. K.A. Heidenreich, S.P. Toledo, L.L. Brunton, M.J. Watson, S. Daniel-Issakani, and B. Strulovici, Insulin stimulates the activity of a novel protein kinase C, PKC ε, in cultured fetal chick neurons, J. Biol. Chem. 265:15076 (1990)
21. D. Schapp and P.J. Parker, Expression, purification, and characterization of protein kinase C-epsilon, J. Biol. Chem. 265:7301 (1990)

A REVIEW OF INSULIN/INSULIN-LIKE PEPTIDE IN THE CENTRAL NERVOUS SYSTEM

Sherin U. Devaskar

Department of Pediatrics, St. Louis University School of Medicine, The Pediatric Research Institute and Cardinal Glennon Children's Hospital, St. Louis, Mo.

Insulin has been demonstrated within the brain of various animal species (1). Investigations undertaken in the chick demonstrated the presence of insulin in embryonic heads prior to the development of pancreatic insulin (2). These observations prompted a search for an extra-pancreatic source of insulin production within the brain. A surge of reports were subsequently noted (3-7) including the recent observations of brain insulin-like transcripts (8-12), thereby defending a central nervous system origin of insulin. On the other hand, there has been accumulating evidence supporting the fact that circulating insulin crosses the blood-cerebro-spinal fluid (CSF) barrier and enters the CSF (13-14). Having entered the CSF, insulin was then observed to be taken up by the brain parenchyma at the circumventricular organs which lack a blood-brain barrier (15-17). Thus, there continues to be an ongoing controversy as to the exact origin of insulin within the CNS. Is it pancreatic or extra-pancreatic?

Prior to undertaking the formidable task of attempting to resolve this controversy, it is essential to demonstrate the biological importance of insulin and the significance of the hormone entering the CNS.

Most of insulin's biological effects in the brain are mediated by insulin receptors that are widely distributed in the adult rat brain (18). A maximal concentration of these receptors have been observed in the choroid plexus, external plexiform layer of olfactory bulb, followed next in structures of the limbic system and hypothalamus, particularly the lateral septum, amygdala, subiculum, hippocampal CA1 region, mammillary body, and arcuate nucleus and, lastly, in the cerebral cortex and cerebellum (19). Generally the highest concentration of insulin receptors seem localized to regions concerned with olfaction, appetite and autonomic function (19). These areas contain dendritic fields that receive a rich synaptic afferent input (19). These brain regions also demonstrate relatively high concentrations of phosphotyrosine containing proteins providing the basis for these tyrosine kinase-insulin receptor dependent intracellular signalling pathways (20). This regional distribution of receptors and tyrosine phosphorylation of proteins coincides with the demonstration of a role for insulin in neurotransmission (21,22) and modulation of feeding behavior and obesity in the adult animal (23,24).

At the other end of the spectrum, in the early avian embryonic state, insulin receptors have been localized to neural folds, neural tube and optic vesicles during neurulation (25). In fact, at this early developmental stage, there are more abundant brain IGF-I receptors than insulin receptors (26); however, the insulin-dependent phosphorylation is equal to the IGF-I dependent phosphorylation (25), suggesting that, in the brain, insulin is at least equipotent to IGF-I. Overall insulin induced protein, DNA and RNA synthesis in the post-neurula chick embryo is far greater than that accomplished by IGF-I (27). In pre-implantation 8-cell rat embryos, insulin stimulates DNA, RNA and protein synthesis (28) and, in post-implantation rat embryos, the absence of insulin results in severe growth retardation (29,30). In fact, in xenopus laevis stage IV oocytes, extracellular insulin increases the synthesis of RNA, protein and glycogen (31). Further, there is experimental evidence to support the fact that maternal insulin is internalized by pre-implantation mouse embryos via a receptor-mediated endocytosis (32) and rat extra-placental membranes have been determined to express insulin II mRNA (33). These mechanisms appear to be in place to ensure the growth and differentiation of fertilized oocytes, pre- and post-implantation embryos, well before pancreatic insulin appears developmentally.

In specific reference to the CNS, then, investigations employing cultured 7 day old chick neurons demonstrated insulin to rapidly stimulate the overall protein synthesis (34), neuronal growth and differentiation (35-37) while failing to acutely stimulate glucose transport, glucose oxidation or glycogen synthase (35). Further, insulin by increasing tubulin and neurofilament mRNA perhaps augments the neurite outgrowth (38,39). More recently, insulin has been observed to increase the activity of neuronal cytosolic and membranous protein kinase-C by a mechanism that does not involve translocation of the enzyme from cytosol to membrane (40). This insulin-stimulated enzyme activity is independent of calcium and solely involves the protein kinase-C epsilon isoform (40). Additionally, insulin has been shown to activate a cytosolic protein kinase (distinct from protein kinase-C and cAMP dependent protein kinase) that phosphorylates ribosomal S6 in chick neurons (41). Thus, the mechanisms involved in brain insulin receptor signalling seem to involve conventional (42) and unconventional pathways that are distinct from those of other growth factors (tyrosine kinase) in other cell systems. Further, at stages of maximal brain growth and development, insulin receptor numbers peak as seen in the late gestation rat (43) and rabbit (44) fetal brain, declining gradually through the neonatal phases to reach the adult values. In addition, the degree of insulin-stimulated autophosphorylation parallels the developmental changes in brain insulin binding (43).

Post-translational N-linked glycosylation of the brain (neuronal) insulin receptor (45), specifically the sialic acid moiety, is highest in the late gestation fetal stages, followed by a decline resulting in a decrease of the α-subunit molecular size (43). A deficiency in the sialic acid moiety results in a potential for an aborted conventional insulin action (augmentation of glucose uptake) in the mature neuronal cells (46). Thus, in keeping with this theory, insulin has been observed to enhance the neonatal rat neuronal re-uptake of norepinephrine (21) and increased stimulus-evoked transmission at synapses formed in culture by cholinergic fetal rat retinal neurons (22). Whether insulin exerts its neuronal effects mainly via the insulin receptors (which are most abundant in these cells) (47) or via both the insulin and IGF-I receptors remains an enigma (47,48,49). To interact with the latter, supraphysiological concentrations of insulin may have to access the CNS (μM).

Unlike the neuronal cells, insulin stimulates glucose uptake by glial cells (50). Both insulin and IGF-I enhance the glial cell glucose transport by a kinase stimulated (51) increase in the insulin-insensitive (Glut 1) glucose transporter mRNA and protein (52). Additionally, insulin enhances glial cell DNA and RNA synthesis (53) and regulates glial cellular growth (54). However, generally glial cells, specifically astrocytes, have very few (if any) insulin receptors when compared to the neurons (55). Thus, in these cells, it is potentially feasible that insulin's biological effects are mediated by the IGF-I receptors (48,49), thereby requiring greater than physiological concentrations of insulin. However, we have preliminarily determined that insulin interacts with the brain IGF-II receptors (this is unlike what has been described in other tissues) in a manner that is kinetically similar to the authentic IGF-II peptide (56). In support of these observations is the previous finding of abundant IGF-II receptors in rat glial cells (48). Thus the additional possibility exists, particularly in the CNS, that insulin (rather than interacting with IGF-I receptors) interacts with the glial IGF-II receptors (at physiological concentrations of insulin).

Thus overall, regardless of the exact receptor-mediated pathway, insulin has important biological effects in the CNS; certain effects that are present from the early embryonic stages of development. In the early embryonic/fetal stages of development, insulin potentiates brain cell growth (35-37, 53,54), differentiation and neurotransmission (21,22). On the other hand, in the adult, insulin in the brain continues to exhibit a regulatory effect, more on appetite control, feeding behavior and adiposity (23,24).

If insulin is so important to the brain, both in the immature and mature stages of development, where does it originate? In addition to the neuronal cells (9-11, 57), the endothelial cells lining brain vasculature are rich in insulin receptors (58). Studies employing isolated brain microvessels have demonstrated that circulating insulin traverses the blood-brain barrier by these insulin receptor mediated transcytosis (59) and, in fact, circulating insulin gains access to the neonatal rabbit brain parenchyma via the brain vascular endothelial insulin receptors (16). These brain vascular endothelial insulin receptors are more abundant in the neonate when compared to the adult (60).

Based on these previous reports, we (61) and others (62) observed the presence of an increased concentration of immunoassayable insulin (distinct from IGF-I and II) in the late gestation fetal and early neonatal rabbit (61) and rat brains (62). This developmental increase appears to follow the levels of circulating insulin concentrations by one developmental time period (1-2 days of age) (61,62). This time lag, perhaps secondary to the process of transcytosis, is highly suggestive of brain insulin concentrations reflecting circulating insulin concentrations in a slow but integrated manner. Additionally, in order to regulate brain growth and differentiation, it stands to reason that insulin (during the critical fetal and neonatal stages of brain development) should easily access the brain.

There have been an independent set of investigations, including our own, that point towards the "neuron" as an additional site of insulin synthesis (9,11) and secretion (11,57). Originally, in isolated rat brain cell cultures maintained in a serum-containing (insulin-containing) medium, insulin immunoreactivity was observed within a small number of neurons (3). Whether this insulin immunoreactivity was secondary to local synthesis of insulin or a receptor-mediated endocytosis of insulin from the medium was not clear. Additionally, radioimmunoassayable insulin was observed to be secreted by both cultured neurons (57) and glial cells (63) into the conditioned medium. However, both these cultures were maintained in a serum-containing (insulin-containing) medium 24 hours prior to collection of conditioned medium and analysis. Later we observed the presence of insulin mRNA within a certain subset of cultured neonatal rabbit neurons by *in-situ* hybridization (10). Others observed a similar phenomenon in cultured rat and hamster pituitary cells (64). More specifically, we noted an autocrine regulation of neuronal insulin mRNA, insulin immunoreactivity and immunogenic insulin secreted by rabbit cultured neurons that were maintained for 7 days in a serum-free, insulin-free medium (11). In fact, the absence of exogenous insulin in the culture medium resulted in an increased synthesis and secretion of insulin or an insulin-like peptide into the conditioned medium (11). Peak secretion occurred at 24 hours and the cellular secretion of insulin appeared to be cyclical, based on the cellular need for this growth-maintaining peptide (11). The immunoassayable insulin secreted into the culture medium was reliably insulin or an insulin-like peptide, as anti-insulin antibodies employed in the experiments, and otherwise generally used, fail to detect IGF-I and II (61). Additionally, we observed that human brain IGF-I variant (truncated form) (65) was not detected by anti-insulin antibodies.

On the contrary, *in-situ* hybridization may result in the detection of closely homologous transcripts such as the IGF-I or II mRNAs. However, previous investigations in rat brain cells employing the RNAse protection assay have determined that IGF-II mRNAs are mainly detected in cultured glial cells with none found in neuronal cells (66). IGF-I mRNA is present in greater abundance in glial cells with a significantly lesser abundance noted in neuronal cells (66). Thus a cross-hybridization between insulin probes and neuronal IGF-I mRNA is of concern. We have previously observed a 3.2 kb insulin mRNA in rabbit fetal brain poly (A+) RNA. This transcript size, which is larger than authentic pancreatic insulin mRNA (0.6 kb), resembled the rat insulin II transcripts observed in separate extra-pancreatic sites of insulin synthesis; namely, the extra-placental membranes (3) and the fetal liver (67). This large size (3.2 kb) may be secondary to a long poly A+ tail (68) which, in turn, has been observed in other cases to exert a translational control (69). This 3.2 kb brain insulin mRNA failed to parallel the various sizes of the different IGF-I mRNA species that have previously been described by Northern blot analysis (7.5 kb, 4.7 kb, and 1.7 kb, the most prominent one being the 1.7 kb) (70). These differences in mRNA sizes do not necessarily rule out a cross-hybridization of insulin probes with IGF-I transcripts.

RNAse protection experiments, reported previously by us, demonstrated that insulin mRNA sequences in the brain were distinctly different from rat and rabbit pancreatic insulin probe sequences (71), suggesting the presence of an insulin-like peptide in the brain. Thus, by *in-situ* hybridization, the same possibility exists that the insulin cDNA or cRNA probe may be detecting the closely homologous insulin-like peptide mRNA or IGF-I mRNA sequences. We have, however, previously failed to detect IGF-I or II mRNAs in cultured neuronal cells by *in-situ* hybridization, employing the human IGF-I and II probes (10). *In-situ* hybridization experiments, undertaken by others, in the adult rat brain demonstrated the presence of insulin II transcripts in paraventricular structures after prolonged autoradiographic exposure (12).

Despite the presence of insulin-like transcripts in the brain, various attempts *in vivo*, employing brain sections, have failed to convincingly detect the insulin protein (above the non-specific staining) within the adult brain (15). This may be related to the fact that minute amounts of insulin are present within the brain parenchymal cells, possibly at levels below the sensitivity of the different detection techniques or a different kind of a brain insulin-like peptide exists that cannot be detected with conventional anti-insulin antibodies. In support of our hypothesis is the observation in invertebrates, such as the Lymnaea stagnalis, where an insulin-like peptide (related to the IGFs and insulin) mRNA was detected specifically in the growth controlling neurons (9). Whether a similar mRNA and, thereby the peptide, exists in the mammalian brain evading detection by conventional insulin cDNAs (even by the sensitive RNAse protection assay which relies on the presence of complete sequence homology between the cRNA and mRNA) remains to be investigated.

Recent evidence employing hybrid genes in transgenic mice revealed that the SV 40 nuclear non-secreted large T antigen (Tag), under the transcriptional control of insulin II 5'-flanking regulatory sequences, was expressed transiently in regions of the embryonic CNS (72) where future catecholamine and monoamine cells were to develop (73,74). These studies, while not being confirmatory of insulin synthesis within the brain, suggest a common cell lineage for certain cells within the CNS and the endocrine pancreas. Thus a common origin of cell lineage may be the explanation behind a common function, namely the synthesis and secretion of certain hormones. After all, other gut and pancreatic hormones have been described to be produced in the CNS as well (75,76). More recently, in the adult rat brain, preliminary studies have revealed the presence of pre-proinsulin I and II, employing reverse transcription and amplification of the synthesized cDNA products by polymerase chain reaction (77).

Until this recent preliminary evidence, the detection of insulin synthesis or its mRNA in the brain (under *in-vivo* conditions) has remained evasive. The reason for this may be that the brain insulin is indeed an insulin-like peptide (separate from authentic pancreatic insulin, IGF-I or brain IGF-I variant) and present in small, almost undetectable quantities. These low concentrations may be adequate for neurotransmittory function but not fulfill the global need for brain cellular growth and differentiation. An alternative explanation may be that insulin mRNA is rather stable and the translation rates may be low, resulting in little to no insulin peptide which may be produced only episodically based on the local need. A third explanation may be that neuronally synthesized and secreted insulin is utilized immediately in an evanescent fashion (with no cellular storage of this peptide) making its detection difficult. Under *in-vitro* conditions, however, in the absence of the easily available exogenous insulin (circulating insulin *in vivo*) there may be an increased need for endogenous production of insulin (to ensure cellular well being), making the detection and quantitation of the peptide easier. In fact, while late fetal and early neonatal rabbit neuronal cells are capable of synthesizing insulin in the absence of exogenous insulin (78), preliminary investigations examining the early fetal neuronal cells (less than 18 day gestation: term ~31 days) demonstrated the absence of exogenous insulin to be detrimental, resulting in cell death (78). These observations are highly suggestive of the fact that these early immature neurons are incapable of synthesizing insulin to the same degree as the late gestation fetal neurons, thereby making them highly dependent on exogenous insulin for sustenance (78). Whether a similar phenomenon exists *in vivo* is unknown at the present time.

Assuming that endogenous insulin forms a small fraction of total brain insulin content, what remains intriguing is the need for endogenous production of insulin in the developing CNS when exogenous insulin gains access to the brain in sufficient amounts (so as to interact with insulin receptors and sustain the various biological functions of the hormone). Due to a common endocrine cellular lineage, the *in-vivo* transient synthesis of insulin by neurons may reflect a vestigial function (72). Alternatively, the endogenous hormone/peptide may serve as a modulator of neurotransmission (21,22), justifying the need for minute amounts of hormone stored intracellularly, while exogenous circulating hormone ensures brain cell growth and differentiation. One can draw an analogy from the experiments undertaken in xenopus laevis oocytes (as described above) (31) and extrapolate them to the specialized neuronal cell. Thus, akin to the oocytes, intracellular insulin (endogenous) may target its action on the neuronal nucleus exerting transcriptional and translational control, and thereby exhibit an additive effect along with extracellular insulin (exogenous) in enhancing cellular DNA, RNA synthesis and cell growth (31). Regardless of the reason for endogenous insulin production in the CNS, what remains a puzzle is the exact contribution of the neuronal insulin towards the total pool of brain insulin and the overall biological function in the CNS. Similar to the brain IGF-I variant (truncated form) (65), there may be a family of brain insulin-like peptides that exist which, along with the IGFs, may modulate various aspects of hormonal function in the developing CNS.

ACKNOWLEDGEMENT

Supported in part by NIH (HD-25024).

REFERENCES

1. D. Le Roith, S.A. Hendricks, M.A. Lesniak, S. Rishi, K.L. Becker, J. Havrankova, J.L. Rosenweig, M.J. Brownstein, and J. Roth, Insulin in brain and other extra-pancreatic tissues of vertebrates and non-vertebrates, *Adv Metab Dis* 10:303-340 (1983).
2. F. de Pablo, J. Roth, E. Hernandez, and R.M. Pruss, Insulin is present in chicken eggs and early chick embryos, *Endocrinology* 111:1909-1914 (1982).
3. M.K. Raizada, Localization of Insulin-like immunoreactivity in the neurons from primary cultures of rat brain, *Exp Cell Res* 143:351-357 (1983).
4. J. Havrankova, D. Schmechel, J. Roth, and M. Brownstein, Identification of insulin in rat brain, *Proc Natl Acad Sci USA* 75:5737-5741 (1978).
5. J.L. Rosenzweig, M.A. Havrankova, M.A. Lesniak, M. Brownstein, and J. Roth, Insulin is ubiquitous in extrapancreatic tissues in rats and humans, *Proc Natl Acad Sci USA* 77:572-576 (1980).
6. S.U. Devaskar, L. Karycki, and U.P. Devaskar, Varying brain insulin concentrations differentially regulate the fetal brain insulin receptors, *Biochem Biophys Res Commun* 136:208-219 (1986).
7. N.P. Birch, D.L. Christie, and A.G.C. Renwick, Immunoreactive insulin from mouse fetal brain cells in culture and whole rat brain, *Biochem J* 218:19-27 (1984).
8. L. Villa Komaroff, A. Gonzales, H.Y. Song, B. Wentworth, and P. Dobnes, Novel insulin related sequences in fetal brain, *Adv Exp Med Biol* 181:65-86 (1984).
9. A.B. Smit, E. Vreugdenhil, R.H.M. Ebberink, W.P.M. Geraerts, J. Klootwijk, and J. Joosse, Growth-controlling molluscan neurons produce the precursor of an insulin-related peptide, *Nature* 331:535-538 (1988).
10. R. Schechter, L. Holtzclaw, F. Sadiq, A. Kahn, and S. Devaskar, Insulin synthesis by isolated neurons, *Endocrinology* 123:505-513 (1988).
11. R. Schechter, H.F. Sadiq, and S.U. Devaskar, Insulin and insulin mRNA are detected in neuronal cell cultures maintained in an insulin-free/serum-free medium, *J Histochem Cytochem* 38:829-836 (1990).
12. W.S. Young, Periventricular hypothalamic cells in the rat brain contain insulin mRNA, *Neuropeptides* 8:93-97 (1986).
13. L.J. Stein, D.M. Dorsa, D.G. Baskin, D.P. Figlewicz, H. Ikeda, S. Frankmann, M.R.C. Greenwood, D. Porte, Jr., and S.C. Woods, Immunoreactive insulin levels are elevated in the cerebrospinal fluid of genetically obese Zucker rats, *Endocrinology* 113:2299-2301 (1983).

14. B.J. Wallum, J.G. Taborsky, Jr., D. Porte, Jr., D.P. Figlewicz, L. Jacobson, J.C. Beard, W.K. Ward, and D.K. Dorsa, Cerebrospinal fluid insulin levels increase during infusions in man, *J Clin Endocrinol Metab* 64:190-194 (1987).
15. D.G. Baskin, S.C. Woods, D.B. West, M. van Houten, B.I. Posner, D.M. Dorsa, and D. Porte, Jr., Immunocyto-chemical detection of insulin in rat hypothalamus and its possible uptake from cerebrospinal fluid, *Endocrinology* 113:1818-1825 (1983).
16. K.R. Duffy and W.M. Pardridge, Blood-brain barrier trancytosis of insulin in developing rabbits, *Brain Res* 420:32-38 (1987).
17. M. van Houten, B.I. Posner, B.M. Kopriwa, and J.R. Brawer, Insulin binding sites in the rat brain: *In-vivo* localization to the circumventricular organs by quantitative autoradiography, *Endocrinology* 105:666-673 (1979).
18. J. Havankova and J. Roth, Insulin receptors are widely distributed in the central nervous system of the rat, *Nature* 272:827-829 (1978).
19. G.A. Werther, A. Hogg, B.J. Oldfield, M.J. McKinley, R. Figdor, A.M. Allen, and F.A.O. Mendelsohn, Localization and characterization of insulin receptors in rat brain and pituitary gland using *in-vitro* autoradiography and computerized densitometry, *Endocrinology* 121:1562-1570 (1987).
20. A.M. Moss, J.W. Unger, R.T. Moxley, and J.N. Livingston, Location of phosphotyrosine-containing proteins by immunocytochemistry in the rat forebrain corresponds to the distribution of the insulin receptor, *Proc Natl Acad Sci USA* 87:4453-4457 (1990).
21. F.T. Boyd, Jr., D.W. Clarke, T.F. Muther, and M.K. Raizada, Insulin receptors and insulin modulation of norepinephrine uptake in neuronal cultures from rat brain, *J Biol Chem* 260:15880-15885 (1985).
22. D.G. Puro and E. Agardh, Insulin mediated regulation of neuronal maturation, *Science* 225:1170-1172 (1984).
23. S.C. Woods, D. Porte, Jr., E. Bobbioni, E. Ionescu, and J.F. Sautes, Insulin: Its relationship to the central nervous system and to the control of food intake and body weight, *Am J Clin Nutr* 42:1063-1071 (1985).
24. D.P. Figlewicz, S.C. Woods, D.G. Baskin, D.M. Dorsa, L. Wilcox, J. Stein, and D. Porte, Jr., Insulin in the central nervous system: A regulator of appetite and body weight, in: *Insulin, Insulin-like Growth Factors, and Their Receptors in the Central Nervous System,* M.K. Raizada, M.I. Phillips, D. LeRoith, eds., Plenum Press (NY and Lond.) pp 151-162 (1987).
25. M. Girbau, L. Bassas, J. Alemany, and F. de Pablo, *In situ* autoradiography and ligand-dependent tyrosine kinase activity reveal insulin receptors and insulin-like growth factor I receptors in prepancreatic chicken embryos, *Proc Natl Acad Sci USA* 86:5868-5872 (1989).
26. L. Bassas, F. de Pablo, M.A. Lesniak, and J. Roth, Ontogeny of receptors for insulin-like peptides in chick embryo tissues: Early dominance of insulin-like growth factor over insulin receptors in brain, *Endocrinology* 117:2321-2329 (1985).
27. M. Girbau, J.A. Gomez, M.A. Lesniak, and F. de Pablo, Insulin and Insulin-like growth factor I both stimulate metabolism, growth and differentiation in the postneurula chick embryo, *Endocrinology* 121:1477-1482 (1987).

28. M.B. Harvey and P.L. Kaye, Insulin stimulates protein synthesis in compacted mouse embryos, *Endocrinology* 122:1182-1184 (1988).
29. J.P. Travers, M.K. Pratten, and F. Beck, Effects of low insulin levels on rat embryonic growth and development, *Diabetes* 38:773-778 (1989).
30. F. de Pablo, M. Girbau, J.A. Gomez, E. Hernandez, and J. Roth, Insulin antibodies retard and insulin accelerates growth and differentiation in early embryos, *Diabetes* 34:1063-1067 (1985).
31. D.S. Miller, Stimulation of RNA and protein synthesis by intracellular insulin, *Science* 240:506-509 (1988).
32. S. Heyner, L.V. Rao, L. Jarett, and R.M. Smith, Pre-implantation mouse embryos internalize maternal insulin via receptor-mediated endocytosis: Pattern of uptake and functional correlations, *Dev Biol* 134:48-58 (1989).
33. S.J. Giddings and L. Carnaghi, Rat insulin II gene expression by extraplacental membranes: A non-pancreatic source for fetal insulin, *J Biol Chem* 264:9462-9469 (1989).
34. K.A. Heidenreich and S.P. Toledo, Insulin receptors mediate growth effects in cultured fetal neurons. I. Rapid stimulation of protein synthesis, *Endocrinology* 125:1451-1457 (1989).
35. K.A. Heidenrich, G. de Vellis, and P.R. Gilmore, Functional properties of the subtype of insulin receptor found on neurons, *J Neurochem* 51:878-887 (1988).
36. E. Recio-Pinto, F.F. Lang, and D.N. Ishii, Insulin and insulin-like growth factor II permit nerve growth factor binding and the neurite formation response in cultured human neuroblastoma cells, *Proc Natl Acad Sci USA* 81:2562-2566 (1984).
37. E. DiCiccio-Bloom and I.B. Black, Insulin growth factors regulate the mitotic cycle in cultured rat sympathetic neuroblasts, *Proc Natl Acad Sci USA* 85:4066-4070 (1989).
38. J.F. Mill, M.V. Chao, and D.N. Ishii, Insulin, insulin-like growth factor Ii, and nerve growth factor effects on tubulin mRNA levels and neurite formation, *Proc Natl Acad Sci USA* 82:7126-7130 (1985).
39. P. Fernyhough, J.F. Mill, J.L. Roberts, and D.D. Ishii, Stabilization of tubulin mRNAs by insulin and insulin-like growth factor I during neurite formation, *Mol Brain Res* 6:109-120 (1989).
40. K.A. Heidenreich, S.P. Toledo, L.L. Brunton, M.J. Watson, S. Daniel-Issakani, and B. Strulovici, Insulin stimulates the activity of a novel protein kinase C, PKC-e, in cultured fetal chick neurons, *J Biol Chem* 265:15076-15082 (1990).
41. K.A. Heidenreich and S.P. Toledo, Insulin receptors mediate growth effects in cultured fetal neurons. II. Activation of a protein kinase that phosphorylates ribosomal protein S6, *Endocrinology* 125:1458-1463 (1989).
42. R.W. Rees-Jones, S.A. Hendricks, M. Quarum, and J. Roth, The insulin receptor of rat brain is coupled to tyrosine kinase activity, *J Biol Chem* 259:3470-3474 (1984).
43. W.A. Brennan, Developmental aspects of the rat brain insulin receptor: Loss of sialic acid and fluctuation in number characterize fetal development, *Endocrinology* 122:2364-2370 (1988).

44. S.U. Devaskar, N. Holekamp, L. Karycki, and U.P. Devaskar, Ontogenesis of the insulin receptor in the rabbit brain, *Hormone Res* 24:319-327 (1986).
45. K.A. Heidenreich and D. Brandenburg, Oligosaccharide heterogeneity of insulin receptors. Comparison of N-linked glycosylation of insulin receptors in adipocytes and brain, *Endocrinology* 118:1835-1842 (1986).
46. A.I. Salhanick and J.M. Amatruda, Role of sialic acid in insulin action and the insulin resistance of diabetes mellitus, *Am J Physiol* 255:E173-E179 (1988).
47. W.L. Lowe, Jr., F.T. Boyd, D.W. Clarke, M.K. Raizada, C. Hart, and D. LeRoith, Development of brain insulin receptors: Structural and functional studies of insulin receptors from whole brain and primary cell cultures, *Endocrinology* 119:25-35 (1986).
48. I. Ocrant, K.L. Valentino, L.F. Eng, R.L. Hintz, D.M. Wilson, and R.G. Rosenfeld, Structural and immunohistochemical characterization of insulin-like growth factor I and II receptors in the murine central nervous system, *Endocrinology* 123:1023-1034 (1988).
49. J. Shemer, M.K. Raizada, B.A. Masters, A. Ota, and D. LeRoith, Insulin-like growth factor I receptors in neuronal and glial cells: Characterization and biological effects in primary culture, *J Biol Chem* 262:7693-7699 (1987).
50. D.W. Clarke, F.T. Boyd, Jr., M.S. Kappy, and M.K. Raizada, Insulin binds to specific receptors and stimulates 2-deoxy-D-glucose uptake in cultured glial cells from rat brain, *J Biol Chem* 259:11672-11675 (1984).
51. L.M. Mudd, H. Werner, Z. Shen-Orr, C.T. Roberts, Jr., D. LeRoith, H.C. Haspel, and M.K. Raizada, Regulation of rat brain/Hep G2 glucose transporter gene expression by phorbol esters in primary cultures of neuronal and astrocytic glial cells, *Endocrinology* 126:545-549 (1990).
52. H. Werner, M.K. Raizada, L.M. Mudd, H.L. Foyt, I.A. Simpson, C.T. Roberts, Jr., and D. LeRoith, Regulation of rat brain/Hep G2 glucose transporter gene expression by insulin and insulin-like growth factor I in primary cultures of neuronal and glial cells, *Endocrinology* 125:314-320 (1989).
53. D.W. Clarke, F.T. Boyd, Jr., M.S. Kappy, and M.K. Raizada, Insulin stimulates macromolecular synthesis in cultured glial cells from rat brain, *Am J Physiol* 249:C484-C489 (1985).
54. R.P. Saneto and J. de Vellis, Hormonal regulation of the proliferation and differentiation of astrocytes and oligodendrocytes in primary culture, in: *Developmental Biology of Cultured Nerve, Muscle, and Glia,* D. Schubert, ed., Vol 4: pp 125. John Wiley and Sons, New York (1984).
55. V.K.M. Han, J.M. Lauder, and J. D'Ercole, Characterization of somatomedin/insulin-like growth factor receptors and correlation with biologic action in cultured neonatal rat astroglial cells, *J Neuroscience* 7:501-511 (1987).
56. S. Devaskar, F. Sadiq, L. Holtzclaw, and M. George, Developmental regulation of rabbit brain insulin and insulin-like growth factor receptors, *J Neurochem* (submitted).
57. D.W. Clarke, L. Mudd, F.T. Boyd, M. Fields, and M.H. Raizada, Insulin is released from rat brain neuronal cells in culture, *J Neurochem* 47:831-836 (1986).

58. H.J.L. Frank and W.M. Pardridge, Insulin binding to brain microvessels, *Adv Metab Disord* 10:291-303 (1983).
59. H.J.L. Frank, W.M. Pardridge, W.L. Morris, R.G. Rosenfeld, and T.B. Choi, Binding and internalization of insulin and insulin-like growth factors by isolated brain microvessels, *Diabetes* 35:654-661 (1986).
60. H.J.L. Frank, W.M. Pardridge, T. Jankovic-Vokes, T.J. Vinters, and W.L. Morris, Enhanced insulin binding to blood-brain barrier *in vivo* and to brain microvessels *in vitro* in newborn rabbits, *Diabetes* 34:728-733 (1985).
61. R. Schechter and S. Devaskar, Developmental regulation of insulin in the mammalian central nervous system *Developmental Brain Research* (submitted 1990).
62. H.G. Bernstein, A. Dorn, M. Reiser, and M. Zeigler, Cerebral insulin-like immunoreactivity in rats and mice: Drastic decline during postnatal ontogenesis, *Acta Histochem* 74:33-36 (1984).
63. R. Kadle, C. Suksang, E.D. Roberson, and R.E. Fellows, Identification of an insulin-like factor in astrocyte conditioned medium, *Brain Res* 460:60-67 (1988).
64. C.G. Budd, B. Pansky, and B. Cordell, Detection of insulin synthesis in mammalian anterior pituitary cells by immunohistochemistry and demonstration of insulin related transcripts by *in-situ* RNA-DNA hybridization, *J Histochem Cytochem* 34:673-678 (1986).
65. V.R. Sara, C. Carlsson-Skwirut, C. Andersson, E. Hall, B. Sjogren, A. Holmgren, and H. Jornvall, Characterization of somatomedins from human fetal brain: Identification of a variant form of insulin-like growth factor I, *Proc Natl Acad Sci USA* 83:4904-4907 (1986).
66. P. Rotwein, S.K. Burgess, J.D. Milbrandt, and J.E. Krause, Differential expression of insulin-like growth factor genes in rat central nervous system, *Proc Natl Acad Sci USA* 85:265-269 (1988).
67. S.J. Giddings and L.R. Carnaghi, Selective expression of developmental regulation of the ancestral rat insulin II gene in fetal liver, *Mol Endocrinol* 4:1363-1369 (1990).
68. K. Rau, L. Muglia, and J. Locker, Insulin-gene expression in extrafetal membranes of rats, *Diabetes* 38:39-43 (1988).
69. J. Paris and J.D. Richter, Maturation-specific polyadenylation and translational control: Diversity of cytoplasmic polyadenylation elements, influence of poly (A) tail size and formation of stable polyadenylation complexes, *Mol Cell Biol* 10:5634-5645 (1990).
70. P.K. Lund, B.M. Moats-Staats, M.A. Hynes, J.G. Simmons, M. Jansen, A.J. D'Ercole, and J.J. Van Wyk, Somatomedin-C/Insulin-like growth factor-I and insulin-like growth factor-II mRNAs in rat fetal and adult tissues, *J Biol Chem* 261:14539-14544 (1986).
71. S. Devaskar and H.F. Sadiq, Regulation of neuronal insulin-like peptide, in: *Molecular and Cellular Aspects of Insulin and IGF I/II,* M.K. Raizada, ed., Plenum Press, New York (1989) 231-235.
72. S. Alpert, D. Hanahan, and G. Teitelman, Hybrid insulin genes reveal a developmental lineage for pancreatic endocrine cells and imply a relationship with neurons, *Cell* 53:295-309 (1988).

73. F. Lauder and F. Bloom, Ontogeny of monamine neurons in the locus coeruleus, raphe nuclei and substantia nigra of the rat. I. Cell differentiation, *J Comp Neurol* 155: 469-482, (1974).
74. L.A. Specht, V.M. Pickel, T.H. Joh, and D.J. Reis, Light microscopic immunocytochemical localization of tyrosine hydroxylase in prenatal rat brain. I. Early ontogeny, *J Comp Neurol* 199:233-253 (1981).
75. D.J. Drucker and S.L. Asa, Glucagon gene expression in vertebrate brain, *J Biol Chem* 263:13475-13478 (1988).
76. V.K.M. Han, M.A. Hynes, C. Jin, J.M. Towle, and P.K. Lund, Cellular localization of proglucagon/glucagon-like peptide I messenger RNAs in rat brain, *J Neuroscience Res* 16:97-(1986).
77. M.T. Rojeski and J. Roth, Messenger RNA for insulin in brain and other extrapancreatic sites, *Clin Res* 38:296A (1990).
78. R. Schechter, C. Bogey, K. Jackson, and J.R. Gavin, III, Insulin support of neuron cell growth by endocrine and paracrine pathways, *Ped Res* 27:52A (1990).

GLUCOSE TRANSPORTERS IN CENTRAL NERVOUS SYSTEM

GLUCOSE HOMEOSTASIS

Bartosz Z. Rydzewski, Magdalena M. Wozniak, and Mohan K. Raizada

Departments of Physiology (B.Z.R., M.K.R.) and Pharmacology & Experimental Therapeutics (M.M.W.), College of Medicine, University of Florida, Gainesville, FL

The brain is considered to be a glucose-obligatory organ, that is neuronal and glial cells from the central nervous system (CNS) are dependent on glucose as a sole energy source[1]. The ability of various anatomical areas of the brain to utilize glucose has been shown to have a major impact on the control of certain physiological and behavioral functions of the brain. This has resulted in an increased desire to elucidate the cellular and molecular mechanisms involved in the transport, metabolism, and sensitivity to the glucose in the brain. Recent identification and characterization of a facilitative glucose transporter family have set a pace for research involving glucose uptake systems present in the brain cells.

Sensitivity of cells in the CNS to changes in blood glucose concentrations can be influenced by the activity of the glucose transport system at different levels. Firstly, the regulation of activity of the glucose transporter system in brain microvessels and chorioid plexus determines the rate of glucose transport across the blood-brain barrier[2-8]. Activity of this transporter system together with the rate of glucose metabolism by the brain tissue determines the concentration of glucose in the CNS extracellular fluid and therefore controls the amount of glucose available for neuronal and glial uptake. The second level of control is associated with the glucose transporter activity of the individual cells present in the CNS, mainly neurons and astrocytic glia.

The predominant class of glucose transporter present at the blood-brain barrier (BBB) is Glut1[9,10]. This species of facilitative glucose transporter has been identified by numerous investigators in the endothelial cells of the brain vasculature. One of the most relevant means of regulation of Glut1 expression

Edited by M.K. Raizada and D. LeRoith, Plenum Press, New York, 1991

at the BBB is glucose concentration itself. It has been observed in a number of tissues that high glucose diminishes Glut1 mRNA and protein levels whereas low glucose concentrations result in elevation of both mRNA and transporter molecules[11,12]. In the cells associated with the BBB, however, hyperglycemia increases Glut1 mRNA steady state levels[13]. This is not accompanied by the parallel stimulation of transporter protein[14]. In contrast, a hyperglycemia-induced decrease in the number of transporters has been demonstrated, which suggests post-transcriptional inhibition of glucose-transporter mRNA translation[15,16]. Thus it would appear that increased blood glucose levels can result in diminished glucose uptake activity of the BBB which, taken together with reduced cerebral blood flow observed in streptozotocin-induced diabetes[16], may alter glucose availability to the CNS. Hyperglycemia-induced reduction in number of active transporters at the BBB can serve as a mechanism protecting brain cells against exposure to abnormally high glucose concentrations.

Heterogeneity of cells present in the brain makes the identification of glucose transporter classes expressed in them difficult. Cell culture techniques allowed for a separation of neurons from astrocytic glia and preliminary characterization of the glucose transport system present in these cells. Astrocytic glial cells express the Glut1 transporter - the same type of glucose transporter as that found in the BBB - and the Glut3 insulin-insensitive transporter originally demonstrated in human adult cerebral tissue[9,17]. No other glucose transporter species have been associated with the astrocytes up to date. Some investigators have shown the Glut1 mRNA in neuronal cells in culture[18,19]. Steady state levels of this mRNA species are, however, much lower than that found in astrocytic glia[18]. Minor quantities of immunoreactive Glut1 in neuronal cells have also been demonstrated[20]. Other reports dispute the presence of Glut1 in neuronal cells[21]. Whatever the case, Glut1 is apparently not the protein responsible for the majority of the glucose uptake in the neurons. There is some indirect evidence to support this notion: i) in spite of neuronal Glut1 mRNA levels, much lower than those seen in astrocytes[18], glucose uptake activity is higher in neuronal than in astrocytic glial cells[9]; ii) glucose uptake in neuronal cells is not regulated by the factors which alter the expression or activity of Glut1 in other cell types. Factors like insulin, insulin-like growth factor I (IGF I), and phorbol esters cause prominent changes in Glut1 mRNA steady state levels, but fail, however, to affect 2-deoxy glucose uptake in neuronal cells[9,19,22,23]. This suggests either post-transcriptional inhibition of Glut1 expression in cells of neuronal origin as has been previously suggested[19], or the presence of a different transporter responsible for the bulk of glucose uptake in neurons. If the latter hypothesis is correct and the contribution of Glut1 to the total glucose transport activity in neuronal cells is minor, then changes in expression of Glut1 would be relatively insignificant and difficult to detect. Pronounced changes in Glut1 expression during rat brain development[24,25] provide evidence for the presence of a glucose transporter species whose activity would offset low Glut1 levels. Another report has suggested that Glut3 may be the transporter species responsible for the majority of glucose uptake in neuronal cells expressing low quantities of Glut1[20]. This is further supported by the detection of high levels of this transporter species in the brain.

It has been demonstrated that diabetic subjects manifest increased sympathetic outflow which ultimately may lead to hypertension and related disorders. What then is a mechanism of sensing of the diabetic state by the brain? Most obvious manifestations of diabetes in the periphery include altered levels of certain hormones and hyperglycemia. The majority of hormones engaged in glucose homeostasis in the periphery have peptide structures and therefore they are unable to effectively penetrate the BBB[26]. In contrast, glucose, unlike insulin or glucagon, crosses the BBB and its concentrations in cerebrospinal and extracellular fluids are in direct proportion to its plasma levels[26,27]. Thus glucose may serve as an ideal messenger between peripheral circulation and the CNS.

The hypothalamus is engaged in peripheral glucose-level homeostasis through sympathetic modulation of pancreatic function and has been shown to contain glucose-sensitive cells[28]. In the periphery, analogous glucose-regulatable cells are located in the endocrine pancreas. The recently demonstrated presence of Glut2 in the pancreatic β-cells is mechanistically understandable[29,30,31]. The glucose K_m of this transporter is 20-40 mM[30,32] as compared with 2-10 nM for other members of the facilitative glucose transporter family[33,34,35]. This characteristic makes Glut2 capable of transporting glucose at a rate directly proportional to its concentration in ECF even in extreme hyperglycemia. Is it then possible that hypothalamic cells which are responsible for glucose-level sensing express Glut2? Studies conducted so far disproved this notion, showing a virtual absence of Glut2 in the brain tissue[30,36]. This may, however, be due to a very low level of this transporter species in whole-brain preparations. The presence of a distinct, unidentified glucose transporter with characteristics similar to those of Glut2 is also conceivable.

It is unclear at this point what mechanism might be responsible for the glucose sensitivity of the hypothalamic cells or whether glial or neuronal cells serve as "glucose-level sensors". Evidence accumulated so far points at astrocytic glial cells as the ones responsible for glucose-level detection.

Glucose starvation results in an increase in both Glut1 mRNA and protein[12]. The effect is much more pronounced in astrocytic glial than in neuronal cultures. This observation suggests that astrocytic glial cells are more sensitive to glucose level fluctuations than neurons and can be actual "glucose sensors".

In spite of Glut1 being "insulin-insensitive," insulin and IGF-I effectively stimulate both Glut1 mRNA and 2-deoxy glucose uptake in astrocytic glial cells[18]. In contrast, an increase in Glut1 mRNA levels is not associated with the stimulation of 2-deoxy glucose uptake in neuronal cells[18]. ED_{50}'s for insulin and IGF-1 indicate that these factors act through specific receptors[18,37]. In astroglial cultures, insulin- and IGF-1-mediated elevation in 2-deoxy glucose uptake reached its maximum within 1-2 h and was preceded by the elevation of Glut1 mRNA levels[18]. These observations suggest that insulin and IGF-1 alter Glut1 expression as opposed to the translocation of insulin-responsive glucose transporters (Glut4) in insulin-responsive tissues[38,39]. It is argued that one of the actions of peripherally produced insulin in the CNS is a stimulation of "satiety centers"[40]. If this hypothesis is correct then the action of insulin could be mediated through an increased number of glucose transporters (Glut1) followed by higher intracellular glucose concentrations. Insulin then would amplify the effect of high ECF glucose levels. Since neurons are relatively insensitive to insulin, once again astroglial cells would be the ones directly responsive to a stimulus.

Another example of the high sensitivity of the astroglial glucose transporter system involves its modulation by thyroid hormones[41]. In rat astroglial cultures, deficiency of thyroid hormones results in impaired 2-deoxy glucose uptake and a 80% decrease in glucose transporter number as measured by cytochalasin binding. This effect was reversed within 60 s by the supplementation of culture medium with 3,5,3'-triiodo thyronine. The rapidity of this reaction suggests translocation of glucose transporters as a mechanism of increase in 2-deoxy glucose uptake. Regulation of the glucose transport system in astroglial cells by thyroid hormones may be another mode of modulation of brain sensitivity to glucose fluctuations.

The intracellular mechanism responsible for glucose sensing is unknown. The effect of glucose concentration on Glut1 expression may involve an alteration in glycosylation capability of the cell as suggested by Elbein[42]. Low glucose concentration is able to increase both Glut1 mRNA and 2-deoxy glucose uptake in L6 myocytes[12]. It is conceivable that altered glycosylation capacity due to low glucose concentrations is responsible for more profound effects, including those ultimately leading to changes in neuronal activity in diabetic patients.

Recently it has been demonstrated that glucose concentrations as well as glycosylation inhibitors affect the expression of the glucose-regulatable proteins (GRP) at both transcriptional and translational levels[43,44,45]. Elevated levels of the GRP78 mRNA were shown in the brains of prediabetic and diabetic rats, thus suggesting that GRPs may mediate some of the glucose effects[46].

All of the presented observations strongly suggest that the neurons' glucose transporter system is relatively insensitive to various stimuli. Given glucose-dependent metabolism of neuronal cells and their physiological significance, it is teleologically understandable. On the other hand, an astroglial glucose transport system capable of responding to many factors involved in glucose homeostasis may be the actual mediator of the glucose homeostasis information between the peripheral circulation and specialized neuronal centers in the brain. If this is the case, then information needs to be further conveyed to neurons, presumably in a paracrine fashion.

In summary, elucidation of mechanisms responsible for the sensitivity of the CNS to glucose is by no means complete. The availability of highly specific antibodies and cDNA probes for all identified members of facilitative glucose transporter family and employment of cell culture techniques resulted in a dynamic increase in the number of publications on glucose transporter function in the CNS in the recent year. These advances should quickly result in a multiplication of the information available on glucose homeostasis in the brain and ultimately lead us to an understanding of the glucose actions mediated through glial and neuronal intermediaries.

REFERENCES

1. Lund-Andersen, H. Transport of glucose from blood to brain. *Phys. Rev.* 59:305-310, 1979.
2. Dick, A. P., Harik, S. I., Klip, A., Walker, D. M. Identification and characterization of the glucose transporter of the blood-brain barrier by cytochalasin B binding and immunological reactivity. *Proc. Natl. Acad. Sci. USA* 81:7233-7237, 1984.
3. Baldwin, S. A., Cairns, M. T., Gardiner, R. M., Ruggier, R. A. D-glucose-sensitive cytochalasin B binding component of cerebral microvessels. *J. Neurochem.* 45:650-652, 1985.
4. Dick, A. P., Harik, S. I. Distribution of the glucose transporter in the mammalian brain. *J. Neurochem.* 46:1406-1411, 1986.
5. Matthaei, S., Olefsky, J. M., Horuk, R. Biochemical characterization and subcellular distribution of the glucose transporter from rat brain microvessels. *Biochim. Biophys. Acta* 905:417-425, 1987.
6. Kasanicki, M. A., Cairns, M. T., Davies, A., Gardiner, R. M., Baldwin, S. A. Identification and characterization of the glucose-transport protein of the bovine blood/brain barrier. *Biochem. J.* 247:101-108, 1987.
7. Kalaria, R. N., Gravina, S. A., Schmidley, J. W., Perry, G., Harik, S. I. The glucose transporter of the human brain and blood-brain barrier. *Ann. Neurol.* 24:757-764, 1988.
8. Gerhart, D. Z., LeVasseur, R. J., Broderius, M. A., Drewes, L. R. Glucose transporter localization in brain using light and electron immunocytochemistry. *J. Neurosci. Res.* 22:464-472, 1989.
9. Devaskar, S. The mammalian brain glucose transport system. In: *Molecular Biology and Physiology of Insulin and Insulin-like Growth Factors*, eds. Raizada, M. K., and LeRoith, D., Plenum Press, New York, 1991 (in press).

10. Pardridge, W. M., Boado, R. J. and Farrel, C. R. Brain-type glucose transporter (GLUT 1) is selectively localized to the blood brain barrier. *J. Biol. Chem.* 265:18035-18040, 1990.
11. Pessin, J. E., Tillotson, L. G., Yamada, K., et al. Identification of the stereospecific hexose hexose transporter from starved and fed chicken embryo fibroblasts. *Proc. Natl. Acad. Sci. USA* 79:2286-2290, 1982.
12. Walker, P. S., Donovan, J. A., Van Ness, B. G., Fellows, R. E. and Pessin, J. E. Glucose-dependent regulation of glucose transport activity, protein, and mRNA in primary cultures of rat brain glial cells. *J. Biol. Chem.* 263:15594-15601, 1988.
13. Choi, T. B., Boado, R. J. and Pardridge, W. M. Blood-brain barrier glucose transporter mRNA is increased in experimental diabetes mellitus. *Biochem. Biophys. Res. Commun.* 164:375-380, 1989.
14. Harik, S. I., Gravina, S. A. and Kalaria, R. N. Glucose transporter of the blood-brain barrier and brain in chronic hyperglycemia. *J. Neurochem.* 51:1930-1934, 1988.
15. Matthaei, S., Horuk, R. and Olefsky, J. M. Blood-brain glucose transfer in diabetes mellitus. Decreased number of glucose transporters at blood-brain barrier. *Diabetes.* 35:1181-1184, 1986.
16. Pardridge, W. M., Triguero, D. and Farrell, C. R. Downregulation of blood-brain barrier glucose transporter in experimental diabetes. *Diabetes.* 39:1040-1044, 1990.
17. Kayano, T., Fukumoto, H., Eddy, R. L., et al. Evidence for a Family of Human Transporter-like Proteins. *J Biol Chem* 263:15245-15248, 1988.
18. Werner, H., Raizada, M. K., Mudd, L. M., et al. Regulation of Rat Brain/HepG2 glucose Transporter Gene Expression by Insulin and Insulin-Like Growth Factor-I in Primary Cultures of Neuronal and Glial Cells. *Endocrinology* 125 No. 1:314-320, 1989.
19. Mudd, L. M., Werner, H., Shen-Orr, Z., et al. Regulation of Rat Brain/HepG2 Glucose Transporter Gene Expression by Phorbol Esters in Primary Cultures of Neuronal and Glial Cells. *Endocrinology* 126 No. 1:545-549, 1990.
20. Sadiq, F., Holtzclaw, L., Chundu, K., Muzzafar, A. and Devaskar, S. The Ontogeny of the Rabbit Brain Glucose Transporter. *Endocrinology* 126 No. 5:2417-2424, 1990.
21. Boado, R. J. and Pardridge, W. M. The Brain-type Glucose Transporter mRNA is Specifically Expressed at the Blood-Brain Barrier. *Biochem. Biophys. Res. Commun.* 166:174-9, 1990.
22. Clarke, D., Ramaswamy, A., Holmes, L., Mudd, L., Poulakos, J. and Raizada, M. K. Phorbol Esters Stimulate 2-deoxyglucose Uptake in Glia, but Not Neurons. *Brain Research* 421:358-362, 1987.
23. Clarke, D. W., Boyd, F. T., Kappy, M. S. and Raizada, M. K. Insulin Binds to Specific Receptors and Stimulates 2-deoxy D-glucose Uptake in Cultured Glial Cells From Rat Brain. *J. Biol. Chem.* 259:11672-11678, 1984.

24. Sivitz, W., DeSautel, S., Walker, P. S. and Pessin J. E. Regulation of the Glucose Transporter in Developing Rat Brain. *Endocrin.* 124:1875-1880, 1989.
25. Werner, H., Adamo, M., Lowe Jr., W. L., Roberts Jr., C. T. and LeRoith, D. Developmental regulation of the Rat Brain/HepG2 Glucose Transporter Gene Expression. *Mol. Endocrin.* 3:273-279, 1989.
26. Steffens, A. B., Sheurink, A. J. W., Porte Jr., D. and Woods, S. C. Penetration of peripheral glucose and insulin into cerebrospinal fluid in rats. *Am. J. Physiol.* 255:R200, 1988.
27. Hertz, M. and Paulson, O. Glucose Transfer Across the Blood-Brai Barrier. *Adv. Met. Disorders.* 10:178-192, 1983.
28. Oomura, Y. Glucose as a Regulator of Neuronal Activity. *Adv. Met. Disorders.* 10:31-65, 1983.
29. Fukumoto, H., Seino, S., Imura, H., et al. Sequence, tissue distribution, and chromosomal localization of mRNA encoding a human glucose transporter-like protein. *Proc. Natl. Acad. Sci. USA* 85:5434-5438, 1988.
30. Thorens, B., Sarkar, H. K., Kaback, H. R. and Lodish, H. F. Cloning nad functional expression in bacteria of a novel glucose transporter present in liver, intestine, kidney, and beta-pancreatic islet cells. *Cell* 55:281-290, 1988.
31. Chen, L., Alam, T., Johnson, J. H., Hughes, S., Newgard, C. B. and Unger, R. H. Regulation of beta-cell glucose transporter gene expression. *Proc. Natl. Acad. Sci. USA* 87:4088-4092, 1990.
32. Axelrod, J. D. and Pilch, P. F. Unique Cytochalasin B Binding Characteristics of the Hepatic Glucose Carrier. *Biochem.* 22:2222-2227, 1983
33. Wheeler, T. J. and Hinkle, P. C. The Glucose Transporter of Mammalian Cells. *Ann. Rev. Physiol.* 47:503-508, 1989.
34. Keller, K., Strube, M. and Mueckler, M. Functional Expression of the Human HepG2 and Rat Adipocyte Glucose Transporters in Xenopus Oocytes. *J. Biol. Chem.* 264:1884-1890 , 1989.
35. Fukumoto, H., Kayano, T., Buse, J. B., et al. Cloning and characterization of the major insulin-responsive glucose transporter expressed in human skeletal muscle and other insulin responsive tissues. *J Biol Chem* 264:7776-7779, 1989.
36. Thorens, B., Charron, M. J. and Lodish, H. F. Molecular physiology of glucose transporters. *Diabetes Care* 13:209-218, 1990.
37. Adamo, M., Raizada, M. K. and Leroith, D. Insulin and Insulin-Like Growth Factor Receptors in the Nervous System. *Mol. Neurobiol.* 3:72-100, 1989.
38. Wardzala, L. J., Cushman, S. W. and Salans L. B. Mechanism of Insulin Action on Glucose Transport in the Isolated Rat Adipose Cell. *J. Biol. Chem.* 253:8002-8005, 1978.
39. Suzuki, K. and Kono, T. Evidence that Insulin Causes Translocation of Glucose Transport Activity to the Plasma Membrane from an Intracellular Storage Site. *Proc. Natl. Acad. Sci. USA.* 77:2542-2545, 1980.

40. Woods, S. C. and Porte Jr., D. The Role of Insulin as a Satiety Factor in the Central Nervous System. *Adv. Met. Disorders* 10:457-468, 1983.
41. Roeder, L. M., Hopkins, I. B., Kaiser, J. R., Hanukoglu, L. and Tildon, J. T. Thyroid Hormone Action on Glucose Transporter Activity in Astrocytes. *Biochem. Biophys. Res. Commun.* 156:275-281, 1988.
42. Elbein, A. D. Inhibitors of the biosynthesis and processing of N-linked oligosaccharide chains. *Annu. Rev. Biochem.* 56:497-534, 1987.
43. Lee, A. S., Delegeane, A. M., Baker, V. and Chow, P. C. Transcriptional regulation of two genes specifically induced by glucose starvation in a hamster mutant fibroblast cell line. *J. Biol. Chem.* 258:597-603, 1983.
44. Attenello, J. W. and Lee, A. S. Regulation of a hybrid gene by glucose and temperature in hamster fibroblasts. *Science* 226:187-190, 1984
45. Chang, S. C., Wooden, S. K., Nakaki, T., Kim, Y. K., Lin, A. T., Kung, L., Attenello, J. W. and Lee, A. S. Rat gene encoding the 78-kDa glucose regulated protein GRP78: its regulatory sequences and the effect of protein glycosylation on its expression. *Proc. Natl. Acad. Sci. USA.* 84:680-684, 1987.
46. Parfett, C. L. J., Brudzynski, K. and Stiller, C. Enhanced accumulation of mRNA for 78-kilodalton glucose-regulated protein (GRP78) in tissues of nonobese diabetic mice. *Biochem. Cell Biol.* 68:1428-1432, 1990.

THE MAMMALIAN BRAIN GLUCOSE TRANSPORT SYSTEM

Sherin U. Devaskar

Department of Pediatrics, St. Louis University School of Medicine, The Pediatric Research Institute, Cardinal Glennon Children's Hospital, St. Louis, Mo.

Glucose is an essential substrate for brain oxidative metabolism (1,2). Circulating glucose crosses the blood-brain barrier and accesses the brain parenchymal cells. Glucose, being a polar substance, crosses lipid bilayers of cell plasma membranes (3,4) by a saturable stereo-specific carrier system (2,5,6,7). This carrier system consists of a family of closely related membrane associated glycoproteins termed the glucose transporters (GTs) (8-15). In most tissues examined, the facilitative type of GTs transport glucose intracellularly. Besides intracellular transport, these GTs are capable of transcellular transport and transport of glucose outside the cell (16). Structurally they consist of 12 transmembraneous domains with the amino and carboxyl termini facing the cytoplasmic surface of the cell, and a glycosylation site on the exofacial domain located between the first and second transmembraneous segments (8,16). While there is considerable primary sequence homology between the different GT isomers, their tissue specific expression and Km varies based on the individual tissue's glucose needs and function of the specific GT (16). Typically the facilitative GTs are classified into five major types: Glut 1 (erythrocyte/Hep G2/rat brain type; Km - 1-2 mM) (8,9) and Glut 3 (fetal skeletal muscle/brain/placenta type; Km - ?) (10) are the insulin-insensitive types, present almost ubiquitously in most tissues examined and are responsible mainly for the basal transport of glucose. Glut 2 (liver/pancreatic beta islet cell type; Km - 15 mM) is capable of bidirectional transfer of glucose and mainly located in hepatocytes and the pancreatic beta islet cells (11,12). Glut 4 (adipocyte/skeletal muscle/heart type; Km - 5 mM) is the insulin-responsive type present mainly in insulin-responsive tissues and is translocatable to the cell membrane in response to insulin (13,14). Glut 5 (jejunum type) was recently cloned and demonstrated to be present mainly in the antilumenal surface of the epithelial cells lining the microvillus of the jejunum (15,16,17). Finally, Glut 6 was detected in multiple tissues and found to be a psuedogene which cannot be translated into a functional protein (15,16).

Molecular Biology and Physiology of Insulin and Insulin-Like Growth Factors
Edited by M.K. Raizada and D. LeRoith, Plenum Press, New York, 1991

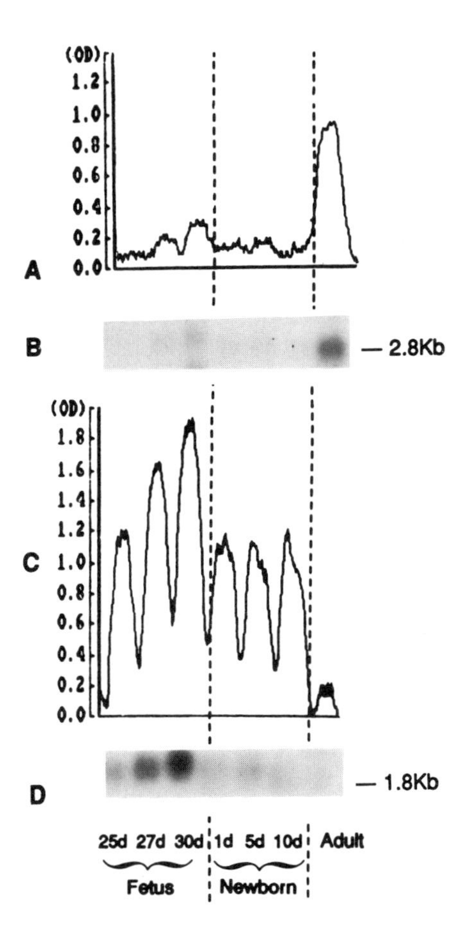

Figure 1. **Northern blot analysis:** *A) and B) demonstrate the densitometric quantitation and the autoradiograph of Glut 1 GT mRNA in rabbit whole brain obtained at various developmental stages. A ~2.8 kb mRNA species with a peak abundance in the adult brain is observed. The Glut 1 mRNA is slightly higher in the late gestation fetal brain when compared to the neonatal stages. C) and D) represent the corresponding densitometric quantitation and autoradiograph for beta actin mRNA. Note a developmental increase in the actin mRNA (1.8 kb) during the late fetal stages. (Reproduced from* Endocrinology *126:2417-2424, 1990, with permission from Williams and Wilkins Publishing, Baltimore, Maryland).*

The presence and developmental regulation of these various types of GTs were explored in the mammalian CNS (18,19,20). Further, brain cell-specific localization of these GTs was undertaken in an attempt to understand the mechanism behind brain glucose transport. We employed rabbit whole brain at various developmental time points ranging from the 25 day gestational age

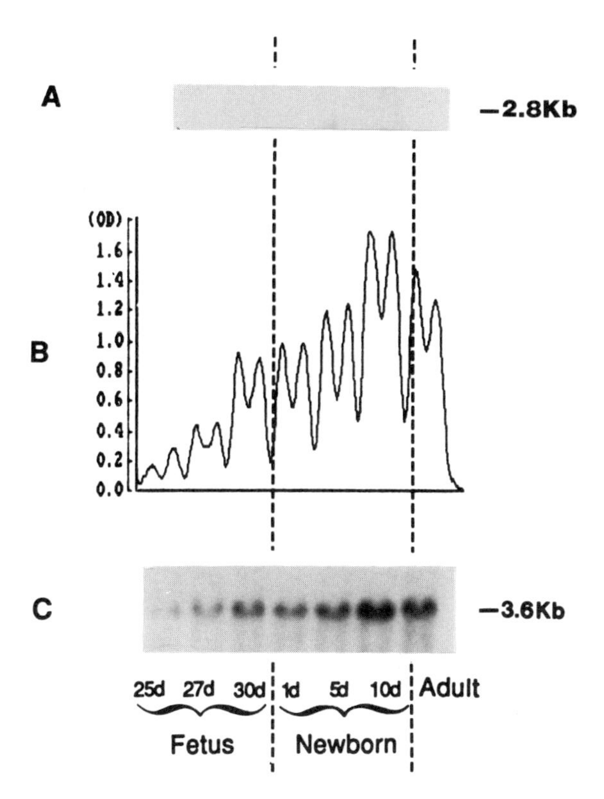

Figure 2. Northern blot analysis: A) represents the negative autoradiograph when rabbit whole brain poly (A+) RNA obtained at different developmental stages was hybridized to Glut 4 cDNA. B) and C) demonstrate the densitometric quantitation and the autoradiograph of the Glut 3 GT mRNA (3.6 kb). Note a gradual increase in the mRNA abundance with age, a peak being observed at 10 day post natal time point. (Reproduced from Endocrinology *126:2417-2424, 1990, with permission from Williams and Wilkins Publishing, Baltimore, Maryland).*

fetus to the adult (25d, 27d, 30d fetus, term being 31 days; 1d, 5d, 10d newborn and adult). The presence of mRNAs for three major types of GTs was explored (Glut 1,3,4) and the protein for Glut 1 and 4 alone was examined. Employing Northern blot analysis and specific human cDNA probes (8,21), we detected both the Glut 1 (2.8 kb) and 3 (3.6 kb) types of GTs (Figure 1). Glut 4 (rat cDNA probe (13) was absent in the CNS (Figure 2). Focusing on the Glut 1 and 3 types, we noted a reciprocal developmental regulation of these two types. The Glut 1 mRNA and the corresponding protein by Western blot analysis (~45 kD) demonstrated a relatively higher abundance during the fetal stages of development, declining to reach a nadir in the 10 day old neonate, increasing again to attain the highest abundance in the adult brain (Figures 1 and 3). Similar observations were made by others in the rat brain (18,19). Glut 3 mRNA, on the other hand, was of a low abundance in the fetal stages, gradually increasing during the newborn stages, and reaching a peak abundance in the 10 day old neonate and adult (Figure 2). Similar to the Glut 4 mRNA, no Glut 4 protein was detected at all stages studied (Figure 3). While no developmental studies exist to date similar to Glut 4, Glut 2 and 5 are not expressed in the adult human brain (11,15).

We then attempted to study the various rabbit brain cell types and the specific localization of Glut 1 and 3. We detected the presence of Glut 1 in isolated rabbit brain microvascular preparation as a 2.8 kb mRNA and a

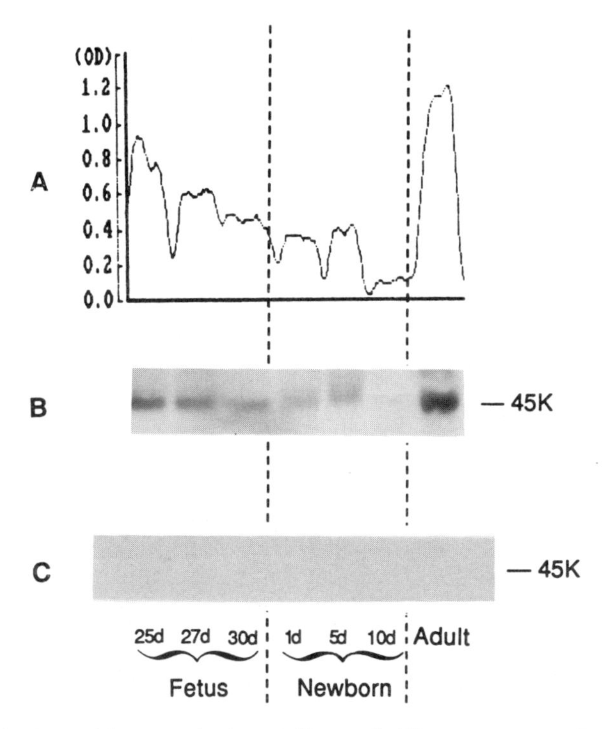

Figure 3. Western blot analysis: A) and B) represent the densitometric quantitation and autoradiograph of the Glut 1 GT protein (~45 K) in the rabbit brain at various developmental stages. Again a higher amount of Glut 1 GT is noted in the late fetal stages when compared to the post natal stages, with peak amounts being present in the adult brain. C) is a representative autoradiograph demonstrating the absence of Glut 4 GT protein in the brain at all stages studied. (Reproduced from Endocrinology *126:2417-2424, 1990, with permission from Williams and Wilkins Publishing, Baltimore, Maryland).*

50-55 kd protein (22,23). This protein represented a glycosylated form of the Glut 1 GT detected in whole brain preparations (~45 kD) as demonstrated recently in the adult rat brain (24). Examination of fetal and neonatal rabbit and rat brain sections by immunohistochemical analysis revealed the presence of a majority of the Glut 1 protein in the endothelial cells of the brain vasculature

A

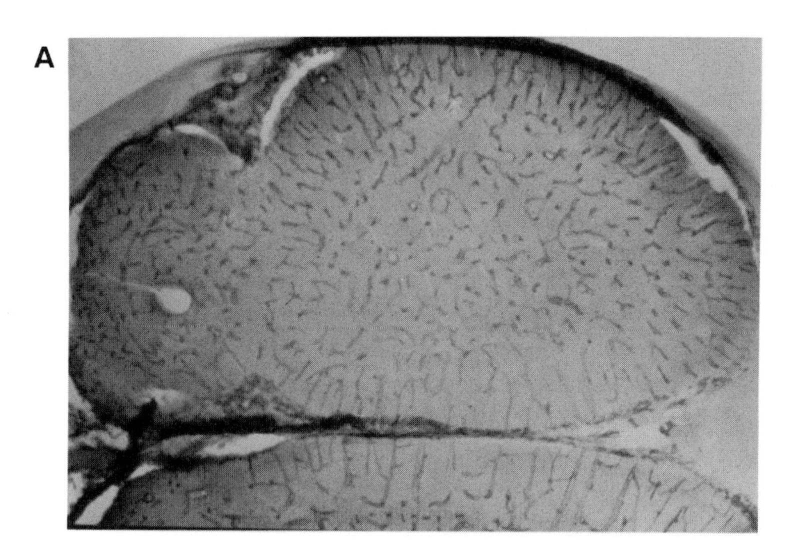

B

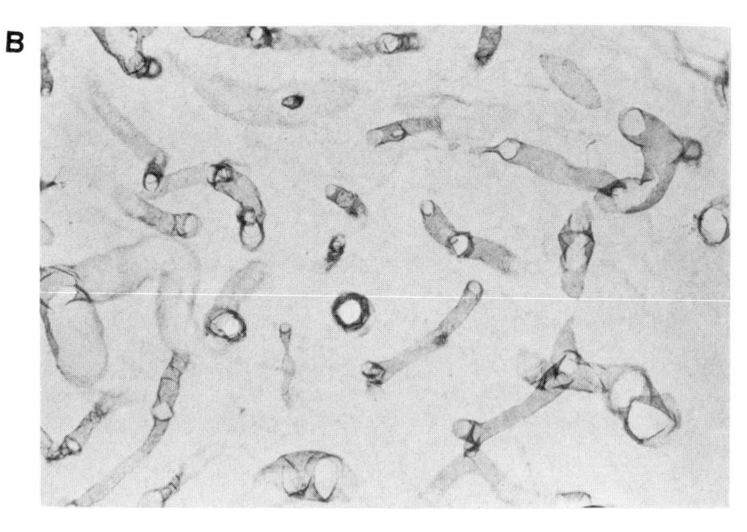

Figure 4. ***Immunohistochemical analysis:*** *A) An 18 day fetal rat brain section (magnification 4X) demonstrating the presence of Glut 1 GT immunoreactivity in the vasculature of the brain. B) demonstrates the same brain section at a higher magnification (20X).*

(Figure 4A and B). More recently, employing isolated adult rat brain microvascular preparations, a group of investigators determined that 100% of the cytochalasin B binding GT sites were secondary to the Glut 1 type of GT alone, ruling out the presence of other types in the brain vasculature (24). Additionally, however, in whole rat brain sections, employing immunohistochemistry, we preliminarily detected the prominent presence of Glut 1 type of GT during certain stages of development in certain brain parenchymal cells of glial origin (23). Although the immunoreaction in glial cells was not as intense as in the vascular endothelial cells, nevertheless, it was present (23). Unlike previous regional brain glucose uptake and utilization studies, (25,26) specific anatomic localization of the GTs proved to be a difficult task. This was based on the fact that these proteins were noted in the vasculature of all brain regions. The additional absence (or minute amounts) of GTs in the neuronal cells did not help with regional localization in the brain.

Attempts at cellular localization were undertaken in-vitro as well. Newborn rabbit brain neuron enriched and glial cell cultures were prepared, and maintained, as previously described (20). Initially, we determined the 2-deoxy-glucose uptake by these two cell types and observed that the neuronal glucose uptake was far higher than that of the glial cell (Figure 5). While insulin augmented the glucose uptake in both cell types, the effect appeared far greater in glial cells than in neurons. Despite some controversy in this regard (27), similar types of observations have been made in rat (28,29) and chick (30) brain cell cultures previously. The total numbers of glucose transporters determined by cytochalasin B binding demonstrated that the rabbit neuronal cells had a relatively equal, if not a higher, number of GTs when compared to glial cells (Figure 6). Equal levels of cytochalasin B binding were observed in primary cultures of rat astrocytes (31) and chick neurons (30).

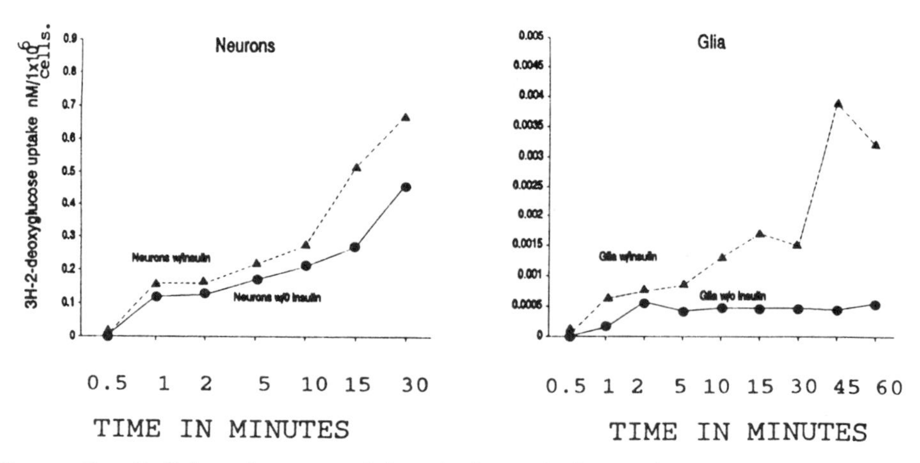

Figure 5. Cellular glucose uptake studies: 2-deoxy-glucose uptake by rabbit neuronal and glial cells in the presence and absence of 100 nM of insulin (for the indicated times) is shown. Glucose uptake was assessed by exposing the cells to 0.5 mM [^{3}H] deoxy-glucose (1 μCi) for 5 min at 37°C.

Studies employing C1300 neuroblastoma and C6 glioma cell lines observed a higher level of cytochalasin B binding in neurally derived cells when compared to the glial counterpart (31). In our present experiments, the Glut 4 type of GT was absent in both the cultured rabbit glial and neuronal cells. Glut 1 mRNA and protein, and Glut 3 mRNA were observed mainly in the glial cells (Figure 7) with little to no GTs in the neuronal cells. Thus it appears that a different type of GT in the neurons meets the cellular metabolic needs.

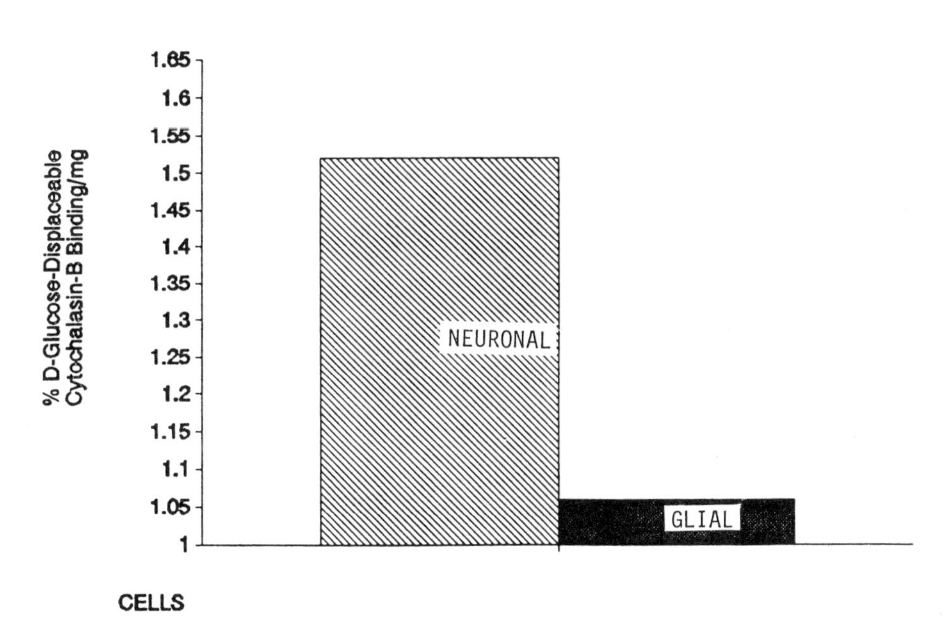

Figure 6. D-glucose (500 mM) displaceable cytochalasin B binding to rabbit neuronal and glial cells is demonstrated.

Immunohistochemical localization of Glut 1 GT protein in cultured glial cells revealed this protein to be associated with the cell membrane and peri-nuclear cytosolic structures (Figure 8) similar to the previously described Glut 1 GT localization in other tissues (32).

Regulation studies of the brain glucose transporter have yielded varying results based on the brain cell type being studied and the concentration of the regulating agent. In-vivo preliminary studies employing the adult rat whole brain have suggested that a streptozotocin-induced diabetic state decreased, while insulin-induced and starvation-induced hypoglycemia increased the Glut 1 GT mRNA (33). We have preliminarily observed that maternal diabetes in the diabetic mouse results in a mild decrease in fetal whole brain Glut 1 GT protein (34). Despite some prevailing controversy (35,36), investigations employing isolated brain microvessels revealed that diabetes (hypoinsulinemic and hyperglycemic) increased the Glut 1 mRNA (37) while decreasing the corresponding protein (36,38). Similar regulation studies were undertaken in-vitro using newborn mammalian brain cell cultures. Extracellular glucose

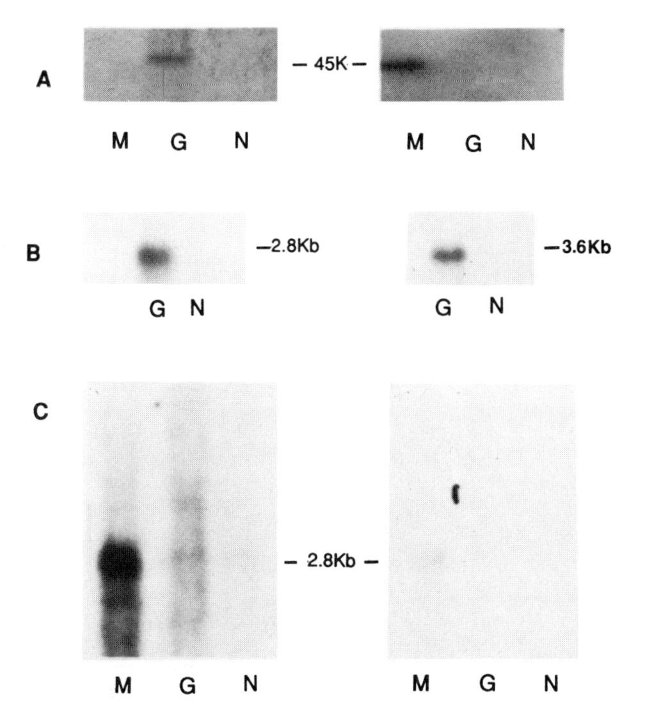

Figure 7. ***Cellular Western and Northern blot analysis:*** *A) Autoradiographs of Western blots demonstrating rabbit skeletal muscle (M), glial (G) and neuronal (N) Glut 1 GT protein (~45 K) in glia alone (left panel) and Glut 4 GT protein (~45 K) in muscle alone (right panel). B) Autoradiographs of Northern blots demonstrating the 2.8 kb Glut 1 GT mRNA in glia (G) alone (left panel) and a 3.6 kb Glut 3 GT mRNA in glia (G) alone (right panel). C) Autoradiographs of Northern blots demonstrating the 2.8 kb Glut 4 mRNA in skeletal muscle (M) alone at relatively low (left panel) and high stringency (right panel) hybridization conditions. (Reproduced from* Endocrinology *126:2417-2424, 1990, with permission from Williams and Wilkins Publishing, Baltimore, Maryland).*

concentrations were observed to reciprocally regulate the rat glial cell Glut 1 GT mRNA and protein (39). Recently, starvation of rat neuronal cells revealed an increase in glucose transport as well (40). Additionally, cellular growth promoting agents such as insulin (41), IGF-I (41) and phorbol esters (42) were observed to increase the glial cell Glut 1 GT mRNA in a dose-dependent manner. In brain microvessels, as well as in glial cells, there is accumulating

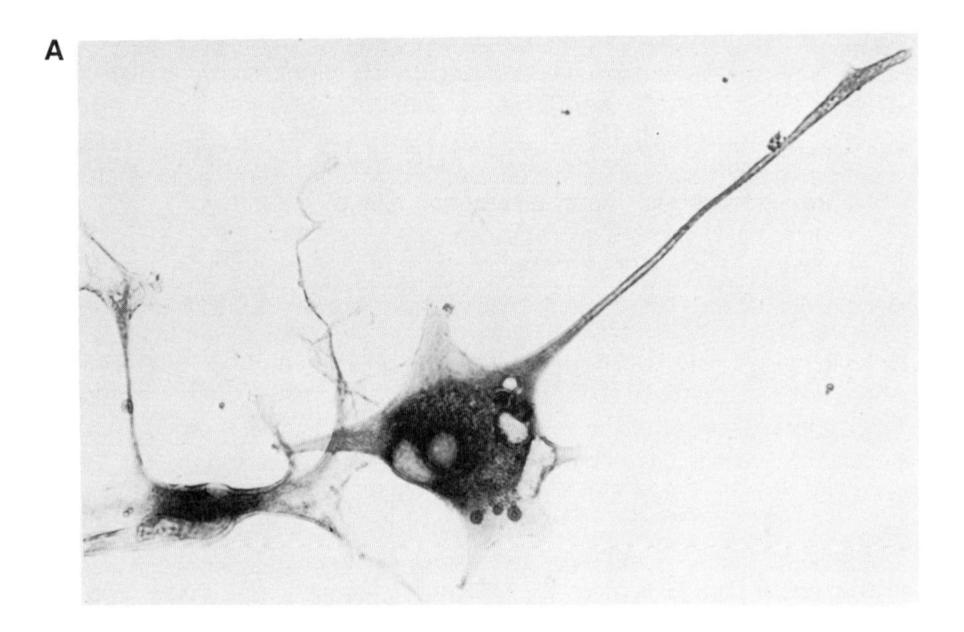

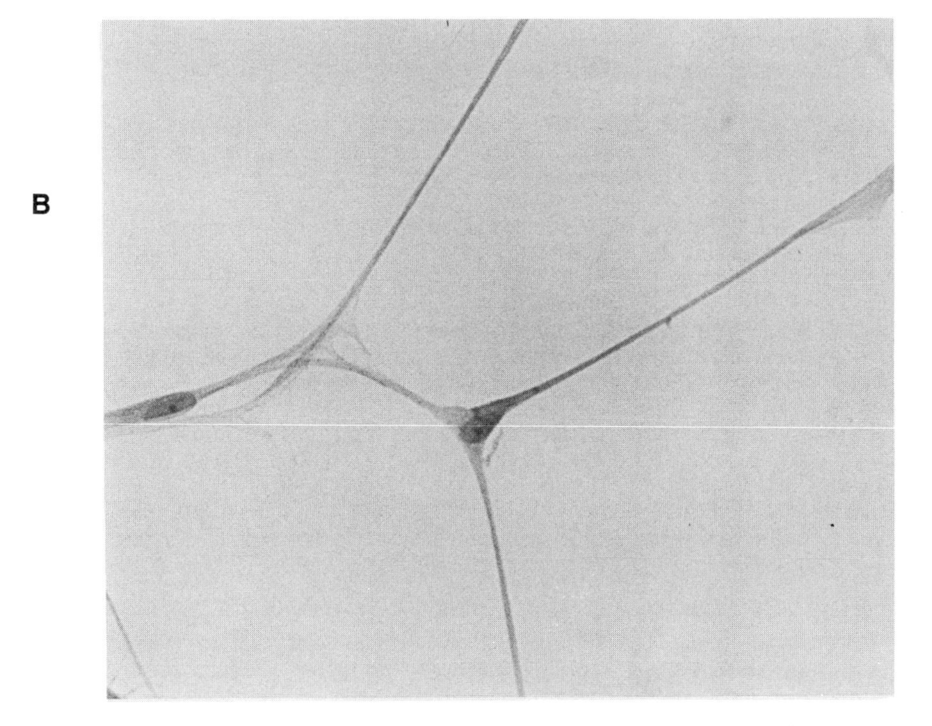

*Figure 8. **Cellular immunohistochemical analysis:*** *A) Glut 1 GT immunoreactivity is observed as dark staining of the perinuclear cytoplasmic region and the plasma membrane of glial fibrillary acidic protein-positive glial cells in culture (magnification 400X). B) An absence of Glut 1 GT immunoreactivity is observed in neuron-specific enolase-positive neuronal cells in culture (magnification 400X).*

evidence to suggest that glucose and/or insulin exert a translational control that is separate from the transcriptional control on the expression of Glut 1 GT (37,38,41,42). Similar regulation studies for the brain glial cell Glut 3 type of GT are not available at the present time. Studies along those lines, among others, may help shed light on the reasons for the need for two types (Glut 1 and 3) of transporters within the non-neuronal cells of the brain.

In summary, glucose is transported across the blood-brain barrier via Glut 1 GTs that are abundantly present in the brain vascular endothelial cells. Having gained access to the brain parenchyma, glucose is transported into glial cells by two types of transporters; namely, Glut 1 and 3. The exact contribution of each of these GTs towards total glial cell glucose transport is unknown at the present time. Further, it appears both *in situ* (brain sections) and in cultured cells, the neuronal glucose transport is mediated by an unconventional transport system.

ACKNOWLEDGEMENTS

I wish to acknowledge Drs. D.E. James and M. Mueckler, Washington University School of Medicine, St. Louis, MO., for generous gifts of the Glut 1 and 4 GT cDNAs and antibodies. I wish to thank Dr. J.S. Flier (Beth Israel Hospital and Harvard Medical School, Boston, MA) for the GT25S (Hep G2 GT cDNA) used in some of these experiments and Dr. G.I. Bell, (Howard Hughes Institute, University of Chicago, Chicago, IL) for the human Glut 3 cDNA.

Supported in part by NIH (HD-25024).

REFERENCES

1. O.E. Owen, A.P. Morgan, H.G. Kemp, J.M. Sullivan, M.G. Herrera, and G.F. Cahill, Brain metabolism during fasting, *J Clin Invest* 46:1589-1959 (1967).
2. H. Lund-Anderson, Transport of glucose from blood to brain, *Physiol Rev* 59:305-352 (1979).
3. T.J. Wheeler and P.C. Hinkle, The glucose transporter of mammalian cells, *Ann Rev Physiol* 47:503-517 (1985).
4. J.E. Pessin and M.P. Czech, *The Enzymes of Biological Membranes,* Plenum Publishing Co., New York (1985) (A.N. Matoosi, ed) Vol 2, pp 497-522.
5. C. Crone, Facilitated transfer of glucose from blood to brain tissue, *J Physiol* 181:103-113 (1965).
6. A. Gjedde, Modulation of substrate transport to the brain, *Acta Neurol Scand* 67:3-25 (1983).
7. W.M. Pardridge, Brain metabolism: A perspective from the blood-brain barrier, *Physiol Rev* 63:1481-1535 (1983).
8. M. Mueckler, C. Caruso, S.A. Baldwin, M. Panico, I. Blench, H.R. Morris, W.J. Allard, G.E. Leinhard, and H.F. Lodish, Sequence and structure of a human glucose transporter, *Science* 229:941-945 (1985).

9. M.J. Birnbaum, H.C. Haspel, and O.M. Rosen, Cloning and characterization of a cDNA encoding the rat brain glucose-transporter protein, *Proc Natl Acad Sci USA* 83:5784-5788 (1986).
10. T. Kayano, H. Fukumoto, R.L. Eddy, Y-S Fan, M. Byers, T.B. Shows, and G.I. Bell, Evidence for a family of human glucose transporter-like proteins: Sequence and gene localization of a protein expressed in fetal skeletal muscle and other tissues, *J Biol Chem* 263:15245-15248 (1988).
11. H. Fukumoto, S. Seino, H. Imura, Y. Seino, R.L. Eddy, Y. Fukushima, M.G. Byers, T.B. Shows, and G.I. Bell, *Proc Natl Acad Sci USA* 85:5434-5438 (1988).
12. M.A. Permutt, L. Koranyi, K. Keller, P.E. Lacy, D.W. Scharp, and M. Mueckler, Cloning and functional expression of a human pancreatic islet glucose transporter cDNA, *Proc Natl Acad Sci USA* 86:8688-8692 (1989).
13. D.E. James, M. Strubbe, and M. Mueckler, Molecular cloning and characterization of an insulin-regulatable glucose transporter, *Nature* (Lond) 338:83-87 (1989).
14. H. Fukumoto, T. Kayano, J.B. Buse, Y. Edwards, P.F. Pilch, G.I. Bell, and S. Seino, Cloning and characterization of the major insulin-responsive glucose transporter expressed in human skeletal muscle and other insulin responsive tissues, *J Biol Chem* 264:7776-7779 (1989).
15. T. Kayano, C.F. Burant, H. Fukumoto, G.W. Gould, Y-S Fan, R.L. Eddy, M.G. Byers, T.B. Shows, S. Seino, and G.I. Bell, Human facilitative glucose transporters, *J Biol Chem* 265:13276-13282 (1990).
16. G.I. Bell, T. Kayano, J.B. Buse, C.F. Burant, J. Takeda, D. Lin, H. Fukumoto, and S. Seino, Molecular biology of mammalian glucose transporters, *Diabetes Care* 13:198-208 (1990).
17. J.B. Meddings, D. deSouza, M. Goel, and S. Thiesen, Glucose transport and microvillus membrane physical properties along the crypt-villus axis of the rabbit, *J Clin Invest* 85:1099-1107 (1990).
18. H. Werner, M. Adamo, W.L. Lowe, Jr., C.T. Roberts, Jr., and D. LeRoith, Developmental regulation of the rat brain/Hep G2 glucose transporter gene expression, *Mol Endocrinology* 3:273-279 (1989).
19. W. Sivitz, S. DeSautel, P.S. Walker, and J.E. Pessin, Regulation of the glucose transporter in developing rat brain, *Endocrinology* 124:1875-1880 (1989).
20. F. Sadiq, L. Holtzclaw, K. Chundu, A. Muzzafar, and S. Devaskar, The ontogeny of the rabbit brain glucose transporter, *Endocrinology* 126: 2417-2424 (1990).
21. J.S. Flier, M. Mueckler, A.L. McCall, and H.F. Lodish, Distribution of glucose transporter mRNA transcripts in tissues of rat and man, *J Clin Invest* 79:657-661 (1987).
22. S. Devaskar, K. Chundu, D.S. Zahm, and L. Holtzclaw, The neonatal rabbit brain glucose transporter, *Dev Brain Res* (submitted 1990).
23. S. Devaskar, D.S. Zahm, L. Holtzclaw, K. Chundu, and B.E. Wadzinski, Developmental regulation of the distribution of rat brain insulin-insensitive glucose transporter (in preparation 1990).

24. W.M. Pardridge, R.J. Boado, and C.R. Farrell, Brain-type glucose transporter (Glut 1) is selectively localized to the blood-brain barrier, *J Biol Chem* 265:18035-18040 (1990).
25. L. Sokoloff, M. Reivich, C. Kennedy, M.H. Des Rosiers, C.S. Patak, K.D. Pettigrew, O. Sakura, and M. Rhinohara, The (^{14}C) deoxy-glucose method for the measurement of local cerebral glucose utilization: Theory, procedure, and normal values in the conscious and anesthetized albino rat, *J Neurochem* 28:897-916 (1977).
26. R.C. Vannucci, M.A. Christensen, and D.T. Stein, Regional cerebral glucose utilization in the immature rat: Effect of hypoxia-ischemia, *Ped Res* 26:208-214 (1989).
27. M. Hara, Y. Matsuda, K. Hirai, N. Okumura, and H. Nakagawa, Characteristics of glucose transport in neuronal cells and astrocytes from rat brain in primary culture, *J Neurochem* 52:902-908 (1989).
28. D.W. Clarke, F.T. Boyd, Jr., M.S. Kappy, and M.K. Raizada, Insulin binds to specific receptors and stimulates 2-deoxy-D-glucose uptake in cultured glial cells from rat brain, *J Biol Chem* 259:11672-11675 (1984).
29. F.T. Boyd, Jr., D.W. Clarke, T.F. Muther, and M.K. Raizada, Insulin receptors and insulin modulation of norepinephrine uptake in neuronal cultures from rat brain, *J Biol Chem* 260:15880-15885 (1985).
30. K. Heidenreich, P.R. Gilmore, and W.T. Garvey, Glucose transport in primary cultured neurons, *J Neuroscience Res* 22:397-407 (1989).
31. K. Keller, K. Lange, and J. Malkewitz, Glucose transporter in plasma membranes of cultured neural cells as characterized by cytochalasin B binding, *J Neurochem* 47:1394-1398 (1986).
32. A. Zorzano, W. Wilkinson, N. Kotliar, G. Thoidis, B.E. Wadzinski, A.E. Ruoho, and P.F. Pilch, Insulin-regulated glucose uptake in rat adipocytes is mediated by two transporter isoforms present in at least two vesicle populations, *J Biol Chem* 264:12358-12363 (1989).
33. L. Koranyi, R. Bourey, F. Fiedorek, and M.A. Permutt, Alterations of brain glucose transporter mRNA in diabetic and chronic glucose starved rats, *Diabetes* 38:65A:260 (abstract).
34. D. Cole, U. Devaskar, M. George, and S. Devaskar, Opposing effects of maternal diabetes on maternal and fetal brain glucose transporter proteins, *Clin Res* 38(3):804A (1990).
35. S.I. Harik, S.A. Gravina, and R.N. Kalaria, Glucose transporter of the blood-brain barrier and brain in chronic hyperglycemia, *J Neurochem* 51:1930-1934 (1988).
36. S. Matthaei, R. Horuk, and J.M. Olefsky, Blood-brain glucose transfer in diabetes mellitus - decreased number of glucose transporters at blood-brain barrier, *Diabetes* 35:1181-1184 (1984).
37. T.B. Choi, R.J. Boado, and Pardridge, Blood-brain barrier glucose transporter mRNA is increased in experimental diabetes mellitus, *Biochem Biophys Research Commun* 164:375-380 (1989).
38. W.M. Pardridge, D. Triguero, and C.R. Farrell, Downregulation of blood-brain barrier glucose transporter in experimental diabetes, *Diabetes* 39:1040-1044 (1990).

39. P.S. Walker, J.A. Donovan, B.G. Van Ness, R.E. Fellows, and J.E. Pessin, Glucose-dependent regulation of glucose transport activity, protein and mRNA in primary cultures of rat brain glial cells, *J Biol Chem* 263:15594-15601 (1988).
40. M. Hara, Y. Matsuda, N. Okumura, K. Hirai, and H. Nakagawa, Effect of glucose starvation on glucose transport in neuronal cells in primary culture from rat brain, *J Neurochem* 52:909-912 (1989).
41. H. Werner, M.K. Raizada, L.M. Mudd, H.L. Foyt, I.A. Simpson, C.T. Roberts, Jr., and D. LeRoith, Regulation of rat Brain/Hep G2 glucose transporter gene expression by insulin and insulin-like growth factor-I in primary cultures of neuronal and glial cells, *Endocrinology* 125:314-320 (1989).
42. L.M. Mudd, H. Werner, Z. Shen-Orr, C.T. Roberts, Jr., D. LeRoith, H.C. Haspel, and M.K. Raizada, Regulation of rat brain/Hep G2 glucose transporter gene expression by phorbol esters in primary cultures of neuronal and astrocytic glial cells, *Endocrinology* 126:545-549 (1990).

REGULATION AND PHYSIOLOGICAL FUNCTION OF INSULIN-LIKE GROWTH FACTORS IN THE CENTRAL NERVOUS SYSTEM

Thomas J. Lauterio

Departments of Internal Medicine and Physiology
Eastern Virginia Medical School
Norfolk, Virginia 23501
and Medical Research Service (151)
Department of Veterans Affairs Medical Center
Hampton, Virginia 23667

BACKGROUND

The function of insulin-like growth factors in the central nervous system (CNS) and the physiological regulators of brain and pituitary IGF synthesis has been studied from many perspectives. Until recently, these efforts have not met with great success. However as more laboratories obtain the means to study IGF mRNA and peptide synthesis the multifaceted approach of investigation has provided new insights on growth factor metabolism.

One line of research was initiated by Tannenbaum and associates (1983) in a study designed to examine the negative feedback mechanism of growth hormone (GH) secretion in the hypothalamus. In a series of experiments, the effect of insulin-like activities (ILAs), a semi-purified preparation rich in insulin-like growth factors (Zapf et al. 1978; Posner et al., 1978; Guyda et al., 1981) on the pulsatile release of GH was examined in conscious rats. Several preparations of the ILAs (Sephadex and Carboxymethyl cellulose chromatographed) were injected into the lateral ventricles of the brain and the effect of these preps along with insulin, bovine serum albumin and saline on peripheral GH concentrations were ascertained. The semi-purified growth factors did suppress GH secretion significantly, but were also found to produce anorexia and body weight loss. Food intake was depressed 66% from saline injected controls and body weight loss was 8% from the previous day's total. Tannenbaum noted these interesting effects of insulin-like growth factors and suggested that the mechanism behind this decrease be explored further. Using purified insulin-like growth factors I and II we pursued the food intake and body weight phenomenon further in our laboratory. Male rats were implanted with lateral ventricle cannulae and allowed to recover from surgery prior to experimentation. All rats increased body weight and food intake by three days following surgery. Animals were handled on a daily basis to reduce stress from injection, which was done utilizing a remote catheter. Food intake and body weight was also recorded on a daily basis before, during and after injection of hormones and/or vehicle. Injections were done over a 30 second interval to reduce pressure of the cerebral spinal fluid and to ensure proper delivery. Food intake and body weight over the 24 hour period following injection was compared between treatment groups as well as to each animal's own food and body weight 24 to 48 hours following

Molecular Biology and Physiology of Insulin and Insulin-Like Growth Factors
Edited by M.K. Raizada and D. LeRoith, Plenum Press, New York, 1991

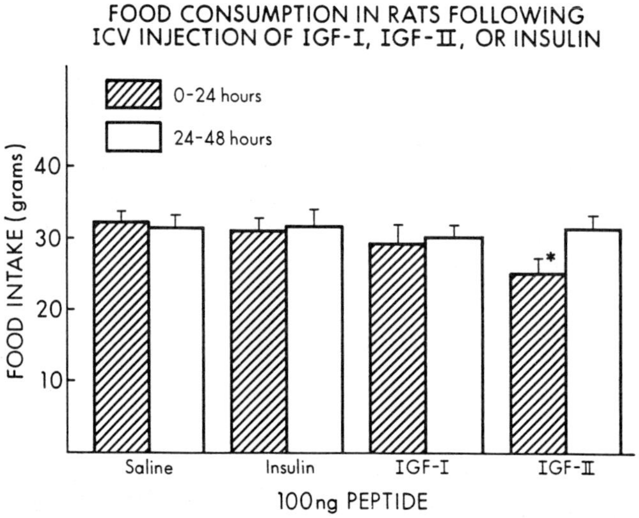

Fig. 1. Effect of intracerebroventricular injection of purified insulin-like growth factors I and II, insulin and vehicle on food intake in male Sprague-Dawley rats. Hormone was administered in 100 ng dose per 10 ul volume by remote catheter. Total food consumed per 24 hr period following injection is represented by hatched bars while food consumption for the 24-48 hr time period following injection is represented by the open bars. Data are expressed as mean food intake (grams) ± standard error of the mean (SEM) N=6. * $P < 0.05$ significance compared to saline control or IGF-II 24-48 hr value. Adapted from Lauterio et al. 1987a.

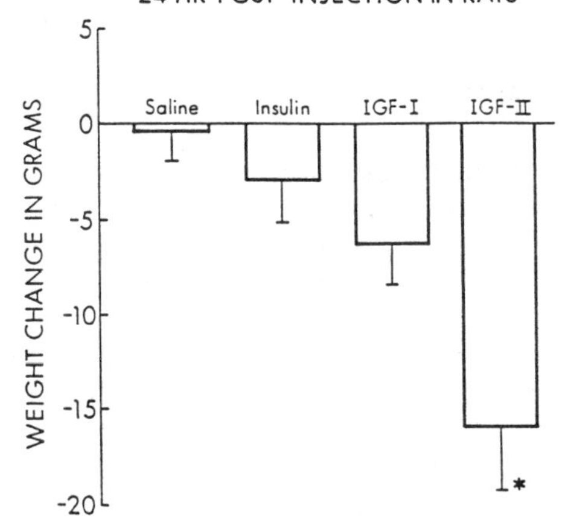

Fig. 2. Body weight changes in rats injected with insulin (100 ng), IGF-I (100 ng) or IGF-II (30, 100 or 300 ng) ICV. Data are presented as grams body weight ± SEM (N=6). * $P < 0.05$ significance or greater compared to saline, insulin or IGF-I injected animals. Adapted from Lauterio et al. 1987a.

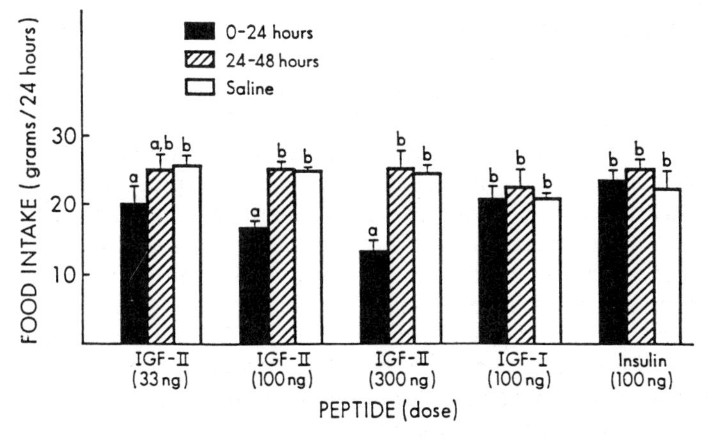

Figure 3 - Food consumption in rats following ICV injection of IGF-I, IGF-II, insulin or vehicle.

injection. Purified IGF-II but not IGF-I or insulin inhibited food intake and depressed body weight in these animals (Figure 1 and 2) (Lauterio et al., 1987a).

Further, this effect was dose dependent and behavioral effects were observed with as little as 33 ng of IGF-II per injection (Figure 3). In this latter study, all rats were injected on 1 day with hormone and on another with vehicle in a random fashion thus reducing individual variation.

Analysis of the hypothalamus for IGF-II peptide concentration, following extraction and removal of binding proteins showed that IGF-II was regionally and differentially distributed in areas concerned with food intake and body weight regulation. A parallel radioimmunological analysis of IGF-I content revealed this hormone was present in much lower concentrations than IGF-II and that distribution was homogeneous.

LOCATION OF INSULIN-LIKE GROWTH FACTORS AND THEIR RECEPTORS WITHIN THE CENTRAL NERVOUS SYSTEM

Both of IGF-II peptide and its mRNA are present in adult human and rat brain (Haselbacher et al., 1985; Lauterio et al., 1987a; Brown et al., 1986; Lund et al. 1986; Murphy et al., 1987; Rotwein et al. 1988; Lauterio et al. 1990) but the function of this hormone is still unknown. The physiological role of IGFs in the central nervous system may be understood better if the anatomical and cellular location of IGFs and their receptors were determined. Some clues may be provided by knowing for example: 1) the regions of the brain that synthesize IGF-II (ie. contain IGF-I or II mRNA); 2) the areas that contain IGF peptides and their respective interneural pathways; 3) the cell types that produce or contain IGFs; and 4) the location of IGF receptors. Recent data from a number of labs also strongly suggest that the types and amounts of binding proteins synthesized or secreted in a particular region be considered when evaluating physiological regulation of the IGFs.

Published reports on the location of IGF-II mRNA in the rat brain have generated controversy. Several investigators have shown IGF-II mRNA to be confined solely to the epithelial cells of the choroid plexus and leptomeninges (Stylianopoulou et al., 1988; Hynes et al. 1988) while others have demonstrated the presence of IGF-II mRNA in the hypothalamus, cortex and other brain regions (Rotwein et al, 1988; Lauterio et al. 1990). The major difference between these studies is the methodology used for RNA detection. In the former two studies, in situ hybridization was performed, while the latter two investigators utilized the solution hybridization method for mRNA detection. Although in situ hybridization provides information as to the cellular location of a particular RNA, solution hybridization is more sensitive and can detect specific mRNA present in very small regions or in low quantities. Rotwein and colleagues found a regional distribution of IGF-II mRNA within the brain and content varied by as much as 5 fold within the brain in some cases. The pons-medulla, cerebellum, olfactory and hippocampus all contained more IGF-II message than did the striatum, hypothalamus, midbrain or cortex. IGF-I mRNA was prevalent only in the olfactory bulb, midbrain and cerebellum. By altering culture conditions of embryonic rat brain cells these investigators were able to selectively establish populations of predominantly glial or neuronal cells which could be probed for IGF mRNA by solution hybridization. While IGF-I mRNA was present in glia and neurons, IGF-II was detected exclusively in glial cells. These findings would suggest that IGF-II has a supportive role in neural growth and differentiation.

Our laboratory (Lauterio et al., 1987a; Lauterio et al., 1990) found a regional distribution of both IGF-II mRNA and peptide within the hypothalamus. In the control animals, IGF-II mRNA content in the ventromedial hypothalamic region was 10 fold greater than that of the dorsal hypothalamic area. These studies emphasize the heterogeneous nature of the brain and the need to examine smaller physiologically relevant regions rather than treating the organ as a whole homogeneous entity. Changes in peptide or message content may be not be detected unless a more focussed approach is taken. For example, insulin treatment decreased IGF-II transcription and translation in the ventral region of the hypothalamus, but increased IGF-II levels in the dorsal and lateral hypothalamic area. Thus the net change observed if one were to process the whole hypothalamus for RNA and peptide determinations would be zero. Physiologically meaningful observations would have been obscured and an erroneous conclusion would be drawn about the ability of insulin to modulate hypothalamic IGF synthesis.

Although the mRNA for IGF-II has been localized to glial and epithelial cells within the central nervous system, there is evidence that the protein hormone may actually be taken up by neurons. Immunohistochemical staining of rat adult brain sections utilizing a monoclonal antisera and the peroxidase-antiperoxidase (PAP) method of visualization revealed the presence of IGF-II in neurons (Lauterio et al., 1987b, figure 4).

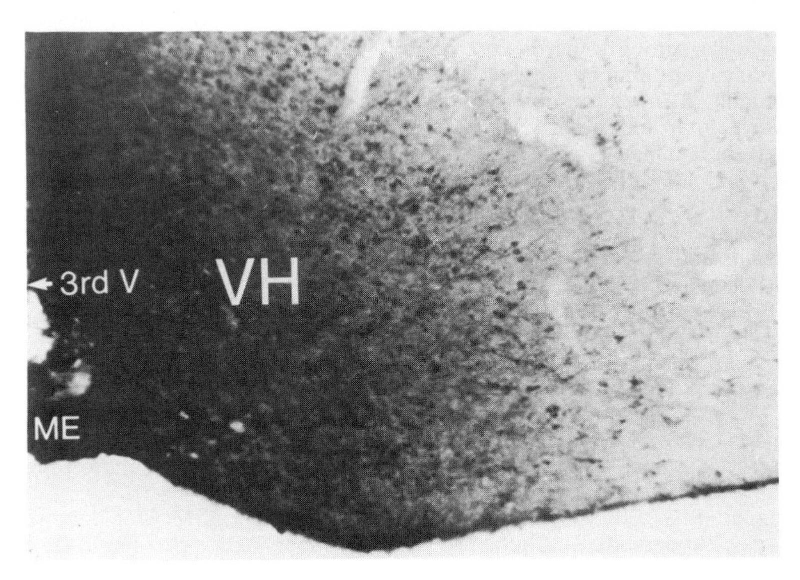

Figure 4 - Immunohistochemical staining of IGF-II neurons and fibers in the rat adult hypothalamus (ventral hypothalamic region) utilizing PAP for visualization. Note particularly intense staining in the median eminence (ME) region of the hypothalamus. Location of the third ventricle (3rd V) is designated by arrow.

The precedent for glial synthesized proteins acting on nervous tissue has been established previously. One such example of this arrangement is illustrated by apoprotein E which is synthesized by the astrocytes but which binds to neuronal low-density lipoprotein (LDL) receptors following secretion (Boyles et al., 1985; Pitas et al., 1987; Mahley, 1988). Apoprotein E or Apo-E (34,000 Mr protein) is a constituent of very-low density lipoproteins (VLDLs) as well as being a component of high-density lipoproteins (HDLs). VLDLs act to transport liver triglycerides to peripheral tissues while a subclass of Apo-E containing HDLs are thought to help redistribute cholesterol among cells. Either one of these functions may account for Apo-E synthesis and neuronal binding by brain tissue (Mahley, 1988). Triglycerides and cholesterol are both required by nervous tissue for regeneration and repair and LDL receptors are expressed at high levels in regenerating nerves (Boyles et al, 1987). IGF-II could feasibly be taken up by the neurons following glial secretion and neuronal receptor binding in an analogous manner.

IGF-II receptors have been localized by immunohistochemistry on rat hypothalamic neurons but to a lower degree than on astrocytes (Ocrant et al, 1988). While this suggests a more significant role for IGF-II in astrocyte metabolism, a neuronal function for the IGFs may still exist. Both type 1 and type 2 receptors are present in mammalian brain (Gammeltoft et al., 1985) and the brain type 1 receptor can bind to IGF-I or II with equal affinity. A subtype of this receptor, present in fetal brain neuronal cultures, has only slightly less affinity for IGF-II than IGF-I (Burgess et al., 1987). Since the brain IGF-II concentration is much greater than that of IGF-I, it is possible that IGF-II exerts biological effects through the type 1 receptor in addition to the type 2 receptor. The biochemical analysis of receptor binding sites in the hypothalamus, anterior pituitary and brain (minu hypothalamus) provides additional evidence for a role of for the IGFs in modulating brain function (Goodyer et al., 1984). IGF-II binding to membrane preparations in the anterior pituitary is 4 fold greater than that of IGF-I and 8 times that of insulin. Hypothalamic and brain minus hypothalamus specific binding of IGF-II also exceeded IGF-I and insulin specific binding by at least 3 fold in the same study. The above experiments serve not only as important initial characterizations of insulin-like growth factors in the CNS but also to stress the complexity of the system. This complexity is further amplified when one considers the physiological role of IGF binding proteins present in the CNS which can act to: a) reduce IGF peptide effects; b) enhance IGF peptide effects; c) exert effects independent of those exerted by IGF peptides.

While studies on IGF action in the CNS are few in number, there are a few important observations which may help sort out which IGFs exert effects in which tissues. Lattemann et al. (1989) investigated the effects of IGF-I and IGF-II on incorporation of ^{3}H-myoinositol into IP1 and membrane lipids in both the hippocampus and the olfactory bulb. IGF-I stimulated inositol incorporation into IP1 in the olfactory bulb, but not the hippocampus whereas IGF-II inhibited inositol incorporation into hippocampus but not olfactory bulb IP1 and membrane lipids. These interesting findings show the diversity of biological effects that two similar peptides can have and emphasizes the need to examine biologically relevant systems to obtain definitive answers on IGF action. Both regions of the brain contain type 1 and type 2 receptors, so these data suggest that the two peptides are working through their respective receptors. These experiments also provide insight as to why IGF-II but not IGF-I decreases food intake when injected intracerebroventricularly.

PHYSIOLOGICAL REGULATION OF IGF-II IN THE CENTRAL NERVOUS SYSTEM

While the in vitro effects of insulin like growth factors have been studied extensively, the in vivo effects of these factors are only now being investigated intensely. This is particularly true of IGF-II which has always been viewed as a fetal growth factor which has no particular importance in the adult animal. Comparison its effects to those of IGF-I as a post-natal growth factor has only dampened enthusiasm for IGF-II research. The fact that IGF-II is present in the brain and may exert some function there has generated some interest, but the brain is a difficult and complex system to study. Our lab has tried to focus on the factor or factors that regulate IGF-II peptide, gene expression and receptor kinetics in the central nervous system but in particular the hypothalamus. The rationale for this line of research stems from previously reported findings that IGF-II inhibits food intake and body weight. Questions we are actively pursuing include: 1) Is IGF-II involved in the long term or short term food intake and body weight control? 2) What are the cues that stimulate IGF-II synthesis and/or secretion? 3) Are these signals peripheral or central in origin? 4) Do certain nutritional components alter IGF-II metabolism and if so which? 5) What role might insulin have in altering brain IGF-I synthesis, secretion or function?

The last question may be the most straightforward to address. Insulin has been proposed to be a satiety factor for a number of years. Insulin resistance has been forwarded as a causative factor in obesity (Kolterman et al., 1979) and a correlation between insulin levels and obesity has been observed for man and other species (Bagdade et al., 1967; Woods et al., 1974). Long term insulin administration into the brain inhibits food intake in a number of species (Porte and Woods, 1981) while short term injections have been reported to have little effect on control of food intake. Thus the question of whether insulin is exerting a direct versus indirect effect on these parameters is not resolved. In addition, insulin is not synthesized in the brain but there are insulin receptors present throughout the CNS. The theory we are testing in our laboratory is that insulin acts to modulate food intake and body weight by altering IGF-II metabolism. The rationale for this hypothesis is based on the relationship between peripheral insulin and IGF status as well as direct effects of insulin on IGF-II receptor binding. A positive relationship appears to exist between insulin and IGF-II peptide levels as reported by our lab and others (Lauterio et al., 1990; Phillips et al., 1985). Further, insulin has been shown to directly up-regulate the number of IGF-II receptors on adipocytes by binding to its own receptor (Wardzala et al., 1984; Oka et al., 1984). These data taken together raises the possibility that insulin acts by altering IGF-II metabolism which in turn regulates food intake and body weight. Our lab has shown that peripheral insulin status can modulate hypothalamic IGF-II peptide and mRNA synthesis in a region specific manner (Lauterio et al., 1990). We are currently investigating the origin of the insulin signal necessary for this modulation by examining the effect of ICV injections and infusions on the synthesis and secretion of IGF-II in hypothalamic regions known to be important for control of body weight and food intake.

In addition to insulin levels, nutritional status may play a part in the regulation of brain IGF-II content and synthesis. This would be expected if there is indeed a role for IGF-II in food intake and body weight regulation. We have reported previously that brain IGF-II content is differentially altered in rats within 20 minutes of a meal compared to fasted counterparts (Lauterio et al. 1987b). Bohannon et al. (1988) have shown IGF-I receptors in the median eminence to be increased by food restriction. While this increase in receptors may reflect decreased IGF-I ligand synthesis and secretion due to food deprivation, it can also be expected to occur if hypothalamic levels of IGF-II were altered by some metabolic parameter. The relationship between IGF levels and those of insulin confounds the interpretation of the above experiments as insulin status is altered with fasting. Specific effects due to insulin need to be separated from those of general nutritional nature.

One approach taken to resolve this problem has been to determine the effect of insulin or 2-deoxy-D-glucose (2DG) on brain IGF-II metabolism utilizing the same experimental paradigm. Administration of 2DG _in vivo_ induces a glucoprivic state and its effect on brain glucose concentrations will essentially be the same as that of insulin administration. If both insulin and 2DG injections produce similar biological responses, it is generally interpreted to mean that the biological response observed is due to insulin's ability to alter glucose concentrations and not due to a direct effect of insulin on the cell or system affected. If on the other hand, insulin elicits a biological effect but 2DG fails to cause a similar response, a direct or at least non-glucose related effect is assumed. Our laboratory has examined the effect of insulin and 2DG on brain IGF-II peptide synthesis in just such a manner. Male Sprague Dawley rats were injected with 2DG, saline or insulin and groups of rats were killed and brains removed for IGF-II analysis 15, 30, 45, 60 or 120 minutes after injection.

In the above study, 2DG increased peripheral blood glucose levels while insulin lowered blood glucose from 15 minutes on to the end of the study. The results suggest that the effects of insulin on brain IGF-II content may be due

TABLE 1. The effect of acute peripheral insulin (INS, 2 U/ rat) or 2-Deoxy-D-Glucose (2DG, 500 mg/ kg body wt) injections on [a]brain IGF-II content. Results are present as significant increase (+), decrease (-) or no effect (0) compared to saline injected animals at the same time point. N=8 per group.

Brain	Time Following Injection (Min)									
	15		30		45		60		120	
Region	2DG	INS	2DG	INS	2DG	INS	2DG	INS	2DG	INS
SCN	0	0	0	0	0	0	-	-	-	0
SON	-	-	0	0	-	-	0	0	0	0
ARC n.	0	0	0	0	+	0	0	-	0	0
DMH	0	0	0	0	0	0	-	-	0	-
PVN	0	0	0	0	+	0	0	+	0	0
VMH	0	0	0	0	0	0	-	0	0	0
VC	0	0	0	0	0	0	0	0	0	+

a - Specific regions within the brain were dissected out, extracted and filtered to remove binding proteins prior to radioimmunoassay as previously described (Lauterio et al. 1987a). SCN- Suprachiasmatic nucleus; SON - Supraoptic nucleus; ARC n. - arcuate nucleus; DMH - dorsomedial hypothalamus; PVN - paraventricular nucleus; VMH - ventromedial hypothalamus; VC - vagal complex.

in part to alteration of glucose concentrations. This would be true in the SCN, SON, DMH and PVN. The vagal complex appears to be affected by insulin but not glucose, while the VMH appears to be more responsive to fluctuations in glucose concentrations than to insulin. An interesting area to note is the arcuate nucleus which was positively and negatively affected by 2DG and insulin respectively at different time points. This time dependent response to insulin and glucose concentrations may be important in the release of hypothalamic peptide and pituitary response to nutrient status. An alternative explanation is that peripheral blood glucose levels trigger one response while central glucose levels elicits a different effect on brain IGF-II content.

In another study assessing the effect of nutrients on IGF-II status in the brain, male Sprague-Dawley rats were placed either on a standard laboratory rat chow diet or a high fat diet consisting of 47% rat chow, 44% sweetened condensed milk, 8% corn oil and 1% corn starch. This high fat diet has been used previously to produce obesity in approximately half of the rats fed the diet (Triscari et al., 1985; Levin et al., 1987). The other half of the rats fed this diet do not become obese. These divergent groups are described as diet-induced obese (DIO) and Diet-resisters (DR). The rats that exhibit obesity are not hyperphagic compared to the chow fed rats and it is generally viewed that these rats are metabolically different from the "resisters" whose body weights remain equal to those in the chow fed group. Obese and resister groups are determined following three months of the dietary treatment which starts at 3 months post-birth. Results showed differential effects of obesity and diet on brain IGF-II concentrations in specific regions of the hypothalamus. For example, IGF-II content in rats fed a high fat diet was reduced by over 60% in the lateral (LH) and dorsomedial hypothalamic (DMH) regions compared to chow fed rats. This result was consistent regardless of whether animals were examined in the fasted state or following a meal. On the other hand, IGF-II levels were elevated in the paraventricular nucleus (PVN, by 25%) and neurointermediary pituitary (NIP, by 125%) in obese rats only. Resisters had IGF-II concentrations in those areas of the brain that were equivalent to those of the chow fed rats. The simplest explanation for these results is that dietary fat drives the changes observed in the LH and DMH whereas some physiological change in the obese rats (ie. insulin levels) could be responsible for altering IGF-II in the PVN and NIP.

In summary of the data thus far, it is apparent that dietary influences can alter IGF-II levels in the brain in a region specific manner. The mechanism(s) by which these changes occur is(are) unknown to date but a good starting point for investigation would be regulators of energy metabolism. Insulin, glucose, and dietary fat all seem to have a role in brain IGF-II metabolism and all of these factors play an important part in the control of food intake and body weight. Whether IGF-II is acting as a neurotransmitter or neuromodulator for regulation of food intake and body weight remains to be determined. Also the nature of the signal for IGF-II synthesis and secretion is not elucidated. A possibility exists that insulin that has gained access to the CNS stimulates IGF-II release, but it is also possible that peripheral insulin changes may trigger an IGF-II response through stimulation of the vagus nerve. How dietary fat and circulating glucose concentrations affect brain peptide metabolism is currently being investigated by this laboratory.

The physiological regulation of brain IGF-II has not been extensively studied in vivo to date and while the above studies represent a start in that direction, much work needs to be done. The task is complicated to undertake for a number of reasons. The first is that the brain is a very complex and heterogeneous organ and the physiology of its' numerous peptides and neurotransmitters is not well understood. Secondly, more comprehensive studies need to be completed on the mapping of brain IGF-II content, areas of synthesis and receptor binding. Determining the cell types that produce and bind IGF-II would help clarify the situation considerably as well as the interconnections between those regions. If IGF-II is synthesized in the ventromedial hypothalamus for example, is it then transported to another brain region or to the pituitary where it performs a function? In lieu of physiological data, anatomical localization studies may provide substantial clues as to the function of IGF-II in the brain.

It is interesting to note that high concentrations of IGF-II and receptors are located in the ventral hypothalamus (Lauterio et al., 1987a; Lesniak et al., 1988; Sara et al., 1982). This region is especially important in regulation of pituitary hormone secretion and it comes under the influence of most neurotransmitters and bioactive peptides. The opioid peptides shown to be involved in food intake and body weight regulations also are secreted in this area. Thus numerous hypothalamic-pituitary systems and connections may be controlling IGF-II synthesis and/or action. Conversely many systems must be considered potential sites for IGF-II influence. The task of narrowing down a particular system which affects IGF-II metabolism or vice versa will be a formidable one indeed.

ACKNOWLEDGEMENTS

This research was supported by the American Diabetes Association, National Institutes of Health grant #DK40982 and the Department of Veterans Affairs.

REFERENCES

Bagdade, J.D., Bierman, E.L. and Porte, D. Jr. (1967) The significance of basal insulin in the evaluation of the insulin response to glucose in diabetic and non-diabetic subjects. Journal of Clinical Investigation 46:1549-1557.

Bohannon, N.J., Corp, E.S., Wilcox, B.J., Figlewicz, D.P., Dorsa, D.M. and Baskin, D.G. (1988) Characterization of insulin-like growth factor I receptors in the median eminence of the brain and their modulation by food restriction. Endocrinology 122:1940-1947.

Boyles, J., Pitas, R.E., Wilson, E., Mahley, R.W. and Taylor, J.M. (1985) Apolipoprotein E associated with astrocyte glia of the central nervous system and with nonmyelinating glia of the peripheral nervous system. J. Clinical Investigation 76:1501-1513.

Boyles, J.K., Hui, D.Y., Weisgraber, K.H., Pitas, R.E. and Mahley, R.W. (1987) Expression of apolipoprotein B,E(LDL) receptors and uptake of apolipoprotein E-containing lipoproteins by the regenerating rat sciatic nerve. Neuroscience 13(1):294.

Brown, A.L., Graham, D.E., Nissley, S.P., Hill, D.J., Strain, A.J. and Rechler, M.M. (1985) Developmental regulation of insulin-like growth factor II mRNA in different rat tissues. Journal of Biological Chemistry 261(28):13144-13150.

Burgess, S.K., Jacobs, S., Cuatrecasas, P. and Sahyoun, N. (1987) Characterization of a neuronal subtype of insulin-like growth factor I receptor. Journal of Biological Chemistry 262(4):1618-1622.

Gammeltoft, S., Haselbacher, G.K., Humbel, R.E., Fehlmann, M. and Van Obberghen, E. (1985) Two types of receptor for insulin-like growth factors in mammalian brain. The EMBO Journal 4(13A):3407-3412.

Goodyer, C.G., De Stephano, L., Lai, W.H., Guyda, H.J. and Posner, B.I. (1984) Characterization of insulin-like growth factor receptors in rat anterior pituitary, hypothalamus, and brain. Endocrinology 114(4):1187-1195.

Guyda, H., Posner, B.I. and Rappaport, R. (1981) Insulin-like growth factors. pp 205-229 in: Pediatric Endocrinology (R. Collu, J. Ducharme and H. Guyda, eds.) Raven Press, New York.

Haselbacher, G.K., Schwab, M.E., Pasi, A. and Humbel, R.E. (1985) Insulin-like growth factor II (IGF-II) in human brain: regional distribution of IGF-II and higher molecular mass forms. Proceedings of the National Academy of Science USA 82:2153-2157.

Hynes, M.A., Brooks, P.J., Van Wyk, J.J. and Lund, P.K. (1988) Insulin-like growth factor II messenger ribonucleic acids are synthesized in the choroid plexus of the rat brain. Molecular Endocrinology 2:47-54.

Kolterman, O.G, Reaven, G.M. and Olefsky, J.M. (1979) Relationship between in vivo insulin resistance and decreased insulin receptors in obese man. Journal of Clinical Endocrinology and Metabolism 48:487-494.

Lattemann, D.F., King, M.G., Szot, P. and Baskin, D.G. (1989) Insulin-like growth factors as regulatory peptides in the adult rat brain. In: LeRoith, D. and Raizada, M.K. (eds.) Molecular and Cellular Biology of Insulin-like Growth Factors and Their Receptors. pp. 427-434, Plenum Press, New York.

Lauterio, T.J., Marson, L., Daughaday, W.H. and Baile, C.A. (1987) Evidence for the role of insulin-like growth factor II (IGF-II) in the control of food intake. Physiology and Behavior 40:755-758.

Lauterio, T.J., Marson, L., Della-Fera, M.A. and Baile, C.A. (1987b) Insulin-like growth factor in rat brain: distribution of IGF-II neurons and fibers and peptide concentration changes with fed state. Neuroscience 13(1):611 Abstract

Lauterio, T.J., Aravich, P.F. and Rotwein, P. (1990) Divergent effects of insulin on insulin-like growth factor-II gene expression in the rat hypothalamus. Endocrinology 126(1):392-398.

Lesniak, M.A., Hill, J.M., Kiess, W., Rojeski, M., Pert, C.B. and Roth, J. (1988) Receptors for insulin-like growth factors I and II: autoradiographic localization in rat brain and comparison to receptors for insulin. Endocrinology 123(4):2089-2099.

Levin, B.E., Triscari, J., Hogan, S. and Sullivan, .C. (1987) Resistance to diet-induced obesity: food intake, pancreatic sympathetic tone, and insulin. American Journal of Physiology 252:R471-R478.

Lund, P.K., Moats-Staats, B.M., Hynes, M.A., Simmons, J.G., Jansen, M., D'Ercole, A.J. and Van Wyk, J.J. (1986) Somatomedin-C/Insulin-like growth factor-I and insulin-like growth factor-II mRNAs in rat fetal and adult tissues. Journal of Biological Chemistry 261(31):14539-14544.

Mahley, R.W. (1988) Apoprotein E: cholesterol transport protein with expanding role in cell biology. Science 240:622-630.

Murphy, L.J., Bell G.I. and Friesen, H.G. (1987) Tissue distribution of insulin-like growth factor I and II messenger ribonucleic acid in the adult rat. Endocrinology 120(4):1279-1282.

Ocrant, I., Valentino, K.L., Eng, L.F., Hintz, R.L., Wilson, D.M. and Rosenfeld, R.G. (1988) Structural and immunohistochemical characterization of insulin-like growth factor I and II receptors in the murine central nervous system. Endocrinology 123(2):1023-1034.

Oka, Y., Mottola, C., Oppenheimer, C.L., Czech, M.P. (1984) Insulin activates the appearance on insulin-like growth factor II receptors on the adipocyte cell surface. Proceedings of the National Academy of Science USA 81:4028-4032.

Phillips, L.S., Fusco, A.C. and Unterman, T.G. (1985) Nutrition and somatomedin. XIV. Altered levels of somatomedins and somatomedin inhibitors in rats with streptozotocin-induced diabetes. Metabolism 34(8):765-770.

Pitas, R.E., Boyles, J.K., Lee, S.H., Foss, D. and Mahley, R.W. (1987) Astrocytes synthesize apolipoprotein E and metabolize apolipoprotein E-containing lipoproteins. Biochimica Biophysica Acta 917:148-161.

Porte, D. Jr. and Woods, S.C. (1981) Regulation of food intake and body weight by insulin. Diabetologia 20:274-281.

Posner, B.I., Guyda, H.J., Corvol, M.T., Rappaport, R., Harley, C. and Goldstein, S. (1978) Partial purification, characterization and assay of a slightly acidic insulin-like peptide (ILAs) from human plasma. Journal of Clinical Endocrinology and Metabolism 47:1240-1250.

Rotwein, P., Burgess, S.K., Milbrandt, J.D. and Krause, J.E. (1988) Differential expression of insulin-like growth factor genes in rat central nervous system. Proceedings of the National Academy of Science, USA 85:265-269.

Sara, V.R., Hall, K., Von Holtz, H., Humbel, R., Sjogren, B., Wetterberg, L. (1982) Evidence for the presence of specific receptors for insulin-like growth factors 1 (IGF-1) and 2 (IGF-2) and insulin throughout the adult human brain. Neuroscience Letters 34:39-44.
Stylianopoulou, F., Herbert, J., Soares, M.B. and Efstratiadis, A. (1988) Expression of insulin-like growth factor II gene in the choroid plexus and the leptomeninges of the adult rat central nervous system. Proceedings of the National Academy of Science USA 85:141-145.
Tannenbaum, G.S., Guyda, H.J. and Posner, B.I. (1983) Insulin-like growth factors: a role in growth hormone negative feedback and body weight regulation via brain. Science 220:77-79.
Triscari, J.C., Nauss-Karol, C., Levin, B.E. and Sullivan, A.C. (1985) Changes in lipid metabolism in diet-induced obesity. Metabolic and Clinical Experimentation 34:580-587.
Wardzala, L.J., Simpson, I.A., Rechler, M.M. and Cushman, S.W. (1984) Potential mechanism of the stimulatory action of insulin on insulin-like growth factor II binding to the isolated rat adipose cell. Apparent redistribution of receptors cycling between a large intracellular pool and the plasma membrane. Journal of Biological Chemistry 259:8378-8383.
Woods, S.C., Decke, E and Vasselli, J.R. (1974) Metabolic hormones and regulation of body weight. Psychological Reviews 81:26-43.
Zapf, J., Rinderknecht, E., Humbel, R.E., and Froesch, E.R. (1978) Nonsuppressible insulin-like activity (NSILA) from human serum: recent accomplishments and their physiologic implications. Metabolism 27:1803-1828.

IGF-I mRNA LOCALIZATION IN TRIGEMINAL AND SYMPATHETIC NERVE TARGET ZONES DURING RAT EMBRYONIC DEVELOPMENT

Carolyn Bondy and Edward Chin

Developmental Endocrinology Branch
National Institute of Child Health and Human Development
National Institutes of Health, Bethesda, MD

In recent years evidence has accumulated suggesting that insulin-like growth factors may have significant autocrine or paracrine roles in regulating the rate of growth or state of differentiation of various types of normal and tumorous tissue (1,2). A paracrine/autocrine role for insulin-like growth factor-I (IGF-I) in embryonic development has been suggested by the following observations. IGF-I is secreted by cultured fetal cells and explants and binds to specific receptors in fetal tissue (3,4). IGF-I immunoreactivity is low in fetal serum (5) but both IGF-I immunoreactivity and mRNA are detected in fetal tissues during the course of gestation (6-11). The presence of the type-I IGF receptor during embryogenesis has been demonstrated by binding studies (12,13) and by detection of receptor mRNA (11,14).

In situ hybridization histochemistry has proven to be a valuable tool for the investigation of potential paracrine/autocrine roles of growth factors. Recent improvements in hybridization and histochemical techniques make it possible to detect growth factor gene expression by a small number of cells or even a single cell in a context of preserved histological structures and relationships, thus providing information about the specific local settings in which growth factor action may be important. We have recently reported that IGF-I mRNA is present in a very distinctive pattern in the mid-gestation rat embryo (11). It was observed, in that study, that IGF-I mRNA is localized in the target zones of the trigeminal nerve, and in regions of sympathetic and spinal ganglion development (11). The present paper further analyzes what appears to be a significant relationship between the pattern of IGF-I gene expression and the development of the peripheral nervous system in rat embryogenesis.

IGF-I mRNA is prominently localized in three discrete regions of the developing face during embryonic days 12 through 15. These are the brow

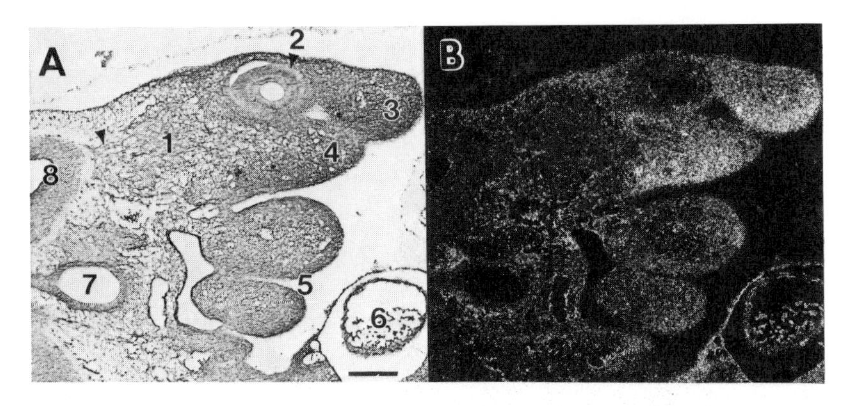

Figure 1. IGF-I mRNA localization in trigeminal fields on embryonic day 12 (E12), shown by in situ *hybridization. Panel A shows a bright field micrograph of a parasagittal section through an E12 rat embryo, and panel B shows the dark field view of the same section. The trigeminal ganglion (1) is very large in proportion to other structures at this early stage. The arrowhead shows its fibers reaching into the rhombencephalon (8). Neurite outgrowth into the trigeminal target zones is just beginning at this time, and IGF-I mRNA (shown by the white grains in the dark field picture) is concentrated in these zones. 1 - trigeminal ganglion, 2 - optic cup, 3 - maxillary process, 4 - mandibular process, 5 - branchial arches, 6 - heart, 7 - otic vesicle, 8 - rhombencephalon. Scale bar = 200 μ. Tissue preparation, probe synthesis and* in situ *hybridization procedures have been described in detail elsewhere (11).*

and the maxillary and the mandibular processes, as shown in Figs. 1 & 2. These regions are the target zones of the ophthalmic, maxillary and mandibular branches of the trigeminal nerve, which is innervating the face during this developmental period (15). Prior to embryonic day 12 (E12), no IGF-I mRNA is detected in the head (not shown). On E12, the trigeminal ganglion first appears as a distinct aggregate of spindle-shaped neurons in the region between the hind brain and the optic cup, and the first processes are seen forming in the direction of the target zones (15). At this time, IGF-I mRNA is very abundant throughout the target zones, but not in the ganglion itself, or in surrounding areas where its processes will not be directed (Fig. 1).

On subsequent days, the IGF-I mRNA concentration recedes from the ganglion towards the distal regions of the target areas. On E14, the densest concentration of mRNA is seen in the superficial dermal areas of the target zones (Fig. 2).

By E15, the trigeminal fibers have fully extended into their target regions, and the pattern of IGF-I gene expression is significantly altered. Clusters of IGF-I mRNA containing cells are found in close association with the body of the trigeminal ganglion itself, and are seen coalescing around the trigeminal fibers. A similar distribution of IGF-I mRNA is associated with developing spinal and sympathetic ganglia. Loosely aggregated, IGF-I mRNA containing cells surround the growing spinal and sympathetic ganglia and their nerve fibers, as shown in

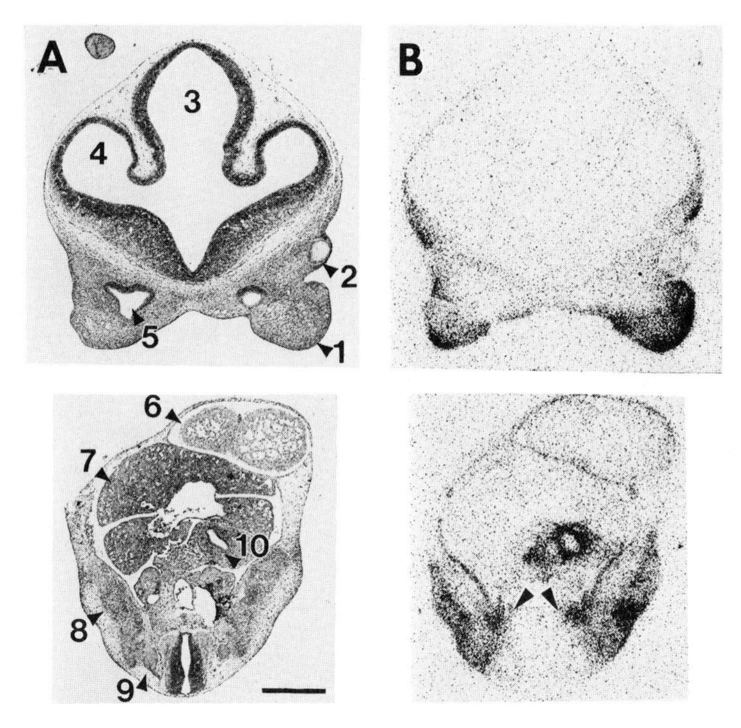

Figure 2. ***Pattern of IGF-I mRNA distribution in an E14 rat embryo in coronal sections through the head (top) and lower trunk (bottom).*** *On the left (A) is the hematoxylin and eosin stained tissue section and on the right (B) the corresponding film autoradiograph. The hybridization signal appears as black grains in the film autoradiograph. Hybridization is not seen in the brain, spinal cord or ganglia, but is present in the discrete supraorbital and maxillary target zones of the trigeminal nerve (B, top), and in areas surrounding the spinal nerves (B, bottom, arrowheads). The paraspinal region indicated by #9 in A and the arrowheads in B is shown in high power micrographs in Fig. 3. A strong hybridization signal is also apparent in the submucosal stomach wall (B, bottom). 1 - maxillary process, 2 - eye, 3 - 3rd ventricle of the diencephalon, 4 - lateral ventricle of the telencephalon, 5 - nasal passage, 6 - heart, 7 - liver, 8 - myotome, 9 - dorsal root ganglion just to the left of the spinal cord, 10 - stomach. Scale bar = 1 cm.*

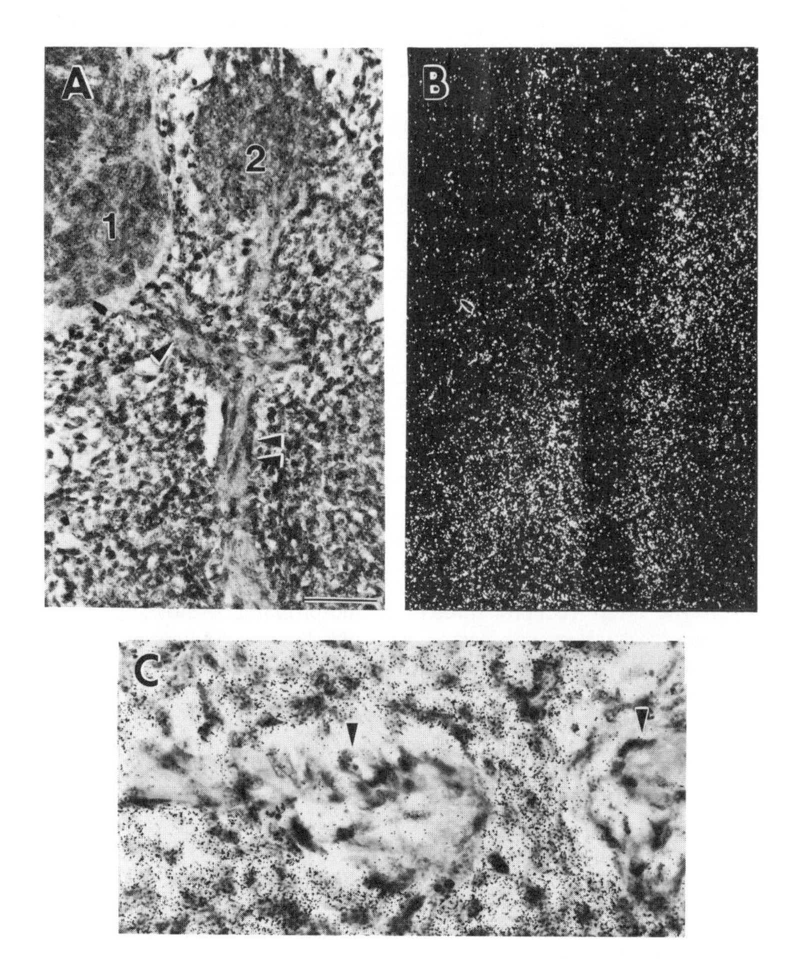

Figure 3. ***IGF-I mRNA localization in tissues surrounding developing dorsal root ganglion and spinal nerve roots (A&B) and sympathetic ganglia (C).*** *Panels A & B are bright and dark field micrographs, respectively, of the paraspinal area indicated by #9 and arrowheads in Fig. 2. 1 - ventral horn of the spinal cord, 2 - dorsal root ganglion. The single arrowhead indicates the ventral, or motor root and the double arrowheads indicate the spinal nerve. Hybridization of IGF-I mRNA, signified by white grains, is at background levels over the spinal cord, ganglion and nerves, but is intense in the undifferentiated tissue adjoining the nascent nerves and ganglion. Scale bar = 50 μ. Panel C is a high power micrograph of the sympathetic ganglia chain from the thoracic region of an E15 rat embryo. The ganglia (arrowheads) themselves are negative, but surrounded cells show abundant exposed silver grains, indicating high levels of IGF-I mRNA. Scale bar in C = 10 μ.*

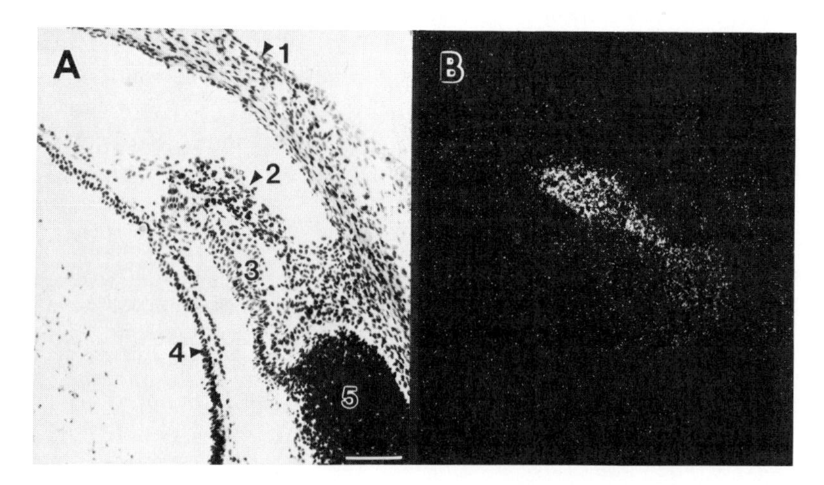

Figure 4. IGF-I mRNA is abundant in the developing ocular iris. *The panel on the left (A) shows a bright field micrograph of the forming iris and ciliary body in an E20 rat. On the right (B) is the dark field view of the same section, showing a very intense level of IGF-I mRNA hybridization in the iris alone. 1 - cornea, 2 - iris, 3 - ciliary body, 4 - lens, 5 - retina. Scale bar = 100 μ.*

Figs. 2 & 3. These loose cells, which appear to be condensing around the forming ganglia and nerves, could be fibroblasts involved in the formation of perineurial sheaths, or they could be primitive Schwann cells joining nerves during this time of development.

IGF-I mRNA is also localized in specific target zones of sympathetic innervation. We have previously described the focal concentration of IGF-I mRNA in the cardiac atria (11), a tissue notable for abundant sympathetic innervation. Another target tissue of dense sympathetic innervation is the ocular iris. As shown in Fig. 4, IGF-I mRNA is extremely abundant in this tissue. Traces of IGF-I mRNA hybridization are first seen in the developing iris on E16 (not shown); levels are most abundant perinatally and decline significantly several weeks after birth.

The transient presence of high levels of IGF-I mRNA concentrated in focal target zones of trigeminal and sympathetic innervation during the developmental time periods in which their innervation is accomplished, provides circumstantial evidence that IGF-I may be involved in that process. Insulin-like growth factors have been shown to stimulate neurite outgrowth in a variety of neuronal cell types in culture (reviewed in ref. 16) and therefore it might be suggested that IGF-I induces the process of neurite outgrowth into these particular regions *in vivo*. The observation that there is a topographic gradient of IGF-I gene expression in these innervation fields (the maxillary processes, for example) which actually recedes during the course of trigeminal fiber outgrowth, raises the possibility of a chemoattractant role for IGF-I. Alternatively, IGF-I may act to promote the survival of neurons and their processes, in a manner analogous to the mode of action of nerve-growth factor (NGF, 17). There are a number of interesting parallels between IGF-I and NGF mRNA localizations; for example, both are elevated in trigeminal target zones (17) and in the heart and iris (18).

Much work remains to be done to clarify the role of IGF-I in the development of the peripheral nervous system. The relationship of mRNA to peptide levels must be established, and it will be of interest to determine whether IGF-I is transported from peripheral target sites back to ganglion cell bodies, as appears to be the case for NGF (19). There is evidence for axoplasmic transport of IGF-I (20). The role of IGF-I in neural development should be illuminated by the *in situ* neutralization of IGF-I action by antibodies or antagonists, as has proven informative for NGF. The engineering of transgenic animals lacking IGF-I gene expression may also provide insight into the role of IGF-I in embryonic development.

REFERENCES

1. Underwood, L.E., D'Ercole, A.J., Clemmons, D.R. and Van Wyk, J.J. (1986) Paracrine functions of somatomedins. *Clinics in Endocrinology and Metabolism* 15:59-77.
2. Daughaday, W. (1990) Editorial: The possible autocrine/paracrine and endocrine roles of insulin-like growth factors of human tumors. *Endocrinology* 127:1-4.
3. Underwood, L.E. and D'Ercole, A.J. (1984) Tissue growth factors. *Clinics in Endocrinology and Metabolism* 13:69-89.
4. D'Ercole, A.J. and Underwood, L.E. (1986) in: *Human Growth* (Faulkner, F. & Tanner, J.M., eds) Vol. I, pp. 327-338, Plenum Publishing Co, New York.
5. Daughaday, W.H., Parker, K.A., Borowsky, S., Trivedi, B. and Kapadia, M. (1982) Measurement of somatomedin-related peptides in fetal, neonatal and maternal rat serum. *Endocrinology* 110:575-581.
6. D'Ercole, A.J., Hill, D.J., Strain, L.E. and Underwood, L.E. (1986) *Pediatric Res.* 20:253-258.
7. Lund, P.K., Moats-Staats, B.M., Hynes, M.A., Simmons, J..G, Jansen, M., D'Ercole, A.J .and Van Wyk, J.J. (1986) Somatomedin-C/insulin-like growth factor-I and insulin-like growth factor-II mRNAs in rat fetal and adult tissues. *J. Biol. Chem.* 261:14539-14544.
8. Han, V.K., D'Ercole, A.J., and Lund, P.K. (1987) Cellular localization of somatomedin (insulin-like growth factor) mRNA in the human fetus. *Science* 286:193-196.
9. Rotwein, P., Burgess, S.K., Milbrandt, J.D. and Krause, J.E. (1988) Differential expression of insulin-like growth factor genes in rat central nervous system. *Proc. Natl. Acad. Sci.* 85:265-268.

10. Beck, F., Samani, N.J., Penschow, J.D., Thorley, B., Tregear, G.W. and Coghlan, J.P. (1987) Histochemical localization of IGF-I and -II mRNA in the developing rat embryo. *Dev.* 101:175-184.
11. Bondy, C.A., Werner, H., Roberts, C.T., Jr., and LeRoith, D. (1990) Cellular pattern of IGF-I and Type I IGF receptor gene expression in organogenesis: comparison with IGF-II gene expression. *Molec. Endocrinol.* 4:1386-1398.
12. Girbau, M., Bassas, L., Alemany, J. and de Pablo, F. (1989) *In situ* autoradiography and ligand-dependent tyrosine kinase activity reveal insulin receptors and insulin-like growth factor receptors in prepancreatic chicken embryos. *Proc. Natl. Acad. Sci. USA* 86:5868-5872.
13. Mattson, B.A., Rosenblum, I.Y., Smith, R.M. and Heyner, S. (1988) Autoradiographic evidence for insulin and insulin-like growth factor binding to early mouse embryos. *Diabetes* 37:585-589.
14. Werner, H., Woloschek, M., Adamo, M., Shen-Orr, Z., Roberts, C.T., Jr., and LeRoith, D. (1989) Developmental regulation of the rat IGF-I receptor gene. *Proc. Natl. Acad. Sci.* 86:7451-7455.
15. Erzurumlu, R.S. and Killackey, H.P. (1983) Development of order in the rat trigeminal system. *J. Comp. Neurol.* 213:365-380.
16. Recio-Pinto, E. and Ishii, D.N. (1988) Insulin and related growth factors: Effects on the nervous system and mechanism for neurite growth. *Neurochem. Int.* 12:397-414.
17. Davies, A.M., Bandtlow, C., Heumann, R., Korsching, S., Roher, H. and Thoenen, H. (1987) Timing and site of nerve growth factor synthesis in developing skin in relation to innervation and expression of the receptor. *Nature* 326:353-358.
18. Shelton, D.L. and Reichardt, L.F. (1984) Expression of the beta-nerve growth factor gene correlates with the density of sympathetic innervation in effector organs. *Proc. Natl. Acad. Sci.* 81:7951-7955.
19. Heumann, R., Korsching, S., Scott, J. and Thoenen, H. (1984) Relationship of NGF and its mRNA in sympathetic ganglia and peripheral target tissues. *EMBO* 3:3183-3189.
20. Hansson, H.-A., Rozell, B. and Skottner, A. (1987) Rapid axoplasmic transport of insulin-like growth factor I in the sciatic nerve of adult rats. *Cell Tissue Res.* 247:241-247.

NEUROACTIVE PRODUCTS OF IGF-I AND IGF-2 GENE EXPRESSION IN THE CNS

Vicki R. Sara [A], Ann-Christin Sandberg-Nordqvist [A], Christine Carlsson-Skwirut [A], Tomas Bergman [B] and Christianne Ayer-LeLievre [C]

[A] Karolinska Institute's Department of Pathology and [B] Department of Medical Chemistry, Karolinska Hospital, Stockholm Sweden

[C] Institute d'Embryologie du CNRS Nogent sur Marne, France

INTRODUCTION

The insulin-like growth factors (IGFs) consist of IGF-1 and IGF-2, as well as variants arising from either alternative RNA processing or post-translational modification of the IGF precursor. In the extracellular fluid or circulating, the IGFs are associated with their carrier proteins which are believed to function as transporters, directing the IGFs to their target cells (35). The IGFs act as both endocrine hormones on distal target cells, as well as locally as paracrine or autocrine hormones. While the IGFs have long been recognized as growth and anabolic factors for a wide variety of tissues and cell types, such as cartilage, muscle, and fibroblasts, their role within the nervous system has only been widely recognized over the last several years. However, historically this can be traced back to the experiments of Stephan Zamenhof in the 1940's, who demonstrated that crude pituitary extracts of growth hormone stimulated the growth of tadpole and rat brains. This brain growth-promoting activity was later shown to be due to a growth hormone dependent growth factor, later identified as truncated IGF-1 (32). Both IGF-1 and IGF-2 are synthesized within the central nervous system where they are believed to fulfill different functions mediated via their receptors.

BIOSYNTHESIS OF IGF-1 IN THE CNS

IGF-1 Gene Expression - Characterization, Localization and Regulation

The IGF-1 gene is expressed within the CNS in a developmentally and regionally specific manner. This has been demonstrated in both rats (30) and man (31) where IGF-1 mRNA is far more abundant during fetal life than in the adult. In the adult, regional specificity has only been examined extensively in the rat, where the major expression was found in the olfactory bulb and spinal cord (30).

The primary transcript from the IGF-1 gene can be alternatively spliced to result in either IGF-1a or IGF-1b mRNA which encode prohormones differing in the length and structure of their carboxylterminal E domains (29). The IGF-1 gene transcripts have recently been characterized in the human brain. Using PCR

Molecular Biology and Physiology of Insulin and Insulin-Like Growth Factors
Edited by M.K. Raizada and D. LeRoith, Plenum Press, New York, 1991

amplification of cDNA followed by molecular cloning and direct sequencing, Sandberg et al (in preparation) have obtained the nucleotide sequence of both IGF-1a and IGF-1b cDNA in human fetal brain. The nucleotide sequences of the brain IGF-1a and IGF-1b cDNAs were identical to that obtained in other tissues, such as human liver, with the exception of a base change in position 270 of IGF-1a cDNA. Although this base change may have arisen from the techniques employed, it was repeatedly found using either cloning or direct sequencing. Thus, the possibility of a mutation in this position, which does not influence the amino acid encoded, may be considered. Both IGF-1a and IGF-1b mRNA have been identified in the rat brain by solution hybridization/RNAse protection assay (22). However, unlike in man where the presence of exon 4 or 5 is mutually exclusive, both are present in the IGF-1b mRNA of the rat and also the mouse, which leads to a change in the translational reading frame (39). Thus, the carboxylterminal peptides of the IGF-1b in murine and human vary considerably. In addition, in both the mouse and the rat, transcription appears to be initiated at different sites in the IGF-1 gene. The expression of these 5' untranslated regions is tissue specific, with the class C 5' untranslated region predominating in rat brain (23).

The IGF-1 gene is expressed by both isolated neuronal and glial cells in culture (30). While IGF-1 mRNA has been identified in preparations from whole brain and even various CNS regions, it has only recently been possible, using *in situ* hybridization histochemical techniques, to localize the sites of IGF-1 synthesis within the CNS. In certain areas of the embryonic rat brain, such as the cortex, thalamus, striatum and tectum, the expression of the IGF-1 gene is low and widespread (3). In other areas it appears to be expressed in specific restricted cell groups in a tightly regulated developmental manner, suggesting a specific function during development. In the adult rat brain, IGF-1 mRNA is found in the olfactory bulb, hippocampus and cerebellum (43). As in the embryonic rat (3), intense IGF-1 hybridization in the olfactory bulb is restricted to the glomerular and mitral cells. In the hippocampus, hybridization was to pyramidal cells of Ammon's horn in CA1 and CA2 layers and dentate gyrus, whereas in the cerebellum, it was located to the granular cell layer. These sites of IGF-1 synthesis are adjacent to, or overlap, IGF-1 receptors (42).

The regulation of IGF-1 gene expression within the CNS is poorly understood. In contrast to many other tissues and cells, particularly in the adult where GH is a major stimulator o IGF-1 gene transcription (23), GH does not appear to directly stimulate neuronal or glial cell IGF-1 production. Similarly to the peripheral nervous system (13), IGF-1 synthesis may respond to local tissue injury, however, the signal eliciting this response remains elusive at this stage. Glucocorticoids which are well established inhibitors of brain cell proliferation, have been demonstrated to reduce IGF-1 mRNA in primary cultures of both neuronal and glial cells (1).

Protein Products

The protein products of expression of the IGF-1 gene in the human brain have been isolated and their amino acid sequences determined. Post-translational processing of the IGF-1 prohormone results in two peptides which are proposed to fulfill distinct functions within the CNS (Fig. 1). A truncated form of IGF-1, which lacks the aminoterminal tri-peptide GPE (gly-pro-glu), has been characterized in both fetal and adult human brain (7,33). The truncated IGF-1 appears to be the major gene product in the CNS since no evidence for the presence of intact IGF-1 could be obtained. The truncated IGF-1 similarly appears as the major peptide in bovine colostrum (11), human platelets (19), porcine uterus (26) and has been proposed to represent the locally acting autocrine or paracrine form of IGF-1 (35). The nucleotide sequence of human fetal brain cDNA confirms that the aminoterminal truncation represents the only sequence modification and suggests that this cleavage occurs as a post-translational modification of the IGF-1 prohormone. The second product of proteolytic cleavage of the IGF-1 prohormone is the tripeptide, GPE (32). Although both these products have now been identified within the human brain, the

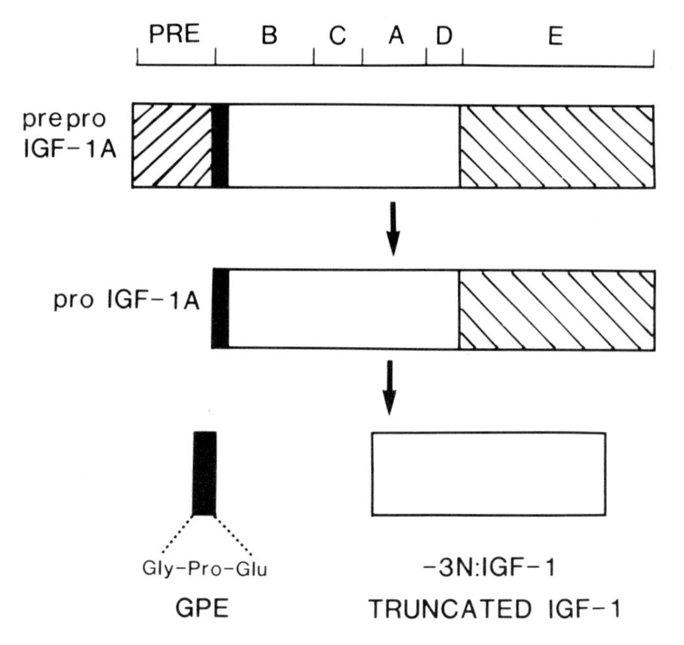

Fig. 1 GPE and truncated IGF-1 are two identified neuroactive products of IGF-1 gene expression in the CNS. These peptides result from post-translational processing of the IGF-1 prohormone, which in the CNS, is proposed to be pro-IGF-1a.

biosynthetic pathway and stages in this proteolytic process have yet to be completely defined. As illustrated in Figure 1, the predominance of the IGF-1a mRNA in the rat brain has led to the suggestion that proteolytic processing of the IGF-1a prohormone results in the production of truncated IGF-1 and GPE (32, 35).

BIOLOGICAL ACTIVITY

The peptide products from expression of the IGF-1 gene in the brain, namely truncated IGF-1 and GPE, appear to induce biological responses via two separate mechanisms. The action of truncated IGF-1 is mediated via the IGF-1 receptor. GPE does not cross-react in the IGF-1 receptor, but rather in the NMDA receptor, and possibly an additional, as yet undefined, mechanism (34).

The IGF-1 receptor appears to be present on both neurones and glial cells *in vitro*, with a structural subtype displaying altered glycosylation of the hormone-binding a-subunit being present on neurones (6). The expression of the receptor is enhanced during rapid growth phases (36,43). In the adult, the receptor is widely distributed. In the human brain for example, the highest densities of IGF-1 receptor are found in the hippocampus, amygdala and parahippocampal gyrus, followed by cerebellum, cerebral cortex and caudate nucleus (2). The developmental and regional expression of the IGF-1 receptor suggests a role in growth regulation during early development, as well as metabolic regulation in the adult as the distribution in the adult brain occurs in areas of high metabolic activity. Additionally, the presence of both IGF-1 mRNA and immunoreactivity is found to coincide with the distribution and occurrence of the receptors, supporting a paracrine or autocrine role for IGF-1 within the CNS (42).

Over the last decade much evidence has accumulated to demonstrate that IGF-1 and also IGF-2 have a potent growth-promoting action of neuronal and glial cell precursors *in vitro*. Additionally, a role in differentiation has been suggested. For example, IGF-1 has been reported to induce the differentiation of oligodendrocytes from their bipotential precursors (25). The growth-promoting actions of both IGF-1 and IGF-2 appear to be mediated via interaction in the IGF-1 receptor where both peptides cross-react almost equipotently. The biological actions of the IGFs in the CNS have been recently reviewed and will not be detailed here (32). Truncated IGF-1 displays enhanced neurotrophic activity both *in vitro* and *in vivo*, when compared to intact IGF-1 and IGF-2. Enhanced biological activity *in vitro* can be mainly attributed to failure to bind to the IGF-BPs which regulate IGF bioavailability to the target cells (8). Whereas the addition of BP1 to the incubation medium, blocks the action of intact IGF-1, it fails to bind truncated IGF-1 and has no influence on its stimulation of neuronal and glial cell proliferation (8). Failure to be bound by the IGF BPs results in rapid degradation and shorter half-life of truncated IGF-1 in the circulation (10). Thus, systemic administration of truncated IGF-1, as opposed to intact IGF-1, fails to induce a significant growth response in neonatal rats. However, the reverse is found following local application where truncated and not intact IGF-1 displays potent growth-promoting activity. For example, Giacobini et al (12), investigated the effects of both intact and truncated IGF-1 on intraocular grafts of embryonic brain tissue. Truncated IGF-1 displayed a potent neurotrophic action on cortex and spinal cord grafts, whereas intact IGF-1 had no significant effect, presumably due to its binding to BPs present in synovial fluid which prevented bioavailability to the target cells.

GPE is an additional protein product from expression of the IGF-1 gene within the CNS. GPE fails to cross-react in any IGF receptor and does not display growth-promoting activity either *in vitro* or *in vitro* (34). Instead, GPE cross-reacts in the NMDA (N-methyl-D-aspartate) receptor which is a subtype of receptors for the major excitatory amino acid neurotransmitter glutamate (34). GPE interaction appears to be specific for the NMDA receptor subtype as the tripeptide fails to cross-react in either the kainate or quisqualate receptors. The carboxylterminal glutamate residue of GPE is necessary for NMDA receptor binding while the aminoterminal glycine residue potentiates this binding, suggesting the model shown in Fig. 2. GPE is proposed to cross-react in both the glutamate recognition site as well as the glycine allosteric site of the NMDA receptor. GPE potentiates the release of dopamine via interaction in the NMDA receptor. However, GPE has an additional action which is not mediated via the NMDA receptor, namely the facilitation of acetylcholine release. GPE potentiates the potassium evoked release of acetylcholine from rat striatal slices at concentrations far less than those interacting in the NMDA receptor and this action cannot be inhibited by the use of specific NMDA receptor blockers (34). The mechanism for GPE's potent facilitation of ACh release has not yet been clarified.

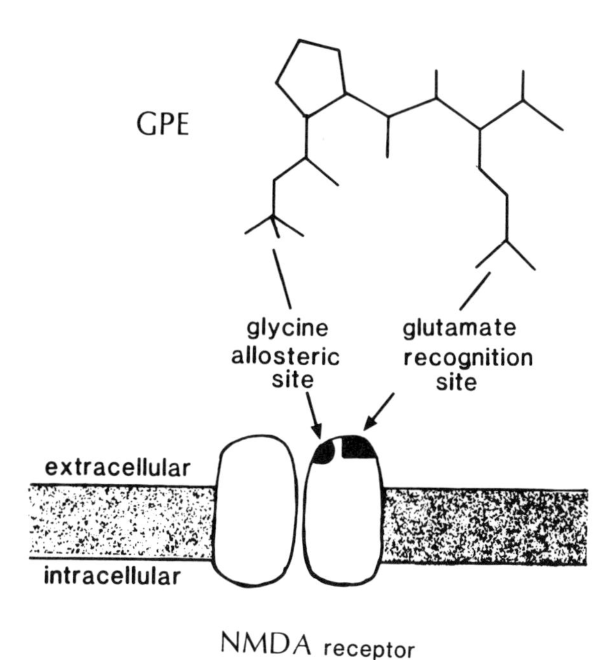

Figure 2. Model of GPEs interaction in the NMDA receptor. It is proposed that GPE cross-reacts in the glycine allosteric site as well as the glutamate recognition site.

In conclusion, there are two identified neuroactive products of IGF-1 gene expression in the CNS, namely truncated IGF-1 and the tripeptide, GPE. Based upon evidence available today, these peptides appear to fulfill quite distinct functions within the CNS. Truncated IGF-1 is proposed to function as an autocrine or paracrine anabolic factor, involved in regulation of proliferation and differentiation and possibly also metabolic regulation in the adult, whereas GPE is believed to act as a neuromodulator regulating neurotransmission. GPE is the first example of the product of a growth factor gene having a specific role in neurotransmission.

BIOSYNTHESIS OF IGF-2 IN THE CNS

IGF-1 Gene Expression - Characterization, Localization and Regulation

The expression of the IGF-2 gene in the nervous system has been well characterized, especially in the rat where the gene continues to be expressed in adult brain (5,24,30). Multiple IGF-2 transcripts, which contain identical coding regions but differ in their untranslated regions, are produced by the initiation of transcription at several different promoter sites in the IGF-2 gene (39). The 4.Okb IGF-2 transcript is greatest in the rat brain, suggesting initiation of transcription at the third promoter site. IGF-2 mRNA is most abundant in the brain during early development, but in contrast to most other tissues in the rat, is also found to be widely distributed in various brain regions in the adult. The explanation for the widespread occurrence of IGF-2 mRNA in brain extracts has become clear with its localization by *in situ* hybridization. The IGF-2 gene is not expressed in neurones or glia but rather in the meninges, choroid plexus, as well as mesenchymal cells surrounding the blood vessels in the adult brain (16,38). Contamination of brain regions by these cells is thus, unavoidable. Similarly in the fetus, apart from the choroid plexus and leptomeninges, IGF-2 mRNA has been detected to be present in a developmentally dependent way in hypothalamus, the floor of the third ventricle, pineal primordium and the pars intermedia of the pituitary (3,4,37). Thus, IGF-2 is synthesized at highly vascularized sites within the CNS, suggesting a role in the production of extracellular fluids and supply of substrates to neural tissue. However, an IGF BP is also synthesized in the choroid plexus (40) and the possibility must also be considered that IGF-2 associates with its BP and is transported to interact in distal IGF-2 receptors throughout the brain.

In the human brain, IGF-2 mRNA is most abundant in the fetus and barely detectable in adult tissue with little membrane contamination (31). A single 6.Okb transcript is found in the fetal brain (31) and adult hypothalamus (17), indicating that transcription in the human brain is initiated at the third promoter site in the IGF-2 gene. A similar transcript has been identified in brain tumors where there is an over-expression of the IGF-2 gene (31).

Protein Products

The protein products of expression of the IGF-2 gene have been identified in the human brain as IGF-2 identical to that first isolated by Rinderknecht and Humbel from serum (7,33), as well as a higher molecular weight form (15). The latter has yet to be sequenced but presumably represents a partially processed form of proIGF-2 which, similar to that isolated from serum, consists of IGF-2 with a carboxylterminus extension peptide. In contract to IGF-1, IGF-2 is found in the cerebrospinal fluid (14). The higher molecular weight form of IGF-2 predominates in human CSF (14) where both forms of IGF-2 are associated with an IGF-2 specific BP (27). Thus, the IGF-2-BP complex may circulate via the CSF to reach the widely distributed IGF-2 receptors.

BIOLOGICAL ACTIVITY

The biological activity of IGF-2 may be mediated via two mechanisms, namely the IGF-1 receptor and the IGF-2/Man-6-P receptor. In purified preparations of human fetal brain, IGF-2 cross-reacts almost equipotently with IGF-1 in the IGF-1 receptor, whereas only IGF-2 cross-reacts in the IGF-2/Man-6-P receptor (28). Based upon studies using blocking receptor antibodies in non-neural cells, it is most likely that IGF-2 induces neuronal and glial cell precursor proliferation by interaction in the IGF-1 receptor. A biological role mediated via the IGF-2/Man-6-P receptor in the brain has yet to be demonstrated. The IGF-2/Man 6-P receptor is widely distributed throughout the brain, and in contract to the IGF-1 receptor, is found in choroid plexus and cerebral vasculature (21,41). Studies in non-neural cells, have implicated the IGF-2/Man-6-P receptor in intracellular protein trafficking and in protein catabolism (20).

Recently, a trophic role of IGF-2 in synaptogenesis has been suggested. Ishii (18) has reported a marked correlation between the expression of the IGF-2 gene in muscle and the rate of neuromuscular synapse formation during synaptogenesis, as well as during muscle reinnervation. Nerve sprouting has been observed following exposure of adult rat gluteus muscle to IGF-2 *in vivo* (9). These studies suggest that IGF-2 may act as a trophic factor from the target cells to induce their innveration. However, further studies to determine the mechanism of this action and the specificity of IGF-2 involvement remain to be performed.

CONCLUSION

The IGFs are synthesized within the CNS to fulfill distinct functions. It has been proposed that the IGF-1 gene is expressed in neurones and glial cells where the protein products, namely truncated IGF-1 and GPE, have an autocrine/paracrine action to regulate growth and modulate neurotransmission, respectively. In contract, IGF-2 is synthesized in choroidal epithelial cells and vascular endothelial cells and in addition to a possible local action on substrate transport, may circulate as the IGF-2-BP complex via the CSF to distal targets within the CNS.

ACKNOWLEDGEMENTS

Supported by the Swedish Medical Research Council, the Cancer Fund in Stockholm, Nordic Insulin Foundation, and the Osterman Fund.

REFERENCES

1. Adamo, M., Werner, H., Farnsworth, W., Roberts, C.T., Jr., Raizada, M. and LeRoith, D. Dexamethasone reduces steady state insulin-like growth factor I messenger ribonucleic acid levels in rat neuronal and glial cells in primary culture. *Endocrinology*, 123:2565, 1988.

2. Adem, S., Jossan, S.S., d'Argy, R., Gillberg, P.G., Nordberg, A., Winblad, B. and Sara, V.R. Insulin-like growth factor I (IGF-1) receptors in the human brain: quantitative autoradiographic localization. *Brain Res*. 503:299, 1989.

3. Ayer-Lievre, A., Ståhlbom, P-A. and Sara, V.R. Expression of IGF-1 and II mRNA in the brain and cranio-facial region of the rat fetus. *Development* (In Press).

4. Beck, F., Samani, N.J., Byrne, S., Morgan, K., Gebhard, R. and Brammar, W.J. Histochemical localization of IGF-1 and IGF-11 mRNA in the rat between birth and adulthood. *Development* 104:29, 1988.

5. Brown, A.L., Graham, D.E., Nissley, S.P., Hill, D.J., Strain, A.J. and Rechler, M.M. Developmental regulation of insulin-like growth factor II mRNA in different rat tissues. *J. Bio. Chem*. 261:13144, 1986.

6. Burgess, S.K., Jacobs, S., Cuatrecasas, P. and Sahyoun, N. CHaracterization of a neuronal subtype of insulin-like growth factor I receptor. *J. Bio. Chem.* 262:1618, 1987.

7. Carlsson-Skwirut, C., Jörnvall, H., Holmgren, A., Andersson, C., Bergman, T., Lundquist, G., Sjögren, B. and Sara, V.R. Isolation and characterization of variant IGF-1 as well as IGF-2 from adult human brain. *FEBS Lett*. 201:3656, 1986.

8. Carlsson-Skwirut, C., Lake, M., Hartmanis, M., Hall, K. and Sara, V.R. A comparison of the biological activity of the recombinant intact and truncated insulin-like growth factor 1 (IGF-1). *Biochem. Biophys. Acta*. 1011:192, 1989.

9. Caroni, P. and Grandes, P. Nerve sprouting in innervated adult skeletal muscle induced by exposure to elevated levels of insulin-like growth factors. *J. Cell. Biol.* 110:1307, 1990.

10. Drakenberg, K., Östenson, C-G, and Sara, V.R. Circulating forms and biological activity of intact and truncated insulin-like growth factor 1 (IGF-1) in adult and neonatal rat. *Acta. Endocrinol.* 123:43, 1990.

11. Francis, L.G., Read, L.C., Ballard, F.J., Bagley, C.J., Upton, F.M. Gravestock, P.M. and Wallace, J.C. Purification and partial sequence analysis of insulin-like growth factor-1 from bovine colostrum. *Biochem. J.* 233:207, 1986.

12. Giacobini, MB, M.J., Olson, L., Hoffer, B.J. and Sara, V.R. Truncated IGF-1 exerts tropich effects on fetal brain tissue grafts. *Exper. Neurol.* 108:33, 1990.

13. Hansson, H.A., Dahlin, L.B., Danielsen, N., Fryklund, L., Nachemson, A.K., Polleryd, P., Rozell, B., Skottner, A., Stemme, S. and Lundborg, G. Evidence indicating trophic importance of IGF-1 in regenerating peripheral nerves. *Acta. Physiol. Scand.* 126:609, 1986.

14. Haselbacher, G. and Humbel, R. Evidence for two species of insulin-like growth factor II (IGF II and "big" IGF II) in human spinal fluid. *Endocrinol.* 110:1822, 1982.

15. Haselbacher, G.K., Schwab, M.E., Pasi, A. and Humbel, R. Insulin-like growth factor II (IGF II) in human brain: Regional distribution of IGF II and of higher molecular mass forms. *Proc. Nat'l. Acad. Sci.* 82:2153, 1985.

16. Hynes, M.A., Brooks, P.J., Van Wyk, J.J. and Lund, P.K. Insulin-like growth factor II messenger ribonucleic acids are synthesized in the choroid plexus of the rat brain. *Mol. Endocrinol.* 2:47, 1988.

17. Irminger, J-C., Rosen, K.M., Humbel, R.E. and Villa-Komaroff, L. Tissue-specific expression of insulin-like growth factor II mRNAs with distinct 5' untranslated regions. *Proc. Nat'l Acad. Sci.* 84:6330, 1987.

18. Ishii, D.N. Relationship of insulin-like growth factor II gene expression in muscle to synaptogenesis. *Proc. Nat'l. Acad. Sci.* 86:2998, 1989.

19. Karey, K.P., Marquardt, H. and Sirbasku, D.A. Human platelet-derived mitogens. 1. Identification of insulin-like growth factors I and II by purification and N amino acid sequence analysis. *Blood* 74:1084, 1989.

20. Kovacina, K.S., Steele-Perkins, G. and Roth, R.A. A role for the insulin-like growth factor II/Mannose-6-phosphate receptor in the insulin-induced inhibition of protein catabolism. *Mol. Endocrinol.* 3:901, 1989.

21. Lesniak, M.A., Hill, J.M., Kiess, W., Rojeski, M., Pert, C.B. and Roth, J. Receptors for insulin-like growth factors I and II: Autoradiographic localization in rat brain and comparison to receptors for insulin. *Endocrinol.* 123:2089, 1988.

22. Lowe, W.L., Jr., Lasky, S.R., LeRoith, D. and Roberts, C.T., Jr. Distribution and regulation of rat insulin-like growth factor I messenger ribonucleic acids encoding alternative carboxyterminal E-peptides: Evidence for differential processing and regulation in liver. *Mol. Endocrinol.* 2:528, 1988.

23. Lowe, W.L., Roberts, C.T., Jr., Lasky, S.R. and LeRoith, D. Differential expression of alternative 5' untranslated regions in mRNAs encoding rat insulin-like growth factor I. *Proc. Nat'l. Acad. Sci.* 84:8946, 1987.

24. Lund, P.K., Moats-Staats, B.M., Hynes, M.A., Simmons, J.G., Jansen, M., D'Ercole, A.J. and Van Wyk, J.J. Somatodemin-C/insulin-like growth factor I and insulin-like growth factor II mRNAs in rat fetal and adult tissues. *J. Biol. Chem.* 261:14539, 1986.

25. McMorris, F.A., Smith, T.M., DeSalvo, S. and Furlanetto, R.W. Insulin-like growth factor I/somatomedin C: A potent inducer of oligodendrocyte development. *Proc. Nat'l. Acad. Sci.* 83:822, 1986.

26. Ogasawara, M., Karey, K.P., Marquardt, H. and Sirbasku, D.A. Identification and purification of truncated insulin-like growth factor I from porcine uterus. Evidence for high biological potency. *Biochem.* 28:2710, 1989.

27. Roghani, M., Hossenlopp, P., Lepage, P., Balland, A. and Binoux, M. Isolation from human cerebrospinal fluid of a new insulin-like growth factor binding protein with a selective affinity for IGF-II. *FEBS Lett.* 2:253, 1989.

28. Roth, R.A., Steele-Perkins, G., Hari, J., Strover, C., Pierce, S., Turner, J., Edman, J.C. and Rutter, W.J. Insulin and insulin-like growth factor receptors and responses. Cold Spring Harbor Symposia on Quantitative Biology. Vol. LIII, Cold Harbor Laboratory:537, 1988.

29. Rotwein, P. Two insulin-like growth factor I messenger RNAs are expressesd in human liver. *Proc. Nat'l. Acad. Sci.* 83:77, 1986.

30. Rotwein, P., Burgess, S.K., Milbrandt, J.D. and Krause, J.E. Differential expression of insulin-like growth factor genes in rat central nervous system. *Proc. Nat'l. Acad. Sci.* 85:265, 1988.

31. Sandberg, A-C., Engberg, C., Lake, M., von Holst, H. and Sara, V.R. The expression of insulin-like growth factor I and insulin-like growth factgor II genes in the human fetal and adult brain and in glioma. *Neurosci. Lett.* 93:114, 1988.

32. Sara, V.R. and Carlsson-Skwirut, C. Insulin-like growth factors in the central nervous system: Biosynthesis and biological role. In: Growth Factors: From Genes to Clinical Application. V.R. Sara, K. Hall and H. Löw (Eds.). Raven Press, N.Y. 179, 1990.

33. Sara, V.R., Carlsson-Skwirut, C., Andersson, C., Hall, E., Sjögren, B., Holmgren, A. and Jörnvall, H. Characterization of somatomedins from human fetal brain: Identification of a variant form of insulin-like growth factor I. *Proc. Nat'l. Acad. Sci.* 83:4904, 1986.

34. Sara, V.R., Carlsson-Skwirut, C., Bergman, T., Jörnvall, H., Roberts, P.J., Crawford, M., Nilsson Håkanson, L., Civalero, I. and Nordberg, A. Identification of gly-pro-glu (GPE), the aminoterminal tripeptide of insulin-like growth factor I which is truncated in brain, as a novel neuroactive peptide. *Biochem. Biophys. Res. Comm.* 165:766, 1989.

35. Sara, V.R. and Hall, K. Insulin-like growth factors and their binding proteins. *Physiol. Rev.* 70:591, 1990.

36. Sara, V.R., Hall, K. Misaki, M., Fryklund, L., Christensen, N. and Wetterberg, L. Ontogensis of somatomedin and insulin receptors in the human fetus. *J. Clin. Invest.* 71:1084, 1983.

37. Stylianopoulou, F., Efstratiadis, A., Herbert, J. and Pintar, J. Pattern of the insulin-like growth factor II gene expression during rat embryogenesis. *Development* 103:497, 1988.

38. Stylianopoulou, F., Herbert, J., Soares, M.B. and Efstratiadis, A. Expression of the insulin-like growth factor II gene in the choroid plexus and the leptomeninges of the adult rat central nervous system. *Proc. Nat'l. Acad. Sci.* 85:141, 1988.

39. Sussenbach, J.S. The gene structure of the insulin-like growth factor family. *Prog. Growth Fact. Res.* 1:33, 1989.

40. Tseng, L. Y-H., Brown, A.L., Yang, Y. W-H., Romanus, J.A., Orlowski, C.C., Taylor, T. and Rechler, M.M. The fetal rat binding protein for insulin-like growth factors is expressed in the choroid plexus and cerebrospinal fluid of adult rats. *Mol. Endocrinol.* 3:1559, 1989.

41. Valentino, K.I., Ocrant, I. and Rosenfeld, R.G. Developmental expression of insulin-like growth factor II receptor immunoreactivity in the rat central nervous system. *Endocrinol.* 126:914, 1989.

42. Werther, G.A., Abata, M., Hoff, A., Cheesman, H., Oldfield, B., Hards, D., Hudson, P., Power, B., Freed, K., and Herington, A.C. Localization of insulin-like growth factor I mRNA in rat brain by *in situ* hybridization - Relationship to IGF-I receptors. *Mol. Endocrinol.* 4:773, 1990.

43. Werner, H., Woloschak, M., Adamo, M., Shen-Orr, Z., Roberts, C.T., Jr. and LeRoith, D. Developmental regulation of the rat insulin-like growth factor I receptor gene. *Proc. Nat'l. Acad. Sci.* 86:7451, 1989.

DISTRIBUTION OF INSULIN-LIKE GROWTH FACTOR 1 (IGF-1) AND 2 (IGF-2) RECEPTORS IN THE HIPPOCAMPAL FORMATION OF RATS AND MICE

Charles R. Breese, Anselm D'Costa, Rosemarie M. Booze, William E. Sonntag

Department of Physiology and Pharmacology, Bowman Gray School of Medicine, Wake Forest University, Winston-Salem, North Carolina 27103

Since the discovery of specific receptors in the brain for insulin-like growth factor-1 and 2 (IGF-1 and IGF-2), there has been an increased interest in the role of these peptides as neurotrophic factors and/or neuromodulators in the central nervous system. Receptors for IGF-1 and IGF-2 have been characterized in brain membrane homogenates[1-6] and localized to specific brain regions utilizing *in vitro* receptor autoradiographic and immunological techniques[7-14]. Competitive binding and affinity cross-linking studies have demonstrated that IGFs bind to two distinct receptors. The IGF-1 (type 1) receptor is structurally similar to the insulin receptor, consisting of 2 extracellular alpha subunits and 2 transmembranous beta subunits (approximately 130 and 95 kDa respectively)[15-17]. The IGF-2 (type 2) receptor consists of a single chain glycoprotein of approximately 250 kDa, which also binds mannose-6-phosphate[18-22]. Radioligand binding studies have demonstrated crossreactivity of the IGFs and insulin at the receptor level. The type I IGF receptor has higher affinity for IGF-1 than IGF-2 and insulin, and the type II IGF receptor binds IGF-2 with considerably higher affinity than IGF-1, and does not recognize insulin[2,23].

There have been extensive *in vitro* studies on the consequences of IGF receptor activation in neuronal tissue. Both IGF-1 and IGF-2 have been shown to have direct actions on both cultured neurons and glia, and exhibit mitogenic and growth promoting actions through the enhancement of RNA and DNA synthesis[24-31]. IGF-1 has been shown to support differentiation of cortical neurons in culture[32], and induce oligodendrocytes, suggesting an important role in the myelination of neurons derived from the central nervous system[33-37]. IGF-1 decreases K^+ induced acetylcholine release suggesting that IGF-1 may have a neuromodulatory effect on hippocampal neurons[38]. This peptide may also have additional effects by stimulating choline acetyl transferase activity and release of acetylcholine in septal and pons cultured neurons and cortical slices, as well as dopamine release from mesencephalic cultured neurons[39,40]. IGF-1 administration has also been shown to increase somatostatin secretion from the median eminence resulting in a decrease in growth hormone release[41-43]. IGF-2 has been reported to stimulate neurite outgrowth in

Molecular Biology and Physiology of Insulin and Insulin-Like Growth Factors
Edited by M.K. Raizada and D. LeRoith, Plenum Press, New York, 1991

neuroblastoma cells[44-47], and increase survival of sympathetic and sensory neurons in culture[48]. Collectively, these studies suggest that IGFs regulate metabolic pathways, growth and differentiation, and possibly behavioral processes through the activation of specific receptors. These results have led to various hypotheses on the function of IGFs in neural tissue, however the consequences of *in vivo* receptor activation in the fetal and adult brain have yet to be demonstrated.

The data reported to date strongly suggest that IGF-1 and IGF-2 may have important roles as trophic factors and/or modulators of neurotransmission in the central nervous system. In the present study, we have utilized receptor autoradiographic analysis to localize and compare IGF-1 and IGF-2 receptor populations in the brains of rats and mice. The hippocampus has been utilized as a model for central nervous system function, as it represents a well-defined neuroanatomical and neurochemical region in which to study species differences. In addition, the hippocampus is an area of the brain which has previously been shown to exhibit highly specific localization of IGF receptor binding in the rat[7,8,10,12]. Both the rat and mouse exhibit similar hippocampal cytoarchitecture, however, alterations in specific neurotransmitter systems have been reported[49-53]. Comparison of IGF receptor populations between these species may 1) provide insight into the mechanism of IGF neurochemistry and functional anatomy; 2) indicate the consequences of neuropeptide modulation in the central nervous system; and 3) provide important information in the development of appropriate models for the assessment of neurological function in humans.

MATERIALS AND METHODS

Male Fischer-344 rats and C57/BL6 mice (obtained from the National Center for Toxicological Research, Jefferson, AR) were fed *ad libitum* food and water and allowed to habituate for 1 week. Animals were sacrificed by rapid decapitation, the brains were rapidly removed, frozen on dry ice and stored at -80°C. Coronal cryostat cut sections (20μm) were mounted on 5% gelatin-coated slides and analyzed for IGF-1 and IGF-2 receptor binding. Recombinant (Thr59)IGF-1 and IGF-2 (Amgen, Thousand Oaks, CA) were radio-iodinated using a lactoperoxidase-glucose oxidase method. Mounted sections were incubated with ^{125}I-IGF-1 and ^{125}I-IGF-2 in 0.5M Tris-HCl pH 7.5, and 1% BSA overnight at 4°C. Receptor binding assays were performed on slide-mounted sections with varying concentrations of unlabeled peptides and incubated as above. Sections were then washed and removed from the slide and counted on an LKB gamma counter. Receptor autoradiography was performed on slide-mounted sections which were incubated, washed, dried and apposed directly to ^{3}H-Hyperfilm (Amersham). Non-specific binding was determined by the addition of 2μg/ml of unlabeled IGF-1 and IGF-2 to the appropriate incubation solution. Microscales (Amersham) were utilized to quantitate ligand binding. Optical densities of discrete anatomical brain regions were determined using a computer-assisted digitizing system (CARP system, Biographics, Inc, Dallas, TX). Sections were counter-stained with cresyl violet to identify the regions of receptor binding in relation to the lamellar cytoarchitecture of the region.

RESULTS

Previous studies in this and other laboratories[7,8,10,12] have demonstrated a high level of binding of both IGF-1 and IGF-2 in the hippocampus of the rat. However, there have been limited reports on IGF-1 or IGF-2 binding in the mouse brain. Both species exhibit a well-developed hippocampal cytoarchitecture, and analogous cellular arrangement and lamellar structure within CA1, CA3, and the dentate gyrus (Fig. 1). Therefore, it was hypothesized that IGF binding in this region would be similar in both species. Examination of receptor autoradiographs of both IGF-1 and IGF-2 binding in the rat and mouse demonstrated species differences in receptor distribution and density of these peptides in the hippocampus, as well as other brain structures.

IGF Binding Studies

Initially, competitive receptor binding assays were performed to examine the possibility of alterations in the relative affinities of these receptors between species. Receptor binding assays were performed on slide-mounted sections with varying concentrations of unlabeled peptide. As demonstrated in figure 2, the receptor binding curves for the rat and mouse are similar, suggesting that there were no significant differences between the binding affinities for IGF-1 or IGF-2 receptors between these species. These results indicate that differences observed by receptor autoradiography result from changes in the total receptor numbers (B_{max}) within the hippocampus of the rat and mouse, rather than from alterations in receptor affinity.

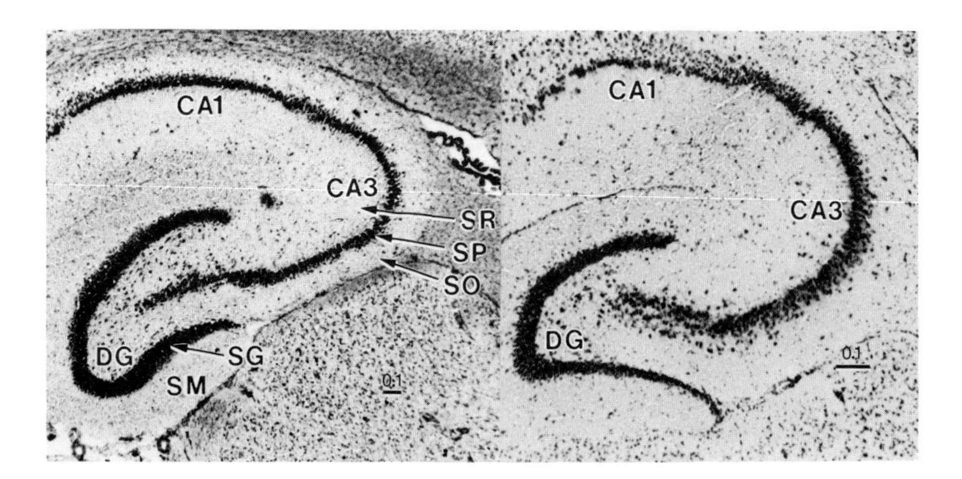

Figure 1. Nissl stained horizontal sections of rat (left) and mouse (right) hippocampal formation. Both species have analogous laminar structure and cell layers. Abbreviations: DG, dentate gyrus; SG, s. granulosum of dentate gyrus; SM, s. moleculare of dentate gyrus; SO, s. oriens of CA1/CA3; SP, s. pyramidale of CA1/CA3; SR, s. radiatum of CA1/CA3. (bar = 0.1mm)

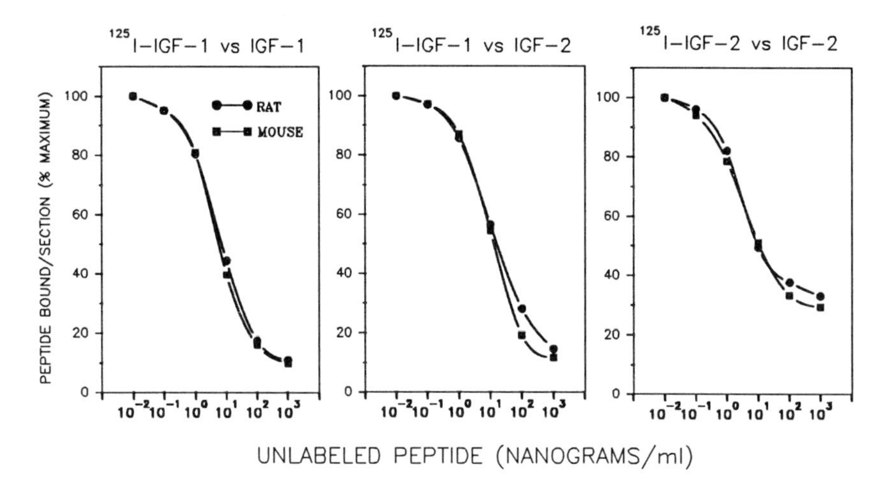

Figure 2. Competition curves for IGF-1 and IGF-2. Slide-mounted sections were incubated with ^{125}I radiolabeled peptide in the absence and presence of varying concentrations of unlabeled peptide. Vertical axis represents percent of total binding. Each point represents the mean of four determinations. Both species exhibited similar relative binding affinities for each peptide.

IGF-1 Receptor Distribution

Receptor autoradiographic analysis of ^{125}I-IGF-1 receptor binding in the rat hippocampus (Fig. 3, left photograph) demonstrated high levels of binding in the stratum oriens and s. radiatum of CA3 as well as the s. moleculare of the dentate gyrus. Moderate to low binding was observed in the s. lacunosum-moleculare, s. lucidum, and s. pyramidale of CA1-CA3. A complete absence of binding was observed in the s. oriens and s. radiatum of CA1. These results suggest that IGF-1 receptors are primarily located on the dendrites of pyramidal cells originating in CA3 and the dentate granular cell layer. In contrast, mouse IGF-1 receptor binding (fig. 3, right photograph) exhibited a reduction in IGF-1 binding in the hippocampus relative to the rat. The greatest density of IGF-1 binding in the mouse was observed on CA1/CA3 pyramidal cells and s. oriens and s. radiatum of CA3. Other regions of CA1 displayed an absence of binding. The most striking difference observed was an absence of IGF-1 binding to the s. moleculare of the dentate gyrus in mice. These results demonstrated an alteration in both receptor density and localization of IGF-1 receptors in the hippocampal formation of rats and mice.

IGF-2 Receptor Distribution

Studies utilizing ^{125}I-IGF-2 confirmed alterations in the binding of IGF-2 in the hippocampus of the rat and mouse. In the rat hippocampus (Fig. 4, left photograph), receptor binding was localized to the s. pyramidale of CA1-CA3 and s. granulosum of the dentate gyrus. There was an absence of binding in other regions of the hippocampus. The mouse hippocampus (Fig. 4, right

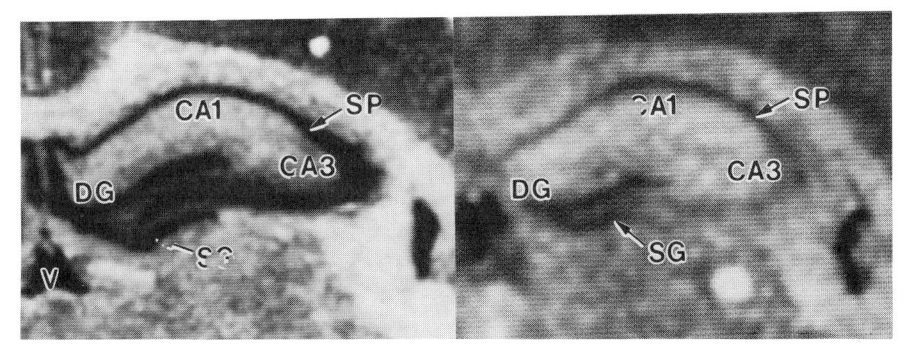

Figure 3. Receptor autoradiography of IGF-1 binding sites in the hippocampal formation of rats and mice. Receptor autoradiography was performed on cryostat-cut brain sections incubated with ^{125}I-IGF-1 and apposed to Hyperfilm ^{3}H. The rat hippocampus (left) demonstrated high levels of binding in the s. oriens and s. radiatum of CA3 as well as the s. moleculare of the dentate gyrus. A complete absence of binding was observed in the s. oriens and s. radiatum of CA1. The mouse exhibited IGF-1 receptor binding (right) on CA1/CA3 pyramidal cells and the s. oriens and s. radiatum of CA3. Abbreviations: DG, dentate gyrus; SM, s. moleculare of dentate gyrus; SO, s. oriens of CA1/CA3; SP, s. pyramidale of CA1/CA3; SR, s. radiatum of CA1/CA3. (bar = 1.0mm)

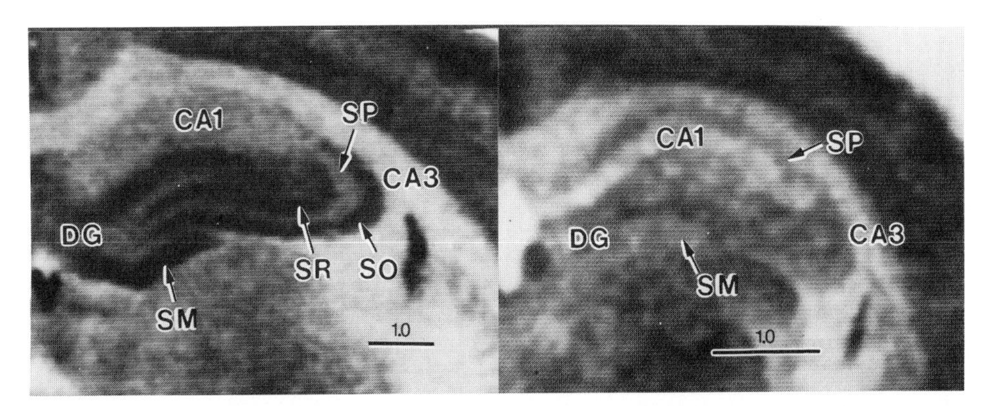

Figure 4. Receptor autoradiography of IGF-2 binding sites in the rat and mouse hippocampus. Receptor autoradiography was performed as previously described with ^{125}I-IGF-2. In the rat (left), high levels of receptor binding were observed in s. pyramidale of CA1/CA3 and the s. granulosum of the dentate gyrus. There is an overall absence of binding in other regions of the hippocampus. The mouse (right) displayed similar localization but less overall binding in the hippocampus than the rat. Abbreviations: DG, dentate gyrus; SG, s. granulosum of dentate gyrus; SP, s. pyramidale of CA1/CA3. (calibration bar in fig. 3)

photograph) displayed a similar anatomical localization of IGF-2, however, a reduction in binding density was observed in the hippocampal cell layers relative to that observed in the rat, with the majority of binding localized to the s. granulosum of the dentate gyrus. These studies suggest that IGF-2 receptors are primarily located on the soma of pyramidal and granular cells in the hippocampus, but do not exclude possible binding on other cell types within these lamina.

SUMMARY and CONCLUSIONS

This study demonstrated species differences in IGF-1 and IGF-2 receptor binding and localization in the hippocampus of the rat and mouse. Competition binding studies indicated that there were no differences in the relative binding affinities for the type 1 or type 2 receptors between the brains of these animals. These results suggested that the observed species differences were not attributable to alterations in IGF receptor kinetics. Receptor autoradiographic analyses demonstrated that IGF-1 binding differed in both the localization and overall receptor densities observed, with the rat demonstrating more specific localization and greater receptor density in the hippocampus than the mouse. The rat also exhibited a greater density of IGF-2 receptors in the hippocampus than the mouse. Despite differences in IGF receptor populations, both species exhibit similar hippocampal structure and lamination. Therefore, these results demonstrate a disparity in the localization of IGF receptor binding in the rat and mouse, suggesting that IGFs in these species are differentially regulated, with distinct neuromodulatory, neurotrophic, and/or developmental roles in this region of the brain.

Previous comparative anatomical studies of the hippocampal formation of rats and mice fail to offer an explanation for the absence or reduction of binding of IGF-1 in the mouse. Although the mouse has a greater cell density in the s. granulosum than the rat[54], and both species exhibit similar glia and synaptic contact densities in the s. moleculare of the dentate gyrus[54,55], the mouse exhibits a complete absence of IGF-1 binding in this region. The lack of anatomical differences in the hippocampal formation of these species suggests that the patterns observed in IGF binding result from alterations in either neurochemical modulation of these neurons or specific neurotrophic requirements of the cells in this region. Differences have been reported on the concentrations and binding of various neurotransmitters in the hippocampus of these species[49-53], however these differences do not easily account for the variations observed in IGF binding in this study. IGFs are known to influence acetylcholine neurotransmission in the hippocampus as well as other brain areas in the rat[38-40]. Recently, a truncated form of IGF-1, in which a tripeptide is cleaved from the N-terminus of the peptide, has been reported in brain[56,57]. The cleaved tripeptide has been shown to activate glutamate receptors, which may dramatically influence excitatory neurotransmission in this region[58]. Therefore, in addition to the possible neurotrophic actions of the peptide itself, subsequent processing of IGF-1 may be an important aspect of IGF-1 activity in the brain. These studies suggest that in the rat, IGFs may have an important role in the modulation of excitatory neurotransmission in the hippocampus, or act as a neurotrophic factor contributing to maintenance of these neurons.

Differences in IGF-2 binding between these species were not as extensive as that noted for IGF-1. Binding of IGF-2 was localized to the cellular layers in the hippocampus in both species, however, IGF-2 receptor density was reduced in the mouse in these regions relative to the rat. As previously mentioned, differences in cellular density could not account for the differences in IGF-2 binding[54,55]. However, since both species exhibit similar receptor localization, IGF-2 may retain an important role in both species. Further studies will be required to assess the actions of IGF-2 in the brain, as well as the specific cellular localization of these receptors.

Rodents are generally utilized as an animal model for human nervous system function, and findings in rodents have broad implications in human development, aging, and neurological disorders. Previous studies have shown that humans exhibit a large quantity of IGF receptors in the hippocampus[11]. Due to the high receptor concentrations found in the hippocampus of the rat, the present study suggests that rat hippocampus may be a preferable model for assessment of IGF neuromodulatory or neurotrophic roles in this region of the brain. However, only further research will clarify the actions of IGF-1 and IGF-2 in the brain of the developing, adult, and aged animal. Further electophysiological and *in vivo* functional studies will be necessary to elucidate the role of receptor activation in the hippocampus, assess the neurochemical function of these peptides in the central nervous system, and explain the exact nature of the differences found in the rat and mouse hippocampus.

ACKNOWLEDGEMENTS: Animals used in this study were obtained from the Project on Caloric Restriction at the National Center for Toxicological Research. This research was supported by grant AG07752 to WES.

REFERENCES

1. Sara, V.R., K. Hall, H. Von Holtz, R. Humbel, B. Sjogren, L. Wtterberg (1982) Evidence for the presence of specific receptors for insulin-like growth factors 1 (IGF-1) and 2 (IGF-2) and insulin throughout the adult human brain. Neuroscience letters 34:39-44.
2. Goodyer, C.G., L. De Stephano, W.H. Lai, H.J. Guyda, and B.I. Posner (1984) Characterization of insulin-like growth factor receptors in rat anterior pituitary, hypothalamus, and brain. Endocrinology 114:1187-1195.
3. Gammeltoft, S., G.K. Haselbacher, R.E. Humbel, M. Fehlmann, and E. Van Obberghen (1985) Two types of receptor for insulin-like growth factors in mammalian brain. Embo Journal 4:3407-3412.
4. Heidenreich, K.A., G.R. Freidenberg, D.P. Figlewicz, and P.R. Gilmore (1986) Evidence for a subtype of insulin-like growth factor I receptor in brain. Regulatory Peptides 15:301-310.
5. Burgess, S.K., S. Jacobs, P. Cuatrecasas, N. Sahyoun (1987) Characterization of a neuronal subtype of insulin-like growth factor I receptor. Journal of Biological Chemistry 62:1618-1687.
6. Rosenfeld, R.G., H. Pham, B.T. Keller, R.T. Borchardt, and W.M. Pardridge (1987) Demonstration and structural comparison of receptors for insulin-like growth factor-I and -II (IGF-I and -II) in brain and blood-brain barrier. Biochemical & Biophysical Research Communications 149:159-166.
7. Bohannon, N.J., E.S. Corp, B.J. Wilcox, D.P. Figlewicz, D.M. Dorsa, and D.G. Baskin (1988) Localization of binding sites for insulin-like growth factor- I (IGF-I) in the rat brain by quantitative autoradiography. Brain Research 444:205-213.
8. Lesniak, M.A., J.M. Hill, W. Kiess, M. Rojeski, C.B. Pert, J. Roth, (1988) Receptors for insulin-like growth factors I and II: Autoradiographic localization in rat brain and comparison to receptors for insulin. Endocrinology 123:2089-2099.

9. Mendelsohn, L.G., M.C. Smith, V.L. Lucaites, G.A. Kerchner, and B. Ghetti (1988) Autoradiographic localization of insulin-like growth factor II receptors in cerebellar cortex of weaver and Purkinje cell degeneration mutant mice. Brain Research 458:361-366.
10. Smith, M., J. Clemens, G.A. Kerchner, and L.G. Mendelsohn (1988) The insulin-like growth factor-II (IGF-II) receptor of rat brain: regional distribution visualized by autoradiography. Brain Research 445:241-246.
11. Adem, A., S.S. Jossan, R. d'Argy, P.G. Gillberg, A. Nordberg, B. Winblad, and V. Sara (1989) Insulin-like growth factor 1 (IGF-1) receptors in the human brain: quantitative autoradiographic localization. Brain Research 503:299-303.
12. Matsuo, K., M. Niwa, M. Kurihara, K. Shigematsu, S. Yamashita, M. Ozaki, and S. Nagataki (1989) Receptor autoradiographic analysis of insulin-like growth factor-I (IGF-I) binding sites in rat forebrain and pituitary gland. Cellular.&.Molecular.Neurobiology. 9:357-367.
13. Sklar, M.M., W. Kiess, C.L. Thomas, and S.P. Nissley (1989) Developmental expression of the tissue insulin-like growth factor II/mannose 6-phosphate receptor in the rat. Measurement by quantitative immunoblotting. Journal of Biological Chemistry 264:16733-16738.
14. Valentino, K.L., I. Ocrant, and R.G. Rosenfeld (1990) Developmental expression of insulin-like growth factor- II receptor immunoreactivity in the rat central nervous system. Endocrinology 126:914-920.
15. Chernausek, S.D., S. Jacobs, and J.J. Van Wyk (1981) Structural similarities between human receptors for somatomedin C and insulin: analysis by affinity labeling. Biochemistry 20:7345-7350.
16. Czech, M.P. (1982) Structural and functional homologies in the receptors for insulin and the insulin-like growth factors. Cell 31:8-10.
17. McElduff, A., P. Poronnik, R.C. Baxter, and P. Williams (1988) A comparison of the insulin and insulin-like growth factor I receptors from rat brain and liver. Endocrinology 122:1933-1939.
18. Roth, R.A., C. Stover, J. Hari, D.O. Morgan, M.C. Smith, V. Sara, and V.A. Fried (1987) Interactions of the receptor for insulin-like growth factor II with mannose-6-phosphate and antibodies to the mannose-6-phosphate receptor. Biochemical & Biophysical Research Communications 149:600-606.
19. Kiess, W., G.D. Blickenstaff, M.M. Sklar, C.L. Thomas, S.P. Nissley, and G.G. Sahagian (1988) Biochemical evidence that the type II insulin-like growth factor receptor is identical to the cation-independent mannose 6-phosphate receptor. Journal of Biological Chemistry 263:9339-9344.
20. MacDonald, R.G., S.R. Pfeffer, L. Coussens, M.A. Tepper, C.M. Brocklebank, J.E. Mole, J.K. Anderson, E. Chen, M.P. Czech, and A. Ullrich (1988) A single receptor binds both insulin-like growth factor II and mannose-6-phosphate. Science 239:1134-1137.
21. Tong, P.Y., S.E. Tollefsen, and S. Kornfeld (1988) The cation-independent mannose 6-phosphate receptor binds insulin-like growth factor II. Journal of Biological Chemistry 263:2585-2588.
22. Waheed, A., T. Braulke, U. Junghans, and K. von Figura (1988) Mannose 6-phosphate/insulin like growth factor II receptor: the two types of ligands bind simultaneously to one receptor at different sites. Biochemical & Biophysical Research Communications 152:1248-54
23. Rechler, M.M., and S.P. Nissley (1985) The nature and regulation of the receptors for insulin- like growth factors. Annual Review of Physiology 47:425-442.
24. Lenoir, D., and P. Honegger (1983) Insulin-like growth factor I (IGF I) stimulates DNA synthesis in fetal rat brain cell cultures. Brain Research 283:205-213.
25. Enberg, G., A. Tham, and V.R. Sara (1985) The influence of purified somatomedins and insulin on foetal rat brain DNA synthesis in vitro. Acta Physiologica Scandinavica 125:305-308.
26. Yang, H.C., and A.B. Pardee (1986) Insulin-like growth factor I regulation of transcription and replicating enzyme induction necessary for DNA synthesis. Journal of Cellular Physiology 127:410-416.

27. Han, V.K., J.M. Lauder, and A.J. D'Ercole (1987) Characterization of somatomedin/insulin-like growth factor receptors and correlation with biologic action in cultured neonatal rat astroglial cells. Journal of Neuroscience 7:501-511.
28. Shemer, J., M.K. Raizada, B.A. Masters, A. Ota, and D. LeRoith (1987) Insulin-like growth factor I receptors in neuronal and glial cells Characterization and biological effects in primary culture. Journal of Biological Chemistry 262:7693-7699.
29. Surmacz, E., L. Kaczmarek, O. Rnning, and R. Baserga (1987) Activation of the ribosomal DNA promoter in cells exposed to insulin-like growth factor I. Molecular & Cellular Biology 7:657-663.
30. DiCicco-Bloom, E., and I.B. Black (1988) Insulin growth factors regulate the mitotic cycle in cultured rat sympathetic neuroblasts. Proc.Natl.Acad.Sci.USA 85:4066-4070.
31. Ota, A., Z. Shen-Orr, and D. LeRoith (1989) Insulin and IGF-I receptors in neuroblastoma cells: increases in mRNA and binding produced by glyburide. Neuropeptides 14:171-175.
32. Aizenman, Y., J. de Vellis (1987) Brain neurons develop in a serum and glial free environment: Effects of transferrin, insulin, insulin-like growth factor-1 and thyroid hormone on neuronal survival, growth, and differentiation. Brain Research 406:32-42.
33. McMorris, F.A., T.M. Smith, S. DeSalvo, and R.W. Furlanetto (1986) Insulin-like growth factor I/somatomedin C: a potent inducer of oligodendrocyte development. Proc.Natl.Acad.Sci.USA 83:822-826.
34. McMorris, F.A., and M. Dubois-Dalcq (1988) Insulin-like growth factor I promotes cell proliferation and oligodendroglial commitment in rat glial progenitor cells developing in vitro. Journal of Neuroscience Research 21:199-209.
35. Mozell, R.L., and F.A. McMorris (1988) Insulin-like growth factor-I stimulates regeneration of oligodendrocytes in vitro. Annals of the New York Academy of Sciences 540:430-432.
36. Saneto, R.P., K.G. Low, M.H. Melner, and J. de Vellis (1988) Insulin/insulin-like growth factor I and other epigenetic modulators of myelin basic protein expression in isolated oligodendrocyte progenitor cells. Journal of Neuroscience Research 21:210-219.
37. van der Pal R.H., J.W. Koper, L.M. van Golde, and M. Lopes-Cardozo (1988) Effects of insulin and insulin-like growth factor (IGF- I) on oligodendrocyte-enriched glial cultures. Journal of Neuroscience Research 19:483-490.
38. Araujo, D.M., P.A. Lapchak, B. Collier, J.G. Chabot, and R. Quirion (1989) Insulin-like growth factor-1 (somatomedin-C) receptors in the rat brain: distribution and interaction with the hippocampal cholinergic system. Brain Research 484:130-138.
39. Nilsson, L., V.R. Sara, and A. Nordberg (1988) Insulin-like growth factor 1 stimulates the release of acetylcholine from rat cortical slices. Neuroscience Letters 88:221-226.
40. Knusel, B., P.P. Michel, J.S. Schwaber, and F. Hefti (1990) Selective and Nonselective Stimulation of Central Cholinergic and Dopaminergic Development Invitro by Nerve Growth Factor, Basic Fibroblast Growth Factor, Epidermal Growth Factor, Insulin and the Insulin-Like Growth Factor- I and Factor-II. Journal of Neuroscience 10:558-570.
41. Tannenbaum, G.S., H.J. Guyda, and B.I. Posner (1983) Insulin-like growth factors: a role in growth hormone negative feedback and body weight regulation via brain. Science 220:77-79.
42. Ceda, G.P., R.G. Davis, R.G. Rosenfeld, and A.R. Hoffman (1987) The growth hormone (GH)-releasing hormone (GHRH)-GH-somatomedin axis: evidence for rapid inhibition of GHRH-elicited GH release by insulin-like growth factors I and II. Endocrinology 120:1658-1662.
43. Yamashita, S., and S. Melmed (1987) Insulinlike growth factor I regulation of growth hormone gene transcription in primary rat pituitary cells. Journal of Clinical Investigation 79:449-452.

44. Bothwell, M. (1982) Insulin and somatomedin MSA promote nerve growth factor-independent neurite formation by cultured chick dorsal root ganglionic sensory neurons. Journal of Neuroscience Research 8:225-231.
45. Recio-Pinto, E., and D.N. Ishii (1984) Effects of insulin, insulin-like growth factor-II and nerve growth factor on neurite outgrowth in cultured human neuroblastoma cells. Brain Research 302:323-334.
46. Mill, J.F., M.V. Chao, and D.N. Ishii (1985) Insulin, insulin-like growth factor II, and nerve growth factor effects on tubulin mRNA levels and neurite formation. Proc.Natl.Acad. Sci.USA 82:7126-7130.
47. Recio-Pinto, E., and D.N. Ishii (1988) Insulin and insulinlike growth factor receptors regulating neurite formation in cultured human neuroblastoma cells. Journal of Neuroscience Research 19:312-320.
48. Recio-Pinto, E., Rechler, M.M., Ishii, D.N., Effects of insulin-like growth factor-II, and nerve growth factor on neurite formation and survival in cultured sympathetic and sensory neurons. Journal of Neuroscience 6:1211-1219, (1986).
49. Marchand, C.M-F., S.P. Hunt, J. Schmidt (1979) Putative acetylcholine receptors in hippocampus and corpus striatum of rat and mouse. Brain Research 160:363-7.
50. Marks, M.J., J.A. Stitzel, E. Romm, J.M. Wehner, A.C. Collinis (1986) Nicotinic binding sites in rat and mouse: comparison of acetylcholine, nicotine and α-bungarotoxin. Molecular Pharmacology 30:427-36.
51. Fredens, K., K. Stengaard-Pedersen, M.N. Wallace (1987) Localization of cholecystokinin in the dentate commissural-associational system of the rat and mouse. Brain Research 401:68-78.
52. Palacios, J.M., D. Hoyer, R. Cortes (1987) α1-Adrenoceptors in the mammalian brain: similar pharmacology but different distribution in rodents and primates. Brain Research 419:65-75.
53. Dietl, M.M., J.M. Palacios (1989) Distribution of cholecystokinin receptors in vertebrate brain: Species differences studied by receptor autoradiography. J. Chemical Neuroanatomy 2:149-61.
54. West, M.J., A.H. Andersen (1980) An allometric study of the area dentata in the rat and mouse. Brain Res. Reviews 2:317-48.
55. Kishi, K., B.B. Stanfield, W.M. Cowan (1979) A note on distribution of glial cells in the molecular layer of the dentate gyrus. Brain Res. Bull. 4:35-41. p73
56. Sara, V.R., C. Carlsson-Skwirut, C. Andersson, E. Hall, B. Sjogren, A. Holmgren, and H. Jornvall (1986) Characterization of somatomedins from human fetal brain: identification of a variant form of insulin-like growth factor I. Proc.Natl.Acad.Sci.USA 83:4904-4907.
57. Carlsson-Skwirut, C., M. Lake, M. Hartmanis, K. Hall, and V.R. Sara (1989) A comparison of the biological activity of the recombinant intact and truncated insulin-like growth factor 1 (IGF- 1). Biochimica Et Biophysica Acta 1011:192-197.
58. Sara, V.R.,C. Carlsson-Skwirut,T. Bergman,H. Jornvall, P.J. Roberts, M. Crawford, L.N. Hakansson, I. Civalero, and A. Nordberg (1989) Identification of Gly-Pro-Glu (GPE), the aminoterminal tripeptide of insulin-like growth factor 1 which is truncated in brain, as a novel neuroactive peptide. Biochemical & Biophysical Research Communications 165:766-771.

LOCALIZATION OF INSULIN AND TYPE 1 IGF RECEPTORS IN RAT BRAIN BY IN VITRO AUTORADIOGRAPHY AND IN SITU HYBRIDIZATION

Jonathan L. Marks, Michael G. King, and Denis G. Baskin

Depts. of Biological Structure and Medicine, University of Washington

and V.A. Medical Center, Seattle WA

Introduction

Insulin and the insulin-like growth factors (IGFs) are structurally related, circulating hormones that regulate growth and intermediary metabolism in virtually all tissues. In many organs their function has been well defined but in the adult CNS no general understanding of their action and importance is available. For many years it has been known that circulating insulin can affect the CNS directly to regulate peripheral metabolism (1). The hypothalamus is considered the most likely site of these actions (2). Our laboratory has been particularly interested in the ability of insulin to reduce food intake and act as a regulator of body weight. This has been demonstrated by peripheral infusions of insulin (3) or by injecting insulin into the third ventricle (4) or into the hypothalamus (5). Electrophysiological studies in hypothalamus (6) and hippocampus (7) support this hypothesis by demonstrating that insulin can affect the firing rate of neurons and therefore can act as a neuromodulator or possibly neurotransmitter. Insulin's inhibition of norepinephrine uptake in cultured neonatal neurones supports a neuromodulatory role for insulin (8). Recently, Figlewicz has confirmed this finding in hippocampal slices from adult rats (9). A role for IGFs as neuromodulators in the adult brain has also been suggested. IGF-1 can affect the release of acetylcholine from hippocampal slices (10) while IGF-2 but not IGF-1 reduces food intake when injected into the third ventricle and may be a physiological satiety factor (11). In the hypothalamus, IGF-1 may modulate release of somatostatin and thereby regulate growth hormone release from the pituitary (12).

Insulin-like immunoactivity has been found in the adult brain by radioimmunoassay (13) and immunohistochemistry (14). It is still controversial whether insulin is synthesized by neurons in the adult brain (15), although insulin mRNA may have been found in very low concentration in a few nuclei (16) and insulin is synthesized in cultured neonatal neurons (17). Circulating insulin may be taken up across brain capillaries into brain intestitium or across the choroid plexus into the CSF (18). Very low levels of IGF-1 are found in the adult brain by radioimmunoasay (19) but discreet production by a few neuronal cell types has been suggested (20). Recently, the mRNA for IGF-1 has been identified in the rat brain by in situ hybridization (21), confirming previous reports of low levels by solution or Northern blot hybridization of whole brain RNA (22). IGF-2 levels are high in CSF (23) and IGF-2 mRNA has been found in high concentrations in choroid plexus and meninges (24) and in low concentrations in hypothalamus by solution hybridization (25).

Numerous studies have demonstrated specific insulin (26, 27), type 1 and type 2 IGF receptors (28) in membrane and solubilized preparations from CNS. All three receptors have a slightly reduced molecular weight on gel electrophoresis compared to receptors in peripheral organs due to different sugar moieties (28,29,30). Nevertheless, their affinities for insulin and the IGFs are similar to peripheral receptors. CNS insulin receptors bind insulin>> IGF-1>IGF-2, while type 1 IGF receptors bind IGF-1>IGF-2>>insulin and type 2 receptors bind IGF-2>IGF-1 but not insulin (28). Therefore binding studies with low concentrations of labelled insulin and IGF-1 recognize almost exclusively the insulin and type 1 IGF receptors respectively. Unlike in peripheral organs, insulin receptors in the adult CNS are not regulated by circulating insulin levels (31,32), but are developmentally (33) and by nutritional state (32,34). Type 1 IGF receptors in the CNS are also regulated during development (10, 35,36) and by nutritional state (37). It is not known if they are regulated by circulating or local brain IGFs.

We have approached the problem of defining the role of insulin and IGFs in the adult brain by localizing their receptors. This involves the cell type expressing the receptors and whether they are pre or postsynaptic. In view of the complicated structure of the adult brain, binding to intact brain slices followed by film autoradiography has been used in our and other laboratories to show the pattern of insulin and IGF binding in the different brain regions (10,38,39,40). Because the role of the type 2 IGF receptor is currently unclear in IGF action, we have concentrated on the type 1 IGF receptor. Studies have shown marked differences in the concentrations of receptors in different anatomical layers and nuclei among various brain regions. With some exceptions, high levels of insulin binding are not associated with high levels of IGF-1 binding in individual layers or nuclei. These results suggest that insulin and IGFs are often acting on different neural pathways. The resolution of film autoradiographs has not allowed the localization of receptors at the cellular level. To overcome this problem, we have used two methods to localize receptors at the cellular level: in situ hybridization and neuronal lesioning studies. We have concentrated on brain regions that have a relatively simple structure and contain to a large extent the processes or cell bodies of relatively few neuronal types. In these areas, binding should be on pre or postsynaptic processes and terminals or on cell bodies of only a few neuron types.

In situ hybridization

Neurons expressing insulin and type 1 IGF receptors have been determined using in situ hybridization with oligonucleotide probes for insulin and IGF receptors (41). Probes have been 3' end labelled with 35S, hybridized overnight on brain slices and washed under stringent conditions. It was found that mixing three labelled oligonucleotide probes, directed to different parts of the same receptor mRNA, generated an adequate hybridization signal. The specificity of the probes for the appropriate receptor mRNA has been confirmed by several methods. First, northern blot hybridizations using the oligonucleotide probes showed single bands of 9-10 kilobases. Second, receptor mRNA signals were abolished by performing hybridizations in the presence of 100 fold excess unlabelled probe. Third, the distribution of signal was identical but weaker when individual probes for each receptor mRNA were used alone. The subregional distribution of receptor mRNA was determined by contact film autoradiography. The expression of specific receptor mRNA by individual cells was detected by liquid emulsion autoradiography, allowing silver grains to be measured over individual cell bodies. These results were correlated with the expression of receptor binding on film autoradiographs.

Neuronal lesioning

Lesioning studies of particiular cell types were performed in the hippocampus of animals sacrificed after neuronal cell bodies and their processes were allowed to degenerate. Injection of kainic acid in the lateral ventricle was used to lesion preferentially CA3 pyramidal cells in Ammon's horn. Ischemia induced by four vessel cerebral artery occlusion was used to lesion CA1 pyramidal cells.

Localization of receptors in brain regions

In the olfactory bulb, IGF-1 binding was found in all layers but was highest on the combined mitral cell, inner external plexiform layers and on the glomerular layer (Fig. 1A) In situ hybridization signal demonstrated by film autoradiography, showed that mRNA for the type 1 IGF receptor was expressed in the granule cell, mitral cell and glomerular layers (Fig. 1B). The cellular localization of type 1 IGF receptor mRNA was determined on emulsion dipped slides, using the characteristic appearance of olfactory bulb neurons, stained with cresyl violet. Silver grains depicting the site of IGF-1 receptor mRNA were found over granule cells in the granule cell layer, mitral and granule cells in the mitral cell layer, tufted

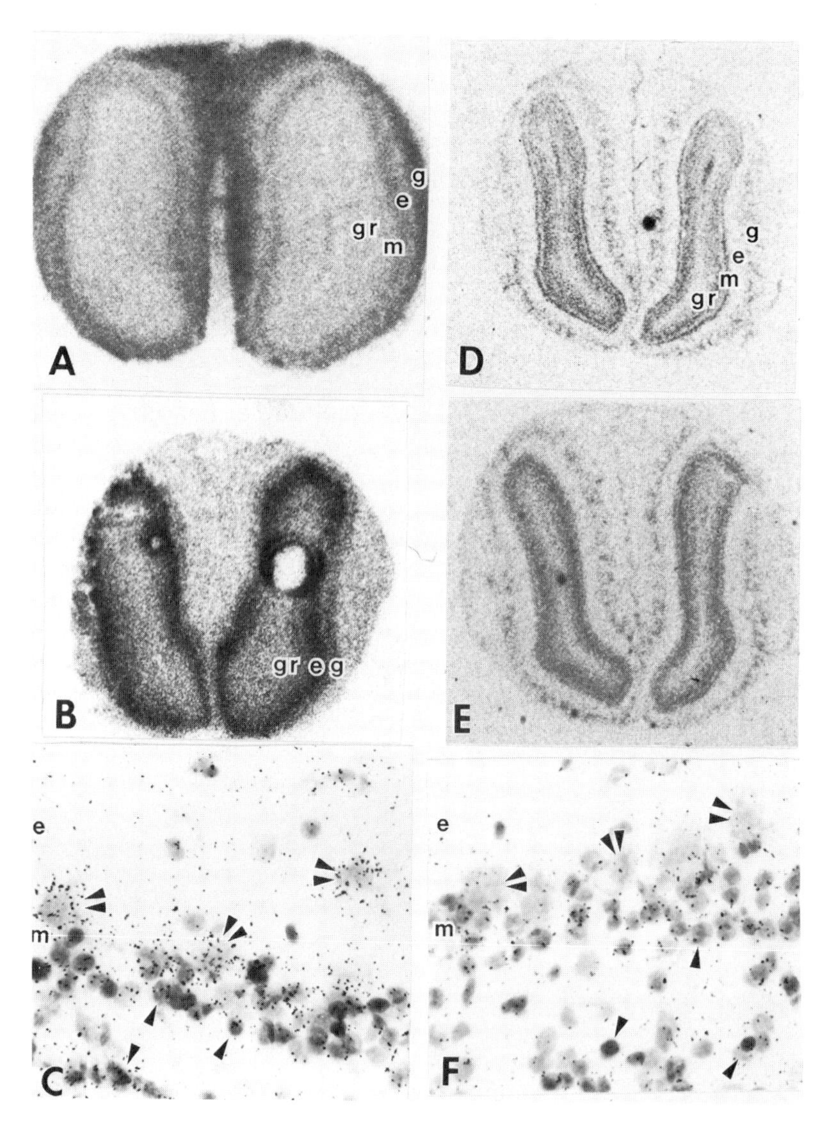

Figure 1. Insulin and type 1 IGF receptor binding and mRNA in the olfactory bulb. A IGF-1 binding by in vitro autoradiography. B Distribution of type 1 IGF receptor mRNA by in situ hybridization. C Cellular localization of type 1 IGF-1 mRNA by in situ hybridization. D Insulin binding by in vitro autoradiography. E Distribution of insulin receptor mRNA by in situ hybridization. F Cellular localization of insulin mRNA by in situ hybridization. Abbreviations: g, glomerular layer; e, external plexiform layer; m, mitral cell layer; gr, granule cell layer. Single arrows indicate granule cells, double arrows indicate mitral and tufted cells.

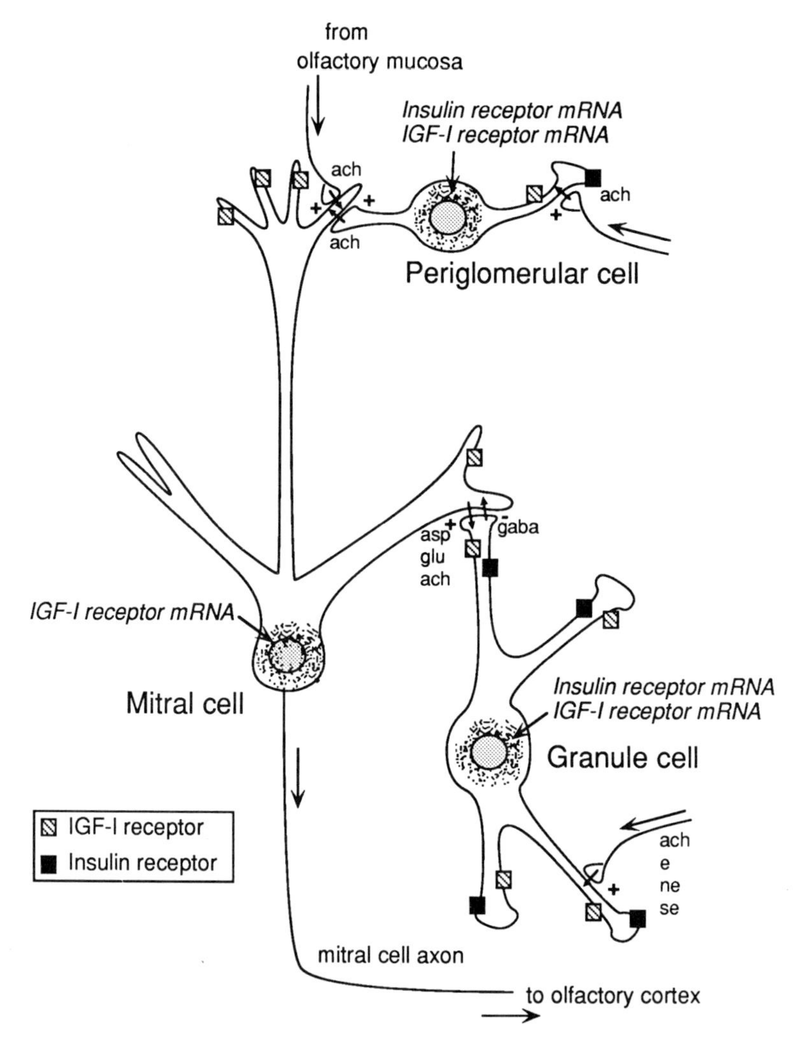

Figure 2. Cellular and subcellular localization of insulin and type 1 IGF receptors in the olfactory bulb.

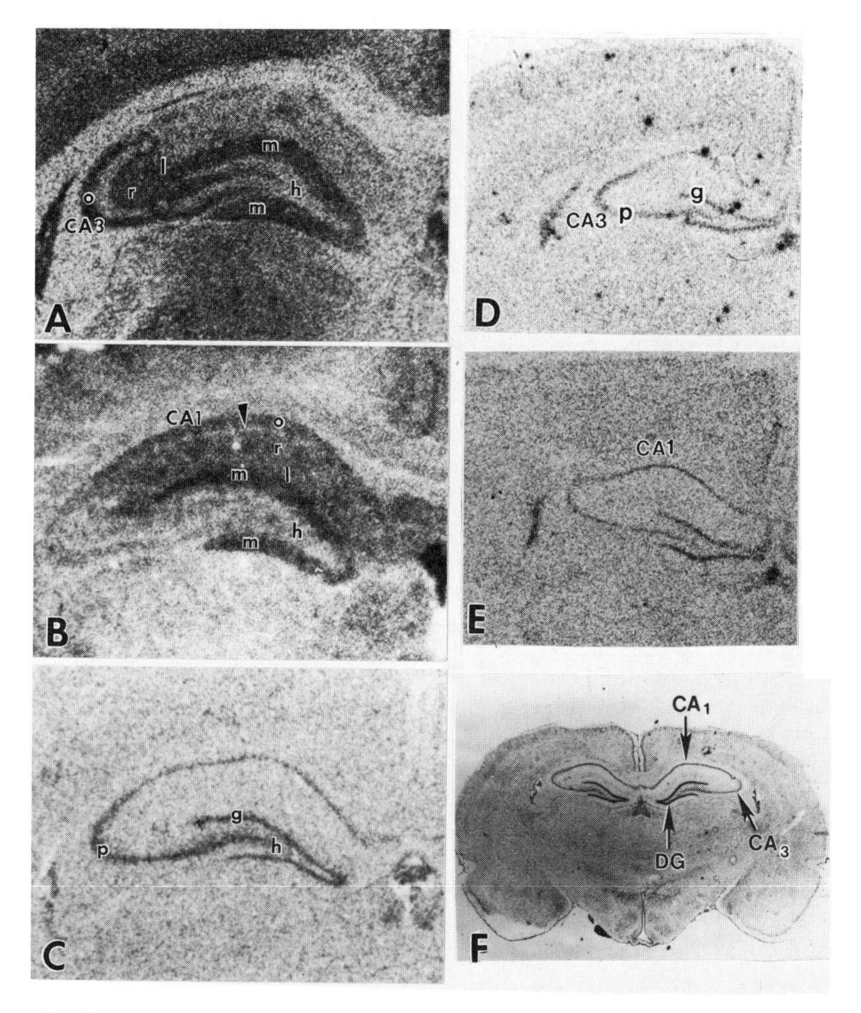

Figure 3. IGF and insulin binding and insulin and type 1 IGF receptor mRNA. in the hippocampus.
A IGF-1 binding by in vitro autoradiography. B Insulin binding by in vitro autoradiography. C IGF-2 binding by in vitro autoradiography. D Distribution of type1 IGF receptor mRNA by in situ hybridization. E Distribution of insulin receptor mRNA by in situ hybridization. F Nissl stain of a coronal section through the hippocampus and hypothalamus. Abbreviations: dg, dentate gyrus; g, granule cell layer of dentate gyrus; h, hilus of dentate gyrus; l, stratum lacunosum-moleculare; m, molecular layer of the dentate gyrus; o, stratum oriens; p, pyramidal cell layer; r, stratum radiatum.

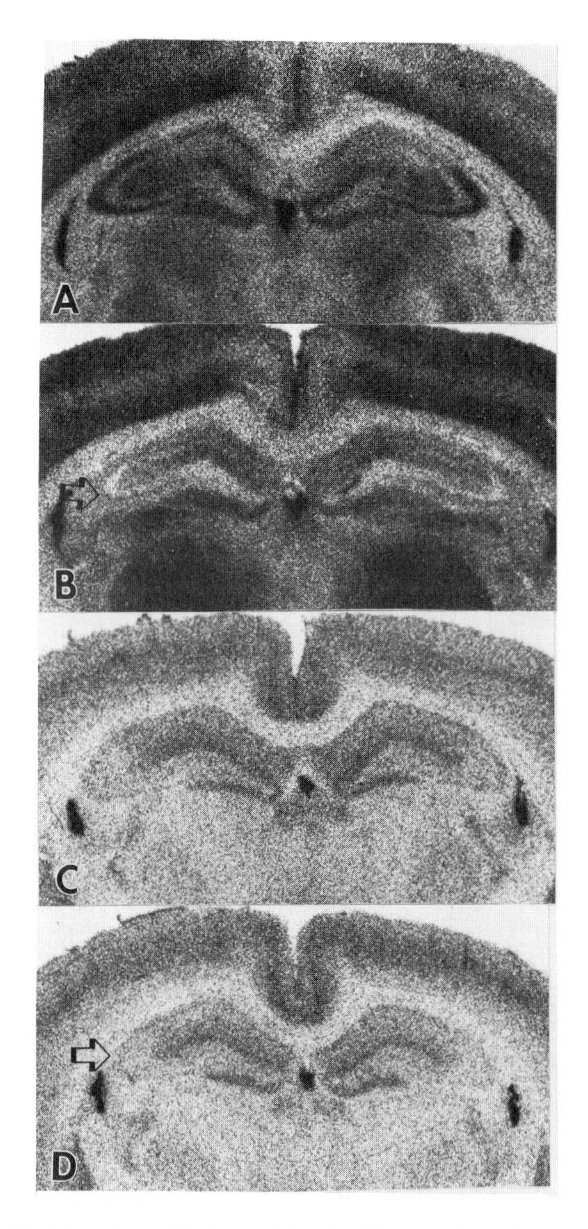

Figure 4. Effect of kainic acid on IGF-1 and insulin binding in the hippocampus. A IGF-1 binding by in vitro autoradiography. B IGF-1 binding in a kainic acid treated animal. C Insulin binding by in vitro autoradiography. D Insulin binding in a kainic acid treated animal. Arrows indicate the effect of kainic acid on binding in CA3.

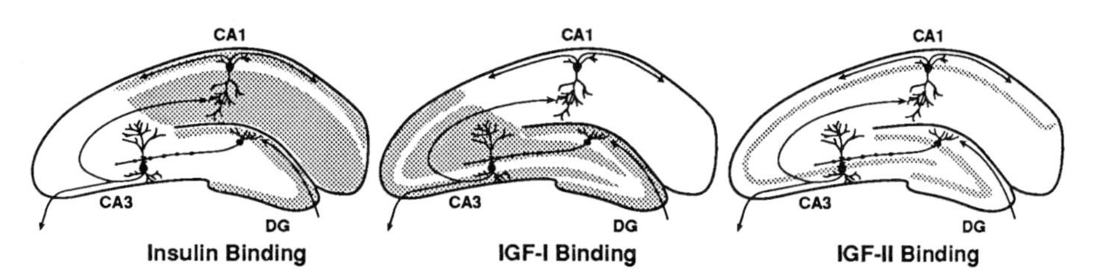

Figure 5. Localization of insulin, type 1 and type 2 IGF receptors in the hippocampus.

cells in the external plexiform layer and periglomerular cells in the glomerular layer (Fig. 1C). Heaviest concentrations of silver grains were found over mitral and tufted cells. Although film autoradiographs consistently showed dense signal for type 1 IGF receptor mRNA over the granule and glomerular layers, it was apparent from cellular localization of signal that this was due to a dense concentration of cell bodies with moderate associated mRNA.

A comparison of IGF-1 binding and mRNA allow us to describe the likely sites of type 1 receptors in the olfactory bulb (Fig. 2). Mitral and tufted cell bodies which express a high concentration of type 1 IGF receptor mRNA are the probable site of the prominent IGF-1 binding in the mitral cell and inner external plexiform layers. As these cells send dendrites that synapse with olfactory nerve terminals and interneurons in the glomerular layer, they may contribute to the high level of binding in the glomerular layer. As periglomerular cells, the interneurons of the glomerular layer also produce type 1 receptor mRNA, IGF-1 binding in this layer is also likely on their cell bodies and/or processes. Granule cell bodies are the probable source of IGF-1 binding in the granule cell layer and as these cells send dendrites to the external plexiform layer, may also contribute to binding in this layer.

Insulin binding in the olfactory bulb was also found in all layers but was particularly heavy in the external plexiform layer (Fig. 1D). Insulin receptor mRNA was found on the granule cell, glomerular and mitral cell layers (Fig 1E). We found that insulin receptor mRNA was expressed by granule cells in the granule and mitral cell layers and by periglomerular cells (not shown) but little or none by mitral and tufted cells (Fig. 1F).

The high number of insulin receptors in the external plexiform layer should be on granule cell dendrites because mitral/tufted cells, whose dendrites are the other major component of this layer, express little insulin receptor mRNA. Granule cell dendrites form reciprocal synapses with mitral cell dendrites and release GABA at these sites. We have prepared isolated synapses or synaptosomes from rat olfactory bulbs and confirmed that they contain a high concentration of insulin binding, compared to other brain regions. For example, when solubilized in Triton X100, insulin binding in olfactory bulb synaptosomes was 70.0±8.0 % total counts/mg protein (mean ±SEM for 4 separate experiments) compared to 35.0±2.1 % total counts/mg protein (n=4) for synaptosomes from cerebral cortex. The difference was significant, $p<0.02$. Other workers have found insulin-like immunoactivity in rat brain synaptosomes, which could be released by depolarization (42). This suggests that locally released insulin could modulate release of GABA at olfactory bulb synapses.

In the hippocampus, IGF-1 binding was found in the molecular layer of the dentate gyrus and the stratum oriens, stratum radiatum and stratum lacunosum-moleculare of Ammon's horn (Fig. 3A). Within Ammon's horn, IGF-1 binding was much more prominent in the CA3 region than in CA1. The mRNA for the type 1 receptor was found over the granule and pyramidal cell body layers of the dentate gyrus and Ammon's horn respectively (Fig. 3C). Commensurate with the binding data, the type 1 receptor mRNA was more abundant in CA3

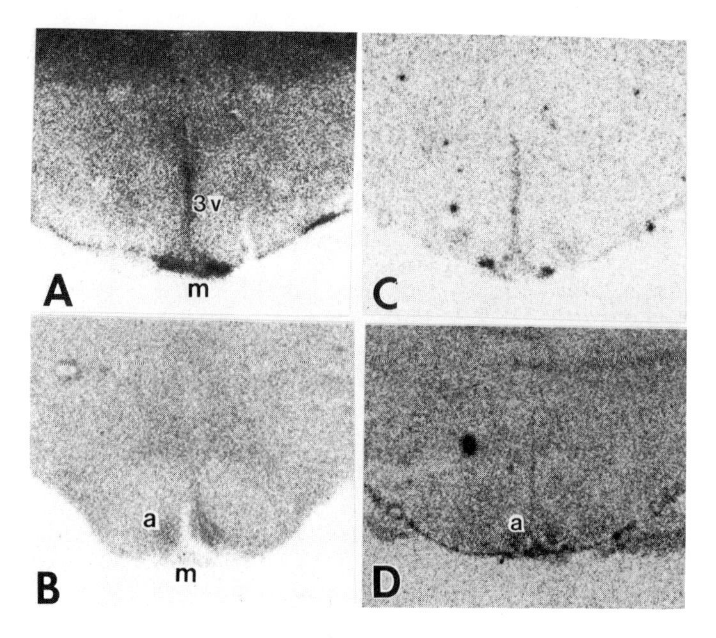

Figure 6. Insulin and type 1 IGF receptor binding and mRNA in the hypothalamus. A IGF-1 binding. B Insulin binding. C Distribution of type 1 IGF receptor mRNA. D Distribution of insulin receptor mRNA by in situ hybridization. Abbreviations: 3V, third ventricle; a, arcuate nucleus; m, median eminence.

than CA1. Cellular localization showed a large numbers of silver grains associated with both granule and pyramidal cells Figs. 7A,B).

Insulin binding was also concentrated in the molecular layer of the dentate gyrus, but in Ammon's horn binding was heavier in the strata of CA1 than those of CA3 (Fig. 3B). Insulin receptor mRNA was associated with the granule and pyramidal cells bodies, with more signal in CA1 than CA3 (Fig. 3E). The molecular layer of the dentate gyrus and the strata of Ammon's horn are hippocampal areas with most concentrated insulin and IGF-1 binding and also conform to the distribution of dendrites of dentate gyrus granule cells and Ammon's horn pyramidal cells (Fig. 5). Since insulin and type 1 IGF receptors are expressed in these cell bodies, the receptors are presumably transported to the dendrites and synaptic terminals of these neurones following their synthesis. By comparison, IGF-2 binding was confined to the cell body layers of Ammon's horn and the dentate gyrus (Fig. 3C).

When lesioned with kainic acid, the primary result was a dramatic reduction in IGF-1 and insulin binding in the CA3 stratum oriens and stratum radiatum (Fig. 4). With ischemic lesioning, insulin and IGF-1 binding were decreased in stratum oriens and stratum radiatum of CA1(43). Thus lesioning studies also show that insulin and type 1 IGF receptors in CA3 and CA1 are associated principally with pyramidal cells.

Insulin and type1 IGF receptor mRNAs were found in all other brain regions studied. Because these hormones reportedly affect hypothalamic function, we were interested in correlating the distribution of their receptor mRNAs with their binding. IGF-1 binding was found at low to moderate levels in many hypothalamic areas but was particularly dense in the median eminence (Fig. 6A). We found that the type 1 IGF receptor mRNA was also concentrated in this region (Fig. 6C). Silver grains were found over ependymal cells lining

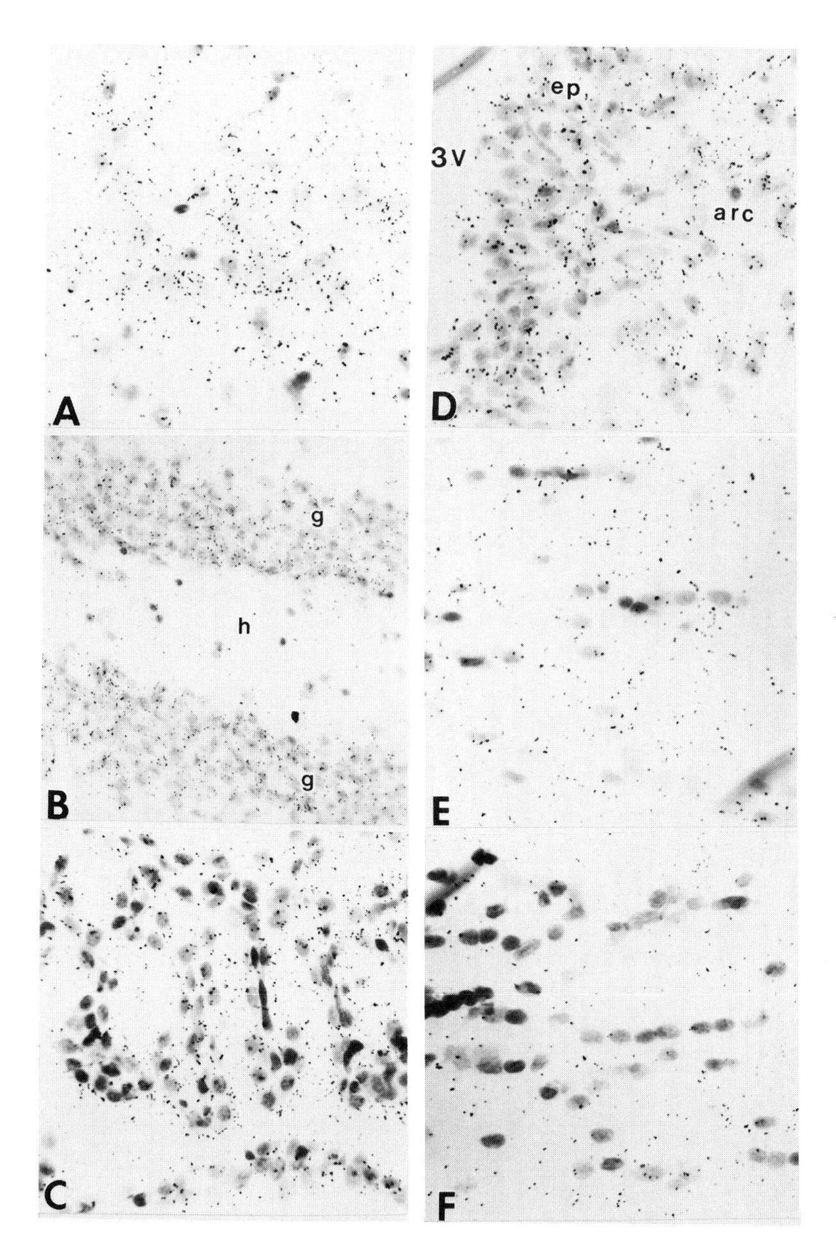

Figure 7. Cellular distribution of insulin and type 1 IGF receptor mRNAs in brain regions. Photomicrographs of type 1 IGF receptor mRNA in A CA3, B dentate gyrus, C the choroid plexus, E the corpus callosum and F in corpus callosum with labelled and 100 fold excess unlabelled probes. D Photomicrograph of insulin receptor mRNA in the arcuate nucleus. Abbreviations: 3V, third ventricle; arc, arcuate nucleus; ep, ependymal layer; g, granule cell layer; h, hilus.

the third ventricle and other cells of undefined nature (not shown). The presence of type 1 IGF receptors in high concentration in the median eminence suggests that this may be a site where circulating IGF-1 regulates the release of neuropeptides into pituitary portal blood.

Insulin binding in the hypothalamus was concentrated over the arcuate nucleus (Fig. 6B), a region of low IGF-1 binding. Insulin receptor mRNA was also concentrated on neuronal cell bodies in the arcuate nucleus (Figs. 6D, 7D), suggesting that considerable insulin binding in this region derives from insulin receptors on local neurons. Lesioning studies have suggested that a minority of insulin receptors in the arcuate nucleus are on noradrenergic terminals (44). These data suggest that insulin receptors in the the arcuate nucleus may be a site where pancreatic insulin modulates hypothalamic function. Circulating insulin might gain access to insulin receptors at this site via the adjacent median eminence, which lacks a blood-brain barrier or by uptake from the third ventricle. Two recently postulated mechanisms for an insulin effect on hypothalamic function are the modulation of IGF-2 (25) and NPY expression (45) in the arcuate nucleus. Hypothalamic NPY is a potent stimulant of food intake (46) and regulator of reproductive function (47). IGF-2 may be a physiological satiety factor (11).

Over areas containing white matter such as the corpus callosum, both insulin and IGF-1 binding and receptor mRNA were very low. Micrographs of white matter hybridized with type 1 IGF receptor probe showed that oligodendrocyte nuclei were associated with few or no silver grains, while similar results were obtained when hybridization was performed with excess unlabelled probe (Figs. 7E,F). We conclude that very little type 1 IGF receptor mRNA was expressed by oligodendrocytes. On the other hand, the choroid plexus was associated with very high levels of insulin and IGF-1 binding and receptor mRNA. Type 1 IGF and insulin (not shown) receptor mRNAs were associated with epithelial cells of the choroid plexus (Fig. 7C).

Conclusions

1) Insulin and type 1 IGF receptor mRNAs are expressed in many different neuronal types in the rat CNS. The receptor mRNAs were also expressed in ependymal cells and epithelial cells of the choroid plexus.

2) By comparing the distibution of binding sites and the neurons expressing the receptor mRNAs, it has been possible to localize the receptors for IGF-1 and insulin in some regions. Binding for each receptors was found on the cell body layers of neurons synthesizing the receptor mRNA. Insulin receptors were localized to the dendrites and postsynaptic terminals of olfactory bulb and hippocampal granule cells, Ammon's horn pyramidal cells. Type1 IGF receptors were localized to the dendrites of hippocampal granule cells, Ammon's horn pyramidal cells and probably to the dendrites of olfactory bulb mitral cells.

Acknowledgements

The authors wish to thank John Breininger for technical assistance and Agnes Koltai for secretarial assistance. Our research has been supported by Merit Review Research Programs of the Medical Research Service, Department of Veterans Affairs; by NIH grants AM17047 and NS24809 and by a Fellowship (J.L.M.) from the American Diabetes Association.

References

1 Szabo AJ and Szabo O. Insulin injected into CNS structures or into the carotid artery: effect on carbohydrate homeostasis of the intact animal. In: Szabo AJ (ed) CNS Regulation of Carbohydrate Metabolism, Advances in Metabolic Disorders vol 10, Academic Press, New York, p 385. 1983.

2 Debons AF, Krimsky I and From A. A direct effect of insulin on the hypothalamic satiety center. Am J Physiol. 219:938, 1970.
3 Vanderweele DA, Pi-Sumner IX, Novin D and Bush J. Chronic insulin infusion suppresses food ingestion and body weight gain in rats, Brain Res Bull. 5:suppl.1:7-11, 1980.
4 Woods SC, Lotter EC, McKay LD and Porte D Jr. Chronic intracerebroventricular infusion of insulin reduces food intake and body weight in baboons. Nature282:503, 1979.
5 McGowan MK, Andrews KM, Kelly J and Grossman SP. Effects of chronic intrahypothalamic infusion of insulin on food intake and diurnal meal patterning in the rat. Behav Neurosci. 104:371, 1990.
6 Oomura Y and Kita H. Insulin acting as a modulator of feeding through the hypothalamus. Diabetologia 20:290, 1981.
7 Palovcik R, Phillips MI, Kappy MS and Raizada MK. Insulin inhibits pyramidal neurons in hippocampal slices. Brain Res. 309:187, 1984.
8 Boyd FT, Clarke DW, Muther TF, et al. Insulin receptors and insulin modulation of norepinephrine uptake in neuronal cultures from rat brain. J Biol Chem. 260:15880, 1985.
9 Figlewicz DP, personal communication.
10 Araujo DM, Lapchak PA, Collier B, Chabot J-G and Quirion R. Insulin-like growth factor-1 (somatomedin-C) rteceptors in the rat brain:distribution and interaction with the hippocampal cholinergic system. Brain Res. 484:130, 1989.
11 Lauterio TJ, Marson L, Daughaday WH and Baile CA. Evidence for the role of insulin-like growth factor II(IGF II) in the control of food intake. Physiol Behav. 40:755, 1987.
12 Berelowitz M, Szabo M, Frohman LA, Firestone SL, Chu L and Hintz RL. Somatomedin-C mediates growth hormone negative feedback by effects on both the hypothalamus and pituitary. Science 212:1279, 1981.
13 Baskin DG, Porte D Jr, Guest K and Dorsa DM. Regional concentrations of insulin in the rat brain. Neuroendocrinology 112:898, 1983.
14 Dorn A, Bernstein H-G, Hahn H-J, Ziegler M and Rummelfanger H. Insulin immunohistochemistry of rodent CNS: apparent species differences but good correlation with radioimmunological data. Histochemistry 71:60, 1981.
15 Giddings SJ, Chirgwin J and Permutt MA. Evaluation of rat insulin messenger RNA in pancreatic and extrapancreatic tissues. Diabetologia 28:343, 1985.
16 Young WS. Periventricular hypothalamic cells in the rat brain contain insulin mRNA. Neuropeptides 8:93, 1986.
17 ClarkeDW, Mudd L, Boyd FT Jr, Fields M and Raizada MK. Insulin is released from rat brain neuronal cells in culture. J. Neurochem. 47:831, 1986.
18 Schwarz MW, Kahn SE, Taborsky GJ Jr, Bergman RN and Porte D Jr. Saturable transport of plasma insulin into the central nervous system (CNS). Soc Neurosci Abstracts 16:979, 1990.
19 Carlsson-Skwirut C, Jornvall H, Holmgren A et al. Isolation and characterization of variant IGF-1 as well as IGF-2 from adult human brain. FEBS Lett. 201:46, 1986.
20 Andersson IK, Edwall D, Norstedt G, Rozell B, Skottner A and Hansson HA. Differing expression of insulin-like growth factor I in the developing and in the adult rat cerebellum. Acta Physiol Scand. 132:167, 1988.
21 Werther GA, Abate M, Hogg A et al. Localization of insulin-like growth factor-1 mRNA in rat brain by in situ hybridization-relationship to IGF-1 receptors. Mol Endocrinology 4:773, 1990.
22 Rotwein P, Burgess SK, Milbrandt JD and Krause JE. Differential expression of insulin-like growth factor genes in rat central nervous system. Proc Natl Acad Sci USA. 85:265, 1988.
23 Hasselbacher GK, Schwab ME, Pasi A and Humbel RE. Insulin-like growth factor II(IGF II) in human brain: regional distribution of IGF II and of higher molecular mass forms. Proc Natl Acad Sci USA. 82:2153, 1985.
24 Stylianopoulou F, Herbert J, Soares MB and Efstratiadis A. Expression of the insulin-like growth factor II gene in the choroid plexus and leptomeninges of the rat central nervous system. Proc Natl Acad Sci USA.85:141, 1988.
25 Lauterio TJ, Aravich PF and Rotwein P. Divergent effects of insulin and insulin-like growth factor-II gene expression in the rat hypothalamus. Endocrinology 126:392, 1990

26 Havrankova J and Roth J. Insulin receptors are widely distributed in the central nervous system of the rat. Nature 272:827, 1978.
27 Marks JL, Maddison J and Eastman CJ. Subcellular localization of rat brain insulin binding sites. J Neurochem 50:774, 1988.
28 Gammeltoft S, Haselbacker GK,Humbel RE, Fehlmann M and Van Obberghen E. Two typers of receptors for insulin-like growth factors in mammalian brain. EMBO J. 4:3407, 1985.
29 Heidenreich KA, Zahniser NR, Berhanu P, Brandenburg D and Olefsky JM. Structural differences between insulin receptors in the brain and peripheral tissues. J Biol Chem. 258:8527, 1983.
30 McElduff A, Poronnik P and Baxter RC. The insulin-like growth factor-II (IGF II) receptor from rat brain is of lower apparent molecular weight than the IGF II receptor from rat liver. Endocrinology 121:1306, 1987.
31 Havrankova J, Roth J and Brownstein M. Concentrations of insulin and of insulin receptors in the brain are independent of peripheral insulin levels. J Clin Invest. 64:636, 1979.
32 Marks JL and Eastman CJ. Effect of starvation on insulin binding in rat brain. Neurosci. 30:551, 1989.
33 Marks JL and Eastman CJ. Ontogeny of insulin binding in different regions of rat brain. Developmental Neurosci. in press.
34 Figlewicz DP, Ikeda H, Hunt T, Stein R et al. Brain insulin binding is decreased in Wistar-Kyoto rats carrying the "fa" gene. Peptides 7:61, 1986.
35 Bondy CA, Werner H, Roberts CT Jr and LeRoith D. Cellular pattern of insulin-like growth factor-1 (IGF-I) and type I IGF receptor gene expression in early organogenesis: comparison with IGF II gene expression. Mol Endocrinology 4:1386, 1990.
36 Pomerance M, Gavaret J-M, Jacquemin C, Matricon C, Toru-Delbauffe D and Pierre M. Insulin and insulin-like growth factor 1 receptors during postnatal development of rat brain. Dev Brain Res. 42:77, 1988.
37 Bohannon NJ, Corp ES, Wilcox BJ, Figlewicz DP, Dorsa DM and Baskin DG. Characterization of insulin-like growth factor-1 receptors in the median eminence of the rat brain and their modulation by food restriction. Endocrinology 122:1940, 1988.
38 Bohannon NJ, Corp ES, Wilcox BJ, Figlewicz DP, Dorsa DM and Baskin DG. Localization of binding sites for insulin-like growth factor-1 (IGF-1) in the rat brain by quantitative autoradiography. Brain Res. 444:205, 1988.
39 Werther GA, Hogg A, Oldfield BJ, et al. Localization and characterization of insulin receptors in the rat brain and pituitary gland using in vitro autoradiography and computerized densitometry. Endocrinology 121:1562, 1987.
40 Lesniak MA, Hill JM, Kiess W, Rojeski M, Pert CB and Roth J. Receptors for Insulin-like growth factors I and II: autoradiographic localization in rat brain and comparison to receptors for insulin. Endocrinology 123:2089, 1988.
41 Marks JL, Porte D Jr, Stahl WL and Baskin DG. Localization of insulin receptor mRNA in rat brain by in situ hybridization. Endocrinology in press.
42 Wei L, Matsumoto H and Rhoads DE. Release of immunoreactive insulin from rat brain synaptosomes under depolarizing conditions. J Neurochem. 54:1661, 1990
43 King MG, Franck JE and Baskin DG. Localization of insulin and insulin-like growth factor (IGF) receptors in rat hippocampus following kainate and ischemic lesions. Soc Neurosci Abstracts 16:431, 1990.
44 Wilcox BJ, Matsumoto AM, Dorsa DM and Baskin DG. Reduction of insulin binding in the arcuate nucleus of the rat hypothalamus after 6-hydroxydopamine treatment. Brain Res. 500:149, 1989.
45 Williams G, Steel JH, Cardoso H et al. Increased hypothalamic neuropeptide Y concentrations in diabetic rat. Diabetes 37:763, 1988.
46 Stanley BG, Kyrkouli SE and Lampert S. Neuropeptide Y chronically injected into the hypothalamus: a powerful neurochemical inducer of hyperphagia and obesity. Peptides 7:1189 1986.
47 Kalra SP, Clark JT, Sahu A, Kalra PS and Crowley WR. Hypothalamic NPY: a local circuit in the control of reproduction and behaviour. In: Mutt V et al (eds) Neuropeptide Y. Raven Press, New York, p 229, 1989.

INSULIN-LIKE GROWTH FACTOR BINDING PROTEINS IN THE NERVOUS SYSTEM

Ian Ocrant

Brown University Department of Biology and Medicine
Rhode Island Hospital Department of Pediatrics
Providence, RI

INTRODUCTION

Insulin-like growth factor binding proteins (IGFBPs) are a group of discrete but structurally related polypeptides,[1-3] which bind and modulate the biological activity of insulin-like growth factors (IGFs) I and II.[4-9] Since available evidence suggests that IGF-I and -II participate in growth, metabolism, differentiation, and injury repair,[10-24] to name just a few biologic processes in which IGFs are implicated, IGFBPs may have important regulatory actions in these and other diverse biological processes.

At least 3 IGFBPs have been cloned, sequenced, and designated IGFBP-1, -2, and -3.[1-3] Available evidence suggests that there may be several more IGFBPs which, because of the rapid advancement of this field, may by cloned and sequenced by the time this review is published. IGFBP-1 is the major IGFBP in human amniotic fluid and has a molecular weight of 25,274 in the human.[1] The rat counterpart of IGFBP-1 has been tentatively identified in the H35 rat hepatoma cell line and migrates at M_r=30,000.[25] IGFBP-2 has a core protein molecular weight of 29,564 in the rat and 31,300 in the human, and is not glycosylated.[2] IGFBP-2 has been shown to be the predominant IGFBP of neonatal rat serum and rat amniotic fluid.[26] IGFBP-3 is a glycoprotein with a core polypeptide molecular weight of 28,500 in the human, and migrates at M_r=40,000 to 53,000, depending on glycosylation and the presence or absence of reducing agent.[3] IGFBP-3 is the major IGF carrier protein in adult human and rat serum and forms a complex with a non-binding, acid labile subunit yielding a circulating complex with M_r=150,000.[27]

IGFBPs appear to be secreted by many, if not most mammalian cells, and are found in virtually all physiologic fluids including plasma, cerebrospinal fluid (CSF), amniotic fluid, seminal plasma, lymph, milk, and follicular fluid.[25-35] Homologues for each IGFBP thus far sequenced have been found in a variety of mammalian systems indicating evolutionary conservation.[25,36] The quantity and type of IGFBP secreted appears to be tissue- or cell-specific,[25] and is regulated by a variety of factors including hormonal,[37] biochemical,[38] and ontogenic influences.[26] Structurally different IGFBPs also differ in their affinity for IGF-I and -II.[29,30] The diversity of the IGFBP family adds yet another layer of complexity to IGF physiology which is already characterized by complex genetic, hormonal, and developmental regulation of IGF-I and -II synthesis and secretion; homology and shared biochemical characteristics between insulin and the IGFs; homology between insulin receptors and the type 1 IGF receptors; and cross-reactivity of insulin, IGF-I and IGF-II for insulin and IGF receptors.[39]

IGF BINDING PROTEINS IN NERVOUS TISSUE

Cerebrospinal Fluid, Choroid Plexus and Brain

IGFBPs have been characterized in human,[28,29,31] rat,[26,40] and bovine CSF.[41] When studied as an aggregate, human CSF has preferential affinity for IGF-II, and one study reported that rat CSF has equal affinity for IGF-I and -II.[42] However, it is clear that CSF is a heterogenous mixture of IGFBPs consisting mainly of IGFBP-2,[28,40] with lesser quantities of other IGFBPs; these include a high molecular weight IGFBP which, in the rat, migrates in polyacrylamide gels identically with IGFBP-3.[40] One group has reported sequence data on a human CSF IGFBP with an apparent M_r=34,000 in unreduced polyacrylamide gels, which is different from IGFBP-2 or any other known IGFBP, and which has preferential affinity for IGF-II.[29,30] The relative abundance of this IGFBP compared to IGFBP-2 in human CSF is not known. The concentration of IGFBPs in rat CSF appears to be developmentally regulated, with much increased binding activity in neonatal rats, especially that corresponding to IGFBP-2.[40]

Recently, primary dispersed cultures of rat choroid plexus cells were found to secrete only one IGFBP which was immunoprecipitable with an antisera that recognizes rat IGFBP-2.[40] Therefore, the other IGFBPs found in CSF must come from other sources; possibly from astrocytes and neurons, or from the blood by transport across the blood-brain barrier. Since astrocytes and neurons are capable of secreting other IGFBPs including IGFBP-3 (see below),[4,43] many, if not all CSF IGFBPs are probably derived locally from brain tissue. Additionally, since the relative amounts of other IGFBPs in CSF is small compared with the amounts of IGFBP-2, available data suggest that the contribution from brain parenchyma is small despite its larger volume.

IGFBP synthesis in the brain has been studied by in situ hybridization histochemistry and Northern blotting. IGFBP-3 messenger RNA is below the level of detection by in situ hybridization in rat hypothalamus and cortex;[44] a finding which is difficult to reconcile with the results of studies in cultured neurons and astrocytes (see below). However, a large IGFBP-2 mRNA signal is present in choroid plexus.[42] No in situ hybridization studies using IGFBP-1 probes in central nervous system (CNS) tissues have been reported, however, immunohistochemical analysis disclosed no IGFBP-1 immunoreactivity in human fetal brain tissue.[45] Northern blot analysis demonstrated only vanishingly low levels of IGFBP-1 mRNA in CNS tissue.[46,47] On the other hand, Northern analysis easily detects IGFBP-2 mRNA in rat brain, hypothalamus, and pituitary.[48]

Cultured Astrocytes and Neurons

Cultured fetal rat neurons and neonatal rat astrocytes secrete IGFBPs into culture medium.[4,43] Fetal rat neurons secrete predominately an IGFBP corresponding to IGFBP-2 and lesser amounts of an unidentified low molecular weight IGFBP (M_r=23,000). Astrocytes secrete 2 IGFBPs corresponding to IGFBP-2 and -3, with IGFBP-3 predominating. Contrary to our findings,[43] one study reported that the high molecular weight IGFBP corresponding to IGFBP-3 in neonatal rat astrocyte cultures was not glycosylated.[4] The reason for this discrepancy is unclear, but it could be related to differences in handling and potency of the Endoglycosidase F used in the different studies. IGFBP-2 mRNA has been detected by Northern blot analysis in cultured fetal rat neuronal and astroglial cells.[48]

Retinal Pigment Epithelium

Virtually nothing is known about IGFBPs in neuroretinal tissues. Recently, we studied the production of IGFBPs by cultured bovine retinal pigment epithelium, a neuroectodermal derivative that is embryologically, morphologically, and functionally related to choroid plexus epithelium. We found that, like choroid plexus, these cells preferentially secrete IGFBP-2.[41] Additionally, we found that IGFBP-2 is the major IGFBP in bovine and rat vitreous, and its levels are increased in juvenile rat vitreous relative to that of adult rats.

Pituitary Gland

IGFBPs are produced by primary cultures of adult rat anterior and neurointermediate (NI) pituitary cells (see references in it for Binoux and Shiu and Paterson).[49-51] Anterior pituitary cells secrete IGFBPs with apparent M_r=35,000, 27,000, and 24,000 while NI pituitary cells secrete primarily the 27,000 variety.[49] While no direct comparison has been made between NI IGFBP and IGFBP-1, -2, or -3, NI IGFBP is similar to IGFBP-2 in size, lack of glycosylation, and relatively higher affinity for IGF-II.[49] Anterior pituitary IGFBP also had greater affinity for IGF-II than IGF-I.

Interestingly, the IGFBP produced by rat anterior pituitary, though readily detectable by affinity cross-linking, was difficult to detect by Western ligand blotting, a technique for detection of IGFBPs that is similar to Western immunoblotting, except that the former uses [^{125}I]IGF-I or -II as a reporter molecule rather than immunoglobulin.[52] Detection was enhanced by Endoglycosidase F treatment of the rat anterior pituitary IGFBP, suggesting that the glycosylation state of anterior pituitary IGFBP affects the binding of the IGFBP to the nitrocellulose filter, or that the binding site on the glycoprotein form is somehow more susceptible to denaturation than on the deglycosylated protein. Western ligand blotting should not, therefore, be relied upon as the sole method for screening conditioned media, physiological fluids, or tissue extracts for IGFBPs.

Recently, IGFBP-2 mRNA was detected by in situ hybridization in fetal rat infundibulum.[53] This embryonic tissue is the progenitor anlage for the posterior pituitary gland.

Neural Cell Lines

Three neuroectodermal cell lines have been reported to secrete IGFBPs. B104 cells are derived from an artificially induced rat central nervous system neuroblastoma, which secrete a partially characterized IGFBP that migrates at M_r=23,000 to 27,000 which is lower than that IGFBP-1, -2, or -3.[43,54] C6 cells, a rat astroglial cell line secretes primarily IGFBP-3, and trace quantities of smaller molecular weight IGFBPs.[25,40] A human glioblastoma cell line designated 1690, was recently reported to secrete IGFBP-3 in sufficient quantities to markedly affect the partitioning of IGF-I between cell surface binding sites (cell surface IGFBP and IGF receptors) and assay buffer (containing free IGF-I and IGFBP-bound IGF-I).[55]

Biological Actions

Like insulin, IGF-I, IGF-II, and their receptors, IGFBPs are found in nervous tissue, primary dispersed cultures of nervous tissue, and in cell lines derived from nervous tissue (see above). Available evidence suggests that IGFs participate in growth and differentiation of neurons and astrocytes,[10-17] synapse formation,[18] repair processes,[19,20] modulation of satiety,[21-23] and feedback regulation of pituitary somatotropin secretion.[24] However, little is known about the mechanisms by which IGFBPs may modulate these effects. Recently it was determined that IGFBPs may be stimulatory (to IGF action) in some experimental systems[7-9] and inhibitory in others[4-6] depending on the parameters of the specific model system. Only one study has thus far reported experimental results bearing on the biological actions of IGFBPs in nervous tissue.[4] This study suggested that the IGFBP purified from BRL-3A cells, a rat liver cell line, which has been designated IGFBP-2,[36] reduced IGF-I stimulated [^{3}H]thymidine incorporation in cultured rat astroglial cells. However, in interpreting these results, one must consider the results of a recent study by Conover[56] which indicated that the biological activity of IGFBPs in vitro depended on whether cells are exposed to IGFBP and IGF-I at the same time, or if cells are pre-incubated with IGFBP prior to exposure to IGF-I. In this study, IGFBP-3 potentiated IGF-I stimulated-aminoisobutyric acid uptake in bovine fibroblasts preincubated with IGFBP-3, but was IGFBP-3 was inhibitory if added to cells simultaneously with IGF-I. Caution is, therefore, indicated when extrapolating from in vitro studies to in vivo physiology.

An additional conundrum concerns variant IGF-I, which is found in human brain. This variant lacks three amino terminal amino acids and is reported to be the predominant form of IGF-I in the CNS.[57] Variant IGF-I reportedly lacks the ability to bind to IGFBP-1 and -2.[58,59] Whether or not this is true for other types of IGFBPs is unknown, as is the biological significance of desamino-(1-3)-IGF-I since IGF-II mRNA and peptide are much more abundant in CNS and fetal tissues.[60-63]

To summarize, data on the role of IGFBPs in nervous tissue are virtually nonexistent. Therefore, the assumption that IGFBPs participate in growth, differentiation, repair, and differentiated functions of nervous tissue is based upon extrapolation of the known effects of the IGFs in several experimental systems, and on the ability of nervous tissue to produce IGFBPs which can, in turn, bind IGFs.

CONCLUSIONS AND FUTURE DIRECTIONS

IGFBP-2 and -3 are secreted by neuroectodermal tissues, but IGFBP-1 is secreted in vanishingly low quantities, if at all. IGFBP-3 appears to be preferentially secreted by glial cells; IGFBP-2 is secreted by neuroepithelial and neuronal cells, and is the predominant IGFBP in CSF and vitreous. Other species of IGFBP secreted by neuroectodermal cells are likely to be quantified and characterized in the future.

IGFBP secretion is developmentally regulated in the CNS. Biochemical and hormonal regulatory influences of CNS IGFBP secretion remain to be explored.

Virtually nothing is known about the biological actions of IGFBPs in nervous tissue. Advances in this area are will require careful in vivo and in vitro neurobiologic studies with highly purified IGFBPs, once these become available. Further studies will also be required to understand the relationship of variant IGF-I and IGFBPs in CNS physiology.

REFERENCES

1. M.T. Brewer, G.L. Stetler, C.H. Squires, R.C. Thompson, W.H. Busby, and D.R. Clemmons, Cloning, characterization, and expression of a human insulin-like growth factor binding protein, Biochem Biophys Res Commun 152:1289 (1988).
2. J.B. Margot, C. Binhert, J.-L. Mary, J. Landwehr, G. Heinreich, and J. Schwander, A low molecular weight insulin-like growth factor binding protein from the rat: cDNA cloning and tissue distribution of its messenger RNA, Mol Endocrinol 3:1053 (1989).
3. S. Shimisaki, A. Koba, M. Mercado, M. Shimonaka, and N. Ling, Complementary DNA structure of the high molecular weight rat insulin-like growth factor binding protein (IGF-BP3) and tissue distribution of its mRNA, Biochem Biophys Res Commun 165:907 (1989).
4. V.K.M. Han, J.M. Lauder, and A.J. D'Ercole, Rat astroglial somatomedin/insulin-like growth factor binding proteins: characterization and evidence of biologic function, J Neurosci 8:3135 (1988).
5. A.G. Frauman, S. Tsuzaki, and A.C. Moses, The binding characteristics and biological effects in FRTL5 cells of placental protein-12, an insulin-like growth factor-binding protein purified from human amniotic fluid, Endocrinology 124:2289 (1989).
6. W.M.Burch, J. Correa, J.E. Shively, and D.R. Powell, The 25-kiloDalton insulin-like growth factor (IGF)-binding protein inhibits both basal

and IGF-I-mediated growth of chick embryo pelvic cartilage in vitro, J Clin Endocrinol Metab 70:173 (1990).

7. R.S. Bar, B.A. Booth, M. Boes, and B.L. Dake, Insulin-like growth factor-binding proteins from vascular endothelial cells: purification, characterization, and intrinsic biological activities, Endocrinology 125:1910 (1989).

8. W.H. Busby, P. Hossenlopp, M. Binoux, and D.R. Clemmons, Purified preparations of the amniotic fluid-derived insulin-like growth factor-binding protein contain multimeric forms that are biologically active, Endocrinology 125:773 (1989).

9. W.F. Blum, E.W. Jenne, F. Reppin, K. Kietzmann, M.B. Ranke, and J.R. Bierich Insulin-like growth factor I (IGF-I)-binding protein complex is a better mitogen than free IGF-I, Endocrinology 125:766 (1989).

10. E. Recio-Pinto, F.F. Lang, and D.N. Ishii, Insulin and insulin-like growth factor II permit nerve growth factor binding and the neurite formation response in cultured human neuroblastoma cells, Proc Natl Acad Sci USA 81:2562 (1984).

11. E. Recio-Pinto, M.M. Rechler, and D.N. Ishii, Effects of insulin, insulin-like growth factor-II, and nerve growth factor on neurite formation and survival in cultured sympathetic and sensory neurons, J Neurosci 6:1211 (1986).

12. Y. Aizenman and J. deVellis, Brain neurons develop in a serum and glial free environment: effects of transferrin, insulin, insulin-like growth factor I and thyroid hormone on neuronal survival, growth and differentiation, Brain Res 406:32 (1987).

13. J. Shemer, M.K. Raizada, B.A. Masters, A. Ota, and D. LeRoith, Insulin-like growth factor I receptors in neuronal and glial cells. Characterization and biological effects in primary culture, J Biol Chem 262:7693 (1987).

14. S.K. Burgess, S. Jacobs, P. Cuatrecasas, and N. Sahyoun, Characterization of a neuronal subtype of insulin-like growth factor I receptor, J Biol Chem 262:1618 (1987).

15. R. Ballotti, F.G. Nielsen, N. Pringle, A. Kowalski, W.D. Richardson, E. Van Obberghen, and S. Gammeltoft, Insulin-like growth factor I in cultured rat astrocytes: expression of the gene, and receptor tyrosine kinase, EMBO J 6:3633 (1987).

16. E. DiCicco-Bloom and I.B. Black, Insulin-like growth factors regulate the mitotic cycle of cultured rat sympathetic neuroblasts, Proc Natl Acad Sci USA 85:4066 (1988).

17. F.A. McMorris and M. Dubois-Dalcq, Insulin-like growth factor I promotes Cell Proliferation and oligodendroglial commitment in rat glial

progenitor cells developing in vitro, J Neurosci Res 21:199 (1988).

18. D.N. Ishii, Relationship of insulin-like growth factor II gene expression in muscle synaptogenesis, Proc Natl Acad Sci USA 86:2898 (1989).
19. H.A. Hansson, B. Rozell, and A. Skottner, Rapid axoplasmic transport of insulin-like growth factor I in the sciatic nerve of adult rats, Cell Tiss Res 247:241 (1987).
20. J. Sjoberg and M. Kanje, Insulin-like growth factor (IGF-1) as a stimulator of regeneration in the freeze-injured rat sciatic nerve, Brain Res 485:102 (1989).
21. D.P. Figlewicz, S.C. Woods, D.G. Baskin, D.M. Dorsa, B.J. Wilcox, L.J. Stein, and D. Porte Jr., Insulin in the central nervous system: a regulator of appetite and body weight, in: "Insulin, Insulin-like Growth Factors, and Their Receptors in the Central Nervous System," M. K. Raizada, M.I. Phillips, D. LeRoith, eds., Plenum Press, New York (1987).
22. G.S. Tannenbaum, H.J. Guyda, and B.I. Posner, Insulin-like growth factors: a role in growth hormone negative feedback and body weight regulation via brain, Science 220:77 (1983).
23. T.J. Lauterio TJ, L. Marson, W.H. Daughaday, and C.A. Baile, Evidence for the role of insulin-like growth factor II (IGF-II) in the control of food intake, Physiol Behav 40:755 (1987).
24. R.G. Rosenfeld and A.R. Hoffman, Insulin-like growth factors and their receptors in the pituitary and hypothalamus, in: "Insulin, Insulin-like Growth Factors, and Their Receptors in the Central Nervous System," M.K. Raizada, M.I. Phillips, and D. LeRoith, eds., Plenum Press, New York (1987).
25. Y.W.-H. Yang, A.L. Brown, C.C. Orlowski, D.E. Graham, L.Y.-H. Tseng, J.A. Romanus, and M.M. Rechler, Identification of rat cell lines that preferentially express insulin-like growth factor binding proteins rIGFBP-1,2, or 3, Mol Endocrinol 4:29 (1990).
26. S.M. Donovan, Y. Oh, H. Pham, and R.G. Rosenfeld, Ontogeny of serum insulin-like growth factor binding proteins in the rat, Endocrinology 125:2621 (1989).
27. Baxter R.C. Characterization of the acid-labile subunit of the growth hormone-dependent insulin-like growth factor binding protein complex, J Clin Endocrinol Metab 67:265 (1988).
28. J.A. Romanus, L.Y.-H. Tseng, Y.W.-H. Yang, and M.M. Rechler, The 34 kiloDalton insulin-like growth factor binding protein in human cerebrospinal fluid and the A673 rhabdomyosarcoma cell line are human

homologues of the rat BRL-3A binding protein, Biochem Biophys Res Commun 163:875 (1989).

29. P. Hossenlopp, D. Seurin, B. Segovia-Quinson, and M. Binoux, Identification of an insulin-like growth factor-binding protein in human cerebrospinal fluid with a selective affinity for IGF-II, FEBS Lett 208:439 (1986).

30. M. Roghani, P. Hossenlopp, P. Lepage, A. Ballard, and M. Binoux, Isolation from human cerebrospinal fluid of a new insulin-like growth factor-binding protein with a selective affinity for IGF-II, FEBS Lett 255:253 (1989).

31. R.G. Rosenfeld, H. Pham, C.A. Conover, R.L. Hintz, and R.C. Baxter, Structural and immunological comparison of insulin-like growth factor binding proteins of cerebrospinal fluid and amniotic fluid, J Clin Endocrinol Metab 68:638 (1989).

32. R.G. Rosenfeld, H. Pham, Y. Oh, G. Lamson, and L.C. Guidice, Identification of insulin-like growth factor-binding protein-2 (IGF-BP-2) and a low molecular weight IGF-BP in human seminal plasma, J Clin Endocrinol Metab 70:551 (1989).

33. M. Binoux and P. Hossenlopp, Insulin-like growth factor (IGF) and IGF-binding proteins: comparison of human serum and lymph, J Clin Endocrinol Metab 67:509 (1988).

34. A.N. Corps, K.D. Brown, L.H. Rees, J. Carr, and C.G. Prosser, The insulin-like growth factor I content in human milk increases between early and full gestation, J Clin Endocrinol Metab 67:25 (1988).

35. M. Seppala, T. Wahlstrom, A.I. Koskimies, A. Tenhunen, E.-M. Rutanen, R. Koistinen, I. Huhtaniemi, H. Bohn, and U.-H. Stenman, Human preovulatory follicular fluid, luteinized cells of hyperstimulated preovulatory follicles, and corpus luteum contain placental protein 12, J Clin Endocrinol Metab 58:505 (1984).

36. S.L.S. Drop, On the nomenclature of the insulin-like growth factor binding proteins, Mol Cell Endocrinol 67:243 (1989).

37. C.A. Conover, Regulation of insulin-like growth factor (IGF)-binding protein synthesis by insulin and IGF-I in cultured bovine fibroblasts, Endocrinology 126:3139 (1990).

38. M.S. Lewitt and R.C. Baxter, Inhibitors of glucose uptake stimulate the production of insulin-like growth factor-binding protein (IGFBP-1) by human fetal liver, Endocrinology 126:1527 (1990).

39. Reviewed in: "Molecular and cellular biology of insulin-like growth factors and their receptors," D. LeRoith and M.K. Raizada, eds., Plenum Press, New York (1990).

40. I. Ocrant, C.T. Fay, and J.T. Parmelee, Characterization of insulin-like growth factor binding proteins produced in the rat central nervous system, Endocrinology 127:1260 (1990).
41. I. Ocrant, C.T. Fay, and J.T. Parmelee, Expression of insulin and insulin-like growth factors and binding proteins by retinal pigment epithelium, Exp Eye Res, in press.
42. L.Y.-H. Tseng, A.L. Brown, Y.W.-H. Yang, J. A. Romanus, C.C. Orlowski, T. Taylor, and M.M. Rechler, The fetal rat binding protein for insulin-like growth factors is expressed in the choroid plexus and cerebrospinal fluid of adult rats, Mol Endocrinol 3:1559 (1989).
43. I. Ocrant, H. Pham, Y. Oh, and R.G. Rosenfeld, Characterization of insulin-like growth factor binding proteins of cultured rat astroglial and neuronal cells, Biochem Biophys Res Commun 159:1316 (1989).
44. S. Shimasaki, A. Kobu, M. Mercado, M. Shimonaka, and N. Ling, Complementary DNA structure of the high molecular weight rat insulin-like growth factor binding protein (IGF-BP3) and tissue distribution of its mRNA, Biochem Biophys Res Commun 165:907 (1989).
45. D.J. Hill, D.R. Clemmons, S. Wilson, V.K.M. Han, A.J. Strain, and R.D.G. Milner, Immunological distribution of one form of insulin-like growth factor (IGF)-binding protein and IGF peptides in human fetal tissues, J Mol Endocrinol 2:31 (1989).
46. L.J. Murphy, C. Seneviratne, G. Ballejo, F. Croze, and T.G. Kennedy, Identification and characterization of a rat decidual insulin-like growth factor-binding protein complementary DNA, Mol Endocrinol 4:329 (1990).
47. G.T. Ooi, C.C. Orlowski, A.L. Brown, R.E. Becker, T.G. Unterman, and M.M. Rechler, Differential tissue distribution and hormone regulation of messenger RNAs encoding rat insulin-like growth factor-binding proteins-1 and -2, Mol Endocrinol 4:321.
48. G. Lamson, H. Pham, Y. Oh, I. Ocrant, J Schwander, and R.G. Rosenfeld, Expression of the BRL-3A insulin-like growth factor binding protein (rBP-30) in the rat central nervous system, Endocrinology 123:1100 (1989).
49. R.G. Rosenfeld, H. Pham, Y. Oh, and I. Ocrant, Characterization of insulin-like growth factor-binding proteins in cultured rat pituitary cells, Endocrinology 124:2867 (1989).
50. M. Binoux, P. Hossenlopp, C. Lassarre, and N. Hardouin, Production of insulin-like growth factors and their carrier by rat pituitary gland and brain explants in primary culture, FEBS Lett 124:178 (1981).

51. R.P.C. Shiu and J.A. Patterson, Characterization of insulin-like growth factor II peptides secreted by explants of neonatal brain and adult pituitary from rats, Endocrinology 123:1456 (1988).

52. P. Hossenlopp, D. Seurin, B. Segovia-Quinson, S. Hardouin, and M. Binoux, Analysis of serum insulin-like growth factor binding proteins using Western blotting: use of the method for titration of the binding proteins and competitive binding studies, Anal Biochem 154:138 (1986).

53. T.L. Wood, A.L. Brown, M.M. Rechler, and J.E. Pintar, The expression pattern of an insulin-like growth factor (IGF)-binding protein gene is distinct from IGF-II in the midgestational rat embryo, Mol Endocrinol 4:1257 (1990).

54. M.A. Sturm, C.A. Conover, H. Pham, and R.G. Rosenfeld, Insulin-like growth factor receptors and binding protein in rat neuroblastoma cells, Endocrinology 124:388 (1989).

55. R.H. McCusker, C. Camacho-Hubner, M.L. Bayne, M.A. Cascieri, and D.R. Clemmons, Insulin-like growth factor (IGF) binding to human fibroblast and glioblastoma cells: the modulating effect of cell released IGF binding proteins (IGFBPs), J Cell Physiol 144:244 (1990).

56. C.A. Conover, Biological actions of insulin-like growth factor binding protein-3 in cultured bovine fibroblasts, Endocrine Society Abstracts 72:186A (1990).

57. V.R. Sara, C. Carlsson-Skwirut, C. Andersson, E. Hall, B. Sjogren, A. Holmgren, and H. Jornvall, Characterization of somatomedins from human fetal brain: identification of a variant form of insulin-like growth factor I, Proc Natl Acad Sci USA 83:4904 (1986).

58. C. Carlsson-Skwirut, M. Lake, M. Hartmanis, K. Hall, and V.R. Sara, A comparison of the biological activity of the recombinant intact and truncated insulin-like growth factor 1 (IGF-1), Biochim Biophys Acta 1011:192 (1989).

59. C.J. Bagley, B.L. May, L. Szabo, P.J. McNamara, M. Ross, G.L. Francis, F.J. Ballard, and J.C. Wallace, A key functional role for the insulin-like growth factor 1 N-terminal pentapeptide, Biochem J 259:665 (1989).

60. F. Beck, N.J. Samani, S. Byrne, K. Morgan, R. Gebhard, and W.J. Brammar, Histochemical localization of IGF-I and IGF-II mRNA in the rat between birth and adulthood, Development 104:29 (1988).

61. P. Rotwein, S.K. Burgess, J.D. Milbrandt, and J.E. Krause, Differential expression of insulin-like growth factor genes in rat central nervous system, Proc Natl Acad Sci USA 85:265 (1988).

62. L.J. Murphy, G.I. Bell, and H.G. Friesen, Tissue distribution of insulin-like growth factor I and II messenger ribonucleic acid in the adult rat, Endocrinology 120:1279 (1987).

63. G.K. Haselbacher, M.E. Schwab, A. Pasi, and R.E. Humbel, Insulin-like growth factor II (IGF II) in human brain: regional distribution of IGF II and of higher molecular mass forms, Proc Natl Acad Sci USA 82:2153 (1985).

BINDING OF [^{125}I]-INSULIN-LIKE GROWTH FACTOR-1 (IGF-1) IN BRAINS OF ALZHEIMER'S AND ALCOHOLIC PATIENTS

Fulton T. Crews, R. McElhaney, G. Freund, W.E. Ballinger, Don W. Walker, Bruce E. Hunter, and Mohan K. Raizada

Departments of Pharmacology and Therapeutics (F.T.C., R.M.), Medicine (G.F.), Pathology (W.E.B.), Neuroscience (D.W.W., B.E.H.), and Physiology (M.K.R.), University of Florida College of Medicine, Gainesville, FL 32610

ABSTRACT

Patients with chronic alcoholism and/or Alzheimer's disease suffer from degenerative changes in the cerebral cortex and hippocampus. To investigate possible changes in IGF-1 receptor binding sites in brain tissue of patients with these pathological conditions, the binding of [^{125}I]-IGF-1 was determined in tissues obtained from control, Alzheimer's and/or patients with a history of alcoholism. The four experimental groups examined consisted of patients from similar age groups. Specific binding of [^{125}I]-IGF-1 to cerebral cortical membranes from Alzheimer's patients had significantly more binding sites than age-matched controls, alcoholic patients and alcoholic patients with Alzheimer's disease. Regression analyses indicated that there were no significant differences in [^{125}I]-IGF-1 binding in cerebral cortex with regard to age of patients (1.1% of total variance with a range of 52 to 92 years). Likewise, the time interval between death and autopsy contributed only 1.4% to the total variance in IGF-1 binding. No statistical differences in [^{125}I]-IGF-1 binding were noted in hippocampal tissue from the various patient groups. Thus, human IGF-1 binding sites in cerebral cortex and hippocampus appear to be relatively stable for a number of variables. The increase in cerebral cortical [^{125}I]-IGF-1 binding sites could be due to upregulation of IGF-1 receptors resulting from a decrease in IGF-1 levels in Alzheimer's patients.

INTRODUCTION

Insulin-like growth factor-1 (IGF-1) is a neurotrophic peptide hormone that stimulates glial proliferation and differentiation, and neuronal growth (1,3,4).

Molecular Biology and Physiology of Insulin and Insulin-Like Growth Factors
Edited by M.K. Raizada and D. LeRoith, Plenum Press, New York, 1991

IGF-1 binds to specific receptors and stimulates intrinsic tyrosine kinase activity in mammalian and human brains (1-6). Activation of tyrosine kinase is proposed to be the first step in the action of this hormone. IGF-1 appears to have important actions on neuronal survival, glial differentiation, maintenance of the blood-brain barrier and other processes that may be affected in Alzheimer's disease and chronic alcoholism. Various hypotheses on the mechanisms of Alzheimer's disease overlap with the possible disruption of IGF-1 actions in the CNS (7,8). Furthermore, chronic alcohol abuse is known to cause cerebral cortical and hippocampal pathology (9,10). IGF-1 binding sites in postmortem samples from the cerebral cortex and hippocampus of patients with Alzheimer's disease and/or a history of chronic alcohol consumption were investigated. [^{125}I]-IGF-1 bound to human brain membranes with high affinity similar to that reported for nonhuman mammalian brain. We report here that membranes prepared from cerebral cortex (Brodman area #4), but not hippocampus, of Alzheimer's patients have more IGF-1 binding sites than those from age-matched controls, alcoholic patients or alcoholic patients with Alzheimer's disease.

MATERIALS AND METHODS

Tissue Collection

The methods for tissue collection have been reported in detail previously (9-11). In brief, samples of frontal cortex and hippocampus were obtained at autopsy from patients over 18 years of age with absence of coma and localized brain disease such as neoplasia and stroke. Blocks of tissue were frozen and adjacent sections fixed for histological diagnosis of Alzheimer's disease. Detailed microscopic examinations of each brain region were performed to document presence, location and degree of lipofuscin accumulation, neurofibrillary changes, senile plaque formation, granulovacuolar degeneration, neuronal death and gliosis. Histopathological and gross pathological changes including distribution and location of cortical atrophy, ventricular dilatation and the extent of atherosclerotic disease in the vasculature of the circle of Willis were recorded. In addition to pathological findings, extensive computer data files on samples included clinical and postmortem diagnoses, cause of death, death-to-autopsy time interval, records of all medications administered in the 10 days preceding death, and alcohol consumption histories.

Alcohol consumption histories were obtained from hospital admission charts including medical histories, social workers' database interviews and detailed questionnaires on alcohol use from relatives or close friends of the deceased. Patients were classified as described previously (9-11), with alcohol abuse defined as daily consumption of more than 80 grams of absolute alcohol/day for more than 10 years.

Membrane Preparation

Approximately 100 mg each of human frontal cerebral cortex and hippocampus were thawed following storage at -70°C. Wet weights were obtained and the tissues homogenized with a Polytron homogenizer in 20 volumes (wt/vol) of 50 mM TRIS-HCl containing 1 mM phenylmethylsulfonyl fluoride (PMSF) and 210 mM leupeptin. The homogenates were centrifuged at 40,000xg for 15 minutes at 4°C, supernatants were discarded and the pellets were resuspended in incubation buffer consisting of 50 mM TRIS-HCl with 1 mg/ml bacitracin. Protein concentrations were determined using the method of Lowry et al. (12), and membranes were frozen at -70°C. On the day of the experiment, membranes were thawed and diluted with the incubation buffer to yield a protein concentration of 1 mg/ml.

Membranes from the brains of adult rats were also prepared essentially as described above.

Measurement of [^{125}I]-IGF-1 Binding

Binding of [^{125}I]-IGF-1 to membranes prepared from various brain areas was carried out as described previously (13) with a few modifications. [^{125}I]-IGF-1 binding was determined in 250 μl of binding buffer (50 mM TRIS-HCl, 1 mg/ml bacitracin and 10% [wt/vol] insulin-free BSA) containing 25 pM [^{125}I]-IGF-1. Binding was initiated by addition of 50 μg of membrane protein. Triplicate determinations of total binding were made for each tissue sample, while nonspecific binding (duplicate determinations) was determined by the addition of 3 nM unlabeled IGF-1. Typically, specific binding was 80% of total binding. Samples were incubated for 90 minutes at 25°C and the reaction was stopped by high-speed centrifugation at 4°C in a Beckman microfuge. The supernatant was aspirated, the pellet washed with an additional 1 ml of binding buffer and the samples were centrifuged again. Following aspiration of the final supernatant, the tips of the microfuge tubes containing the pellets were cut off, placed in 12×75 mm tubes and counted in a gamma counter with approximately 78% counting efficiency. Specific binding was determined by subtracting nonspecific from total DPMs bound.

To minimize nonspecific binding of [^{125}I]-IGF-1 to incubation tubes, incubations were conducted in microfuge tubes precoated using 3% BSA in 50 mM TRIS-HCl. The BSA solution was pipetted into tubes, allowed to sit for 30 minutes at 25°C, aspirated and the tubes allowed to air-dry prior to use.

RESULTS

Rat brains were used initially to characterize the binding of [^{125}I]-IGF-1 and to determine the distribution of IGF-1 receptors in the brain. Table 1 shows that the highest binding was present in the olfactory bulb, cortex and hippocampus.

In view of the significance for memory and other related functions, combined with the presence of relatively high levels of [^{125}I]-IGF-1 binding, further characterization of receptors was carried out in frontal cortex and hippocampus from human brain. Figure 1 shows a competition-inhibition experiment of [^{125}I]-IGF-1 binding to membranes prepared from cortex.

Table 1. [^{125}I]-IGF-1 Binding to Membranes from Rat Brain.

REGION	[^{125}I]-IGF-1 BOUND (fmoles/mg protein) ± SEM
OLFACTORY BULB	10.3 ± 0.8
CORTEX	6.9 ± 0.3
HIPPOCAMPUS	5.1 ± 0.2
BRAIN STEM	4.8 ± 1.0
CEREBELLUM	3.1 ± 0.2
HYPOTHALAMUS	4.4 ± 0.5

Binding of [^{125}I]-IGF-1 in membranes prepared from various brain regions of adult rat was determined essentially as described in the methods for human brain. Data are presented as specific binding (difference between the binding in the absence and presence of 100 ng/ml unlabeled IGF-1).

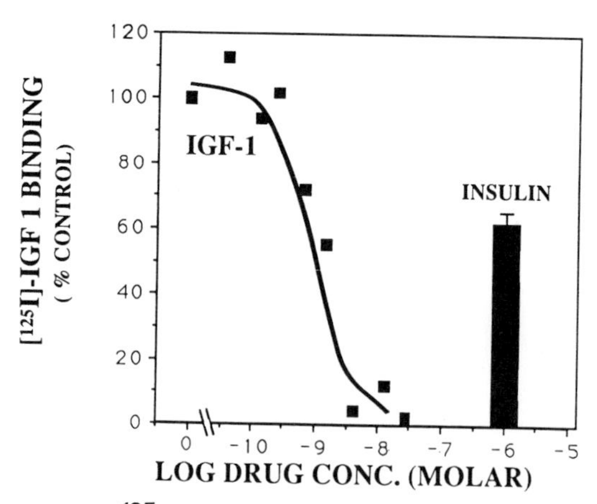

Figure 1. Inhibition of [^{125}I]-IGF-1 binding by IGF-1 and insulin in normal human cerebral cortical membranes.
Data shown represent the mean of triplicate binding determinations using 250 pM [^{125}I]-IGF-1 and 200 μg of membrane protein. Specific binding was approximately 45-50% of total binding, which was 26,000 dpm/200 μg protein.

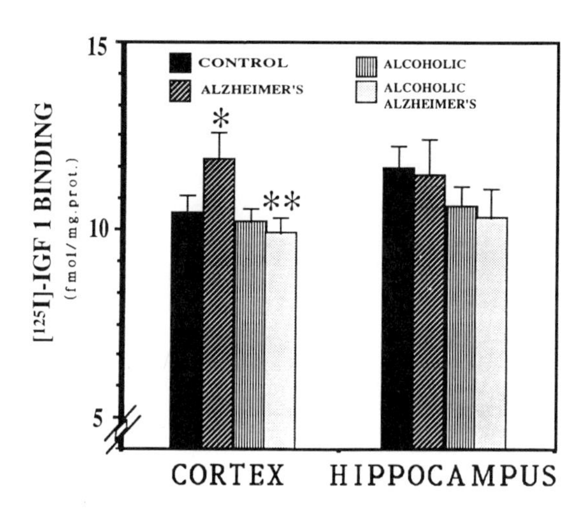

Figure 2. Specific binding of [^{125}I]-IGF-1 to membranes of control, Alzheimer's, alcoholic, and alcoholic Alzheimer's patients.

Binding of [^{125}I]-IGF-1 was determined using homogenates of frozen sections of human pre-central gyrus, frontal cortex Brodman area #4 or hippocampus as described in the methods. Shown is the specific binding of [^{125}I]-IGF-1. The age of patients at autopsy and the time interval between death and autopsy did not contribute to the differences shown. The numbers of patients in the groups were: control (20), Alzheimer's (10), alcoholic (20) and alcoholic Alzheimer's (5). No statistical differences in patient groups were noted in hippocampal tissue from the various treatment groups. Regression analyses for age revealed that age contributed only 1.1% (cerebral cortex) or 3.3% (hippocampus) of the total variance of patients ranging from 58 to 92 years of age. There were no differences in the effect of age in alcoholics compared to controls. No differences in time interval to autopsy were noted. Regression analysis suggested that time interval contributed only 1.4% (cerebral cortex) or 1.9% (hippocampus) of the variance with post-mortem time intervals up to 36 hours. Significant differences of cerebral cortical membranes were: Alzheimer's vs. controls ($P < 0.05$), and alcoholic Alzheimer's vs. Alzheimer's (**$P < 0.05$).*

Unlabeled IGF-1 competed with $[^{125}I]$-IGF-1 for binding sites in a dose-dependent manner with an IC_{50} of 1 nM. Insulin had 1/1000th the affinity of IGF-1 for competition for $[^{125}I]$-IGF-1 binding.

Cortical membranes were prepared from controls, alcoholics, Alzheimer's patients and Alzheimer's patients who were alcoholics to quantitate their ability to bind $[^{125}I]$-IGF-1. Figure 2 shows that membranes from Alzheimer's patients bound 14% more $[^{125}I]$-IGF-1 compared with controls or the other two groups. This increase in binding was significant ($P<0.05$) and could not be accounted for by age, time interval to autopsy, medications or any other variables that were analyzed using regression analyses.

Hippocampal membranes were prepared from the same four groups of patients to determine whether an increase in the binding of $[^{125}I]$-IGF-1 could also be seen in that brain region or whether it was specific for cortex. Hippocampal membranes contained slightly higher numbers of IGF-1 receptor sites than did cerebral cortical membranes. In contrast to the finding in cerebral cortex, there was no significant change in $[^{125}I]$-IGF-1 binding in hippocampal membranes prepared from Alzheimer's patients (Figure 2). Alcoholic and alcoholic Alzheimer's patients tended to have slightly lower IGF-1 binding than controls, but this decrease was not statistically significant. Similar detailed analyses of other potential variables again suggested that age, time interval from death to autopsy, cause of death, and CNS-active drugs had little or no significant effect on $[^{125}I]$-IGF-1 receptor binding.

Several medications were found to have no significant effects on $[^{125}I]$-IGF-1 binding in either cerebral cortical or hippocampal membranes. Patients who received opiates (n=25), anticonvulsants (n=2), amitriptyline (n=1), phenothiazine (n=3), or barbiturates (n=1) showed no changes in hippocampal or cortical IGF-1 binding outside a 2-standard-deviation range. These findings suggest that a variety of medications known to affect the central nervous system do not significantly affect IGF-1 receptor binding in either cerebral cortex or hippocampus.

Since ethanol is known to cause hippocampal degeneration in a chronic rat model of alcoholism, rats were tested for changes in IGF-1 receptors after 5 months of treatment. Similar to the human alcoholic patient data, no remarkable changes in IGF-1 receptors were found (Table 2).

DISCUSSION

IGF-1 is known to have important actions on neuronal and glial growth and differentiation in primary cultures from rat brain (1,2,13-15). In addition, recently it has been hypothesized that IGF-1 may play an important role in modulation of intercellular communication among the various cell types of the brain (16). The findings presented here demonstrate that specific high-affinity binding sites for IGF-1 are present in human brain and substantiate the findings of others (6). Consistent with studies performed with rodent brain, the human IGF-1 receptor binding sites are slightly greater in the hippocampus than in the

Table 2. [^{125}I]-IGF-1 Binding (specific fMol bound/mg protein)

TISSUE	SUCROSE (AVG ± SEM)	ETHANOL (AVG ± SEM)
CORTEX	6.84 ± 0.33	7.51 ± 0.42
HIPPOCAMPUS	5.11 ± 0.26	5.86 ± 0.37
CEREBELLUM	3.13 ± 0.19	3.18 ± 0.27
ADRENAL GLAND	1.88 ± 0.21	1.70 ± 0.16

Binding of [^{125}I]-IGF-1 was determined using membranes of rat tissues as described in the methods. Rats were treated with ethanol or sucrose liquid diets for 5 months as described previously (26).

cerebral cortex. The IGF-1 receptor has intrinsic tyrosine kinase activity, has a high affinity for IGF-1, and is likely to be involved in brain function. The finding that the IGF-1 receptor binding site is resistant to changes in different time intervals from death to autopsy and age is consistent with other types of receptors, including the muscarinic cholinergic receptor and the benzodiazepine receptor (9-11). The hippocampus is an area with extensive pathology in Alzheimer's disease and in alcoholic patients. However, no significant change in IGF-1 receptor sites in hippocampal membranes from Alzheimer's or alcoholic patients was found. Freund and Ballinger (10) have previously shown that muscarinic cholinergic receptors are decreased 30% in the hippocampus and 40% in the cerebral cortex of alcoholic patients. Furthermore, benzodiazepine receptors are decreased 30% in the hippocampus and 25% in the cerebral cortex of alcoholic patients compared to matched controls (9). In the hippocampus of patients with Alzheimer's disease, there are decreases in glutamate receptor sites, particularly the NMDA receptor subtype (23). Thus, there are receptor changes occurring in the hippocampus of patients with Alzheimer's disease and alcoholism where no changes in IGF-1 binding sites were found.

The increased IGF-1 binding seen in the cerebral cortex of Alzheimer's patients is quite interesting, since Brodman area #4 of cerebral cortex, unlike the hippocampus, is not an area normally associated with marked pathology in Alzheimer's disease. IGF-1 has been shown to have a number of effects on cholinergic neuronal function, including stimulation of acetylcholine release from rat brain slices (17), and increased choline acetyltransferase activity in rat neuronal cultures (18). Furthermore, IGF-1 has been shown to stimulate neuronal and glial growth and differentiation (14,15,19). The increase in IGF-1 receptors in the cerebral cortex of Alzheimer's patients may represent a compensatory mechanism for neuronal dysfunction and/or that cells responsive to IGF-1 grow and thrive, whereas receptor-negative cells are lost, resulting in a relative increase in IGF-1 receptor density. Studies of nerve growth factor receptors have suggested that receptor loss in the basal forebrain of Alzheimer's patients is correlated with the loss of the cholinergic neurons, which have nerve-growth-factor receptors (20-22). Decreases in serotonergic, glutaminergic and nicotinic-cholinergic receptors have been reported in the cerebral cortex of Alzheimer's patients (23). Similar to the hippocampus, no change in muscarinic M1 receptors occurs in Alzheimer's patients' cerebral cortex. Our finding of IGF-1 receptor increases represents the first example of an increase in receptors in Alzheimer's patients' cerebral cortex and may be particularly significant due to the known actions of IGF-1 on cellular growth and differentiation.

Although IGF-1 in the central nervous system is generally thought to be independent of peripheral IGF-1, both Alzheimer's and alcoholic patients do have changes in circulating levels of IGF-1. Alzheimer's patients have higher circulating IGF-1 levels than age-matched controls (6,24). In contrast, chronic ethanol has been reported to decrease IGF-1 levels in plasma (25). A complete understanding of the interaction of IGF-1 receptors in Alzheimer's disease and in alcoholic and/or Alzheimer's patients will require further experimentation. The findings presented clearly indicate that IGF-1 receptor binding studies are feasible in human brain, and that the receptor binding site itself is stable in postmortem studies and not markedly affected by a number of medications.

REFERENCES

1. Adamo, M., Raizada, M.K. and LeRoith, D. (1989) Insulin and insulin-like growth factor receptors in the nervous system. *Mol. Neurobiol.* 3:71-100.
2. Masters, B.A., Shemer, J., LeRoith, D. and Raizada, M.K. (1989) Insulin-like growth factor receptors in the central nervous system: Phosphorylation events and cellular mediators of biological function. In: *Molecular and Cellular Biology of IGFs and Their Receptors*, D. LeRoith and M.K. Raizada, eds., Plenum Press, New York, pp. 341-358.
3. Giacobini, M.M., Olson, L., Hoffer, B.J. and Sara, V.R. (1990) Truncated IGF-1 exerts trophic effects on fetal brain tissue grafts. *Exp. Neurol.* 108(1):33-37.
4. Recio, P.E. and Ishii, D.N. (1988) Insulin and insulin-like growth factor receptors regulating neurite formation in cultured human neuroblastoma cells. *J. Neurosci. Res.* 19(3):312-320.
5. Gammeltoft, S., Haselbacher, G.K., Humbel, R.E., Fehlmann, M. and Van, O.E. (1985) Two types of receptor for insulin-like growth factors in mammalian brain. *Embo. J.* 4(13A):3407-3412.
6. Sara, V.R., Hall, K., Von, H.H., Humbel, R., Sjogren, B. and Wetterbergn, L. (1982) Evidence for the presence of specific receptors for insulin-like growth factors 1 (IGF-1) and 2 (IGF-2) and insulin throughout the adult human brain. *Neurosci. Lett.* 34(1):39-44.
7. Thienhaus, O.J., Hartford, J.H., Skelly, M.F. and Bosmann, H.B. (1985) Biologic markers in Alzheimer's disease. *J. Am. Geriatr. Soc.* 33(10):715-726.

8. Reubi, J.C. and Palacios, J. (1986) Somatostatin and Alzheimer's disease: A hypothesis. *J. Neurol.* 233(6):370-372.
9. Freund, G. and Ballinger, W.E., Jr. (1988) Decrease of benzodiazepine receptors in frontal cortex of alcoholics. *Alcohol* 5:275-282.
10. Freund, G. and Ballinger, W.E., Jr. (1989) Loss of muscarinic cholinergic receptors from temporal cortex of alcohol abusers. *Metabolic Brain Disease* 4:121-141.
11. Freund, G. and Ballinger, W.E., Jr. (1988) Loss of cholinergic muscarinic receptors in the frontal cortex of alcohol abusers. *Alcoholism: Clinic. Exp. Research* 12:630-638.
12. Lowry, O.H., Rosebrough, N.J., Fan, A.J. and Randall, R.J. (1951) Protein measurement with the folin phenol reagent. *J. Biol. Chem.* 193:265-275.
13. Shemer, J., Raizada, M.K., Masters, B.A., Ota, A. and LeRoith, D. (1987) Insulin-like growth factor 1 receptors in neuronal and glial cells: Characterization and biological effects in primary culture. *J. Biol. Chem.* 262(16):7693-7699.
14. Masters, B.A., Werner, H., Roberts, C.T., Jr., LeRoith, D. and Raizada, M.K. (1991) Developmental regulation of insulin-like growth factor 1 stimulated glucose transporter in rat brain astrocytes. *Endocrinology* (in press).
15. Masters, B.A., Werner, H., Roberts, C.T., Jr., LeRoith, D. and Raizada, M.K. (1991) Insulin-like growth factor 1 receptors and IGF-1 actions in oligodendrocytes from rat brain. *Regulatory Peptides* (in press).
16. Raizada, M.K. (1991) Insulin-like growth factor 1 in the brain: A possible modulator of intercellular communication. In: *Molecular Biology and Physiology of Insulin and Insulin-like Growth Factors*, M.K. Raizada and D. LeRoith, eds., Plenum Press, New York (in press).
17. Nilsson, L., Sara, V.R., and Nordberg, A. (1988) Insulin-like growth factor 1 stimulates the release of acetylcholine from rat cortical slices. *Neurosci. Lett.* 88:221-226.
18. Knusel, B., Michel, P.P., Schwaber, J.S., and Hefti, F. (1991) Selective and non-selective stimulation of central cholinergic and dopaminergic development *in vitro* by nerve growth factor, basic fibroblast growth factor, epidermal growth factor, insulin and the insulin-like growth factors I and II. *Dev. Neurosci.* (in press).
19. LeRoith, D., Shemer, J., Adamo, M., Raizada, M.K., Heffez, D. and Zick, Y. (1989) Insulin and IGF-I stimulate phosphorylation of their respective receptors in intact neuronal and glial cells in primary culture. *J. Mol. Neurosci.* 1(1):3-8.
20. Hefti, F., and Mash, D. (1989) Localization of nerve growth factor receptors in the normal human brain and in Alzheimer's disease. *Neurobiol. of Aging* 10:75-87.
21. Higgins, G., and Mufson, E.J. (1989) NGF receptor gene expression is decreased in the nucleus basalis in Alzheimer's disease. *Exp. Neurology* 106:222-236.
22. Mufson, E.J., Bothwell, M., and Kordower, J.H. (1989) Loss of nerve growth factor receptor-containing neurons in Alzheimer's disease: A quantitative analysis across subregions of the basal forebrain. *Exp. Neurology* 105:221-232.

23. Whitehouse, P.J. (1987) Neurotransmitter receptor alterations in Alzheimer's disease: A review. *Alzheimer Dis. Assoc. Disord.* 1(1):9-18.
24. Sara, V.R., Hall, K., Enzell, K., Gardner, A., Morawski, R. and Wetterberg, L. (1982) Somatomedins in aging and dementia disorders of the Alzheimer type. *Neurobiol. of Aging* 3:117-120.
25. Sonntag, W.E. and Boyd, R.L. (1988) Chronic ethanol feeding inhibits plasma levels of insulin-like growth factor-1. *Life Sciences* 43:1325-1330.
26. Gonzales, R.A., Ganz, N. and Crews, F.T. (1987) Variations in membrane sensitivity of brain region synaptosomes to the effects of ethanol *in-vitro* and chronic *in-vivo* treatment. *J. Neurochem.* 49:158-162.

INSULIN-LIKE GROWTH FACTOR I: A POSSIBLE MODULATOR OF INTERCELLULAR COMMUNICATION IN THE BRAIN

Mohan K. Raizada

Department of Physiology, Box J-274, JHMHC, University of Florida, College of Medicine, Gainesville, FL 32610.

INTRODUCTION

Insulin-like growth factor I (IGF-I) is produced by the liver in response to growth hormone (GH), and plays a key mediator in its physiological responses. In addition, IGF-I has also been implicated in a wide variety of trophic and metabolic actions in the peripheral tissues. In recent years it has become evident that the liver, although the major site, is not the sole site of IGF-I synthesis and secretion. In fact, a number of peripheral tissues as well as the central nervous system (CNS) has been documented to locally synthesize this hormone. This has led to the proposal that locally produced IGF-I may exert physiological effects in an autocrine and paracrine fashion.

The primary objective of this review is to discuss IGF-I, its receptors and actions in the CNS, and to provide evidence in support of the hypothesis that IGF-I may be an important chemical messenger in intra- and intercellular communication in the brain. In addition, we will also review the possible role of IGF-I and their receptors in developmentally related events in the brain. The peripheral actions of IGF-I, its receptors, and IGF-I gene expression are not within the scope of this review and readers are advised to consult elsewhere (1-6).

A physiologically relevant autocrine/paracrine hormonal system should possess the following properties: (i) the tissue should be able to synthesize or should have access to the hormone. In addition, the levels of the hormone should be influenced by various physiological, pathophysiological, and nutritional states; (ii) the tissue should have specific receptors for the hormone and their distribution should be consistent with the physiological actions of this hormone; (iii) there should exist a mechanism by which physiological and pathophysiological changes are reflected and consistent with levels of the available hormone to activate its receptors. The brain IGF-I system fulfills many of these criteria and will be discussed below.

IGF-I in the CNS

Immunoreactive IGF-I has been shown in the brain extract and in the CSF (8-9), although its concentrations are significantly lower than in the liver and plasma. This raises the question if IGF-I in the CNS is produced there or is transported from the blood through areas where the blood brain barrier (BBB) is not tight (i.e. circumventricular organs). Sara and her associates have demonstrated that post-translational processing of IGF-I prohormone results in two peptides, a truncated form of IGF-I and a tripeptide (8, 10-11). Their experiments suggest that the truncated IGF-I is the major gene product in the CNS and a few other tissues including bovine colostrum, porcine uterus, and human platelets (12-14). Specific IGF-I cDNA probes have been used to demonstrate the presence of specific IGF-I transcripts in the brain supporting the contention that the IGF-I found in the brain is synthesized there (15-17). In adult rat brain, IGF-I mRNA has been shown to be present in olfactory bulb, hippocampus, and cerebellum (15, 18-19). There appears to be a developmentally related change in the levels of IGF-I mRNA. IGF-I mRNA transcripts are far more abundant in the brains of fetuses compared with brains from adults animals (9). In fetal brain, high levels of IGF-I mRNA are observed in olfactory bulb, hypothalamus, and cortex, whereas it is relatively low and wide spread in thalamus, striatum, and tectum (9, 18-22). In olfactory bulb the hybridization with IGF-I probe is restricted to the glomerular and mitral cells. In the hippocampus, hybridization was seen in the pyramidal cells of Ammon's horne in CA1 and CA2 layers, and dentate gyrus, whereas in the cerebellum it was localized on granule cell layers (9, 18-22). These studies indicate relatively widespread IGF-I gene expression in the brain whose expression may be altered during development and differentiation.

IGF-I gene expression has been studied in neuronal and astroglial cells in primary culture to determine which cell types in the brain are responsible for IGF-I production. It appears that relatively similar levels of IGF-I mRNA transcripts are present in both neuronal and glial cells prepared from the brain (15,23). Class C transcripts account for most of IGF-I mRNA in neuronal and glial cells which is consistent with its predominance in all other tissues (24). Consistent with this is the observation that Class C transcripts contribute the majority of IGF-I mRNA in brains of 2-day-old rats (24). These studies raise an interesting question concerning the origin of IGF-I. It could be argued that astroglial cells may be the predominant cell type in the brain responsible for IGF-I production. The existence of IGF-I mRNA in neuronal cultures could be explained by the contamination of 15-20% astrocytic glial cells in neuronal cultures (25-26).

Physiological and biochemical studies indicate that IGF-I gene expression is highly regulated. In the peripheral system the major stimulator of IGF-I gene expression and translation is the GH. However, the influence of GH on the brain IGF-I production is not clearly understood. GH does not directly stimulate IGF-I production in neuronal and glial cells. However, there is evidence of its action on the brain. Mathews, et.al. (26) showed that the brain IGF-I levels are significantly lower in GH deficient mouse. In addition, hypophysectomy-induced GH deficiency results in a decrease in IGF-I mRNA in the brain which can be induced by ICV injection of GH (27). It is proposed that changes in the brain IGF-I levels by GH are mediated by local negative feedback systems rather than changes in the serum IGF-I mediated feedback. Berelowitz (28) proposed that locally synthesized IGF-I could act within the hypothalamus to regulate somatostatin and growth hormone-releasing hormone (GHRH), thus inducting the negative feedback loop. This hypothesis is further supported by the fact that centrally administered IGF-I inhibits GH secretion, and IGF-I stimulates secretion of somatostatin *in vitro* (28).

Regulation of IGF-I mRNA by glucocorticoids has been demonstrated in both astroglial and neuronal cells (23). IGF-I acts as a mitogen in these cells via an autocrine/paracrine mechanism. It is reasonable to postulate that glucocorticoids inhibit actions on neuronal cell proliferation, and brain growth may be mediated via its inhibition of IGF-I synthesis and secretion.

IGF-I Receptors in the CNS

The second component of IGF-I action is the expression of IGF-I specific receptors. Binding and other pharmacological studies have established that the CNS contains specific IGF-I receptors (9, 29-35). Binding of [^{125}I]-IGF-I to membranes prepared from various brain regions exhibit a high affinity for IGF-I and relatively lower affinities for IGF-II and insulin. Affinity crosslinking and other structural studies have shown that IGF-I receptor in the brain, like its counterpart in the peripheral system, is a hetero-tetramer consisting of two identical α-subunits and two identical β-subunits (9,29). The α-subunit is extracellular and contains IGF-I binding sites, whereas the β-subunit is transmembrane protein and is a protein tyrosine kinase. In contrast to these similarities, IGF-I receptor in the brain is significantly smaller than its counterpart from the periphery (29, 30, 31). This difference in molecular weight is primarily due to a result of differences in the glycosylation of both α- and β-subunits (29-35). This suggests that the processing of this receptor in the CNS may be different compared with its peripheral counterpart.

Neuronal, astroglial, and oligodendrocytes in primary culture have been used to determine cell type specificity of IGF-I receptor and IGF-I action in the CNS. Binding studies indicate that IGF-I receptor is found in all three cell types. Both IGF-I binding α-subunit and β-subunit expressing protein tyrosine kinase activity constitute the receptor. Although kinetic and structural similarities exist in IGF-I receptors from neurons, astrocytes, and oligodendrocytes, it appears that functionally these receptors may induce distinct physiological events. IGF-I receptor stimulation is associated with proliferation and differentiation of progenitor cells into mature oligodendrocytes (36-38). Induction of enzymes associated with oligodendrocytes such as 3-cyclic nucleotide, 3-phosphodiesterase, sulfate incorporation into sulfolipids, and glycerol-3-phosphate dehydrogenase are also stimulated by IGF-I (38-41). Expression of specific antigenic markers including the myelin basic protein demonstrating the maturation and differentiation of oligodendrocytes are induced by IGF-I (42). Recently, a transgenic mouse model has been developed to determine the significance of IGF-I on oligodendrocyte function under *in vivo* conditions (43). These studies show that mouse which carried IGF-I transgene expressed 2-fold higher levels of IGF-I in the brain, which is associated with an increase in brain weight and myelin content compared with their nontransgene mouse (43). This suggests that the production of more myelin per oligodendrocyte in transgenic mouse is probably associated with increased IGF-I levels in the brain.

IGF-I receptors on astrocytes have been associated with the metabolic and proliferative actions of IGF-I in the brain. Evidence for such an action of IGF-I are based on studies indicating that IGF-I causes stimulation of sugar uptake and expression of Glut-1 gene both at the levels of transcription and translation in neonatal rat brain (44-45). In addition, stimulation of thymidine incorporation by IGF-I in neonatal astrocytes is also shown (30).

IGF-I appears to be neurotrophic in neurons maintained in cell culture. Figure 1 shows the effect of IGF-I on neuronal cells prepared from 1-day-old

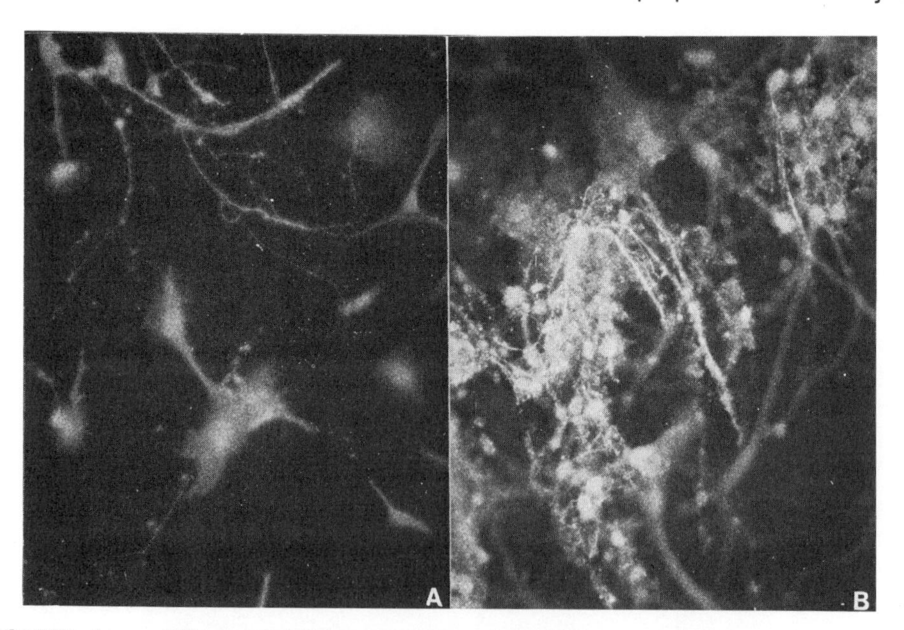

FIGURE 1. <u>**Effect of IGF-I on Neuronal Cells in Primary Culture**</u>.
Cells were dissociated from one-day-old rat brains and plated on to poly-L-lysine precoated tissue culture dishes. After removal of ARC, cultures were incubated in DMEM (A) or DMEM with 100 nM IGF-I (B) for 3 days. They were fixed and stained with antineurofilament antibody essentially as described previously (25-26).

rat brains. Clearly, IGF-I causes a significant increase in the development of neuronal processes, extensive neurite outgrowth, and the number of neuronal cell bodies. In addition, the degree of neurite branching and interconnections are significantly induced by IGF-I. Trophic actions of IGF-I on neurons have been demonstrated by its ability to stimulate DNA and RNA synthesis (30,46) in whole brain neuronal cultures, and in cholinergic and dopaminergic neurons (47-50). These observations suggest that in primary cultures, IGF-I not only stimulates multiplication of neuronal precursor cells, but also the expression of certain differentiated functions of neurons including neurofilament specific proteins, acetylcholine activity, and dopaminergic system. It will be interesting to determine if IGF-I exerts such neurotrophic effects in the CNS *in vivo*.

In situ hybridization techniques have been used in recent years to demonstrate the distribution of IGF-I receptor mRNA and to correlate this with the distribution of IGF-I receptors. High concentrations of IGF-I receptor mRNA transcripts are localized in olfactory bulb, hippocampus, and hypothalamic areas of the brain, although its distribution is widespread at low levels throughout the brain. The distribution of the receptor mRNA is similar with the distribution of IGF-I specific receptors as evidenced by autoradiography (51). In addition, binding sites and mRNA for IGF-I are also observed in ependymal and epithelial cells of the choroid plexues (51). Such an unique distribution of IGF-I receptors and its mRNA hypothesizes its role in behaviors and memory modification. However, these views are highly speculative at the present time.

IGF Binding Proteins (IGFBPs) in the CNS

It is well established that in the periphery IGFs actions are not only regulated by the levels of IGF and its receptors, but also the synthesis and secretion of a set of IGFBPs. At least 3 BPs (BP-1, BP-2, and BP-3) have been cloned and their physiological role in IGF-I actions in normal and pathological states are currently being deduced (52-54). However, the role of IGFBPs in control of IGF actions in the brain is not as clear and is the subject of intense investigation at the present time. For example, how many BPs are found in the brain, which cell types (neurons, glia, epithelial cells) synthesize them, and how do they participate in the control of IGF-I actions in the brain?

Cerebrospinal fluid (CSF) of both humans and animals have been shown to contain IGFBPs (55-59). IGFBP-2 is the predominant BP, although low concentrations of IGFBP-3 have also been seen (56,59). This raises questions as to which cells in the CNS synthesize and secrete these BPs. Dispersed cultures of choroid plexus, primary cultures of neurons and glial cells in conjunction with *in situ* hybridization have been used in an attempt to answer this. It appears that both fetal neurons and astrocytes predominantly secrete IGFBP-2 with astrocytes also secreting BP-3, indicating that CSF IGFBPs originate from both neuronal and glial cells (60-62). These observations are not consistent with the *in situ* hybridization results. For example, IGFBP-2 mRNA is easily seen by Northern blot analysis in the brain (62), but its predominance is seen in choroid plexus (59). Also, observations of the presence of IGFBP-3 as observed in CSF are not corroborated by *in situ* hybridization and Northern blot analyses (54). Thus, it appears further experiments with cells in culture, as well as in whole brains will be required to clear these inconsistencies.

Attempts to understand the role of IGFBPs in brain cell functions have been limited. A study by Han, et.al. showed that IGFBP-2 exerts inhibitory effects on IGF-I stimulated thymidine incorporation in astrocytes (60). It appears that the presence of IGFBP-2 attenuated IGF-I's stimulatory response on astroglial DNA synthesis. Along the same lines our group has investigated IGF-I receptors and IGF-I actions in astrocytes from neonatal and 21-day-old rat brains (44-45). These studies show that IGF-I stimulates glucose uptake and Glut-1 gene expression in neonate astrocytes. Such an effect of IGF-I is completely attenuated in astrocytes from older rats, suggesting that a developmental control of IGF-I system (IGF-I, IGF-I receptors, and IGF BPs) may be involved in the lack of IGF-I's stimulatory effect. This could not be due to a lack of IGF-I receptors or their coupling with the intrinsic tyrosine kinase activity (9,45). Figure 2 shows that lack of response of IGF-I on glucose uptake could be attenuated by preincubating the IGF-I nonresponsive astrocytes from 21-day-old rat brains with IGFBP-2 antibody. This indicates that astrocytes

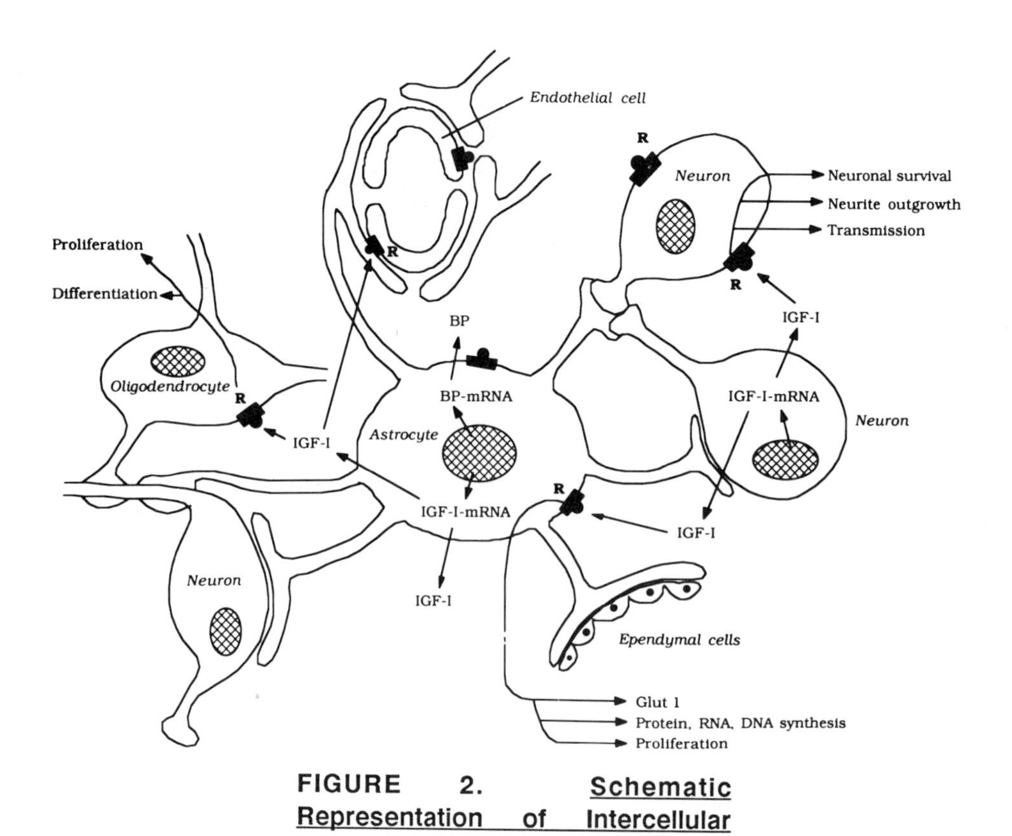

FIGURE 2. Schematic Representation of Intercellular Interactions Among Various Brain Cells.

may undergo an increase in the expression of IGFBP-2 during brain development. It is an attractive hypothesis suggesting a regulatory role of IGFBPs in brain development and growth.

THE HYPOTHESIS

I would like to propose that IGF-I plays a key modulator in the intercellular communication among various cell types in the brain, and that astrocytic glial cells play a central role in such communication. There are anatomical and biochemical evidence to support this contention. Figure 3 is a diagramatic representation of proposed anatomical interactions among the brain

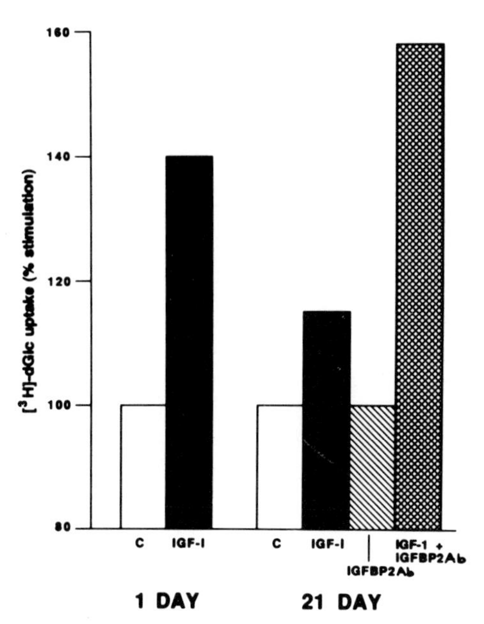

FIGURE 3. **Effect of IGFBP-2 Antibody on IGF-I Stimulation of [^{3}H] 2-deoxy-D-Glucose Uptake in Astroglial Cells From 1-day and 21-day-old Rat Brains.**

Astrocytes from 21-day-old rat brains were preincubated with IGFBP-2 antiserum for 30 minutes at 37°C. They were rinsed and incubated with 100 nM IGF-I for 30 minutes at 37°C. Their ability to take up [^{3}H]-dGlc uptake was determined essentially as described previously (44). IGF-I stimulates dGlc uptake in astrocytes from 1-day-old brains (44), whereas 21-day-old astrocytes do not respond. However, their ability to take up dGlc uptake in response to IGF-I is stimulated by the presence of IGFBP-2 antibodies in the medium. This suggests that a lack of responsiveness of 21-day-old astrocytes may be an inability of IGF-I to stimulate its receptors due to its inactivation by IGFBP-2.

cells. Astrocytes extend their processes called "astrocytic feet" and establish physical connections with neurons, oligodendrocytes, endothelial and ependymal cells. Astrocytic connections are not only present on the neuronal cell body, but they also wraparound certain synapses. This type of interaction may allow immediate availability of various neurotransmitters released from neurons to stimulate astrocytic transmitter receptors including α- and β-adrenergic receptors. Biochemical support for such an interaction has recently been provided by demonstrating that norepinephrine (NE) induced c-AMP production, which was previously thought to be a neuronal event, is in fact a result of activation of β-adrenergic receptors on astrocytes (63). In addition, the presence of many transmitter receptors on astrocytes further support this view (64-67). Finally, it has been suggested that neuronal effects of many hormones including angiotensin II, epidermal growth factor, and VIP are suggested to be mediated by their interaction with astrocytic receptors (68-71). Astrocytic feet have also been suggested to wraparound the nonmyelinated areas of the axon. Such an anatomical interaction suggests a communication between the astrocytes and axons on one hand, and oligodendrocytes on the other, in the regulation of myelin formation. Astrocytes also extend their processes and completely surround the capillaries of the brain consisting of endothelial cells. This provides a tight blood brain barrier where molecules and hormones can be separated from the brain. Moreover, such an arrangement can provide transfer of chemical signals from peripheral hormones to the brain by activation of receptors on astrocytes. These observations strongly suggest that anatomically, astrocytes are in excellent position to serve as a center for intercellular communication.

All these cell types (astrocytes, oligodendrocytes, and neurons) in the brain express IGF-I receptors and IGF-I-induced physiological responses. In neurons, IGF-I is neurotrophic and also stimulates neuronal precursor cells to undergo a limited degree of differentiation. In astrocytes and oligodendrocytes, IGF-I is a proliferative hormone and stimulates glial progenitors to differentiate into oligodendrocytes. IGF-I is also produced in neuronal as well as astroglial cells, whereas the major site of IGFBP's has been proposed to be astroglial cells. Thus, IGF-I produced in astrocytes and/or neurons is released into the extracellular space where it acts on various cells of the brain in an autocrine/paracrine fashion to stimulate unique effects on neurons, astrocytes, oligodendrocytes, and possibly cereberoendothelial cells. The actions of IGF-I on these cells could not only be dependent on synthesis and release of IGF-I, and its interactions with receptors, but also on the regulation of IGFBP synthesis and secretion.

The hypothesis that IGF-I may be important in intercellular communication is supported by the following examples:

1. Hypoglycemia induced under many physiological and pathophysiological conditions has been shown to stimulate sympathetic activity and changes in brain catecholamine levels. This leads to an increased release of NE and stimulation of β-adrenergic receptors in astrocytes (64). This may lead to

changes in IGF-I levels and IGF-I-induced Glut-1 gene expression, which will ultimately influence the metabolism of glycogen predominantly found in astrocytes. Support for such a cascade of events is provided by the fact that; 1) β-adrenergic receptors are present in astrocytes (64), 2) NE stimulates β-adrenergic receptors which leads to a many-fold increase in cyclic AMP levels (64), 3) Cyclic AMP has been shown to stimulate Glut-1 gene expression in astrocytes (44-45), and 4) Glycogen is predominantly found in astrocytes (73). Other similar physiological stimuli could also mediate changes in the IGF-I system. For example, stress activates noradrenergic neurons in hypothalamus resulting in the release of NE which influences secretion of GH and IGF-I. This may result in a feedback control at the level of hypothalamus. How does hypothalamic astrocytes and neurons play coordinating roles in this action is not clear at the present time.

2. Corticosteroids have also been shown to exert diverse effects on the brain including decrease in cerebral glucose utilization (74-76). In view of the observation that glucocorticoids also inhibit IGF-I gene expression, it is likely that inhibitory effects of these steroids may be mediated primarily by their actions on IGF-I system. These observations combined with other studies suggesting that primary targets for glucocorticoid action may be astrocytes (77-78), indicate that these steroids interact with type II steroid receptors in astrocytes and inhibit IGF-I synthesis. This may result in a decrease in IGF-I induced transport and Glut-1 gene expression in other brain cells. Support for such a sequence of events is provided by observations indicating the presence of type II steroid hormone receptors in astrocytes (78), glucocorticoid induced biological changes in enzymes of astroglial origin (79-81), changes in cell shape, microfilament contents, and suppression of cell growth (82-84).

3. Our preliminary studies have indicated that insulin increases IGF-I mRNA levels in astrocytes and not in neurons. This suggests that astrocytes around the CVO may be stimulated to produce IGF-I under hyperinsulinemia. Thus, IGF-I may be responsible for maintaining some degree of glucopenia in the CNS by stimulation of glucose uptake and Glut-1 gene expression in astrocytes during hyperinsulinemia and insulin resistant states. This is of particular importance since insulin does not appear to cross the BBB in any significant amounts.

In summary, it appears that astrocytic glial and neuronal IGF-I system may be a key in the intercellular communication of the brain. Above are a few examples of such a proposed role of IGF-I. Obviously, further experimental evidence is needed to support such a function of brain IGF-I. Answers to the following several critical questions will be an important step in this direction:

1. Do hyperglycemia and high insulin selectively influence the expression of IGF-I in neurons and glia?

2. Does stress induce changes in brain IGF-I?

3. How does IGF-I participate in injury induced gliosis and in other neurodegenerative diseases?

4. How does IGFBPs participate in the control of IGF-I action in the brain?

These and other questions need further investigation in order to define IGF-I's role in the brain. In addition, studies are continuing with the use of selective brain cell culture techniques to deduce which components of IGF-I system are glial, neuronal, or of endothelial origin.

REFERENCES

1. VanWyk, J.J. (1984). In: *Hormonal Proteins and Peptides*, Eds. Li, C.H., Academic Press, New York, pp. 81.
2. Czech, M.P. (1989). *Cell* 59:235.
3. Gammeltoft, S. (1989). In: *Peptide Hormones as Prohormones*. Ellis Harwood Limited, pp. 176.
4. dePablo, F., Scott, L.A. and Roth, J. (1990). *Endocrine Rev*. Vol. 11, 558.
5. Hansson, H-A (1990). In: *Current Medical Literature: Growth and Growth Factors*. 5(1), pp. 3, Royal Society of Medicine.
6. Daughaday, W.H. (1989). In: *Molecular and Cellular Biology of Insulin-Like Growth Factors and Their Receptors*. Eds. LeRoith, D. and Raizada, M.K. Plenum Press, New York, pp. 1.
7. Sara, V.R. and Hall, K. (1990). *Physiol. Rev*. 70:591.
8. Sara, V.R. and Carlsson-Skwirut, C. (1990). In: *Growth Factors: From Genes to Clinical Application*. Eds. Sara, V.R., Hall, K. and Low, H. Raven Press, New York, pp. 179.
9. LeRoith, D., Roberts, C.T., Werner, H., Raizada, M.K. and Adamo, M. (1991). In: *Neurotrophic Factors*. Eds. Loughlin, S.E. and Fallon, J.H., Academic Press, Orlando, FL., 1991.
10. Carlsson-Skwirut, C., Jornvall, H., Holmgreu, A., Andersson, C., Bergman, T., Lundquist, G., Sjogren, B. and Sara, V.R. (1986). *FEBS Lett.* 201:3656.
11. Sara, V.R., Carlsson-Skwirut, C., Andersson, C., Hall, E., Sjogren, B., Holmgren, A. and Jornvall, H. (1986). *Proc. Natl. Acad. Sci., USA* 83:4904.
12. Francis, L.G., Reed, L.C., Ballard, F.J., Bagley, C.J., Upton, F.M., Gravestock, P.M. and Wallace, J.C. (1986). *Biochem. J.* 233:203.
13. Karey, K.P., Marquardt, H. and Sirbasku, (1989). *Blood* 74:1084.
14. Ogasawara, M., Karey, K.P., Marquardt, H. and Sirbasku, (1989). *Biochem.* 28:2710.
15. Rotwein, P., Burgess, S.K., Milbrandt, J.D. and Krause, J.E. (1988). *Proc. Natl. Acad. Sci., USA* 85:265.
16. Sandberg, A-C, Engberg, C., Lake, M., vonHolst, H. and Sara, V.R. (1988). *Neurosci. Lett.* 93:114.

17. Wood, T.L., Berelowitz, M. and McKelvy, J.F. (1989). In: *Molecular and Cellular Biology of Insulin-Like Growth Factors and Their Receptors*. Eds. LeRoith, D. and Raizada, M.K., Plenum Press, New York, pp. 209.
18. Werther, G.A., Abate, M., Hogg, A., Cheesman, H., Oldfield, B., Hards, D., Hudson, P., Power, B., Freed, K. and Herington, A.C. (1990). *Mol. Endocrinol.* 4:773.
19. Bondy, C.A., Werner, H., Roberts, C.T. and LeRoith, D. (1990). *Mol. Endocrinol.* 4:1386.
20. Werner, H., Woloschak, M., Adamo, M., Shen-orr, Z., Roberts, C.T. and LeRoith, D. (1989). *Proc. Natl. Acad. Sci., USA* 86:7451.
21. Sara, V.R., Sandberg-Nordquvist, A-C, Carlsson-Skwirut, C., Bergman, T. and Ayer-LeLievre, (1991). In: *Molecular Biology and Physiology of Insulin and Insulin-Like Growth Factors.* Eds. Raizada, M.K. and LeRoith, D., Plenum Press, New York, pp.
22. Andersson, I.K., Edwall, D., Norstedt, G., Rozell, B., Skottner, A. and Hansson, H-A (1988). *Acta. Physiol. Scand.* 132:167.
23. Adamo, M., Werner, H., Fransworth, W., Roberts, C.T., Raizada, M.K. and LeRoith, D. (1989). *Endocrinol.* 123:2565.
24. Lowe, Jr. W.L., Roberts, C.T., Lasky, S.R. and LeRoith, D. (1987). *Proc. Natl. Acad. Sci., USA* 84:8946.
25. Raizada, M.K. (1983). *Exp. Cell. Res.* 133:261.
26. Mathews, L-S, Norstedt, G. and Palmiter, R.D. (1986). *Proc. Natl. Acad. Sci., USA* 83:9343.
27. Hynes, M.A., VanWyk, J.J., Brooks, P.J., D'Ercole, A.J., Jansen, M. and Lund, P.K. (1987). *Mol. Endocrinol.* 1:233.
28. Berelowitz, M. (1989). In: *Molecular and Cellular Biology of Insulin-Like Growth Factors and Their Receptors*. Eds. LeRoith, D. and Raizada, M.K., Plenum Press, New York, pp. 25.
29. Adamo, M., Raizada, M.K. and LeRoith, D. (1989). *Mol. Neurobiol.* 3:71.
30. Shemer, J., Raizada, M.K., Masters, B.A., Ota, A. and LeRoith, D. (1987). *J. Biol. Chem.* 262:6793.
31. Shemer, J., Adamo, M., Raizada, M.K., Heffez, D., Zick, Y. and LeRoith, D. (1989). *J. Mol. Neurosci.* 1:3.
32. McElduff, A., Poronnik, P., Baxter, R.C. and William, P. (1988). *Endocrinol.* 122:1933
33. Sara, V.R., Hall, K., VonHoltz, H., Humbel, R., Sjogren, B. and Wtterberg, (1982). *Neurosci. Lett.* 34:39.
34. Gammeltoft, S., Haselbacher, G.K., Humbel, R.E., Fehlman, M. and VanObberghen, E. (1985). *EMBO Journal,* 4:3407.
35. Heidenreich, K.A., Freidenberg, G.R., Figlewicz, D.P. and Gilmore, P.R. (1986). *Reg. Peptides,* 15:301.
36. McMorris, F.A., Smith, T.M., Desalvo, S. and Furlane, R.W. (1986). *Proc. Natl. Acad. Sci., USA* 83:822.
37. McMorris, F.A. and Dubois-Dalegn, M. (1988). *J. Neurosci. Res.*, 21:199.
38. Masters, B.A., Werner, H., Roberts, C.T., LeRoith, D. and Raizada, M.K. (1991). *Reg. Peptides* (In Press).
39. Lenoir, D. and Honegger, P. (1983). *Dev. Brain Res.* 7:380.
40. Debbage, P.L. (1986). *J. Neurological Sci.* 72:319.

41. Van der Pal, R.H.M., Koper, J.W., VanGolde, L.M.G. and Lopes-Cardozo, M. (1988). *J. Neurosci. Res.* 19:483.
42. Saneto, R.P., Low, K.G., Melner, M.H. and deVellis, J. (1988). *J. Neurosci. Res.* 21:210.
43. Carson, M. (1990). PhD Dissertation, University of Pennsylvania.
44. Werner, H., Raizada, M.K., Mudd, L.M., Foyt, H.L., Simpson, I.A., Roberts, C.T. and LeRoith, D. (1989). *Endocrinol.* 123:314.
45. Masters, B.A., Werner, H., Roberts, C.T., LeRoith, D. and Raizada, M.K. (1991). *Endocrinol.* (In Press).
46. Nielsen, F.C., Wang, E. and Gammeltoft, S. (1991). *J. Neurochem.* 56:12.
47. Knusel, B., Michel, P.P., Schwaber, J.S. and Hefti, F. (1990). *J. Neurosci.* 10:558.
48. Ishii, D.N. and Recio-Pinto, E. (1987). In: *Insulin, Insulin-Like Growth Factors, and Their Receptors in the CNS.* Eds. Raizada, M.K., Phillips, M.I. and LeRoith, D., Plenum Press, New York, pp. 315.
49. Nilsson, L., Sara, V.R. and Nordberg, (1988). *Neurosci. Lett.* 88:221.
50. Knusel, B. and Hefti, (1991). In: *Molecular Biology and Physiology of Insulin and Insulin-Like Growth Factors.* Eds. Raizada, M.K. and LeRoith, D., Plenum Press, New York.
51. Marks, J.L., King, M.G. and Baskin, D. (1991). In: *Molecular Biology and Physiology of Insulin and Insulin-Like Growth Factors.* Eds. Raizada, M.K. and LeRoith, D., Plenum Press, New York.
52. Brewer, M.T., Stetler, G.L., Squires, C.H., Thompson, R.C., Busby, W.H. and Clemmons, D.R. (1988). *Biochem. Biophys. Res. Commun.* 152:1289.
53. Margot, J.B., Binhert, C., Mary, J.L., Landwehr, J., Heinreich, G. and Schwander, J. (1989). *Mol. Endocrinol.* 3:1053.
54. Shimisaki, S., Koba, A., Mercado, M., Shimonaka, M. and Ling, N. (1989). *Biochem. Biophys. Res. Commun.* 165:905.
55. Hossenlopp, P., Seurin, D., Segovia-Quinson, B. and Binoux, M. (1986). *FEBS Lett.* 208:439.
56. Romanus, J.A., Tseng, L.Y.H., Yang, Y.W.H. and Rechler, M.M. (1989). *Biochem. Biophys. Res. Commun.* 163:875.
57. Roghani, M., Hossenlopp, P., Lepage, P., Ballard, A. and Binoux, M. (1989). *FEBS Lett.* 255:253.
58. Donovan, S.M., Oh, Y., Pham, H. and Rosenfeld, R. (1989). *Endocrinol.* 125:2621.
59. Ocrant, I., Fay, C.T. and Parmelee, J.T. (1990). *Endocrinol.* 127:1260.
60. Han, V.K.M., Lauder, J.M. and D'Ercole, A.J. (1988). *J. Neurosci.* 8:3135.
61. Ocrant, I., Pham, H., Oh, Y. and Rosenfeld, R.G. (1989). *Biochem. Biophys, Res. Commun.* 159:1316.
62. Lamson, G., Pham, H., Oh, Y., Ocrant, I., Schwander, J. and Rosenfeld, R.G. (1989). *Endocrinol.* 123:1100.
63. Rosenberg, P.A. and Dichter, M.A. (1989). *J. Neurosci.* 9:2654.
64. Baker, S.P., Sumners, C., Pitha, J. and Raizada, M.K. (1986). *J. Neurochem.* 46:1318.
65. Bowman, C.L. and Kimelberg, H.K. (1987). *Brain Res.* 423:403.

66. Richards, E.M., Sumners, C., Chou, Y-C, Raizada, M.K. and Phillips, M.I. (1989). *J. Neurochem.* 53:287.
67. Puig, J.F., Pacitti, A.J., Guzman, N.J., Crews, F.T., Sumners, C. and Raizada, M.K. (1990). *Brain Res.* 527:318.
68. Olson, J.A., Shiverick, K.T., Ogilvie, S., Buhi, W.C. and Raizada, M.K. (1991). *Proc. Natl. Acad. Sci., USA* 88: (In Press).
69. Morrison, R.S., Kornblum, H.I., Leslie, F.M. and Bradshaw, R.A. (1987). *Science* 238:72.
70. Wang, S-L, Shiverick, K.T., Ogilvie, S., Dunn, R. and Raizada, M.K. (1989). *Endocrinol.* 124:240.
71. Brenneman, D.E., Neale, E.A., Foster, G.A., D'Autremont, S.W. and Westbrook, G.L. (1987). *J. Cell Biol.* 104:1603.
72. Clemmons, D.R., Thrailkill, K.M., Handwerger, S. and Busby, W.H., Jr. (1990). *Endocrinol.* 127:643.
73. Saneto, R.P. and deVellis, J. (1987). In: *Neurochemistry, A Practical Approach.* Eds. Turner, A.J. and Bachelard, H.S., IRL Press, Washington, DC, pp. 27.
74. Sapolsky, R.M. (1985). *J. Neurosci.* 5:1228.
75. Sapolsky, R.M., Krey, L.M. and McEwen, B.S. (1986). *Endocrine Reviews*, 7:284.
76. Horner, H.C., Packan, D.R. and Sapolsky, R.M. (1990). *Neuroendocrinol.* 52:57.
77. Rydzewski, B., Roberts, C.T., LeRoith, D., Sumners, C. and Raizada, M.K. (1991). Submitted.
78. Chou, Y-C (1990). Ph.D. Dissertation, University of Florida.
79. Holbrook, N.J., Grasso, R. and Hackney, J. (1981). *J. Neurosci. Res.* 6:75.
80. Meyer, J., Leveille, P.J., deVellis, J. and McEwen, B.S. (1982). *J. Neurochem.* 39:423.
81. Kimar, S. and deVellis, J. (1988). In: *Glial Cell Receptors* Ed. Kimelberg, H.K., Raven Press, New York, pp. 243.
82. Berlinger, J.A., Bennett, K. and deVellis, J. (1978). *J. Cell Physiol.* 94:321.
83. Morrison, R.S., deVellis, J., Lee, Y.L., Bradshaw, R.A. and Eng, L.F. (1985). *J. Neurosci. Res.* 14:167.
84. Armelin, M.C.S., Stocco, R.C. and Armelin, H.A. (1983). *J. Cell Biol.* 97:455.

CONTRIBUTORS

Domenico Accili
Diabetes Branch
NIDDK
National Institutes of Health
Bethesda, MD 20892

Martine Aggerbeck
Hormone Research Institute and
Dept. of Biochemistry and Biophysics
University of California,
San Francisco
San Francisco, CA 94143

Alaric Arenander
Dept. of Anatomy and Cell Biology
Mental Retardation Research Center,
NPI
UCLA School of Medicine
Los Angeles, CA

Kumiko Asakawa
Dept. of Medicine
Tokyo Women's Medical College
Tokyo 162
JAPAN

Christianne Ayer-LeLievre
Institut d'Embryologie du CNRS
Nogent sur Marne
FRANCE

Mark A. Bach
Section of Molecular and Cellular
Physiology
Diabetes Branch
NIDDK
National Institutes of Health
Bethesda, MD 20892

W. E. Ballinger
Dept. of Pathology
College of Medicine, Box J-275
University of Florida
Gainesville, FL 32610

Denis G. Baskin
Depts. of Biological Structure
and Medicine
University of Washington, and
V. A. Medical Center (151)
1660 S. Columbian Way
Seattle, WA 98108

Robert C. Baxter
Endocrinology Department
Royal Prince Alfred Hospital
Camperdown, NSW 2050
AUSTRALIA

Marvin L. Bayne
ept. of Growth Factor Research
Merck Sharp & Dohme Research
Laboratories
P. O. Box 2000
Rahway, NJ 07065

Per Belfrage
Dept. of Medical and Physiological
Chemistry
University of Lund
Lund
SWEDEN

Tomas Bergman
Dept. of Medical Chemistry
Karolinska Hospital
Stockholm
SWEDEN

Philip J. Bilan
Division of Cell Biology
The Hospital for Sick Children
555 University Avenue
Toronto, Ontario M5G 1X8
CANADA

Michel Binoux
INSERM U. 142
Hôpital Saint-Antoine
75012 Paris
FRANCE

Carolyn Bondy
Developmental Endocrinology Branch
NICHHD
National Institutes of Health
Bethesda, MD

Rosemarie M. Booze
Dept. of Physiology and
Pharmacology
Bowman Gray School of Medicine
Wake Forest University
Winston-Salem, NC 27103

Charles R. Breese
Dept. of Physiology and
Pharmacology
Bowman Gray School of Medicine
Wake Forest University
Winston-Salem, NC 27103

Sharron Northcutt Brown
Laboratory of Molecular Biology
Jerome H. Holland Laboratory for
the Biomedical Sciences
American Red Cross
15601 Crabbs Branch Way
Rockville, MD 20855

Alexandra L. Brown
Growth and Development Section
Molecular, Cellular and Nutritional
Endocrinology Branch
NIDDK
National Institutes of Health
Bethesda, MD 20892

Carmelo B. Bruni
Dipartimento di Biologia e Patologia
Cellulare e Molecolare
II Facoltà di Medicina e Chirurgia
Università degli Studi di Napoli
ITALY 80122

Alessandro Cama
Diabetes Branch
NIDDK
National Institutes of Health
Bethesda, MD 20892

Christine Carlsson-Skwirut
Department of Pathology
Karolinska Institute
Stockholm
SWEDEN

Margaret A. Cascieri
Dept. of Biochemical Endocrinology
Merck Sharp & Dohme Research
Laboratories
P. O. Box 2000
Rahway, NJ 07065

Stefano Casola
Dipartimento di Biologia e Patologia
Cellulare e Molecolare
II Facoltà di Medicina e Chirurgia
Università degli Studi di Napoli
ITALY 80122

Janet Cheng
Dept. of Anatomy and Cell Biology
Mental Retardation Research Center,
NPI
UCLA School of Medicine
Los Angeles, CA

Gary G. Chicchi
Dept. of Biochemical Endocrinology
Merck Sharp & Dohme Research Laboratories
P. O. Box 2000
Rahway, NJ 07065

Edward Chin
Developmental Endocrinology Branch
NICHHD
National Institutes of Health
Bethesda, MD

Jan Christiansen
Dept. of Clinical Chemistry
Bispebjerg Hospital
DK 2400 Copenhagen NV
DENMARK

David R. Clemmons
Division of Endocrinology
University of North Carolina
Chapel Hill, NC

Vittorio Colantuoni
Dipartimento di Biochimica e Biotecnologie Mediche
II Facoltà di Medicina e Chirurgia
Università degli Studi di Napoli
ITALY 80122

Fulton T. Crews
Dept. of Pharmacology and Therapeutics
College of Medicine, Box J-267
University of Florida
Gainesville, FL 32610

Anselm D'Costa
Dept. of Physiology and Pharmacology
Bowman Gray School of Medicine
Wake Forest University
Winston-Salem, NC 27103

Eva Degerman
Dept. of Medical and Physiological Chemistry
University of Lund
Lund
SWEDEN

Hiroshi Demura
Dept. of Medicine
Tokyo Women's Medical College
Tokyo 162
JAPAN

Sherin U. Devaskar
Department of Pediatrics
St. Louis University School of Medicine
St. Louis, MO

Jean de Vellis
Dept. of Anatomy and Cell Biology
Mental Retardation Research Center, NPI
UCLA School of Medicine
Los Angeles, CA

Alana L. Dudley
Research Division
Joslin Diabetes Center
Harvard Medical School
1 Joslin Place
Boston, MA 02215

D. Z. Ewton
Biology Department
Syracuse University
Syracuse, NY 13244

Raffaella Faraonio
Dipartimento di Biochimica e Biotecnologie Mediche
II Facoltà di Medicina e Chirurgia
Università degli Studi di Napoli
ITALY 80122

J. R. Florini
Biology Department
Syracuse University
Syracuse, NY 13244

Beverly Foster
Laboratory of Molecular Biology
Jerome H. Holland Laboratory for
the Biomedical Sciences
American Red Cross
15601 Crabbs Branch Way
Rockville, MD 20855

G. Freund
Dept. of Medicine
College of Medicine, Box J-277
University of Florida
Gainesville, FL 32610

E. Rudolph Froesch
Metabolic Unit
Dept. of Medicine
University Hospital
8091 Zurich
SWITZERLAND

Susan C. Frost
Dept. of Biochemistry and Molecular
Biology
University of Florida
Gainesville, FL 32610

Izumi Fukuda
Dept. of Medicine
Tokyo Women's Medical College
Tokyo 162
JAPAN

Steen Gammeltoft
Dept. of Clinical Chemistry
Bispebjerg Hospital
DK 2400 Copenhagen NV
DENMARK

T. Gloudemans
Laboratory for Physiological
Chemistry
State University of Utrecht
Utrecht
THE NETHERLANDS

Barry J. Goldstein
Research Division
Joslin Diabetes Center
Harvard Medical School
1 Joslin Place
Boston, MA 02215

Linda K. Gowan
Electro-Nucleonics, Inc.
7101 Riverwood Drive
Columbia, MD 21046

Hans-Peter Guler
Ciba-Geigy Corp.
Summit, NJ 07901

Joyce F. Haskell
Dept. of Obstetrics and Gynecology
University of Alabama at Birmingham
Birmingham, AL

Leif Havton
Dept. of Anatomy and Cell Biology
College of Physicians and Surgeons
Columbia University
New York, NY 10032

Franz Hefti
Ethel Percy Andrus Gerontology
Center
University of Southern California
Los Angeles, CA 90089-0191

K. A. Heidenreich
Dept. of Medicine; M-023E
University of California, San Diego
La Jolla, CA 92093

Naomi Hizuka
Dept. of Medicine
Tokyo Women's Medical College
Tokyo 162
JAPAN

P. Holthuizen
Laboratory for Physiological
Chemistry
State University of Utrecht
Utrecht
THE NETHERLANDS

Paul Hossenlopp
INSERM U. 142
Hôpital Saint-Antoine
75012 Paris
FRANCE

Eileen C. Hoyt
Dept. of Pediatrics
University of North Carolina
at Chapel Hill
Chapel Hill, NC 27599

Ming-Sing Hsu
Dept. of Anatomy and Cell Biology
College of Physicians and Surgeons
Columbia University
New York, NY 10032

W. Robert Hudgins
Laboratory of Molecular Biology
Jerome H. Holland Laboratory for
the Biomedical Sciences
American Red Cross
15601 Crabbs Branch Way
Rockville, MD 20855

Bruce E. Hunter
Dept. of Neuroscience
College of Medicine, Box J-244
University of Florida
Gainesville, FL 32610

Eiichi Imano
Diabetes Branch
NIDDK
National Institutes of Health
Bethesda, MD 20892

Douglas N. Ishii
Physiology Department
Colorado State University
Fort Collins, CO 80523

E. Jansen
Laboratory for Physiological
Chemistry
State University of Utrecht
Utrecht
THE NETHERLANDS

Hiroko Kadowaki
Diabetes Branch
NIDDK
National Institutes of Health
Bethesda, MD 20892

Takashi Kadowaki
Diabetes Branch
NIDDK
National Institutes of Health
Bethesda, MD 20892

Hisanori Kato
Dept. of Agricultural Chemistry
Faculty of Agriculture
The University of Tokyo
Bunkyo-ku, Tokyo 113
JAPAN

K. A. Kenner
Dept. of Medicine; M-023E
University of California, San Diego
La Jolla, CA 92093

Wieland Kiess
Children's Hospital
University of Munich
Munich
GERMANY

Michael G. King
Dept. of Medicine
University of Washington, and
V. A. Medical Center (151)
1660 S. Columbian Way
Seattle, WA 98108

Amira Klip
Division of Cell Biology
The Hospital for Sick Children
555 University Avenue
Toronto, Ontario M5G 1X8
CANADA

Beat Knusel
Ethel Percy Andrus Gerontology
Center
University of Southern California
Los Angeles, CA 90089-0191

Otakar Koldovsky
Dept. of Pediatrics and Anatomy
AHSC, The Children's Research Center
1501 N. Campbell Avenue
Tucson, AZ 85724

Thomas J. Lauterio
Dept. of Internal Medicine
Eastern Virginia Medical School
Norfolk, VA 23501

Derek LeRoith
Section of Molecular and Cellular Physiology
Diabetes Branch
NIDDK
National Institutes of Health
Bethesda, MD 20892

Yi Li
Biochemistry Department
Colorado State University
Fort Collins, CO 80523

P. Kay Lund
Dept. of Physiology
University of North Carolina at Chapel Hill
Chapel Hill, NC 27599

Jiangming Luo
Dept. of Internal Medicine
Faculty of Medicine
University of Manitoba
Winnipeg R3E 0W3
CANADA

K. A. Magri
Biology Department
Syracuse University
Syracuse, NY 13244

Vincent C. Manganiello
Laboratory of Cellular Metabolism
National Heart, Lung, and Blood Institute
National Institutes of Health
Bldg. 10, Room 5N-307
Bethesda, MD 20892

Jonathan L. Marks
Endocrine Unit
West Mead Hospital
West Mead, NSW 2145
AUSTRALIA

Janet L. Martin
Endocrinology Department
Royal Prince Alfred Hospital
Camperdown, NSW 2050
AUSTRALIA

David M. McCracken
Dept. of Pediatrics and Anatomy
AHSC, The Children's Research Center
1501 N. Campbell Avenue
Tucson, AZ 85724

R. McElhaney
Dept. of Pharmacology and Therapeutics
College of Medicine, Box J-267
University of Florida
Gainesville, FL 32610

D. Meinsma
Dept. of Pediatrics
State University of Utrecht
Utrecht
THE NETHERLANDS

Thomas J. Merimee
Dept. of Medicine
Division of Endocrinology
Box J-226, JHMHC
University of Florida
Gainesville, FL 32610-0226

Yutaka Miura
Dept. of Agricultural Chemistry
Faculty of Agriculture
The University of Tokyo
Bunkyo-ku, Tokyo 113
JAPAN

Liam J. Murphy
Dept. of Physiology
Faculty of Medicine
University of Manitoba
Winnipeg R3E 0W3
CANADA

Russell B. Myers
Dept. of Obstetrics and Gynecology
University of Alabama at Birmingham
Birmingham, AL

Finn C. Nielsen
Dept. of Clinical Chemistry
Bispebjerg Hospital
DK 2400 Copenhagen NV
DENMARK

Peter Nissley
Metabolism Branch
National Cancer Institute
National Institutes of Health
Bethesda, MD 20892

Tadashi Noguchi
Dept. of Agricultural Chemistry
Faculty of Agriculture
The University of Tokyo
Bunkyo-ku, Tokyo 113
JAPAN

Ian Ocrant
Dept. of Biology and Medicine
Brown University
Providence, RI

Guck T. Ooi
Growth and Development Section
Molecular, Cellular and Nutritional Endocrinology Branch
NIDDK
National Institutes of Health
Bethesda, MD 20892

Craig C. Orlowski
Growth and Development Section
Molecular, Cellular and Nutritional Endocrinology Branch
NIDDK
National Institutes of Health
Bethesda, MD 20892

James F. Perdue
Laboratory of Molecular Biology
Jerome H. Holland Laboratory for the Biomedical Sciences
American Red Cross
15601 Crabbs Branch Way
Rockville, MD 20855

Jeffrey E. Pessin
Dept. of Physiology & Biophysics
Bowen Science Building
The University of Iowa College of Medicine
Iowa City, IA 52242

Anthony F. Philipps
Dept. of Pediatrics and Anatomy
AHSC, The Children's Research Center
1501 N. Campbell Avenue
Tucson, AZ 85724

Sarah Pierce
Dept. of Pharmacology
Stanford University School of Medicine
Stanford, CA 94305

John E. Pintar
Dept. of Anatomy and Cell Biology
College of Physicians and Surgeons
Columbia University
New York, NY 10032

Suzanne Quinn
Dept. of Medicine
Division of Endocrinology
Box J-226, JHMHC
University of Florida
Gainesville, FL 32610-0226

Mohan K. Raizada
Dept. of Physiology
College of Medicine, Box J-274
University of Florida
Gainesville, FL 32610

Toolsie Ramlal
Division of Cell Biology
The Hospital for Sick Children
555 University Avenue
Toronto, Ontario M5G 1X8
CANADA

Radhakrishna Rao
Dept. of Pediatrics and Anatomy
AHSC, The Children's Research Center
1501 N. Campbell Avenue
Tucson, AZ 85724

Matthew M. Rechler
Growth and Development Section
Molecular, Cellular and Nutritional Endocrinology Branch
NIDDK
National Institutes of Health
Bethesda, MD 20892

Jeanne M. Richardson
Dept. of Physiology & Biophysics
Bowen Science Building
The University of Iowa College of Medicine
Iowa City, IA 52242

William Riley
Dept. of Medicine
Division of Endocrinology
Box J-226, JHMHC
University of Florida
Gainesville, FL 32610-0226

Robert Risch
Dept. of Biochemistry and Molecular Biology
University of Florida
Gainesville, FL 32610

Charles T. Roberts, Jr.
Section of Molecular and Cellular Physiology
Diabetes Branch
NIDDK
National Institutes of Health
Bethesda, MD 20892

Monireh Roghani
INSERM U. 142
Hôpital Saint-Antoine
75012 Paris
FRANCE

Leslie Rogler
Dept. of Anatomy and Cell Biology
College of Physicians and Surgeons
Columbia University
New York, NY 10032

Elena Rossi
Dipartimento di Biologia e Patologia Cellulare e Molecolare
II Facoltà di Medicina e Chirurgia
Università degli Studi di Napoli
ITALY 80122

Richard A. Roth
Dept. of Pharmacology
Stanford University School of Medicine
Stanford, CA 94305

Peter Rotwein
Depts. of Medicine and Genetics
Washington University School of Medicine
Box 8127
St. Louis, MO 63110

Betty Russell
Dept. of Medicine
Division of Endocrinology
Box J-226, JHMHC
University of Florida
Gainesville, FL 32610-0226

William J. Rutter
Hormone Research Institute and
Dept. of Biochemistry and Biophysics
University of California, San Francisco
San Francisco, CA 94143

Bartosz Rydzewski
Dept. of Physiology
College of Medicine, Box J-274
University of Florida
Gainesville, FL 32610

Ann-Christin Sandberg-Nordqvist
Department of Pathology
Karolinska Institute
Stockholm
SWEDEN

Vicki R. Sara
Department of Pathology
Karolinska Institute
Stockholm
SWEDEN

Joan Scheuermann
Laboratory of Molecular Biology
Jerome H. Holland Laboratory for the Biomedical Sciences
American Red Cross
15601 Crabbs Branch Way
Rockville, MD 20855

Werner Schurr
Ciba-Geigy AG
4002 Basle
SWITZERLAND

Charita Seneviratne
Dept. of Internal Medicine
Faculty of Medicine
University of Manitoba
Winnipeg R3E 0W3
CANADA

Kazuo Shizume
Research Laboratory
Foundation for Growth Science
Tokyo 162
JAPAN

Maria de la Luz Sierra
Diabetes Branch
NIDDK
National Institutes of Health
Bethesda, MD 20892

William I. Sivitz
Dept. of Internal Medicine
Bowen Science Building
The University of Iowa College of Medicine
Iowa City, IA 52242

Carolyn J. Smith
Laboratory of Cellular Metabolism
National Heart, Lung, and Blood Institute
National Institutes of Health
Bldg. 10, Room 5N-307
Bethesda, MD 20892

William E. Sonntag
Dept. of Physiology and Pharmacology
Bowman Gray School of Medicine
Wake Forest University
Winston-Salem, NC 27103

Bethel Stannard
Section of Molecular and Cellular Physiology
Diabetes Branch
NIDDK
National Institutes of Health
Bethesda, MD 20892

P. H. Steenbergh
Laboratory for Physiological Chemistry
State University of Utrecht
Utrecht
THE NETHERLANDS

Cynthia Stover
Dept. of Pharmacology
Stanford University School of Medicine
Stanford, CA 94305

Randal D. Streck
Dept. of Anatomy and Cell Biology
College of Physicians and Surgeons
Columbia University
New York, NY 10032

Izumi Sukegawa
Dept. of Medicine
Tokyo Women's Medical College
Tokyo 162
JAPAN

J. S. Sussenbach
Laboratory for Physiological Chemistry
State University of Utrecht
Utrecht
THE NETHERLANDS

Gyorgyi Szebenyi
Depts. of Medicine and Genetics
Washington University School of Medicine
Box 8127
St. Louis, MO 63110

Kazue Takano
Dept. of Medicine
Tokyo Women's Medical College
Tokyo 162
JAPAN

Simeon I. Taylor
Diabetes Branch
NIDDK
National Institutes of Health
Bethesda, MD 20892

S. P. Toledo
Dept. of Medicine; M-023E
University of California, San Diego
La Jolla, CA 92093

Hans Tornqvist
Dept. of Medical and Physiological Chemistry
University of Lund
Lund
SWEDEN

Lucy Y.-H. Tseng
Growth and Development Section
Molecular, Cellular and Nutritional Endocrinology Branch
NIDDK
National Institutes of Health
Bethesda, MD 20892

Tsutomu Umezawa
Dept. of Agricultural Chemistry
Faculty of Agriculture
The University of Tokyo
Bunkyo-ku, Tokyo 113
JAPAN

M. A. van Dijk
Laboratory for Physiological Chemistry
State University of Utrecht
Utrecht
THE NETHERLANDS

Judson J. Van Wyk
Dept. of Pediatrics
University of North Carolina at Chapel Hill
Chapel Hill, NC 27599

Valeria Vasta
Dept. of Biochemistry
University of Florence
Florence
ITALY

Sten Verland
Dept. of Clinical Chemistry
Bispebjerg Hospital
DK 2400 Copenhagen NV
DENMARK

Don W. Walker
Dept. of Neuroscience
College of Medicine, Box J-244
University of Florida
Gainesville, FL 32610

Chiang Wang
Physiology Department
Colorado State University
Fort Collins, CO 80523

Haim Werner
Section of Molecular and Cellular Physiology
Diabetes Branch
NIDDK
National Institutes of Health
Bethesda, MD 20892

Katharina Wettstein
Metabolic Unit
Dept. of Medicine
University Hospital
8091 Zurich
SWITZERLAND

Odile Whitechurch
Transgène S.A.
67082 Strasbourg
FRANCE

Jean M. Wilson
Dept. of Pediatrics and Anatomy
AHSC, The Children's Research Center
1501 N. Campbell Avenue
Tucson, AZ 85724

Teresa L. Wood
Dept. of Anatomy and Cell Biology
College of Physicians and Surgeons
Columbia University
New York, NY 10032

Magdalena M. Wozniak
Dept. of Pharmacology & Experimental Therapeutics
College of Medicine, Box J-267
University of Florida
Gainesville, FL 32610

Yvonne W.-H. Yang
Growth and Development Section
Molecular, Cellular and Nutritional Endocrinology Branch
NIDDK
National Institutes of Health
Bethesda, MD 20892

Kazuyoshi Yonezawa
Dept. of Pharmacology
Stanford University School of Medicine
Stanford, CA 94305

Jurgen Zapf
Metabolic Unit
Dept. of Medicine
University Hospital
8091 Zurich
SWITZERLAND

Raffaele Zarrilli
Dipartimento di Biologia e Patologia Cellulare e Molecolare
II Facoltà di Medicina e Chirurgia
Università degli Studi di Napoli
ITALY 80122

INDEX